W9-BWB-624

BERGEY'S MANUAL® OF

Systematic Bacteriology

Second Edition

Volume Two

The *Proteobacteria*

Part A

Introductory Essays

BERGEY'S MANUAL® OF

Systematic Bacteriology

Second Edition

Volume Two

The *Proteobacteria*

Part A

Introductory Essays

Don J. Brenner
Noel R. Krieg
James T. Staley
EDITORS, VOLUME TWO

George M. Garrity
EDITOR-IN-CHIEF

WITH CONTRIBUTIONS FROM 339 COLLEAGUES

George M. Garrity, Sc.D.
Bergey's Manual Trust
Department of Microbiology and Molecular Genetics
Michigan State University
East Lansing, MI 48824-4320
USA

ISBN-10: 0-387-24143-4 Printed on acid-free paper.
ISBN-13: 978-0387-24143-2

Printed in the United States of America. (IMP/MVY)

9 8 7 6 5 4 3 2 1 SPIN 10993712

springeronline.com

This volume is dedicated to our colleagues,
David R. Boone, Don J. Brenner,
Richard W. Castenholz, and Noel R. Krieg, who
retired from the Board of Trustees of Bergey's Manual
Trust as this edition was in preparation. We deeply
appreciate their efforts as editors and authors; they
have devoted their time and many years in helping
the Trust meet its objectives.

Preface to Volume Two of the Second Edition of *Bergey's Manual® of Systematic Bacteriology*

There is a long-standing tradition for the Editors of *Bergey's Manual* to open their respective editions with the observation that the new edition is a departure from the earlier ones. As this volume goes to press, however, we recognize a need to deviate from this practice, by offering a separate preface to each volume within this edition. In part, this departure is necessary because the size and complexity of this edition far exceeded our expectations, as has the amount of time that has elapsed between publication of the first volume of this edition and this volume.

Earlier, we noted that systematic procaryotic biology is a dynamic field, driven by constant theoretical and methodological advances that will ultimately lead to a more perfect and useful classification scheme. Clearly, the pace has been accelerating as evidenced in the super-linear rate at which new taxa are being described. Much of the increase can be attributed to rapid advances in sequencing technology, which has brought about a major shift in how we view the relationships among *Bacteria* and *Archaea*. While the possibility of a universally applicable natural classification was evident as the First Edition was in preparation, it is only recently that the sequence databases became large enough, and the taxonomic coverage broad enough to make such an arrangement feasible. We have relied heavily upon these data in organizing the contents of this edition of *Bergey's Manual of Systematic Bacteriology*, which will follow a phylogenetic framework based on analysis of the nucleotide sequence of the small ribosomal subunit RNA, rather than a phenotypic structure. This departs from the First Edition, as well as the Eighth and Ninth Editions of the *Determinative Manual*. While the rationale for presenting the content of this edition in such a manner should be evident to most readers, they should bear in mind that this edition, as in all preceding ones represents a progress report, rather than a final classification of procaryotes.

The Editors remind the readers that the *Systematics Manual* is a peer-reviewed collection of chapters, contributed by authors who were invited by the Trust to share their knowledge and expertise of specific taxa. Citation should refer to the author, the chapter title, and inclusive pages rather than to the Editors. The Trust is indebted to all of the contributors and reviewers, without whom this work would not be possible. The Editors are grateful for the time and effort that each expended on behalf of the entire scientific community. We also thank the authors for their good grace in accepting comments, criticisms, and editing of their manuscripts. We would also like to thank Drs. Hans Trüper, Brian Tindall, and Jean Euzéby for their assistance on matters of nomenclature and etymology.

We would like to express our thanks to the Department of Microbiology and Molecular Genetics at Michigan State University for housing our headquarters and editorial office and for providing a congenial and supportive environment for microbial systematics. We would also like to thank Connie Williams not only for her expert secretarial assistance, but also for unflagging dedication to the mission of Bergey's Manual Trust and Drs. Julia Bell and Denise Searles for their expert editorial assistance and diligence in verifying countless pieces of critical information and to Dr. Timothy G. Lilburn for constructing many of the phylogenetic trees used in this volume. We also extend our thanks to Alissa Wesche, Matt Chval and Kristen Johnson for their assistance in compilation of the bibliography.

A project such as the *Systematics Manual* also requires the strong and continued support of a dedicated publisher, and we have been most fortunate in this regard. We would also like to express our gratitude to Springer-Verlag for supporting our efforts and for the development of the Bergey's Document Type Definition (DTD). We would especially like to thank our Executive Editor, Dr. William Curtis for his courage, patience, understanding, and support; Catherine Lyons for her expertise in designing and developing our DTD, and Jeri Lambert and Leslie Grossberg of Impressions Book and Journal Services for their efforts during the pre-production and production phases. We would also like to acknowledge the support of ArborText, Inc., for providing us with state-of-the-art SGML development and editing tools at reduced cost. Lastly, I would like to express my personal thanks to my fellow trustees for providing me with the opportunity to participate in this effort, to Drs. Don Brenner, Noel Krieg, and James Staley for their enormous efforts as volume editors and to my wife, Nancy, and daughter, Jane, for their continued patience, tolerance and support.

Comments on this edition are welcomed and should be directed to Bergey's Manual Trust, Department of Microbiology and Molecular Genetics, 6162 Biomedical and Physical Sciences Building, Michigan State University, East Lansing, MI, USA 48824-4320. Email: garrity@msu.edu

George M. Garrity

Preface to the First Edition of *Bergey's Manual*® of *Systematic Bacteriology*

Many microbiologists advised the Trust that a new edition of the *Manual* was urgently needed. Of great concern to us was the steadily increasing time interval between editions; this interval reached a maximum of 17 years between the seventh and eighth editions. To be useful the *Manual* must reflect relatively recent information; a new edition is soon dated or obsolete in parts because of the nearly exponential rate at which new information accumulates. A new approach to publication was needed, and from this conviction came our plan to publish the *Manual* as a sequence of four subvolumes concerned with systematic bacteriology as it applies to taxonomy. The four subvolumes are divided roughly as follows: (a) the Gram-negatives of general, medical or industrial importance; (b) the Gram-positives other than actinomycetes; (c) the archaeobacteria, cyanobacteria and remaining Gram-negatives; and (d) the actinomycetes. The Trust believed that more attention and care could be given to preparation of the various descriptions within each subvolume, and also that each subvolume could be prepared, published, and revised as the area demanded, more rapidly than could be the case if the *Manual* were to remain as a single, comprehensive volume as in the past. Moreover, microbiologists would have the option of purchasing only that particular subvolume containing the organisms in which they were interested.

The Trust also believed that the scope of the *Manual* needed to be expanded to include more information of importance for systematic bacteriology and bring together information dealing with ecology, enrichment and isolation, descriptions of species and their determinative characters, maintenance and preservation, all focused on the illumination of bacterial taxonomy. To reflect this change in scope, the title of the *Manual* was changed and the primary publication becomes *Bergey's Manual of Systematic Bacteriology*. This contains not only determinative material such as diagnostic keys and tables useful for identification, but also all of the detailed descriptive information and taxonomic comments. Upon completion of each subvolume, the purely determinative information will be assembled for eventual incorporation into a much smaller publication which will continue the original name of the *Manual*, *Bergey's Manual of Determinative Bacteriology*, which will be a similar but improved version of the present *Shorter Bergey's Manual*. So, in the end there will be two publications, one systematic and one determinative in character.

An important task of the Trust was to decide which genera should be covered in the first and subsequent subvolumes. We were assisted in this decision by the recommendations of our Advisory Committees, composed of prominent taxonomic authorities to whom we are most grateful. Authors were chosen on the basis of constant surveillance of the literature of bacterial systematics and by recommendations from our Advisory Committees.

The activation of the 1976 Code had introduced some novel problems. We decided to include not only those genera that had been published in the Approved Lists of Bacterial Names in January 1980 or that had been subsequently validly published, but also certain genera whose names had no current standing in nomenclature. We also decided to include descriptions of certain organisms which had no formal taxonomic nomenclature, such as the endosymbionts of insects. Our goal was to omit no important group of cultivated bacteria and also to stimulate taxonomic research on "neglected" groups and on some groups of undoubted bacteria that have not yet been cultivated and subjected to conventional studies.

The invited authors were provided with instructions and exemplary chapters in June 1980 and, although the intended deadline for receipt of manuscripts was March 1981, all contributions were assembled in January 1982 for the final preparations. The *Manual* was forwarded to the publisher in June 1982.

Some readers will note the consistent use of the stem -var instead of -type in words such as biovar, serovar and pathovar. This is in keeping with the recommendations of the Bacteriological Code and was done against the wishes of some of the authors.

We have deleted much of the synonymy of scientific names which was contained in past editions. The adoption of the new starting date of January 1, 1980 and publication of the Approved Lists of Bacterial Names has made mention of past synonymy obsolete. We have included synonyms of a name only if they have been published since the new starting date, or if they were also on the Approved Lists and, in rare cases with certain pathogens, if the mention of an old name would help readers associate the organism with a clinical problem. If the reader is interested in tracing the history of a name we suggest he or she consult past editions of the *Manual* or the *Index Bergeyana* and its *Supplement*. In citations of names we have used the abbreviation AL to denote the inclusion of the name on the Approved Lists of Bacterial Names and VP to show the name has been validly published.

In the matter of citation of the *Manual* in the scientific literature we again stress the fact that the *Manual* is a collection of authored chapters and the citation should refer to the author, the chapter title and its inclusive pages, not the Editor.

To all contributors, the sincere thanks of the Trust is due; the Editor is especially grateful for the good grace with which the authors accepted comments, criticisms and editing of their manuscripts. It is only because of the voluntary and dedicated efforts of these authors that the *Manual* can continue to serve the science of bacteriology on an international basis.

A number of institutions and individuals deserve special acknowledgment from the Trust for their help in bringing about the publication of this volume. We are grateful to the Department of Biology of the Virginia Polytechnic Institute and State University for providing space, facilities and, above all, tolerance for the diverted time taken by the Editor during the preparation of the book. The Department of Microbiology at Iowa State University of Science and Technology continues to provide a welcome home for the main editorial offices and archives of the Trust and we acknowledge their continued support. A grant (LM-03707) from the National Library of Medicine, National Institutes of Health to assist in the preparation of this and the next volume of the *Manual* is gratefully acknowledged.

A number of individuals deserve special mention and thanks for their help. Professor Thomas O. McAdoo of the Department of Foreign Languages and Literatures at the Virginia Polytechnic Institute and State University has given invaluable advice on the etymology and correctness of scientific names. Those assisting the Editor in the Blacksburg office were R. Martin Roop II, Don D. Lee, Eileen C. Falk and Michael W. Friedman and their help is sincerely appreciated. In the Ames office we were ably assisted by Gretchen Colletti and Diane Triggs during the early period of preparation and by Cynthia Pease during the major portion of the editing process. Mrs. Pease has been responsible for the construction of the List of References and her willingness to handle the cumbersome details of text editing on a big computer is gratefully acknowledged.

John G. Holt

Preface to the First Edition of *Bergey's Manual® of Determinative Bacteriology*

The elaborate system of classification of the bacteria into families, tribes and genera by a Committee on Characterization and Classification of the Society of American Bacteriologists (1911, 1920) has made it very desirable to be able to place in the hands of students a more detailed key for the identification of species than any that is available at present. The valuable book on "Determinative Bacteriology" by Professor F. D. Chester, published in 1901, is now of very little assistance to the student, and all previous classifications are of still less value, especially as earlier systems of classification were based entirely on morphologic characters.

It is hoped that this manual will serve to stimulate efforts to perfect the classification of bacteria, especially by emphasizing the valuable features as well as the weaker points in the new system which the Committee of the Society of American Bacteriologists has promulgated. The Committee does not regard the classification of species offered here as in any sense final, but merely a progress report leading to more satisfactory classification in the future.

The Committee desires to express its appreciation and thanks to those members of the society who gave valuable aid in the compilation of material and the classification of certain species. . . .

The assistance of all bacteriologists is earnestly solicited in the correction of possible errors in the text; in the collection of descriptions of all bacteria that may have been omitted from the text; in supplying more detailed descriptions of such organisms as are described incompletely; and in furnishing complete descriptions of new organisms that may be discovered, or in directing the attention of the Committee to publications of such newly described bacteria.

David H. Bergey, *Chairman*
Francis C. Harrison
Robert S. Breed
Bernard W. Hammer
Frank M. Huntoon
Committee on Manual.
August, 1923.

Contents

Contributors

Sharon L. Abbott
Microbial Diseases Laboratory, Communicable Disease Control, California Department of Health Services, Berkeley, CA 94704-1011, USA

Wolf-Rainer Abraham
Chemical Microbiology Group, GBF-National Research Centre for Biotechnology, Mascheroder Weg 1, D-38124 Braunschweig, Germany

Paula Aguiar
Portland State University, Portland, OR 97207-0751, USA

Azeem Ahmad
Dept. Organismic and Evolutionary Biology, Harvard University Biolabs., Cambridge, MA 02144, USA

Raymond J. Akhurst
Division of Entomology, Commonwealth Scientific and Industrial Research Organization (CSIRO), Canberra Australian Cap. Terr. 2601, Australia

Serap Aksoy
Department of Epidemiology and Public Health, Yale University School of Medicine, New Haven, CT 06520-8034, USA

Milton J. Allison
Department of Microbiology, Iowa State University, Ames, IA 50011-3211, USA

Rudolf Amann
Nachwuchsgruppe Molekulare Ökologie, Max Planck-Institute für Marine Mikrobiologie, Celsiusstrasse 1, D-28359 Bremen, Germany

Øystein Angen
Danish Veterinary Laboratory, Bülowsvej 27, 1790 Copenhagen V, Denmark

Jacint Arnau
Meat Technology Centre, Institut for Food & Agricultural Res. & Tech., Granja Camps i Armet s/n, 17121 Monells, Spain

Georg Auling
Institute für Mikrobiologie, Universität Hannover, Schneiderberg 50, D-30167 Hannover, Germany

Dawn A. Austin
Department of Biological Sciences, School of Life Sciences, Heriot-Watt University, Riccarton, Edinburgh EH14 4AS, United Kingdom

Hans-Dietrich Babenzien
Department of Limnology of Stratified Lakes, Inst. of Freshwater Ecology & Inland Fisheries, Alte Fischerhütte 2, D-16775 Neuglobsow, Germany

Marcie L. Baer
Biology Department, Shippensburg University, Shippensburg, PA 17257, USA

Simon C. Baker
Birkbeck College, Malet Street, Bloomsbury, London WC1E 7HX, United Kingdom

José Ivo Baldani
Centro Nacional de Pesquisa de Agrobiologia, Empresa Brasileira de Pesquisa Agropecuária, Room 247-23851-970 Seropédica, Caixa Postal 74.505, Rio de Janeiro 465, Brazil

Vera Lúcia Divan Baldani
Centro Nacional de Pesquisa de Agrobiologia, Empresa Brasileira de Pesquisa Agropecuária, Room 247-23851-970 Seropédica, Caixa Postal 74.505, Rio de Janeiro 465, Brazil

David L. Balkwill
Department of Biological Science, Florida State University, Tallahassee, FL 32306-4470, USA

Menachem Banai
Ministry of Agriculture, Veterinary Services & Animal Health, Kimron Veterinary Institute, P.O. Box 12, Bet Dagan 50 250, Israel

Claudio Bandi
Dipartimento di Patologia Animale, Igiene e Sanità Pubblica Veterinaria, Sezione di Patologia Generale e Parassitologia, Università degli Studi di Milano, Via Celoria 10 20133 Milano, Italy

Ellen Jo Baron
Clinical Microbiology/Virology Laboratory, Stanford University Medical Center, Stanford, CA 94305-5250, USA

Linda Baumann
Department of Microbiology #0875, College of Letters and Science, University of California, Davis, CA 95616-8665, USA

Paul Baumann
Department of Microbiology #0875, College of Letters and Science, University of California, Davis, CA 95616-8665, USA

Janiche Beeder
Section for Biotechnology, Novsk Hydro ASA Research Centre, P. O. Box 2560, N-3901 Porsgruun, Norway

Julia A. Bell
Dept. of Microbiology and Molecular Genetics, Michigan State University, East Lansing, MI 48824-4320, USA

Hervé Bercovier
Hadassah Medical School, The Hebrew University, Jerusalem, Israel

Karen M. Birkhead
Foodborne & Diarrheal Diseases Lab. Section, Division of Bacterial and Mycotic Diseases, Centers for Disease Control and Prevention, Atlanta, GA 30333, USA

Magne Bisgaard
Department of Veterinary Microbiology, The Royal Veterinary & Agricultural University, Bulowsvej 13, DK-1870 Frederiksberg C, Denmark

Judith A. Bland
Merck and Company, Inc., WS2F-45, Whitehouse Station, NJ 08889-0100, USA

Nancy M.C. Bleumink-Pluym
Dept. of Bacteriology, Inst. of Infectious Diseases & Immunology, Vet. Medicine, Universität Utrecht, Yalelaan 1, 3584 CL Utrecht, The Netherlands

Eberhard Bock
Inst. für Allgemeine Botanik und Botanischer Garten, Universität Hamburg, Ohnhorststrasse 18, D-22609 Hamburg, Germany

Noël E. Boemare
Laboratoire de Pathologie comparée, C.P. 101, Laboratoire associé, Université Montpellier II, NRA-CNRS-UM II, 34095 Montpellier Cedex 05, France

David R. Boone
Department of Environmental Biology, Portland State University, Portland, OR 97207-0751, USA

Edward J. Bottone
Department of Infectious Diseases, The Mount Sinai Hospital, New York, NY 10029-6574, USA

Kjell Bøvre (Deceased)
Kaptein W. Wilhelmsen og Frues Mikrobiologiske Institutt, University of Oslo, Rikshospitalet, N-0027 Oslo, Norway

John P. Bowman
School of Agricultural Science, University of Tasmania, Antartic CRC, Private Bag 54, Hobart 7001, Tasmania, Australia

John F. Bradbury
CABI Bioscience, Bakeham Lane, Egham, Surrey TW20 9TY, United Kingdom

Kristian K. Brandt
Section of Genetics and Microbiology, Department of Ecology, Royal Veterinary and Agricultural University, DK-1871 Frederiksberg, Denmark

Don J. Brenner
Meningitis & Special Pathogens Branch Laboratory Section, Centers for Disease Control & Prevention, Atlanta, GA 30333, USA

Frances W. Brenner
Foodborne & Diarrheal Diseases Lab. Section, Division of Bacterial and Mycotic Diseases, Centers for Disease Control and Prevention, Atlanta, GA 30333, USA

Thorsten Brinkhoff
Inst. für Chemie und Biol. des Meeres (ICBM), Carl von Ossietzky Universität Oldenburg, D-26111 Oldenburg, Germany

Thomas D. Brock
Department of Bacteriology, University of Wisconsin, Madison, WI 53706, USA

George H. Brownell
Department of Biochemistry & Molecular Biology, Medical College of Georgia, Augusta, GA 30912-2100, USA

Marvin P. Bryant (Deceased)
Department of Animal Science, University of Illinois, Urbana, IL 61801-3838, USA

Hans-Jürgen Busse
Institut für Bakteriologie, Mykologie und Hygiene, Veterinärmedizinische Universität Wien, Veterinärplatz 1, A-1210 Wien, Austria

Douglas E. Caldwell
Dept. of Applied Microbiology and Food Science, University of Saskatchewan, Saskatoon, 51 Campus Drive, Saskatchewan S7N 5A8 SK, Canada

Daniel N. Cameron
Foodborne & Diarrheal Diseases Lab. Section, Division of Bacterial and Mycotic Diseases, Centers for Disease Control and Prevention, Atlanta, GA 30333, USA

Pierre Caumette
Department Debiologie L.E.M, Universite de Pau, Av de L'Universite BP1155, Pau F-64013, France

Wen Xin Chen
Department of Microbiology, Biology College, Beijing Agricultural University, Beijing, P.R. China

Henrik Christensen
Department of Veterinary Microbiology, Stigbøjlen 4, Frederiksberg C 1870, Denmark

Penelope Christensen
National Institute for Genealogical Studies, Faculty of Information Studies, University of Toronto, Toronto, Ontario, Canada

John D. Coates
Plant and Microbial Biology, University of California, Berkeley, Berkeley, CA 94720-3102. USA

Matthew D. Collins
Department of Food Science and Technology, University of Reading, Earley Gate-White-knights Rd., Reading RG6 6AP, United Kingdom

Michael J. Corbel
National Institute for Biol. Standards & Control, Blanche Lane, South Mimms, Potters Bar, Hertfordshire EN6 3QG, United Kingdom

Heribert Cypionka
Inst. für Biol. und Chemie des Meeres (ICBM), Universitat Oldenburg, Oldenburg, PFS 2503, D-26111, Germany

Milton S. da Costa
Departamento de Zoologia, Centro de Neurociências, Universidade de Coimbra, Apartado 3126, P-3004-517 Coimbra, Portugal

Colin Dale
Botany and Microbiology, Auburn University, Auburn, AL 36849-5407, USA

Subrata K. Das
Institute of Life Sciences, Nalco square, Bhubaneswar 751 023, India

Gregory A. Dasch
Division of Viral and Rickettsial Diseases, Viral and Rickettsial Zoonoses Branch, National Center for Infectious Diseases, Centers for Disease Control and Prevention, Atlanta, GA 30333, USA

Catherine Dauga
Génopole de l'Institut Pasteur, Plateau Technique 4, Bât Le Pasteur, Institut Pasteur, 28 rue du Docteur Roux, 75724 Paris Cedex 15, France

Frank B. Dazzo
Department of Microbiology and Molecular Genetics, Michigan State University. East Lansing, MI 48824-4320, USA

Jody W. Deming
School of Oceanography, University of Washington, Seattle, WA 98195-0001, USA

Ewald B.M. Denner
Abteilung Mikrobiologie und Biotechnologie, Institut für Mikrobiologie und Genetik, Dr. Bohr-Gasse 9, A-1030 Wein, Austria

Richard Devereux
NHEERL, Gulf Ecology Division, U.S.E.P.A., Gulf Breeze, FL 32561, USA

Paul De Vos
Dept. Biochem., Physiology & Micro. (WE 10V), University of Gent, K.L. Ledeganckstraat 35, B-9000 Gent, Belgium

Kim A. DeWeerd
Department of Chemistry, State University of New York, University at Albany, Albany, NY 12222, USA

Floyd E. Dewhirst
Department of Molecular Genetics, The Forsyth Institute, 140 The Fenway, Boston, MA 02115-3799, USA

Johanna Döbereiner (Deceased)
Centro Nacional de Pesquisa de Agrobiologia, Empresa Brasiliera de Pesquisa Agropecuária, Room 247, 23851-970 Seropédica, Caixa Postal 74.505, Rio de Janeiro 465, Brazil

Nina V. Doronina
Inst. of Biochemistry & Physiology of Microorganisms RAS, Laboratory of Methylotrophy, Russian Academy of Sciences, Pushchino-on-the-Oka, Moscow Region 142290, Russia

Michel Drancourt
Faculté de Médecine, Unité des Rickettsies, 27 Boulevard Jean Moulin, 13385 Marseille Cedex 05, France

Galina A. Dubinina
Institute of Microbiology, Russian Academy of Sciences, Prospect 60-let. Oktyabrya 7/2, Moscow, Russia

J. Stephen Dumler
Division of Microbiology, Department of Pathology, The Johns Hopkins Hospital, Univ. School of Medicine, Baltimore, MD 21287-7093, USA

Jürgen Eberspächer
Institut für Mikrobiologie (250), Universität Hohenheim, Garbenstrasse 30, D-70599 Stuttgart, Germany

Thomas W. Egli
Department of Microbiology, EAWAG, Überlandstrasse 133, CH 8600 Düebendorf, Switzerland

Matthias A. Ehrmann
Lehrstuhl für Mikrobiologie, Technische Universität München, Weihenstephan, Freising 85350, Germany

Stefanie J.W.H. Oude Elferink
ID TNO Animal Nutrition, P.O. Box 65, 8200 AB Lelystad, The Netherlands

Takayuki T. Ezaki
Department of Microbiology and Bioinformatics, Regeneration and Advanced Medical Science, Gifu University School of Medicine, 40 Tsukasa-machi, Gifu 500 8705, Japan

J.J. Farmer III
Foodborne & Diarrheal Diseases Lab. Section, Division of Bacterial and Mycotic Diseases, Centers for Disease Control and Prevention, Atlanta, GA 30333, USA

Mark Fegan
Coop. Research Centre for Tropical Plant Protection, Dept. of Micro. & Parasitology, The University of Queensland, St. Lucia, Brisbane, Queensland 4072, Australia

Andreas Fesefeldt
Geibelallee 12a, 24116 Kiel, Germany

Kai W. Finster
Department of Microbial Ecology, Institute of Biological Sciences, University of Aarhus, Building 540, Ny, Munkegade, DK-8000 Åarhus C, Denmark

Carmen Fischer-Romero
Institut für Medizinische Mikrobiologie, Universität Zürich, Gloriastrasse 30/32, CH-8028 Zürich, Switzerland

Geoffrey Foster
Veterinary Division, Scottish Agricultural College, Drummondhill, Stratherrick Road, Inverness IV2 4JZ, United Kingdom

Pierre-Edouard Fournier
Faculté de Médecine, Unité des Rickettsies, 27, Boulevard Jean Moulin, 13385 Marseille Cedex 05, France

James G. Fox
Department of Comparative Medicine, Massachusetts Institute of Technology, Cambridge, MA 02139, USA

Wilhelm Frederiksen
Dept. of Diagnostic Bacteriology and Antibiotics, Statens Seruminstitut, DK-2300 Copenhagen S, Denmark

Michael Friedrich
Abteilung Biogeochemie, Max Planck-Institut für Terrestrische Mikrobiologie, Karl-von-Frisch-Strasse, D-35043 Marburg, Germany

John L. Fryer
Dept. of Microbiology Ctr./Salmon Disease Research, Oregon State University, Corvallis, OR 97331-3804, USA

Georg Fuchs
Mikrobiologie, Institut für Biologie II, Albert-Ludwigs-Universität Freiburg, D-79104 Freiburg, Germany

John A. Fuerst
Center for Bacterial Diversity and Identification, Department of Microbiology, University of Queensland, Brisbane, Queensland 4072, Australia

Tateo Fujii
Department of Food Science and Technology, Tokyo University of Fisheries, 4-5-7 Konan, Minato-ku, Tokyo 108-8477, Japan

Jean-Louis Garcia
Laboratoire de Microbiologie, ORSTOM-ESIL-Case 925, Université de Provence, 163, Avenue de Luminy, 13288 Marseille, Cédex 9, France

Monique Garnier (Deceased)
Institut National de la Recherche Agronomique et Université Victor Ségalen, Laboratoire de Biologie Cellulaire et Moléculaire, Bordeaux 2, 33883, BP 81, Villenave d'Ormon Cedex, France

Margarita Garriga
Centro de Tecnologia de la Carne, Inst. de Recerca i Tecnologia Agroalimentàries, Granja Camps i Armet s/n, 17121 Monells (Girona) España, Spain

George M. Garrity
Dept. of Microbiology and Molecular Genetics, Michigan State University, East Lansing, MI 48824, USA

Rainer Gebers
Depenweg 12, D-24217 Schönberg/Holstein, Germany

Connie J. Gebhart
Division of Comparative Medicine, University of Minnesota Health Center, Minneapolis, MN 55455, USA

Allison D. Geiselbrecht
Floyd Snider McCarthy, Inc, Seattle, WA 98104-2851, USA

Barbara R. Sharak Genthner
Center for Environmental Diagnostics and Bioremediation, University of West Florida, Pensacola, FL 32514, USA

Peter Gerner-Smidt
Department of Gastrointestinal Infections, Statens Serum Institut, Artillerivej 5, DK-2300 Copenhagen S, Denmark

Monique Gillis
Laboratorium voor Microbiologie Vakgroep WE 10V, Universiteit Gent, K.L. Ledeganckstraat 35, B-9000 Gent, Belgium

Christian Gliesche
Institut für Ökologie, Ernst-Moritz-Arndt-Universität, Greifswald Schwedenhagen 6, D-18565 Kloster/Hiddensee, Germany

Frank Oliver Glöckner
Max Planck-Institute for Marine Microbiology, Celsuisstrasse 1, Bremen D-28359, Germany

Peter N. Golyshin
Division of Microbiology, GBF-Natl. Research Centre for Biotechnology, Mascheroder Weg 1, 38124 Braunschweig, Germany

José M. González
Departamento de Microbiologia y Biologia Celular, Facultad de Farmacia, Universidad de La Laguna, 38071 La Laguna. Tenerife, Spain

Yvonne E. Goodman
Department of Medical Bacteriology, University of Alberta, Medical Services Building, Edmonton, Alberta, Canada

Vladimir M. Gorlenko
Institute of Microbiology, Russian Academy of Sciences, Prospect 60-letiya, Oktyabrya 7, korpus 2, Moscow 117312, Russia

Hans-Dieter Görtz
Department of Zoology, Biologisches Institut, Universität Stuttgart, Pfaffenwaldring 57, D-70550 Stuttgart, Germany

John J. Gosink
Amgen, Inc., Seattle, WA 98101, USA

Jennifer Gossling
8401 University Drive, St. Louis, MO 63105-3641, USA

Masao Goto
Plant Pathology Laboratory, Faculty of Agriculture, Shizuoka University, 836 Ohya, Shizuoka 422-8017, Japan

Peter N. Green
National Collection of Industrial & Marine Bacteria, 23 St. Machar Drive, Aberdeen AB24 3RY, United Kingdom

Francine Grimont
Unité des Entérobactéries, Inst. Natl. de la Santé et de la Recherce Médicale, Institut Pasteur, 28 rue du Docteur Roux, Unité 389 75724 Paris Cedex 15, France

Patrick A.D. Grimont
Unité des Entérobactéries, Inst. Natl. de la Santé et de la Recherce Medicale, Institut Pasteur, 28 rue du Docteur Roux, Unité 389, F-75724 Paris Cedex 15, France

Rémi Guyoneaud
Institut d'Ecologica Aquatica-Microbiologia, Campus de Montilivi, E-17071 Girona, Spain

Lotta E-L. Hallbeck
Department of Cell and Molecular Biology, Göteborg University, Medicinaregatan 9 C, Box 462, S-405 30 Göteborg, Sweden

Theo A. Hansen
Department of Microbial Physiology, Groningen Biomolecular Sci. & Biotech. Inst., University of Groningen, P. O. Box 14, 9750 AA Haren, The Netherlands

Shigeaki Harayama
Marine Biotechnology Institute, 3-75-1 Heita, Kamaishi, Ivate 026-001, Japan

Anton Hartmann
Institute of Soil Ecology, Rhizosphere Biology Division, GSF Research Center, PO Box 1129, D-85764 Neuherberg, München, Germany

Fawzy M. Hashem
Sustainable Agriculture Laboratory, Animal and Natural Resources Institute, Beltsville Agricultural Research Institute,USDA-ARS, Beltsville, MD 20705, USA

Lysiane Hauben
Applied Maths BVBA, Keistraat 120, B-9830 Sint Martens-Latem, Belgium

Ian M. Head
Fossil Fuels & Environ. Geochem. Postgraduate Inst. (NRG), University of Newcastle-upon-Tyne, Newcastle-upon-Tyne NE1 7RU, United Kingdom

Brian P. Hedlund
Department of Biological Sciences, University of Nevada, Las Vegas, Las Vegas, NV 89154-4004, USA

Johann Heider
Mikrobiologie, Institut für Biologie II, Universität Freiburg, Schänzlestrasse 1, D-79104 Freiburg, Germany

Robert B. Hespell (Deceased)
Natl. Center of Agricultural Utilization Research, Agricultural Research Service, United States Department of Agriculture, Peoria, IL 61604-3902, USA

Karl-Heinz Hinz
Klinik für Geflügel der Tierärztlichen Hochschule, Bünteweg 17, D-30559 Hannover, Germany

Akira Hiraishi
Department of Ecological Engineering, Toyohashi University of Technology, Tempaku-cho, Toyohashi 441-8580, Japan

Peter Hirsch
Institut für Allgemeine Mikrobiologie der Biozentrum, Universität Kiel, Am Botanischen Garten 1-9, D-24118 Kiel, Germany

Becky Hollen
Department of Biological Sciences, Louisiana State University, Baton Rouge, LA 70803, USA

Barry Holmes
Public Health Laboratory Service, Central Public Health Laboratory, National Collection of Type Cultures, 61 Colindale Avenue, London NW9 5HT, United Kingdom

John Holt
Department of Microbiology and Molecular Genetics, Michigan State University, East Lansing, MI 48824-1101, USA

Marta Hugas
Meat Technology Centre, Inst. for Food & Agricultural Research & Tech., Granja Camps i Armet s/n, 17121 Monells, Spain

Philip Hugenholtz
Ecosystem Sciences Division, Department of Environmental Science, Policy, and Management, University of California, Berkeley, Berkeley, CA 94720-3110, USA

Thomas Hurek
Arbeitsgruppe Symbioseforschung, Planck-Institut für Terrestrische Mikrobiologie, Karl-von-Frisch-Strasse, D-35043 Marburg, Germany

Johannes F. Imhoff
Institut für Meereskunde, Abt. Marine Mikrobiologie, Universität Kiel, Düsternbrooker Weg 20, D-24105 Kiel, Germany

Kjeld Ingvorsen
Department of Microbial Ecology, Institute of Biological Sciences, University of Aarhus, Building 540, Ny Munkegade, DK-8000 Aarhus C, Denmark

Francis L. Jackson
Medical Microbiology and Immunology, University of Alberta, 1-41-Medical Sciences Building, Edmonton, Alberta AB T6G 2H7, Canada

J. Michael Janda
Microbial Diseases Laboratory, Communicable Disease Control, California Department of Health Services, Richmond, CA 94804, USA

Holger W. Jannasch (Deceased)
Department of Biology, Woods Hole Oceanographic Institution, Woods Hole, MA 02543, USA

Cheryl Jenkins
Department of Microbiology, University of Washington, Seattle, WA 98195-0001, USA

Bo Barker Jorgensen
Max Planck-Institute, Celsuisstrasse 1, Bremen 28359, Germany

Samuel W. Joseph
Microbiology Department, University of Maryland, College Park, MD 20742, USA

Karen Junge
School of Oceanography, University of Washington, Seattle, WA 98195-0001, USA

Elliot Juni
Department of Microbiology and Immunology, University of Michigan Medical School, Ann Arbor, MI 48109-0620, USA

Sibylle Kalmbach
Studienstiftung des Deutschen Volkes, Mirbachstrasse 7, D-53173 Bonn, Germany

Peter Kämpfer
Institut für Angewandte Mikrobiologie, Justus-Liebig-Universität Giessen, Heinrich-Buff-Ring 26-32, IFZ, D-35392 Giessen, Germany

Yoshiaki Kawamura
Department of Microbiology, Gifu University School of Medicine, 40 Tsukasa-machi, Gifu 500 8705, Japan

Donovan P. Kelly
Department of Biological Sciences, University of Warwick, Coventry CV4 7AL, United Kingdom

Suzanne V. Kelly
Professor of Biology, Scottsdale Community College, Scottsdale, AZ 85250, USA

Christina Kennedy
Department of Plant Pathology, College of Agriculture, The University of Arizona, Tucson, AZ 85721-0036, USA

Allen Kerr
Waite Agricultural Research Institute, The University of Adelaide, Glen Osmond 5064, South Australia

Karel Kersters
Lab. voor Microbiologie, Vakgroep Biochemie, Fysiologie en Microbiologie, Rijksuniversiteit Gent, K.L. Ledeganckstraat 35, B-9000 Gent, Belgium

Mogens Kilian
Dept. of Medical Microbiology & Immunology, The University of Aarhus, DK-8000 Aarhus C, Denmark

Bon Kimura
Department of Food Science and Technology, Tokyo University of Fisheries, 4-5-7 Konan, Minato-ku, Tokyo 108-8477, Japan

Hans-Peter Klenk
VP Genomics, Epidauros Biotechnology Inc., Am Neuland 1, D-82347 Bernried, Germany

Oliver Klimmek
Biozentrum Niederursel, Institut für Mikrobiologie der Johann Wolfgang Goethe-Universität, Marie-Curie-Strasse 9, D-60439 Frankfurt am Main, Germany

Allan E. Konopka
Department of Biological Science, Purdue University, West Lafayette, IN 47907-2054, USA

Hans-Peter Koops
Abteilung Mikrobiologie, Inst. für Allgemeine Botanik und Botanischer Garten, Universität Hamburg, Ohnhorststrasse 18, D-22609 Hamburg, Germany

Yoshimasa Kosako
The Institute of Physical and Chemical Research, Japan Collection of Microorganisms, RIKEN, Wako-shi, Saitama 351-0198, Japan

Julius P. Kreier
Department of Microbiology, The Ohio State University, Columbus, OH 43201, USA

Noel R. Krieg
Department of Biology, Virginia Polytechnic Institute & State University, Blacksburg, VA 24061-0406, USA

Achim Kröger (Deceased)
Biozentrum Niederursel, Institut für Mikrobiologie der Johann Wolfgang Goethe-Universität, Marie-Curie-Strasse 9, D-60439 Frankfurt am Main, Germany

J. Gijs Kuenen
Faculty of Chemical Tech. & Materials Science, Kluyver Laboratory for Biotechnology, Delft University of Technology, 2628 BC Delft, The Netherlands

Jan Kuever
Department of Microbiology, Institute for Material Testing, Foundation Institute for Materials Science, D-28199 Bremen, Germany

Hiroshi Kuraishi
1-29-10 Kamiikebukuro, Toshima-ku, Tokyo 170-0012, Japan

L. David Kuykendall
Molecular Plant Pathology Laboratory, Plant Sciences Institute, United States Department of Agriculture, Beltsville, MD 20705-2350, USA

David P. Labeda
Natl. Ctr. For Agricultural Utilization Research, Microbial Properties Research, U.S. Department of Agriculture, Peoria, IL 61604-3999, USA

Matthias Labrenz
Institut für Allgemeine Mikrobiologie, Biologiezentrum, University of Kiel, Am Botanischen Garten 1-9, 24118 Kiel, Germany

Catherine N. Lannan
Department of Microbiology, Ctr./Salmon Disease Research, Oregon State University, Corvallis, OR 97331-3804, USA

Bernard La Scola
CNRS UMR6020, Unité des Rickettsies, 27 Boulevard Jean Moulin, 13385 Marseille Cedex 05, France

Adrian Lee
School of Microbiology and Immunology, University of New South Wales, Kensington, Sydney, Australia

Léon E. Le Minor
Entérobactéries, Institut Pasteur, 28 Rue du Docteur Roux, 75724 Paris Cedex 15, France

Werner Liesack
Max Planck-Institut für Terrestrische Mikrobiologie, Karl-von-Frisch-Strasse, D-35043 Marburg, Germany

Timothy Lilburn
ATCC Bioinformatics, Manassas, VA 20110-2209, USA

John A. Lindquist
Department of Bacteriology, University of Wisconsin, Madison, WI 53706, USA

André Lipski
Abteilung Mikrobiologie, Fachbereich Biologie/Chemie, Universität Osnabrück, 49069 Osnabrück, Germany

Niall A. Logan
School of Biological and Biomedical Sciences, Glasgow Caledonian University, Cowcaddens Road, Glasgow G4 0BA, United Kingdom

Derek R. Lovley
Department of Microbiology, University of Massachusetts, Physiology & Ecology of Anaerobic Micro., Amherst, MA 01003, USA

Wolfgang Ludwig
Lehrstuhl für Mikrobiologie, Technische Universität München, Am Hochanger 4, D-85350 Freising, Germany

Melanie L. MacDonald
Guilford College, Greensboro, NC 27410, USA

Barbara J. MacGregor
Max Planck-Institute for Marine Microbiology, Celsiusstrasse 1, D-28359 Bremen, Germany

Michael T. Madigan
Department of Microbiology, Life Science II, Southern Illinois University, Carbondale, IL 62901-6508, USA

Åsa Malmqvist
ANOX AB, Klosterangsvagen 11A, S-226 47 Lund, Sweden

Henry Malnick
Laboratory of Hospital Infection, Central Public Health Laboratory, London NW9 5HT, United Kingdom

Werner Manz
Section G3, Ecotoxicology and Biochemistry, German Federal Institute of Hydrology, Kaiserin-Augusta-Anlagen 15-17, P. O. Box 20 02 53, D-56002 Koblenz, Germany

Amy Martin-Carnahan
Dept. of Epidemiology and Preventive Medicine, University of Maryland School of Medicine, Baltimore, MD 21201, USA

Esperanza Martínez-Romero
Centro de Investigación sobre Fijación de Nitrógeno, UNAM, Ap Postal 565–A, Cuernavaca, Morelos, México

Abdul M. Maszenan
Environmental Engineering Research Centre, School of Civil and Structural Engineering, Nanyang Technological University, Block N1, #1a-29, 50 Nanyang Avenue, Singapore 639798

Ian Maudlin
Sir Alexander Robinson Ctr. for Trop. Vet. Med., Royal Dick School of Vet. Stud., University of Edinburgh, Easter Bush, Roslin, Midlothian EH25 9RG, United Kingdom

Anthony T. Maurelli
Department of Microbiology and Immunology, Uniformed Services Univ. of the Health Sciences, F. Edward Hébert School of Medicine, Bethesda, MD 20814-4799, USA

Michael J. McInerney
Department of Botany and Microbiology, The University of Oklahoma, Norman, OK 73019-6131, USA

Thomas A. McMeekin
Inst. for Antartic and Southern Ocean Studies, University of Tasmania, Antartic CRC, GPO Box 252-80, Hobart, Tasmania 7001, Australia

Steven McOrist
Department of Biomedical Sciences, Tufts University College of Veterinary Medicine, North Grafton, MA 01536, USA

Thoyd T. Melton (Deceased)
North Carolina A&T State University, Greensboro, NC 27411, USA

Roy D. Meredith (Deceased)

Joris Mergaert
Laboratorium voor Microbiologie Vakgroep Biochemie, Fysiologie en Microbiol., Universiteit Gent, K.L. Ledeganckstraat 35, B-9000 Gent, Belgium

Ortwin D. Meyer
Lehrstuhl für Mikrobiologie, Universität Bayreuth, Universitätsstrasse 30, D-95440 Bayreuth, Germany

Henri H. Mollaret
Institut Pasteur, 28 Rue du Docteur Roux, 75724 Paris Cedex 15, France

Kristian Møller
Department of Microbiology, Danish Veterinary Laboratory, Bulousvej 27, DK-1790 Copenhagen V, Denmark

Edward R.B. Moore
Programme of Soil Quality and Protection, The Macaulay Research Institute, Macaulay Dr., Craigiebuckler, AB15 8QH Aberdeen, United Kingdom

Nancy A. Moran
Dept. of Ecology and Evolutionary Biology, University of Arizona, Tucson, AZ 85721-0088, USA

Maurice O. Moss
Department of Microbiology, School of Biological Sciences, University of Surrey, Guildford, Surrey GU2 5XH, United Kingdom

R.G.E. Murray
Department of Microbiology and Immunology, The University of Western Ontario, London, Ontario N6A 5C1, Canada

Reinier Mutters
Institut für Medizinische Mikrobiologie und Krankenhaushygiene, Klinikum der Philipps-Universität Marburg, D-35037 Marburg, Germany

Gerard Muyzer
Kluyver Laboratory for Biotechnology, Department of Microbiology, Delft University of Technology, 2628 BC Delft, The Netherlands

Yasuyoshi Nakagawa
Biological Resource Center (NBRC), Department of Biotechnology, National Institute of Technology and Evaluation, 2-5-8, Kazusakamatari, Kisarazu, Chiba 292-0818, Japan

Hirofumi Nishihara
School of Agriculture, Ibaraki University, 3-21-1 Chu-ou, Ami-machi, Inashiki-gun, Ibaraki 300-0393, Japan

M. Fernanda Nobre
Departmento de Zoologia, Universidade de Coimbra, Apartado 3126, P-3000 Coimbra, Portugal

Caroline M. O'Hara
Diagnostic Microbiology Section, Division of Healthcare Quality Promotion, Centers for Disease Control and Prevention, Atlanta, GA 30333, USA

Tomoyuki Okamoto
Research and Development Center, Kirin Brewery Company, Ltd., 100-1 Hagiwara-machi, Takasaki-shi, Gunma 370-0013, Japan

Frans Ollevier
Laboratorium voor Aquatische Ecologie, Zoological Institute, Ch. de Bériotstraat 32, Leuven B-3000, Belgium

Bernard Ollivier
Laboratoirede Microbiologie—LMI, ORSTOM, Case 925, Université de Provence, ESIL, 163 Avenue de Luminy, Marseille 13288 Cedex 09, France

Ingar Olsen
Det Odontologiske Fakultet, Institutt for oral biologi, Moltke Moesvei 30/32, Universitetet I Oslo, Postboks 1052 Blindern, N-0316 Oslo, Norway

Stephen L.W. On
Danish Veterinary Institute, Bülowsvej 27, DK-1790, Copenhagen V, Denmark

Ronald S. Oremland
Water Research Division, U.S. Geological Survey, Menlo Park, CA 94025-3591, USA

Aharon Oren
Division of Microbial and Molecular Ecology, The Institute of Life Science, and the Moshe Shilo Minerva Center for Marine Biogeochemistry, The Hebrew University of Jerusalem, Givat Ram, Jerusalem 91904, Israel

Jani L. O'Rourke
School of Microbiology and Immunology, University of New South Wales, Kensington, Sydney, Australia

Ro Osawa
Division of Bioscience, Grad. Sch. of Science & Tech., Kobe University, Rokko-dai 1-1, Nada-ku, Kobe City 657-8501, Japan

Dr. Jörg Overmann
Inst. für Chemie & Biologie des Meeres (ICBM), Universität Oldenburg, Postfach 25 03, D026111, Oldenburg, Germany

Norberto J. Palleroni
Rutgers, North Caldwell, NJ 07006-4146, USA

Bruce J. Paster
Department of Molecular Genetics, The Forsyth Institute, 140 The Fenway, Boston, MA 02115-3799, USA

Bharat K.C. Patel
Microbial Discovery Research Unit, School of Biomolecular Sciences, Griffith University, Nathan Campus, Kessels Road, Brisbane, Queensland 4111, Australia

Dominique Patureau
Laboratoire de Biotechnologie de l'Environnement, INRA Narbonne, avenue des étangs, 11 100 Narbonne, France

Karsten Pedersen
Department of Cell and Molecular Biology, Göteborg University, Medicinaregatan 9 C, Box 462, S-405 30 Göteborg, Sweden

John L. Penner
Dept. of Medical Genetics & Microbiology Grad. Dept./Mol. & Med. Genet., University of Toronto, Toronto, Ontario M5S 3E2, Canada

Jeanne S. Poindexter
Department of Biological Sciences, Barnard College, Columbia University, New York, NY 10027-6598, USA

Andreas Pommerening-Röser
Abteilung Mikrobiologiem, Inst. für Allgemeine Botanik und Botanischer Garten, Universität Hamburg, Ohnhorststrasse 18, D-22609 Hamburg, Germany

Michel Y. Popoff
Unite de Génétique des Bactéries Intracellulaires, Institut Pasteur, 28 rue du Docteur Roux, F-75724 Paris Cedex 15, France

Bruno Pot
Science Department, Yakult Belgium, Joseph Wybranlaan 40, B-1070 Brussels, Belgium

Fred A. Rainey
Department of Biological Sciences, Louisiana State University, Baton Rouge, LA 70803, USA

Didier Raoult
Faculté de Médecine, CNRS, Unité des Rickettsies, 27 Boulevard Jean Moulin, 13385 Marseille Cedex 05, France

Christopher Rathgeber
Department of Microbiology, The University of Manitoba, Winnipeg, Manitoba R3T 2N2, Canada

Gavin N. Rees
Murray-Darling Freshwater Research Centre, CRC Freshwater Ecology, Ellis Street, Thurgoona, PO Box 921, Albury NSW 2640, Australia

Hans Reichenbach
Arbeitsgruppe Mikrobielle Sekundärstoffe, Gesellschaft für Biotechnologische Forschung mbH, Mascheroder Weg 1, D-38124 Braunschweig, Germany

Barbara Reinhold-Hurek
Universität Bremen, Fachbereich 2, Allgemeine Mikrobiologie, P. O. Box 330440, D-28334 Bremen, Germany

Anna-Louise Reysenbach
Department of Environmental Biology, Portland State University, Portland, OR 97207, USA

Yasuko Rikihisa
Department of Veterinary Biosciences, The Ohio State University, 1925 Coffey Road, Columbus, OH 43210-1093, USA

Lesley A. Robertson
Kluyver Laboratory for Biotechnology, Delft University of Technology, Julianalaan 67, P. O. Box 5057, 2628BC Delft, The Netherlands

Julian I. Rood
Monash University, Bacterial Pathogenesis Research Group, Department of Microbiology, Clayton 3168, Australia

Ramon A. Rosselló-Mora
Inst. Mediterrani d'Estudis Avançats (CSIC-UIB), C/Miquel Marque's 21, E-07290 Esporles, Mallorca, Spain

Paul Rudnick
Maryland Technology Development Center, SAIC, Rockville, MD 20850, USA

Gerard S. Saddler
Scottish Agricultural Science Agency, 82 Craigs Road, East Craigs, Edinburgh EH12 8NJ, United Kingdom

Takeshi Sakane
Institute for Fermentation, Osaka, Yodogawa-ku, Osaka 532-8686, Japan

Riichi Sakazaki (Deceased)
Nippon Institute of Biological Sciences, 9-2221-1 Sinmachi, Oume, Tokyo 198-0024, Japan

Abigail A. Salyers
Department of Microbiology, University of Illinois, Urbana-Champaign, Urbana, IL 61801-3704, USA

Antonio Sanchez-Amat
Faculty of Biology, Department of Genetics and Microbiology, University of Murcia, Murcia 30100, Spain

Gary N. Sanden
Epidemic Investigations Laboratory, Meningitis and Special Pathogens Branch, Division of Bacterial and Mycotic Diseases, Centers for Disease Control and Prevention, Atlanta, GA 30333, USA

Masataka Satomi
National Research Institute of Fisheries Science, 2-12-4 Fukuura, Knazawa-ku, Yokohama, Kanagawa 236-8648, Japan

Hiroyuki Sawada
National Institute of Agro-Environmental Sciences, 3-1-1 Kannondai, Tsukuba, Ibaraki 305-8604, Japan

Flemming Scheutz
WHO, The Int. *Escherichia* & *Klebsiella* Centre, Statens Serumininstitut, Artillerivej 5, DK-2300 Copenhagen S, Denmark

Jiri Schindler, Sr.
Clinical Microbiology Group and Natl. Ctr. of Surveillance of Antibiotic Resistance, National Institute of Public Health, Prague 10 10042, Czech Republic

Bernhard H. Schink
Fakultät für Biologie, Lehrstuhl für Mikrobielle Ökologie, Universität Konstanz, Postfach 55 60, D-78457 Konstanz, Germany

Karl-Heinz Schleifer
Lehrstuhl für Mikrobiologie, Technische Universität München, Am Hochanger 4, D-85350 Freising, Germany

Heinz Schlesner
Institut für Allgemeine Mikrobiologie, Universität Kiel, Am Botanischen Garten 1-9, Biologiezentrum, D-24118 Kiel, Germany

Helmut J. Schmidt
Biological Faculty, University of Kaiserslautern, Building 14, Pf 3049, D-67653 Kaiserslautern, Germany

Jean M. Schmidt
Department of Microbiology, Arizona State University, Tempe, AZ 85287-2701, USA

Dirk Schüler
Max Planck-Institute for Marine Microbiology, Celsiusstrasse 1, D-28359 Bremen, Germany

Heide N. Schulz
Section of Microbiology, University of California, Davis, Davis, CA 95616, USA

Paul Segers
Lab. voor Microbiologie Vakgroep WE 10V, Universiteit Gent, K.L. Ledeganckstraat 35, B-9000 Gent, Belgium

Robert J. Seviour
Biotechnology Research Centre, La Trobe University, P.O. Box 199, Bendigo VIC 3550, Australia

Richard Sharp
School of Applied Sciences, South Bank University, 103 Borough Road, London SE1 0AA, United Kingdom

Tsuneo Shiba
Shimonoseki University of Fisheries, Dept. of Food Science and Technology, Yoshimi-Nagatahoncho Shimonose, Yamaguchi 759-65, Japan

Martin Sievers
University of Applied Sciences, Department of Biotechnology, Molecular Biology, CH 8820 Wädenswil, Switzerland

Anders B. Sjöstedt
Department of Microbiology, National Defense Research Establishment, Cementvagen 20, S-901 82 Umeå, Sweden

Lindsay I. Sly
Centre for Bacterial Diversity and Identification, Department of Microbiology and Parasitology, University of Queensland, St. Lucia, Brisbane, Queensland 4072, Australia

Peter H.A. Sneath
Department of Microbiology and Immunology, School of Medicine, University of Leicester, P.O. Box 138, Leicester LE1 9HN, United Kingdom

Martin Sobieraj
Department of Environmental Biology, Portland State University, P. O. Box 751, Portland, OR 97207-0751, USA

Francisco Solano
Dept. of Biochemistry and Molecular Biology B, School of Medicine, University of Murcia, Murcia 30100, Spain

Dimitry Y. Sorokin
S.N. Winogradsky Inst. of Microbiology, Russian Academy of Sciences, Prospect 60-let. Oktyabrya, 7/2, 117312, Moscow, Russia and Department of Environmental Biotechnology, Delft University of Technology, Julianalaan 67, 2628 BC, Delft, The Netherlands

Eva Spieck
Inst. für Allgemeine Botanik und Botanischer Garten, Universität Hamburg, Ohnhorststrasse 18, D-22609 Hamburg, Germany

Georg A. Sprenger
Forschungszentrum Jülich GmbH, Institut für Biotechnologie 1, P. O. Box 1913, D-52425 Jülich, Germany

Stefan Spring
DSM-Deutsche Sammlung von Mikrooorganismen und Zellkulturen, GmbH, D-38124 Braunschweig, Germany

Erko S. Stackebrandt
Deutsche Sammlung von Mikroorganismen und Zellkulturen, GmbH, and GBF, Forschung GmbH2, Mascheroder Weg 1b, D-38124 Braunschweig, Germany

David A. Stahl
Civil and Environmental Engineering, University of Washington, Seattle, WA 98195-2700, USA

James T. Staley
Department of Microbiology, University of Washington, Seattle, WA 98195-0001, USA

Alfons J.M. Stams
Department of Microbiology, Wageningen Agricultural University, Hesselink Van Suchtelenweg 4, NL-6703 CT Wageningen, The Netherlands

Patricia M. Stanley
Minntech Corporation, North, Minneapolis, MN 55447-4822, USA

David J. Stewart
CSIRO, Australian Animal Health Laboratory, Private Bag 24, 5 Portarlington Road, Geelong Victoria 3220, Australia

John F. Stolz
Department of Biological Sciences, Duquesne University, Pittsburgh, PA 15282-2504, USA

Adriaan H. Stouthamer
Dept. of Molecular Cell Physiology/Molecular Microbial Ecology, Vrije Universiteit, De Boelelaan 1087, NL-1081 HV Amsterdam, The Netherlands

Nancy A. Strockbine
Foodborne and Diarrheal Diseases Branch, Division of Bacterial and Mycotic Diseases, Centers for Disease Control and Prevention, Atlanta, GA 30333, USA

William R. Strohl
Merck Research Laboratories, West Point, PA 19486, USA

Joseph M. Suflita
Environmental and General Applied Microbiology, Department of Botany & Micro., The University of Oklahoma, Norman, OK 73019-0245, USA

Jörg Süling
Institut für Meereskunde, Abt. Marine Mikrobiologie, Universität Kiel, Düsternbrooker Weg 20, D-24105 Kiel, Germany

Jean Swings
Laboratorium voor Microbiologie Vakgroep WE10V, Fysiologie en Microbiologie, Universiteit of Gent, K.L. Ledeganckstraat 35, B-9000 Gent, Belgium

Ulrich Szewzyk
Department of Microbial Ecology, Technical University Berlin, Franklinstrasse 29, Secr. OE 5, D-10587 Berlin, Germany

Zhiyuan Tan
Department of Microbiology and Molecular Genetics, College of Agronomy, South China Agricultural University, 510642, China

Ralph S. Tanner
Department of Botany and Microbiology, University of Oklahoma, Norman, OK 73019-6131, USA

Anders Ternström
ANOX AB, Klosterangsvagen 11A, S-226 47 Lund, Sweden

Andreas Teske
Department of Biology, Woods Hole Oceanographic Institution, Woods Hole, MA 02543, USA

An Thyssen
GCPCP, Johnson & Johnson Pharm. Res. & Develop., Turnhoutseweg 30, B-4320 Beerse, Belgium

Kenneth N. Timmis
National Research Centre for Biotechnology, Division of Microbiology, Gesellschaft/Biotechnologische Forschung mbH, Mascheroder Weg 1b, D-38124 Braunschweig, Germany

Brian J. Tindall
Deutsche Sammlung von Mikroorganismen und Zellkulturen, GmbH, Mascheroder Weg 1b, D-38124 Braunschweig, Germany

Tone Tønjum
Institute of Microbiology, Section of Molecular Microbiology A3, Rikshospitalet (National Hospital), Pilestredet 32, N-0027 Olso, Norway

G. Todd Townsend
University of Oklahoma, Norman, OK 73072, USA

Yuri A. Trotsenko
Institute of Biochemistry and Physiology of Microorganisms RAS, Laboratory of Methylotrophy, Prospekt Nauki, 5, Moscow Region 142290, Russia

Hans G. Trüper
Institut für Mikrobiologie und Biotechnologie, Universität Bonn, Mechenheimer Allee 168, W-53115 Bonn, Germany

John J. Tudor
Department of Biology, St. Joseph's University Philadelphia, PA 19131-1308, USA

Richard F. Unz
Department of Civil Engineering, The Pennsylvania State University, University Park, PA 16802-1408, USA

Teizi Urakami
Biochemicals Development Div., Mitsubishi Building, Mitsubishi Gas Chemical Company, 5-2, Marunouchi 2-chome, Chiyoda-ku, Tokyo 100-8324, Japan

Marc Vancanneyt
Laboratorium voor Microbiologie, Universiteit Gent, K.L. Ledeganckstraat 35, B-9000 Gent, Belgium

Peter Vandamme
Lab. voor Microbiologie en Microbiele Genetica, Univeristeit of Gent, Faculteit Wetenschappen, K.L. Ledeganckstraat 35, B-9000 Gent, Belgium

Bernard A.M. van der Zeijst
National Institute of Public Health and Environ., Antonie van Leeuwenhoeklaan 9, P.O. Box 1, P.O. Box 80.165, 3720 BA Bilthoven, The Netherlands

Frederique Van Gijsegem
Laboratorium Moleculaire Genetica, Universiteit Gent, K.L. Ledeganckstraat 35, B-9000 Gent, Belgium

Rob J.M. van Spanning
Department of Molecular Cell Physiology/Molecular Microbial Ecology, Vrije Universiteit, De Boelelaan 1087, NL-1081 HV Amsterdam, The Netherlands

Henk W. van Verseveld
Dept. of Molecular Cell Physiology, Molecular and Microbial Ecology, Vrije Universiteit, De Boelelaan 1087, NL-1081 HV Amsterdam, The Netherlands

Leana V. Vasilyeva
Institute of Microbiology RAN, 117811, Russian Academy of Sciences, 60-let. Oktyabrya 7 build. 2, Moscow, Russia

Jill A. Vaughan
CSIRO, Australian Animal Health Laboratory, Private Bag 24, 5 Portarlington Road, Geelong Victoria 3220, Australia

Antonio Ventosa
Departamento de Microbiologia y Parasitologia, Facultad de Farmacia, Universidad de Sevilla, Apdo. 874, 41080 Sevilla, Spain

Rudi F. Vogel
Lehrstuhl für Mikrobiologie, Technische Universität München, Freising-Wihen 85350, Germany

Russell H. Vreeland
Department of Biology, West Chester University, West Chester, PA 19383, USA

David H. Walker
Department of Pathology, University of Texas Medical Branch, 301 University Boulevard, Galveston, TX 77555-0609, USA

En Tao Wang
Departamento de Microbiologia, Escuela Nacional de Ciencias Biológicas, Instituto Politécnico Nacional, Carpio y Plan de Ayala S/N, México D.F. 11340, México

Naomi L. Ward
The Institute for Genomic Research, Rockville, MD 20850, USA

Richard I. Webb
Department of Microbiology, University of Queensland, Brisbane, Queensland 4072, Australia

Ronald M. Weiner
Cell Biology Cluster, Division of Molecular and Cellular Biosciences, National Science Foundation, Arlington, VA 22230, USA

Susan C. Welburn
Sir Alexander Robinson Ctr. for Trop. Vet. Med., Royal Dick School of Vet. Stud., University of Edinburgh, Easter Bush, Roslin, Midlothian EH25 9RG, United Kingdom

David F. Welch
Laboratory Corporation of America, Dallas, Texas 75230, USA

Aimin Wen
Food Science and Technology Program, Pacific Agri-Food Research Centre, Summerland BC V0H 1Z0, Canada

John H. Werren
Department of Biology, University of Rochester, Rochester, NY 14627-0211, USA

Hannah M. Wexler
Department of Veterans Affairs, West Los Angeles Medical Ctr., UCLA School of Medicine, 11301 Wilshire Boulevard, Los Angeles, CA 90073, USA

Robbin S. Weyant
Meningitis & Special Pathogens Branch, Centers for Disease Control and Prevention, Atlanta, GA 30333, USA

Anne M. Whitney
Meningitis & Special Pathogens Branch Lab. Section, MS D-11, Centers for Disease Control & Prevention, Atlanta, GA 30303, USA

Friedrich W. Widdel
Abteilung Mikrobiologie, Max Planck-Institut für Marine Mikrobiologie, Celsiusstrasse 1, D-28359 Bremen, Germany

Jürgen K.W. Wiegel
Department of Microbiology, University of Georgia, Athens, GA 30602-2605, USA

Anne Willems
Laboratorium voor Microbiologie, Universiteit Gent, K.L. Ledeganckstraat 35, B-9000 Gent, Belgium

Henry N. Williams
Department of OCBS, Dental School, University of Maryland at Baltimore, Baltimore, MD 21201-1510, USA

Washington C. Winn, Jr.
Microbiology Laboratory, Medical Center Hospital of Vermont DVE, Fletcher Allen Health Care, UHC Campus, Burlington, VT 05401-3456, USA

Ann P. Wood
Microbiology Research Group, King's College, London Div. of Life Sciences, Franklin-Wilkins Building, 150 Stamford Street, London SE1 8WA, United Kingdom

Eiko Yabuuchi
Aichi Medical University, Omiya 4-19-18, Asahi-ku, Osaka 535-0002, Japan

Michail M. Yakimov
Istituto Sperimentale Talassografico-CNR, Spianata S. Raineri, 86, 98122 Messina, Italy

Kazuhide Yamasato
Department of Fermentation Science, Faculty of Applied Bioscience, Tokyo University of Agriculture, Sakuragaoka, Setagaya-ku, Tokyo 158-0852, Japan

Akira Yokota
Institute of Molecular and Cellular Biosciences, The University of Tokyo, Yayoi 1-1-1, Bunkyo-ku, Tokyo 113-0032, Japan

John M. Young
Mt. Albert Research Centre, Landcare Research New Zealand Ltd., Private Bage 92 170, Auckland, New Zealand

Xue-jie Yu
Department of Pathology, University of Texas Medical Branch, 301 University Boulevard, Galveston, TX 77555-0609, USA

Vladimir V. Yurkov
Department of Microbiology, The University of Manitoba, Winnipeg, Manitoba R3T 2N2, Canada

George A. Zavarzin
Institute of Microbiology, Russian Academy of Sciences, Building 2, Prospect 60-let. Oktyabrya 7a, Moscow 117312, Russia

Tatjana N. Zhilina
Institute of Microbiology, Russian Academy of Sciences, Prospect 60-let. Oktyabrya 7a, Moscow 117312, Russia

Stephen H. Zinder
Department of Microbiology, Cornell University, Ithaca, NY 14853-0001, USA

The History of *Bergey's Manual*

R.G.E. Murray and John G. Holt

Introduction

Bergey's Manual of Determinative Bacteriology has been the major provider of an outline of bacterial systematics since it was initiated in 1923 and has provided a resource ever since to workers at the bench who need to identify bacterial isolates and recognize new species. It originated in the Society of American Bacteriologists (SAB) but it has since become a truly international enterprise directed by an independent Trust which was founded in 1936. It has gone through nine editions and has generated, as a more comprehensive resource, a unique compendium on bacterial systematics, *Bergey's Manual of Systematic Bacteriology* (Holt et al., 1984–1989), which now enters its second edition.

A number of dedicated bacteriologists (Table 1) have formed, guided the development of, and edited, each edition of *Bergey's Manual*. Many of these individuals have been well known for activity in their national societies and devotion to encouraging worldwide cooperation in bacteriology and particularly bacterial taxonomy. Some of them worked tirelessly on the international stage towards an effective consensus in taxonomy and common approaches to classification. This led to the formation in 1930 of an International Association of Microbiological Societies (IAMS) holding regular Congresses. The regulation of bacterial taxonomy became possible within IAMS through an International Committee on Systematic Bacteriology (ICSB), thus recognizing the need for international discussions of the problems involved in bacterial systematics. Eventually, the need for a Code of Nomenclature of Bacteria was recognized and was published in 1948 (Buchanan et al., 1948), and a Judicial Commission (JC) was formed by ICSB to adjudicate conflicts with the Rules. Despite these efforts, an enormous number of synonyms and illegitimate names had accumulated by the 1970s and were an evident and major problem for the Editor/Trustees of *Bergey's Manual* and for all bacteriologists (Buchanan et al., 1966; Gibbons et al., 1981). A mechanism for recognizing useful, and abandoning useless, names was accomplished by the ICSB and the JC largely due to the insistent arguments of V.B.D. Skerman. Lists were made based on the names included in the Eighth edition of *Bergey's Manual of Determinative Bacteriology* (Buchanan and Gibbons, 1974), because they had been selected by expert committees and individual author/experts, together with the recommendations of sub-committees of ICSB. The results were (1) the published Approved Lists of Bacterial Names (Skerman et al., 1980); (2) a new starting date for bacterial names of January 1, 1980 to replace those of May 1, 1753; (3) freeing of names not on the Approved Lists for use in the future; and (4) definition in the Bacteriological Code (1976 revision; Lapage et al., 1975) of the valid and invalid publication of names. It is now evident that the care and thought of contributors to *Bergey's Manual* over the years played a major part in stimulating an orderly nomenclature for taxonomic purposes, in the development of a useful classification of bacteria often used as a basal reference, and in providing a continuing compendium of descriptions of known bacteria.

The *Manual* started as a somewhat idiosyncratic assembly of species and their descriptions following the interests and prejudices of the editor/authors of the early editions. Following the formation of the Bergey's Manual Trust in 1936 and the international discussions of the ICSB at Microbiological Congresses, the new editions became more and more the result of a consensus developed by advisory committees and specialist authors for each part or chapter of the volumes. This did not happen all at once; it developed out of practice and trials, and it is still developing as the basic sciences affecting taxonomy bring in new knowledge and new understanding of taxa and their relationships.

Antecedents of *Bergey's Manual*

Classification of named species of bacteria did not arise quickly or easily (Buchanan, 1948). The Linnaean approach to naming life forms was adopted in the earliest of systems, such as Müller's use of *Vibrio* and *Monas* (Müller, 1773, 1786), for genera of what we would now consider bacteria. There were few observations, and there was insufficient discrimination in the characters available during most of the nineteenth century to allow any system, even the influential attempts by Ehrenberg (1838) and Cohn (1872, 1875), to provide more than a few names that still survive (e.g. *Spirillum*, *Spirochaeta*, and *Bacillus*). Most descriptions could rest only on shape, behavior, and habitat since microscopy was the major tool.

Müller's work was the beginning of the descriptive phase of bacteriology, which is still going on today because we now realize that the majority of bacteria in nature have not been grown or characterized. Early observations such as Müller's were made by cryptogamic botanists studying natural habitats, usually aquatic, and who usually gave Linnaean binomials to the objects they described microscopically. The mycologist H.F. Link (1809) described the first bacterium that we still recognize today, which he named *Polyangium vitellinum* and is now placed with the fruiting myxobacteria. Bizio (1823) attempted to explain the occurrence of red pigment formation on starchy foods such as polenta as the result of microbial growth and named the organism he found there *Serratia marcescens*, a name now associated with the prodigiosin-producing Gram-negative rod. Perhaps one of the most significant observers of infusoria in the early nineteenth

TABLE 1. Members of the Board of Trustees

David H. Bergey	1923–1937
David R. Boone	1994–
Robert S. Breed	1923–1957 (Chairman 1937–1956)
Don J. Brenner	1979–2001
Marvin P. Bryant	1975–1986
R.E. Buchanan	1951–1973 (Chairman 1957–1973)
Richard W. Castenholz	1991–2001
Harold J. Conn	1948–1965
Samuel T. Cowan	late 1950s–1974
Paul De Vos	2002–
Geoffrey Edsall	late 1950s–1965
George M. Garrity	1997–
Norman E. Gibbons	1965–1976
Michael Goodfellow	1999–
Bernard W. Hammer	1923–1934
Francis C. Harrison	1923–1934
A. Parker Hitchens	1939–1950
John G. Holt	1973–2000
Frank M. Huntoon	1923–1934
Noel R. Krieg	1976–1991, 1996–2002
Stephen P. Lapage	1975–1978
Hans Lautrop	1974–1979
John Liston	1965–1976 (Chairman 1973–1976)
A.G. Lochhead	late 1950s–1960
James W. Moulder	1980–1989
E.G.D. Murray	1934–1964
R.G.E. Murray	1964–1990 (Chairman 1976–1990)
Charles F. Niven, Jr.	Late 1950s–1975
Norbert Pfennig	1978–1991
Arnold W. Ravin	1962–1980
Fred A. Rainey	1999–
Karl-Heinz Schleifer	1989–
Nathan R. Smith	1950–1964
Peter H.A. Sneath	1978–1994 (Chairman 1990–1994)
James T. Staley	1976– (Chairman 2000–)
Roger Y. Stanier	1965–1975
Joseph G. Tully	1991–1996
Jan Ursing	1991–1997
Stanley T. Williams	1989–2000 (Chairman 1994–2000)

century was C.G. Ehrenberg, who described many genera of algae and protozoa and, coincidentally, some bacteria (Ehrenberg, 1838). He named genera such as *Spirochaeta* and *Spirillum*, still recognized today, and *Bacterium*, which became a catch-all for rod-shaped cells, and was made *nomen rejiciendum* in 1947.

Logical classifications were attempted throughout the nineteenth century and that of Ferdinand Cohn (1872, 1875), with his attempts to classify the known bacteria, was most influential. In his 1872 paper Cohn recognized six genera of bacteria (*Micrococcus, Bacterium, Bacillus, Vibrio, Spirillum,* and *Spirochaeta*) and later (1875) expanded the classification to include the cyanobacteria while adding more bacterial genera (*Sarcina, Ascococcus, Leptothrix, Beggiatoa, Cladothrix, Crenothrix, Streptococcus* [not those recognized today], and *Streptothrix*). Buchanan (1925) suggested that Cohn's 1875 classification could be the starting date for bacterial nomenclature instead of Linnaeus' *Species Plantarum* of 1753 and discussed various ideas for the proper starting date for bacterial nomenclature, anticipating by a quarter of a century the actual change in starting date proposed in the revised Bacteriological Code (Lapage et al., 1975). The realization that cultivation was possible, and the development of pure culture techniques, extended enormously the capability to recognize and describe species by adding their growth characteristics and effects on growth media. The vague possibilities of pleomorphism gave way to a concept of fixity of species. All this was aided by the human preoccupation with health, the seriousness of infectious diseases, and the growing awareness of the association of particular kinds of bacteria with particular diseases. The result was a rapid increase in the number of taxonomic descriptions and the recognition that similar but not identical species of bacteria were to be found both associated with higher life forms and more generally distributed in nature.

Between 1885 and 1910 there were repeated attempts at classification and arrangements based on perceived similarities, mostly morphological. There were genuine attempts to bring order out of chaos, and a preliminary publication often stimulated subsequent and repeated additions and revisions, but all these authors neglected the determinative requirements of bacteriology. Some notable examples were Zopf (1885), Flügge (1886), Schroeter (1886), and Trevisan (1887, 1889). Migula produced his first outline in 1890 and new versions in 1894, 1895, 1897, and 1900; others followed, notably Fischer (1895), and importantly, because of a degree of nomenclatural regularity, Lehmann and Neumann published their atlas in 1896. The latter was probably the most successful of the systems and was used in successive editions until 1930, especially in Europe. All these were important in their time. However, a major influence in the subsequent development of *Bergey's Manual* in the environment of the Society of American Bacteriologists (SAB) was the work of F.D. Chester, who produced reports in 1897 and 1898 of bacteria of interest in agriculture, to be followed in 1901 by his *Manual of Determinative Bacteriology*. Chester had recognized that the lack of an organized assembly of descriptions and a scheme of classification made the identification of isolates as known species and the recognition of new species an insurmountable task. Another classification provided by Orla-Jensen (1909, 1919) was influential because it represented an interpretation of "natural relationships", reflecting a more physiological approach to description based on his own studies of the lactic acid bacteria encountered in dairy bacteriology. He delimited genera and species on the basis of characteristics such as metabolic byproducts, fermentation of various sugars, and temperature ranges for growth, in addition to morphology. Most classifications to that time reflected the idiosyncrasies of the authors and their areas of experience. What was yet to come was the ordering of assemblies of all known bacteria, arranged with properties documented to facilitate determination and presenting continuing trials of hierarchical arrangements; it was in that format that *Bergey's Manual* started.

Steps Leading to the First Edition of the *Manual*

Bergey's Manual of Determinative Bacteriology arose from the interest and efforts of a group of colleagues in the Society of American Bacteriologists, who were fully aware of previous attempts to systematize the information available on bacterial species and who recognized that the determination of bacterial identity was difficult and required extensive experience. A committee was formed with C.-E.A. Winslow as chairman and J. Broadhurst, R.E. Buchanan, C. Krumweide Jr., L.A. Rogers, and G.H. Smith as members. Their discussions at the meetings of the SAB and their reports, which were published in the Journal of Bacteriology (Winslow et al., 1917, 1920), were signposts for future efforts in systematics. There were two "starters" for a *Manual*: R.E. Buchanan (Fig. 1a), a rising star in the bacteriological firmament, and President of the SAB in 1918, working at Iowa State College, and D.H. Bergey (Fig. 1b), a senior and respected bacteriologist and President of the SAB for 1915, working at the University of Pennsylvania.

Between 1916 and 1918 Buchanan wrote ten papers entitled "Studies on the nomenclature and classification of the bacteria"

FIGURE 1. *A*, Robert Earle Buchanan, 1883–1973; *B*, David Henricks Bergey, 1860–1937; *C*, Robert Stanley Breed, 1877–1956; *D*, Everitt G.D. Murray, 1890–1964. (Fig. 1C courtesy of American Society for Microbiology Archives Collection.)

(Buchanan, 1916; 1917a, b, c; 1918a, b, c, d, e, f) which provided substance for the Winslow Committee (Buchanan was a member), and was intended to be the basis of a systematic treatise. These papers were revolutionary, in the sense that they included all the bacteria (except the cyanobacteria) that were described at that time. Buchanan included, and named the higher groupings of, bacteria such as the actinomycetes, myxobacteria, phototrophs, and chemolithotrophs, along with the other bacteria included in the classifications of the day. This classification had a logical and aesthetic appeal that helped launch the systematic efforts that followed. No doubt Buchanan was driven by dissatisfaction with sloppy and confusing nomenclature as well as inadequate descriptions of "accepted" bacteria (indeed, much of his later work on *Bergey's Manual* and the *Index Bergeyana* reflected his preoccupation with names and illegitimacy, and had much to do with getting a bacteriological code of nomenclature started.) He must have known of Bergey's book and, perhaps because of increasing academic responsibilities, publication of his concepts in his *General Systematic Bacteriology* was delayed until 1925 (Buchanan, 1925). The book did not try to duplicate *Bergey's Manual*, but rather presented a history of bacterial classification and nomenclature, followed by a discussion of the history of all the bacterial genera and higher ranks, listed alphabetically.

R.E. Buchanan was the key player in the renewal of concern for a sensible (not necessarily "natural") classification of bacteria, with a well-regulated nomenclature, working continuously and firmly to those ends from 1916 to the end of his life. He was a man of his times developing his own priorities and prejudices, yet he recognized in the end that new science was needed for a significant phylogeny to develop. Furthermore, he was more influential in gaining support for the initiation and progress of the first few editions of *Bergey's Manual* under the slightly reluctant aegis of the SAB than is obvious in the *Manual's* pages and prefaces. He also played a dominant role in international efforts (representing the SAB) concerning the regulation and codification of classification and nomenclature. As a member of the "Winslow Committee" of the SAB directed to report on the classification of bacteria, he furnished much of the basis for discussion through his series of papers in the Journal of Bacteriology. He provided voluminous detailed suggestions for the revision of Dr. Winslow's drafts for their reports to the SAB (1917 and 1920). He was also in a powerful position to influence decisions, being elected President of SAB for 1918–1919 when critical discussions were taking place.

The Winslow Committee was engaged in protecting ("conserving") the generic names for well-established species by listing them as *genera conservanda*, together with type species for discussion at the 1918 SAB meeting. The intention was to provide a basis for recommendations for formal action at the next International Botanical Congress, since they were working under the general rules of the Botanical Code. They went further by classifying the genera within higher taxa and providing a key to assist recognition. They intended seeking formal approval of the whole report by the SAB. At this stage, R.S. Breed (Fig. 1c) wrote many letters of objection to having any society ratify the concepts involved in contriving a classification, because it would suggest that it is "official", and he attempted unsuccessfully to gain a postponement of the report's presentation. This polemical correspondence with Committee members, including Buchanan, ended in Breed's withdrawing his name from the report despite his evident interest in a workable classification and a more stable nomenclature. Winslow read the report to the SAB meeting on December 29, 1919. Although it emphasized that its listings were not to be considered as a standard or official classification, it did ask "that the names be accepted as definite and approved genera". The report was then published in the *Journal of Bacteriology*. The Committee was discharged and a new Committee on Taxonomy was appointed with R.E. Buchanan as Chairman. In 1920 Breed was added as a member of the new committee, with the responsibility of making the representations at the Botanical Congress because of his membership on the Botanical Code Revision Committee.

It was at this time and in this climate of opinion that Dr. David Bergey decided to put his own studies of bacteria together with the current views on their classification. To do this required more than one person and he assembled a like-minded group to form a Committee of the SAB for the production of a *Manual of Determinative Bacteriology* (F.C. Harrison, R.S. Breed, B.W. Hammer, and F.M. Huntoon). There is no direct evidence that Buchanan was ever asked to participate or, equally, that he raised any formal objections; it seems more likely that there could have been none of the formal encouragement to go ahead evident in 1921 and 1922 without his support. Indeed he seems to have thought it a good enterprise (Preface in his 1925 book). However, he did find it difficult to work with Breed (letter of January 8, 1951 to J.R. Porter) and in expressing this stated "I have ... always refused to become a member of the Editorial Board of the *Manual*". One wonders if his experiences with Breed between 1918 and 1951 ("Scarcely a month passes in which we do not have some disagreement ... but he has a good many excellent qualities") had kept him at arm's length but not out of touch with what was going on with the *Manual*.

The Winslow Committee had put before the SAB the possibility of a major compilation on bacterial systematics. No doubt Buchanan was in a position, as a Past President, to reinforce the value of that project in principle and David Bergey, likewise a Past President, must have been aware of all the discussions. At the time of the last report (Winslow et al., 1920) Bergey must have started on his book, because R.S. Breed reported to the 1922 SAB Council meeting that the work was approaching completion. A more formal proposal was made to the same Council meeting that Bergey's book be published under the aegis of the Society. The SAB agreed to this with the proviso that it go to a substantial publishing house and, following a discussion of the disposition of royalties, *Bergey's Manual of Determinative Bacteriology* was published in 1923 by the Williams & Wilkins Co., Baltimore (Bergey et al., 1923). It was a group effort from the start, with the authors listed as D.H. Bergey, F.C. Harrison, R.S. Breed, B.W. Hammer, and F.M. Huntoon, and there was an acknowledgment of the assistance of six other colleagues on special groups.

One can imagine that Buchanan was upset by this turn of events, for which the only evidence is his sending Bergey a long list of errors he found in the published book (personal communication). However, he was quite generous in his preface to his 1925 book, with his assessment of *Bergey's Manual* as a step towards reducing chaos and confusion in the classification, phylogeny, and naming of bacteria. He writes: "The most hopeful sign of importance in this respect probably has been the work of the committee on taxonomy of bacteria of the Society of American Bacteriologists under the chairmanship of Dr. Winslow and of the more recent work of a committee on classification of bacteria under the chairmanship of Dr. Bergey.... It is to be expected that, as a result of their work, eventually a practical system of nomenclature which will be satisfactory and applicable to all fields of bacteriology will be evolved" (Buchanan, 1925). Furthermore, he emphasized the differences between practical (medical) and academic attitudes towards individual species and the requirements of a classification. He was then, as later, concerned that bacterial nomenclature was not regulated by an appropriate Code. He writes: "It seems to be self-evident that until the bacteriologists can agree upon a code and follow it consistently, there is little hope or remedy for our present chaos". So it is not surprising that he contributed a section to the Fourth Edition (Bergey et al., 1934) discussing the International Botanical Code as a basis for a bacteriological code with modifications to make it more appropriate.

The committee that organized the First Edition stated that they did not regard their classification of species "as in any case final, but merely a progress report leading to more satisfactory classifications in the future". Clearly there was some feeling in the UK and Europe that this classification was an imposition on the part of the SAB*. As a counter, the Third Edition (Bergey et al., 1930) included a box opposite the title page which declares that it is *"Published at the direction of the Society"* which "disclaims any responsibility for the system of classification followed"; and states further that it "has not been formally approved by the Society and is in no sense official or standard" (italics are in the original). This shows that there had been, as indicated by the article by I.C. Hall in 1927 (Hall, 1927), some degree of contention among members of the SAB with the decisions of the Committee.

Hall's objections to the presentations of the Committee of the SAB on characterization and classification of bacterial types starts with the final report (Winslow et al., 1920) being "presented only to a small minority of the members of the Society who happened to return from lunch in time to attend a business session of the twenty-first annual meeting, which was held in Boston more than four months before the publication of the report". He regrets lack of opportunity for scientific consideration and "practically no discussion because only a few knew what was coming". He evidently objected to physiological criteria and believed that morphology should define genera, families, and orders; furthermore he disputed the validity of habitat and believed that serological characterization was futile. He was prepared to use cultural and physiological properties as criteria for species. He sought "unambiguous criteria". He quotes others who disagreed with the *Bergey's Manual* approach including W.W.C. Topley, who also expressed his distaste in his famous textbook ("Topley and Wilson") that was published in 1929.

Bergey's Manual was launched and successful enough for the publisher to encourage further editions with corrections and additions in 1925 and 1930, for which Bergey had the support of the same four co-authors. There were problems ahead. By 1930 Bergey was aging and becoming somewhat frail so that he was concerned about the *Manual*'s governance and future. He turned to Breed to an increasing degree for the overall editing and as a major contributor, but also to fight for financial support and for a degree of independence. The agreement co-signed by Bergey and Breed with the Society in 1922 had recommended that royalties " . . . be accumulated in a separate fund to be used to stimulate further work in this field" and Bergey himself felt that he had "donated" this fund to the Society for that purpose.

*As can be gathered from skeptical sentiments in the famous textbook by W.W.C Topley and G.S. Wilson, *Principles of Bacteriology and Immunity*, 1st ed. (1929), Edward Arnold Ltd., London, and continued in large part to the Fifth Edition (1964) but not thereafter.

The Struggle for Financial and Editorial Independence

Breed's correspondence after 1930 with the powerful Secretary-Treasurers of the SAB (J.M. Sherman 1923–1934; I.L. Baldwin 1935–1942) seeking funds to assist the business of producing new editions became increasingly sharp and argumentative because this assistance was almost uniformly refused. The royalties were small and the publisher did not pay any until the costs were covered; the result was that the Society felt they were exposed to risk with a property that they considered not likely to go on much longer. Sherman, in particular, strongly objected to Breed's rhetoric and proprietary attitude, yet he reluctantly agreed in 1933 to cede $900 (half the accumulated royalties) for Fourth Edition purposes. The Society felt that the funds were theirs (the contract was between the Society and Williams & Wilkins) and there might be others deserving of support from the fund. A request for funds by A.T. Henrici in 1935 brought the whole matter of ownership back into contention and into Baldwin's more diplomatic hands. At the same time Breed was asking for $1000 (essentially the remainder of royalties plus interest) and decisions had to be made during a flurry of correspondence with a repetitive *non placet obligato* from Sherman. There was also a *Bergey's Manual* Committee (Winslow, Buchanan and Breed) reporting to the Council in support of a mechanism for funding the *Manual*. In the end, and agreeably to all parties for different reasons, it was decided between Sherman and Baldwin that the SAB should cede the rights to the *Manual*, the royalties to come, and the accumulated fund to Dr. Bergey to do with as he would wish, and the Council agreed (December 28, 1935). In large part it was a gesture of respect for Dr. Bergey because both of them stated in letters that they did not expect the *Manual* to go through more editions, in which respect they were mistaken.

In preparation for the Fourth Edition, and recognizing that Bergey was not well and that Harrison, Hammer, and Huntoon would not stay for long, Breed added E.G.D. Murray (Fig. 1d) to his corps of editors/authors, so that with Harrison still enlisted there were two Canadian members. With the Fourth edition published in 1934, from late 1935 until early 1936 was a time of negotiation. It is clear that Bergey, Breed, and Murray wanted an independent entity, while Buchanan with his own ideas was presenting a plan to Baldwin involving sponsorship by the Society, and Breed was trying unsuccessfully to make peace with Buchanan. Bergey, for his part, was (January, 1936) consulting with the SAB and advisors in preparation for developing a deed of trust for the future development of the *Manual*, and asking that there be no further controversy. His feeling about the whole sad tale was voiced on January 29, 1936: "The arrangement I have made will be without hindrance from a group of persons who appear to have no kindly feeling toward advances in bacteriology in which they could not dictate every step". The *Bergey's Manual* Trust was indentured on January 2, 1936 in Philadelphia, Pennsylvania, and the Trustees were Bergey, Breed, and Murray. The only concession to the SAB, that continues to the ASM today, is that one of the Trustees is chosen as a representative who reports annually to the Society on the state of the Trust and its work.

Mr. R.S. Gill, the representative of the Williams & Wilkins Co., informed Breed in December, 1934 that copies of the Fourth Edition were exhausted and sought agreement for a new edition; Breed prevaricated because the situation was not yet clear. However, by 1937 he was seeking contributions from a number of colleagues for a future volume. Sadly, D.H. Bergey died on September 5, 1937 at age 77, but the trustees retained his name on the masthead of the Fifth Edition published in 1939. Breed was now Chairman of the Trust and remarked in a letter to E.B. Fred and I.L. Baldwin (January 26, 1938) that Dr. Bergey, who was so interested in seeing the *Manual* revised, would have liked "... to know how well his plans are developing and how ... interested specialists are cooperating with us in making this new edition much better than anything we have had before". So a new way of producing the *Manual* with many contributors was now in place for elaboration in future editions. The first printing of 2000 copies of the Fifth Edition (Bergey et al., 1939) was sold out before the end of the year and 1000 more copies were printed. It was obvious that the *Manual* was needed and served a useful purpose, vindicating the optimism Bergey and Breed had maintained in the face of opposition. Breed, Murray, and A.P. Hitchens (who was appointed to the Board of Trustees in 1939) had to organize a Sixth Edition, which needed to be completely revised and required much to be added. There were 1335 species descriptions in the Fifth Edition and the Sixth, when accomplished, would have 1630. They were faced not only with the need to make changes in the outline classification but also to make decisions about the inclusion or exclusion of large numbers of dubious and inadequately described bacteria. Furthermore, the exigencies of World War II took some of the trustees and many of their contributors out of contention for the duration. Nevertheless, the Sixth Edition was published in 1948 (Breed et al., 1948a) and acknowledged the assistance of 60 contributors. Some of the incompletely described species appeared in appendices following the listings in genera and the book included an index of sources and habitats as an attempt to be helpful. A novelty, and an approach not to be fully realized until 35 years later in the *Systematic Manual*, was a section on the *Myxobacterales* containing a preliminary discussion of the nomenclature and biological characteristics of members of that Order. For this, credit is given to J.M. Beebe, R.E. Buchanan, and R.Y. Stanier; it seems likely to those who knew all of them that this approach originated with Stanier. Additions to the Sixth Edition were sections on the classification of *Rickettsiales* prepared by I.A. Bengston and on the *Virales* or Filterable Viruses prepared by F.O. Holmes. The former was appropriate but the latter pleased very few, certainly preceded an adequate understanding that would have allowed for a rational classification, and never appeared again.

The original Board of Trustees went through changes due to death and the enlargement of the Board. H.J. Conn, a colleague of Breed's at Cornell, was added in 1948 to join Breed, Murray, and Hitchens. The next year A.P. Hitchens died and was replaced by N.R. Smith, an expert on *Bacillus* species. R.E. Buchanan was added as a member in 1951 and began to take an active role in the affairs of the Trust. In 1952 Breed expressed a desire to step down as Editor-in-Chief, he was 75, and the Board debated about his successor. Among those considered were E.G.D. Murray, who was about to retire from McGill University, L.S. McClung of Indiana University, and C.S. Pederson of Cornell, but no decision was made. In correspondence to Breed, Smith wrote that "No doubt, Dr. Buchanan would like to take over when you step aside . . . In fact one can read between the lines that 'no one besides Buchanan is capable of editing the *Manual*' ". This change, however, did not come to pass as Breed stayed on until his death in 1956.

Breed pursued actively the production of a Seventh Edition in the 1950s with the active support of Murray and Smith (Breed et al., 1957). The task was no less formidable, and there were many new authorities mounting increasingly pointed discussions about shortcomings in bacterial taxonomy in the dinner sessions that Breed arranged at the annual SAB meetings. It was to be the last edition in which the bacteria are classified as Schizomycetes within a Division of the Plantae, the Protophyta, primordial plants. In fact, the Preface tells us, the opening statement describing the Schizomycetes as "typically unicellular plants", was hotly debated without attaining a change, yet there were some concessions to cytology in the rest of that description, particularly concerning nucleoids. Ten Orders were recognized, adding to the five in the Sixth Edition, and these now included *Mycoplasmatales* and considerable division of the Order *Eubacteriales*. The keys to the various taxa were improved for utility and, recognizing the many difficulties involved in determination, an inclusive key to the genera described in the book was devised by V.B.D. Skerman and appended. This key, which was referred to as a comprehensive key, was designed to lead the user by alternative routes to a diagnosis of a genus when a character might be variable. It proved to be extremely popular and useful with readers and was repeated as an updated version in the Eighth Edition. Overall, the substance of the Seventh Edition of the *Manual* was due to the efforts of 94 contributors from 14 different countries. The *Manual* was becoming an international effort; however, Breed complained that the slowness of communication between the USA and Europe hampered their efforts.

Breed did not see the fruits of his labors as Editor-in-Chief; he died February 10, 1956, with many of the contributions arranged and the form of the book decided, but leaving a serious problem of succession. The position of Chairman of the Board of Trustees and Editor-in-Chief was decided, appropriately, and given to R.E. Buchanan whose interest in bacterial nomenclature and taxonomy, with direct and indirect involvement in the *Manual*, dated back to its origins. There was the immediate problem of finishing the editorial work on the Seventh Edition after Breed's death. E.F. Lessel Jr. had been working as a graduate student with Breed in Geneva, NY on the *Manual*, but was called into military service before the job was finished, and was stationed at a camp in Texas. Upon taking over the Chairmanship, Buchanan contacted W. Stanhope Bayne-Jones, of the Army's Office of the Surgeon General and Lessel's superior, to ask that Lessel be assigned to work on the completion of the *Manual* while in the service. Bayne-Jones agreed and assigned Lessel to the Walter Reed Hospital in Washington, DC. Thus the last editorial polishing of the book could take place without undue delays. After his service commitments were fulfilled Lessel went to Iowa State and finished his Ph.D. under Buchanan's direction and acted on occasion as recording secretary for Trust meetings.

R.E. Buchanan for many years had held three important administrative posts at Iowa State (Bacteriology Department Head since 1912; Dean of Graduate College since 1919; and Director of the Agricultural Experiment Station since 1936), retiring from all three in 1948. After 1948 some of his energies went to compiling and annotating the text for the 1952 publication of the Bacteriological Code and starting the *International Bulletin of Bacteriological Nomenclature and Taxonomy*. The *International Bulletin* received its initial monetary start in 1950 with a $150 gift from the *Bergey's Manual* Trust, to which Murray objected, saying "the *Journal* would be ephemeral." Fortunately he was wrong because the *Bulletin* later changed its name to the *International Journal of Systematic Bacteriology* and is still being published by ICSB (IAMS) with about 1200 pages in the 1997 volume. When Buchanan became Editor-in-Chief of the *Manual*, he induced the Department of Bacteriology at Iowa State to provide him an office suite and the title of Research Professor, from which position he obtained grants from the National Library of Medicine to support the office. This support continued until his death in 1973 at the age of 89 years.

Buchanan's twenty-year involvement in the Trust was to see, near its end, the start of a new era, despite his many objections to change. The chief change to come arose from a growing lack of confidence in the sanctity of higher taxa, there being few and often no objective tests of correctness. In the production of the Seventh Edition, it was recognized that an expanding synonymy and the ever-growing list of species that were unrecognizable or inadequately described provided a burden that made for wasted space and unreasonably extensive appendices. The addition of Breed's collection of reprints to Buchanan's considerable collection formed an extensive taxonomic archive in the Trust headquarters. With this resource in mind the Trust decided that a separate publication was needed to assemble as complete a listing as possible of the names and references of all the taxa included in the *Manual*, as well as "species formerly found as appendices or indefinitely placed as *species incertae sedis*" that might or should have appeared in the *Manual*. These, together with an assessment of whether or not each name was validly published and legitimate, formed a monster book of nearly 1500 pages, published as *Index Bergeyana* (Buchanan et al., 1966). Each and every reference was checked for accuracy, for Buchanan rightly stated that there "was a lot of gossip about the description of each name." These labors were a personal interest of R.E. Buchanan, who directed several years of effort by J.G. Holt (then at Iowa State), E.F. Lessel Jr., and a number of graduate students and clerks in the undertaking. The lists served as a finder mechanism, an alphabetical listing of the names of the bacteria, and of special use as a reference after the new starting date for nomenclature, January 1, 1980, mandated by the revised Code (Lapage et al., 1975). Addenda were inevitable and more names were collected as a *Supplement to Index Bergeyana* published in 1981 under the direction of N.E. Gibbons, K.B. Pattee and J.G. Holt. These substantial reference works assisted the refining of the content of the Seventh and Eighth Editions of the *Manual* and allowed concentration on effectively described and legitimate taxa.

There were seemingly interminable discussions about what needed to be done for an effective new edition. This was particularly true in the period 1957–1964 after Breed's death, when the Trust membership changed and new ideas and new scientific approaches to taxonomy became available. In the late 1950s the Board of Trustees was enlarged with the addition of S.T. Cowan, C.F. Niven Jr., G. Edsall, and A.G. Lochhead (the record is unclear on the exact date of their appointment). The election of Cowan from the UK added a European member and continued the internationalization of the Board (Fig. 2). Each of these new members brought expertise in different areas of bacteriology and that policy of diversity of interest among members has continued to this day. Later, in 1962, Arnold Ravin, a bacterial geneticist, was added to replace the retiring Lochhead. Of primary concern in the late 1950s and early 1960s was the position of Editor-in-Chief and location of Trust headquarters. An arrangement with Iowa State University to have a candidate assume a professorship at the University and house the headquarters there was made. The position was offered to P.H.A. Sneath, who had gained renown with his invention of numerical taxonomy and production of a masterful monograph on the genus *Chromobacterium*. By 1963, however, Sneath chose to stay in England and Buchanan stayed on as Editor-in-Chief. All other efforts to find a new editor failed until Buchanan's death in 1973. As he grew older, more difficult, and more autocratic, progress on a new edition slowed considerably. Even the replacement of E.G.D. Murray, Conn, Smith, and Edsall by R.G.E. Murray, J. Liston, R.Y. Stanier, and N.E. Gibbons did not change the speed of Board actions. It became a war of wills between Buchanan and the others on what was important and where progress could be made. Until decisions on the taxa to be included and their circumscriptions were made, there was slow progress in naming and putting to work the 20 or more advisory committees needed to direct the authors of the final texts on genera and species. One novel (to the Trust) approach for obtaining consensus on taxonomic matters was the organization of a conference of advisory committee members and trustees held in May, 1968 at Brook Lodge in Augusta, MI, under the auspices of the Upjohn Co. and chaired by R.G.E. Murray. Fifteen advisory committee members joined in discussions with the Trust to assess the status of current knowledge on the major groups of bacteria to be included in the Eighth Edition. Despite this helpful preliminary, it brought no agreement between Buchanan as Chairman, whose main focus was then on nomenclature, and the rest of the Trustees, whose interests mostly focused on biological, functional, and eco-physiological attributes. It was clear that many of the higher taxa rested on shaky ground and were hard to assess on strict taxonomic terms. Accordingly, there was a long argument over abandoning formal names above family level wherever possible, agreeing that a large number of genera were of uncertain affiliation or, at least, could only be related on the basis of some diagnostic characters, such as gliding motility, shape and Gram reaction, and methane production, all of which might or might not have phylogenetic significance. All former ideas about phylogeny and relationships were discarded. The Eighth Edition was planned as a book divided into "Parts", each with a vernacular descriptor. The Advisory Committee for each part (some needed more than one) was assigned a member of the Trust who was responsible for action and who, eventually, had to see that each genus had an assigned author (131 in the end) who was willing to write.

Molecular/genetic technology was well established by 1974 when the Eighth Edition was published, but was not yet widely applied to play a role in broad decisions in taxonomy. The procaryotic nature of bacteria and all cells related to them (i.e. including the Cyanobacteria) could be recognized and used to define the Kingdom *Procaryotae*. *Monera* was the old and partially applicable higher taxon but the description was not cytologically based. The molecular composition of DNA was useful for separating phenotypically similar but genetically distinct groups (e.g., *Micrococcus* and *Staphylococcus*) and many descriptions could include mol% G + C as a character. Genetic and subsequent biochemical-molecular data told us that species were only relatively "fixed" in their expressed characters. This concept needed to be addressed in the circumscriptions and aids to identification. Greater use was made of diagnostic tables and wherever possible there were indications regarding uncertainties and the percentages of positive or negative reactions for tests. The value of the Eighth Edition for identification purposes was increased by the emphasis of both the Trustees and the authors on refining descriptions (in terms as up-to-date as possible), tables, keys, and illustrations. As in previous editions, many old names of dubious or unrecognizable entities were discarded and synonymy was reduced to essentials; the old information and its location was not lost because it was available in the *Index Bergeyana* (Buchanan et al., 1966), or later in the *Supplement to Index Bergeyana* (Gibbons et al., 1981).

The Eighth Edition was a long time in gestation—17 years—

FIGURE 2. Photograph of Trustees meeting at Iowa State University, Ames, November, 1960. *L. to R.*, G. Edsall, R.E. Buchanan, C.F. Niven, Jr., N.R. Smith, and E.G.D. Murray.

but its success (40,000 copies over the next 10 years, and more than half outside of North America) was a testament to its necessity and utility. Most of the primary journals involved in publishing microbiological papers suggested or required the *Manual* as the nomenclatural resource for bacterial names, all this despite the treatment of some groups (e.g., the *Enterobacteriaceae*) not being universally accepted. But it was truly an international enterprise, with authors from 15 countries who could, at last, be named in literature citations as authors.

The editing of the Eighth Edition became a major operation requiring sharing of responsibilities and some redirection of effort. This was in part due to the age and increasing infirmities of R.E. Buchanan who had been both Chairman of the Trust and Editor, directing his efforts to nomenclature, synonymy, and etymology. It became evident that a Co-Editor was required and fortunately N.E. Gibbons, recently retired from the National Research Council of Canada, agreed to undertake the task. Shortly thereafter Gibbons became the *de facto* editor, due to Buchanan's illness and death in January, 1973, and did all the general technical editing from his home in Ottawa with help from his wife, Alice Gibbons (who handled the Index of Names), and a number of Trustees, especially S.T. Cowan. The book was published in 1974 (Buchanan and Gibbons, 1974).

With publication of the Eighth Edition the Board of Trustees went through another major change of membership, and over a period of two years Niven, Ravin, Liston, Gibbons, and Stanier left the Board. At the first meeting after Buchanan's death, held in October, 1973, J.G. Holt, who had served as Secretary to the Board from 1963–1966 and co-edited the *Index Bergeyana*, was elected member and Secretary. In 1974–1975, H. Lautrop, S. Lapage, and M. Bryant were added, and later in 1976 N.R. Krieg and J.T. Staley joined the Board. In 1975 Holt was appointed Editor-in-Chief. With the publication and healthy sales of the Eighth Edition and increasing international profile, it was decided to meet at locations separate from the ASM venue and to meet every other year outside North America, and the 1975 meeting was held in Copenhagen, Denmark, at the Statens Seruminstitut. From then on a segment of each meeting was devoted to consultation with taxonomically inclined colleagues in that area.

The Trust had recognized, in the process of deciding the format of the book, that students and technologists were important users, with primary interests in identification and a lesser need for the extensive descriptions of individual species. An abridged edition of the Sixth Edition of the *Manual* had been produced (Breed et al., 1948b), but was only a modest success and not carried forward to the Seventh Edition. In 1974 the need seemed to be greater, so preparations were made to assemble an outline classification; the descriptions of genera, families and such higher taxa as were recognized; all the keys and tables for the identification of species; the glossary; all the illustrations; and two informative introductory chapters. It was recognized that there were both deletions and additions (new keys and synopses as well as new genera) to the material from the parent edition, so that at the most the abridged version would be considered an abstract of the work of the authors of the larger text. Therefore, citation could only be made to the complete Eighth Edition. It was published as *The Shorter Bergey's Manual of Determinative Bacteriology* in 1977 (Holt, 1977). It too was a great success, selling 20,000 copies over a span of 10 years. A few years later it was translated into Russian and sold throughout the USSR, with royalties accruing to the Trust.

The development of bacteriology, as we now appreciate, required the recognition and differentiation of the various groups of microbes as taxonomic entities. At the time that *Bergey's Manual* started, the nature of bacterial cells was not known. Bacteria were classified and named under the Botanical Code of Nomenclature as Schizomycetes and no one could then have substantiated present understanding that *Cyanophyceae* are really bacteria. The international discussions of bacterial classification were minimal and took place at Botanical Congresses, as befitted the view that the Schizomycetes and the *Schizophyceae* within the Phylum Schizophyta (later Protophyta) belonged in the plant kingdom. This interpretation was maintained in *Bergey's Manuals* up to and including the Seventh Edition (1957); however, it was stated in an introductory chapter that E.G.D. Murray "... felt most strongly that the bacteria and related organisms are so different from plants and animals that they should be grouped in a kingdom equal in rank with these kingdoms". As expressed by Stanier and Van Niel (1962) in their seminal paper "The concept of a bacterium" it is "... intellectually distressing (for a biologist) to devote his life to the study of a group that cannot be readily and satisfactorily defined in biological terms ...". This marked the beginning of the useful and directive description of bacteria as cells of unique nature. With this approach it was clear that the cyanobacteria were included and there was, at last, a satisfactory unity. This was to be slowly elaborated in the next three decades by the recognition of phylogenetic information recorded in molecular sequences of highly conserved macromolecules, but in the meantime the Eighth Edition (1974) subscribed to the view based on cytological data that the bacteria (all the procaryotes) belong in a separate kingdom, the *Procaryotae*. This was not a surprising decision because two Trustees, Stanier and R.G.E. Murray, were then involved in the description of bacteria as cells with unique features.

International Efforts to Regulate Taxonomy

The founders of *Bergey's Manual* were fully aware of the substratum of opinion, albeit not supported then by strong data, that the bacteria were a special form of life, requiring special methods and a different approach to classification, not necessarily the same as that required by the Botanical Code. In fact, between 1927 and 1930 there was a considerable international correspondence between bacteriologists interested in taxonomy in the varied fields of application in agriculture, medicine, soil science, etc, expressing their concerns. The correspondence also concerned what should be done about discussing bacteria at the forthcoming Botanical Congress to be held in Cambridge, England, in 1930, and about resolutions adopted by the Bacteriological Section of the Botanical Congress, of which J.M. Sherman had been Secretary, held in Ithaca, NY, in 1926. The resolutions were (1) exclusion of the requirement for a Latin diagnosis in bacteriological nomenclature; (2) greater emphasis on the "type concept"; (3) a special international and representative committee was needed to coordinate the special nomenclatural interests of bacteriologists; and (4) that a permanent International Commission on Bacteriological Nomenclature should be formed. Sherman, then Secretary-Treasurer of SAB, wrote to Prof. J. Briquet of the Permanent International Committee on Botanical Nomenclature pointing out that the past two Congresses had authorized a bacteriological committee on nomenclature, that it should be organized, and that the Bacteriological Section had prepared a distinguished list of nominations for membership. The list included three of the major contributors to discussions of systematics in the SAB (Buchanan, Breed, and Harrison) and two of them were intimately involved with *Bergey's Manual*.

A lively correspondence among the authorities resulted and much of it was stimulated by Breed writing to bacteriologists in Europe as well as America. He sums up an impression of the responses in a letter to the Secretary of the Botanical Congress, as follows: ". . . there is a general feeling that unless the Congress welcomes us into the ranks of botanists with the recognition of our peculiar and perplexing problems in the taxonomic field, we must organize an independent international group". At the same time he recognized the value of the work of Congresses in maintaining useful rules of nomenclature and reiterating the list of resolutions. The British correspondents were generally agreeable to bacteriological discussions but expressed sharp divisions as to associating or not with the botanists. Other players namely the newly formed International Society of Microbiology, and the Cambridge committee charged with organizing the bacteriological component of the 1930 Botanical Congress came on the scene in 1927. The former encouraged some thoughts of an independent base for microbiological congresses and taxonomy committees, while the latter questioned whether or not a Section of Bacteriology was desirable or even feasible, and asked H.R. Dean (Professor of Pathology at Cambridge University) to seek interest and act on it. Dean's correspondents in this matter were numerous and mostly British, but also included Breed, Buchanan, B. Issatchenko (USSR), and K.B. Lehmann (Germany) (letters regarding this information are now filed in The American Society for Microbiology Archives). The responses generally supported a Section at the Congress but the overall opinions on continuing association with the botanists varied from the enthusiastic (mostly general microbiologists) to outright contrary opinion (mostly medical bacteriologists). Paul Fildes wrote: "Personally I am of the opinion that bacteriology has nothing to gain by a close association with botany." And Sir John Ledingham, while agreeing with having general bacteriological discussions, thought in the future "If the botanists will not have us, maybe that is all to the good". J.W. McLeod wrote: "Frankly, I am not very enthusiastic about a Section of Bacteriology at an International Botanical Congress especially if we are going to have an International Association of Microbiology". Other views crept into letters such as one from F. Löhnis: "I know that there exists within ... (the SAB) ... a small but very active minority extremely eager to advance a scheme of classification and nomenclature that seems to me as to others quite contrary to international usage ... this minority has advanced its ideas in the U.S.A. and will probably try the same scheme at Cambridge in 1930 if there should be a separate Section of Bacteriology". Breed wrote Dean that there would be support in the SAB for a delegation and added a few remarks on differences with the botanists, including: "Our troubles, for example, do not concern type specimens kept in a herbarium. They are intimately concerned with the maintenance of type culture collections such as the English bacteriologists have been able to establish so splendidly at the Lister Institute". There were more meetings in 1929 of a subcommittee appointed to settle a program for the Bacteriology Section (Dean as Chairman, with Boycott, Topley, Ledingham, Paine, Thornton, Thaysen, and Murray) and charged to keep Briquet (Botanical Nomenclature Committee) informed of any discussion of bacteriological nomenclature that might take place.

Attitudes to studying and naming bacteria were rather different in the UK and Europe in the 1920s than was evident in the USA and Canada. The influential members of the SAB involved in *Bergey's Manual* seemed to be able to muster support for their views and seek consensus even if there were rumblings of dissent (q.v. Hall, 1927). In Europe many, like Orla-Jensen, believed that individual bacteriologists of substance should prevail because they were the ones who knew their groups of bacteria and he objected to imposition from outside. Internationalism did not and does not come easily.

The International Society for Microbiology (ISM), formed during an international conference on rabies sponsored by the Institute Pasteur in April, 1927, elected Prof. J. Bordet as President and R. Kraus as Secretary-General. It was stated in the brochure that: "It will not only compose the Science of Bacteriology but all the sciences associated with Microbiology" and the concept was based on "the unanimous conviction that Science should unite Nations...". The idea that all Societies of Microbiology may join, and that National Committees may present individual microbiologists as members, was expressed. So, the concept of an international association was born in Europe without anyone from North America among the founding members from 14 countries. There was interest: Harrison wrote to Dean suggesting that contact should be established between the ISM and the Bacteriological Section meeting at the Botanical Congress. Ledingham wrote to Dean in June, 1928, to support a meeting of the Nomenclature Committee of the Pathological Society of Great Britain and Ireland with Breed and others who were visiting, "particularly with regard to joint action on this matter by the botanical bacteriologists and the new International Society for Microbiology. Possibly they might consent to turn the matter over entirely to the new International Society (if adequate guarantees given)". It is not clear what group meeting resulted although hints were made.

1930 was the year of change because the First International Congress for Microbiology was held in Paris and by a vote agreed to follow the rules of nomenclature accepted by the International Congresses of Botany and Zoology *"in so far as they may be applicable and appropriate"* (italics as given by Breed, 1943). This opened the doors for a dedicated committee which would be in action at the following Congress (1936, in London, England), and set in train the development of an International Committee for Systematic Bacteriology, the regulatory mechanisms that were to be so important to taxonomic decisions in years to come, and a bacteriological code of nomenclature. The Microbiology Congress and the Botanical Congress, prompted by its Bacteriology Section (and probably by a questionnaire circulated by Breed), both approved in plenary session that the starting date for bacteriological nomenclature should be May 1, 1753, the date of publication of *Species Plantarum* by Linnaeus.

No doubt, there was much going on behind the scenes and some degree of consensus about the ever contentious matters involved in bacterial taxonomy. However, it was clear that bacterial taxonomy would be a matter of international concern from then on.

The Enlargement of the Scope of the *Manual*

In the period following the death of R.E. Buchanan, John Liston took over as Chairman until 1976 when he retired and was replaced by R.G.E. Murray. It was during this subsequent period, in the late 1970s, that plans were laid to expand the informational coverage of the *Manual*. What started as a discussion of a new edition of the determinative manual developed into a plan to include much more information on the systematics, biology, and cultivation of each genus covered. Hans Lautrop had analyzed the content of the Eighth Edition and suggested a format that would allow authors to expound on further descriptive information, isolation and maintenance, and taxonomic problems. Other planned departures from past editions included the profuse use of high quality illustrations and allowing publication of new names and combinations in the *Manual*. It was also decided to preface the book with essays on general aspects of bacterial systematics such as modern genetic techniques, culture collections, and nomenclature. This expanded coverage meant a large increase in the number of pages and it was decided to publish the book in four volumes, each containing a set of taxa divided along somewhat practical lines. The final arrangement consisted of volumes covering the Gram-negatives of medical importance, the Gram-positives of medical importance, the other Gram-negatives (including the *Archaea* and, for the first time, the Cyanobacteria), and lastly, the Actinomycetes. This division allowed users to purchase separate volumes that suited their special professional requirements. This expansion demanded a more descriptive title and it was decided to call the book *Bergey's Manual of Systematic Bacteriology*. Production of each volume was set up on a cascading schedule with completion planned for the mid 1980s. Trust members were chosen to edit each sub-volume, with the final editing being done in the Ames office. Obviously, such an undertaking was an expensive endeavor, beyond royalty income, and extra funding was provided by a grant from the National Library of Medicine of the US National Institutes of Health for volumes 1 and 2, and an advance on royalties from the publisher. In the end the complete project cost around $400,000. Volume 1 was published in 1984 (Krieg and Holt, 1984), Volume 2 in 1986 (Sneath et al., 1986), and Volumes 3 and 4 in 1989 (Staley et al., 1989; Williams et al., 1989).

The book was a truly international project in which 290 scientists from 19 countries (and 6 continents) participated, and as much of a success as the Trust and its authors could have expected. Each of the volumes sold between 10 and 23 thousand copies in the 1984–1996 period and more than half of the sales were outside of the USA. The total royalties add up to in excess of $450,000, making the *Systematic Manual* both a scientific and business success. The challenge now is to find the finances, energies, and means to keep the *Manual* up to date, affordable and reasonably current.

One of the mandates of the Trust is to further bacterial taxonomy, and the modern Board of Trustees has taken other initiatives besides the publication of books to promote the field. There has been monetary support, however small, for worthwhile causes, such as the aforementioned gift to launch the *International Bulletin of Bacteriological Taxonomy and Nomenclature*. Also in 1980, the Trust contributed $3000 towards the publication of the Approved Lists of Bacterial Names (Skerman et al., 1980). Two ways have been found to honor people who have made important contributions to the field of bacterial systematics. In 1978 the Bergey Award was instituted as a joint effort by Williams & Wilkins and the Trust; the first award went to R.Y. Stanier and is an annual event. Table 2 lists the recipients of this award, which consists of $2,000 and expenses to allow travel to a meeting of the recipient's choice to receive the award. In the 1990s the Trust commissioned a medal, the Bergey Medal (Fig. 3), to be given to individuals who have made significant lifetime contributions to bacterial sys-

TABLE 2. Recipients of the Bergey Award

Recipient	Year
Roger Y. Stanier	1979
John L. Johnson	1980
Morrison Rogosa	1981
Otto Kandler	1982
Carl R. Woese	1983
W. E. C. Moore	1984
Jozef De Ley	1985
William H. Ewing	1986
Patrick A. D. Grimont	1987
Lawrence G. Wayne	1988
Hubert A. Lechevalier	1989
M. David Collins	1990
Erko Stackebrandt	1991
Wolfgang Ludwig	1992
Wesley E. Kloos	1993
Friedrich Widdel	1994
Michael Goodfellow	1995
Karel Kersters	1996
Rosmarie Rippka	1997
Barry Holmes	1998
David A. Stahl	1999
William B. Whitman	2000
Lindsay I. Sly	2001
Peter Vandamme	2002
Peter Kämpfer	2003
Rudolf Amann	2004

FIGURE 3. Obverse view of the Bergey Medal, 3 in. diam., See Table 3 for a list of recipients.

tematics and to recognize the service of Trustees (Table 3). In 1982, the Board of Trustees decided to stimulate the involvement of more people in the affairs of the Trust, beyond the legal limit of nine regular members set in the By-Laws. It instituted the appointment of *Bergey's Manual* Associates for five-year terms to contribute their scientific expertise to the needs of the *Manuals*, the Trust and its Editors (Table 4).

The *Systematic Manual* was produced during a time of significant advances in our understanding of relationships between bacterial taxa based on the comparison of molecular sequences in highly conserved nucleic acids and proteins. The work of Carl Woese and others dating from the 1970s began to provide solid, initially sparse but now burgeoning, information on the phylogenetic relationships of the bacteria and, indeed, all life forms. This new information had a potential impact on the organization of the taxa in the *Manual*, however, the Trust and its advisors decided to continue to organize the book on phenotypic grounds. First, because the bench workers needing to identify isolates have to use these characters and, secondly, because the phylogenetic data were accumulating slowly during the early 1980s. The Trust decided to continue with a phenotypic arrangement and indicate, where appropriate and data were sufficient, the phylogenetic placement of the taxon being discussed. Finally, enough progress has been made in the last 20 years for this Second edition to be phylogenetically organized, although there are still gaps and uncertainties in our knowledge.

In the 1980s and early 1990s there was a large turnover in Board membership and leadership. New Board members included D.J. Brenner, J.W. Moulder, S.T. Williams, K.-H. Schleifer, N. Pfennig, P.H.A. Sneath, R.W. Castenholz, J.G. Tully, and J. Ursing, some of whom have since retired (Table 1 and Fig. 4). In 1990 Board Chairman R.G.E. Murray retired after a long and fruitful tenure and was replaced by P.H.A. Sneath, who served until 1994 when S.T. Williams took over the helm. It should be explained that the Board of Trustees has a retirement age of 70 (members call it the "Buchanan Amendment"), which is no reflection on the quality of service of retired Board members. See Fig. 5 for the current membership of the Board of Trustees and Editors of sub-volumes of this Second Edition.

TABLE 3. Recipients of the Bergey Medal

Eyvind A. Freundt	1994
R.G.E. Murray	1994
Riichi Sakazaki	1994
V.B.D. Skerman	1994
Dorothy Jones	1995
Norberto Palleroni	1995
Norbert Pfennig	1995
Thomas D. Brock	1996
Marvin P. Bryant	1996
John G. Holt	1996
Emilio Weiss	1996
Lillian H. Moore	1997
Ralph S. Wolfe	1997
George A. Zavarzin	1997
Kjell Bøvre	1998
Holger Jannasch	1998
Juluis P. Kreier	1998
Peter H.A. Sneath	1998
Wilhelm Frederiksen	1999
James W. Moulder	1999
Karl O. Stetter	1999
Hans G. Trüper	1999
Peter Hirsch	2000
Hans Reichenbach	2000
Stanley T. Williams	2000
Eiko K. Yabuuchi	2000
Floyd E. Dewhirst	2001
E. Imre Friedmann	2001
Joseph G. Tully	2001
Don J. Brenner	2002
Rita R. Colwell	2002
Noel R. Krieg	2002
Monique Gillis	2003
Hans Hippe	2003

One important change in the Trust operations has been the establishment of a permanent headquarters. In the late 1980s the Trust decided to move from Iowa State University where it had resided since 1958, and set out to find a permanent home for the Editorial Office that was not tied to the tenure of the Editor. After an active search such a home was eventually found at Michigan State University which has a large, active Department of Microbiology and is the base for the NSF-funded Center for Microbial Ecology. In December, 1990, Holt and the Trust office and archives moved to East Lansing, Michigan. Holt subsequently retired as Editor-in-Chief in 1996 and a replacement was found who continued as a faculty member in the Department. The new Editor-in-Chief, George M. Garrity, assumed his duties in 1996.

All of these changes were accompanied by an increasingly active publishing program. After publication of the last two volumes of the *Systematic Manual* in 1989, plans were made to produce the Ninth Edition of the *Determinative Manual*. Based on a concept of N.R. Krieg, the format of the book was changed to a style between the Eighth Edition and the *Shorter Manual*; the species descriptions are summarized in extensive tables. It was published in 1994 (Holt et al., 1994) in softcover and contained the determinative information from the *Systematic* book plus descriptions of new genera and species named since publication of the larger book. This *Manual* is intended to be a prime resource for bench workers and all who are engaged in diagnostic bacteriology and the identification of isolates. The Trust published other books in the early 1990s, notably *Stedman's/Bergey's Bacteria Words* (Holt et al., 1992) (one of a series of wordbooks compiled

TABLE 4. Past and Present Bergey's Manual Associates (1982–2003)

Martin Altwegg	Wolfgang Ludwig
Paul Baumann	Thomas McAdoo
David R. Boone	W.E.C. Moore
Richard W. Castenholz	Aharon Oren
Jongsik Chun	Norberto J. Palleroni
Rita R. Colwell	Fred A. Rainey
Gregory A. Dasch	Anna-Louise Reysenbach
Floyd E. Dewhirst	Morrison Rogosa
Paul De Vos	Abigail Salyers
Karin Everett	Juri Schindler
Takayuki Ezaki	Karl -Heniz Schleifer
Monique Gillis	Haroun N. Shah
Michael Goodfellow	Lindsay I. Sly
Peter Hirsch	Robert M. Smibert
Lillian Holdeman-Moore	Erko Stackebrandt
Barry Holmes	Karl O. Stetter
J. Michael Janda	James M. Tiedje
Dorothy Jones	Hans G. Trüper
Lev V. Kalakoutskii	Anne Vidaver
Peter Kämpfer	Naomi Ward
Otto Kandler	Lawrence G. Wayne
Karel Kersters	Robbin S. Weyant
Helmut König	William B. Whitman
Micah I. Krichevsky	Friedrich Widdel
L. David Kuykendall	Annick Wilmotte
David P. Labeda	Stanley T. Williams
Mary P. Lechevalier	George A. Zavarzin

for medical transcriptionist use), and provided the general editing of the Second Edition of the CDC manual on the *Identification of Unusual Pathogenic Gram-negative Aerobic and Facultatively Anaerobic Bacteria* (Weyant et al., 1996).

The Publication Process

It is no mean task to produce and get into print a taxonomic compendium; it is a major and complex project for authors, editors, and not least the publisher. The Williams & Wilkins Co. of Baltimore was the publisher of the *Manuals* from 1923 to 1998, and over those years there was an extraordinarily effective partnership between the Trust and the publisher which was mutually advantageous. The various editions of the *Determinative Manual* have been very successful in both the scientific and the commercial sense. The confidence of the publisher allowed them to provide financial support for the preparation of other ventures such as the *Systematic Manual*, which required some years of work and several editorial offices, adding to the up-front expenses. The great success of the published volumes vindicated and more than repaid the publisher's generous support of the enterprise. After major changes in the management of Williams & Wilkins and the merger of the company with another publisher, the Trust reexamined its publishing arrangements and entertained offers from other firms. In late 1998 a new publishing agreement was signed with Springer-Verlag of New York to publish this edition of the *Systematic Manual*, ushering in a new era of cooperation between the Trust, representing the microbiological community, and its publisher, who is committed to disseminating high-quality and useful books to that community.

Because of the number and complexity of the entries, the number of scientists involved in generating the text (or revising it, as is now more often the case), and the sheer number of indexable items, it has been obvious for years that some form of computer assistance would become essential. One of the long-term goals of the Trust and its publishers has been to produce an electronic version of the *Manual*. There were a number of objectives associated with this project. One was the obvious provision of a searchable CD-ROM version of the data contained in the *Manual*. The other, not so obvious, was the ability to streamline the process of updating new editions by supplying the phenotypic data of each taxon in a database that can be easily updated by authors and to which new information (which is accru-

FIGURE 4. Trustees at their meeting in Stamford, England, September, 1985. *L. to R.*, D. Brenner, P. Sneath, N. Krieg, J. Holt, J. Moulder, N. Pfennig, J. Staley, S. Williams, M. Bryant, and R. Murray.

FIGURE 5. Current Trustees (with Emeritus Chairman P.H.A. Sneath) taken at Sun River, OR, August, 1997. *L. to R.*, J. Staley, S. Williams, G. Garrity, J. Holt, K. Schleifer, D. Brenner, N. Krieg, R. Castenholz, D. Boone, and P. Sneath.

ing at an alarming rate) can be added. The Trust editorial office is now using the latest computer technology in producing this and subsequent versions of its manuals, utilizing the power of Standard Generalized Markup Language (SGML) to facilitate the storage, retrieval, typesetting, and presentation of the information in both print and electronic form. Planning for this new edition of the *Systematic Manual* has been underway for the past four years and two major problems have faced the Board and its Advisory Committees. One is the rapid rate of description of new taxa, many of which are not adequately differentiated by phenotypic characteristics. The other is the requirement that the book reflect the best of current science, including a phylogenetic classification based on semantides, particularly 16S rRNA. The phylogeny is incomplete but the gaps are being slowly filled. Problems occur when there is little correlation between the phylogenetic classification and the phenotypic groupings that prove essential to the initiation of identification. Therefore, broadly based and informational descriptions remain an essential feature of the *Manual* as well as a text that stimulates research.

We were most fortunate over the years to enjoy not only a cooperative and productive relationship with Williams & Wilkins, but also the friendly assistance of a series of liaison officers who have represented the Company and its interests and concerns. Among these most helpful people were Robert S. Gill, Dick Hoover, Sara Finnegan, and William Hensyl, whose abilities as facilitators and as interpreters of the disparate requirements of Trust and Publisher were essential. We look forward to our new relationship with Springer-Verlag which should be productive and benefit the entire microbiological community.

The concept of the *Bergey's Manuals*, i.e. encyclopedic taxonomic treatments of the procaryotic world that aid microbiologists at all levels and in all sub-disciplines, is alive and well. The vision of Bergey and Breed is being carried on by their successors and will continue well into the next millennium.

Acknowledgments

A consideration of the history of the *Manuals* and publications of the Bergey's Manual Trust would be incomplete without an acknowledgment of the contributions of a large number of individuals. One such person that we wish to thank for assistance is The American Society of Microbiology Archivist, Jeff Karr, who sought and found correspondence and minutes that were of great use in preparing this manuscript. Many other people were often involved in complex operations going on in their place of work with no or limited formal recognition of their contribution, and frequently without recompense as in the case of wives of editors. Some individuals performed major tasks (e.g., Alice Gibbons, the whole index for the Eighth Edition). A long succession of helpers were involved over the 32 years the headquarters was at Iowa State, and their contributions were invaluable. Of special note was the long service to R.E. Buchanan of Elsa Zvirbulis, Mildred McConnell, and Vlasta Krakowska in Ames. J.G. Holt has been ably assisted by a series of excellent secretary/editorial assistants, especially Cynthia Pease in Ames, and Betty Caldwell and Constance Williams in East Lansing. Taxonomy and the production of useful compilations and classifications are "labors of love" involving both dedication and unremitting effort of those so inclined AND the people around them.

On Using the *Manual*

Noel R. Krieg and George M. Garrity

Arrangement of the Manual

One important goal of the *Manual* has always been to assist in the identification of procaryotes, but another goal, equally important, is to indicate the relatedness that exists among the various groups of procaryotes. This goal seemed elusive until the late 1950s and early 1960s, with the realization that the DNA of an organism makes it what it is. Initially, overall base compositions of DNAs (mol% G + C values) were used to compare procaryotic genomes, and organisms for which mol% G + C values differed markedly were obviously not of the same species. If, however, two organisms had the same mol% G + C value, they might or might not belong to the same species, and thus a much more precise method of comparison was needed. The development of DNA–DNA hybridization techniques fulfilled this need. The continua that often blurred the separation between groups defined by phenotypic characteristics did not usually occur with DNA–DNA hybridization. Organisms tended to be either closely related or not, because DNA–DNA duplex formation did not even occur if base pair mismatches exceeded 10–20%. Thus DNA–DNA hybridization solved many of the problems that had long plagued bacterial taxonomy at the species level of classification. It was almost useless, however, for estimating more distant relationships among procaryotes, i.e., at the generic level, family level, or above. An important development was the discovery by Doi and Igarashi (1965) and by Dubnau et al. (1965) that the ribosomal RNA (rRNA) cistrons in bacterial species were conserved (slower to change) to a greater extent than the bulk of the genome, probably because of their critical function for the life of a cell. This function would allow only slow changes in nucleotide sequence to occur over long periods of time, relative to other genes that were not so critical for the cell. This in turn led to the idea that rRNA–DNA hybridization might be useful for deducing the broader relationships that DNA–DNA hybridization could not reveal. For instance, in 1973 a monumental study by Palleroni et al. showed that the genus *Pseudomonas* consisted of five rRNA groups (tantamount to five different genera).

In the 1970s, the idea—based on cellular organization—that there were only two main groups of living organisms, the procaryotes and the eucaryotes, was challenged by Woese and Fox, who compared oligonucleotide catalogs of the 16S rRNA (and the analogous eucaryotic 18S rRNA) from a broad spectrum of living organisms. These comparisons indicated that there were two fundamentally different kinds of procaryotes: the *Archaea* (also called archaebacteria or archaeobacteria), and the *Bacteria* (also called eubacteria). Urcaryotes—i.e., that portion of eucaryotes exclusive of mitochondria or chloroplasts, these being endosymbionts undoubtedly derived from procaryotes—differed from both the *Bacteria* and the *Archaea*. These findings led Woese, Kandler, and Wheelis (1990) to the view that the *Archaea*, the *Bacteria*, and the eucaryotes (now called the *Eucarya*) evolved by separate major evolutionary pathways from a common ancestral form, although just where the deepest branchings occur is still not clear.

Improvements in the methodologies of molecular biology have now made it possible to determine and compare sequences of the rDNA cistrons from a great number of procaryotes, and a comprehensive classification of procaryotes, based on relatedness deduced from 16S rDNA sequences, is underway. It is hoped that such a phylogenetic classification scheme will lead to more unifying concepts of bacterial taxa, to greater taxonomic stability and predictability, to the development of more reliable identification schemes, and to an understanding of how bacteria have evolved. However, sequencing of the complete genomes of a number of procaryotes and comparison of various genes among the organisms has led to some reservations about whether 16S rDNA sequence analyses are completely reliable for reconstructing evolutionary phylogenies. The study of genes other than rRNA genes has sometimes led to different phylogenetic arrangements. Some bacteria have been found to contain certain genes of the archaeal type, and some archaea have been reported to have certain genes of the bacterial type. These discrepancies might be due in part to a lateral transfer of genes by transformation, transduction, or conjugation from one present-day species to another, as distinguished from vertical transfer from ancestral forms. Thus the location of the deep evolutionary branchings deduced from 16S rRNA gene sequences may not be as firm as once thought. On the other hand, acquisition of an eclectic assortment of genes might have occurred in very primitive life-forms that existed prior to the divergence of the three major evolutionary pathways. In any event, the present edition of the *Manual* provides the best available phylogenetic scheme based on 16S rRNA gene sequencing. Figure 1 shows the major groups of procaryotes and their relatedness to one another. The deeper branching points are not shown because of their uncertainty, and some crossover of branch points beneath the plane of projection is likely. Each branch is the equivalent of either a class or an order. The arrangement shown is reasonably firm and is unlikely to change very much as more information is gathered.

Within the *Archaea*, two phyla are recognized: *"Crenarchaeota"* and *"Euryarchaeota"*. In the present classification, one class is accommodated in the *"Crenarchaeota"*: *"Thermoprotei"*, and seven in the *"Euryarchaeota"*: *"Methanobacteria"*, *"Methanococci"*, *"Halobacte-*

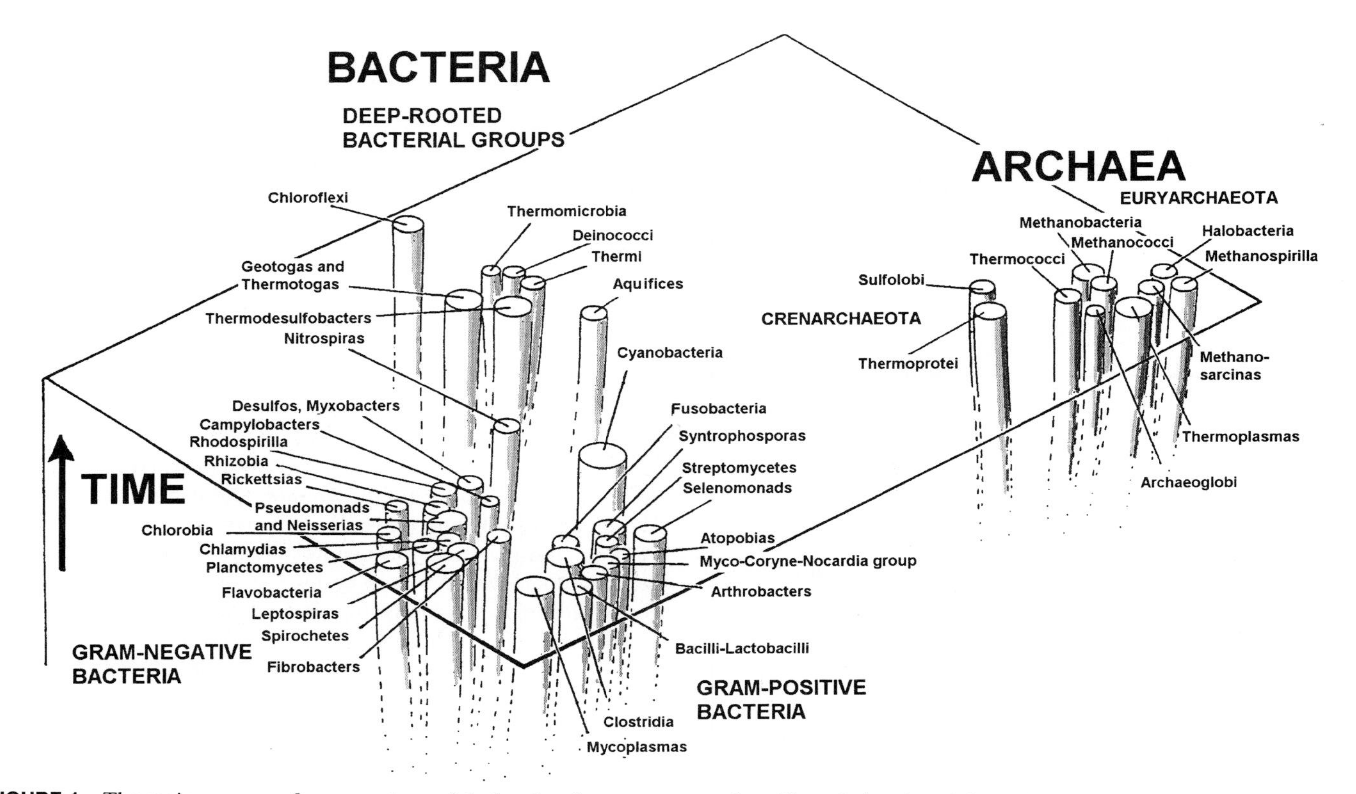

FIGURE 1. The major groups of procaryotes and their relatedness to one another. The relative size of the oval discs is an approximate indicator of the number of species in each group. The deeper origin of these groups, i.e., their evolution from more primitive forms, is still debatable and therefore is represented only by dashed lines. (Courtesy of Peter H.A. Sneath.)

ria", *"Thermoplasmata"*, *"Thermococci"*, *"Archaeoglobi"*, and *"Methanopyri"*.

The *Bacteria* have been grouped into 23 phyla, which are further subdivided into 28 classes. The deep-rooted *Bacteria* encompass nine phyla: the *"Aquificae"*, *"Thermotogae"*, *"Thermodesulfobacteria"*, the *"Deinococcus–Thermus"* phylum, the single-species phylum *"Chrysiogenetes"* (not shown), *"Chloroflexi"*, *"Thermomicrobia"*, *"Nitrospirae"*, and *"Deferribacteres"*.

Within the Gram-negatives, the *"Proteobacteria"* have been elevated to a phylum and subdivided into five classes corresponding to the *"Alphaproteobacteria"* (Rhodospirilla, Rhizobia, Rickettsias), the *"Betaproteobacteria"* (Neisserias), the *"Gammaproteobacteria"* (Pseudomonads), the *"Deltaproteobacteria"* (Desulfos and Myxos), and the *"Epsilonproteobacteria"* (Campylobacters); other Gram-negative phyla include the *"Planctomycetes"* (Planctomyces), the *"Chlamydiae"* (Chlamydias), the *"Chlorobi"* (Chlorobia), the *"Spirochaetes"* (Spirochetes and Leptospiras), the *"Fusobacteria"* (Fusiforms), the *"Verrucomicrobia"* (not shown), the *"Bacteroidetes"* (Bacteroides, Flavobacteria, Sphingobacteria), the *"Acidobacteria"* (not shown), *"Fibrobacteres"* (Fibrobacters), the *Cyanobacteria* (Cyanobacteria), and *"Dictyoglomi"* (not shown).

Traditionally, the Gram-positive bacteria have been separated on the basis of mol% G + C. The low G + C Gram-positives have been assigned to the phylum *"Firmicutes"* and include the Mycoplasmas, Clostridia, Bacilli–Lactobacilli, and Syntrophospora branches shown in Fig. 1. The high G + C Gram-positive bacteria have been assigned to the *"Actinobacteria"* and encompass the Arthrobacters, the Myco/Coryne/Nocardia group, the Atopobias, and the Streptomycetes. The *Cyanobacteria* represent the last of the phyla depicted in the figure and consistently appear in close proximity to the Gram-positive bacteria, but represent a distinct phylogenetic lineage. These major groups are also shown in Fig. 2, in which boxes have been used to enclose the related groups. Only a few representative groups are shown in the figures because of space considerations.

THE PHYLA

The use of a phylogenetic schema for the organization and presentation of the contents of the *Manual* represents a departure from the first edition, in which genera were grouped together based on a few readily determined phenotypic criteria. All accepted genera have now been placed into a provisional taxonomic framework based upon the best available 16S rDNA sequence data (>1000 nts and <0.01% ambiguities). At the close of 1999, 16S rDNA sequences for approximately two-thirds of the validly published type strains were publicly available. In those instances where 16S rDNA data for the type species were not available, placement of a genus into the framework was based on either phenotypic characteristics or data derived from a closely related species. Some validly named genera are known to be either para[chphyletic or polyphyletic. In such cases, allocation of the genus within the taxonomic framework was determined by the phylogenetic position of the type species. The order in which taxa are presented is based, in part, on the topology of the RDP tree (Release 6.01, November 1997) with some notable exceptions. This is discussed in more detail in the article entitled "Roadmap to the Manual".

It is generally agreed that procaryotes fall into two major lines

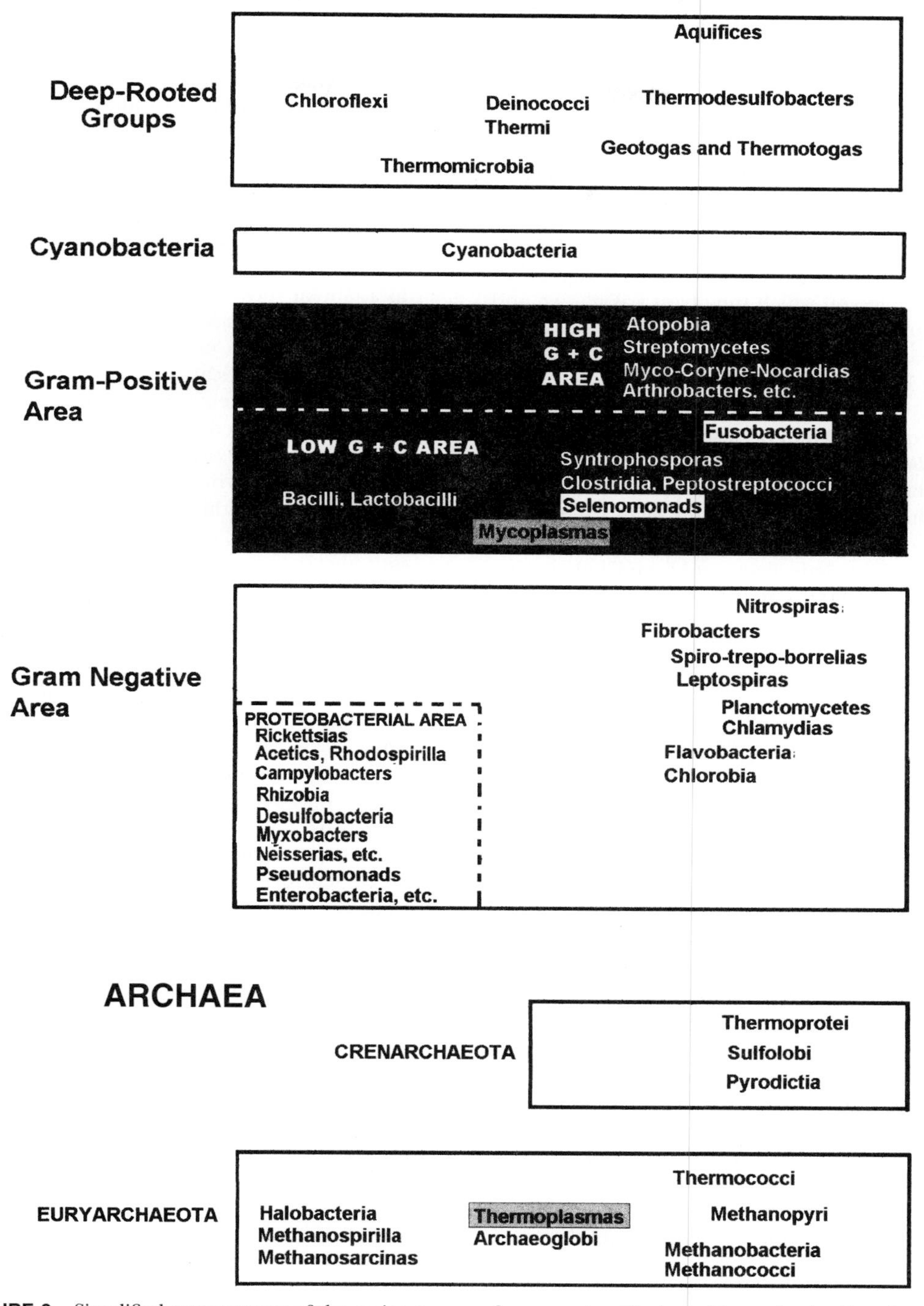

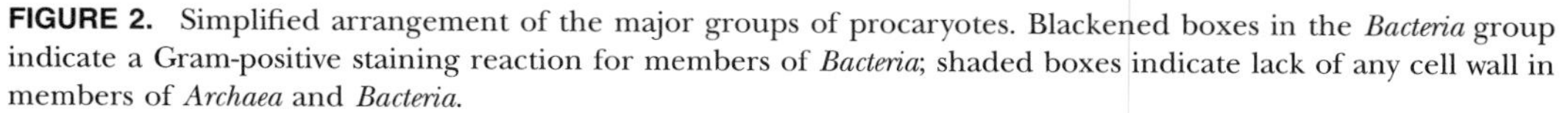

FIGURE 2. Simplified arrangement of the major groups of procaryotes. Blackened boxes in the *Bacteria* group indicate a Gram-positive staining reaction for members of *Bacteria*; shaded boxes indicate lack of any cell wall in members of *Archaea* and *Bacteria*.

of evolutionary descent: the ***Archaea*** and ***Bacteria***. These will be dealt with as **domains**. The domains have been further subdivided into phyla that represent the major procaryotic lineages and will serve as the main organizational unit in this edition of the *Manual*. At present, the *Archaea* have been subdivided into two phyla and the *Bacteria* into 23 phyla. With the exception of the *Cyanobacteria* and the *"Actinobacteria"*, phyla are further subdivided into classes, orders, and families. In the case of the former, families are replaced by subsections, which may be further divided into subgroups. In addition, species generally do not appear as discrete entities. Rather, these are represented by strain designations. Some genera are also referred to as Form-genera. In the case of *"Actinobacteria"*, the taxonomic hierarchy is slightly modified and includes subclasses and suborders. Readers should note that names above the class level are not covered by the *Bacteriological Code* and should be regarded as informal or colloquial names. Furthermore, as additional sequence data become available, some phyla are likely to be combined while others may be split.

Articles

Each article dealing with a bacterial genus is presented wherever possible in a definite sequence as follows:

a. *Name of the Genus.* Accepted names are in **boldface**, followed by the authority for the name, the year of the original description, and the page on which the taxon was named and described. The superscript AL indicates that the name was included on the Approved Lists of Bacterial Names, published in January 1980. The superscript VP indicates that the name, although not on the Approved Lists of Bacterial Names, was subsequently validly published in the *International Journal of Systematic Bacteriology.* Names given within quotation marks have no standing in nomenclature; as of the date of preparation of the *Manual* they had not been validly published in the *International Journal of Systematic Bacteriology,* although they had been "effectively published" elsewhere. Names followed by the term "gen. nov." are newly proposed but will not be validly published until they appear in the *International Journal of Systematic Bacteriology*; their proposal in the *Manual* constitutes only "effective publication", not valid publication.
b. *Name of Author(s).* The person or persons who prepared the Bergey article are indicated. The address of each author can be found in the list of Contributors at the beginning of the *Manual.*
c. *Synonyms.* In some instances a list of some synonyms used in the past for the same genus is given. Other synonyms can be found in the *Index Bergeyana* or the *Supplement to the Index Bergeyana.*
d. *Etymology of the Genus Name.* Etymologies are provided as in previous editions, and many (but undoubtedly not all) errors have been corrected. It is often difficult, however, to determine why a particular name was chosen, or the nuance intended, if the details were not provided in the original publication. Those authors who propose new names are urged to consult a Greek and Latin authority before publishing, in order to ensure grammatical correctness and also to ensure that the name means what it is intended to mean.
e. *Capsule Description.* This is a brief resume of the salient features of the genus. The most important characteristics are given in **boldface**. The name of the type species of the genus is also indicated.
f. *Further Descriptive Information.* This portion elaborates on the various features of the genus, particularly those features having significance for systematic bacteriology. The treatment serves to acquaint the reader with the overall biology of the organisms but is not meant to be a comprehensive review. The information is presented in a definite sequence, as follows:
   i. Colonial morphology and pigmentation
   ii. Growth conditions and nutrition
   iii. Physiology and metabolism
   iv. Genetics, plasmids, and bacteriophages
   v. Phylogenetic treatment
   vi. Antigenic structure
   vii. Pathogenicity
   viii. Ecology
g. *Enrichment and Isolation.* A few selected methods are presented, together with the pertinent media formulations.
h. *Maintenance Procedures.* Methods used for maintenance of stock cultures and preservation of strains are given.
i. *Procedures for Testing Special Characters.* This portion provides methodology for testing for unusual characteristics or performing tests of special importance.
j. *Differentiation of the Genus from Other Genera.* Those characteristics that are especially useful for distinguishing the genus from similar or related organisms are indicated here, usually in a tabular form.
k. *Taxonomic Comments.* This summarizes the available information related to taxonomic placement of the genus and indicates the justification for considering the genus a distinct taxon. Particular emphasis is given to the methods of molecular biology used to estimate the relatedness of the genus to other taxa, where such information is available. Taxonomic information regarding the arrangement and status of the various species within the genus follows. Where taxonomic controversy exists, the problems are delineated and the various alternative viewpoints are discussed.
l. *Further Reading.* A list of selected references, usually of a general nature, is given to enable the reader to gain access to additional sources of information about the genus.
m. *Differentiation of the Species of the Genus.* Those characteristics that are important for distinguishing the various species within the genus are presented, usually with reference to a table summarizing the information.
n. *List of the Species of the Genus.* The citation of each species is given, followed in some instances by a brief list of objective synonyms. The etymology of the specific epithet is indicated. Descriptive information for the species is usually presented in tabular form, but special information may be given in the text. Because of the emphasis on tabular data, the species descriptions are usually brief. The type strain of each species is indicated, together with the collection(s) in which it can be found. (Addresses of the various culture collections are given in the article entitled Culture Collections: An Essential Resource for Microbiology.) The 16S rRNA gene sequence used in phylogenetic analysis and for placement of the genus into the taxonomic framework is given, along with the GenBank accession number and RDP identifier for the aligned sequence. Additional comments may be provided to point the reader to other well-characterized strains of the species and any other known DNA sequences that may be relevant.
o. *Species Incertae Sedis.* The List of Species may be followed in some instances by a listing of additional species under the heading "Species Incertae Sedis". The taxonomic placement or status of such species is questionable and the reasons for the uncertainty are presented.
p. *Literature Cited.* All references given in the article are listed alphabetically at the end of the volume rather than at the end of each article.

Tables

In each article dealing with a genus, there are generally three kinds of tables: (a) those that differentiate the genus from similar or related genera, (b) those that differentiate the species within the genus, and (c) those that provide additional information about the species (such information not being particularly useful for differentiation). Unless otherwise indicated, the meanings of symbols are as follows:

+: 90% or more of the strains are positive
d: 11–89% of the strains are positive
−: 90% or more of the strains are negative
D: different reactions occur in different taxa (e.g., species of a genus or genera of a family)
v: strain instability (NOT equivalent to "d")

Exceptions to the use of these symbols, as well as the meaning of additional symbols, are clearly indicated in footnotes to the tables.

Use of the *Manual* for Determinative Purposes

Each chapter has keys or tables for differentiation of the various taxa contained therein. Suggestions for identification may be found in the article on Polyphasic Taxonomy. For identification of species, it is important to read both the generic and species descriptions because characteristics listed in the generic descriptions are not usually repeated in the species descriptions.

The index is useful for locating the names of unfamiliar taxa and discovering what has been done with a particular taxon. Every bacterial name mentioned in the *Manual* is listed in the index. In addition, an up-to-date outline of the taxonomic framework, along with an alphabetized listing of genera, is provided in the Roadmap to the Manual. The table also provides the reader with an indication to which section a genus either was, or would have been, assigned in the first edition.

Errors, Comments, Suggestions

As indicated in the Preface to the first edition of *Bergey's Manual of Determinative Bacteriology*, the assistance of all microbiologists in the correction of possible errors in the text is earnestly solicited. Comments on the presentation will also be welcomed, as well as suggestions for future editions. Correspondence should be addressed to:

Editorial Office,
Bergey's Manual Trust,
Michigan State University
East Lansing, MI 48824–4320
Telephone 517–432–2457;
fax 517–432–2458;
e-mail: garrity@msu.edu

Procaryotic Domains

Noel R. Krieg

Procaryotes can be described as follows:

Single cells or simple associations of similar cells (usually 0.2–10.0 μm in smallest dimension, although some are much larger) forming a group defined by cellular, not organismal, properties (i.e., by the structure and components of the cells of an organism rather than by the properties of the organism as a whole). The nucleoplasm (genophore) is, with a few exceptions, not separated from the cytoplasm by a unit-membrane system (nuclear membrane). Cell division is not accompanied by cyclical changes in the texture or staining properties of either nucleoplasm or cytoplasm; a microtubular (spindle) system is not formed. The plasma membrane (cytoplasmic membrane) is frequently complex in topology and forms vesicular, lamellar, or tubular intrusions into the cytoplasm; vacuoles and replicating cytoplasmic organelles independent of the plasma membrane system (chlorobium vesicles, gas vacuoles) are relatively rare and are enclosed by nonunit membranes. Respiratory and photosynthetic functions are associated with the plasma-membrane system in those members possessing these physiological attributes, although in the cyanobacteria there may be an independence of plasma and thylakoid membranes. Ribosomes of the 70S type (except for one group—the *Archaea*—with slightly higher S values) are dispersed in the cytoplasm; an endoplasmic reticulum with attached ribosomes is not present. The cytoplasm is immobile; cytoplasmic streaming, pseudopodial movement, endocytosis, and exocytosis are not observed. Nutrients are acquired in molecular form. Enclosure of the cell by a rigid wall is common but not universal. The cell may be nonmotile or may exhibit swimming motility (mediated by flagella of bacterial type) or gliding motility on surfaces.

In organismal terms, these ubiquitous inhabitants of moist environments are predominantly unicellular microorganisms, but filamentous, mycelial, or colonial forms also occur. Differentiation is limited in scope (holdfast structures, resting cells, and modifications in cell shape). Mechanisms of gene transfer and recombination occur, but these processes never involve gametogenesis and zygote formation.

Although procaryotic organisms can usually be readily differentiated from eucaryotic microorganisms, in some instances it may be difficult, especially with procaryotes that exhibit some attributes similar to those of microscopic eucaryotes. For instance, the hyphae formed by actinomycetes might be confused with the hyphae formed by molds; a fascicle of bacterial flagella could give the misleading impression of being a single eucaryotic flagellum; the ability of spirochetes to twist and contort their shape is suggestive of the flexibility exhibited by certain protozoa; some eucaryotic cells are as small as bacteria, and some bacteria are as large as eucaryotic cells (see footnote to Table 1). The most reliable approach is probably the demonstration of the absence of a nuclear membrane in procaryotes, but this involves electron microscopy of thin sections. Other procaryotic cell features range from those that are relatively easy to determine to the molecular characteristics that require sophisticated methods. Fluorescent- labeled gene probes that can easily distinguish between procaryotic and eucaryotic cells have been developed.

Some characteristics that may help to differentiate between procaryotes and eucaryotes are listed in Table 1.

Archaea vs. *Bacteria*

As shown in Figs. 1 and 2 in "On Using the Manual", the two fundamentally different groups (domains) that comprise the procaryotes are the *Bacteria* and the *Archaea*. Recent phylogenetic analyses of the *Bacteria*, *Archaea*, and *Eucarya* using conserved protein sequences have shown that the majority of trees support a closer relationship between the *Archaea* and *Eucarya* than between *Archaea* and *Bacteria*. Although the possibility of interdomain horizontal gene transfer complicates this picture, the apparent *Archaea–Eucarya* sisterhood raises interesting questions about the phylogenetic relationships between procaryotes and eucaryotes and the root of the universal tree of life. Table 2 provides some characteristics differentiating these two procaryotic groups. A general description of each group follows.

Bacteria For practical purposes the *Bacteria* may be divided into three phenotypic subgroups: (1) those that are Gram-negative and have a cell wall, (2) those that are Gram-positive and have a cell wall, and (3) those that lack any cell wall. (See chapters by Garrity and Holt and Ludwig and Klenk, this *Manual*, for a discussion of the phylogenetic relationships between Gram-positive and Gram-negative *Bacteria*.)

Gram-negative *Bacteria* that have a cell wall These have a Gram-negative type of cell-wall profile consisting of an outer membrane and an inner, relatively thin peptidoglycan layer (which contains muramic acid and is present in all but a few organisms that have lost this portion of wall; see footnote to Table 1) as well as a variable complement of other components outside or between these layers. They usually stain Gram-negative, although the presence of a thick exopolysaccharide layer around the outer membrane may result in a Gram-positive staining reaction, as seen in the cyst-like forms of some *Azospirillum* species. Cell shapes may be spheres, ovals, straight or curved rods, helices, or filaments; some of these forms may be sheathed or capsulated. Reproduction is by binary fission, but some groups show budding, and a rare group (Subsection II of the cyanobacteria) shows

TABLE 1. Some differential characteristics of procaryotes and eucaryotes[a]

Characteristic	Procaryotes	Eucaryotes
Cytological features		
Nucleoplasm (genophore, nucleoid) separated from the cytoplasm by a unit-membrane system (nuclear membrane)	–	+
Size of smallest dimension of cells (width or diameter):		
Usually 0.2–2.0 µm	+[b]	–
Usually 2.0 µm	–	+
Mitochondria present	–	Usually +
Chloroplasts present in phototrophs	–	+
Vacuoles, if present, enclosed by unit membranes	–	+
Gas vacuoles present[c]	D	–
Golgi apparatus present	–	D
Lysosomes present	–	D
Microtubular systems present	–[d]	D
Endoplasmic reticulum present	–	+
Ribosome location:		
Dispersed in the cytoplasm	+	–
Attached to endoplasmic reticulum	–	+
Cytoplasmic streaming, pseudopodial movement, endocytosis, and exocytosis	–	D
Cell division accompanied by cyclical changes in the texture or staining properties of either nucleoplasm or cytoplasm	–	+
Diameter of flagella, if present:		
0.01–0.02 µm	+	–
~0.2 µm	–	+
In cross-section, flagella have a characteristic "9 + 2" arrangement of microtubules	–	+
Endospores present[e]	D	–
Antibiotic susceptibility		
Susceptible to penicillin, streptomycin, or other antibiotics specific for procaryotes	D	–
Susceptible to cycloheximide or other antibiotics specific for eucaryotes	–	D
Features based on chemical analysis		
Poly-β-hydroxybutyrate present (as a storage compound in cytoplasmic inclusions)	D	–
Teichoic acids present (in cell walls)	D	–
Polyunsaturated fatty acids possibly present (in membranes)	Rare	Common
Branched-chain *iso-* or *anteiso-*fatty acids and cyclopropane fatty acids present (in membranes)	Common	Rare
Sterols present (in membranes)	–[f]	Common
Diaminopimelic acid present (in cell walls)	D[g]	–
Muramic acid present (in cell walls)	D[h]	–
Peptidoglycan (containing muramic acid) present in cell walls	D[h]	–
Nutrition		
Nutrients acquired by cells as soluble small molecules; to serve as sources of nutrients, particulate matter or large molecules must first be hydrolyzed to small molecules by enzymes external to the plasma membrane	+	D
Metabolic features		
Respiratory and photosynthetic functions and associated pigments and enzymes (e.g., chlorophylls, cytochromes), if present, are associated with the plasma membrane or invaginations thereof	+[i]	–
Chemolithotrophic type of metabolism occurs (inorganic compounds can be used as electron donors by organisms that derive energy from chemical compounds)	D	–
Ability to fix N_2	D	–
Ability to dissimilate NO_3^- to N_2O or N_2	D	–
Methanogenesis	D	–
Ability to carry out anoxygenic photosynthesis	D	–
Enzymic features		
Type of superoxide dismutase:		
Cu-Zn type	Rare	+
Mn and/or Fe type	+	–[j]
Reproductive features		
Cell division includes mitosis, and a microtubular (spindle) system is present	–	+
Meiosis occurs	–	D
Mechanisms of gene transfer and recombination, if they occur, involve gametogenesis and zygote formation	–	+
Molecular biological properties		
Number of chromosomes present per nucleoid	Usually 1	Usually 1
Chromosomes circular	+	–
Chromosomes linear	–[k]	+
Sedimentation constant of ribosomes:		
70S	+	–[l]
80S	–	+
Sedimentation constants of ribosomal RNA:		
16S, 23S, 5S	+	–
18S, 28S, 5.85S, 5S	–	+

(continued)

TABLE 1. *(continued)*

Characteristic	Procaryotes	Eucaryotes
First amino acid to initiate a polypeptide chain during protein synthesis:		
Methionine	D	+
N-Formylmethionine	D	−
Messenger-RNA binding site at AUCACCUCC at 3′ end of 16S or 18S ribosomal RNA	+	−

[a]Symbols: +, positive; −, negative; D, differs among organisms.

[b]A few bacteria (e.g., certain treponemes, mycoplasmas, *Haemobartonella*) may have a width as small as 0.1 μm; a few bacteria (e.g., *Achromatium, Macromonas*) may have a width greater than 10 μm. The largest known procaryote is a spherical, sulfur bacterium provisionally named *"Thiomargarita namibiensis"* and has a diameter of 100–750 μm. It is a member of the γ *Proteobacteria* and is related to the genus *Thioploca*. Its cytoplasm occurs as a narrow layer surounding a large, central, liquid vacuole that contains nitrate. The organism has not yet been isolated. Another large procaryote is *Epulopiscium fishelsoni*, a noncultured cigar-shaped bacterium that inhabits the intestinal tract of surgeonfish from the Red Sea; it can be larger than 80 × 600 μm. This organism is also viviparous, producing two live daughter cells within the mature cell.

[c]Gas vacuoles are not bounded by a unit membrane but by a protein. The vesicles composing the vacuoles can be caused to collapse by the sudden application of hydrostatic pressures, a feature essential to identify them. They can also be identified by electron microscopy.

[d]However, certain intracellular fibrils that may be microtubules have been reported in *Spiroplasma*, certain spirochetes, the cyanobacterium *Anabaena*, and in bacterial L forms.

[e]Bacterial endospores are usually resistant to a heat treatment of 80°C or more for 10 min, however, some types of endospores may be killed by this heat treatment and may require testing at lower temperatures.

[f]Except in membranes of most mycoplasmas.

[g]Present in virtually all Gram-negative bacteria and in many Gram-positive bacteria.

[h]Present in walled *Bacteria* except chlamydiae and planctomycetes; absent in *Archaea*.

[i]However, in cyanobacteria there may be an independence of cytoplasmic membrane and thylakoid membranes.

[j]Except in mitochondria, in which the Mn type occurs.

[k]A few bacteria such as some *Borrelia* species have linear chromosomes.

[l]Except in mitochondria and chloroplasts, which have 70S ribosomes.

multiple fission. Fruiting bodies and myxospores may be formed by the myxobacteria. Swimming motility, gliding motility, and nonmotility are commonly observed. Members may be phototrophic or nonphototrophic (both lithotrophic and heterotrophic) bacteria and include aerobic, anaerobic, facultatively anaerobic, and microaerophilic species; some members are obligate intracellular parasites.

GRAM-POSITIVE *BACTERIA* THAT HAVE A CELL WALL These *Bacteria* have a cell-wall profile of the Gram-positive type; there is no outer membrane and the peptidoglycan layer is relatively thick. Some members of the group have teichoic acids and/or neutral polysaccharides as components of the wall. A few members of the group have cell walls that contain mycolic acids. Reaction with Gram's stain is generally, but not always, positive; exceptions such as *Butyrivibrio*, which has an unusually thin wall and stains Gram-negative, may occur. Cells may be spheres, rods, or filaments; the rods and filaments may be nonbranching, but many show true branching. Cellular reproduction is generally by binary fission; some produce spores as resting forms (endospores or spores on hyphae). The members of this division include simple asporogenous and sporogenous bacteria, as well as the actinomycetes and their relatives. Gram-positive *Bacteria* are generally chemosynthetic heterotrophs and include aerobic, anaerobic, facultatively anaerobic, and microaerophilic species; some members are obligate intracellular parasites. Only one group, the heliobacteria, is photosynthetic and although these have a Gram-positive type of cell wall they nevertheless stain Gram- negative.

Table 3 provides some characteristics that help to differentiate the Gram-positive *Bacteria* from the Gram-negative *Bacteria*.

BACTERIA LACKING A CELL WALL These *Bacteria* are commonly called the mycoplasmas. They do not synthesize the precursors of peptidoglycan and are insensitive to β-lactam antibiotics or other antibiotics that inhibit cell wall synthesis. They are enclosed by a unit membrane, the plasma membrane. The cells are highly pleomorphic and range in size from large deformable vesicles to very small (0.2 μm), filterable elements. Filamentous forms with branching projections are common. Reproduction may be by budding, fragmentation, and/or binary fission. Some groups show a degree of regularity of form due to the placing of internal structures. Usually, they are nonmotile, but some species show a form of gliding motility. No resting forms are known. Cells stain Gram-negative. Most require complex media for growth (high-osmotic-pressure surroundings) and tend to penetrate the surface of solid media forming characteristic "fried egg" colonies. The organisms resemble the naked L-forms that can be generated from many species of bacteria (notably Gram-positive *Bacteria*) but differ in that the mycoplasmas are unable to revert and make cell walls. Most species are further distinguished by requiring both cholesterol and long-chain fatty acids for growth; unesterified cholesterol is a unique component of the membranes of both sterol-requiring and nonrequiring species if present in the medium. The mol% G + C content of rRNA is 43–48 (lower than the 50–54 mol% of walled Gram-negative and Gram-positive *Bacteria*); the mol% G + C content of the DNA is also relatively low, 23–46, and the genome size of the mycoplasmas at 0.5–1.0 × 10^9 Da is less than that of other procaryotes. The mycoplasmas may be saprophytic, parasitic, or pathogenic, and the pathogens cause diseases of animals, plants, and tissue cultures.

Archaea The *Archaea* are predominantly terrestrial and aquatic microbes, occurring in anaerobic, hypersaline, or hydrothermally and geothermally heated environments; some also occur as symbionts in animal digestive tracts. They consist of aerobes, anaerobes, and facultative anaerobes that grow chemolithoautotrophically, organotrophically, or facultatively organotrophically. *Archaea* may be mesophiles or thermophiles, with some species growing at temperatures up to 110°C.

The major groups of the *Archaea* include (a) the methanogenic *Archaea*, (b) the sulfate-reducing *Archaea*, (c) the extremely

TABLE 2. Some characteristics differentiating *Bacteria* from *Archaea*[a]

Characteristic	*Bacteria*	*Archaea*
General morphological and metabolic features		
Strict anaerobes that form methane as the predominant metabolic end product from H_2/CO_2, formate, acetate, methanol, methylamine or H_2/methanol. Cells exhibit a blue-green epifluorescence when excited at 420 nm	−	D
Strict anaerobes that form H_2S from sulfate by dissimilatory sulfate reduction. Extremely thermophilic (some grow up to 110°C). Exhibit blue-green epifluorescence when excited at 420 nm	−	D
Cells stain Gram negative or Gram positive and are aerobic or facultatively anaerobic chemoorganotrophs. Rods and regular to highly irregular cells occur. Cells require a high concentration of NaCl (1.5 M or above). Neutrophilic or alkaliphilic. Mesophilic or slightly thermophilic (up to 55°C). Some species contain the red-purple photoactive pigment bacteriorhodopsin and are able to use light for generating a proton motive force	−	D
Thermoacidophilic, aerobic, coccoid cells lacking a cell wall	−	D
Obligately thermophilic, aerobic, facultatively anaerobic, or strictly anaerobic Gram-negative rods, filaments, or cocci. Optimal growth temperature between 70°C and 105°C. Acidophiles and neutrophiles. Autotrophic or heterotrophic. Most species are sulfur metabolizers	−	D
Cell walls (if present)		
Contain muramic acid	+[b]	−
Antibiotic susceptibility		
Susceptible to β lactam antibiotics	D	−
Lipids		
Membrane phospholipids consist of:		
long chain alcohols (phytanols) that are ether linked to glycerol to form C_{20} diphytanyl glycerol diethers or C_{40} dibiphytanyl diglycerol tetraethers	−	+
Long chain aliphatic fatty acids that are ester linked to glycerol	+	−
Pathway used in formation of lipids:		
Mevalonate pathway	−	+
Malonate pathway	+	−
Molecular biological features		
Ribothymine is present in the "common arm" of the tRNAs	Usually +	−
Pseudouridine or 1-methylpseudouridine is present in the "common arm" of the tRNAs	−	+
First amino acid to initiate a polypeptide chain during protein synthesis:		
Methionine	−	+
N-Formylmethionine	+	−
Aminoacyl stem of the initiator tRNA terminates with the base pair "AU"	−	+
Protein synthesis by ribosomes inhibited by:		
Anisomycin	−	+
Kanamycin	+	−
Chloramphenicol	+	−
ADP-Ribosylation of the peptide elongation factor EF-2 is inhibited by diphtheria toxin	−	+
Elongation factor 2 (EF-2) contains the amino acid diphthamide	−	+
Some tRNA genes contain introns	−	+
DNA-dependent RNA polymerases are:		
Multicomponent enzymes	−	+
Inhibited by rifampicin and streptolydigin	+	−
Replicating DNA polymerases are inhibited by aphidicolin or butylphenyl-dGTP	+	−

[a] Symbols: +, positive; −, negative; D, differs among organisms.

[b] Except planctomycetes and chlamydiae, which have a protein wall.

halophilic *Archaea*, (d) the *Archaea* lacking cell walls, and (e) the extremely thermophilic S^0-metabolizing *Archaea*.

A unique biochemical feature of all *Archaea* is the presence of glycerol isopranyl ether lipids. The lack of murein (peptidoglycan-containing muramic acid) in cell walls makes *Archaea* insensitive to β-lactam antibiotics. The "common arm" of the tRNAs contains pseudouridine or 1-methylpseudouridine instead of ribothymidine. The nucleotide sequences of 5S, 16S, and 23S rRNAs are very different from their counterparts in *Bacteria* and *Eucarya*.

Archaea share some molecular features with *Eucarya*: (a) the elongation factor 2 (EF-2) contains the amino acid diphthamide and is therefore ADP-ribosylable by diphtheria toxin, (b) amino acid sequences of the ribosomal "A" protein exhibit sequence homologies with the corresponding eucaryotic (L-7/L12) protein, (c) the methionyl initiator tRNA is not formylated, (d) some tRNA genes contain introns, (e) the aminoacyl stem of the initiator tRNA terminates with the base pair "AU," (f) the DNA-dependent RNA polymerases are multicomponent enzymes and are insensitive to the antibiotics rifampicin and streptolydigin, (g) like the α-DNA polymerases of eucaryotes, the replicating DNA polymerases of *Archaea* are not inhibited by aphidicolin or butylphenyl-dGTP, and (h) protein synthesis is inhibited by anisomycin but not by chloramphenicol.

Autotrophic *Archaea* do not assimilate CO_2 via the Calvin cycle. In methanogens, autotrophic CO_2 is fixed via a pathway involving the unique coenzymes methanofuran, tetrahydromethanopterin, coenzyme F_{420}, HS-HTP, coenzyme M, HTP-SH, and coenzyme F_{430} whereas in *Acidianus* and *Thermoproteus*, autotrophic CO_2 is fixed via a reductive tricarboxylic acid pathway. Some methanogenic *Archaea* can fix N_2.

Gram stain results may be positive or negative within the same subgroup because of very different types of cell envelopes. Gram-positive-staining species possess pseudomurein, methanochondroitin, and heteropolysaccharide cell walls; Gram-negative-staining cells have (glyco-) protein surface layers. The cells may have a diversity of shapes, including spherical, spiral, plate or rod; unicellular and multicellular forms in filaments or aggregates also occur. The diameter of an individual cell may be 0.1–15 μm, and the length of the filaments can be up to 200 μm. Multiplication is by binary fission, budding, constriction, fragmentation, or unknown mechanisms. Colors of cell masses may be red, purple, pink, orange-brown, yellow, green, greenish black, gray, and white.

TABLE 3. Some characteristics differentiating Gram-positive bacteria having a cell wall from Gram-negative bacteria having a cell wall[a]

Characteristic	Gram-negative bacteria	Gram-positive bacteria
Cytological features		
An outer membrane is present (in the cell wall) in addition to the plasma (cytoplasmic) membrane	+	–
Acid-fast staining	–	D[b]
Endospores present	–[c]	D[d]
Filamentous growth with hyphae that show true branching	–	D
Locomotion		
Gliding motility occurs	D	–
Chemical features		
Percentage of the dry weight of the cell wall that is represented by lipid	Usually 11–22%	Usually 4%[e]
Teichoic or lipoteichoic acids present	–	D
Lipopolysaccharide (LPS)[f] occurs (in the outer membrane of the cell wall)	+	–
2-Keto-3-deoxyoctonate (KDO) present[g]	D	–
Percentage of the dry weight of the cell wall represented by peptidoglycan	Usually <10%	Usually 10%
Mycolic acids present	–	D[h]
Phosphatidylinositol mannosides present	–	D[i]
Phosphosphingolipids present	D[j]	–
Metabolic features		
Energy derived by the oxidation of inorganic iron, sulfur, or nitrogen compounds	–	D
Enzymic features		
Citrate synthases:		
Inhibited by reduced nicotinamide adenine dinucleotide (NADH)	Usually +[k]	Usually –
Molecular weight:		
~250,000	Usually +[l]	Usually –
~100,000	Usually –	Usually +
Succinate thiokinases, molecular weight of:		
70,000–75,000	Usually –	Usually +
140,000–150,000	Usually +	Usually –

[a] Symbols: +, positive; –, negative, D, differs among organisms.

[b] Acid-fast staining occurs in the genus *Mycobacterium*, and in some *Nocardia* species.

[c] An exception may be the genus *Coxiella*.

[d] Endospores occur in the genera *Bacillus, Clostridium, Desulfotomaculum, Sporosarcina, Thermoactinomyces, Sporomusa, Metabacterium*, and *Polyspora*.

[e] Except for *Mycobacterium, Corynebacterium, Nocardia*, and other genera whose walls contain mycolic acids.

[f] LPS consists of Lipid A (a β-linked D-glucosamine disaccharide to which phosphate residues are linked at positions 1 and 4 and fatty acids are linked to both the amino and hydroxyl groups of the glucosamines), a core polysaccharide (a short acidic heteropolysaccharide), and O antigens (side chains that are polysaccharide composed of repeating units).

[g] In many but not all Gram-negative bacteria, the core polysaccharide contains KDO which, if present, can serve as an indicator of the presence of LPS.

[h] Mycolic acids occur in *Corynebacterium, Nocardia, Mycobacterium, Bacterionema, Faenia*, and *Rhodococcus*.

[i] Present in certain actinomycete and coryneform bacteria.

[j] For example, in *Bacteroides*.

[k] Known exceptions include *Acetobacter, Thermus*, and cyanobacteria.

[l] One known exception is *Thermus*.

Classification of Procaryotic Organisms and the Concept of Bacterial Speciation

Don J. Brenner, James T. Staley and Noel R. Krieg

Classification Nomenclature and Identification

Taxonomy is the science of classification of organisms. Bacterial taxonomy consists of three separate, but interrelated areas: classification, nomenclature, and identification. Classification is the arrangement of organisms into groups (taxa) on the basis of similarities or relationships. Nomenclature is the assignment of names to the taxonomic groups according to international rules (*International Code of Nomenclature of Bacteria* [Sneath, 1992]). Identification is the practical use of a classification scheme to determine the identity of an isolate as a member of an established taxon or as a member of a previously unidentified species.

Some 4000 bacterial species thus far described (and the tens of thousands of postulated species that remain to be described) exhibit great diversity. In any endeavor aimed at an understanding of large numbers of entities it is practical, if not essential, to arrange, or classify, the objects into groups based upon their similarities. Thus classification has been used to organize the bewildering and seemingly chaotic array of individual bacteria into an orderly framework. Classification need not be scientific. Mandel said that "like cigars,... a good classification is one which satisfies" (Mandel, 1969). Cowan observed that classification is purpose oriented; thus, a successful classification is not necessarily a good one, and a good classification is not necessarily successful (Cowan, 1971, 1974).

Classification and adequate description of bacteria require knowledge of their morphologic, biochemical, physiological, and genetic characteristics. As a science, taxonomy is dynamic and subject to change on the basis of available data. New findings often necessitate changes in taxonomy, frequently resulting in changes in the existing classification, in nomenclature, in criteria for identification, and in the recognition of new species. The process of classification may be applied to existing, named taxa, or to newly described organisms. If the taxa have already been described, named, and classified, new characteristics may be added or existing characteristics may be reinterpreted to revise existing classification, update it, or formulate a new one. If the organism is new, i.e., cannot be identified as an existing taxon, it is named and described according to the rules of nomenclature and placed in an appropriate position in an existing classification, i.e., a new species in either an existing or a new genus.

Taxonomic ranks Several levels or ranks are used in bacterial classification. The highest rank is called a Domain. All procaryotic organisms (i.e., bacteria) are placed within two Domains, *Archaea* and *Bacteria*. Phylum, class, order, family, genus, species, and subspecies are successively smaller, non-overlapping subsets of the Domain. The names of these subsets from class to subspecies are given formal recognition (have "standing in nomenclature"). An example is given in Table 1. At present, neither the kingdom nor division are used for *Bacteria*. In addition to these formal, hierarchical taxonomic categories, informal or vernacular groups that are defined by common descriptive names are often used; the names of such groups have no official standing in nomenclature. Examples of such groups are: the procaryotes, the spirochetes, dissimilatory sulfate- and sulfur-reducing bacteria, the methane-oxidizing bacteria, methanogens, etc.

Species The basic and most important taxonomic group in bacterial systematics is the species. The concept of a bacterial species is less definitive than for higher organisms. This difference should not seem surprising, because bacteria, being procaryotic organisms, differ markedly from higher organisms. Sexuality, for example, is not used in bacterial species definitions because relatively few bacteria undergo conjugation. Likewise, morphologic features alone are usually of little classificatory significance because the relative morphologic simplicity of most procaryotic organisms does not provide much useful taxonomic information. Consequently, morphologic features are relegated to a less important role in bacterial taxonomy in comparison with the taxonomy of higher organisms.

The term "species" as applied to bacteria has been defined as a distinct group of strains that have certain distinguishing features and that generally bear a close resemblance to one another in the more essential features of organization. (A strain is made up of the descendants of a single isolation in pure culture, and usually is made up of a succession of cultures ultimately

TABLE 1. Taxonomic ranks

Formal rank	Example
Domain	*Bacteria*
Phylum	*Proteobacteria*
Class	*Alphaproteobacteria*
Order	*Legionellales*
Family	*Legionellaceae*
Genus	*Legionella*
Species	*Legionella pneumophila*
Subspecies	*Legionella pneumophila* subsp. subsp. *pneumophila*

derived from an initial single colony). Each species differs considerably and can be distinguished from all other species.

One strain of a species is designated as the type strain; this strain serves as the name-bearer strain of the species and is the permanent example of the species, i.e., the reference specimen for the name. (See the chapter on Nomenclature for more detailed information about nomenclatural types). The type strain has great importance for classification at the species level, because a species consists of the type strain and all other strains that are considered to be sufficiently similar to it as to warrant inclusion with it in the species. Any strain can be designated as the type strain, although, for new species, the first strain isolated is usually designated. The type strain need not be a typical strain.

The species definition given above is one that was loosely followed until the mid-1960s. Unfortunately, it is extremely subjective because one cannot accurately determine "a close resemblance", "essential features", or how many "distinguishing features" are sufficient to create a species. Species were often defined solely on the basis of relatively few phenotypic or morphologic characteristics, pathogenicity, and source of isolation. The choice of the characteristics used to define a species and the weight assigned to these characteristics frequently reflected the interests and prejudices of the investigators who described the species. These practices probably led Cowan to state that "taxonomy... is the most subjective branch of any biological science, and in many ways is more of an art than a science" (Cowan, 1965).

Edwards and Ewing (1962, 1986) were pioneers in establishing phenotypic principles for characterization, classification and identification of bacteria. They based classification and identification on the overall morphologic and biochemical pattern of a species, realizing that a single characteristic (e.g., pathogenicity, host range, or biochemical reaction) regardless of its importance was not a sufficient basis for speciation or identification. They employed a large number of biochemical tests, used a large and diverse strain sample, and expressed results as percentages. They also realized that atypical strains, when adequately studied, are often perfectly typical members of a given biogroup (biovar) within an existing species, or typical members of a new species.

Numerical taxonomic methods further improved the validity of phenotypic identification by further increasing the number of tests used, usually to 100–200, and by calculating coefficients of similarity between strains and species (Sneath and Sokal, 1973). Although there is no similarity value that defines a taxospecies (species determined by numerical taxonomy), 80% similarity is commonly seen among strains in a given taxospecies. Despite the additional tests and added sensitivity of numerical taxonomy, even a battery of 300 tests would assess only between 5–20% of the genetic potential of bacteria.

It has long been recognized that the most accurate basis for classification is phylogenetic. Kluyver and van Niel (1936) stated that "many systems of classification are almost entirely the outcome of purely practical considerations . . . (and) are often ultimately impractical . . . " They recognized that "taxonomic boundaries imposed by the intuition of investigators will always be somewhat arbitrary—especially at the ultimate systematic unit, the species. One must create as many species as there are organisms that differ in sufficiently fundamental characters" and they realized that "the only truly scientific foundation of classification is in appreciating the available facts from a phylogenetic view". The data necessary to develop a natural (phylogenetic) species definition became available when DNA hybridization was utilized to determine relatedness among bacteria.

DNA hybridization is based upon the ability of native (double-stranded) DNA to reversibly dissociate or be denatured into its two complementary single strands. Dissociation is accomplished at high temperature. Denatured DNA will remain as single strands when it is quickly cooled to room temperature after denaturation. If it is then placed at a temperature between 25 and 30°C below its denaturation point, the complementary strains will reassociate to again form a double-stranded molecule that is extremely similar, if not identical, to native DNA (Marmur and Doty, 1961). Denatured DNA from a given bacterium can be incubated with denatured DNA (or RNA) from other bacteria and will form heteroduplexes with any complementary sequences present in the heterologous strand–DNA hybridization. This is the method used to determine DNA relatedness among bacteria.

Perfectly complementary sequences are not necessary for hybridization; the degree of complementary required for heteroduplex formation can be governed experimentally by changing the incubation temperature or the salt concentration. Increasing the incubation temperature and/or lowering the salt concentration in the incubation mixture increases the stringency of heteroduplex formation (fewer unpaired bases are tolerated), whereas decreasing the temperature and/or increasing the salt concentration decreases the stringency of heteroduplex formation. The percentage of unpaired bases within a heteroduplex is an indication of the degree of divergence present. One can approximate the amount of unpaired bases by comparing the thermal stability of the heteroduplex to the thermal stability of a homologous duplex. This is done by stepwise increases in temperature and measuring strand separation. The thermal stability is calculated as the temperature at which 50% of strand separation has occurred and is represented by the term "$T_{m(e)}$".

The ΔT_m values of heteroduplexes range from 0 (perfect pairing) to ~20°C, with each degree of instability indicative of approximately 1% divergence (unpaired bases). As DNA relatedness between two strains decreases, divergence usually increases.

A number of different DNA–DNA and DNA–RNA hybridization methods have been used to determine relatedness among bacteria (Johnson, 1985). Two of these, free solution reassociation with separation of single- and double-stranded DNA on hydroxyapatite (Brenner et al., 1982) and the S-1 endonuclease method (Crosa et al., 1973) are currently the most widely used for this purpose. These methods have been shown to be comparable (Grimont et al., 1980). An in-depth discussion of DNA hybridization methods has been presented by Grimont et al. (1980) and by Johnson (1985).

Experience with thousands of strains from several hundred well-established and new species led taxonomists to formulate a phylogenetic definition of a species (genomospecies) as "strains with approximately 70% or greater DNA–DNA relatedness and with 5°C or less ΔT_m. Both values must be considered" (Wayne et al., 1987). They further recommended that a genomospecies not be named if it cannot be differentiated from other genomospecies on the basis of some phenotypic property. DNA relatedness provides a single species definition that can be applied equally to all organisms and is not subject to phenotypic variation, mutations, or variations in metabolic or other plasmids. The major advantage of DNA relatedness is that it measures overall relatedness, and therefore the effects of atypical biochemical reactions, mutations, and plasmids are minimal since they affect only a very small percentage of the total DNA.

Once genomospecies have been established, it is simple to determine which variable biochemical reactions are species specific, and therefore to have an identification scheme that is compatible with the genetic concept of species. The technique is also extremely useful in determining the biochemical boundaries of a species, as exemplified for *Escherichia coli* in Table 2. The use of DNA relatedness and a variety of phenotypic characteristics in classifying bacteria has been called polyphasic taxonomy (Colwell, 1970), and seems to be the best approach to a valid description of species. DNA relatedness studies have now been carried out on more than 10,000 strains representing some 2000 species and hundreds of genera, with, to our knowledge, no instance where other data invalidated the genomospecies definition.

Stackebrandt and Goebel (1994) reviewed new species descriptions published in the *International Journal of Systematic Bacteriology*. In 1987, 60% of species descriptions included DNA relatedness studies, 10% were described on the basis of serologic tests, and 30% did not use these approaches. In 1993, 75% of species descriptions included DNA relatedness data, 8% used serology, and 3% used neither method. In the remaining 14%, 16S rRNA sequence analysis was the sole basis for speciation. As 16S rRNA sequence data have accumulated, the utility of this extremely powerful method for phylogenetic placement of bacteria has become evident (Woese, 1987; Ludwig et al., 1998b). The number of taxonomists using 16S rRNA sequencing is or soon will be greater than the number using DNA hybridization (Stackebrandt and Goebel, 1994), and many of them were creating species solely or largely on the basis of 16S rRNA sequence analysis. It soon became evident, however, that 16S rRNA sequence analysis was frequently not sensitive enough to differentiate between closely related species (Fox et al., 1992; Stackebrandt and Goebel, 1994). Stackebrandt and Goebel (1994) concluded that the genetic definition of 70% relatedness with 5% or less divergence within related sequences continues to be the best means of creating species. They concluded that 16S rRNA sequence similarity of less than 97% between strains indicates that they represent different species, but at 97% or higher 16S rRNA sequence similarity, DNA relatedness must be used to determine whether strains belong to different species.

TABLE 2. Classification of atypical strains that could be *E. coli*

Relatedness of biogroup to typical *E. coli*	Characteristic
80% or more	Urea positive and KCN positive
	Mannitol negative
	Inositol positive
	Adonitol positive
	H_2S positive or H_2S positive and yellow pigmented
	H_2S positive and citrate positive
	Citrate positive
	Phenylalanine deaminase positive
	Lysine and ornithine decarboxylase and arginine dihydrolase negative
	Indol negative
	Methyl red negative
	Methyl red negative and mannitol negative
	Urea positive and mannitol negative
	Anaerogenic, nonmotile, and lactose negative
60% or less	Yellow pigment, cellobiose positive, and KCN positive = *Escherichia hermannii*
	Urea positive, KCN positive, citrate positive, cellobiose positive = *Citrobacter amalonaticus*

The validity and utility of the DNA relatedness based genetic definition of a species has been questioned (Maynard Smith, 1995; Vandamme et al., 1996a; Istock et al., 1996). These criticisms fall into several categories: (a) DNA relatedness (and any other current means of speciation) does not sufficiently sample bacterial diversity by employing large numbers of wild isolates from many different habitats; (b) it employs an arbitrary cutoff for a species whereas evolution is a continuum; (c) the DNA-relatedness based definition does not achieve standardization of species; (d) bacterial species are not real entities—named species are useful but not meaningful from an evolutionary standpoint; (e) DNA relatedness results are not comparable due to different methods; (f) DNA relatedness tests are too difficult and/or tedious to perform. In view of these perceived problems, it has been recommended that the best solution to the species problem in the absence of a "gold standard", which has not been provided by DNA relatedness, is a pragmatic polyphasic (consensus) taxonomy that integrates all available data.

Each of these criticisms has some merit; however each can be addressed, and none, in our opinion, represent fatal flaws nor significantly negate the usefulness of the DNA-relatedness based definition of a species. Large numbers of diverse strains (50–100) have been tested for DNA relatedness in a number of species including *E. coli*, *Legionella pneumophila*, *Enterobacter agglomerans*, *Klebsiella oxytoca*, *Yersinia enterocolitica*. In no case did the sample size or the diversity of sources and/or phenotypic characteristics change the results. For many other species only one or a few strains were tested—usually because that was the total number of strains available.

It is true that the 70% relatedness and 5% divergence values chosen to represent strains of a given species are arbitrary, and that there is a "gray area" around 70% for some species. Nonetheless, these values were chosen on the basis of results obtained from multiple strains, usually 10 or more, of some 600 species studied in a number of different reference laboratories. There are few, if any cases, in which the species defined in this manner have been shown to be incorrect.

The DNA relatedness approach has standardized the means of defining species by providing a single, universally applicable criterion. Since it has been successful, one must believe that it generates species that are compatible with the needs and beliefs of most bacteriologists. There are two areas in which genomospecies have actually or potentially caused problems. One of these is where two or more genomospecies cannot be separated phenotypically. In this case it has been recommended that these genomospecies not be formally named (Wayne et al., 1987). Alternatively, especially if a name already exists for one of the genomospecies, the others can be designated as subspecies. In this way there is no confusion at the species level and, one can, if one wishes, distinguish between the genomospecies using a genetic technique. The other "problem" is with nomenspecies that were split or lumped, usually on the basis of pathogenicity or phytopathogenic host range. These include species in the genera *Bordetella*, *Mycobacterium*, *Brucella*, *Shigella*, *Klebsiella*, *Neisseria*, *Yersinia*, *Vibrio*, *Clostridium*, and *Erwinia*. In some of these cases (*Klebsiella*, *Erwinia*) the classification has been changed and is now accepted. In the others, changes have not yet been proposed or, as in the case of *Yersinia pestis* and *Yersinia pseudotuberculosis*, which are the same genomospecies, the change was rejected by the Judicial Commission because of possible danger to public health if there was confusion regarding *Y. pestis*, the plague bacillus.

If one agrees that a true species definition is not possible, the genomospecies definition is still useful in providing a single, universally applicable basis for designating species.

To criticize DNA relatedness because results obtained using different methods may not be totally comparable seems somewhat unjustified. When compared, the most frequently used methods have given similar results. Obviously, one should be careful in comparing data from various laboratories, especially when different methods are used. However, this is at least equally true for sequence data and phenotypic tests.

It is true that large amounts of DNA are required for the DNA relatedness protocols now used for taxonomic purposes, and that it is necessary to use radioactive isotopes. As for the difficulty involved and the limitations in strains that can be assayed (it is not uncommon to do 40–80 DNA relatedness comparisons daily), surely these are not credible reasons to stop using the method. Efforts can and should be made to automate the system, to miniaturize it, and to substitute nonradioactive compounds for the radioactive isotopes. With these improvements, the method will be available for use in virtually any laboratory. Even without them, one can argue that DNA hybridization is more affordable and practical than a consensus classification system in which several hundred tests must be done on each strain.

It is noteworthy that bacterial species can be compared to higher organisms on a molecular basis using mol% G + C range, DNA–DNA or DNA–rRNA relatedness, and similarity of 16S vs. 18S rDNA sequences (Staley, 1997, 1999). Thus, *E. coli* can be compared with its primate hosts based on the results of DNA–DNA hybridization. When this is done, it is apparent that the bacterial species is much broader than that of its hosts. For example, humans and our closest relative, the chimpanzee (*Pan troglodytes*), show 98.4% relatedness by this technique (Sibley and Ahlquist, 1987; Sibley et al., 1990). Indeed, even lemurs, which exhibit 78% DNA relatedness with humans, would be included in the same species as humans if the definition of a bacterial species was used. Furthermore, none of the primates would be considered to be threatened species using the bacterial definition. Likewise, the range of mol% G + C and the range of small subunit ribosomal RNA within *E. coli* strains shows a similar result, namely, that the bacterial species is much broader than that of animals (Staley, 1999).

One consequence of the broad bacterial species definition is that very few species have been described, fewer than 5000, compared with over a million animals. This has led some biologists to erroneously conclude that bacteria comprise only a minor part of the biological diversity on Earth (Mayr, 1998). In addition, with such a broad definition, not a single free-living bacterial species can be considered to be threatened with extinction (Staley, 1997). Therefore, biologists should realize, as mentioned earlier in this section, that the bacterial species is not at all equivalent to that of plants and animals.

In summary, the genetic definition of a species, if not perfect, appears to be both reliable and stable. DNA relatedness studies have already resolved many instances of confusion concerning which strains belong to a given species, as well as for resolving taxonomic problems at the species level. It has not been replaced as the current reference standard. It should remain the standard, at least until another approach has been compared to it and shown to be comparable or superior.

Subspecies A species may be divided into two or more subspecies based on consistent phenotypic variations or on genetically determined clusters of strains within the species. There is evidence that the subspecies concept is phylogenetically valid on the basis of frequency distribution of ΔT_m values. There are presently essentially no guidelines for the establishment of subspecies, which, although frequently useful, are usually designated at the pleasure of the investigator. Subspecies is the lowest taxonomic rank that is covered by the rules of nomenclature and has official standing in nomenclature.

Infrasubspecific Ranks Ranks below subspecies, such as biovars, serovars, phagovars, and pathovars, are often used to indicate groups of strains that can be distinguished by some special character, such as antigenic makeup, reactions to bacteriophage, etc. Such ranks have no official standing in nomenclature, but often have great practical usefulness. A list of some common infrasubspecific categories is given in Table 3.

Genus All species are assigned to a genus, which can be functionally defined as one or more species with the same general phenotypic characteristics, and which cluster together on the basis of 16S rRNA sequence. In this regard, bacteriologists conform to the binomial system of nomenclature of Linnaeus in which the organism is designated by its combined genus and species names. There is not, and perhaps never will be, a satisfactory definition of a genus, despite the fact that most new genera are designated substantially on the basis of 16S rRNA sequence analysis. In almost all cases, genera can be differentiated phenotypically, although a considerable degree of flexibility in genus descriptions is often needed. Considerable subjectivity continues to be involved in designating genera, and considerable reclassification, both lumping and splitting, is still occurring at the genus level. Indeed, what is perceived to be a single genus by one systematist may be perceived as multiple genera by another.

Higher Taxa Classificatory relationships at the familial and higher levels are even less certain than those at the genus level, and descriptions of these taxa are usually much more general, if they exist at all. Families are composed of one or more genera that share phenotypic characteristics and that should be consistent from a phylogenetic standpoint (16S rRNA sequence clustering) as well as from a phenotypic basis.

Major developments in bacterial classification

A century elapsed between Antony van Leeuwenhoek's discovery of bacteria and Müller's initial acknowledgement of bacteria in

TABLE 3. Infrasubspecific designations

Preferred name	Synonym	Applied to strains having:
Biovar	Biotype	Special biochemical or physiologic properties
Serovar	Serotype	Distinctive antigenic properties
Pathovar	Pathotype	Pathogenic properties for certain hosts
Phagovar	Phage type	Ability to be lysed by certain bacteriophages
Morphovar	Morphotype	Special morphologic features

a classification scheme (Müller, 1786). Another century passed before techniques and procedures had advanced sufficiently to permit a fairly inclusive and meaningful classification of these organisms. For a comprehensive review of the early development of bacterial classification, readers should consult the introductory sections of the first, second, and third editions of *Bergey's Manual of Determinative Bacteriology*. A less detailed treatment of early classifications can be found in the sixth edition of the *Manual*, in which post-1923 developments were emphasized.

Two primary difficulties beset early bacterial classification systems. First, they relied heavily upon morphologic criteria. For example, cell shape was often considered to be an extremely important feature. Thus, the cocci were often classified together in one group (family or order). In contrast, contemporary schemes rely much more strongly on 16S rRNA sequence similarities and physiological characteristics. For example, the fermentative cocci are now separated from the photosynthetic cocci, which are separated from the methanogenic cocci, which are in turn separated from the nitrifying cocci, and so forth; with the 16S rRNA sequences of each group generally clustered together. Secondly, the pure culture technique which revolutionized bacteriology was not developed until the latter half of the 19th century. In addition to dispelling the concept of "polymorphism", this technical development of Robert Koch's laboratory had great impact on the development of modern procedures in bacterial systematics. Pure cultures are analogous to herbarium specimens in botany. However, pure cultures are much more useful because they can be (a) maintained in a viable state, (b) subcultured, (c) subjected indefinitely to experimental tests, and (d) shipped from one laboratory to another. A natural outgrowth of the pure culture technique was the establishment of type strains of species which are deposited in repositories referred to as "culture collections" (a more accurate term would be "strain collections"). These type strains can be obtained from culture collections and used as reference strains to duplicate and extend the observations of others, and for direct comparison with new isolates.

Before the development of computer-assisted numerical taxonomy and subsequent taxonomic methods based on molecular biology, the traditional method of classifying bacteria was to characterize them as thoroughly as possible and then to arrange them according to the intuitive judgment of the systematist. Although the subjective aspects of this method resulted in classifications that were often drastically revised by other systematists who were likely to make different intuitive judgments, many of the arrangements have survived to the present day, even under scrutiny by modern methods. One explanation for this is that the systematists usually knew their organisms thoroughly, and their intuitive judgments were based on a wealth of information. Their data, while not computer processed, were at least processed by an active mind to give fairly accurate impressions of the relationships existing between organisms. Moreover, some of the characteristics that were given great weight in classification were, in fact, highly correlated with many characteristics. This principle of correlation of characteristics appears to have started with Winslow and Winslow (1908), who noted that parasitic cocci tended to grow poorly on ordinary nutrient media, were strongly Gram-positive, and formed acid from sugars, in contrast to saprophytic cocci which grew abundantly on ordinary media, were generally only weakly Gram-positive and formed no acid. This division of the cocci studied by the Winslows (equivalent to the present genus *Micrococcus* (the saprophytes) and the genera *Staphylococcus* and *Streptococcus* (the parasites) has held up reasonably well even to the present day.

Other classifications have not been so fortunate. A classic example of one which has not is that of the genus *"Paracolobactrum"*. This genus was proposed in 1944 and is described in the Seventh Edition of *Bergey's Manual* in 1957. It was created to contain certain lactose-negative members of the family *Enterobacteriaceae*. Because of the importance of a lactose-negative reaction in identification of enteric pathogens (i.e., *Salmonella* and *Shigella*), the reaction was mistakenly given great taxonomic weight in classification as well. However, for the organisms placed in *"Paracolobactrum"*, the lactose reaction was not highly correlated with other characteristics. In fact, the organisms were merely lactose-negative variants of other lactose-positive species; for example *"Paracolobactrum coliform"* resembled *E. coli* in every way except in being lactose-negative. Absurd arrangements such as this eventually led to the development of more objective methods of classification, i.e., numerical taxonomy, in order to avoid giving great weight to any single characteristic.

Phylogenetic Classifications We have already discussed the impact of DNA relatedness at the species level. Unfortunately, this method is of marginal value at the genus level and of no value above the genus level because the extent of divergence of total bacterial genomes is too great to allow accurate assessment of relatedness above the species level. At the genus level and above, phylogenetic classifications, especially as based on 16S rRNA sequence analysis, have revolutionized bacterial taxonomy (see Overview: A Phylogenetic Backbone and Taxonomic Framework for Procaryotic Systematics by Ludwig and Klenk).

Official Classifications A significant number of bacteriologists have the impression that there is an "official classification" and that the classification presented in *Bergey's Manual* represents this "official classification". It is important to correct that misimpression. There is no "official classification" of bacteria. (This is in contrast to bacterial nomenclature, where each taxon has one [and usually only one] valid name, according to internationally agreed-upon rules, and judicial decisions are rendered in instances of controversy about the validity of a name.) The closest approximation to an "official classification" of bacteria would be one that is widely accepted by the community of microbiologists. A classification that is of little use to bacteriologists, regardless of how fine a scheme or who devised it, will soon be ignored or significantly modified. The editors of *Bergey's Manual* and the authors of each chapter make substantial efforts to provide a classification that is as accurate and up-to-date as possible, however it is not and cannot be "official".

It also seems worthwhile to emphasize something that has often been said before, viz. bacterial classifications are devised for microbiologists, not for the entities being classified. Bacteria show little interest in the matter of their classification. For the systematist, this is sometimes a very sobering thought!

Note Added in Proof

Recently a committee of bacterial taxonomists met to re-evaluate the bacterial species definition (Stackebrandt et al., 2002b). The committee recognized that, since the report by Wayne et al. (1987), several new methods have been developed that greatly aid in bacterial taxonomy, including 16S rDNA sequence analyses, restriction enzyme typing methods, multilocus sequencing, whole genome sequence analyses, Fourier-Transformed Infrared Spectroscopy and pyrolysis-mass spectrometry. Special methods noted by the committee that show great promise for taxonomists include sequencing of housekeeping genes, DNA profiling and the application of DNA arrays. Microbiologists were encouraged to develop new methods that would allow data to be compared to DNA–DNA reassociation, which the committee concluded should remain the standard for species circumscription for *Bacteria* and *Archaea.* Other recommendations were made to base the species description on more than a single strain, to follow guidelines established by the subcommittees of ICSP (International Committee on Systematics of Prokaryotes) for minimal characterization of a species, and to recognize the importance of phenotypic properties for species identification. Also, because electronic databases are an immensely important aid for the international community of bacterial systematists, the committee recommended the development of standards for electronic exchange of taxonomic information.

Acknowledgments

This chapter is dedicated to the memory of John L. Johnson, a consummate scientist, trusted colleague and friend, whose search for truth was uncompromising and unhindered by personal ego.

Identification of Procaryotes

Noel R. Krieg

The Nature of Identification Schemes

Identification schemes are not classification schemes, although there may be a superficial similarity. An identification scheme for a group of organisms can be devised only **after** that group has first been classified (i.e., recognized as being different from other organisms). Identification of that group is based on one or more characteristics, or on a pattern of characteristics, which all the members of the group have and which other groups do not have.

The particular pattern of characteristics used for identifying a bacterial group should not be found in any other bacterial group. Following classification of the group, a relatively few characteristics which, taken together, are unique to that group are selected. The identifying characteristics may be phenotypic, such as cell shape and Gram reaction or the ability to ferment certain sugars, or they may be genotypic, such as a particular nucleotide sequence.

Pure Cultures

Although it is possible to identify specific organisms, and even individual cells, in a mixed culture, pure cultures are usually used for identification. Moreover, in most laboratories identification is still being done mainly on the basis of the phenotypic characteristics of the culture, although it may be aided by commercial multitest identification systems, usually involving 96–well microtiter plates, that are capable of determining a variety of characteristics easily and quickly. Phenotypic identification systems work reasonably well with pure cultures, but if the culture is not pure the results will be a composite from all of the different organisms in the culture and thus can be very misleading.

In obtaining a pure culture, it is important to realize that the selection of a single colony from a plate does not necessarily assure purity. This is especially true if selective media are used; live but non-growing contaminants may often be present in or near a colony and can be subcultured along with the chosen organism. It is for this reason that non-selective media are preferred for final isolation, because they allow such contaminants to develop into visible colonies. Even with non-selective media, apparently well-isolated colonies should not be isolated too soon; some contaminants may be slow growing and may appear on the plate only after a longer incubation. Another difficulty occurs with bacteria that form extracellular slime or that grow as a network of chains or filaments; contaminants often become firmly embedded or entrapped in such matrixes and are difficult to remove. In the instance of cyanobacteria, contaminants frequently penetrate and live in the gelatinous sheaths that surround the cells, making pure cultures difficult to obtain.

In general, colonies from a pure culture that has been streaked on a solid medium are similar to one another, providing evidence of purity. Although this is generally true, there are exceptions, as in the case of S → R variation, capsular variants, pigmented or nonpigmented variants, etc., which may be selected by certain media, temperatures, or other growth conditions. Another criterion of purity is morphology: organisms from a pure culture generally exhibit a high degree of morphological similarity in stains or wet mounts. Again, there are exceptions, coccoid body formation, cyst formation, spore formation, pleomorphism, etc., depending on the age of the culture, the medium used, and other growth conditions. For example, examination of a broth culture of a marine spirillum after 2 or 3 days may lead one to believe the culture is highly contaminated with cocci, unless one is previously aware that following active growth such spirilla generally develop into thin-walled coccoid forms.

Universal Systems for Identifying a Pure Culture Although the goal of identification is merely to provide the name of an isolate, most identification systems depend on first determining a number of morphological, biochemical, cultural, antigenic, and other phenotypic characteristics of the isolate before the name can be assigned. An ideal universal system would be one that provides the name without having to determine these characteristics. In a sense, such a system would be a kind of "black box" into which the isolate, or an extract of it, is placed, to be followed some time later by a display of the name of the organism.

One system that has proven extremely useful is automated cellular fatty acid (CFA) analysis (Onderdonk and Sasser, 1995). The system depends on saponifying the fatty acids with sodium hydroxide, converting them to their volatile methyl esters, and then separating and quantifying each fatty acid by gas-liquid chromatography. A computer compares the resulting fatty acid profile with thousands of others in a huge database and calculates the best match or matches for the isolate. The computer can also indicate that an isolate does not closely match any other fatty acid profile, which can lead to discovery of new genera or species. The entire procedure is simple and takes about 2 h, and numerous specimens can be analyzed rapidly each day. One drawback is that the isolate must be cultured under highly standardized conditions of media and temperature in order to provide a valid basis of comparison with other fatty acid profiles. Another drawback is that the system may not be able to differentiate species that are very closely related by DNA–DNA hybridization, for example, *Escherichia coli* and *Shigella.* Still another drawback is that the system is extremely expensive to purchase or lease. Some commercial laboratories will perform the entire identification procedure on an isolate that is sent to them; this is helpful for one or a few isolates but becomes expensive if many isolates are to be identified.

A second universal system, and the one of choice at present, is one in which all or most of the nucleotide sequence of the 16s rRNA gene of an unknown isolate is determined. DNA is isolated from the strain and then universal primers are used to amplify the 16S rDNA by the polymerase chain reaction (PCR). The sequence of the PCR product is compared with other sequences stored in an enormous database. One such database is that used in the Ribosomal Database Project-II (RDP-II), which is a cooperative effort by scientists at Michigan State University and the Lawrence Berkeley National Laboratory. 16S sequencing is rapid enough to handle a large number of isolates in a short time; this service is provided by a number of institutions for medical and other types of isolates. Given a well-equipped sequencing lab, 16S sequences can be obtained and analyzed within 48 hours. The main drawback with sequence-based identification is that of the need for sophisticated equipment, which is present in relatively few microbiology labs. Another drawback is that although sequence-based identification is very effective for assigning an isolate to its most likely genus, it may not be able to identify an isolate to the species level if the sequences for two or more related species have greater than 97% similarity.

Traditional Identification Schemes for Identifying a Pure Culture

Phenotypic characteristics chosen for an identification scheme should be easily determinable by most microbiology laboratories. Such characteristics should not be restricted to research laboratories or special facilities. Characteristics useful for identification are often not those that were involved in classification of the group. Classification might be based on a DNA–DNA hybridization study or on ribosomal RNA gene sequencing, whereas identification might be based on a few phenotypic characteristics that have been found to correlate well with the genetic information. Serological reactions, which generally have only limited value for classification, often have enormous value for identification. Slide agglutination tests, fluorescent antibody techniques, and other serological methods can be performed simply and rapidly and are usually highly specific; therefore, they offer a means for achieving quick, presumptive identification of bacteria. Their specificity is frequently not absolute, however, and confirmation of the identification by additional tests is usually required.

The goal of having easily determinable identifying characteristics may not always be possible, particularly with genera or species that are not susceptible to being identified by traditional phenotypic tests. For instance, the inability of *Campylobacter* species to use sugars makes phenotypic identification of species of this genus much more difficult than, say, the species of the family *Enterobacteriaceae.* In such instances one may need to resort to less common phenotypic characteristics such as the ability to grow at a specific temperature, antibiotic susceptibilities, and the ability to grow anaerobically with various electron acceptors such as trimethylamine oxide. There may even be a requirement for more sophisticated procedures, such as the use of cellular lipid patterns, DNA–DNA hybridization, or nucleic acid probes, in order to achieve an accurate identification. It may even be necessary to send the culture in question to a major reference facility that has the necessary equipment and technical expertise.

Identification of a strain should depend on a pattern of several characteristics, not merely one or a very few characteristics. If one feature is given great importance, it is possible that some strains may be mutants that do not exhibit that particular characteristic yet do have the other identifying features. For instance, hippurate hydrolysis was given great emphasis in differentiating *Campylobacter jejuni* from other *Campylobacter* species, but later it was discovered that hippurate-negative strains may occur. At first, these hippurate-negative strains were incorrectly thought to belong to a different species, *Campylobacter coli,* until DNA–DNA hybridization experiments showed that this was not correct.

Identification should rely on relatively few characteristics compared to classification schemes. Classification may involve hundreds of characteristics, as in a numerical taxonomy study but the prospect of inoculating hundreds of tubes of media in order to identify a strain is daunting. It may be possible, however, to use a large number of characteristics if they can be determined easily. To alleviate the need for inoculating large numbers of tubed media, a variety of convenient and rapid multitest systems have been devised and are commercially available for use in identifying particular groups of bacteria, particularly those of medical importance. A summary of some of these systems has been given by Smibert and Krieg (1994) and Miller and O'Hara (1995) but new systems are being developed continually. Each manufacturer provides charts, tables, coding systems, and characterization profiles for use with the particular multitest system being offered. It is important to realize that each system is for use in identifying only certain taxa and may not be applicable to other taxa. For instance, the commercial systems for identifying members of the family *Enterobacteriaceae* would give results that would be meaningless for identifying *Campylobacter* species.

Determination of the characteristics chosen for an identification scheme should be relatively inexpensive. Ordinary microbiology laboratories may not be able to afford expensive apparatus such as those required for cellular fatty acid profiles, 16S rRNA gene sequencing, or DNA probes. In regard to the latter, commercial kits for using DNA probes to identify particular taxa may be simple to use but may also be quite expensive. In general, such probes are best reserved for situations where it is essential to make a definitive identification because no other method will suffice.

The identification scheme should give results rapidly. This is especially true in clinical microbiology laboratories, where the treatment of a patient often depends on a rapid (but accurate) identification, and sometimes even a presumptive identification, of a pathogen. Serological methods have long been used for rapid detection of antigens associated with a particular species. For instance, a swab of the throat of a person with suspected case of streptococcal pharyngitis can be treated to extract the Lancefield Group A polysaccharide indicative of *Streptococcus pyogenes.* Anti-Group A antibodies can then be used in various ways, such as an ELISA test, to identify this antigen. Fluorescent antibodies can be used to obtain presumptive identification of individual cells in a mixture. For instance, cells of *Streptococcus pyogenes* can be seen in a swab from streptococcal pharyngitis by using fluorescent Lancefield Group A antiserum, and cells of *Vibrio cholerae* can be seen in diarrheic stools of cholera patients by using fluorescent O Group I antiserum. Antibodies are not always completely specific, however, and definitive identification usually requires isolation of the organism and determination of various identifying features.

Need for standardized test methods. One difficulty in devising identification schemes based on phenotypic characteristics is that the results of characterization tests may vary depending on the size of the inoculum, incubation temperature, length of the incubation period, composition of the medium, the surface-to-volume ratio of the medium, and the criteria used to define a "positive" or "negative" reaction. Therefore, the results of characterization tests obtained by one laboratory often do not match exactly those obtained by another laboratory, although the results within each laboratory may be quite consistent. The blind acceptance of an identification scheme without reference to the particular conditions employed by those who devised the scheme can lead to error (and, unfortunately, such conditions are not always specified). Ideally, it would be desirable to standardize the conditions used for testing various characteristics, but this is easier said than done, especially on an international basis. The use of commercial multitest systems offers some hope of improving standardization among various laboratories because of the high degree of quality control exercised over the media and reagents, but no one system has yet been agreed on for universal use with any given taxon. **It is therefore advisable to always include strains whose identity has been firmly established** (type or reference strains, available from national culture collections) **for comparative purposes when making use of an identification scheme,** to make sure that the scheme is valid for the conditions employed in one's own laboratory.

Need for definitions of "positive" and "negative" reactions. Some tests may be found to be based on plasmid- or phage-mediated characteristics; such characteristics may be highly mutable and therefore unreliable for identification purposes. Even with immutable characteristics, certain tests are not well suited for use in identification schemes because they may not give highly reproducible results (e.g., the catalase test, oxidase test, Voges-Proskauer test, and gelatin liquefaction are notorious in this regard). Ideally, a test should give reproducible results that are clearly either positive or negative, without equivocal reactions. In fact, no such test exists. The Gram reaction of an organism may be "Gram variable," the presence of endospores in a strain that makes only a few spores may be very difficult to determine by staining or by heat-resistance tests, acid production from sugars may be difficult to distinguish from no acid production if only small amounts of acid are produced, and a weak growth response may not be clearly distinguishable from "no growth". A precise (although arbitrary) definition of what constitutes a "positive" and "negative" reaction is often important in order for a test to be useful for an identification scheme.

Sequence of tests used in identifying an isolate. In identifying an isolate, it is important to determine the **most general features first**. For instance, it would not be wise to begin by determining that melibiose is fermented, gelatin is liquefied, and that nitrate is reduced. Instead, it is better to begin with more general features such as the Gram staining reaction, morphology, and general type of metabolism. It is important to establish whether the new isolate is a chemolithotrophic autotroph, a photosynthetic organism, or a chemoheterotrophic organism. Living cells should be examined by phase-contrast microscopy and Gram-stained cells by light microscopy; other stains can be applied if this seems appropriate. If some outstanding morphological property, such as endospore production, sheaths, holdfasts, acid-fastness, cysts, stalks, fruiting bodies, budding division, or true branching, is obvious, then further efforts in identification can be confined to those groups having such a property. Whether or not the organisms are motile, and the type of motility (swimming, gliding) may be very helpful in restricting the range of possibilities. Gross growth characteristics, such as pigmentation, mucoid colonies, swarming, or a minute size, may also provide valuable clues to identification. For example, a motile, Gram-negative rod that produces a water-soluble fluorescent pigment is likely to be a *Pseudomonas* species, whereas one that forms bioluminescent colonies is likely to belong to the family *Vibrionaceae.*

The **source** of the isolate can also help to narrow the field of possibilities. For example, a spirillum isolated from coastal sea water is likely to be an *Oceanospirillum* species, whereas Gram-positive cocci occurring in grape-like clusters and isolated from the human nasopharynx are likely to belong to the genus *Staphylococcus.*

The relationship of the isolate to oxygen (i.e., whether it is aerobic, anaerobic, facultatively anaerobic, or microaerophilic) is often of fundamental importance in identification. For example, a small microaerophilic vibrio isolated from a case of diarrhea is likely to be a *Campylobacter* species, whereas a Gram-negative anaerobic rod isolated from a wound infection may well be a member of the genera *Bacteroides, Prevotella, Porphyromonas,* or *Fusobacterium.* Similarly, it is important to test the isolate for its ability to dissimilate glucose (or other simple sugars) to determine if the type of metabolism is oxidative or fermentative, or whether sugars are catabolized at all.

Above all, common sense should be used at each stage, as the possibilities are narrowed, in deciding what additional tests should be performed. There should be a reason for the selection of each test, in contrast to a "shotgun" type of approach where many tests are used but most provide little pertinent information for the particular isolate under investigation. As the category to which the isolate belongs becomes increasingly delineated, one should follow the specific tests indicated in the particular diagnostic tables or keys that apply to that category.

The following summary is taken from "The Mechanism of Identification" by S.T. Cowan and J. Liston in the eighth edition of the *Manual,* with some modifications:

1. Make sure that you have a pure culture.
2. Work from broad categories down to a smaller, specific category of organism.
3. Use all the information available to you in order to narrow the range of possibilities.
4. Apply common sense at each step.
5. Use the minimum number of tests to make the identification.
6. Compare your isolate to type or reference strains of the pertinent taxon to make sure the **identification** scheme being used is actually valid for the conditions in your particular laboratory.

If, as may well happen, you cannot identify your isolate from the information contained in the *Manual,* neither despair nor immediately assume that you have isolated a new genus or species; many of the problems of microbial classification are the result of people jumping to this conclusion prematurely. When you fail to identify your isolate, check (a) its **purity**, (b) that you have carried out the **appropriate tests**, (c) that your **methods are reliable**, and (d) that you have used correctly the various keys and tables of the *Manual.* It has been said that the most frequent cause of mistaken identity of bacteria is error in the determination of shape, Gram-staining reaction, and motility. In most cases, you should have little difficulty in placing your isolate into a genus; allocation to a species or subspecies may need the help of a specialized reference laboratory.

On the other hand, it is always possible that you have actually isolated a new genus or species. A comparison of the present edition of the *Manual* with the previous edition indicates that many new genera and species have been added. Undoubtedly, there exist in nature a great number of bacteria that have not yet been classified and therefore cannot yet be identified by existing schemes. However, before describing and naming a new taxon, one must be **very sure that it is really a new taxon** and not merely the result of an inadequate identification.

Use of Probes for for Identification of a Particular Species

DNA probes have made it possible to identify an isolate definitively without relying on phenotypic tests. A probe is a single-stranded DNA sequence that can be used to identify an organism by forming a "hybrid" with a unique complementary sequence on the DNA or rRNA of that organism. Using probes as a "shotgun" approach to identification of an isolate, however, is costly and time-consuming. In general, probes are mainly used to verify the identification of an isolate after the microbiologist already has fairly good clues as to its identity.

Whether a probe consists of only a few nucleotides or many nucleotides, it must be specific for the particular species and must not bind to the DNA of other species. Also, the probe must have a label attached to it so that if it forms a hybrid duplex with a complementary sequence, that duplex can be readily detected.

Labeling can be accomplished by incorporating a radioactive isotope such as ^{32}P into the probe so that the hybrid duplex will be detectable by exposure to a photographic film. Because working with radioisotopes is dangerous and requires safe radioactive waste disposal, nonradioactive labeling of probes has become popular. One commonly used method is to chemically link digoxigenin to the probe. After the probe hybridizes to its target DNA an anti-digoxigenin antibody that has been chemically linked to alkaline phosphatase is used. After the antibody-enzyme conjugate binds to the digoxigenin on the probe, adamantyl-1,2-dioetane is applied as a substrate for the enzyme. The chemical reaction emits light which can be detected with photographic film.

Some of the more convenient procedures for identifying an isolate depend on the use of two probes for a particular organism, a detector (or reporter) probe and a capture probe, which bind to different regions of the same target DNA or RNA. (RNA is preferable because a bacterial cell has much more of it than DNA.) The detector probe has an antigenic group attached to it whereas the capture probe has a "tail" composed of a chain of similar nucleotides such as polyA or polyG. This tail allows the probe/target DNA hybrids to be removed by attachment to beads or plastic rods to which are bound the appropriate complementary chains of nucleotides (i.e., polyT or polyC). Detection of the removed hybrids is then done by means of an antibody/enzyme conjugate for the antigenic group on the detector probe, in which the enzyme attached to the antibody catalyzes a color-yielding reaction.

DNA probes can even be used to identify individual cells in mixed cultures under a microscope. A specific DNA probe is conjugated to a fluorescent dye (e.g., see DeLong et al., 1989; Amann et al., 1990b; Angert et al., 1993) and applied to cells on a slide. If hybridization occurs between the probe DNA and the DNA or rRNA of an appropriate cell, the cell will become fluorescent when viewed under a fluorescence microscope. Methods have even been developed for rapid, nonradioactive, *in situ* hybridization with bacteria in paraffin-embedded tissues (Barrett et al., 1997).

Use of Probes for Identification of Multiple Species in Mixed Cultures

The methods of molecular biology have now made possible the definitive identification of many different organisms in a mixed culture, as in a sample of feces, soil, or water. The basis for this is the fact that approximately 70% of the 16S rRNA genes (i.e., 16S rDNA) of all procaryotes is highly conserved (identical in sequence) whereas other regions are unique to particular genera or species. This has made possible the construction of "universal primers" that can bind to any rDNA so that the various 16S rDNAs present in a mixed culture can be amplified by the polymerase chain reaction (PCR). The resulting PCR products are cloned and the unique rDNAs separated. These are sequenced and the corresponding organism is identified by comparing the sequence to a large database of 16S rDNA sequences (for examples, see Wise et al., 1997; Hugenholtz et al., 1998b). The technique of denaturing gradient gel electrophoresis (DGGE) has been found useful for separating the PCR products derived from mixed cultures (e.g., see Teske et al., 1996; Fournier et al., 1998). When applied to mixed cultures from environmental sources such as soil and water, analysis of the 16S rDNA sequences has indicated that many of the sequences cannot be matched with those from any known organisms (i.e., are not identifiable as any cultured, described organism). The results indicate that even the present edition of the *Manual*, large as it is, probably describes less than 1% of existing procaryotic species.

Identification of a Particular Strain of a Species

It is often necessary to identify one strain among the various strains of a species. One example is the need to identify a particular pathogenic strain so that the source of an outbreak of disease can be determined. For instance, one may wish to determine whether a strain of *Legionella pneumophila* isolated from an air conditioning system is the same strain as that isolated from a patient with Legionnaire's disease. As another example, in an ecological study one might be interested in learning whether a particular strain of *Bacillus sphaericus* that has been isolated from one soil sample is present in soil samples from other areas. The following are various methods for differentiating one bacterial strain from another. Some are traditional methods; others are DNA fingerprinting methods based on the techniques of molecular biology. DNA fingerprinting is the most specific way available to identify individual strains of a species.

Traditional methods

Antigenic Typing (Serotyping). Different strains of a species may have different antigens. The antigens present in a particular strain can be determined by the use of specific antisera. As examples, *Streptococcus pyogenes* is divided into >70 antigenic types based on M-proteins, *Streptococcus pneumoniae* is divided into >80 antigenic types based on capsular polysaccharides, and salmonellas are divided into >2000 serotypes based on O and H antigens.

Phage Typing. Strains of a bacterial species may be subject to attack and lysis by numerous bacteriophages. Some phages may attack a particular strain while others do not. The pattern of lysis by various bacteriophages constitutes the phage type of a strain. For example, *Salmonella typhi* can be divided into 33 phage types.

Antibiograms. Which of a large spectrum of antibiotics can inhibit growth of a strain and which cannot constitutes a specific identifying pattern.

DNA fingerprinting This method of identifying a bacterial strain can be done in various ways, as follows.

DNA fingerprinting using a probe and agarose gel electrophoresis. The DNA is treated with a restriction endonuclease to cleave it into many small pieces of differing molecular weight, which are then separated on an agarose gel according to their molecular weight. The gel is treated with an alkali to convert the double-stranded DNA fragments into single-stranded fragments. The pattern of DNA fragments on the gel is transferred to a nitrocellulose membrane and a labeled DNA probe is added. The probe binds only to DNA fragments containing a base sequence complementary to that of the probe. After removing any unbound probe, the location of the bound probe is determined by overlaying the membrane with photographic film, which will be exposed to either radiation or chemiluminescence.

Ribotyping. Ribotyping is a variation of the DNA fingerprinting method in which the DNA probe that is applied to the membrane is complementary to the gene for rRNA. A bacterial chromosome contains genes for three kinds of rRNA (23S, 16S, and 5S rRNA). These genes are transcribed from an *rrn* operon to yield a single large 30S precursor RNA molecule, which then undergoes a maturation process to yield the three different kinds of rRNA. Most operons in procaryotes occur only once on a chromosome but *rrn* operons occur more than once—from 2 to 14 per genome, depending on the species (Rainey et al., 1996). Ribotyping depends on the fact that the sequence of the DNA *between* the *rrn* operons varies from strain to strain in a bacterial species and, consequently, the sites for cleavage of this DNA by a restriction endonuclease will vary from one strain to another. If the DNA from each of two strains is treated with an appropriate endonuclease, the size of the resulting DNA fragments that contain an *rrn* operon will differ between the two strains and can be visualized by agarose gel electrophoresis. A universal DNA probe for 16S rDNA can be used to detect *only* the *rrn*-containing fragments on a membrane blot, and the pattern of these particular fragments will be unique for each strain.

DNA Fingerprinting Using Pulsed Field Gel Electrophoresis (PFGE). No probe is used in this method. When bacterial DNA is treated with a rare-cutting restriction endonuclease, many short fragments but only a few fragments of 1,000,000 bp or more are formed. Long DNA fragments cannot be separated on conventional agarose gel in the same way as short fragments but instead they "worm" their way through the matrix, as if they were going through a narrow, winding tube, and all migrate at a similar rate away from the cathode. Consequently, no banding pattern can be formed that could be used to characterize the bacterial strain. However, if the angle of the electric field suddenly changes, these DNAs must reorient their long axes along the new direction of the field before they can continue to migrate. The higher the molecular weight of the fragment, the longer the time it takes for this reorientation to occur. Thus the longer the fragment, the longer it takes to migrate through the gel. A PFGE apparatus causes a periodic switching of the angle of the electric field and thus allows the long fragments to become well separated and form distinct bands. The bands can be visualized merely by soaking the gel in a solution of ethidium bromide, which binds to the fragments and fluoresces under ultraviolet light.

One problem with PFGE, however, is that the DNA must be treated *very gently* to avoid random mechanical breakage, because the only breakage must be that caused by the restriction endonuclease. Therefore, the intact bacterial cells are embedded in small blocks of low melting point agarose and lysed *in situ* before being treated with the restriction endonuclease. The blocks are then placed into a gel slab and subjected to pulsed-field gel electrophoresis at a low temperature for several hours.

Randomly amplified polymorphic DNA (RAPD) strain typing. RAPD strain identification is based on the PCR technique and the use of a single 10-base primer. Because the primer is short, there are usually many complementary sequences on the genomic DNA to which the primer will bind. DNA polymerase adds other bases to the primer, creating short pieces of double-stranded DNA. The PCR technique then creates millions of copies of these pieces. The various sizes of DNA pieces are then separated electrophoretically on an agarose gel and viewed by staining with ethidium bromide.

Further Reading

Amann, R.I., L. Krumholz and D.A. Stahl. 1990. Fluorescent-oligonucleotide probing of whole cells for determinative, phylogenetic, and environmental studies in microbiology. J. Bacteriol. *172*: 762–770.

Angert, E.R., K.D. Clements and N.R. Pace. 1993. The largest bacterium. Nature *362*: 239–241.

Barrett, D.M., D.O. Faigel, D.C. Metz, K. Montone and E.E. Furth. 1997. *In situ* hybridization for *Helicobacter pylori* in gastric mucosal biopsy specimens: quantitative evaluation of test performance in comparison with the CLO test and thiazine stain. J. Clin. Lab. Anal. *11*: 374–379.

Barrow, G.I. and R.K.A. Feltham (Editors). 1993. Cowan and Steel's Manual for the Identification of Medical Bacteria, 3rd Ed., Cambridge University Press, Cambridge.

Board, R.G., D. Jones and F.E. Skinner (Editors). 1992. Identification Methods in Applied and Environmental Microbiology, Blackwell Scientific Publications, Oxford.

DeLong, E.F., G.S. Wickham and N.R. Pace. 1989. Phylogenetic stains: ribosomal RNA-based probes for the identification of single cells. Science *243*: 1360–1363.

Forbes, B.A., D.F. Sahn and A.S. Weissfeld. 1998. Bailey and Scott's Diagnostic Microbiology, 10th Ed., Mosby, St. Louis.

Fournier, D., R. Lemieux and D. Couillard. 1998. Genetic evidence for highly diversified bacterial populations in wastewater sludge during biological leaching of metals. Biotechnol. Lett. *20*: 27–31.

Hugenholtz, P., C. Pitulle, K.L. Hershberger and N.R. Pace. 1998. Novel division level bacterial diversity in a Yellowstone hot spring. J. Bacteriol. *180*: 366–376.

Logan, N.A. 1994. Bacterial systematics, Blackwell Scientific Publications, Oxford.

Miller, J.M. and C.M. O'. 1995. Substrate utilization systems for the identification of bacteria and yeasts, *In* Murray, Baron, Pfaller, Tenover and Yolken (Editors), Manual of Clinical Microbiology, 6th Ed., American Society for Microbiology, Washington, D.C. pp. 103–109.

Murray, P.R., E.J. Baron, M.A. Pfallen, F.C. Tenover and R.H. Yolken (Editors). 1995. Manual of Clinical Microbiology, 6th Ed., American Society for Microbiology, Washington, D.C.

Onderdonk, A.B. and M. Sasser. 1995. Gas-liquid and high-performance chromatographic methods for the identification of micoorganisms, *In* Murray, Baron, Pfaller, Tenover and Yolken (Editors), Manual of Clinical Microbiology, 6th Ed., American Society for Microbiology, Washington, D.C. pp. 123–129.

Podzorski, R. and D.H. Persing. 1995. Molecular detection and identification of microorganisms, *In* Murray, Baron, Pfaller, Tenover and Yolken (Editors), Manual of Clinical Microbiology, 6th Ed., American Society for Microbiology, Washington, D.C. pp. 130–157.

Rainey, F.A., N.L. Ward-Rainey, P.H. Janssen, H. Hippe and E. Stackebrandt. 1996. *Clostridium paradoxum* DSM 7308^T contains multiple 16S rRNA genes with heterogeneous intervening sequences. Microbiology (Reading) *142*: 2087–2095.

Smibert, R.M. and N.R. Krieg. 1995. Phenotypic characterization, *In* Gerhardt, Murray, Wood and Krieg (Editors), Methods for General and Molecular Bacteriology, American Society for Microbiology, Washington, D.C. pp. 607–654.

Teske, A., P. Sigalevich, Y. Cohen and G. Muyzer. 1996. Molecular identification of bacteria from a coculture by denaturing gradient gel electrophoresis of 16S ribosomal DNA fragments as a tool for isolation in pure cultures. Appl. Environ. Microbiol. *62*: 4210–4215.

Wise, M.G., TV. Matchers and LC. Shanties. 1997. Bacterial diversity of a Carolina bay as determined by 16S rRNA gene analysis: confirmation of novel taxa. Appl. Environ. Microbiol. *63*: 1505–1514.

Numerical Taxonomy

Peter H.A. Sneath

Numerical taxonomy (sometimes called **taxometrics**) developed in the late 1950s as part of multivariate analyses and in parallel with the development of computers. Its aim was to devise a consistent set of methods for classification of organisms. Much of the impetus in bacteriology came from the problem of handling the tables of data that result from examination of their physiological, biochemical, and other properties. Such tables of results are not readily analyzed by eye, in contrast to the elaborate morphological detail that is usually available from examination of higher plants and animals. There was thus a need for an objective method of taxonomic analyses, whose first aim was to sort individual strains of bacteria into homogeneous groups (conventionally species), and that would also assist in the arrangement of species into genera and higher groupings. Such numerical methods also promised to improve the exactitude in measuring taxonomic, phylogenetic, serological, and other forms of relationship, together with other benefits that can accrue from quantitation (such as improved methods for bacterial identification; see the discussion by Sneath of Numerical Identification in this *Manual*).

Numerical taxonomy has been broadly successful in most of these aims, particularly in defining homogeneous **clusters** of strains, and in integrating data of different kinds (morphological, physiological, antigenic). There are still problems in constructing satisfactory groups at high taxonomic levels, e.g., families and orders, although this may be due to inadequacies in the available data rather than any fundamental weakness in the numerical methods themselves.

The application of the concepts of numerical taxonomy was made possible only through the use of computers, because of the heavy load of routine calculations. However, the principles can easily be illustrated in hand-worked examples. In addition, two problems had to be solved: the first was to decide how to weight different variables or characters; the second was to analyze similarities so as to reveal the taxonomic structure of groups, species, or clusters. A full description of numerical taxonomic methods may be found in Sneath (1972) and Sneath and Sokal (1973). Briefer descriptions and illustrations in bacteriology are given by Skerman (1967), Lockhart and Liston (1970), Sneath (1978a), Priest and Austin (1993), and Logan (1994). A thorough review of applications to bacteria is that of Colwell (1973).

It is important to bear in mind certain definitions. Relationships between organisms can be of several kinds. Two broad classes are as follows.

Similarity on Observed Properties. Similarity, or resemblance, refers to the attributes that an organism possesses today, without reference to how those attributes arose. It is expressed as proportions of similarities and differences, for example, in existing attributes, and is called the **phenetic relationship**. This includes similarities both in phenotype (e.g., motility) and in genotype (e.g., DNA pairing).

Relationship by Ancestry, or Evolutionary Relationship. This refers to the **phylogeny** of organisms, and not necessarily to their present attributes. It is expressed as the time to a common ancestor, or the amount of change that has occurred in an evolutionary lineage. It is not expressed as a proportion of similar attributes, or as the amount of DNA pairing and the like, although evolutionary relationship may sometimes be deduced from phenetics on the assumption that evolution has indeed proceeded in some orderly and defined way. To give an analogy, individuals from different nations may occasionally look more similar than brothers or sisters of one family; their phenetic resemblance (in the properties observed) may be high though their evolutionary relationship is distant.

Numerical taxonomy is concerned primarily with phenetic relationships. It has in recent years been extended to phylogenetic work, by using rather different techniques; these seek to build upon the assumed regularities of evolution so as to give, from phenetic data, the most probable phylogenetic reconstructions. A review of the area is given by Sneath (1974).

The basic taxonomic category is the species. It is noted in the chapter on "Bacterial Nomenclature" that it is useful to distinguish a **taxospecies** (a cluster of strains of high mutual phenetic similarity) from a **genospecies** (a group of strains capable of gene exchange), and both of these from a **nomenspecies** (a group bearing a binomial name, whatever its status in other respects). Numerical taxonomy attempts to define taxospecies. Whether these are justified as genospecies or nomenspecies turns on other criteria. One may also distinguish a **genomospecies**, a group of strains that have high DNA–DNA relatedness. It should be emphasized that groups with high genomic similarity are not necessarily genospecies: genomic resemblance is included in phenetic resemblance; genospecies are defined by gene exchange.

Groups can be of two important types. In the first, the possession of certain invariant properties defines the group without permitting any exception. All triangles, for example, have three sides, not four. Such groupings are termed **monothetic**. Taxonomic groups, however, are not of this kind. Exceptions to the most invariant characters are always possible. Instead, taxa are **polythetic**, that is, they consist of assemblages whose members share a high proportion of common attributes, but not necessary any invariable set. Numerical taxonomy produces polythetic groups and thus permits the occasional exception on any character.

Logical Steps in Classification

The steps in the process of classification are as follows:

1. Collection of data. The **bacterial strains** that are to be classified have to be chosen, and they must be examined for a number of relevant properties (**taxonomic characters**).
2. The data must be coded and scaled in an appropriate fashion.
3. The **similarity** or **resemblance** between the strains is calculated. This yields a table of similarities (**similarity matrix**) based on the chosen set of characters.
4. The similarities are analyzed for **taxonomic structure**, to yield the groups or clusters that are present, and the strains are arranged into **phenons** (phenetic groups), which are broadly equated with taxonomic groups (**taxa**).
5. The properties of the phenons can be tabulated for publication or further study, and the most appropriate characters (**diagnostic characters**) can be chosen on which to set up **identification systems** that will allow the best identification of additional strains.

It may be noted that those steps must be carried out in the above order. One cannot, for example, find diagnostic characters before finding the groups of which they are diagnostic. Furthermore, it is important to obtain complete data, determined under well-standardized conditions.

Data for numerical taxonomy The data needed for numerical taxonomy must be adequate in quantity and quality. It is a common experience that data from the literature are inadequate on both counts; most often it is necessary to examine bacterial strains afresh by an appropriate set of tests.

Organisms Most taxonomic work with bacteria consists of examining individual strains of bacteria. However, the entities that can be classified may be of various forms—strains, species, genera—for which no common term is available. These entities, t in number, are therefore called **operational taxonomic units (OTUs)**. In most studies OTUs will be strains. A numerical taxonomic study, therefore, should contain a good selection of strains of the groups under study, together with type strains of the taxa and of related taxa. Where possible, recently isolated strains, and strains from different parts of the world, should be included.

Characters A **character** is defined as any property that can vary between OTUs. The values it can assume are **character states.** Thus, "length of spore" is a character and "1.5 µm" is one of its states. It is obviously important to compare the same character in different organisms, and the recognition that characters are the same is called the **determination of homology**. This may sometimes pose problems, but in bacteriology these are seldom serious. A single character treated as independent of others is called a **unit character**. Sets of characters that are related in some way are called **character complexes**.

There are many kinds of characters that can be used in taxonomy. The descriptions in the *Manual* give many examples. For numerical taxonomy, the characters should cover a broad range of properties: morphological, physiological, biochemical. It should be noted that certain data are not characters in the above sense. Thus the degree of serological cross-reaction or the percent pairing of DNA are equivalent, not to character states, but to similarity measures.

Number of Characters Although it is well to include a number of strains of each known species, numerical taxonomies are not greatly affected by having only a few strains of a species. This is not so, however, for characters. The similarity values should be thought of as estimates of values that would be obtained if one could include a very large number of phenotypic features. The accuracy of such estimates depends critically on having a reasonably large number of characters. The number, n, should be 50 or more. Several hundred are desirable, though the taxonomic gain falls off with very large numbers.

Quality of Data The quality of the characters is also important. Microbiological data are prone to more experimental error than is commonly realized. The average difference in replicate tests on the same strain is commonly about 5%. Efforts should be made to keep this figure low, particularly by rigorous standardization of test methods. It is very difficult to obtain reasonably reproducible results with some tests, and they should be excluded from the analysis. As a check on the quality of the data, it is useful to reduplicate a few of the strains and carry them through as separate OTUs; the average test error is about half the percentage discrepancy in similarity of such replicates (e.g., 90% similarity implies about 5% experimental variation).

Coding of the Results The test reactions and character states now need coding for numerical analysis. There are several satisfactory ways of doing this, but for the present purposes of illustration only one common scheme will be described. This is the familiar process of coding the reactions or states into positive and negative form. The resulting table, therefore, contains entries + and − (or 1 and 0, which are more convenient for computation), for t OTUs scored for n characters. Naturally, there should be as few gaps as possible.

The question arises as to what weight should be given to each character relative to the rest. The usual practice in numerical taxonomy is to give each character equal weight. More specifically, it may be argued that unit characters should have unit weight, and if character complexes are broken into a number of unit characters (each carrying one unit of taxonomic information), it is logical to accord unit weight to each unit character. The difficulties of deciding what weight should be given *before* making a classification (and hence in a fashion that does not prejudge the taxonomy) are considerable. This philosophy derives from the opinions of the eighteenth-century botanist Adanson, and therefore numerical taxonomies are sometimes referred to as Adansonian.

Similarity The $n \times t$ table can then be analyzed to yield similarities between OTUs. The simplest way is to count, for any pair of OTUs, the number of characters in which they are identical (i.e., both are positive or both are negative). These **matches** can be expressed as a percentage or a proportion, symbolized as S_{SM} (for simple matching coefficient). This is the most common coefficient in bacteriology. Other coefficients are sometimes used because of particular advantages. Thus the Gower coefficient S_G accommodates both presence–absence characters and quantitative ones, the Jacquard coefficient S_J discounts matches between two negative results, and the Pattern coefficient S_P corrects for apparent differences that are caused solely by differences between strains in growth rate and hence metabolic vigor. These coefficients emphasize different aspects of the phenotype (as is quite legitimate in taxonomy) so one cannot regard one or another as necessarily the correct coefficient, but fortunately this makes little practical difference in most studies. Various special similarity coefficients can also be employed for electrophoretic and chemotaxonomic data.

The similarity values between all pairs of OTUs yields a checkerboard of entries, a square table of similarities known as a **similarity matrix** or ***S* matrix**. The entries are percentages, with 100% indicating identity and 0% indicating complete dissimilarity between OTUs. Such a table is symmetrical (the similarity of *a* to *b* is the same as that of *b* to *a*), so that usually only one half, the left lower triangle, is filled in.

These similarities can also be expressed in a complementary form, as dissimilarities. Dissimilarities can be treated as analogs of distances, when "taxonomic maps" of the OTUs are prepared, and it is a convenient property that the quantity $d = (1 - S_{SM})^{1/2}$ is equivalent geometrically to a distance between points representing the OTUs in a space of many dimensions (a **phenetic hyperspace**).

Taxonomic structure A table of similarities does not of itself make evident the **taxonomic structure** of the OTUs. The strains will be in an arbitrary order that will not reflect the species or other groups. These similarities therefore require further manipulation. It will be seen that a table of serological cross-reactions, if complete and expressed in quantitative terms, is analogous to a table of percentage similarities, and the same is true of a table of DNA pairing values. Such tables can be analyzed by the methods described below, though in serological and nucleic studies there are some particular difficulties on which further work is needed.

There are two main types of analyses to reveal the taxonomic structure: **cluster analysis** and **ordination**. The result of the former is a treelike diagram or **dendrogram** (more precisely a **phenogram**, because it expresses phenetic relationships), in which the tightest bunches of twigs represent clusters of very similar OTUs. The result of the latter is an **ordination diagram** or **taxonomic map**, in which closely similar OTUs are placed close together. The mathematical methods can be elaborate, so only a nontechnical account is given here.

In cluster analysis, the principle is to search the table of similarities for high values that indicate the most similar pairs of OTUs. These form the nuclei of the clusters and the computer searches for the next highest similarity values and adds the corresponding OTUs onto these cluster nuclei. Ultimately all OTUs fuse into one group, represented by the basal stem of the dendrogram. Lines drawn across the dendrogram at descending similarity levels define, in turn, phenons that correspond to a reasonable approximation to species, genera, etc. The most common cluster methods are the **unweighted pair group method with averages** (**UPGMA**) and **single linkage**.

In ordination, the similarities (or their mathematical equivalents) are analyzed so that the phenetic hyperspace is summarized in a space of only a few dimensions. In two dimensions this is a scattergram of the positions of OTUs from which one can recognize clusters by eye. Three-dimensional perspective drawings can also be made. The most common ordination methods are **principal components analysis** and **principal coordinates analysis**.

A number of other representations are also used. One example is a similarity matrix in which the OTUs have first been rearranged into the order given by a clustering method and then the cells of the matrix have been shaded, with the highest similarities shown in the darkest tone. In these "shaded *S* matrices", clusters are shown by dark triangles. Another representation is a table of the mean similarities between OTUs of the same cluster and of different clusters (**inter-** and **intragroup similarity table**); if based on S_{SM} with UPGMA clustering, this table expresses the positions and radii of clusters (Sneath, 1979a) and consequently the distance between them and their probable overlap—properties of importance in numerical identification, as discussed later.

For general purposes, a dendrogram is the most useful representation, but the others can be very instructive, since each method emphasizes somewhat different aspects of the taxonomy.

The analysis for taxonomic structure should lead logically to the establishment or revision of taxonomic groups. We lack, at present, objective criteria for different taxonomic ranks, that is, one cannot automatically equate a phenon with a taxon. It is, however, commonly found that phenetic groups formed at about 80% *S* are equivalent to bacterial species. Similarly, we lack good tests for the statistical significance of clusters and for determining how much they overlap, though some progress is being made here (Sneath, 1977, 1979b). The fidelity with which the dendrogram summarizes the *S* matrix can be assessed by the **cophenetic correlation coefficient**, and similar statistics can be used to compare the **congruence** between two taxonomies if they are in quantitative form (e.g., phenetic and serological taxonomies). Good scientific judgment in the light of other knowledge is indispensable for interpreting the results of numerical taxonomy.

Descriptions of the groups can now be made by referring back to the original table of strain data. The better diagnostic characters can be chosen—those whose states are very constant within groups but vary between groups. It is better to give percentages or proportions than to use symbols such as +, (+), *v*, *d*, or − for varying percentages, because significant loss of statistical information can occur with these simplified schemes. It would, however, be superfluous to list percentages based on very few strains. As systematic bacteriology advances, it will be increasingly important to publish the actual data on individual strains or deposit it in archives; such data will show their full value when test methods become very highly standardized.

It is evident that numerical taxonomy and numerical identification place considerable demands on laboratory expertise. New test methods are continually being devised. New information is continually being accumulated. It is important that progress should be made toward agreed data bases (Krichevsky and Norton, 1974), as well as toward improvements in standardization of test methods in determinative bacteriology, if the full potential of numerical methods is to be achieved.

Numerical Identification

The success of numerical taxonomy has in recent years led to the development of a new diagnostic method based upon it, called **numerical identification**. The rapidly growing field is well reviewed by Lapage et al. (1973), and Willcox et al. (1980). The essential principles can be illustrated geometrically (Sneath, 1978b) by considering the columns of percent positive test reactions in a new table, a table of *q* taxa for *m* diagnostic characters. If an object is scored for two variables, its position can be represented by a point on a scatter diagram. Use of three variables determines a position in a three-dimensional model. Objects that are very similar on the variables will be represented by clusters of points in the diagram or the model, and a circle or sphere can be drawn round each cluster so as to define its position and radius. The same principles can be extended to many variables or tests, which then represent a multidimensional space or "hyperspace". A column representing a species defines, in effect, a region in hyperspace, and it is useful to think of a species as being represented by a hypersphere in that space, whose position and radius are specified by the numerical values of these percentages. The tables form a reference library, or database, of properties of the taxa.

The operation of numerical identification is to compare an unknown strain with each column of the table in turn, and to calculate a distance (or its analog) to the center of each taxon hypersphere. If the unknown lies well within a hypersphere, this will identify it with that taxon. Further, such systems have important advantages over most other diagnostic systems. The numerical process allows a likelihood to be attached to an identification, so that one can know to some order of magnitude the certainty that the identity is correct. The results are not greatly affected by an occasional aberrant property of the unknown, or an occasional experimental mistake in performing the tests. Furthermore, the system is robust toward missing information, and quite good identifications can be obtained if only a moderate proportion of the tests have been performed.

Numerous applications of numerical identification are now being made. Most commercial testing kits or automatic instruments for microbial identification are based on these concepts, and they require the comparison of results on an unknown strain with a database using computer software or with printed material prepared by such means. Research sponsored by the Bergey's Manual® Trust (Feltham et al., 1984) shows that these concepts can be extended to a very wide range of genera.

Further Reading

Feltham, R.K.A., P.A. Wood and P.H.A. Sneath. 1984. A general-purpose system for characterizing medically important bacteria to genus level. J. Appl. Bacteriol. *57*: 279–290.

Lapage, S.P., S. Bascomb, W.R. Willcox and M.A. Curtis. 1973. Identification of bacteria by computer.I. General aspects and perspectives. J. Gen. Microbiol. *77*: 273–290.

Logan, N.A. 1994. Bacterial Systematics, Blackwell Scientific Publications, Oxford.

Sneath, P.H.A. 1978. Identification of microorganisms, *In* Norris and Richmond (Editors), Essays in Microbiology, John Wiley, Chichester. 10/1–10/32.

Willcox, W.R., S.P. Lapage and B. Holmes. 1980. A review of numerical methods in bacterial identification. Antonie van Leeuwenhoek J. Microbiol. Serol. *46*: 233–299.

Polyphasic Taxonomy

Monique Gillis, Peter Vandamme, Paul De Vos, Jean Swings and Karel Kersters

INTRODUCTION

Bacterial taxonomy comprises the interrelated areas of classification, nomenclature, and identification and is supposed to reflect phylogeny and evolution. When looking back over the changes in bacterial systematics during the last 25 years, it is clear that the most spectacular changes occurred mainly in the areas of characterization and phylogeny. Characterization changed from simple procedures, in which a limited number of features of the bacterial cell (mainly morphological and physiological aspects) were studied, to a multidisciplinary approach using phenotypic, genotypic, and chemotaxonomic techniques. Determination of phylogenetic relationships (which is at this time essentially synonymous with 16S and/or 23S rRNA gene sequence similarities) became a routine procedure in bacterial taxonomy.

While the rules of bacterial nomenclature remain largely unchanged (Lapage et al., 1992; Stackebrandt and Goebel, 1994; Murray and Stackebrandt, 1995), the tools for identification diversified with the multidisciplinary approach to bacterial characterization. The names of bacteria, and certainly the number of named taxa, have also changed and/or increased drastically as a result of the application of this conceptual approach to bacterial taxonomy.

The term "polyphasic taxonomy" was introduced 30 years ago by Colwell (1970) to refer to a taxonomy that assembles and assimilates many levels of information, from molecular to ecological, and incorporates several distinct, and separable, portions of information extractable from a nonhomogeneous system to yield a multidimensional taxonomy. Nowadays, polyphasic taxonomy refers to a consensus type of taxonomy and aims to utilize all the available data in delineating consensus groups, decisive for the final conclusions.

The species is the basic unit of bacterial taxonomy, and the first recommendation for a polyphasic consensus delineation of a bacterial species is based on "the phylogenetic species definition" of Wayne et al. (1987). These authors defined a species as a group of strains, including the type strain, sharing at least 70% total genome DNA–DNA hybridization and less than 5°C ΔT_m.* Phenotypic features should agree with this genotypic definition and should override the "phylogenetic" concept of species only in a few exceptional cases. Total genome DNA–DNA hybridization values are the key parameter in this species delineation.

Considering the perception of a bacterial species, taxonomists either sustain a coherent species definition without questioning if this corresponds with a biological reality, or they try to visualize bacterial species as condensed nodes in a cloudy and confluent taxonomic space. Genera and families represent mostly agglomerates of nodal species and internodal strains, and agglomerates of genera, respectively. Although in the present chapter most of the attention will be focused on the species level, the hierarchical structure of all current taxonomic classification requires us to consider higher taxa, such as genera and families, as well. Compared to the bacterial species, the higher taxa are much more difficult to delineate and phylogenetic divergence is not necessarily supported by phenotypic, chemotaxonomic, or polyphasic data. At present, no clear-cut genus definition is available and this has led to the creation of genera in which the genotypic and phenotypic divergence varies with the individual concepts of taxonomists. Therefore, delineation of genera by a consensus approach, including simple differential parameters and an accompanying polyphasic definition, is highly desirable if the present concept of bacterial classification is to be retained.

DIFFERENT TYPES OF INFORMATION USED IN POLYPHASIC TAXONOMY

In principle both genotypic and phenotypic information may be incorporated into polyphasic taxonomic studies. Sources of information and diverse techniques available to retrieve this information are represented schematically in Fig. 1. The ultimate characterization on the genomic level is the determination of the sequence and the organization of the total bacterial genome. As long as this cannot be performed routinely, the polyphasic approach is the most obvious strategy to collect a maximum amount of direct and indirect information about the total genome. It is not our intention to describe all the available techniques here. Our aim is to discuss the major categories of taxonomic techniques required to obtain a useful polyphasic characterization. Practical and theoretical aspects of the different techniques listed in Fig. 1 can be found in various papers (see Vandamme et al., 1996a) and handbooks. Of paramount importance is the level of taxonomic resolution of the different methods. Fig. 2 presents the discriminatory taxonomic power of the techniques summarized in Fig. 1. On the basis of this parameter, different categories of techniques can be distinguished: (i) those with a broad taxonomic resolution, of which the rRNA gene-based techniques are the best known for their impact on phylogenetic conclusions; (ii) those revealing differences on the species and/or genus level and (iii) various typing methods that are not necessarily relevant on the species level but can be used to screen for groups of similar strains. The various techniques differ

*T_m is the melting temperature of the hybrid as measured by stepwise denaturation; ΔT_m is the difference in T_m in degrees Celsius between the homologous and the heterologous hybrids formed under standard conditions.

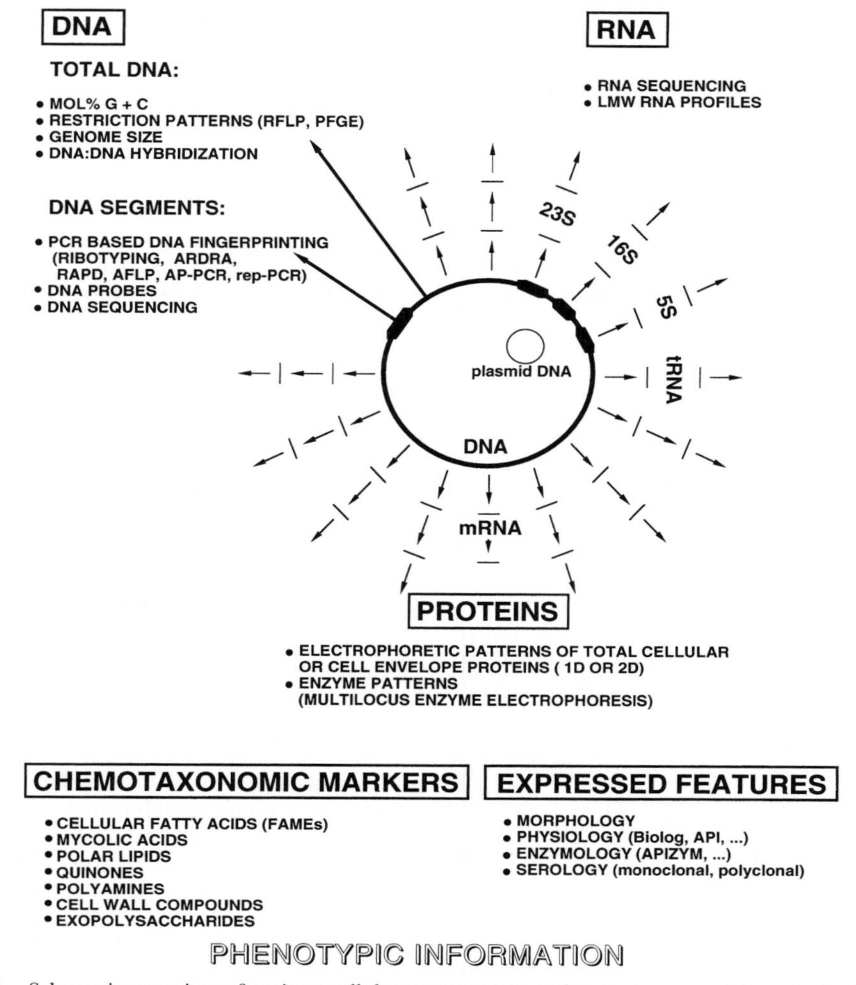

FIGURE 1. Schematic overview of various cellular components and techniques used in polyphasic bacterial taxonomy (adapted from Vandamme et al., 1996a). Abbreviations: AFLP, amplified fragment length polymorphism; AP-PCR, arbitrarily primed PCR; ARDRA, amplified rDNA restriction analysis; FAMEs, fatty acid methyl esters; LMW, low molecular weight; PFGE, pulsed-field gel electrophoresis; RAPD, randomly amplified polymorphic DNA; rep-PCR, repetitive element sequence-based PCR; RFLP, restriction fragment length polymorphism; 1D, 2D, one- and two-dimensional, respectively.

in the amount of effort required. Some have been automated, and some are relatively fast and cheap. It is obvious that fast, cost-effective, and preferentially automated methods with a fine taxonomic resolution (below the species level) are among those to be used for primary screening purposes, while total genome DNA–DNA hybridization, being a more laborious technique, can be restricted to a minimum number of strains representing groups defined using other appropriate methods. Techniques based on rRNA genes of representative strains are the most suitable to determine the phylogenetic position of bacterial groups.

In practice, it is nearly impossible to gather all the information that could possibly be used in a polyphasic study. The strategy in modern polyphasic taxonomy is to first estimate the different levels of taxonomic discrimination to be covered and then to choose the techniques accordingly. The total number of strains to be studied will also significantly affect the final choice.

The consensus polyphasic approach starts with making a choice of complementary techniques to be used simultaneously or stepwise in order to characterize and classify an individual strain or any group of strains. The goal is to evaluate all the results in relation to each other and to obtain a consensus view of the data with a minimum number of inconsistencies. Nomenclatural implications complete the evaluation, together with the search for adequate identification procedures. Each taxon should be described and, preferably, differentiated from related or similar taxa by its phenotypic, genotypic, and chemotaxonomic characteristics.

The minimal requirements for obtaining useful polyphasic data are: (i) a preliminary screening for groups of similar strains; (ii) determination of the phylogenetic placement of these groups; (iii) measurement of the relationships between the groups and their closest neighbors, and (iv) collection of various descriptive data, preferentially on different aspects of the cell.

Polyphasic Strategy

There is no universal strategy that can be employed in all polyphasic studies. The taxonomic levels to be covered vary with the

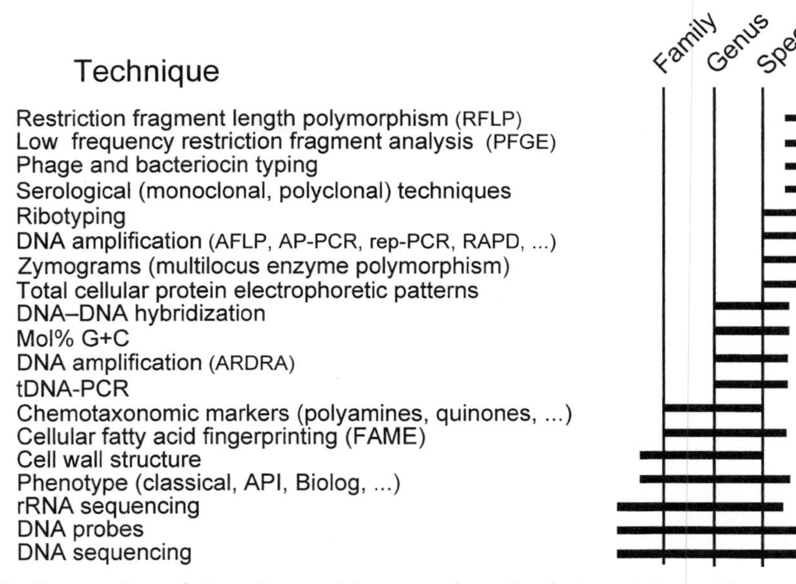

FIGURE 2. Taxonomic resolution of some of the currently used techniques in bacterial taxonomy (adapted from Vandamme et al., 1996a). Abbreviations: see legend of Figure 1.

objective of each study. The choice of the techniques to be used also depends on the number of strains to be studied. The more strains to be screened, the more one needs a fast and preferentially automated screening technique. On the other hand, the requirement for special analytical methods for a given technique is generally less important when only a few strains are under investigation. The taxonomic resolution of many techniques can differ depending on the bacterial group studied. The choice of methods can also be taxon-dependent when for any reason the preparation of particular cell constituents is very difficult or inefficient.

Development of a strategy for a polyphasic taxonomy can be illustrated by a theoretical example: suppose that one has to classify 50 bacterial isolates for which a minimal characterization (e.g., Gram reaction, origin, morphology, growth conditions) is available. According to the minimal requirements mentioned above, the 50 strains should initially be screened to identify groups of similar strains, preferably by at least two non-overlapping methods. The choice of techniques will certainly also be affected by the availability of the required instrumentation and the knowledge of each research group. For the delineation of groups, a thorough knowledge of the resolving power of each technique is necessary. An awareness of the limitations of the methods used to analyze and cluster the results is also essential. Armed with this knowledge, the main consensus groups obtained by the various techniques can then be determined. The second goal is to determine the phylogenetic position of the consensus groups by sequencing the 16S rRNA genes of representative strains.

Different theoretical possibilities for studying the relationships among the consensus groups and providing an emended or a new description of taxa exist. These possibilities are listed below:

Case 1 If the 16S rRNA gene sequence similarity between the representative strains under study and those found in GenBank for a particular genus exceeds 97%, it can be assumed that these strains are members of that genus. DNA–DNA hybridizations can then be performed between several representative strains of each consensus group and all known species of that genus, to find out if the new consensus group belongs to one of the known species or constitutes a new species, as recommended by Stackebrandt and Goebel (1994). Members of a particular consensus group can be identified as a known species, the polyphasic consistency of the species can be verified, and, if needed, the description can be emended. When a new group is identified as belonging to a particular genus but not to one of its described species, the creation of a new species can be planned. Therefore a polyphasic description of the new taxon is required before a new species, with an appropriate name, can be proposed and described. Phenotypic characterization remains an indispensable part of the description allowing differentiation and description of the groups, but in the future genotypic and chemotaxonomic parameters should enhance the description of new taxa and assure the differentiation between the taxa.

Case 2 There are no clearcut recommendations for the delineation of bacterial genera or higher taxonomic levels. If the 16S rRNA gene sequence similarity between representative strains of a consensus group and those found in GenBank is less than 97%, it is often not straightforward to decide whether the particular group belongs to a new or existing genus. It is recommended to evaluate the stability of the phylogenetic position of the group in question and to compare its overall genotypic, chemotaxonomic, and phenotypic profile with that of its closest relatives. When both phenotypic and chemotaxonomic parameters support the phylogenetic group, the creation of a new genus can be considered.

Case 3 Strains not belonging to consensus groups must be further characterized to determine their exact taxonomic status.

Polyphasic taxonomy in practice

Many examples of polyphasic taxonomic studies of diverse bacterial groups are available, and a general evaluation of the results shows that the various conclusions of these studies depend on the bacterial group(s) studied, on the techniques applied, and on the researcher involved (Vandamme et al., 1996a). In many cases the identification of consensus groups and the formulation of conclusions is not simple and may show various inconsistencies either with (i) earlier classification and nomenclature or (ii) within the new conclusions themselves. Some striking examples will be discussed briefly in Part A and the main problems will be summarized in Part B.

Part A

1. Polyphasic classification does not necessarily conform to special purpose classification practiced by specialists in the field, e.g., in the genera *Agrobacterium* (Kersters and De Ley, 1984) and *Xanthomonas* (Vauterin et al., 1995). In the former example the polyphasic groups did not at all correspond with the named species, of which the type species *A. tumefaciens* has a conserved status; this has led to the use of a "biovar" system to indicate the polyphasic groups. In *Xanthomonas* 14 new species corresponding to polyphasic groups have been created, partly replacing the former pathovar system.
2. Certain groups constitute very tight phylogenetic clusters that can be biochemically quite versatile e.g., *Bordetella* (Vandamme, 1998).
3. Occasionally members of very tight polyphasic groups are not classified accordingly e.g., *Escherichia* and *Shigella* species sharing more than 85% total DNA–DNA hybridization. For pragmatic reasons they remain classified in two separate genera (Brenner, 1984).
4. In contrast, biochemically restricted groups can be phylogenetically extremely heterogeneous. For example, *Campylobacter* (Vandamme et al., 1991) and *Capnocytophaga* (Vandamme et al., 1996b) originally included a large number of taxa characterized by a minimal set of common phenotypic features. The ability of various techniques to distinguish between members of these taxa at various taxonomic levels was reexamined leading to the development of molecular diagnostic tests.
5. For the lactic acid bacteria, traditionally applied phenotypic classification schemes do not correspond with the phylogenetic-based classification because of a large amount of sequence variation in the 16S rRNA genes (Vandamme et al., 1996a). The traditional phenotypic analysis remains important for identification purposes because the phylogenetic data have not yet been translated into new identification strategies.
6. In several bacterial lineages, such as *Comamonadaceae*, multiple subbranches (16S rRNA gene sequence similarities of 95–96%) have been identified, and some of them can be considered candidates for separate generic status (Willems et al., 1991a; Wen et al., 1999). However, as with species, it is important that genera exhibit some phenotypic coherence. This discrepancy has resulted either in the combination of multiple subbranches into a single genus or in the creation of separate genera for individual sub-branches.
7. In many examples genotypic groups could not be described phenotypically and therefore remain unnamed within species e.g., in *Comamonas terrigena* (Willems et al., 1991c) or within genera e.g., *Acinetobacter*, containing several unnamed genomic species (Bouvet and Jeanjean, 1989; Nishimura et al., 1987). Additional methods for unambiguous identification are required. It should be stressed that in such cases the requirement for phenotypic differentiation as formulated by Wayne et al. (1987) makes it impossible to properly name genomic groups and can thus hinder the recognition of biological diversity.
8. In some cases it is very difficult to determine consensus groups because many strains occupy separate positions or the clusters are too narrow, due to the techniques used being too discriminating. Supplementary techniques, with slightly different levels of discriminating power, are recommended to improve the delineation of significant groups.
9. On occasion, a new genus has been created on the basis of significant phenotypic and physiological differences despite sharing more than 99% 16S rRNA gene similarity with an existing genus (Yurkov et al., 1997).
10. Molecular tools allow one to obtain various DNA sequences from diverse biotypes, providing an image of the total bacterial populations and consortia. Most uncultured or "unculturable" bacteria are only characterized by their 16S rRNA gene sequence, ITS, or 23S rRNA gene sequence, and many may represent new taxa that cannot yet be characterized polyphasically (Hugenholtz et al., 1998a). Identical sequences may originate from different cells of the same strain, from different strains of the same species, or from strains of closely related species. Therefore, it is not appropriate to propose to classify them like cultured organisms and to propose binomial species names. It is recommended to include such organisms temporarily in a new category, *Candidatus* as proposed by Murray and Schleifer (1994) and Murray and Stackebrandt (1995).
11. During the last years, many new species containing a single strain have been described. Likewise, there are numerous genera consisting of only one species. This does not correspond with an ideal definition nor with the reality of nature. However, if only a single strain representing a new taxon can be isolated, the creation of single strain species illustrates the breadth of bacterial diversity. Attempts to obtain additional isolates should be encouraged.

Part B The major sources of conflict in practicing consensus taxonomy are: (i) characteristics expressing variability among organisms appear to be superior parameters for certain taxonomic ranks (DNA–DNA hybridization on the species level, rRNA gene sequencing on the genus and family level). However, the delineated groups cannot necessarily be revealed by the examination of phenotypic parameters. This has significant impact because of the need for a phenotypic description and differentiating phenotypic features; (ii) the lack of guidelines or minimal standards for description of a (polyphasic) genus.

1. **DNA–DNA hybridizations.** Classical hybridization techniques are laborious, require considerable amounts of DNA, and are consequently not suitable for large scale use. Moreover, several methods that do not necessarily give the same quantitative results are now in use, making quantitative comparisons difficult. New, more rapid, miniaturized, and standardized methods that will quickly delineate species (Adnan et al., 1993; Ezaki et al., 1989) are under development. The use of a single, rapid, and standardized method requiring small amounts of DNA is recommended. Regardless of the method used, it remains difficult to apply the 70% DNA–DNA hybridization rule for species definition because this rule was proposed mainly on the basis of differences among species in the family *Enterobacteriaceae*. These species are phenotypi-

cally well studied and exhibit a high degree of phenotypic heterogeneity that does not always correspond with genotypic heterogeneity. In many other bacterial families, members of a single species share DNA hybridization values of 40–100%, and evaluation of these lower values is often difficult.

2. **rRNA gene sequencing.** The comparison of 16S and 23S rRNA gene sequences is indispensable in polyphasic taxonomy and provides the phylogenetic framework for present day classification. The 16S rRNA gene sequence does not contain enough discriminating power to delineate species within certain groups, and additional DNA–DNA hybridizations are often required (Stackebrandt and Goebel, 1994). Moreover, all 16S rRNA gene sequences employed in classification should be used with care because high levels of sequence variation have been observed, even between strains of the same species. This variation is attributed to inter-operon differences, as well as to other differences (Clayton et al., 1995; Young and Haukka, 1996). For the reconstruction of phylogenetic trees, it is also important to include a wide range of related and unrelated reference organisms. Bootstrap analysis is highly recommended to determine the significance of the branching points.
3. **Phenotypic analysis.** According to Wayne et al. (1987), phenotypic data deserve special attention because of their impact on species delineation and because a description of new species requires a minimum number of phenotypic characteristics. Historically, phenotypic analysis was very important and many conventional tests have been used to describe and differentiate taxa. Taxonomic reports do not always provide much new phenotypic data, and data from older literature do not always reflect possible adaptations of strains or minor changes in test media and conditions. Nowadays few research groups perform extensive conventional phenotypic analysis because it is laborious and sometimes not reproducible. More and more commercialized, automated, and miniaturized methods, mostly conceived for particular bacterial groups are being used, resulting in the analysis of a restricted set of phenotypic properties and creating a dependence on the commercial dealer. A minimal phenotypic description may still be required in the long-term, but in the future genotypic and chemotaxonomic parameters should complete the description of new taxa and facilitate differentiation between the taxa. Phenotypic coherence at the species level does not usually represent a problem. However, at higher (genus) levels phenotypic coherence cannot always be found in the various phylogenetic groups, and clear differentiation of these groups is often doubtful or even impossible.
4. **Genus delineation.** Although there is a rather broad consensus among taxonomists that phylogenetic data are of superior value for the delineation of genera, the goal remains to define genera polyphasically, to describe them, and to differentiate them from their neighbors. However, there are no rules to delineate genera, except that it is generally accepted that genera should reflect phylogenetic relationships. The phylogenetic divergence within genera can differ with the bacterial groups under consideration, although most taxonomists do not accept very large (phylogenetically) heterogenous genera. The level of phylogenetic relatedness, as shown in an rRNA dendrogram, that corresponds to a given hierarchical line showing phenotypic coherence varies considerably (mostly between 4% and 10% 16S rRNA gene sequence difference). Phenotypic coherence does not always correspond to delineated phylogenetic groups and vice versa. Any new genus needs to be described. The goal of phenotypically differentiating genera from other closely related genera is regularly not fulfilled because the phenotypic data are often not complete or not comparable with results obtained from conventional tests described in the literature. We recommend, therefore, to also include chemotaxonomic and genomic data in order to improve the description and differentiation of taxa. Therefore, universal, comprehensive databases containing various kinds of molecular patterns, as well as phenotypic and chemotaxonomic data are required. To obtain reliable, reproducible, and exchangeable profiles which can be consulted, preferably on-line, the standardization of the experimental conditions and the use of tools to correct for inevitable small experimental aberrations becomes extremely important. Software programs for constructing and consulting such databases should be developed. For any new genus, a type species and an appropriate name must be proposed. Named taxa are necessary for the recognition of groups and for the practical use of bacterial classification.

We strongly recommend guidelines or minimal standards for delineating genera including (i) a phylogenetic parameter expressed as percentage 16S rRNA gene sequence similarity and a high bootstrap value for the relevant branching points; (ii) a polyphasic description including phenotypic, genomic, and chemotaxonomic data to provide a comprehensive description and allow differentiation.

Conclusion

The main conclusions concerning polyphasic taxonomy as it has been practiced widely during the last 20 years are as follows:

1. Replacing minimal numbers of characteristics by large numbers of features and characterizing different aspects of bacterial cells has resulted in more stable polyphasic classification systems.
2. Polyphasic species descriptions should (i) reflect phylogenetic relationships, (ii) be based on total genome DNA–DNA hybridization to determine the genomic relationships between representatives of groups and within these groups, and (iii) provide further descriptive genomic, phenotypic and chemotaxonomic information, as well as information on the infraspecific clonal structure as revealed by fine typing methodologies. In principle all methods studying a particular aspect of the cell can be useful as sources of information. In practice a choice of complementary methods has to be made to tackle any taxonomic problem.
3. The polyphasic bacterial species is more complex than the species defined by Wayne et al. (1987) because more aspects are considered. Polyphasic practice differs according to the groups studied, and the final impact of a particular category of characters may vary considerably. From polyphasic taxonomy studies, the bacterial species appears as an assemblage of isolates originating from a common ancestor population, in which the steady generation of genetic diversity has resulted in clones with different degrees of recombination. The polyphasic species is characterized by a degree of phenotypic consistency, by a significant degree of total genome DNA–DNA hybridization, and by over 97% 16S rRNA gene sequence similarity. Usually, a minimal phenotypic description

is required, supported by genomic and/or chemotaxonomic differentiating parameters.

4. Polyphasic taxonomy is purely empirical, follows no strict rules or guidelines, may integrate any type of information, and results in a consensus classification that reflects the phylogenetic relationships, a guarantee for its stability. The aim is to reflect as closely as possible the biological reality.
5. The usefulness of polyphasic characterization is to enable the selection of an appropriate technique to be used for a quick and accurate identification. For many bacteria encountered in routine diagnostic laboratories, monophasic, mostly phenotypic, identification will still be used, but other bacteria may require the utilization of more than one technique e.g., rRNA identification to determine the phylogenetic position and an appropriate fine technique (genomic, chemotaxonomic, and/or phenotypic tests) for the species level. Comparison with a standardized, accessible, universal database is a *conditio sine qua non.*

Perspectives

1. One of the most interesting perspectives in bacterial systematics is the technological progress to be expected in the near future and its enormous impact on polyphasic methodology. Data for large numbers of bacterial strains will be gathered even faster, but the challenge will be the processing of these enormous amounts of data into a classification system. Large sets of data can only be analyzed by computer- assisted techniques, and appropriate software programs are needed to agglomerate the most closely related strains and to represent the agglomerates. The application of fuzzy logic (Kosko, 1994), based on the idea that an isolate does not have to belong to a particular set of strains but can have a partial degree of membership in more than one set, may open new perspectives.
2. The accessibility of standardized genotypic, chemotaxonomic, and phenotypic features via universal databases is another goal, which can only be realized when complete standardization of all techniques is achieved. New methodologies that can fuse different databases are also required.
3. A further goal is the design of an accessible, cumulative, dynamic system allowing continuous recalculation of all existing information into "new synthetic taxonomies of the moment".
4. We will be dependent on these and other developments if we want to perform better in the discovery and description of bacterial biodiversity in nature. In such a system, nonclustered isolates and gene sequences of uncultured bacteria have their place and are available for comparison at the same level as recognized named taxa. In order to streamline labeling of taxa, the simplification of the actual nomenclatural practice might be considered.
5. Other macromolecules potentially useful for phylogenetic comparison e.g., β-subunit of ATPase and elongation factor Tu (Ludwig et al., 1993), chaperonin (Viale et al., 1994), various ribosomal proteins (Ochi, 1995), RNA polymerases (Zillig et al., 1989) and tRNAs (Höfle, 1990, 1991) should be further investigated to allow comparison of the results of phylogenetic analysis with the rRNA gene based dendrograms.
6. More whole-genome sequences and insights into genomic organization are available for a variety of bacterial organisms and will also become accessible to microbial taxonomists. It will be a formidable challenge in the next century to use this information to evaluate our present view on polyphasic classification.
7. Together with the comparison of sequences of particular genes or gene families, a better understanding of the evolution of bacterial genomes will become possible, shining a new light on the present (in)consistencies in bacterial systematics. If horizontal gene transfer is indeed not a marginal phenomenon but an important mechanism of procaryote evolution (Lake et al., 1999), complete genome sequencing may yield major revelations about the evolutionary tree of life.

Acknowledgments

We are indebted to the Fund for Scientific Research (Flanders, Belgium) for research and personnel grants (M.G., P.D.V., K.K. and J.S.) and for funding Postdoctoral Research Fellow (P.V.) and Research Director (P.D.V.) positions.

Overview: A Phylogenetic Backbone and Taxonomic Framework for Procaryotic Systematics

Wolfgang Ludwig and Hans-Peter Klenk

Introduction

Despite its relatively short history, microbial systematics has never been static but rather constantly subject to change. The evidence of this change is provided by many reclassifications in which bacterial taxa have been created, emended, or dissected, and organisms renamed or transferred. The development of a procaryotic systematics that reflects the natural relationships between microorganisms has always been a fundamental goal of taxonomists. However, the task of elucidating these relationships could not be addressed until the development of molecular methods (the analysis of macromolecules) that could be applied to bacterial identification and classification. Determination of genomic DNA G + C content, and chemotaxonomic methods such as analysis of cell wall and lipid composition, in many cases proved superior to classical methods based upon morphological and physiological traits. These tools provide information that can be used to differentiate taxa, but do not allow a comprehensive insight into the genetic and phylogenetic relationships of the organisms. DNA–DNA reassociation techniques provide data on genomic similarity and hence indirect phylogenetic information, but the resolution of this approach is limited to closely related strains. DNA–DNA hybridization is the method of choice for delimiting procaryotic species and estimating phylogeny at and below the species level. The current species concept is based on two organisms sharing a DNA–DNA hybridization value of greater than 70% (Wayne et al., 1987).

With improvement in molecular sequencing techniques, the idea of Zuckerkandl and Pauling (1965) to deduce the phylogenetic history of organisms by comparing the primary structures of macromolecules became applicable. The first molecules to be analyzed for this purpose were cytochromes and ferredoxins (Fitch and Margoliash, 1967). Subsequently, Carl Woese and coworkers demonstrated the usefulness of small subunit (SSU) rRNA as a universal phylogenetic marker (Fox et al., 1977). These studies suggested natural relationships between microorganisms on which a new procaryotic systematics could be based. The aims of this chapter are to provide a brief description of the methods used to reconstruct these phylogenetic relationships, to explore the phylogenetic relationships suggested by 16S rRNA and alternative molecular chronometers, and to present a justification for the use of the current 16S rRNA-based procaryotic systematics as a backbone for the structuring of the second edition of *Bergey's Manual of Systematic Bacteriology*.

Reconstruction and Interpretation of Phylogenetic Trees

Sequence alignment The critical initial step of sequence-based phylogenetic analyses is undoubtedly the alignment of primary structures. Alignment is necessary because only changes at positions with a common ancestry can be used to infer phylogenetic conclusions. These homologous positions have to be recognized and arranged in common columns to create an alignment, which then provides the basis for subsequent calculations and conclusions. Sequences such as SSU rRNA that contain a number of conserved sequence positions and stretches can be aligned using multiple sequence alignment software such as CLUSTAL W (Swofford et al., 1996). Furthermore, these conserved islands can be used a guide for arranging the intervening variable regions. The alignment of variable regions may remain difficult if deletions or insertions have occurred during the course of evolution. In addition, the homologous character of positions in variable regions is not necessarily indicated by sequence identity or similarity and hence can often not be reliably recognized. However, functional homology, if detectable or predictable, can be used to improve the alignment. In the case of rRNAs, functional pressure apparently dictates the evolutionary preservation of a common core of secondary or higher order structure which is manifested by the potential participation of 67% of the residues in helix formation by intramolecular base pairing. The majority of these structural elements are identical or similar with respect to their position within the molecule as well as number and position of paired bases, or internal and terminal loops. The primary structure sequence alignment can be evaluated and improved by checking for potential higher structure formation (Ludwig and Schleifer, 1994). Furthermore, the character of the base pairing, G–C versus non-G–C, Watson–Crick versus non-Watson–Crick, may be used to refine an alignment. The pairing is a byproduct of thermodynamic stability and consequently has an impact on function. Therefore, adjustments to the alignment appear rational from an evolutionary point of view. However, the recognition of homologous positions in regions which are highly variable with respect to primary as well as higher order structure may still be difficult or even impossible.

The principal problems of aligning rRNA sequences can be avoided by the routine user, if they take advantage of comprehensive databases of aligned sequences (including higher order structure information) that can be obtained from the Ribosomal Database Project (Maidak et al., 1999), the compilations of small

TABLE 1. Transformation of measured distances (lower triangle) into phylogenetic distances (upper triangle): applying the Jukes Cantor (Jukes and Cantor, 1969) transformation[a]

	Escherichia coli	*Klebsiella pneumoniae*	*Proteus vulgaris*	*Pseudomonas aeruginosa*	*Bacillus subtilis*	*Thermus thermophilus*	*Geotoga subterranea*
Escherichia coli		3.2	7	15.6	26	28.5	35.8
Klebsiella pneumoniae	3.1		7	15.1	25.8	28.2	36.4
Proteus vulgaris	6.7	6.7		17.6	26.6	29.9	37.8
Pseudomonas aeruginosa	14.1	13.7	15.7		23.5	29.2	34.3
Bacillus subtilis	22	21.8	22.4	20.2		27	30.4
Thermus thermophilus	23.7	23.5	24.7	24.2	22.6		32.4
Geotoga subterranea	28.5	28.8	29.7	27.6	25	26.3	

[a]The uncorrected distances were used for the reconstruction of the tree in Fig. 1. Given that the data are not ultrametric (see Swofford et al, 1996), they do not directly correlate with the branch lengths in the tree.

and large subunit rRNAs at the University of Antwerp (De Rijk et al., 1999; Van de Peer et al., 1999), or the ARB project as a guide to inserting new sequence data. The RDP offers alignment of submitted sequences as a service while the ARB program package contains tools for automated alignment, secondary structure check, and confidence test.*

Treeing methods The number and character of positional differences between aligned sequences are the basis for the inference of relationships. These primary data are then processed using treeing algorithms based on models of evolution. Usually, the phylogenetic analysis is refined by positional selection or weighting according to criteria such as variability or likelihood. The results of these analyses are usually visualized as additive trees. Terminal (the "organisms") and internal (the common "ancestors") nodes are connected by branches. The branching pattern indicates the path of evolution and the (additive) lengths of peripheral and internal branches connecting two terminal nodes indicate the phylogenetic distances between the respective organisms. There are two principal versions of presentation: radial trees or dendrograms (Fig. 1). The advantage of radial tree presentation is that phylogenetic relationships, especially of only moderately related groups, can usually be shown more clearly, and that all of the information is condensed into an area which can be inspected "at a glance". However, the number of terminal nodes (sequences, organisms, taxa) for which the relationships can be demonstrated is limited. This number is not limited in dendrograms.

A number of different treeing methods or algorithms based on sequence data have been developed. Most of them are based on models of evolution. These models describe assumed rules of the evolutionary process concerning parameters such as (overall) base frequencies or (number and weighting of) substitution types. A comprehensive review on methods for phylogenetic analyses, models of evolution, and the mathematical background is given by Swofford et al. (1996). The three most commonly used treeing methods, distance matrix, maximum parsimony, and maximum likelihood, operate by selecting trees which maximize the congruency of topology and branch lengths with the measured data under the criteria of a given model of evolution.

Distance treeing methods such as Neighbor Joining (Saitou and Nei, 1987) or the method of Fitch and Margoliash (Fitch and Margoliash, 1967) rely on matrices of distance values obtained by binary comparison of aligned sequences and calculation of the fraction of base differences. These treeing programs mostly perform modified cluster analyses by defining pairs and, subsequently, groups of sequences sharing the lowest distance values and connecting them into the framework of a growing tree. The tree topology is optimized by maximizing the congruence between the branch lengths in the tree and the corresponding inferred distances of the underlying matrix.

Before treeing, the measured differences are usually transformed into evolutionary distance values according to models of evolution. The underlying assumption is that the real number of evolutionary changes is underestimated by counting the detectable differences in present day sequences. For example, the Jukes Cantor transformation (Jukes and Cantor, 1969) accounts for this underestimation by superelevation of the measured distances (Table 1). Although the theoretical assumptions that provide the basis for transforming the measured distance values into phylogenetic distances are convincing with respect to overall branch lengths, there is a certain risk of misinterpretation or overestimation of local tree topologies. An intrinsic disadvantage of distance treeing methods is that only part of the phylogenetic information, the distances, is used, while the character of change is not taken into account. However, there are methods available to perform more sophisticated distance calculations than simply counting the differences (Felsenstein, 1982).

In contrast to distance methods, maximum parsimony-based treeing approaches use the original sequence data as input. According to maximum parsimony criteria, tree reconstruction and optimization is based on a model of evolution that assumes preservation to be more likely than change. Parsimony methods search for tree topologies that minimize the total tree length. That means the most parsimonious (Edgell et al., 1996) tree topology (topologies) require(s) the assumption of a minimum number of base changes to correlate the tree topology and the original sequence data. In principle, the problem of plesiomorphies (see below) can be handled more appropriately with parsimony than with distance methods, given that the most probable ancestor character state is estimated at any internal node of the tree. Long branch attraction is a disadvantage of the maximum parsimony approach. The parsimony approach does infer branching patterns but does not calculate branch lengths *per se.* To superimpose branch lengths on the most parsimonious tree topologies additional methods and criteria have to be applied. Both PAUP* and ARB parsimony tools are able to combine the

**Editorial Note:* Software available from O. Strunk and W. Ludwig, Department of Microbiology, Technische Universität, München, Munich, Germany. ARB is a software environment for sequence data located at: www.mikro.biologie.tu-muenchen.de/pub/ARB.

**Editorial Note:* Software available from David Swofford at the Laboratory of Molecular Systematics, National Museum of Natural History, Smithonian Institution, Washington, D.C. Available from Sinauer Associates of Sunderland, MS at www.sinauer.com/formpurch.htm.

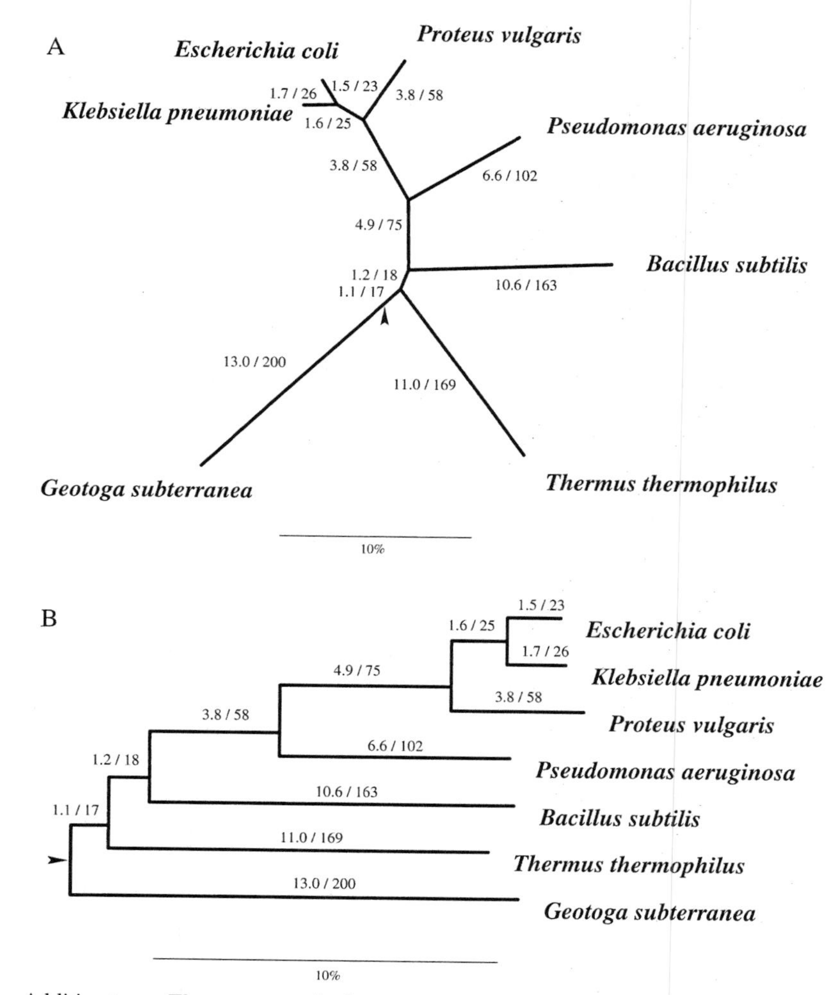

FIGURE 1. Additive trees. The same tree is shown as a radial tree (*A*) and a dendrogram (*B*). The tree was reconstructed by applying the neighbor joining method (Saitou and Nei, 1987) to a matrix of uncorrected binary 16S rRNA sequence differences for the organisms shown in the tree and a selection of archaeal sequences as outgroup references. *Arrowheads* indicate the branching of the archaeal reference sequences and the root of the trees. Bar = 10% sequence difference. The distance between two sequences (organisms) is the sum of all branch lengths directly connecting the respective terminal nodes or the sum of the corresponding horizontal branch lengths in the radial tree or the dendrogram, respectively. The numbers at the individual branches indicate overall percentage sequence divergence, followed by the number of different sequence positions (the length of the *E. coli* 16S rRNA sequence [1542 nucleotides] was used as reference in all calculations). Note: the tree topology was not evaluated by applying different methods and parameters.

reconstruction of topologies and the estimation of branch lengths.

The most sophisticated of the three independent phylogenetic treeing methods is maximum likelihood, where a tree topology is regarded as optimal if it reflects a path of evolution that, according to the criteria of given models of evolution, most likely resulted in the sequences of the contemporary organisms. The corresponding evolutionary models may include parameters such as transition/transversion ratio, positional variability, character state probability per position and many others. Given that the maximum likelihood approach utilizes more of the information content of the underlying sequences, it is considered to be superior to the other two treeing methods. An accompanying disadvantage is the need for expensive computing time and performance. Even if powerful computing facilities are accessible only a limited number of sequences can be handled within a reasonable time. Rapid development in the field of computing hardware suggests that this powerful method may become applicable for larger data sets in the near future.

The use of filters Most commonly used programs for phylogenetic treeing are capable of including filters or weighting masks that remove or weight down individual alignment columns while treeing, thus reducing the influence of highly variable positions. Conservation profiles can be calculated by simply determining the fraction of the most frequent character. More sophisticated approaches define positional variability, the rate of change, or the likelihood of a given character state, with respect to an underlying tree topology according to parsimony criteria, or by using a maximum likelihood approach. The choice of phylogenetic entities for which filters or masks should be generated depends on the group of organisms or the phylogenetic level

TABLE 2. Phylogenetic information content of procaryotic small subunit rRNA[a]

	Bacteria				*Archaea*			
Intra-domain similarity	&mt;67%				&mt;67%			
	Conserved		Variable		Conserved		Variable	
	Pos.	%	Pos.	%	Pos.	%	Pos.	%
Sequence conservation	568	36.8	974	63.2	571	37	971	63
Potential information (bits)			1948				1942	
Number of characters	1	2	3	4	1	2	3	4
Positional variability, %	36.8	23.2	13.5	26.5	37	28.3	15.2	19.5
Corrected information (bits)			1506				1385	

[a]The calculations were performed using the 16S rRNA sequence of *E. coli* (1542 nucleotides) as a reference. To avoid influences of sequencing, database, and alignment errors a 98% similarity criterion was applied to define "invariant" positions. Therefore, the term 'conserved' was used instead "invariant". Bits (of information) were calculated by multiplying the logarithm to the base two of the permissive character states (positional variability: different nucleotides per position) times the number of informative (variable) sites. Potential information was calculated as the maximum information content assuming positional variability of four. These values were corrected according measured positional variability.

(the corresponding area in a tree) of interest. Tools for the generation of profiles, masks, or filters are implemented in the ARB software package or available from other authors (Swofford et al., 1996; Maidak et al., 1999). The removal of positions also means loss of information; therefore it is recommended to perform treeing analyses of a given data set several times applying different filters. This helps to visualize the robustness or weakness of a specific tree topology and to estimate whether or not variable positions have had a substantial influence. Filters or masks should only be calculated using comprehensive data sets of full sequences; then these filters can also be applied to the analysis of partial sequences. The results of many years of tree reconstruction have shown that positions should only be removed up to 60% positional conservation, to avoid the loss of too much information. In most cases use of a 50% conservation filter is appropriate.

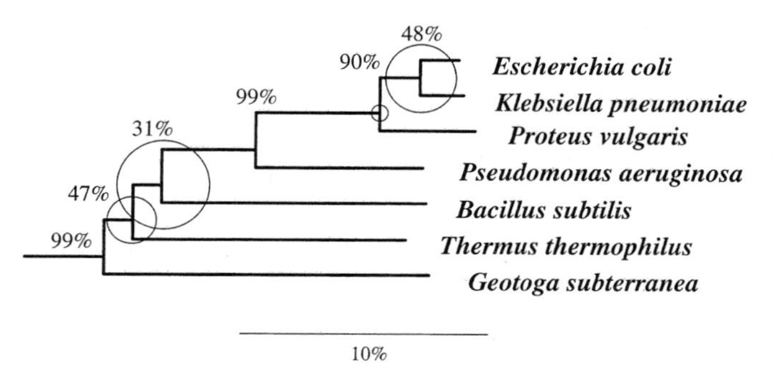

FIGURE 2. Confidence tests on tree topology. 1000 bootstrap operations were performed for evaluation of the tree in Figure 1B. The numbers at the furcations indicate the fraction of (1000 bootstrapped) trees which support the separation of the respective subtree (branches to the right of the particular furcation) from all other branches or groups in the tree. Circles indicating an area of "unsharpness" were calculated as a function of bootstrap values and branch lengths using ARB.

Confidence tests Different treeing methods handle data according to particular assumptions and consequently may yield different results. The many inconsistencies of real sequence data also prevent easy and reliable phylogenetic inference; therefore the careful evaluation of tree topologies is to be recommended. Besides the application of filters and weighting masks and the use of different treeing approaches, resampling techniques can be used to evaluate the statistical significance of branching order. Bootstrapping or jackknifing (Swofford et al., 1996) are procedures that randomly sample or delete columns in sequence data (alignments) or distance values (distance matrix). Usually 100–1000 different artificial data sets are generated as inputs for treeing operations by these methods. For each data set the optimum tree topologies are inferred by the particular treeing method and, finally, a consensus tree topology is generated. In this consensus tree, bootstrap or jackknife values are assigned to the individual branches. These values indicate the number of treeing runs in which the subtree defined by the respective branch appeared as monophyletic with respect to all other groups. An example of a bootstrapped tree is shown in Fig. 2. Besides the bootstrap value, an area of low significance is indicated by circles centered on the individual (internal) nodes. These areas were estimated from the sampling values in relation to the corresponding (internal) branch lengths using the ARB software tools. No convincing significance can be expected if only a few residues provide information supporting the separation of branches or subtrees. Given that in most cases branch lengths indicate the degree of estimated sequence divergence, a subtree separated from the remainder of a phylogenetic tree by a short internal branch is highly unlikely to be assigned a high resampling value.

The resampling techniques can only be used to estimate the robustness of a tree reconstructed by applying a single treeing method and parameter set. Thus for reliable phylogenetic conclusions it is necessary to combine different treeing methods, as well as filters and weighting masks and resampling techniques. Even if appropriate software and powerful hardware are accessible, high quality tree evaluations may get rather expensive in working and computing time. An approach for estimating "upper bootstrap" limits without the need of expensive multiple treeings was developed and implemented in the ARB package. In the absence of resampling values, a critical "reading" of trees allows a rough estimate of the confidence of relative branching orders at a glance, assuming that a short branch length in most cases also indicates low significance of separation.

Why do trees differ? Tree reconstruction can often be a frustrating experience, especially for researchers not familiar with the theoretical principles of phylogenetic treeing, when the application of different treeing methods or parameters to a single data set results in different tree topologies. This is not surprising since different treeing methods are based on different models of evolution, and therefore the data are processed in different ways. Consequently, a perfect match of tree topologies cannot necessarily be expected even if identical data sets are analyzed using identical parameters. None of the models reflect perfectly the reality of the evolutionary process. The assumption of independent evolution of different sequence positions, for example, does not hold true for the many functionally correlated residues such as base paired nucleotides in rRNAs. In addition, none of the treeing methods and software programs can really exhaustively test and optimize all possible tree topologies. For example, with only 20 sequences there would be 10^{20} possible tree topologies to be examined. Other factors, such as data selection (the organisms and sequence positions included in calculations), the order of data addition to the tree, and the presence of positions that have changed at a higher rate than the remainder of the data set, also influence tree topology. These instabilities do not usually concern the global tree topology but rather local branching patterns.

Limitations of Tree Reconstruction

Information content of molecular chronometers The reconstruction of gene or organismal history, based upon the degree of divergence of present day sequences, relies on the number and character of detectable sequence changes that have accumulated during the course of evolution. Thus the maximum information content of molecules is defined by the number of characters (monomers), and the number of potential character states (different residues), per site. With real data, only a fraction of the sites are informative, as a reasonable degree of sequence conservation is needed to demonstrate the homologous character of molecules or genes and to recognize a phylogenetic marker as such. For example, there are 974 (63.2%) variable and hence informative positions in the 16S rRNA genes of members of the *Bacteria*, and 971 (63%) such positions in the *Archaea*. Given that the maximum information content per position is defined by the number of possible character states i.e., the four nucleotides (the potential fifth character state, deletion or insertion, is not considered here), there could be 1948 (*Bacteria*) or 1942 (*Archaea*) bits of information (logarithm to base 2 of the number of possible character states times the number of informative positions) in the SSU rRNA. However, due to functional constraints and evolutionary selective pressure, the number of allowed character states varies from position to position. As shown in Table 2, there are only 407 (26.4%; *Bacteria*) or 301 (19.5%; *Archaea*) positions in the investigated data set at which all four nucleotides are found, whereas only three different residues apparently are tolerated at 209 (13.6%; *Bacteria*) or 233 (15.2%; *Archaea*) positions, and only two character states are realized at 358 (23.2%; *Bacteria*) or 437 (28.3%; *Archaea*) positions. Thus the theoretical information content of 1984 (*Bacteria*) or 1938 (*Archaea*) bits in reality is reduced to 1506 (*Bacteria*) or 1385 (*Archaea*). The reduced information content draws attention to the need for careful sequence alignment and analysis.

The problem of plesiomorphy Any homologous residue in present day sequences can only report one evolutionary event. The higher the number of permitted characters at a particular position, the higher the probability that such an evolutionary event is directly detectable (by a difference). The majority of these events remain obscure since, especially at variable positions, identical residues are probably the result of multiple changes during the course of evolution, simulating an unchanged position (plesiomorphy). The effect of plesiomorphy on the topology of the resulting trees depends on the number of plesiomorphies supporting branch attraction and also on the treeing method used. Such plesiomorphic sites may cause misleading branch attraction, as shown in Fig. 3, where a short stretch of aligned real 16S rRNA sequences is used to visualize branch attraction. Plesiomorphies may also be responsible for the observation that long "naked" branches represented by only one or a few highly similar sequences often "jump" in phylogenetic trees when the reference data set is changed or expanded. The positioning can usually be stabilized when further representatives of different phylogenetic levels of that branch become available. The rooting of trees may also be influenced by identities at plesiomorphic sites when single sequences are used as outgroup references. The influence of plesiomorphic positions can be reduced by using them at a lower weight for tree reconstruction, but is nevertheless still present.

Partial sequence data There are several convincing arguments for the use of only complete sequence data in the reconstruction of phylogenetic trees. These include the limited information content of the molecule, and the fact that different parts of the primary structure carry information for different phylogenetic levels (Ludwig et al., 1998b). Whenever partial sequences are added to a database of complete primary structures and phylogenetic treeing approaches are applied to the new data set, the new sequences may influence the overall tree topology. The inclusion of partial sequence data may impair phylogenetic trees or influence conclusions previously based on full data. Software which allows the addition of new data to a given data set, and placement of the new sequence according to optimality criteria in a validated tree without changing its topology, is now available. The ARB implementation of this software is capable of removing short partial sequences from a tree prior to the integration of a new highly similar but more complete sequence. Thus the more informative sequence is not "attracted" by a probably misplaced partial sequence. After finding the most similar sequences, the ARB tool compares the number of determined characters, removes the shorter version, and reinserts the data in the order of completeness. There are a number of recent publications presenting comprehensive trees based upon data sets which have been truncated to the regions comprised by included partial sequences. This procedure is not acceptable, given all the limitations of partial sequence data and of the methods of analysis.

Partial sequence data of appropriate regions of the gene may contain sufficient information for the identification of organisms. The determination and comparative analysis of partial sequences may be sufficient to reliably assign an organism to a phylogenetic group if the database contains sequences from closest relatives. A fraction of the 5′-terminal region of the SSU rRNA (*Escherichia coli* pos. 60–110) is one the most informative or discriminating regions for closely related organisms. Hence partial sequence data that include this region can be used to find the closest relative of an organism or to indicate a novel species. Short diagnostic regions (15–20 nucleotides) of partial sequences can also be used as targets for taxon-specific probes or PCR primers that are commonly used for the sensitive detection and identification of microorganisms (Schleifer et al., 1993; Amann et al., 1995; Ludwig et al., 1998a).

Bush-like trees The majority of names and definitions of major phylogenetic groups, such as the phylum *"Proteobacteria"* and the corresponding classes (*"Alphaproteobacteria"*, *"Betaproteobacteria"*, *"Gammaproteobacteria"*, *"Deltaproteobacteria"*, and *"Epsilonproteobacteria"*) originated in the early years of comparative rRNA sequence analysis. At that time phylogenetic clusters could easily be delimited, given that the trees contained many long "naked"

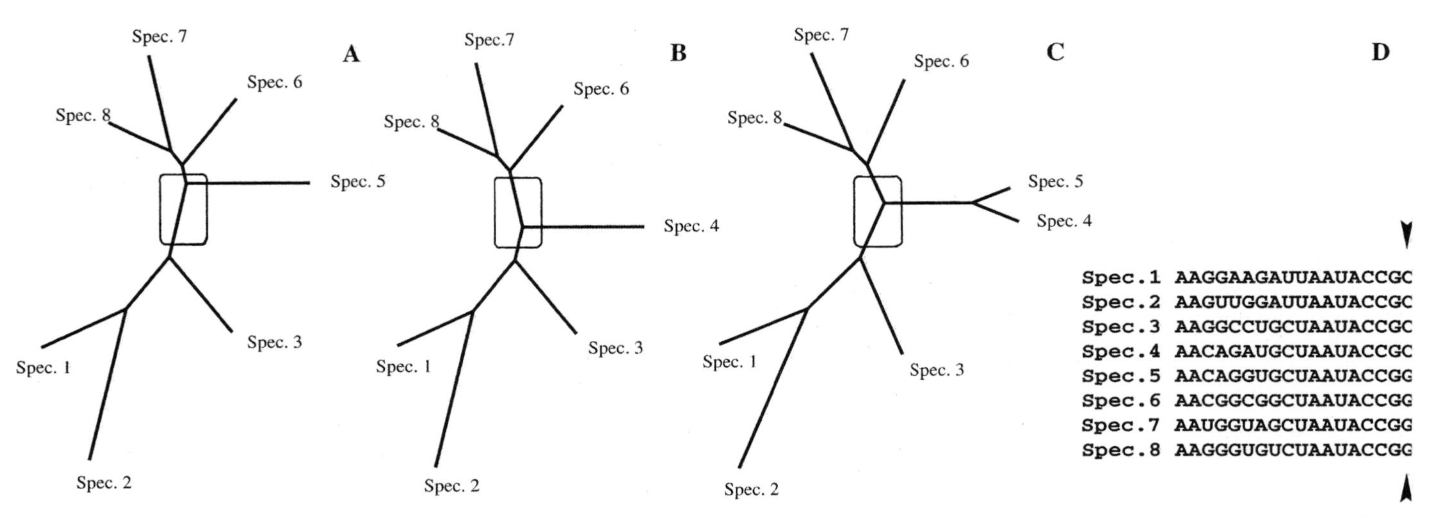

FIGURE 3. "Branch attraction". The trees show the effects of separate (*A*, *B*) or combined (*C*) inclusion of sequences Spec. 4 and Spec. 5 on the tree topology. The *rectangle* highlights the region of the tree where major changes can be seen. The trees were reconstructed using the neighbor joining method on the aligned 16S rRNA sequence fragments shown (*D*). The column of residues responsible for the attraction of Spec. 5 branch and the Spec. 6–8 subtree in *A* as well as the attraction of Spec. 4 branch and the Spec. 1–3 subtree in *B* is marked by arrows.

branches separating subtrees. This "phylogenetic clarity" was mainly an effect of the limited amount of available sequence data and has often been obscured by the rapid expansion in the number of sequence database entries. Most of the long "naked" branches have expanded and the tree-like topology changed to a bush-like topology. It is probably only a matter of time before the missing links will be found for the remaining "naked" branches such as the *Chlamydiales*, the *"Flexistipes"*, or branches assigned to cloned environmental sequences.

In bush-like areas of a tree, the probability that a given branch will exchange positions with a neighboring branch decreases with the distance between the two branches. This indicates that the relative order of closely neighboring branches cannot be reliably reconstructed, although their separation from more distantly located lineages remains robust. As a consequence, delimitation of taxa often cannot be based on individual local branching order. The use of criteria or additional data for the definition of taxonomic units remains the subjective decision of the taxonomist. In some cases, this leads to the definition of taxa that include paraphyletic groups.

Presentation of Phylogenetic Trees

The main purpose of drawing trees is to visualize the phylogenetic relationships of the organisms or markers, and to allow the reader to recognize these relationships at a glance. It is often difficult to combine an easy-to-grasp presentation of phylogenetic relationships and associated information on the significance of branching patterns. There is no optimum solution to the problem of "correct" presentation of trees; however, ways of addressing this problem can be suggested.

One acceptable procedure would be to present all the trees (which may differ locally) obtained from the same data set by the application of different treeing methods and parameters. However, this may prove more confusing than helpful for readers not experienced in phylogenetic treeing. A more user-friendly solution is to present only one tree topology and to indicate the significance of the individual branches or nodes. However, showing multiple confidence values at individual nodes, or depicting areas of confidence by shading or circles around the nodes, may make the tree unreadable, especially in areas of bush-like topology.

In many cases use of a consensus tree is advantageous. Some programs for consensus tree generation are able to present local topologies as multifurcations at which a relative branching order is not significantly supported by the results of tree evaluations. A fairly acceptable compromise is to use a consensus tree, and to visualize both a detailed branching pattern where stable topologies can be validated, and multifurcations that indicate inconsistencies or uncertainties. Such a multifurcation indicates missing information on that particular era of evolution rather than multiple events resulting in a high diversification within a narrow span of evolutionary time. This type of presentation is certainly more informative for the reader than a choice of various tree topologies, each showing low statistical significance (Ludwig et al., 1998b).

None of the modes of presentation described above can be applied to bush-like topologies and yield meaningful results. Although individual branches are likely to change their positions only locally within a large bush-like area, depending on the methods and parameters applied for treeing, there is no way to split up such an area by several multifurcations. No methods are currently available for the calculation of confidence values for the next, second, third and so on neighboring nodes, and highlighting areas of unsharpness makes the tree difficult to read. A legitimate solution is to base the calculations on the full data set but to hide some of the branches for presentation purposes, and show a tree topology containing a smaller number of significantly separated branches. Thus, while the tree would be based on all available information, only that part of the tree topology of interest for the particular phylogenetic problem is shown clearly laid out (Ludwig et al., 1998b).

16S rRNA: The Benchmark Molecule for Procaryote Systematics

In principle, all the requirements of a phylogenetic marker molecule are fulfilled in SSU rRNAs to a greater extent than in almost all other described phylogenetic markers (Woese, 1987; Olsen and Woese, 1993; Olsen et al., 1994; Ludwig and Schleifer, 1994; Ludwig et al., 1998b). Besides functional constancy, ubiquitous distribution, and large size (information content), genes coding for SSU rRNA exhibit both evolutionarily conserved regions and highly variable structural elements. The latter characteristic results from different functional selective pressures acting upon the independent structural elements. This varying degree of sequence conservation allows reconstruction of phylogenies for a broad range of relationships from the domain to the species level. A comprehensive SSU rRNA sequence data set (currently more than 16,000 entries) is available in public databases (Ludwig, 1995*; Maidak et al., 1999; Van de Peer et al., 1999) in plain or processed (aligned) format, and is rapidly increasing in size. A significant fraction of validly described procaryotic species are represented by 16S rRNA sequences from type strains or closely related strains.

As with any new technique in the field of taxonomy, it took time to establish comparative sequencing of SSU rRNA (genes) as a powerful standard method for the identification of microorganisms and defining or restructuring procaryotic taxa according to their natural relationships. Rapid progress in sequencing and *in vitro* nucleic acid amplification technology led to the replacement of an expensive, sophisticated, and tedious methodology, available only to specialists, by rapid and easy-to-apply routine techniques. As a result, analysis of the genes coding for SSU rRNA is one of the most widely used classification techniques in procaryotic identification and systematics. It is widely accepted that SSU rRNA analysis should be integrated into a polyphasic approach for the new description of bacterial species or higher taxa.

Some Drawbacks of 16S rRNA Gene Sequence Analysis

Functional constraints Depending on functional importance, the individual structural elements of rRNAs cannot be freely changed. It is therefore assumed that sequence change in the rRNAs occurs in jumps rather than as a continuous process. The divergence of present day rRNA sequences may document the succession of common ancestors and their present day descendants, but a direct correlation to a time scale cannot be postulated.

Multiple genes It has been known since the early days of comparative rRNA sequence analysis that the genomes of microorganisms may contain multiple copies of some genes or operons. However, until recently it was commonly assumed that there are no remarkable differences between the rRNA gene sequences of a given organism. A significant degree of sequence divergence among multiple homologous genes within the same organism, such as has been found in *Clostridium paradoxum* (Rainey et al., 1996) and *Paenibacillus polymyxa* (Nübel et al., 1996), would call any sequence-based interorganism relationships into question. The underestimation of this problem may be attributed to the fact that such differences are not easy to recognize using sequencing techniques which depend on purified rRNA or amplified rDNA, and can be mistaken for artifacts. Only frame shifts resulting from inserted or deleted residues can be readily recognized. New techniques, such as denaturing gradient gel electrophoresis (DGGE) (Nübel et al., 1996), allow sequence variation in PCR-amplified rDNA fragments to be detected. The rapidly progressing genome sequencing projects have also provided detailed information on the topic of intraorganism rRNA heterogeneities. Different organisms vary with respect to the presence and degree of intercistron primary structure variation, and most differences concern variable positions and affect base-paired positions (Engel, 1999; Nübel et al., 1996). Although some projects to systematically investigate interoperon differences have been initiated, no comprehensive survey of the spectrum of microbial phyla has been performed. Current and future investigations will show whether regularities or hot spots for interoperon differences can be defined in general or in particular for certain phylogenetic groups. This knowledge can then be used to remove or weight such positions for phylogenetic reconstructions.

Interpretation of high 16S rRNA gene sequence similarity Organisms sharing identical SSU rRNA sequences may be more diverged at the whole genome level than others which contain rRNAs differing at a few variable positions. This has been shown by comparison of 16S rRNA sequence and genomic DNA–DNA hybridization data (Stackebrandt and Goebel, 1994). In the interpretation of phylogenetic trees, it is important to note that branching patterns at the periphery of the tree cannot reliably reflect phylogenetic reality. Given the low phylogenetic resolving power at these levels of close relatedness (above 97% similarity), it is highly recommended to support conclusions based on SSU rRNA sequence data analysis by genomic DNA reassociation studies (Stackebrandt and Goebel, 1994).

Comparative Analyses of Alternative Phylogenetic Markers

Other genes have been investigated as potential alternative phylogenetic markers, to determine whether SSU rRNA-based phylogenetic conclusions can describe the relationships of the organisms, or merely reflect the evolutionary history of the respective genes. For sound testing of phylogenetic conclusions based on SSU rRNA data, the sequences used must originate from adequate phylogenetic markers. The principal requirements for such markers are ubiquitous distribution in the living world combined with functional constancy, sufficient information content, and a sequence database which represents diverse organisms, containing at least members of the major groups (phyla and lower taxa) as defined based upon SSU rRNA.

How many alternative phylogenetic markers are out there? Comparative analysis of the completed genome sequences suggests that there are only a limited number of genes that occur in all genomes and which also share sufficient sequence similarity to be recognized as ortho- or paralogous. Analysis of the first eight completely sequenced genomes (six *Bacteria*, one *Archaea*, and one yeast) showed that only 110 clusters of orthologous groups (COGs) were present in all genomes (Tatusov et al., 1997; Koonin et al., 1998; updated in www.ncbi.nlm.nih.gov/COG/) and only eight additional genes were ubiquitous in procaryotes. Another 126 COGs were found in the remaining five microbial genomes, excluding the mycoplasmas, which have a reduced genomic complement. The majority of the universally conserved COGs (65 out of 110) belong to the information storage and processing proteins, which appear to hold more promise for future phylogenetic analysis than the metabolic proteins. However, about half

Editorial Note: Software available from O. Strunk and W. Ludwig, Department of Microbiology, Technische Universität, München, Munich, Germany. ARB is a software environment for sequence data located at: www.mikro.biologie.tu-muenchen.de/pub/ARB.

of these information processing COGs contain ribosomal proteins, which are small and therefore not sufficiently informative for the inference of global phylogenies. This leaves us with about 40–100 genes that fulfill the basic requirements of useful phylogenetic markers.

It has been proposed that many genes involved in the processing of genetic information (components of the transcription and translation systems) exhibit concurrent evolution due to their housekeeping function (Olsen and Woese, 1997). It appears logical that these key systems would be optimized early and then conserved to confer maximum survival and evolutionary benefit on the organism.

Although the databases of alternative phylogenetic markers are small relative to that of the SSU rRNA, some of the other requirements for markers, including representation of phylogenetically diverse organisms, are met by, for example, LSU rRNA, elongation factor Tu/1α, the catalytic subunit of the proton translocating ATPase, *recA*, and the hsp 60 heat shock protein. For some other markers fulfillment of the ubiquity requirement can not be assessed because of the limited state of the sequence databases.

Some Drawbacks of Alternative Phylogenetic Markers

Lateral gene transfer and gene duplication Comparative analyses of the 18 published complete microbial genome sequences does not reveal a consensus picture of the root of the tree of life (Klenk et al., 1997) or of the relative branching order of the early lineages within the domains. This contradicts the marked separation of the primary domains based on morphology, physiology, biochemical characteristics, and overall genome sequence data. A monophyletic origin of the domain *Archaea* has been put in question by some authors (Gupta, 1998), but genomic evidence for monophyly of this group has also been reported (Gaasterland and Ragan, 1999). This contradiction has led to the assumption that lateral gene transfer and/or gene duplications, often followed by the loss of one or more gene variants in different lineages, has occurred in some potential marker molecules, especially genes coding for proteins involved in central metabolism (Brown and Doolittle, 1997). Obviously, such genes or markers cannot be used for testing major phylogenies deduced from SSU rRNA data.

The usefulness of many proteins as potential phylogenetic markers is curtailed by the presence of duplicated genes in certain organisms. The degree of sequence divergence in these duplicated markers ranges from the interdomain level, as shown for the catalytic subunit of vacuolar and F_1F_0-ATPases of *Enterococcus hirae*, to the species level, exemplified by EF-Tu of *Streptomyces ramocissimus*. When conserved proteins are used as phylogenetic markers for inferring intradomain phylogenies, one has to take care that orthologous genes (common origin) rather than paralogous genes (descendants of duplications) are compared. The recognition of paralogous genes is a central problem in phylogenetic analyses, especially when only limited data sets are available as in the case of the catalytic subunit of the proton-translocating ATPase. Although the sequence similarities between bacterial F_1F_0 type, and archaeal and eucaryal vacuolar type, ATPase subunits are rather low (around 20%), it was initially assumed that the corresponding subunits (β and A or α and B) are homologous molecules (Iwabe et al., 1989; Ludwig et al., 1993). The presence of an F_1F_0 type ATPase β-subunit gene has been shown for all representatives of the domain *Bacteria* investigated thus far (Ludwig et al., 1993; Neumaier, 1996). However, the finding that *Thermus* and other members of the *"Deinococcus-Thermus"* phylum contain vacuolar type ATPases (Tsutsumi et al., 1991; Neumaier, 1996) threatened this ATPase-based phylogenetic picture. It was later found that genes for subunits of vacuolar type ATPases exist in many (but not all) bacterial species from different phyla in addition to the corresponding F_1F_0 type ATPase subunit genes (Kakinuma et al., 1991; Neumaier, 1996). It is commonly accepted that F_1F_0 type ATPase subunits α and β resulted from an early gene duplication and should be regarded as paralogous. The same is assumed for the vacuolar type ATPase subunits A and B. The findings described above suggest additional early gene duplications probably leading to the ancestors of F_1F_0 and vacuolar type ATPase (subunits). Whereas α and β, or A and B subunits, coexist in all cases investigated so far, this is not the case for the F_1F_0 and vacuolar type paralogs. The available data indicate that the former would have become the essential energy-gaining version in the bacterial domain, the latter in the archaeal and eucaryal domains. During the course of evolution, the other member of the duplicate pair apparently changed its function (Kakinuma et al., 1991) and may have lost its essential character. Therefore, the nonessential copy could have been lost by many (even closely related) organisms during the course of evolution. The *"Deinococcus–Thermus"* phylum, in which only vacuolar type ATPases have been found, might be an exception. Assuming an early diversification of the bacterial phyla, the functional diversification of the duplicated ATPases could have occurred during this era of evolution. The ancestor of the members of the *"Deinococcus–Thermus"* phylum may have lost the F_1F_0 version early in evolution. However, early lateral gene transfers as postulated by some authors (Hilario and Gogarten, 1993) cannot be excluded.

There are other examples of gene duplications and premature phylogenetic misinterpretations, as documented by the history of glyceraldehyde-3-phosphate dehydrogenase (GAPDH) based phylogenetic investigations (Martin and Cerff, 1986; Brinkmann et al., 1987; Martin et al., 1993; Henze et al., 1995). Besides these early gene duplications, there are also indications of more recent events, such as the EF-Tu of *Streptomyces*, *hsp60* of *Rhizobium*, or *recA* of *Myxococcus*. Paralogous genes occurring as a result of gene duplication or lateral gene transfer can only be recognized as such in organisms which have preserved more than one version of the (duplicated) gene. And even then it may remain difficult or even impossible to decide which genes can be regarded as orthologous. Obviously, only the orthologous gene, which represents the functionally essential compound, can be used for inferring or evaluating phylogenies. Thus, whenever new potential phylogenetic markers are investigated and major discrepancies with rRNA-based conclusions are found, a comprehensive data base should be established, accompanied by an extensive search for potential gene duplications.

Limited information content Based on currently available sequence data, the LSU rRNA is the only marker which carries more phylogenetic information than the small subunit rRNA. There are more than twice as many informative residues in the large subunit rRNA (Ludwig et al., 1998b). In the case of protein markers, the amino acid sequences are preferred over the coding gene sequences for phylogenetic analysis. Proteins provide the function, and consequently the amino acid sequences are the targets of evolutionary selective pressure. In contrast, the DNA sequence differences, especially at third base positions, are under pressure of the codon preferences of the particular organism. Most of the proteins recognized as useful phylogenetic markers comprise less informative primary structure sites than the rRNA markers. For example, EF-Tu/-1α and ATPase catalytic subunit protein primary structures contain 311 and 359 informative residues, respectively. This deficiency could be partly compensated for by the 20 possible character states (amino acids) per position. However, in real data the number of allowed character states—the positional variability—is reduced due to functional constraints. The current data sets (EF-Tu/-1α and ATPase catalytic subunit) do not contain positions at which more than 15 different amino acids occur, and the largest fractions of positions (18%–20%, 11%–12%, 9%–12%) are represented by positions with only 2, 3, or 4 different residues, respectively.

Conflicting tree topologies Identical tree topologies cannot be expected from phylogenetic analysis of different markers. Given the low phylogenetic information content of each of the markers, and the wide grid of resolution, it is unlikely that independently evolving markers have preserved information on the same eras of evolutionary time. In principle, one would expect that this missing phylogenetic information would yield reduced resolution but not change the topology of the tree. However, the latter is often the case, as shown in Fig. 4. A small stretch of aligned real 16S rRNA sequences was used to generate the tree in Fig. 4A. If it is assumed that this tree illustrates the phylogenetic truth and that the information for the common origin of Spec. 4 and Spec. 5 was lost during the course of evolution, one would expect a reduction in resolution. Removing the alignment column (marked by an arrowhead) responsible for this relationship should result in shortening or deleting of the common branch of Spec. 4 and Spec. 5, producing a multifurcation as shown in Fig. 4B. However, due to branch attraction by residues at other alignment positions the branches of Spec. 4 and Spec. 5 are separated as shown in Fig. 4C, misleadingly simulating a different history. Consequently, local differences of resolution and topology in trees derived from alternative phylogenetic markers do not necessarily indicate a different path of evolution.

Alternative Gene Trees

Large subunit rRNA As alluded to above, the LSU rRNA may be the most informative alternative phylogenetic marker. The primary structure of this molecule is at least as conserved as that of the SSU rRNA, and it contains more and longer stretches of informative positions. The spectrum of the LSU rRNA database is superior to that of all other alternative (protein) markers. Given that both rRNAs are involved in the translation process, it can be assumed that a similar selective pressure has been exerted on both genes. Consequently, LSU rRNA should be more useful for supporting rather than evaluating SSU rRNA-based conclusions. The internal structure (branching orders of the major lineages) of the intradomain trees can also be evaluated, given the availability of representative data sets for both molecules. The overall topologies of trees based upon the sequences of small and large subunit rRNA genes are in good agreement (De Rijk et al., 1995; Ludwig et al., 1998b). A 23S rRNA-based bacterial phyla tree is shown in Fig. 5, the corresponding 16S rRNA-based tree in Fig. 6. Slight local differences between trees reconstructed from both genes with the same method and parameters have been documented (De Rijk et al., 1995; Ludwig et al., 1995). This finding does not really cast doubt on the SSU rRNA-based branching patterns but rather underlines the previously mentioned limitations of phylogenetic markers. The LSU rRNA might be the better phylogenetic marker, providing more information and greater resolution, but the major drawback of this molecule is the currently limited database. Unfortunately, this database has not grown as fast as that for the SSU rRNA.

Elongation factors The elongation factors are also intrinsic components of the translation process but are functionally different from the rRNAs. It is generally assumed that the different classes of elongation (and probably initiation) factors are paralogous molecules resulting from early gene duplications. At present, a reasonable data set is available for EF-Tu/1α. In general, EF-Tu/1α-based domain trees (Fig. 7) globally support rRNA-derived branching patterns (Ludwig et al., 1998b). However, some general problems of protein markers are also exhibited by EF-Tu/1α sequences. As with the rRNA markers, no significant relative branching order for the major intradomain lines of descent can be determined. No major contradictions, e.g., members of a given phylum defined by rRNA sequences clustering among representatives of another phylum, were seen between rRNA and EF-Tu/1α tree topologies. However, in detailed

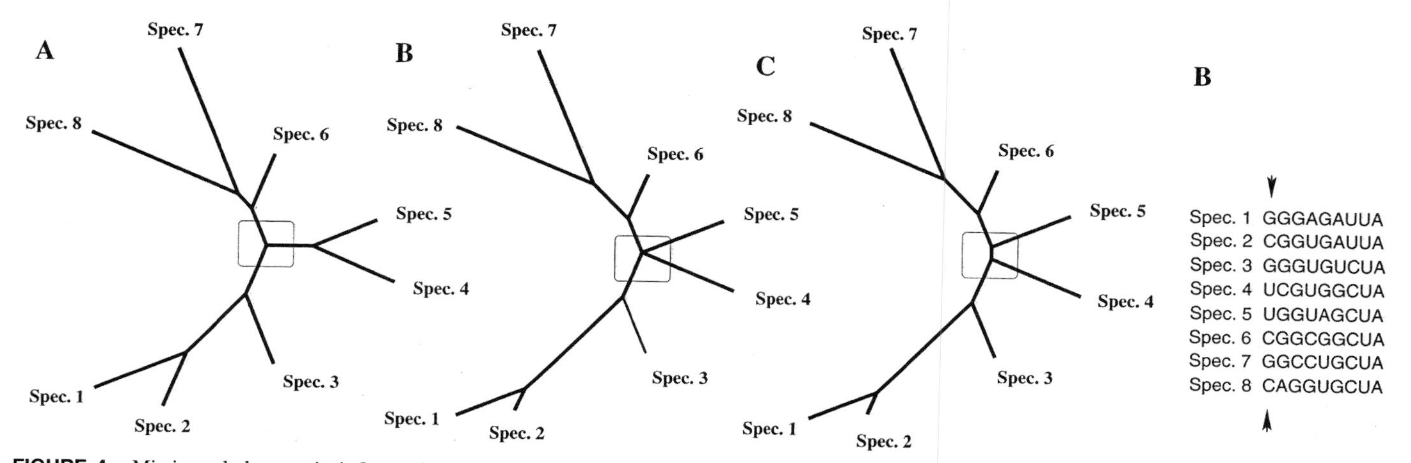

FIGURE 4. Missing phylogenetic information. If the tree in *A* reflects the true phylogeny, a tree topology showing a multifurcation for Spec. 4 and Spec. 5 as well as the other subtrees as shown in *B* would be correct if the phylogenetic information on the monophyletic origin of Spec. 4 and Spec. 5 was not preserved in present day sequences. This can be simulated by exclusion of the column marked by arrowheads in *D*. The loss of this information produces the misleading tree topology of *C* as a result of branch attraction.

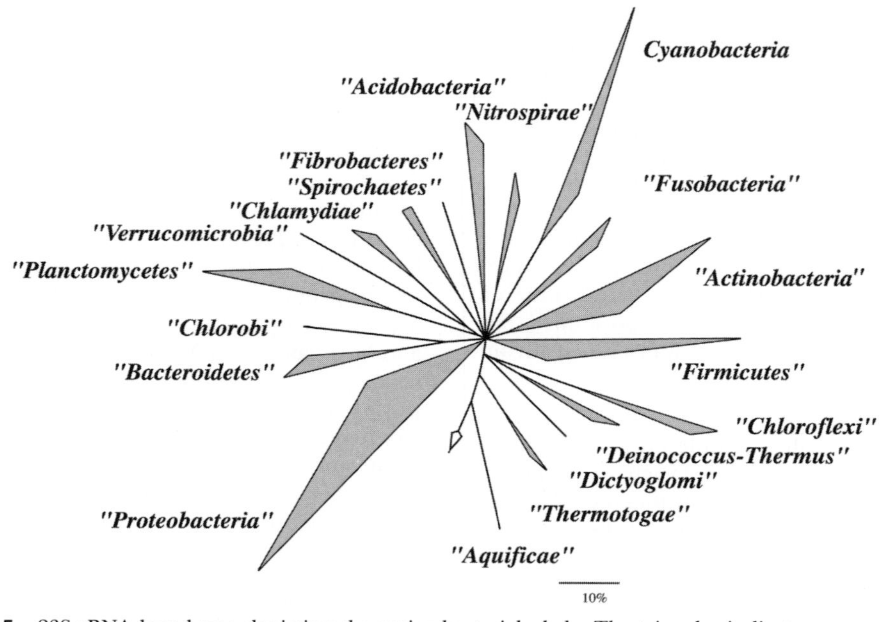

FIGURE 5. 23S rRNA based tree depicting the major bacterial phyla. The triangles indicate groups of related organisms, while the angle at the root of the group roughly indicates the number of sequences available and the edges represent the shortest and longest branch within the group. The tree was reconstructed, evaluated and optimized using the ARB parsimony tool. Only sequence positions sharing identical residues in at least 50% of all bacterial sequences were included in the calculations. All available almost complete homologous sequences from *Archaea* and *Eucarya* were used as outgroup references to root the tree (indicated by the *arrow*). Multifurcations indicate that a relative branching order could not be defined.

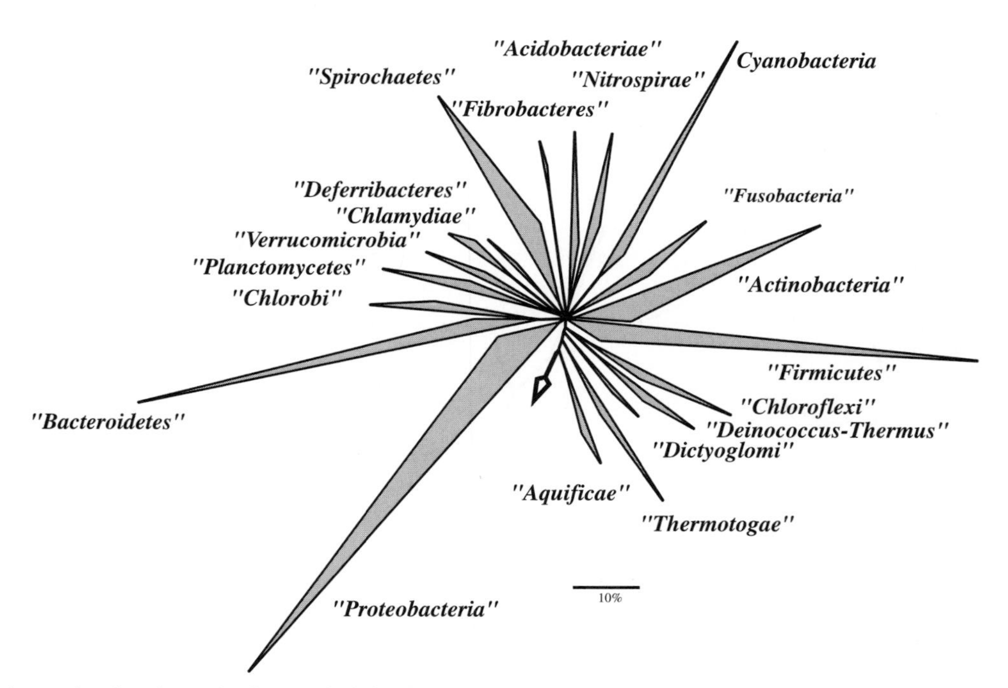

FIGURE 6. 16S rRNA based tree showing the major bacterial phyla. Tree reconstruction was performed as described for Figure 5. Tree layout of this and subsequent trees was according to the description for Figure 5.

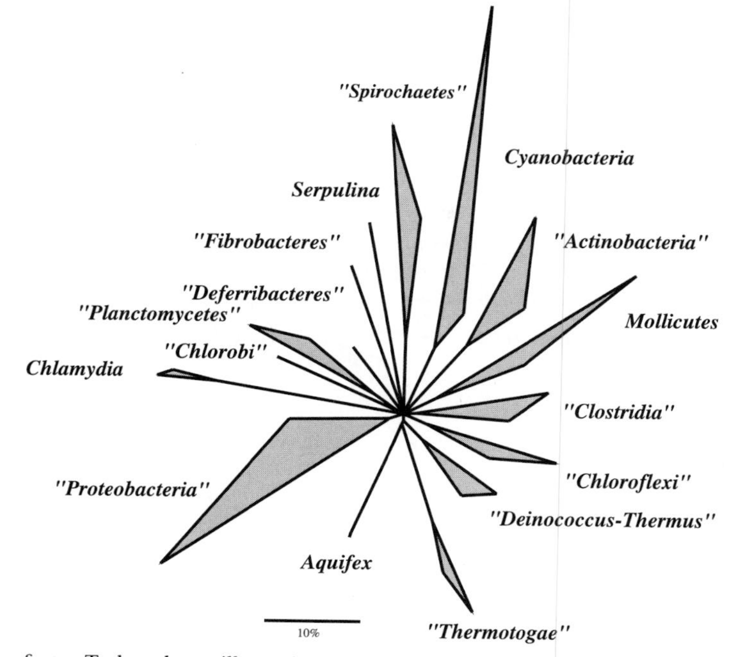

FIGURE 7. Elongation factor Tu based tree illustrating relationships among the major bacterial phyla. The tree was reconstructed from amino acid sequence data, and evaluated and optimized using the ARB parsimony tool. The tree is shown as unrooted, and only positions sharing identical residues in at least 30% of all sequences were included in the calculations.

trees local topological differences have been demonstrated (Ludwig et al., 1993). The reduced phylogenetic information content of EF-Tu (656 bits versus 1506 bits in the SSU rRNA; Ludwig et al., 1998b) may be responsible for the fact that the monophyletic status of some phyla such as the *"Proteobacteria"* is not supported by the protein-based trees. The separation of subgroups such as proteobacterial classes, however, is globally in agreement with the rRNA-based trees.

Interdomain sequence similarities for the rRNAs are 50% and higher, allowing the rooting and (at least to some extent) structuring of the lower branches for a given domain tree versus the other two. The interdomain protein similarities of the elongation factors are low (not more than 30%), making a reliable rooting or structuring of the bacterial tree difficult. The elongation factor database also contains examples of paralogy resulting from gene duplications or lateral gene transfer (Vijgenboom et al., 1994).

RNA polymerases The DNA-directed RNA polymerases (RNAPs) are essential components of the transcription process in all organisms, and the genes for the largest subunits (β and β′ in *Bacteria*; A′, A′′ and B in *"Crenarchaeota"*; B′ and B′′ in *"Euryarchaeota"*) are highly conserved and ubiquitous. The public databases contain RNAP sequences for about 40 species of *Bacteria* and 10 species of *Archaea*. The genes coding for RNAPs are located next to each other on the chromosomes of both *Bacteria* and *Archaea*, and contain 2300 (*Archaea*) to 2400 (*Bacteria*) amino acids that can be clearly aligned for phylogenetic purposes (Klenk et al., 1994). No paralogous genes are known for RNAPs. In general, for the *Bacteria* the intradomain topology of the trees derived from both RNAP large subunits supports the 16S rRNA-based tree in almost all details, with only one major discrepancy: the position of the root of the domain. Intensive rooting experiments with a variety of archaeal and/or eucaryotic outgroups does not place the root of the *Bacteria* close to the extreme thermophiles (*Aquifex* or *Thermotoga* species) as in the rRNA tree, but next to *Mycoplasma* (Klenk et al., 1999). Since the placement of a root within a phylogenetic tree is not critical for most taxonomic purposes, it can be concluded that rRNAs and RNAPs in general support the same intradomain branching pattern for the *Bacteria*.

Proton translocating ATPase The catalytic subunit of proton-translocating ATPase is another example of a protein marker for which a reasonable data set is available, at least with respect to the spectrum of bacterial phyla (Ludwig et al., 1993, 1998b; Ludwig and Schleifer, 1994; Neumaier, 1996). This marker should be more appropriate than elongation factors or RNA polymerases for testing the validity of rRNA-based trees for organismal phylogeny, as the ATPase has nothing in common functionally with transcription or translation except its own synthesis.

In general, the F_1F_0 ATPase β-subunit data support the rRNA-based tree (Fig. 8), but the information content and resolving power is reduced. Again, local differences in branching patterns have been shown, and the monophyletic structure of some phyla, defined by rRNA analysis, is not supported (Ludwig et al., 1993).

A correct rooting of the ATPase β-subunit-based bacterial domain tree with the paralogous catalytic subunit of the vacuolar type ATPase (Hilario and Gogarten, 1998; Ludwig et al., 1998b) is not possible as the overall sequence similarities between the two paralogs are not higher than 23%.

There are not sufficient data available for the F_1F_0 ATPase α-subunit (most likely the paralogous pendant of the β-subunit) to allow effective comparison with the rRNA data. However, the currently available α-subunit data set does not indicate great dif-

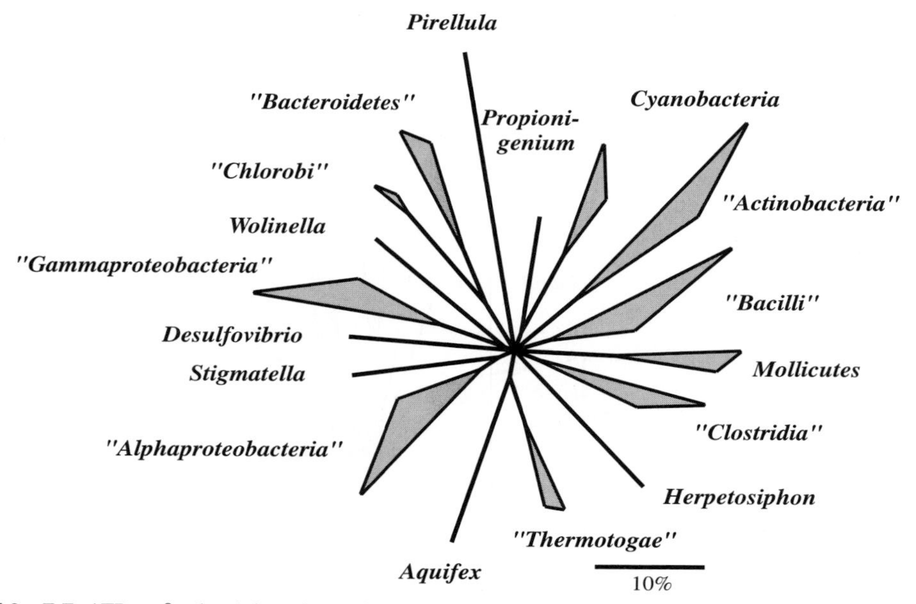

FIGURE 8. F_1F_0 ATPase β-subunit-based tree depicting the major bacterial phyla. The tree is shown as unrooted. Tree reconstruction was performed as described for Figure 7.

ferences in phylogenetic conclusions inferred from the two data sets. There are also insufficient data for the paralogous subunits A and B of the vacuolar type ATPase; however, a clear separation of the *Eucarya* from the *Bacteria* and *Archaea* is seen when the currently available data set is analyzed. The bacterial and archaeal lines appear intermixed at the lowest level of the corresponding subtree. At present, this intermixing cannot be proven or correctly interpreted (Neumaier, 1996). There is low significance for any branching pattern at this level of (potential) relatedness. Furthermore, only a few positions, which currently cannot be tested for plesiomorphy, are responsible for this intermixing. In addition, functional constancy can not be assumed for eucaryal and archaeal versus bacterial vacuolar type ATPases, and lateral gene transfer cannot be excluded (Hilario and Gogarten, 1993).

recA *protein* Most of the bacterial phyla are represented by one or a few sequences in the *recA* protein sequence data base (Wetmur et al., 1994; Eisen, 1995; Karlin et al., 1995). Comparative analysis of these data again supports the rRNA-based view of bacterial phylogeny. A homologous counterpart for the archaeal and eucaryal phyla has not yet been identified. A significant relative branching order of phyla cannot be defined. Although monophyly of the *"Proteobacteria"* or the Gram-positive bacteria with a low DNA G + C content is not observed, no major contradictions to the rRNA-based phylogeny have been reported. The higher phylogenetic groups (*"Proteobacteria"*, *Cyanobacteria*, *"Actinobacteria"*, *Chlamydiales*, *"Spirochaetes"*, *"Deinococcus–Thermus"*, *"Bacteroidetes"*, as well as *"Aquificae"*) are separated from each other as in the rRNA-derived phylogeny. However not surprisingly local differences in detailed branching patterns were found.

There is one major discrepancy: phylogenetic analysis of *Acidiphilium* using *recA* sequence data does not show it to cluster within the *"Alphaproteobacteria"* as is found with rRNA analyses. Two *recA* genes, which differ remarkably in sequence, have been found in *Myxococcus xanthus* and may indicate the occurrence of gene duplications or lateral gene transfer. Therefore it is possible that such phenomena have occurred in the evolution of *recA* in *Acidiphilium*.

hsp60 heat shock proteins Sequences for hsp60 chaperonin have been determined for a wide spectrum of bacterial phyla (Viale et al., 1994; Gupta, 1996; 1998). A distant relationship has been postulated for hsp60, the eucaryotic TCP-1 complex, and the archaeal Tf-55 protein (Brown and Doolittle, 1997). However, given the low similarities, the homologous character of hsp60 and the TCP-1 complex or the Tf-55 protein cannot be demonstrated unambiguously.

Trees based upon the currently available hsp60 sequence data set support rRNA-based trees in that the different phyla are well separated from one another, and in cases where several sequences are available for a given phylum, subclusters resemble the rRNA-derived phylogeny. For example, the *"Gammaproteobacteria"* and *"Betaproteobacteria"* are more closely related to one another than to the *"Alphaproteobacteria"* sister group in both hsp60 and rRNA analyses. However, the use of hsp60 as a phylogenetic marker molecule is again complicated by the existence of duplicated genes as, for example, among *Rhizobium* species.

Other supporting and nonsupporting protein markers The hsp70 (70 kDa heat shock protein)-based tree globally supports rRNA-based clustering. The phyla appear to be separated and even the branching order of the classes of the *"Proteobacteria"* (*"Alphaproteobacteria"*, *"Gammaproteobacteria"*, *"Betaproteobacteria"*) is corroborated. The major concern associated with the hsp70-derived phylogeny is the intermixed rooting of bacterial and archaeal major lines of descent (Brown and Doolittle, 1997; Gupta, 1998). No significant branching order can be defined for the intermixed lines, and, as discussed above for the ATPase phylogeny, these findings may reflect missing resolution at the interdomain level.

At first glance, many other proteins (reviewed by Brown and Doolittle, 1997) seem to support the intradomain tree structures of rRNA-based phylogenies. However, meaningful comparative evaluation is difficult due to limitations in phylogenetic information content and/or databases that are insufficient in size and scope. Examples are provided by family B DNA polymerases which might represent useful markers for all three domains, aminoacyl-tRNA synthetases which differ in size and hence in potential information content, and ribosomal proteins which generally are short polypeptides and thus of very limited phylogenetic use (Brown and Doolittle, 1997). Among the enzymes involved in central metabolism, the usefulness of 3-phosphoglycerate kinase is also curtailed by a limited sequence database.

There are a number of potential protein markers for which deduced trees do not clearly support rRNA-based intradomain phylogenetic conclusions, including DNA gyrases and topoisomerases, some enzymes of the central metabolism, and of amino acid synthesis and degradation. However, as no comprehensive sequence databases are available, careful evaluation of the tree topologies is not possible.

Rationale for a 16S rRNA-derived Backbone for *Bergey's Manual*

The introduction of comparative primary structure analysis of the SSU rRNA by Carl Woese and coworkers was undoubtedly a major milestone in the history of systematic biology. This approach opened the door to the elucidation of the evolutionary history of the procaryotes, and provided the first real opportunity to approach the ultimate goal in taxonomy i.e., systematics based upon the natural relationships between organisms. The rapid development of experimental procedures enabled the scientific community to characterize the majority of described species at the 16S rRNA level. During preparation of the new edition of *Bergey's Manual*, coordinated efforts to close the gaps and to investigate the missing species were initiated. There is a realistic prospect of completing the database with respect to all known validly described species in the near future.

Although the resolving power of the SSU rRNA approach has sometimes been overestimated, it has allowed a tremendous expansion in our knowledge of procaryotic relationships during recent years. This has been accompanied by the recognition of limitations in the existing procaryotic taxonomy, and efforts to redress these limitations. The taxonomic history of the pseudomonads is one impressive example of the "phylogenetic cleaning" of a genus that was phylogenetically heterogeneous in composition (Kersters et al., 1996).

It appears that the SSU rRNA is currently the most powerful phylogenetic marker, in terms of information content, depth of taxonomic resolution, and database size and scope. There is also good congruence between global tree topologies derived from different phylogenetic markers, indicating that SSU rRNA-based phylogenetic conclusions indeed reflect organismal evolution, at least at the global level. Local discrepancies in phylogenetic trees resulting from different information content, different rate or mode of change, or inadequate data analysis do not greatly compromise this general picture. The underlying cause of major tree discrepancies may in some cases be the analysis of paralogous genes, as indicated by multiple genes arising from duplication, loss, or lateral transfer of genes.

The logical consequence of these investigations and observations is to structure the present edition of *Bergey's Manual* according to our current (rRNA-based) concept of procaryotic phylogeny, using the global tree topology as a backbone, and to propose an emended framework of hierarchical taxa.

It should be considered that all phylogenetic conclusions and tree topologies presented here are models that represent the present, imperfect view of evolution. The information content of the SSU rRNA database is rather limited for representation of 3–4 billion years of evolution of cellular life. Furthermore, the methods of data analysis and the software and hardware for deciphering and visualizing this information are far from being optimal. For these reasons, the proposed backbone of the taxonomic scheme might be subject to change in the future. The introduction of new taxonomic tools and methods has always had a major impact on contemporaneous taxonomy. New sequence data and improved methods of data analysis may change our view of procaryotic phylogeny. Comparison of previous editions of *Bergey's Manual*, as well as updates of the Approved Lists of Bacterial Names (Skerman et al., 1980), indicates that the contemporary view of microbial taxonomy is determined mainly by the availability, applicability, and resolving power of the methods used to characterize organisms and elucidate their genetic and phylogenetic relationships.

The Small Subunit rRNA-based Tree

The global SSU rRNA-based intradomain phylogenetic relationships are discussed for the *Archaea* and *Bacteria* below. Given that the relationships of these organisms are described in detail in subsequent chapters, only higher phylogenetic levels are shown here. Reconstruction of general trees was performed using only sequences that were at least 90% complete (in relation to the *E. coli* 16S rRNA reference sequence). Lines of descent or phylogenetic groups containing a single or only a few sequences are (usually) not shown in these trees. Environmental sequences from organisms which have not yet been cultured were included in the calculations but are not depicted in the trees. The trees and discussions are based upon a comparative analysis of the current RDP (Maidak et al., 1999) and ARB trees. The RDP tree was reconstructed by applying a maximum likelihood method combined with resampling, whereas for the ARB tree a special maximum parsimony approach in combination with different optimization methods and upper bootstrap limit determination was used. The RDP tree contains the *Bacteria* and *Archaea*, while the ARB tree also includes the *Eucarya*. In both cases, the rooting and internal structuring of the domain trees was estimated using the full data set of the other domains. Although these trees were reconstructed using different methods, their global topologies are in good agreement.

A statistically significant relative branching order cannot be unambiguously determined for the majority of the phyla in the *Bacteria*, or for many of the intraphylum groups, as indicated by multifurcations within the trees. However, clustering tendencies are common to both trees. It should also be considered that most phyla were defined in the early days of comparative rRNA sequencing (Woese, 1987) when the data set was small and long "naked" branches facilitated clear-cut phylum delimitation. With the rapidly expanding database most of these "naked" branches expanded and in some cases it is no longer possible to demonstrate a monophyletic structure or to clearly delimit traditional phyla and other groups, as exemplified by the *"Proteobacteria"* and the low G + C Gram-positive bacteria (*"Bacilli"*, *"Clostridia"*, *Mollicutes*). The inter- and intra-genus relationships of each group are discussed in detail in subsequent chapters; described below is an overview of the phyla of the bacterial and archaeal domains and their major phylogenetic subclusters (Figs. 9, 10, 11, 12, 13, 14, 15, and 16).

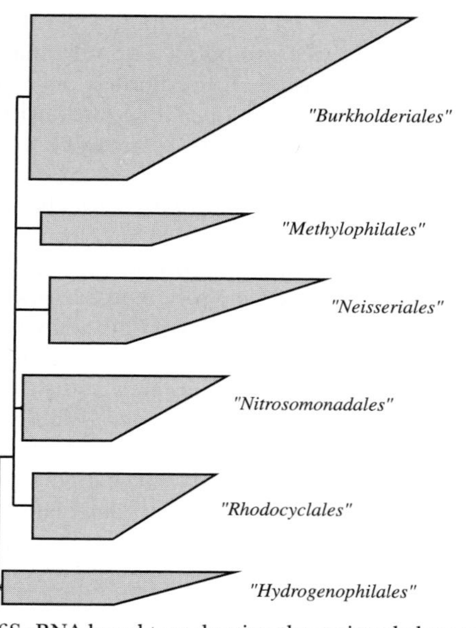

FIGURE 9. 16S rRNA-based tree showing the major phylogenetic groups of the *"Betaproteobacteria"*. Only groups represented by a reasonable number of almost complete sequences are shown. Tree topology is based on the ARB database of 16,000 sequences entries and was reconstructed, evaluated, and optimized using the ARB parsimony tool. A filter defining positions which share identical residues in at least 50% of all included sequences from *"Betaproteobacteria"* was used for reconstructing the tree. The topology was further evaluated by comparison with the current RDP tree, which was generated using a maximum likelihood approach in combination with resampling (Maidak et al., 1999). A relative branching order is shown if supported by both reference trees. Multifurcations indicate that a (statistically) significant relative branching order could not be determined or is not supported by both reference trees.

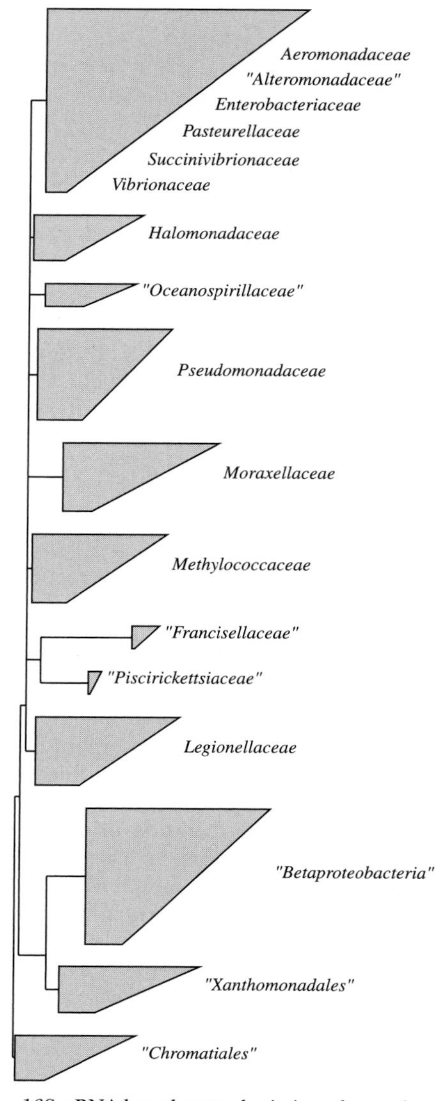

FIGURE 10. 16S rRNA-based tree depicting the major phylogenetic groups within the *"Gammaproteobacteria"*. Tree reconstruction and evaluation was performed as described for Figure 9 with the exception that a 50% filter calculated for the *"Gammaproteobacteria"* was used.

The Bacteria

THE *"Proteobacteria"* The traditional view of the *"Proteobacteria"* as a monophyletic phylum is not completely supported by careful analyses of the current 16S rRNA database. Although there is support for monophyly in the RDP tree, with the *"Deltaproteobacteria"* and *"Epsilonproteobacteria"* forming the deeper branches, a monophyletic structure that includes these two groups is not clearly supported by the ARB tree. Confidence analyses indicate that the significance of a relative branching order within the *"Proteobacteria"* is low in both trees. However, a closer relationship of the *"Gammaproteobacteria"* and *"Betaproteobacteria"*, as well as a common origin of these groups and the *"Alphaproteobacteria"*, is supported by the RDP as well as the ARB tree.

The *"Betaproteobacteria"* (Fig. 9) clearly represents a monophyletic group, comprising the described or proposed higher taxa *"Burkholderiales"*, *"Methylophilales"*, *"Nitrosomonadales"*, *"Neisseriales"*, and *"Rhodocyclales"*. A slightly deeper-branching group comprises the *"Hydrogenophilales"*.

The classical members of the *"Gammaproteobacteria"* (Fig. 10) represent a monophyletic group which includes the *"Betaproteobacteria"* as a major line of descent. In both reference trees the family *"Xanthomonadaceae"* appears to be the most likely sister group of the *"Betaproteobacteria"*. A common clustering of the families *Aeromonadaceae, "Alteromonadaceae", Enterobacteriaceae, Pasteurellaceae,* and *Vibrionaceae* is supported in both trees. A relative branching order of this cluster and other major groups of the *"Gammaproteobacteria"* such as the families *Halomonadaceae, Legionellaceae, Methylococcaceae, Moraxellaceae, "Oceanospirillaceae", Pseudomonadaceae,* and the *"Francisellaceae"-"Piscirickettsiaceae"* group cannot be unambiguously determined. In both trees, these groups branch off higher than the *"Betaproteobacteria"-"Xanthomonadaceae"* branch, whereas the order *"Chromatiales"* forms a deeper branch. The phylogenetic position of the families *Moraxellaceae* and *Cardiobacteriaceae* relative to that of the *"Gammaproteobacteria"-"Xanthomonadaceae"* lineage depends on the treeing method used.

A closer relationship between the families *Rickettsiaceae* and *Ehrlichiaceae* within the *"Alphaproteobacteria"* (Fig. 11) can be seen in both reference trees. The results of tree evaluations indicate branching of this cluster followed by the families *"Sphingomonadaceae"* and the *"Rhodobacteraceae"*. The families *"Bradyrhizobiaceae"*, *Hyphomicrobiaceae*, *"Methylobacteriaceae"*, and *"Methylocystaceae"* represent another subcluster among the *"Alphaproteobac-*

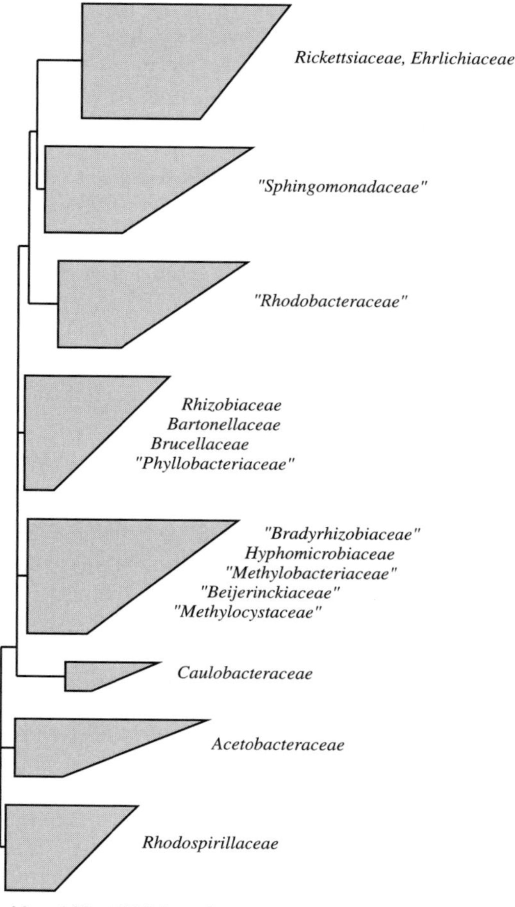

FIGURE 11. 16S rRNA-based tree showing the major phylogenetic groups within the *"Alphaproteobacteria"*. Tree reconstruction and evaluation was carried out as described for Figure 9 with the exception that a 50% filter calculated for the *"Alphaproteobacteria"* was used.

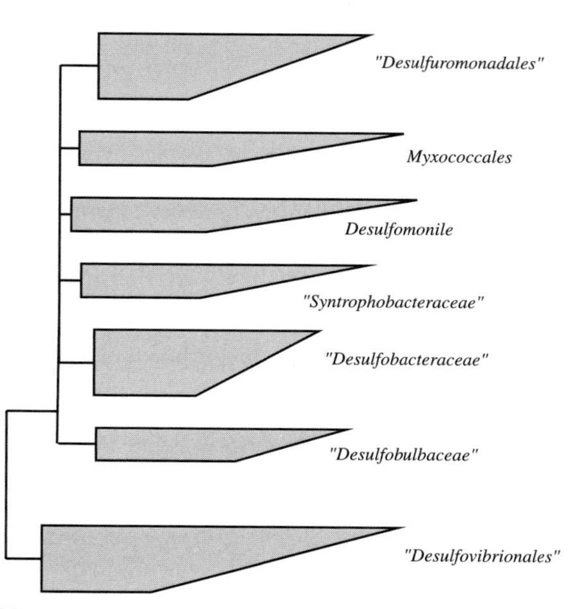

FIGURE 12. 16S rRNA-based tree illustrating the major phylogenetic groups within the *"Deltaproteobacteria"*. Tree reconstruction and evaluation was performed as described for Figure 9 with the exception that a 50% filter calculated for the *"Deltaproteobacteria"* was used.

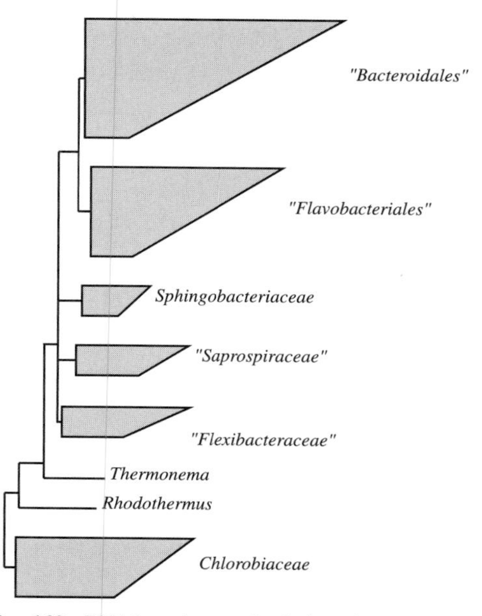

FIGURE 13. 16S rRNA-based tree depicting the major phylogenetic groups within the phylum *"Bacteroidetes"*. Tree reconstruction and evaluation was performed as described for Figure 9 with the exception that a 50% filter calculated for the *"Bacteroidetes"* phylum was used.

teria". A closer interrelated group is formed by the families *Bartonellaceae, Brucellaceae, Rhizobiaceae,* and *"Phyllobacteriaceae"*. No reliable resolution of these major groups and the family *Caulobacteraceae* can be achieved, but it appears that a deeper branching of the families *Acetobacteraceae* and *Rhodospirillaceae* among the *"Alphaproteobacteria"* is indicated.

The order *"Desulfovibrionales"* currently represents the deepest branch of the *"Deltaproteobacteria"* (Fig. 12). Three other major subgroups comprise *Desulfomonile* and relatives, the *"Syntrophobacteraceae"*, as well as the *"Desulfobulbaceae"*. These subgroups are phylogenetically equivalent in depth to the lineages *"Desulfobacteraceae"*, *"Geobacteraceae"*, and *Myxococcales.*

The families *"Helicobacteraceae"* and *Campylobacteraceae* are the two major lines that form the *"Epsilonproteobacteria"*.

THE *"Spirochaetes"* The *"Spirochaetes"* phylum currently comprises three major subgroups: the sister groups of the families *Spirochaetaceae* and *"Serpulinaceae"*, as well as the deeper branching family *Leptospiraceae.*

"Deferribacteres" AND *"Acidobacteria"* PHYLA To date, the *"Deferribacteres"* phylum is represented by only two cultured species, while only three cultured species are found in the *"Acidobacteria"* phylum. However, a comprehensive data set of environmental sequences indicates a phylogenetic depth and diversity within the *"Acidobacteria"* comparable to that of the *"Proteobacteria"* (Ludwig et al., 1997).

THE *Cyanobacteria* The chloroplast organelles comprise a monophyletic subgroup within the *Cyanobacteria* phylum, which also contains a number of other major lines of descent. The current taxonomy of the cyanobacteria is far from being in accordance with the phylogenetic structure of the phylum.

"Verrucomicrobia", *"Chlamydiae"*, AND *"Planctomycetes"* The phylum *"Verrucomicrobia"* comprises a number of environmental sequences as well as a few cultured members of the genera *Verrucomicrobium* and *Prosthecobacter* (Hedlund et al., 1996). Both

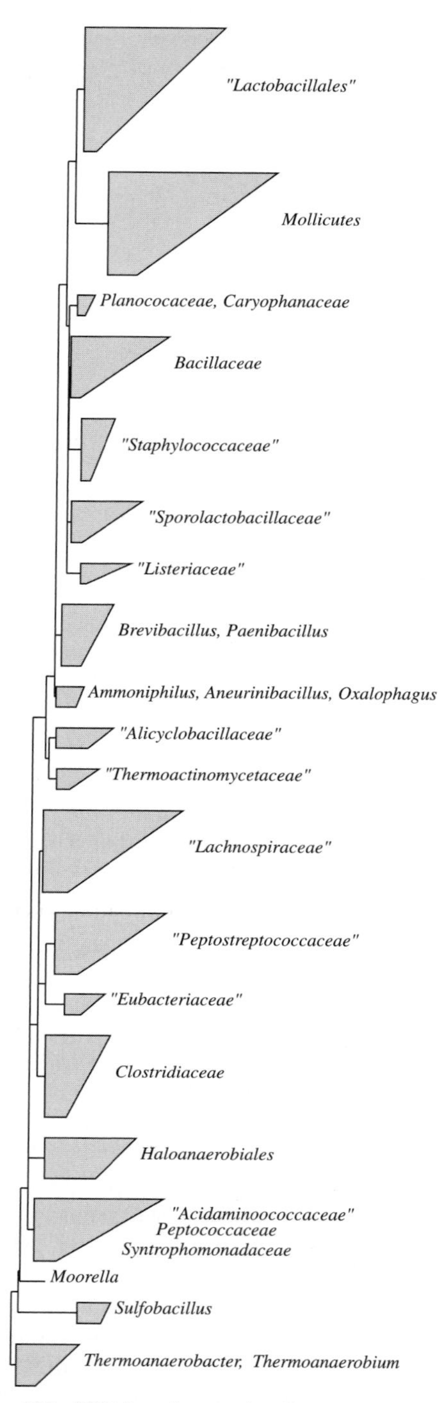

FIGURE 14. 16S rRNA-based tree showing the major phylogenetic groups of the *"Firmicutes"* (Gram-positive bacteria with a low DNA G + C content). Tree reconstruction and evaluation was carried out as described for Figure 9 with the exception that a 50% filter calculated for a core set of sequences (excluding the *Mycoplasmatales* and the deeper groups represented by *Moorella, Sulfobacillus, Thermoanaerobacter,* and *Thermoanaerobium*) was used.

reference trees indicate a moderate degree of relationship between the *"Verrucomicrobia"* and the *Chlamydiales* phylum. However, given the limited number of available sequences for the *"Verrucomicrobia"* and the long naked branch of the *Chlamydiales*, a sister group relationship between these two phyla should be regarded as tentative. A moderate relationship between these two phyla and the *"Planctomycetes"* phylum is also indicated in both the ARB and RDP trees. However, the significance of this branching point is low, and their relationship may not be supported in the future by a growing database. The intraphylum structure of the *"Planctomycetes"* indicates two pairs of sister groups: *Pirellula/Planctomyces* and *Isosphaera/Gemmata.*

"Chlorobi" AND *"Bacteroidetes"* A monophyletic origin of the *"Chlorobi"* (containing the genera *Chlorobium, Pelodictyon, Prosthecochloris,* and some environmental sequences) and the *"Bacteroidetes"* (Gosink et al., 1998) phyla (Fig. 13) can be seen in both trees and is supported by alternative markers such as large subunit rRNA, and β-subunit of F_1F_0 ATPase. The thermophilic genera *Rhodothermus* and *Thermonema* represent the deepest branches of the phylum *"Bacteroidetes"*. A common root of the *"Bacteroidales"* and *"Flavobacteriales"* within the phylum is supported in both reference trees. This cluster seems to be phylogenetically equivalent to the other major groups i.e., the *Sphingobacteriaceae, "Saprospiraceae", "Flexibacteraceae", Flexithrix,* and *Hymenobacter.*

LOW G + C GRAM-POSITIVE BACTERIA Other than for the *"Proteobacteria"*, the most comprehensive 16S rRNA gene sequence database (with more than 1750 almost complete sequences) is available for the Gram-positive bacteria with a low DNA G + C content (*"Bacilli", "Clostridia", Mollicutes*). The common origin of the organisms classically assigned to this group is not significantly supported by all reference trees (see Fig. 14). The *Mollicutes,* comprising the families *Mycoplasmataceae, Acholeplasmataceae,* and their walled relatives, represent a monophyletic unit. The classical lactic acid bacteria are members of the families *"Aerococcaceae", "Carnobacteriaceae", "Enterococcaceae", Lactobacillaceae, "Leuconostocaceae"*, and *Streptococcaceae,* and are unified in the order *"Lactobacillales"*. A clear resolution of the relationships between the families *Bacillaceae, Planococcaceae, "Staphylococcaceae", "Sporolactobacillaceae"*, and *"Listeriaceae"* cannot be achieved. Two slightly deeper branching clusters comprise the genera groups of *Brevibacillus–Paenibacillus* and *Ammoniphilus–Aneurinibacillus–Oxalophagus.* The *"Alicyclobacillaceae"* and *Thermoactinomyces* groups represent a further deeper branch. Another major subbranch unifies the *"Eubacteriaceae", Clostridiaceae, "Lachnospiraceae"*, and *"Peptostreptococcaceae"* . The *"Eubacteriaceae"* and *"Peptostreptococcaceae"* appear to be sister groups. The phylogenetic position of the order *Haloanaerobiales* is strongly influenced by the treeing method applied and should be regarded as tentative. The families *Haloanaerobiaceae* and *Halobacteroidaceae* constitute a well-defined phylogenetic unit in both reference trees. However, the assignment of this unit to the low G + C Gram-positive phylum is not clearly supported when different treeing methods are applied, suggesting that this group may represent its own phylum. A deeper rooting within the phylum is indicated for the *Peptococcaceae–Syntrophomonadaceae* cluster but the phylogenetic position of the genera *Moorella, Sulfobacillus, Thermoanaerobacter,* and *Thermoanaerobium* is uncertain. The latter two genera represent a phylogenetic unit, but this unit and each of the other genera probably represent additional phyla.

"Fusobacteria" PHYLUM The *"Fusobacteriaceae"* phylum so far comprises only three subclusters: *Fusobacterium, Propionigenium–Ilyobacter* and *Leptotrichia–Sebaldella.*

HIGH G + C GRAM POSITIVE BACTERIA (*"Actinobacteria"*) The phylum of the Gram-positive bacteria with a high G + C DNA content (the *"Actinobacteria"*) provides an example of a

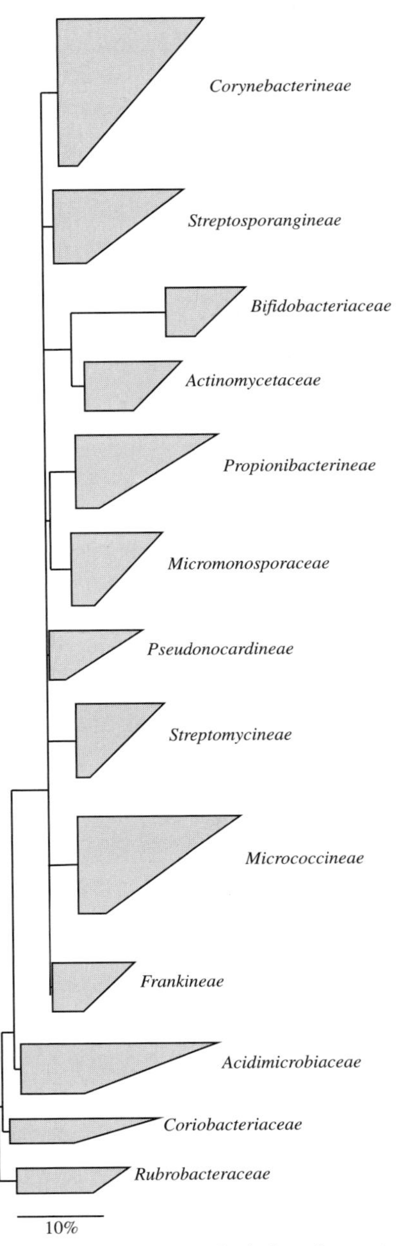

FIGURE 15. 16S rRNA-based tree depicting the major phylogenetic groups within the *"Actinobacteria"* (Gram-positive with a high DNA mol% G + C content). Tree reconstruction and evaluation was performed as described for Figure 9 with the exception that a 50% filter calculated for this phylum was used.

clearly defined and delimited major bacterial line of descent. As seen in Fig. 15, the families *Rubrobacteraceae* and *Coriobacteriaceae* currently represent the deepest branches of the phylum, whereas the family *Acidimicrobiaceae* occupies an intermediate position between the former two and the remaining major subgroups of the phylum. There is some support for a common origin of the *Bifidobacteriaceae* and *Actinomycetaceae*, and for the clustering of the families *Propionibacteriaceae* and *Micromonosporaceae.* No significant or stable branching order for these and other subgroups such as *Corynebacteriaceae, Frankineae, Pseudonocardiaceae, Streptomycetaceae,* and *Streptosporangineae* could be achieved.

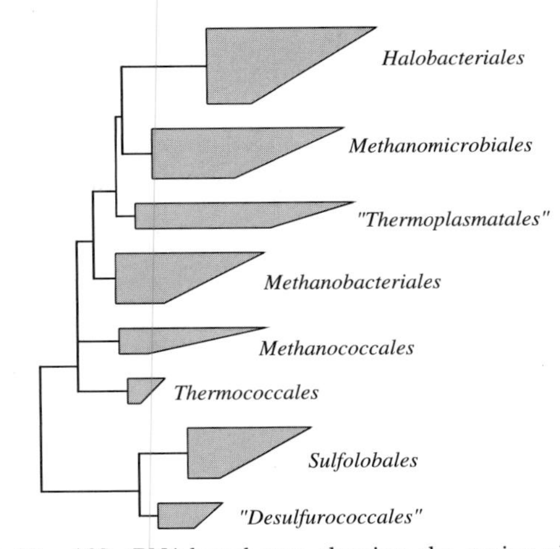

FIGURE 16. 16S rRNA-based tree showing the major phylogenetic groups within the *Archaea.* Tree reconstruction and evaluation was carried out as described for Figure 9 with the exception that tree optimization was performed independently for the *"Euryarchaeota"* and *"Crenarchaeota"* using a 50% filter in each case.

Other phyla The *"Nitrospirae"* phylum contains a limited number of organisms, namely representatives of the genera *Nitrospira, Leptospirillum, Thermodesulfovibrio,* and *Magnetobacterium.* Similarly, only a limited number of organisms and environmental sequences represent the phylum of the green non-sulfur bacteria, which includes the families *"Chloroflexaceae"*, *"Herpetosiphonaceae"*, and *"Thermomicrobiaceae"*. Two major subgroups, the *Deinococcaceae* and the *"Thermaceae"*, have been identified within the *"Deinococcus–Thermus"* phylum. The orders *"Thermotogales"* and *"Aquificales"* constitute two of the deeper branching phyla within the bacterial domain.

The existence of additional phyla is suggested by the phylogenetic position of organisms such as *Dictyoglomus thermophilum* and *Desulfobacterium thermolithotrophum,* and of some environmental sequences. However, the phylum status of these lineages cannot be evaluated at this time, due to the paucity of available sequence data.

The Archaea Two major lines of descent (phyla) have been delineated within the *Archaea*: the *"Euryarchaeota"*, and the *"Crenarchaeota"*. Within the *"Euryarchaeota"*, the orders *Halobacteriales, Methanomicrobiales,* and *"Thermoplasmatales"* share a common root. A relationship between the first two orders is suggested in both reference trees (Fig. 16), and the order *Methanobacteriales* is indicated as the next deepest branch. A stable and significant tree topology resolving the relationship between these four orders and the orders *"Archaeoglobales"*, *Methanococcales, Thermococcales,* and *"Methanopyrales"* cannot be deduced from the current database.

The orders *Sulfolobales* and *"Desulfurococcales"* appear to be sister groups within the *"Crenarchaeota"*, while a monophyletic structure of the *Thermoproteales* is somewhat questionable. The genus *Thermofilum* tends to root outside the *Thermoproteales* group, however the significance of this branching is low and the database does not contain sufficient entries to allow careful evaluation of this outcome.

A third archaeal phylum, *"Korarchaeota"*, has been postulated on the basis of two partial environmental 16S rRNA sequences, but representatives of this lineage have not yet been isolated in pure culture. Consequently, the phylum status as well as phylogenetic position of the lineage can currently not be assessed.

Nucleic Acid Probes and Their Application in Environmental Microbiology

Rudolf Amann and Karl-Heinz Schleifer

I. Introduction

Microbiology has entered the molecular age. Almost every month another complete bacterial genome sequence is published (e.g., Fleischmann et al., 1995; Cole et al., 1998), and it is now routine to start the classification of a newly isolated microorganism with the determination and comparative analysis of at least one nucleic acid sequence. Clearly, the most commonly used molecule for this purpose is the ribonucleic acid of the small subunit of the ribosome, the 16S rRNA of *Bacteria* and *Archaea.* The high information content of nucleic acid sequences can, in principle, be accessed by two techniques. One is sequencing followed by comparative sequence analysis; the second is hybridization with nucleic acid probes.

As a first, rough definition, nucleic acid probes can be described as single-stranded pieces of nucleic acids that have the potential to bind specifically to their counterparts, complementary nucleic acid sequences. By this process, the so-called hybridization, probes facilitate the detection of their respective target molecules based on primary structure. In most cases, this is accomplished by conferring a detectable moiety, the label, to the target molecules (Fig. 1).

Nucleic acid hybridization techniques, such as DNA–DNA reassociation or DNA–rRNA hybridization, have been used by bacterial taxonomists at times when nucleic acid sequencing was still difficult and available only to specialists. Taxonomists adopted the methodology soon after the first hybridization experiments were performed in the early 1960s (Marmur and Lane, 1960; Hall and Spiegelman, 1961). *Bacteria* and *Archaea* generally lack the morphological diversity necessary for microscopic identification to be reliable. The traditional identification methods, based on the phenotypic characterization of pure culture isolates, were slow and often yielded unclear results because of the influence of exogenous and endogenous parameters on the expressed phenotype. Nucleic acid-based, genotypic methods promised a faster and more reliable identification based on stable genetic markers that were independent of the cultivation conditions. Furthermore, genotypic methods should allow bacterial taxonomists to transform an artificial classification system, suitable only for identification, into a more natural one reflecting the phylogeny of the bacteria. Indeed, studies using DNA–rRNA hybridization (Palleroni et al., 1973; De Ley and De Smedt, 1975) in the 1970s and 1980s yielded significant insights into the genotypic relationships of bacteria (Schleifer and Stackebrandt, 1983; De Ley, 1992). A DNA–DNA similarity of $\geq$70% continues to be used as an important determinant for placing bacterial strains into species. Meanwhile, nucleic acid probes have become a standard method of identifying fastidious, slow growing, or even hitherto uncultured bacteria (Amann et al., 1991).

The increasing availability of nucleic acid sequences in databases, the ease of synthesizing oligonucleotide probes, and numerous other methodological advances in molecular biology, such as the polymerase chain reaction (PCR; Saiki et al., 1988), have made nucleic acid probes a routinely used tool in microbiological laboratories. The use of probes is now so common that it has become impossible to review all applications. In this introductory chapter we will, therefore, focus on basic principles behind nucleic acid probing, describe the steps necessary for directed design and reliable application of nucleic acid probes, and discuss selected examples of the use of nucleic acid probes in environmental microbiology. Additional information may be obtained from two earlier reviews on the subject (Stahl and Amann, 1991; Schleifer et al., 1993) and a recently published book on "Molecular Approaches to Environmental Microbiology" (Pickup and Saunders, 1996). Examples for the application of

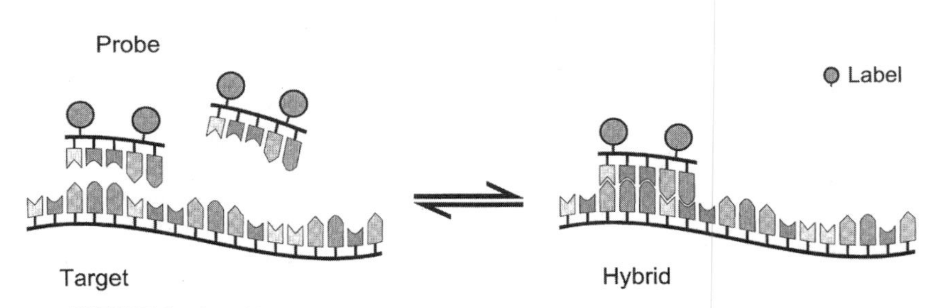

FIGURE 1. Specific hybridization of a nucleic acid probe to a target molecule.

nucleic acid probes in identification of bacteria are given by Amann et al. (1996a) and Schleifer et al. (1995).

II. Basic Principles of Nucleic Acid Probing

This part is intended to provide the groundwork for the rest of this chapter by explaining the principles of nucleic acid hybridization in a coherent way. It may also be used as a glossary for the specialized terminology.

In every cell there are two types of nucleic acids: deoxyribonucleic acid or DNA and ribonucleic acid or RNA. Whereas the former is the storage medium of genetic information, RNA molecules occur as ribosomal RNA (rRNA), messenger RNA (mRNA), or transfer RNA (tRNA) and are involved in the translation of genotypic information into the expressed characters. DNA usually forms a duplex of two antiparallel strands of polynucleotides that are fully complementary to each other. This means the base adenine (A) on one strand is opposing a thymine (T) on the other strand and cytosine (C) base pairs with guanine (G). The non-covalent bonding between the base pairs is mediated by hydrogen bonds which can be broken by physical or chemical means, resulting in the denaturation of the DNA molecule into single strands. The process is fully reversible and upon cooling or neutralization the DNA will reassociate to form the original duplex.

In RNA the base thymine is replaced by uracil (U). Internal base complementarities cause the single stranded RNA to fold into secondary structures that may, in addition to the canonical base pairs G–C and A–U, also be stabilized by non-canonical pairs such as G–U or G–A. It was first shown by Hall and Spiegelman (1961) that RNA may bind to or hybridize with denatured DNA, resulting in DNA–RNA duplex structures. The term hybridization is also used to describe the binding of a labeled single-stranded nucleic acid, the probe, to an unlabeled single-stranded nucleic acid, the target. When the degree of probe binding is plotted against temperature or the concentration of the naturing agent, the resulting profile is sigmoid (Fig. 2). Hybridization may also occur between two strands that are not fully complementary. In this case, canonical base pairs or matches stabilize a certain number of mismatches. The resulting hybrid will be less stable than a fully complementary hybrid and show a lower temperature of dissociation (T_d) of the probe as compared to the fully complementary target nucleic acid. The T_d is defined as that temperature at which 50% of the maximally bound probe has dissociated from the immobilized target. It is similar to the melting temperature, or T_m, of double stranded DNA. Parameters such as temperature and composition of the hybridization buffer have a strong influence on the kinetics and specificity of hybridization. The combined effect of these parameters is often referred to as the stringency of hybridization. As outlined in Fig. 2, determination of the optimal stringency of hybridization for a given probe is of utmost importance for the specificity of a hybridization assay and its capacity to discriminate between target and nontarget nucleic acids.

The optimum hybridization temperature is usually close to but below the T_d. If the stringency of hybridization is too low, the probe specificity may be compromised. If the stringency is too high, the sensitivity of the hybridization assay will be decreased.

III. Development of Nucleic Acid Probes

The development of a new nucleic acid probe for the identification of a given strain, species, or a defined group of microorganisms can be accomplished by two different approaches. In addition to an approach that is based on comprehensive nucleic acid sequence collections, it is possible to generate nucleic acid probes for bacterial identification without prior knowledge of the target nucleic acid. In the following, the two approaches are discussed separately in the sections Empirical Probe Selection and Rational Probe Design.

A. Empirical probe selection In a very simple hybridization format it is possible to use the total chromosomal DNA of a given strain after radioactive or nonradioactive labeling as a probe for the screening of total DNA. Such genomic DNA probes have been referred to as whole-cell probes (Stahl and Amann, 1991; Schleifer et al., 1993). Genomic DNA of the strain of interest is, e.g., labeled with photobiotin (Forster et al., 1985) and hybridized to DNA of reference strains that have been immobilized in microtiter plates (Ezaki et al., 1989). This assay is similar to, but much faster than, the traditional DNA–DNA hybridization, in which DNA must be released and purified from a large number of reference strains before it is subjected to pairwise hybridizations. Whole cell probes are, however, ill-defined and by their nature will always contain a fraction of sequences that are highly conserved, such as the rRNA genes. A considerable amount of nonspecific hybridization with less related species is therefore common (Grimont et al., 1985; Hyypiä et al., 1985) and often can not be prevented, even when highly stringent hybridization conditions are used (Ezaki et al., 1989).

Another straightforward possibility for using a natural nucleic acid fraction is the application of isolated RNA as a hybridization probe. This fraction consists mainly of rRNA, which when compared to the other RNAs (e.g., mRNA, tRNA) is more abundant and less rapidly degraded. It is important to clarify the essential difference between a chromosomal DNA–DNA hybridization and a DNA–(16S or 23S) rRNA hybridization. In the latter case, the probe is well known and characterized and therefore, *sensu stricto*, is not an example of empirical probe selection. With the saturation hybridization method (De Ley and De Smedt, 1975) the differences in the melting temperature of homologous and heterologous pairs of DNA and rRNA are determined as ΔT_m (e) values. These allow for the determination of intra- and intergeneric relationships of bacteria (De Ley, 1992). The limits of this widely accepted taxonomic method originate from the specific nature of the 16S and 23S rRNAs which are usually too conserved to allow for the differentiation of strains within one species or even closely related species. Furthermore, DNA–rRNA hybridization does not allow reconstruction of relationships above the level of families or orders (De Ley, 1992).

Random DNA fragments have the potential to be strain-specific. They can be selected from recombinant genomic DNA libraries by screening randomly chosen clones for specificity. Those clones that do not hybridize with closely related non-target strains are further evaluated. While this approach has been used frequently (for a review see e.g., Schleifer et al., 1993) it is rather time-consuming. Several strategies to enrich for DNA fragments with unique sequences have been developed. In the format of subtractive hybridization (Schmidhuber et al., 1988), DNA restriction fragments of the strain of interest are hybridized with biotinylated DNA fragments of a closely related strain. DNA that is similar in both strains is removed by binding to immobilized avidin, leaving behind a fraction that is enriched for DNA fragments unique to the strain of interest. Subtracter DNA may be obtained from one or several related organisms and is always

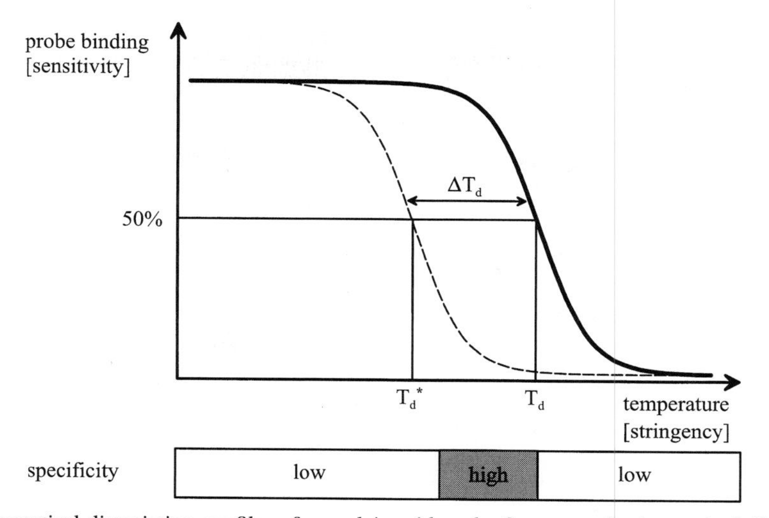

FIGURE 2. Theoretical dissociation profiles of a nucleic acid probe from a perfectly matched (*bold line*) and an imperfectly matched immobilized target nucleic acid (*broken line*). Probe binding (y-axis), which is directly proportional to the sensitivity, is shown over temperature (x-axis) which represents only one parameter that defines the hybridization stringency. Note that T_d, the temperature of dissociation of the perfect hybrid is higher than the T_d* of the imperfect hybrid. It is the primary goal of probe design to maximize ΔT_d, the difference between the temperature of dissociation of a probe from target and non-target nucleic acid. The bar below the dissociation profiles indicates hybridization temperatures/stringencies with high and low specificity of discrimination of target and non-target nucleic acid.

used in excess. The limitations of this technique originate from its reliance on a single removal system which might not be sufficiently effective, and from problems with cloning the small quantities of DNA fragments resulting from this approach. Recently, an improved method that combines subtractive hybridization with PCR amplification and several removal systems and thereby largely overcomes these problems was described (Bjourson et al., 1992). A further simplification of the method was introduced by Wassill et al. (1998) and led to the successful identification of individual strains of lactococci.

After a random DNA probe has been selected for further use, the target sequence should be localized on either the chromosome or a plasmid, and in the latter case, the copy number of the plasmid should be established. Since plasmids are often mobile and might be lost, plasmid-targeted probes have been selected for identification of bacteria in only a few cases (e.g., Totten et al., 1983). Like probes targeted to species-specific repetitive sequences (e.g., Clark-Curtiss and Docherty, 1989), plasmid-targeted probes have the advantage of enhanced sensitivity, attributable to the increased number of target sites per cell. However, a clear disadvantage of randomly selected DNA fragments is that the biological function may remain unknown even after a sequence has been determined. Bacterial genome data clearly show that comparative sequence analysis still fails to assign defined functions to a substantial fraction of the sequences accumulated thus far (e.g., Fleischmann et al., 1995; Cole et al., 1998). Nevertheless, a complete sequence contains plenty of target sites for strain-specific probes. These probes are of importance in the monitoring of biotechnological strains such as starter cultures in food microbiology or production strains in amino acid fermentation.

B. Directed probe design The exponentially increasing number of nucleic acid sequences in various databases has prompted a move to directed probe design, which usually starts with the selection of a target site. Good targets might, for example, be genes coding for well described virulence factors that allow for the differentiation of virulent and avirulent strains (Moseley et al., 1982). Other DNA probes have been targeted to genes specifying surface epitopes (Korolik et al., 1988) or antibiotic resistance (Groot Obbink et al., 1985). Because of rapid advances in molecular techniques, it is now easier and faster to screen for the presence of a gene of interest than to show its function. One has to realize, however, that detection of a gene by hybridization does not necessarily prove that it is present in a functional form, without deleterious mutations, or that it is expressed. The latter problem can be addressed by switching to mRNA as a target.

In the following, an example of a directed probe design that goes one step beyond the use of cloned DNA probes to target known molecules will be discussed in detail. The directed design of short oligonucleotide probes exploits defined signatures, e.g., single point mutations, that are initially detected in sequence databases by comparative analysis. This type of probe design is possible for any gene for which a reasonably large sequence database exists. By far the most commonly used target molecule, in this respect, is the 16S rRNA. The general steps in the design and optimization of an oligonucleotide probe are described in the following sections:

i. generation of a nucleic acid sequence database (see **Nucleic acid sequence databases**)
ii. design of probe (see **Computer-assisted probe design**)
iii. synthesis and labeling of the probe (see **Synthesis and labeling of oligonucleotide probes**)
iv. preparation of the target nucleic acid (see **Some Common formats of oligonucleotide hybridization**)
v. optimization of the hybridization conditions (see **Optimizing the hybridization conditions**)
vi. evaluation of the probe specificity and sensitivity (see **Evaluation of probe specificity and sensitivity**)

Before starting the design of new probes, it is recommended that one checks whether suitable probes have already been developed and published. Rapid growth in the number of such probes precludes the provision of a compilation of available oligonucleotide probes, even if we restrict ourselves to rRNA-targeted probes as was done previously (Amann et al., 1995). Several probes databases are available, but are updated infrequently. Readers interested in rRNA-targeted probes might want to start with the Oligonucleotide Probe Database (OPD) which was accessible via the World Wide Web at the time of publication (http://www.cme.msu.edu/OPD; Alm et al., 1996). In addition to the probe sequences and references, this database also provides information on optimal hybridization conditions and T_d values. Another special feature is the integration of OPD and the Ribosomal Database Project (RDP; Maidak et al., 1999), which allows the end user to reevaluate probe specificity against the constantly increasing rRNA sequence databases.

1. Nucleic acid sequence databases Several large nucleic acid databases exist that are readily accessible via the Internet. For the design of 16S rRNA-targeted oligonucleotide probes specialized databases, offered by the Ribosomal Database Project (RDP) (Maidak et al., 1999) or the University of Antwerp (Van de Peer et al., 1999), are an ideal starting point. Both databases have collected more than 16,000 sequences of small subunit rRNA molecules. This includes the 16S rRNA of approximately 2750 of the validly described species of *Bacteria* and *Archaea* and numerous 18S rRNA sequences of *Eucarya.* It is likely that coverage of the 16S rRNA sequences of the cultured procaryotes will be almost complete in the near future. The value of these databases for the identification of microorganisms can not be overestimated.

For a scientist interested in the design of a new probe, the initial question is the availability of the target sequence. Is there a full or partial 16S rRNA sequence of the microbial strain of interest in the public databases? Have additional strains of the same and closely related species also been sequenced? If a complete detection system, consisting of multiple probes, is to be developed for a genus or an even wider taxonomic entity, how well do the available sequences cover this group? Are corresponding sequences for those organisms that must be discriminated against available? A critical examination of the database will frequently reveal a need to perform additional sequencing. Today, this is largely facilitated by direct sequencing of PCR products. Conserved primers for the 5′ and 3′ end of the 16S rRNA gene exist (Giovannoni, 1991), which enable amplification of almost full length 16S rRNA genes from most, but not all (Marchesi et al., 1998), procaryotes. When starting from a pure culture, the resulting rDNA PCR product can be directly sequenced. Since a high-quality rRNA database is a prerequisite for reliable probe design, double-stranded sequencing should be performed on almost full length 16S rDNA sequences.[0]

Early in the design of a probe it is important to consider the intended application. If it is merely to screen isolates obtained from a specific set of samples, isolated on standardized media, the specificity requirements are more relaxed since one only needs to discriminate among those bacteria that are culturable under the selected conditions. However, if the probe is designed for *in situ* identification of a given microbial population in a complex environmental sample, it must be kept in mind that we have currently cultivated and described only a minority of the extant bacteria (Amann et al., 1995). It might, in such cases, be highly advisable to initially generate a 16S rRNA gene library of the community of interest to get at least a first impression of the natural diversity at the site of interest. Several environmental samples have already been investigated in this manner (e.g., Giovannoni et al., 1990, 1996; DeLong, 1992; Fuhrman et al., 1992; Barns et al., 1996; Snaidr et al., 1997; Hugenholtz et al., 1998b). The results not only indicate a huge microbial diversity but also add important 16S rRNA sequence information to the databases from organisms that have hitherto not been cultured. In the near future such "environmental" rRNA sequences (Barns et al., 1996) will outnumber sequences of well described, pure cultures and contribute significantly to our ability to perform directed probe design based on a reliable database.

2. Computer-assisted probe design The principle behind directed design of oligonucleotide probes is the identification of sites at which all target sequences are identical and maximally different from all nontarget sequences. The process can be best described as a systematic search of a number of aligned sequences. In an alignment, sequences are arranged in such a way that homologous positions are written in columns. In a difference alignment, only those nucleotides that deviate from the uppermost sequence, which is usually the sequence of the target organism, are given as letters. A window of the width of the intended oligonucleotide probe is then shifted from left to right with the aim of identifying a region at which all nontarget sequences contain one or several mismatches (Fig. 3). If several such sites are identified, the number, quality, and location of mismatches provide the basis for a ranking of the potential target sites. The primary goal is to maximize the difference in the temperatures of dissociation of the probe from the target and the nontarget sequences (ΔT_d). In our example (Fig. 3) the third option would be the best. It contains not only more, but also stronger, mismatches. It has been shown by Ikuta et al. (1987) that for 19 base pair oligonucleotide–DNA duplexes containing different single mismatches, e.g., the destabilizing effect of A-A, T-T, C-A is more pronounced than that of the only slightly destabilizing G-T, G-A base pairs.

An important fine tuning of the ΔT_d may be achieved by shifting the probe target position in a way that the strong mismatches are located in the center. It has been shown previously that mismatches at the end of an oligonucleotide are generally less destabilizing than internal mismatches (Szostak et al., 1979). It is very difficult to differentiate a single terminal mismatch, whereas a strongly destabilizing A-A or T-U mismatch at position 11 of a 17mer results in a significant reduction of binding to

0. *The direct sequencing of a 16S rDNA PCR product of a pure culture assumes that it contains only one copy or, in the more frequent case of genomes with several rRNA operons (e.g., *E. coli* has seven, *B. subtilis* ten rRNA operons), identical copies of the gene coding for the 16S rRNA. This, however, is not always the case. Microheterogeneities between the different rRNA operons of bacteria exist (Nübel et al., 1996; Rainey et al., 1996). For the archaeon *Haloarcula marismortui* , which contains two rRNA operons, an exceptionally high sequence heterogeneity of 5% has been shown (Mylvaganam and Dennis, 1992). This clearly demonstrates that the "one organism-one rRNA sequence" hypothesis that applies for many organisms might be an oversimplification in some instances. This has various consequences, one being that a group of closely related sequences recovered from an environment may not represent a group of separate, phylogenetically highly related strains, but rather the sequence heterogeneity of the 16S rDNA contained within one strain. Such small sequence differences can be distinguished, e.g., by denaturing gradient gel electrophoresis (DGGE). Thus, DGGE of PCR-amplified 16S rRNA gene fragments from a single strain can result in multiple bands. Oligonucleotide probes targeted to the rRNA microheterogeneities allow one to analyze the expression of the different operons (Nübel et al., 1996).

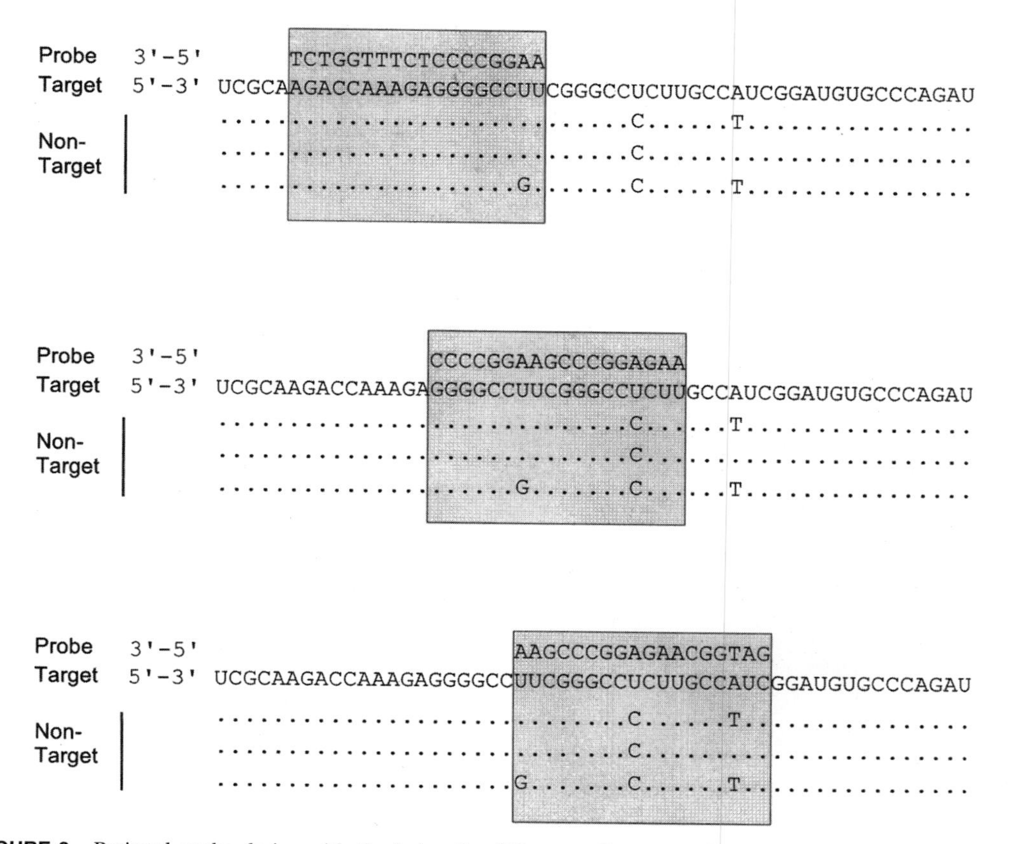

FIGURE 3. Rational probe design with the help of a difference alignment of target and non-target sequences. Note the differences in the number and the location of mismatches between the probe and the non-target sequences.

31% and 22%, respectively, of the binding to target rRNA without a mismatch (Manz et al., 1992). In the same study it was shown that the single mismatch discrimination could be significantly enhanced by competitor oligonucleotides. Competitors are unlabeled derivatives of the probe that are fully complementary to the known nontarget sequence. They are mixed with the labeled probe and efficiently prevent its hybridization to the nontarget sequence without significantly decreasing the homologous hybridization, thereby increasing probe specificity. The advantage of large databases is that such competitors can be designed in a directed way, preventing known potential unspecificities in advance. The variation in destabilizing effect of differently located mismatches during fluorescence *in situ* hybridization (FISH) with 16S rRNA-targeted oligonucleotide probes has also been evaluated (Neef et al., 1996).

Given the high number of 16S rRNA sequences now available, probe design must be performed with the aid of computer programs such as the PROBE_DESIGN tool of ARB* or the DESIGN_PROBE tool of the RDP (Maidak et al., 1999). Since rRNA sequences are patchworks of evolutionarily conserved regions, signature sites can be identified for any taxon between the level of the domains *Archaea, Bacteria, Eucarya,* and single species. In this respect, PROBE_DESIGN has the advantage that designation of target groups is done within a phylogenetic tree, assuring that monophyletic assemblages are targeted. The relatively high degree of conservation of the rRNA molecules usually does not allow for the design of subspecies- and strain-specific rRNA-targeted probes. Modern probe design tools generate an ordered list of potential probe target sites that take into account the above mentioned key parameters, number, quality and location of mismatches. The ranking of target sites should be according to an estimated ΔT_d between the target and nontarget organisms. For group-specific probes, i.e., those targeted to families, orders or classes, it may be necessary to allow for incomplete coverage of the target group and few non-target hits.

It is in the very nature of the evolutionary process that mutations in a sequence site that is characteristic for a particular phylogenetic group may occur. A good signature site might be present in only 95% of all members of the group and, because of high microbial diversity, organisms might exist that are not members of the group but have the identical probe target site (Fig. 4A). A multiple probe approach was developed to address precisely such problems. By an intelligent combination of several probes, identifications can be made with a high degree of confidence. If possible, two or even three probes to separate signature sites of one target sequence are constructed (Amann, 1995a). If they bind to the same cells or colonies or to the same fragment or fraction of DNA, the possibility of false positives is virtually eliminated (Fig. 4B). Multiple probe systems can also be built from nested probe sets (Stahl, 1986) in which the first probe targets, e.g., a signature at the genus level, the second one at the species level, and the third is specific for selected strains

Editorial Note: Software available from O. Strunk and W. Ludwig: ARB: a software environment for sequence data. www.mikro.biologie.tu-muenchen.de (Department of Microbiology, Technische Universität, München, Munich, Germany.)

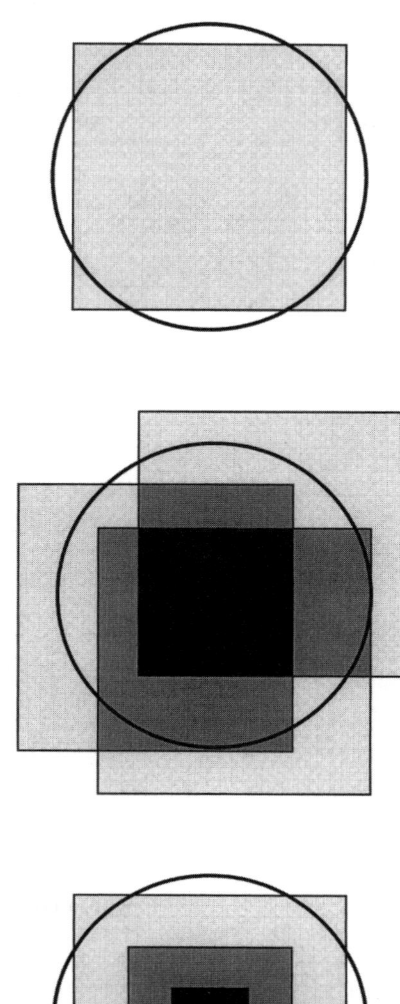

FIGURE 4. The multiple probe approach. The target group is represented by a circle. The specificities of the probes are described as squares. *A*, a probe detects most of the strains in the target group, but also some outside organisms. *B*, simultaneous application of three probes targeted to the same group. Note the different levels of gray indicating those subgroups that are detected by one, two, or three probes. *C*, nested set of probes focusing on increasingly narrow subgroups.

of this species (Fig 4C). If in agreement, the results of such a top- to-bottom approach (Amann et al., 1995) support each other, giving higher confidence to the final identification. For nested sets of probes used to quantify abundance of genera and species in mixed samples, it should also be possible to perform bookkeeping (Devereux et al., 1992) in the sense that the sum of all species-specific probes should be identical to the value obtained with the genus-specific probe. Here, the application of multiple probes allows for the identification of missing species within known genera. This directed screening for hitherto unknown bacteria that are genotypically different from, but related to, known species, might be of considerable interest in biotechnology.

3. SYNTHESIS AND LABELING OF OLIGONUCLEOTIDE PROBES Based on the results of the probe design, oligonucleotides are chemically synthesized on a solid support. Today solid phase synthesis is a fast and reliable technique that is frequently performed by service units or private companies. The cost of oligonucleotide synthesis has become affordable, with a product of 0.2 μmol, sufficient for thousands of hybridization assays usually costing less than $50.

For standard assays the nucleic acid probe must be labeled prior to hybridization. There are two principal types of detection systems: (i) direct systems in which a label that can be directly detected is covalently attached to the probe and (ii) indirect systems in which the initial modification of the probe is detected via the secondary, non-covalent binding of a labeled reporter protein. This second step can, for example, be the specific detection of the vitamin biotin by labeled (strept)avidine (Langer et al., 1981) or an antigen-antibody reaction such as the detection of the hapten digoxigenin (Kessler, 1991). A rather complete compilation of labels and detection systems can be found elsewhere (Kessler, 1994).

The labeling of oligonucleotides during solid phase synthesis by the incorporation of modified nucleotides or the direct attachment of labels to the 5′ end using phosphoamidite chemistry is becoming more commonplace. Alternatively, labels or reporter molecules can be attached after the synthesis by enzymatic or chemical means. T4 polynucleotide kinase catalyzes, e.g., the ^{32}P- or ^{33}P-labeling of oligonucleotides at the 5′-end with γ-labeled nucleotide triphosphate (Maxam and Gilbert, 1980). Terminal transferase may be used to elongate the 3′ end with labeled nucleotide triphosphates (Ratcliff, 1981). Chemical labeling of oligonucleotides after synthesis can be accomplished via primary aliphatic amino groups incorporated into the oligonucleotides during synthesis. Detailed protocols are available for the attachment of activated fluorescent dye molecules such as fluorescein or rhodamine via 5′ aminolinkers (Amann, 1995b) or the covalent labeling of oligonucleotide probes with enzymes such as alkaline phosphatase (Jablonski et al., 1986) or horseradish peroxidase (Urdea et al., 1988).

The choice of detection system and the quality of labeling are of prime importance for the sensitivity of a hybridization assay. It is highly recommended to check any new batch of labeled oligonucleotides for expected length, homogeneity, and completeness of labeling by polyacrylamide gel electrophoresis (Sambrook et al., 1989). If, because of incomplete labeling and purification, a probe batch still contains an equimolar amount of unlabeled oligonucleotide of identical sequence, the resulting competition for target sites would reduce the hybridization signal by one half of the maximum. Another factor that strongly influences the sensitivity of a hybridization assay is, of course, the amount, purity, and accessibility of the target nucleic acid. Since the preparation of target nucleic acids is highly dependent on the hybridization format used, this aspect will be discussed subsequently (Section 4) in the context of some common formats of oligonucleotide hybridization.

4. OPTIMIZING THE HYBRIDIZATION CONDITIONS Unfavorable hybridization conditions may lead to the failure of even a well designed and highly purified probe. These conditions encompass the hybridization buffer, the temperature, and time of hybridization. The two main components of any hybridization

buffer are monovalent cations, added in the form of salts, and a buffer system. Hybridizations require a pH close to neutral. The monovalent cations are important for the speed of hybrid formation and the stability of the resulting duplexes. The time required for hybrid formation is also directly influenced by the complexity of the probe, of which probe length is a good indicator. The higher the probe complexity, the longer the time necessary for hybridization. As a rule of thumb, oligonucleotide hybridizations take 1 h, compared to 5 h or more for polynucleotide probes.

However, the most important characteristic that needs to be determined during the optimization of hybridization conditions is the melting point (T_m) of the probe-target hybrid, or in the case of an oligonucleotide probe, its temperature of dissociation (T_d). For long hybrids the melting point can be quite accurately estimated based on the mol% G + C content of the DNA (Stahl and Amann, 1991). There are also formulae for the estimation of T_d values, the simplest being that of Suggs et al. (1981), which applies for hybridization in 6X SSC (0.9 M sodium chloride; 0.09 M trisodium citrate; pH 7.0):

$T_d = 4N_{G+C} + 2N_{A+T}$ (Suggs et al., 1981)

Here N_{G+C} and N_{A+T} are the numbers of G and C and of A and T which are assumed to add 4 and 2°C, respectively, to the thermal stability. Other more elaborate formulae, such as that of Lathe (1985), are available. They usually include the three parameters previously identified as key determinants of the T_d, the concentration of monovalent cations (M) in the hybridization buffer, and the length (n) and base composition (% G+C) of the oligonucleotide probe:

$T_d = 81.5 + 16.6 \log M + 0.41(\text{mol\% G+C}) - 820/n$ (Lathe, 1985)

All of these relationships have been empirically derived from experimental data. It must be noted that for oligonucleotides, the influence of the exact base sequence on the thermal stability is profound. The formulae should, therefore, only be used during probe design as an attempt to obtain a probe with a T_d within a certain range. As soon as the oligonucleotide has been synthesized, the T_d must be experimentally determined before the probe is used for the identification of microorganisms. For 16S rRNA-targeted oligonucleotide probes, several procedures have been described. In the original protocols, replicates of extracted, filter-immobilized total nucleic acid (Stahl et al., 1988; Raskin et al., 1994b) were hybridized at a relatively low temperature of 40°C with ^{32}P-labeled, rRNA-targeted oligonucleotide probes. The replicates were subsequently washed at successively higher temperatures (e.g., 40, 45, 50... 70°C) in 0.1% SDS–1X SSC for 30 min, and the amount of remaining radioactivity quantified. Alternatively, in the more economical elution technique, the same piece of membrane is transferred after hybridization through a series of washing steps at increasing temperature. Here, the amount of ^{32}P released in each washing step is quantified in scintillation vials and the total amount of released activity is plotted against temperature (Raskin et al., 1994b). In order to determine the optimum wash temperature, dissociation profiles for target and nontarget organisms need to be completed. Only then can conditions that fully discriminate nontarget nucleic acids and simultaneously yield good binding of the probe to the target nucleic acid be defined. Probes are specific only under certain conditions.

Changing the temperature of washing is, of course, not the only way to control the stringency of a hybridization assay. The temperature of hybridization is another obvious possibility, and, even at a constant temperature, the stringency of hybridization can be changed either by the addition of denaturing agents such as formamide or dimethylsulfoxide, or by varying the concentration of the duplex stabilizing monovalent cations. During hybridization of fixed whole microbial cells, high temperatures could have detrimental effects on the morphology. Therefore, formamide has been used to change the stringency of hybridization without altering the hybridization temperature of 46°C. This is done with the assumption that an addition of formamide of 1% is equivalent to an increase in hybridization temperature of 0.5°C (Wahl et al., 1987). The hybridization is followed by a slightly more stringent washing step at 48°C. In order to prevent production of excess amounts of potentially harmful waste, the stringency of the wash buffer is adjusted by lowering the concentration of monovalent cations rather than by the addition of formamide. This adjustment cannot be performed in the hybridization buffer since this would decrease the speed of hybridization. Based on the 0.5°C assumption of Wahl et al. (1987) and the salt term of the formula of Lathe (1985), Table 1 gives formamide concentrations and the salt concentrations that should yield comparable stringency.

By quantifying the fluorescence of the probe of interest after hybridization to selected target and nontarget cells, T_d values and optimum hybridization stringencies for whole cell hybridizations can be determined in a similar way as the optimum wash temperatures for immobilized extracted rRNA (Wagner et al., 1995; Neef et al., 1996). It has also been shown that the T_d values of a probe hybridized against extracted rRNA and against whole fixed cells are very similar (Amann et al., 1990b). One has to keep in mind, however, that the buffers used for T_d determinations are frequently very different. A T_d of 59°C in a buffer containing 2X SSC–0.1% SDS with a concentration of monovalent cations of roughly 390 mM is equivalent to a T_d in 1X SSC–0.1% SDS (195 mM) of 54°C. In the 900 mM NaCl buffer frequently used for *in situ* hybridization (Neef et al., 1996) the same T_d would be 65°C. If the temperature is kept at 46°C and the stringency is increased by adding formamide, half maximal binding would be at a concentration of 38% formamide (65 − 46 = 19°C equivalent to 38% formamide). Long term experience has shown that these correlations are quite robust. However, it should be stressed that it is best to determine T_d values in the

TABLE 1. Hybridization and washing buffers with corresponding stringencies for use in FISH

% Formamide [v/v] in a hybridization buffer containing 900 mM NaCl, 0.01% SDS, 10 mM Tris/HCl; pH 7.4	Concentration [mM] of monovalent cations in a wash buffer containing X mM NaCl, 0.01% SDS, 5 mM Na_2EDTA, 10 mM Tris/HCl; pH 7.4
0	900
5	636
10	450
15	318
20	225
25	159
30	112
35	80
40	56
45	40
50	28
55	20
60	14
65	10

actual format in which the probe is going to be used. If T_d values from separate determinations do not match, one should consider that the thermal stability of sequence identical duplexes increases from DNA–DNA to DNA–RNA and RNA–RNA (Saenger, 1984). As a rule, the T_d of a deoxyoligonucleotide hybridized against rRNA is about 2°C higher than against DNA.

Furthermore, even though each probe has a defined T_d, the optimal hybridization conditions are dependent on the hybridization format and the needs of the researcher. For slot blot hybridizations of total nucleic acids, the wash temperature frequently matches the T_d so as to reduce nonspecific binding. For FISH, in which discrete cells are stained, a weak nonspecific binding does not interfere, as long as it does not exceed the natural background fluorescence. In this format, which tends to be sensitivity-limited, the optimal stringency of hybridization is usually at the highest formamide concentration that still yields full fluorescence.

5. Evaluation of probe specificity and sensitivity Specificity and sensitivity are key aspects of any identification method. In the process of generating a new probe, specificity has already been controlled on two levels: initially during probe design and subsequently in the optimization of hybridization conditions using selected target and nontarget reference strains. However, even when the hybridization conditions have been properly determined, it may still be too early to apply the newly designed probe for determinative purposes. Questions that should first be considered are whether all strains available for a given species are indeed detected by a species-specific probe. Since 16S rRNA sequences are not usually available for all strains of interest, the best approach is to check the newly designed probe against a panel of reference strains. Subsequently, one should consider which bacteria need to be discriminated and whether 16S rRNAs from those bacteria have been sequenced. If this is not the case, it must be demonstrated that these strains do not hybridize at the optimized hybridization conditions.

Finally, hybridization assays should always incorporate proper controls, including at the minimum a positive and a negative control to evaluate the specificity and sensitivity of hybridization.

IV. Formats of Oligonucleotide Hybridization

Numerous hybridization assays exist. A full coverage is beyond the scope of this article and readers interested in a more complete listing are referred to other recent reviews (e.g., Schleifer et al., 1993). We will restrict ourselves to hybridization with oligonucleotide probes and to a few commonly used assays that allow a rapid identification of microorganisms. For the scope of this chapter, it might be sufficient to discriminate, on one hand, standard from reverse formats and, on the other hand, assays that require extracted nucleic acids from those that detect target nucleic acids at their original location within microbial cells.

A. Dot-blot/slot-blot and other membrane-based hybridization formats These assays are all based on the immobilization of target nucleic acids that have been extracted from the samples of interest. Critical steps for these types of assays are the cell lysis, purification of the nucleic acids, denaturation, and immobilization of the target nucleic acid on nitrocellulose or nylon membranes.

1. Quantitative slot-blot hybridization Dot-blot and slot-blot refer to the technique of using a vacuum chamber with round (dot) or longitudinal (slot) holes for the defined application of target nucleic acid solutions to membranes (Kafatos et al., 1979). In contrast to simply spotting samples onto membranes, which is sufficient for qualitative screening of multiple organisms, blotting evenly immobilizes each target nucleic acid on the same, defined area, facilitating quantitation. One particular method, quantitative slot blot hybridization with rRNA-targeted oligonucleotide probes, was introduced to studies in microbial ecology by Stahl et al. (1988).

This assay was designed to be directly applicable to diverse environmental samples without the need to cultivate the populations of interest. The choice of rRNA as a target molecule allows the use of highly disruptive cell lysis methods, which would damage high molecular weight DNA. This is an important advantage, since little is known about the samples *a priori* and quantitation relies on the efficient and representative recovery of nucleic acids from physiologically diverse bacteria (e.g., thin-walled Gram-negative bacteria vs. thick-walled Gram-positive bacteria). DNA is much more sensitive to shearing than RNA, therefore many of the DNA-based methods that are dependent on the retrieval of relatively intact nucleic acids require the use of less harsh lysis protocols. Consequently, those methods might fail to recover certain groups of bacteria present in high abundance in the community under investigation. One such example was the complete absence of 16S rDNA clones of Gram-positive bacteria with a high DNA G + C content in a PCR based gene library obtained from municipal activated sludge known to contain significant numbers of bacteria belonging to this phylogenetic group (Snaidr et al., 1997). In this case the freeze-thaw lysis method applied might have been ineffective in releasing DNA from this important part of the natural microbial community.

For quantitative slot blot hybridizations, total nucleic acid is recovered from the sample of interest by mechanical disruption with zirconium beads. The lysis is performed at low pH in the presence of equilibrated phenol and sodium dodecylsulfate to minimize nucleic acid degradation. Subsequently, nucleic acids are further purified by sequential extraction with phenol/chloroform and chloroform followed by ethanol precipitation. After spectrophotometric quantitation, the RNA is denatured with 2% glutaraldehyde and applied to a nylon membrane using a slot blot device. Air drying and baking is used to further immobilize the nucleic acids. The membranes are prehybridized in a buffer containing Denhardt's solution (Denhardt, 1966) before a synthetic oligonucleotide probe (5′-end labeled with ^{32}P using polynucleotide kinase and [γ-^{32}P]ATP) is applied. Denhardt's solution saturates free nucleic acid binding sites on the membrane that would otherwise increase the background by nonspecifically binding the labeled probe. Membranes are usually hybridized in rotating cylinders to prevent drying during the 40°C incubation, which lasts for several hours. The subsequent 30 min washing step is then performed at, or close to, the T_d determined for each probe. The membranes are dried and the amount of radioactivity bound to each slot is quantified by phospor imaging or autoradiography combined with densitometry. Average signals obtained from triplicates of a particular sample (e.g., with a genus-specific probe) are normalized for differences in the total amount of immobilized rRNA by comparison to the average signal obtained from replicates of the very same sample with a universal probe that binds to the rRNA of all organisms. Several applications of this technique have been published (e.g., Stahl et al., 1988; Raskin et al., 1994a; b).

2. Colony hybridization In the special case of colony hybridization (Grunstein and Hogness, 1975), nucleic acids are released directly onto filters on which colonies were either directly grown or transferred by replica plating or filtration. This method was originally developed for the rapid screening of cloned DNA fragments to search for specific genes. Colony hybridization can also be used for the identification of culturable bacteria (e.g., screening of primary plate isolates obtained from environmental samples) (Sayler et al., 1985; Festl et al., 1986). It must, however, be considered that Gram-positive bacteria need considerably harsher lysis methods than Gram-negative bacteria. Gram-positive bacteria may be resistant to the frequently used alkaline lysis method and may therefore yield false negative results (Jain et al., 1988). It has been shown that pretreatment of cells with 10% sodium dodecylsulfate improves the *in situ* lysis of a variety of Gram-positive bacteria (Betzl et al., 1990; Hertel et al., 1991). Under optimal conditions 1 in 10^6 colonies may be detected (Sayler et al., 1985). However, problems may arise from bacteria that show rapidly spreading growth, such as *Bacillus cereus* subsp. biovar mycoides. In addition, only those bacteria that readily form colonies on the media employed can be identified. Media are always selective and allow the analysis of only a poorly defined subfraction of the microbial cells present in a given environment. These drawbacks are not so important if colony hybridization is used to follow the fate of defined, rapidly growing bacterial strains, so it can be used successfully for these applications. For the sake of brevity, we will provide only one example for several different areas: PCB-degrading bacteria were monitored in soil by detecting specific catabolic genes (Layton et al., 1994). Heavy metal resistant bacteria have been screened and enumerated (Barkay et al., 1985). More recently, colony hybridization was used to differentiate subspecies and biovars of *Lactococcus lactis* with a gene fragment from the histidine biosynthesis operon (Beimfohr et al., 1997). The survival of genetically modified microorganisms has been studied in aquatic environments (Amy and Hiatt, 1989) and in mammalian intestines (Brockmann et al., 1996), and colony hybridization has also been used to monitor the maintenance and transfer of genes (Jain et al., 1987).

B. Reverse hybridization formats In reverse hybridization formats the labeled target nucleic acid is analyzed using an array of immobilized probes. In contrast to the standard hybridization assays, multiple nucleic acid probes, rather than the target, are deposited or even synthesized on a support. Subsequently, the sample of interest, rather than the probe, is labeled and hybridized against the array. This approach was initially used by Saiki et al. (1989) for the genetic analysis of PCR-amplified DNA with immobilized sequence-specific oligonucleotide probes. Meanwhile, it has found numerous application for the identification of bacteria, e.g., *Listeria monocytogenes* (Bsat and Batt, 1993), clostridia (Galindo et al., 1993), and lactic acid bacteria (Ehrmann et al., 1994).

Reverse sample genome probing was introduced to environmental microbiology by Voordouw et al. (1991). This method follows the same concept as reverse hybridization, but uses immobilized genomic DNAs from reference strains to probe environmental DNA that is radioactively labeled by nick translation. Here, the quality of the identification obtained following the incubation of a labeled total chromosomal DNA probe is largely dependent on the number and types of bacterial standards spotted on the master filter, and is therefore restricted by our current ability to retrieve representative pure cultures. Furthermore, since this assay represents essentially a massively parallel but classical DNA–DNA hybridization,the basic principle of this method must be considered. Because of potentially large differences in the mol% G + C content of DNA, which can range from 25–75%, a given incubation temperature might be optimal or relaxed for one type of DNA but highly stringent for another DNA. This directly influences the extent of binding and thereby the potential of this approach for accurate quantitation. The degree of DNA–DNA hybridization is high only between closely related species and quite low between less closely related species. As a consequence, the DNA of fairly closely related reference strains must be immobilized in order to detect even numerically abundant populations. So far the method has been restricted to the well characterized group of sulfate-reducing bacteria (e.g., screening of enrichments and isolates obtained from oil fields; Voordouw et al., 1992).

Supports for probe immobilization range from nylon membranes (e.g., Ehrmann et al., 1994) to microtiter plates (Galindo et al., 1993). Recently, Guschin and coworkers (1997) used oligonucleotide microchips as genosensors for determinative and environmental studies in microbiology. These microchips contain an array of deoxyribonucleotide oligomers that were immobilized after synthesis and purification within a polyacrylamide gel matrix bound to the surface of a glass slide. Oligonucleotide microchips were originally developed for rapid sequence analysis of genomic DNA by hybridization with oligonucleotides (Mirzabekov, 1994) and have proven to be suitable for analysis of mutations and gene polymorphisms (Yershov et al., 1996). Yet another fascinating possibility is the highly parallel synthesis of thousands of oligonucleotides. Here, photolithography is used to generate miniaturized arrays of densely packed oligonucleotides on a glass support (Fodor et al., 1991; Pease et al., 1994). These probe arrays, or DNA chips, are then used in hybridizations in which the analyzed nucleic acid is fluorescently labeled. Subsequently, the fluorescence arising from areas covered by the different oligonucleotides is quantified by laser scanning microscopy. Fluorescence signals from complementary probes were reported to be 5–35 times stronger than those arising from probes with one or two mismatches (Pease et al., 1994).

In the near future, it should be possible to immobilize thousands of species-specific probes or sets of nested probes tailored to the specific needs of microbiologists. Whereas light-generated DNA chips appear perfect for routine applications with a large commercial market (e.g., clinical microbiology), the postsynthesis loading of multiple oligonucleotide probes on suitable supports such as microchips or membranes could also be cost effective for more specialized applications. Along these lines, it is noteworthy that simultaneous transcriptional profiling on all open reading frames of the yeast *Saccharomyces cerevisiae* has been reported recently (Hauser et al., 1998).

C. In situ hybridization *In situ* hybridization, defined in a strict sense, is a localization technique that identifies nucleic acids in cells that remain at the site where they live. In a somewhat wider definition, microbiologists are using the term to describe the detection of target nucleic acids within fixed whole cells, although early attempts were made to discriminate between true *in situ* and whole cell hybridization (Amann et al., 1990b). The *in situ* identification of fixed whole bacterial cells using fluorescently labeled, rRNA-targeted oligonucleotides originally described by DeLong and coworkers (1989) has, over the last decade, found numerous applications in microbiology (for review see Amann et al., 1995; Amann and Kühl, 1998). Ribosomal RNA is not the only target for *in situ* hybridization, but for obvious reasons the most common one. Its stability and high copy number makes rRNA a much easier target than, e.g., mRNA. This does not mean that *in situ* mRNA detection in single cells has not yet been achieved (e.g., Hahn et al., 1993; Hönerlage et al., 1995; Wagner et al., 1998b), but that it has not yet been used for routine applications, as is the case for rRNA-targeted *in situ* hybridization probes.

The basic steps of fluorescence *in situ* hybridization are outlined in Fig. 5.

In principle, all the points that need to be considered for specific and sensitive detection of extracted target nucleic acids also apply to *in situ* hybridization. However, a couple of additional points are critical for *in situ* hybridization, especially for avoiding false negative results.

1. Permeabilization of target cells for nucleic acid probes A prerequisite for successful *in situ* hybridization is that the probe molecules can get to the target molecules. For this, cell components such as the cell wall, membranes, and, if present, capsular material or other extracellular polymeric substances must be permeable for the probe molecules to enter. This is easier when smaller probes are used. Oligonucleotides are, in this regard, better than polynucleotides and small fluorescent labels with a molecular weight below 1 kDa are better than large enzyme labels such as horseradish peroxidase (Amann et al., 1992b; Schönhuber et al., 1997). Furthermore, since intact membranes are generally impermeable to standard oligonucleotides, a fixation step is required. Fixation is usually accomplished by treatment of the sample with crosslinking aldehyde solutions (paraformaldehyde, formalin) and/or denaturing alcohols (for detail see Amann, 1995b). This step also kills the cells. Even though several fairly general fixation protocols have been described (Amann, 1995b; Amann et al., 1995), care should be taken to ensure that the procedure is optimized for the target cells, both so that their morphological integrity is not compromised and so that the cell walls do not become so strongly cross-linked that probe penetration is hindered. Thick-walled Gram-positive bacteria need different fixation protocols than Gram-negative bacteria (Roller et al., 1994; Erhart et al., 1997). Furthermore, diffusion of the probe requires a certain time, which is a function of the distance between the probe and the target. Therefore, larger aggregates need either to be dispersed, e.g., by sonication (Llobet-Brossa et al., 1998) or sectioned to preserve the natural organization (Ramsing et al., 1993; Schramm et al., 1996).

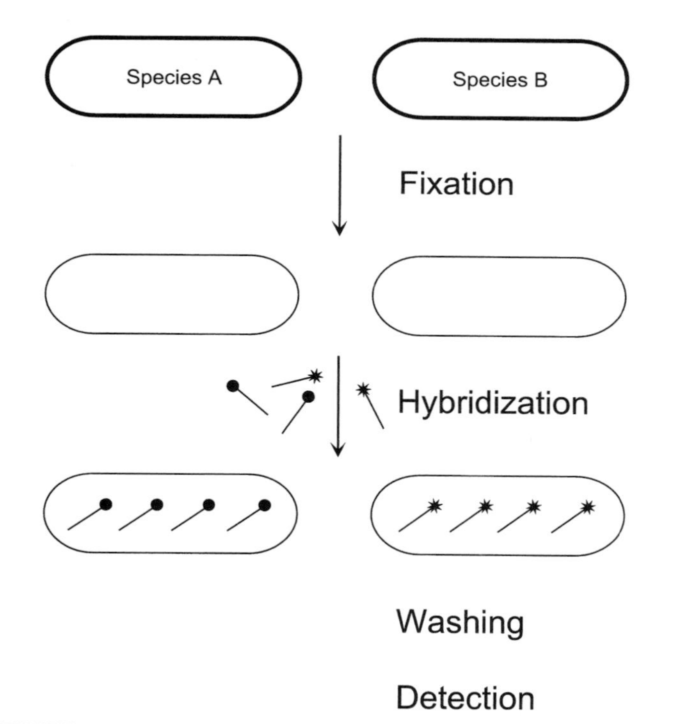

FIGURE 5. Principal steps of fluorescence *in situ* hybridization. The dots and asterisks indicate two different fluorescent dye molecules that are linked to two specific oligonucleotide probes.

The impermeability of the thick peptidoglycan layer of many Gram-positive bacteria to horseradish peroxidase-labeled oligonucleotides has recently been exploited to estimate the state of the cell wall in individual bacteria using FISH (Bidnenko et al., 1998). The authors reasoned that the expression of intracellular, peptidoglycan-hydrolyzing enzymes, such as autolysins or phage-encoded lysins, should permeabilize the cell walls for probe entry and thereby make the cells detectable by this method. The concept worked for strains of *Lactococcus lactis* infected with the virulent bacteriophage bIL66. Whereas only few cells hybridized in an exponentially growing culture, after infection the frequency of hybridizing cells increased sharply to 90%. In contrast, FISH with peroxidase-labeled oligonucleotide probes cannot be used to estimate the state of the cell wall of the Gram-negative *E. coli*, which without further lysis is fully permeable for probes of that size.

2. In situ accessibility of probe target sites Ever since rRNA-targeted FISH was first performed it was obvious that some target sites yield stronger signals than others (Amann et al., 1995; Frischer et al., 1996). For denatured extracted nucleic acids, it is assumed that target molecules are completely single-stranded and that different target sites are equally accessible for different nucleic acid probes. This is not the case for *in situ* hybridization with rRNA-targeted oligonucleotides. Here, the target molecules, which are integral parts of the ribosome, remain in the cell. Consequently, both rRNA-protein and intramolecular rRNA–rRNA interactions may influence the accessibility of the target sites. It is therefore not surprising that 200 fluorescein-labeled oligonucleotides, targeting the 16S rRNA of *Escherichia coli* with a spacing of less than 10 nucleotides (Fuchs et al., 1998), showed large differences in their capacity to fluorescently stain the very same cells (Fig. 6).

A good choice of accessible target sites yielding bright fluorescent signals is of critical importance for the sensitivity of *in situ* identification. Since the higher-order structure of the rRNA molecules and the ribosome are quite conserved, the *in situ* accessibility map of the *E. coli* 16S rRNA should be helpful for the selection of target sites in other organisms. Nevertheless, variations in *in situ* accessibility between different species will exist (Fuchs et al., 1998). Therefore, in the event that a newly designed probe that works on extracted nucleic acid does not yield good signals *in situ*, it is recommended to use one of the well-established, strongly fluorescing, general probes (e.g., the EUB338; Amann et al., 1990a) to determine whether this problem is cell- or probe-related and could possibly be solved by switching to a different target site.

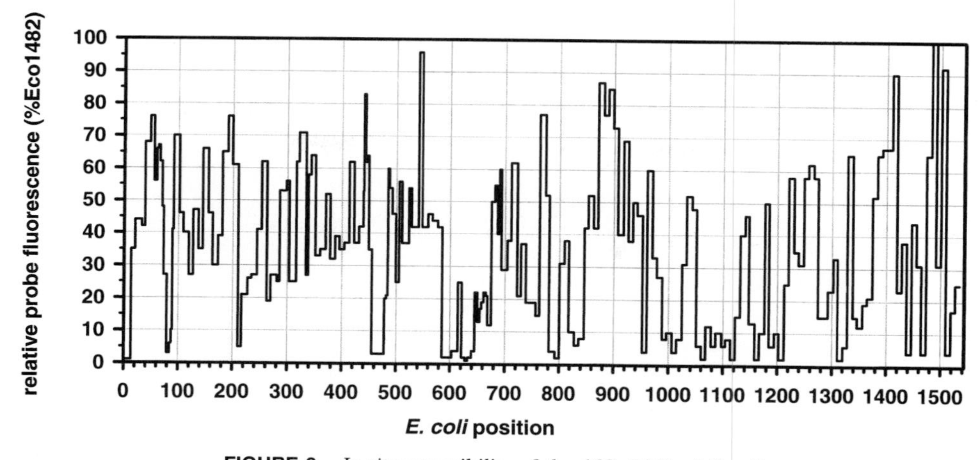

FIGURE 6. *In situ* accessibility of the 16S rRNA of *E. coli*.

3. Improving the sensitivity of in situ hybridization When discussing the sensitivity of *in situ* hybridization, one has first to realize that even though an individual cell can be identified (Fig. 7), this cell first needs to be brought into the microscopic field of observation. In a marine sediment containing $>10^9$ cells/cm^3 and a very high fraction of autofluorescent particles, the detection limit might be no better than 0.1% or 10^6 cells/cm^3. However, given a relatively clean water sample and the right equipment, it should also be possible to detect <1 cell/cm^3.

Another frequently encountered problem is that bacterial cells from the environment have low signals after *in situ* hybridization with fluorescently monolabeled oligonucleotide probes (Amann et al., 1995). The fluorescence conferred by a rRNA-targeted probe will be sensitive to changes in the cellular rRNA content of the target cells. The linear relationship between the growth rate of *Salmonella typhimurium* (Schaechter et al., 1958) and cellular ribosome content is well known. This correlation also applies to other bacteria (Poulsen et al., 1993; Wallner et al., 1993) and might be the reason why small, starving cells with little to no growth are so difficult to detect by FISH with rRNA-targeted probes. On the other hand, if this correlation is really true for cells in the environment, then it should be possible to determine or, at least, to estimate *in situ* growth rates of individual cells based on quantitation of probe-conferred fluorescence. This has been attempted for sulfate-reducing bacteria in a biofilm using digital microscopy (Poulsen et al., 1993). However, there is a large difference between the highly controlled growth conditions in Schaechter's experiments (Schaechter et al., 1958) and those experienced by environmental bacteria which might have to cope with rapid changes in the physical and chemical environment. Since ribosome synthesis is energetically costly, ribosome degradation, as a rapid first response to the slowing of the growth rate, would be very wasteful. Indeed, during periods of starvation of up to several months, bacteria maintain cellular ribosome pools in excess of their current needs (Flärdh et al., 1992; Wagner et al., 1995). Consequently, in strongly fluctuating environments such as, e.g., sediments in the intertidal zone, the cellular ribosome content should not be used to estimate actual growth rates. Nevertheless, the FISH signal of a cell is ecologically meaningful since it reflects the potential of the cell to synthesize protein.

There have been several attempts to combine FISH with short term incubation of environmental samples with nutrients and/or antibiotics, with the aims of increasing the ribosome content of environmental cells or demonstrating viability. Oligotrophic biofilms in potable water were incubated for 8 h with a mixture of carbon sources and an antibiotic preventing cell division (Kalmbach et al., 1997) prior to FISH in a modification of the direct viable count technique (Kogure et al., 1979; 1984). The number of cells detectable by FISH increased from 50% to 80%, clearly demonstrating viability of the majority of the cells. In a similar approach, marine water samples were incubated for approximately one hour with chloramphenicol (Ouverney and Fuhrman, 1997). Again, an increase in detection yield from 75% to almost 100% was observed. It should be noted here that even though both studies described precautions taken to prevent changes in total cell number or microbial composition during the incubation of the samples, the treatments had effects, e.g., on the cellular ribosome content. These methods cannot therefore, in a strict sense, be regarded as *in situ* hybridizations, and it might, in any case, be helpful to also investigate parallel samples after direct fixation.

Recently, technical improvements that result in more sensitive FISH have been reported. These include the use of more sensitive fluorescent dye molecules such as CY3 (Glöckner et al., 1996), dual labeling of oligonucleotide probes (Wallner et al., 1993, Fuhrman and Ouverney, 1998), the application of tyramide signal amplification (Schönhuber et al., 1997), and detection of the probe-conferred fluorescence by highly sensitive cameras (e.g., Ramsing et al., 1996; Fuhrman and Ouverney, 1998). Still, a certain fraction of particles detected by binding of the DNA stain DAPI (Porter and Feig, 1980) and identified as cells by cell morphology cannot be detected by these improved methods. It has been suggested that some of these cells might represent "ghosts" lacking nucleoids (Zweifel and Hagström, 1995). Since many of these DAPI stained spots that remain undetected by FISH are at the limit of resolution of light microscopy, the possibility that they originate from large virus particles can also not be excluded. Furthermore, problems with probe penetration, and the possibility that even the most general 16S rRNA probe target sites contain mutations, should be considered.

Like any other method for the identification of microorganisms, FISH has specific limitations. In addition to those we have

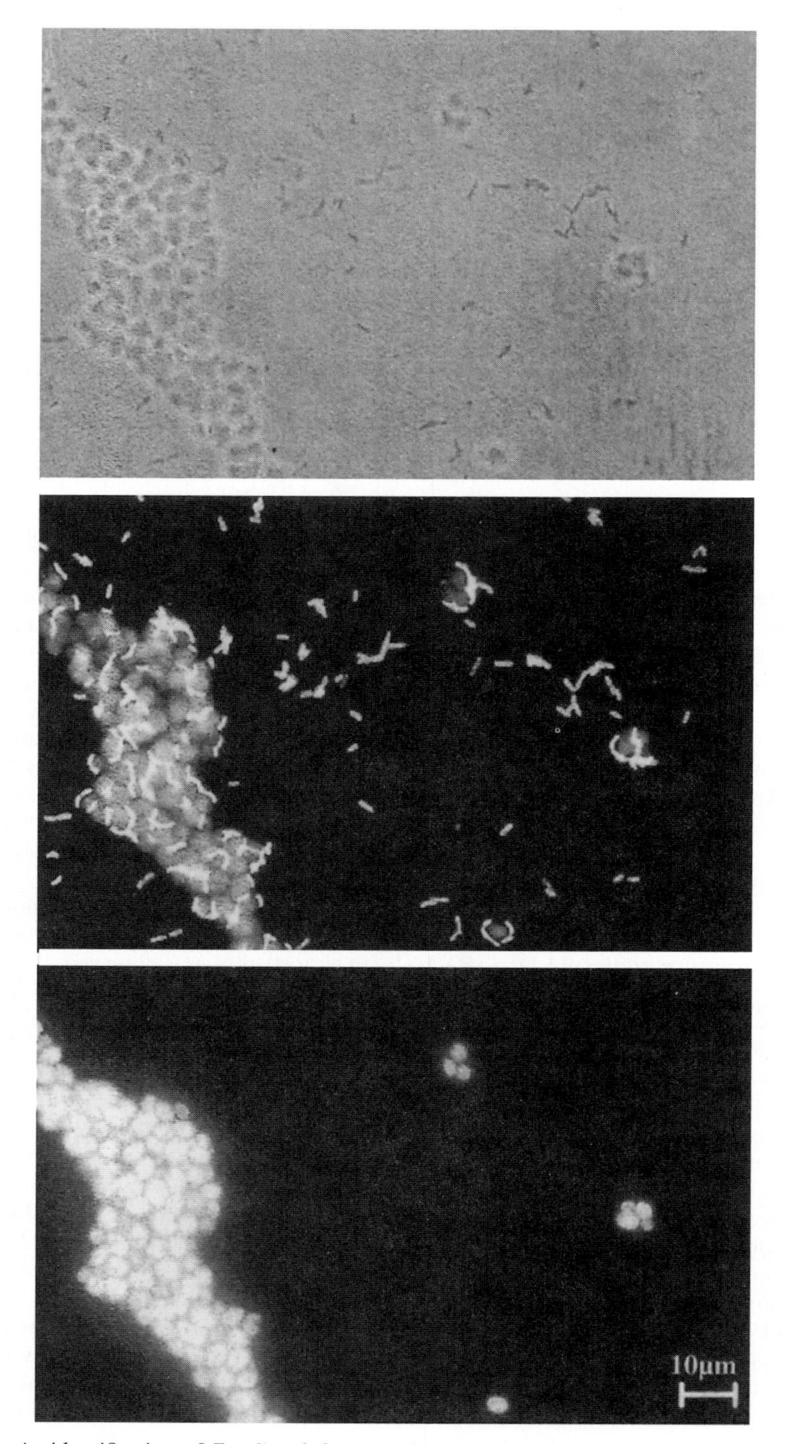

FIGURE 7. *In situ* identification of *E. coli* and the yeast *Saccharomyces cerevisiae* by *in situ* hybridization with two differently labeled rRNA-targeted oligonucleotide probes. Phase contrast and two epifluorescence images are shown for the same microscopic field.

already discussed, a further limitation is the lack of automation, as cells are frequently still counted manually. It should be stressed that FISH is the method of choice for determination of numbers of individual cells or localization of cells. However, for simple yes/no answers or rough estimates, this method may still be too complicated and time-consuming.

V. Applications of Nucleic Acid Probing in Environmental Microbiology

Nucleic acid probes have, in the last decade, revolutionized environmental microbiology. In each of the monthly issues of *Applied and Environmental Microbiology,* there are numerous examples in which hybridization or PCR assays are used to monitor defined strains or specific genes in different environments. This boom is due to the increasingly accepted view that traditional microbiological methods for analyzing environmental samples are less accurate and usually slower than molecular techniques. It also reflects the now-common idea that more than 100 years of scientific microbiology has, in terms of cultivation of representatives of extant microbial diversity, managed to describe only the tip of the iceberg. This does not imply that we should stop our efforts to isolate and study pure cultures. Most of what we know of microbiology and microorganisms is based on studies of pure cultures. Nevertheless, molecular techniques have become a faster and more accurate means to address questions concerning the exact size of a population or the composition of complex microbial communities. Before we review some examples, organized according to habitats, it should be noted that nucleic acid probes do not answer all our questions equally well, especially when the viability and physiology of certain bacteria is the focus of interest. In such instances, cultivation-based methods might still be the method of choice. For reasons of practicability, we will focus on applications of *in situ* hybridization in environmental microbiology. This, however, does not indicate that this method is superior to other assays. If one wants to analyze a relatively poorly studied habitat for microbial diversity or community composition, it is always wise to combine at least two different methods.

A. Soils and sediments Soils and sediments are among the most complex of all microbial habitats. The microbial diversity of soils has always been viewed as high, but the first attempts at quantifying this diversity were not reported until 1990. In their much-cited reassociation study of DNA isolated from a Norwegian forest soil, Torsvik et al. (1990a) suggest that the genetic diversity of DNA extracted from a bacterial cell fraction of a gram of soil corresponds to about 4000 completely different genomes of a size standard for soil bacteria. This was about 200 times the genetic diversity found in the strains isolated from the same soil sample (Torsvik et al., 1990b), which reflects the selectivity of cultivation based methods and the tendency of these methods to underestimate both the absolute number and the diversity of microorganisms already noticed before (Skinner et al., 1952; Sørheim et al., 1989). During the last years, molecular methods including the analysis of mol% G + C profiles (Holben and Harris, 1995; Nüsslein and Tiedje, 1998), the analysis of amplified rDNA by restriction analysis (ARDRA; e.g., Smith et al., 1997; Nüsslein and Tiedje, 1998), and denaturing or thermal gradient gel electrophoresis (Felske et al., 1997; Heuer et al., 1997) have increasingly been applied to soils.

With respect to nucleic acid probing of soils, colony hybridizations have often been performed (e.g., Sayler and Layton, 1990), whereas application of quantitative dot blot hybridization of rRNA–DNA is rather rare, suggesting difficulties with the extraction of good quality nucleic acids from soils (discussed, e.g., in Torsvik et al., 1990a; Holben and Harris, 1995). However, the few recent studies that have been reported (e.g., MacGregor et al., 1997; Rooney-Varga et al., 1997; Sahm et al., 1999) indicate that the method is applicable.

In situ monitoring of bacterial populations in soils and sediments by FISH has also long proven difficult (Hahn et al., 1992). Problems arise from autofluorescence of soil particles, irregular distribution of cells, and low detection yield. With the implementation of improved dyes (Zarda et al., 1997) and microscopic techniques such as confocal laser scanning microscopy (Assmus et al., 1995), the situation is much improved, leading to the application of FISH for studies of microbial community composition both in soil (Zarda et al., 1997; Ludwig et al., 1997; Chatzinotas et al., 1998) and in sediments (Llobet-Brossa et al., 1998; Rosselló-Mora et al., 1999). As an example, Fig. 8 shows detection of a filament of the sulfate-reducing bacterium *Desulfonema* sp. in Wadden Sea sediments. Interestingly, quite unexpected groups of bacteria are found in high numbers in both habitats, e.g., the peptidoglycan-less planctomycetes (Zarda et al., 1997; Chatzinotas et al., 1998), as well as representatives of a thus far uncultured group of Gram-positive bacteria with a high DNA G + C content (Felske et al., 1997) and of the newly described phylum *Holophaga/Acidobacterium* (Ludwig et al., 1997). High abundance of members of the *Cytophaga/Flavobacterium* cluster has been reported for anoxic marine sediments (Llobet-Brossa et al., 1998; Rosselló-Mora et al., 1999).

In the study of Chatzinotas et al. (1998), several molecular methods were compared for their potential to detect broad-scale differences in the microbial community composition of two pristine forest soils. This study highlights one of the many potential methodological pitfalls: FISH of dispersed soil slurries failed to detect Gram-positive bacteria with a high DNA G + C content, even though dot blot hybridization of extracted DNA indicated significant occurrence of members of this group. This was only in part a problem of cell permeability since care had been taken to ensure that at least the vegetative filaments of the actinomycetes were probe-permeable. Filaments could indeed be visualized in nondispersed soil samples. It appears that the filaments were destroyed by the methods used for soil dispersion, vortexing and sonication, which together with the physical effects of inorganic soil particles most likely resulted in the milling of actinomycete filaments. This again shows how important it is not to rely on a single technique for community composition analysis of the highly diverse soil microbiota.

B.* In situ *hybridization of biofilms and aggregates In many natural settings as well as in biotechnological wastewater treatment systems, immobilized communities of bacteria, the so-called biofilms, are the main mediators of biogeochemical reactions rather than free-living bacteria. Since, in addition to determination of community composition, *in situ* localization is of prime importance in the investigation of these systems, biofilms have been studied intensively with FISH. Thicker biofilms, in the mm to cm range, are known as microbial mats. Because of their size these can, unlike biofilms, also be studied with extraction-based molecular techniques, such as slot blot hybridization (e.g., Risatti et al., 1994) or DGGE analysis of amplified rDNA fragments (e.g., Ferris and Ward, 1997; Ferris et al., 1997).

The *in situ* visualization of defined bacterial biofilm populations was first achieved in an anaerobic fixed-bed reactor (Amann et al., 1992a). The initial colonization of a glass surface was monitored using FISH. Two morphologically distinct populations of Gram-negative sulfate reducing bacteria (a thick and a thin vibrio) could be assigned to 16S rDNA sequences related to *Desulfuromonas* and *Desulfovibrio* retrieved from the same reactor by oligonucleotide probing. One of the probes was later used to direct the enrichment and isolation of a sulfate-reducing strain representative for the *Desulfovibrio* sp. population (Kane et al., 1993).

Over the last few years, numerous FISH studies have been performed in systems such as activated sludge plants, trickling filters, or anaerobic sludge digesters (Harmsen et al., 1996a, b). Initial investigations of activated sludge targeted the higher level

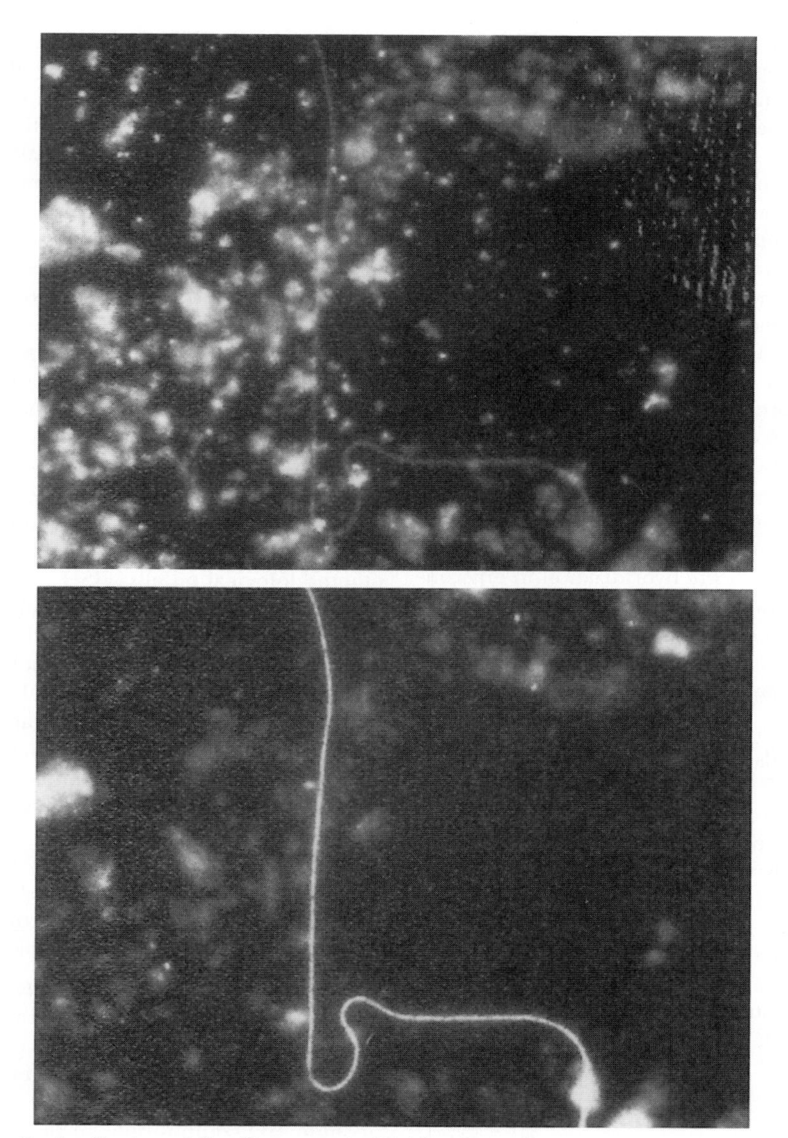

FIGURE 8. *In situ* visualization of *Desulfonema* sp. in Wadden Sea sediments. *Upper panel*, DAPI staining. *Lower panel*, FISH with *Desulfonema* probe.

bacterial taxa by applying, e.g., 16S or oligonucleotide probes for the α, β, and γ *Proteobacteria*, the *Cytophaga-Flavobacterium* cluster, or the *Actinobacteria* (Wagner et al., 1993; Manz et al., 1994; Kämpfer et al., 1996). Probes are now available for functionally important groups such as the ammonia- (Wagner et al., 1995, 1996; Mobarry et al., 1996) and nitrite-oxidizing bacteria (Wagner et al., 1996; Schramm et al., 1998), and for key genera and species in wastewater treatment such as *Acinetobacter* spp. (Wagner et al., 1994), *Zoogloea ramigera* (Rosselló-Mora et al., 1995), and *Microthrix parvicella* (Erhart et al., 1997). The application of these probes has already resulted in some interesting findings. For instance, in contrast to textbook knowledge, *Acinetobacter* spp. seems to play no major role in biological phosphorus removal (Wagner et al., 1994). In the future, important processes such as floc formation and settling will be related to population sizes of defined bacteria. Based on previous indications that β-*Proteobacteria* are a major group in activated sludge, a high genetic diversity within this group was demonstrated by the simultaneous application of three oligonucleotide probes labeled with different fluorochromes (Amann et al., 1996b; Snaidr et al., 1997). Here, the colocalization of two or even three probes within one fixed whole cell results in a better discrimination of closely related β-subclass *Proteobacteria*.

FISH can be combined with microsensor measurements to address both *in situ* structure and activity of biofilms. In the first example of such a study (Ramsing et al., 1993), a mature, thick trickling filter biofilm was first investigated with microsensors for oxygen, sulfide, and nitrate to quantify sulfate reduction before cryosectioning, and FISH was used for the *in situ* localization of SRB. Similarly, the structure/function correlation of nitrification has been analyzed both in a trickling filter treating the ammonia-rich effluent water of an eel farm (Schramm et al., 1996), and in a chemolithoautotrophic fluidized bed reactor (Schramm et al., 1998). Whereas the former contained, as expected, dense clusters of *Nitrosomonas* spp. (Fig. 9) and *Nitrobacter* spp., *Nitrosospira* spp. and *Nitrospira* spp. dominated the more oligotrophic fluidized bed. In another attempt to combine *in situ* activity measurement with *in situ* identification of individual cells, microautoradiography has recently been used in conjunction with FISH (Nielsen et al., 1997). It should not be forgotten that the combination of traditional isolation and physiological characterization of pure cultures with FISH is a very powerful and straightforward way to correlate *in situ* distribution with particular metabolic properties. For example, it has been shown that *Paracoccus* is an important denitrifying genus in a methanol-fed sand filter (Neef et al., 1996). Traditional cultivation resulted in the isolation of numerous strains of *Paracoccus* that had the potential to denitrify using methanol as substrate. Genus-specific probes identified dense clusters of brightly stained paracocci accounting for about 3.5% of all cells in methanol-fed biofilms, whereas almost no cells were detected in a parallel filter that did not receive methanol and therefore showed no denitrification. Here, FISH allows for the assignment of functions studied in pure cultures to defined populations within a complex biofilm. A caveat of this approach may be the known metabolic plasticity of bacteria.

C.* In situ *identification of planktonic bacteria in oligotrophic water samples An early observation in the study of aquatic environments was that the free-living, planktonic bacteria were often more difficult to detect by FISH than those attached to the surfaces of the same water body (Manz et al., 1993). It was pointed out that these surfaces are enriched for nutrients so that immobilized bacteria are not as strongly nutrient-limited as planktonic bacteria. Indeed, the initial applications of FISH to bacterioplankton yielded good results only in highly eutrophic ponds (Hicks et al., 1992) or contaminated coastal water (Lee et al., 1993). The improvement of FISH detection yields by simultaneous application of multiple single labeled (Lee et al., 1993), brighter (Glöckner et al., 1996), or dual-labeled (Fuhrman and Ouverney, 1998) oligonucleotides in combination with image-intensifying CCD cameras (Ramsing et al., 1996; Fuhrman and Ouverney, 1998) indicates that the method remains sensitivity limited, and this is probably a function of small cell size and low cellular ribosome content. Some years ago, Edward DeLong, the pioneer of FISH, needed to apply quantitative dot blot hybridization to determine the abundance of archaeal rRNA in coastal Antarctic surface waters (DeLong et al., 1994). Assuming that rRNA abundance is a good indicator of biomass, it was suggested that as yet uncultured *Archaea* might constitute up to one-third of the total procaryotic biomass. FISH protocols have now been sufficiently improved to allow reliable *in situ* detection of greater than 50% of the bacteria in oligotrophic water samples (Glöckner et al., 1996; Fuhrman and Ouverney, 1998). The microbial community composition of the winter cover and pelagic zone of an Austrian high mountain lake has been described (Alfreider et al., 1996). A seasonal study of microbial community dynamics, that for the first time also included the monitoring of defined bacterial populations based on probes targeted to environmental 16S rDNA retrieved from the same lake, has also

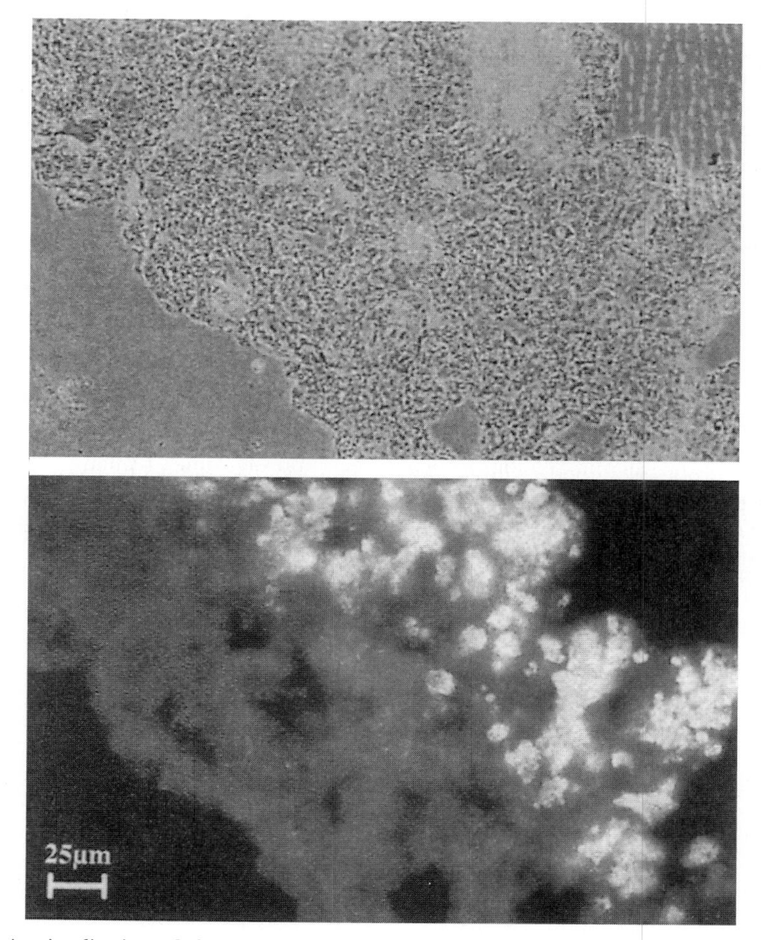

FIGURE 9. *In situ* visualization of clusters of *Nitrosomonas* sp. cells in a nitrifying biofilm. A phase contrast and an epifluorescence image are shown for the identical microscopic field.

been reported (Pernthaler et al., 1998). The combined use of digital microscopy and FISH enabled the determination of biomasses and size distributions, an approach which has recently been extended to studies of morphological and compositional changes in a planktonic bacterial community in response to enhanced protozoan grazing (Pernthaler et al., 1997; Jürgens et al., 1999). While only the environmental applications of FISH have been discussed here, similar techniques may be applied to the analysis of waterborne pathogens such as *Legionella pneumophila* (Manz et al., 1995; Grimm et al., 1998) or bacterial endosymbionts (for a review see Amann et al., 1995).

D. Flow cytometry and fluorescence-activated cell sorting Flow cytometry is a technique for the rapid analysis and sorting of single cells. Up to 10^3 suspended cells per second can pass an observation point where the cells, aligned in a water jet like pearls on a string, interfere with one or several light sources, usually lasers. Several physical and chemical properties of each individual cell can be measured simultaneously on the basis of fluorescence emitted from specifically and stoichiometrically bound dyes, as well as from light scattering. Flow cytometry has a higher throughput than microscopic quantification of specifically stained microbial populations and can more readily be automated, which should facilitate more rapid and frequent monitoring of the composition of microbial communities. It has been demonstrated using pure cultures that FISH of fixed whole bacterial cells can be combined with flow cytometry (Amann et al., 1990a; Wallner et al., 1993). The approach can also be applied to environmental samples as has been shown with samples from a wastewater treatment plant (Wallner et al., 1995). Flow cytometric and microscopic counts were in general agreement, with some discrepancies found for those populations that occurred predominantly in flocs or chains.

Effective cell dispersion is a prerequisite for accurate counting and therefore flow cytometry is better suited for free-living cells than for immobilized microbial communities such as biofilms. Furthermore, application of flow cytometry in microbiology is frequently hindered by the small size and concomitant low scattering and fluorescence of microbial cells.

However, certain features of flow cytometry justify the effort required to change from microscopy to this approach for the analysis of bacteria. There is the above-mentioned high throughput and potential for automation, of which the study by Fuchs et al. (1998) on the quantitation of fluorescence conferred by 200 different 16S rRNA-targeted probes is a good example. An additional attractive feature is that many flow cytometers have a sorting option. It was recently shown that bacteria could be sorted directly from environmental samples,without cultivation, based on differences in light scattering, DNA content, and affiliation to certain phylogenetic groups as revealed by FISH (Wallner et al., 1997). Microscopy of sorted cells showed that populations of originally low abundance could be strongly enriched (up to 1000-fold) by flow sorting (Snaidr et al., 1999). The ultimate purity of the sorted cells also depends on the sample analyzed and the original abundance, but in an optimal case can be close to 100% (Wallner et al., 1997). Gene fragments can subsequently be amplified from the sorted cells for further molecular analysis by PCR. In this way, the combination of flow sorting and FISH with probes targeted to 16S rRNA directly retrieved from the environment without cultivation allows selective access to the genetic information of microorganisms.

VI. Outlook

The prospects for hybridization-based molecular methods in microbiology are bright. The future will likely bring not only even more sensitive methods and automation, but also the massively parallel application of user-friendly probe arrays. These will include nucleic acid probes for taxonomic identification at the species level, but may also allow strain level assignment. Probes for functional genes such as those coding for certain degradative pathways or virulence factors will also be available. It has been pointed out in this introductory chapter that probe design usually relies on knowledge of the target sequence. Therefore, in the near future we will see continued and most likely increased sequencing efforts both for specific genes, such as those coding for the 16S rRNA, and for full genomes. The only threat to the increased use of nucleic acid probes in the future might originate from the very same rapid development of nucleic acid sequencing technology. After all, probes are just tools for determining a short sequence that might in the future be determined as well or even faster by direct sequencing. We expect that in the near future the artificial boundaries between nucleic acid hybridization and sequencing will erode. Sequencing, rather than nucleic acid hybridization may become the standard approach to molecular identification, but this sequencing may be performed via hybridization to nucleic acid arrays.

Acknowledgments

The original work of the authors has been supported by grants from the Deutsche Forschungsgemeinschaft, the Bundesministerium für Forschung und Technologie, the Fonds der Chemischen Industrie, the Max-Planck-Society, and the European Union. We would like to thank Bernhard Fuchs, Frank-Oliver Glöckner, and Wilhelm Schönhuber for artwork and critical reading of the manuscript.

Bacterial Nomenclature

Peter H.A. Sneath

Scope of nomenclature

Nomenclature has been called the handmaid of taxonomy. The need for a stable set of names for living organisms, and rules to regulate them, has been recognized for over a century. The rules are embodied in international codes of nomenclature. There are separate codes for animals, noncultivated plants, cultivated plants, procaryotes, and viruses. But partly because the rules are framed in legalistic language (so as to avoid imprecision), they are often difficult to understand. Useful commentaries are found in Ainsworth and Sneath (1962), Cowan (1978), and Jeffrey (1977). There are proposals for a new universal code for living organisms (see the Proposed BioCode).

The nomenclature of the different kinds of living creatures falls into two parts: (a) informal or vernacular names, or very specialized and restricted names; and (b) scientific names of taxonomic groups (taxon, plural taxa).

Examples of the first are vernacular names from a disease, strain numbers, the symbols for antigenic variants, and the symbols for genetic variants. Thus one can have a vernacular name like the tubercle bacillus, a strain with the designation K12, a serological form with the antigenic formula Ia, and a genetic mutant requiring valine for growth labeled *val*. These names are usually not controlled by the codes of nomenclature, although the codes may recommend good practice for them.

Examples of scientific names are the names of species, genera, and higher ranks. Thus *Mycobacterium tuberculosis* is the scientific name of the tubercle bacillus, a species of bacterium.

These scientific names are regulated by the codes (with few exceptions) and have two things in common: (a) they are all Latinized in form so as to be easily recognized as scientific names, and (b) they possess definite positions in the taxonomic hierarchy. These names are international; thus microbiologists of all nations know what is meant by *Bacillus anthracis*, but few would know it under vernacular names like Milzbrandbacillus or Bactéridie de charbon.

The scientific names of procaryotes are regulated by the *International Code of Nomenclature of Bacteria*, which is also known as the *Revised Code* published in 1975 (Lapage et al., 1975). This edition authorized a new starting date for names of bacteria on January 1, 1980, and the starting document is the Approved Lists of Bacterial Names (Skerman et al., 1980), which contains all the scientific names of bacteria that retain their nomenclatural validity from the past. The operation of these Lists will be referred to later. The *Code* and the Lists are under the aegis of the International Committee on Systematic Bacteriology, which is a constituent part of the International Union of Microbiological Societies. The Committee is assisted by a number of Taxonomic Subcommittees on different groups of bacteria, and by the Judicial Commission, which considers amendments to the *Code* and any exceptions that may be needed to specific Rules. An updated edition of the *Revised Code* was published in 1992 (Lapage et al., 1992).

Latinization

Since scientific names are in Latinized form, they obey the grammar of classic, medieval, or modern Latin (Neo-Latin). Fortunately, the necessary grammar is not very difficult, and the most common point to watch is that adjectives agree in gender with the substantives they qualify. Some examples are given later. The names of genera and species are normally printed in italics (or underlined in manuscripts to indicate italic font). For higher categories conventions vary: in Britain they are often in ordinary roman type, but in America they are usually in italics, which is preferable because this reminds the reader they are Latinized scientific names. Recent articles that deal with etymology and Latinization include that of MacAdoo (1993) and the accompanying article by Trüper on Etymology in Nomenclature of Procaryotes. The latter is particularly valuable because it clarifies the formation of names derived from names of persons.

Taxonomic hierarchy

The taxonomic hierarchy is a conventional arrangement. Each level above the basic level of species is increasingly inclusive. The names belong to successive **categories**, each of which possesses a position in the hierarchy called its **rank**. The lowest category ordinarily employed is that of species, though sometimes these are subdivided into subspecies. The main categories in decreasing rank, with their vernacular and Latin forms, and examples, are shown in Table 1.

Additional categories may sometimes be intercalated (e.g., subclass below class, and tribe below family). There is currently discussion on the best treatment for categories above kingdom; the BioCode (see later) uses the term, domain, above kingdom.

Form of names

The form of Latinized names differs with the category. The species name consists of two parts. The first is the **genus name**. This is spelled with an initial capital letter, and is a Latinized substantive. The second is the **specific epithet**, and is spelled with a lower case initial letter. The epithet is a Latinized adjective in agreement with the gender of the genus name, or a Latin word in the genitive case, or occasionally a noun in apposition. Examples are given in the article by Trüper. Thus in *Mycobacterium*

TABLE 1. The ranking of taxonomic categories

Category	Example
Domain	*Bacteria*
Phylum in zoology or Division in botany and bacteriology	*Actinobacteria*
Class	*Actinobacteria*
Subclass	*Actinobacteridae*
Order	*Actinomycetales*
Suborder	*Actinomycineae*
Family	*Actinomycetaceae*
Genus	*Actinomyces*
Species	*Actinomyces bovis*

tuberculosis, the epithet tuberculosis means "of tubercle", so the species name means the mycobacterium of tuberculosis. The species name is called a **binominal name**, or **binomen**, because it has two parts. When subspecies names are used, a trinominal name results, with the addition of an extra **subspecific epithet**. An example is the subspecies of *Lactobacillus casei* that is called *Lactobacillus casei* subsp. biovar *rhamnosus*. In this name, *casei* is the specific epithet and *rhamnosus* is the subspecific epithet. The existence of a subspecies such as *rhamnosus* implies the existence of another subspecies, in which the subspecific and specific epithets are identical, i.e., *Lactobacillus casei* subsp. biovar *casei*.

One problem that frequently arises is the scientific status of a species. It may be difficult to know whether an entity differs from its neighbors in certain specified ways. A useful terminology was introduced by Ravin (1963). It may be believed, for example, that the entity can undergo genetic exchange with a nearby species, in which event they could be considered to belong to the same **genospecies**. It may be believed the entity is not phenotypically distinct from its neighbors, in which event they could be considered to belong to the same **taxospecies**. Yet, the conditions for genetic exchange may vary greatly with experimental conditions, and the criteria of distinctness may depend on what properties are considered, so that it may not be possible to make clear-cut decisions on these matters. Nevertheless, it may be convenient to give the entity a species name and to treat it in nomenclature as a separate species, a **nomenspecies**. It follows that all species in nomenclature should strictly be regarded as nomenspecies. They are, of course, usually also taxospecies.

Genus names, as mentioned above, are Latinized nouns, and so subgenus names (now rarely used) are conventionally written in parentheses after the genus name; e.g., *Bacillus* (*Aerobacillus*) indicates the subgenus *Aerobacillus* of the genus *Bacillus*. As in the case of subspecies, this implies the existence of a subgenus *Bacillus* (*Bacillus*).

Above the genus level most names are plural adjectives in the feminine gender, agreeing with the word *Procaryotae*, so that, for example, *Brucellaceae* means *Procaryotae Brucellaceae*.

Purposes of the Codes of Nomenclature

The codes have three main aims:

1. Names should be stable,
2. Names should be unambiguous,
3. Names should be necessary.

These three aims are sometimes contradictory, and the rules of nomenclature have to make provision for exceptions where they clash. The principles are implemented by three main devices: (a) priority of publication to assist stability, (b) establishment of nomenclatural types to ensure the names are not ambiguous, and (c) publication of descriptions to indicate that different names do refer to different entities. These are supported by subsidiary devices such as the Latinized forms of names, and the avoidance of synonyms for the same taxon (see Synonyms and Homonyms later in this section).

Priority of Publication

To achieve stability, the first name given to a taxon (provided the other rules are obeyed) is taken as the correct name. This is the **principle of priority**. But to be safeguarded in this way a name obviously has to be made known to the scientific community; one cannot use a name that has been kept secret. Therefore, names have to be published in the scientific literature, together with sufficient indication of what they refer to. This is called **valid publication**. If a name is merely published in the scientific literature, it is called **effective publication**; to be valid it also has to satisfy additional requirements, which are summarized later.

The earliest names that must be considered are those published after an official starting date. For many groups of organisms this is Linnaeus' *Species Plantarum* of 1753, but the difficulties of knowing to what the early descriptions refer, and of searching the voluminous and growing literature, have made the principle of priority increasingly hard to obey.

The *Code* of nomenclature for bacteria, therefore, established a new starting date of 1980, with a new starting document, the Approved Lists of Bacterial Names (Skerman et al., 1980). This list contains names of bacterial taxa that were recognizable and in current use. Names not on the lists lost standing in nomenclature on January 1, 1980, although there are provisions for reviving them if the taxa are subsequently rediscovered or need to be reestablished. To prevent the need to search the voluminous scientific literature, the new provisions for bacterial nomenclature require that for valid publication new names (including new names in patents) must be published in certain official publications. Alternatively, if the new names were effectively published in other scientific publications, they must be announced in the official publications to become validly published. Priority dates from the official publication concerned. At present the only official publication is the *International Journal of Systematic Bacteriology* (now the *International Journal of Systematic and Evolutionary Microbiology*).

Nomenclatural Types

To make clear what names refer to, the taxa must be recognizable by other workers. In the past it was thought sufficient to publish a description of a taxon. This has been found over the years to be inadequate. Advances in techniques and in knowledge of the many undescribed species in nature have shown that old descriptions are usually insufficient. Therefore, an additional principle is employed, that of **nomenclatural types**. These are actual specimens (or names of subordinate taxa that ultimately relate to actual specimens). These type specimens are deposited in museums and other institutions. For procaryotes (like some other microorganisms that are classified according to their properties in artificial culture) instead of type specimens, **type strains** are employed. The type specimens or strains are intended to be typical specimens or strains that can be compared with other material when classification or identification is undertaken, hence the word "type". However, a moment's thought will show that if a type specimen has to be designated when a taxon is first described and named, this will be done at a time when little has yet been found out about the new group. Therefore, it is impossible to be sure that it is indeed a typical specimen. By the time a completely typical specimen can be chosen, the taxon may be so well known that a type specimen is unnecessary; no one would now bother to designate a type specimen of a bird so well known as the common house sparrow.

The word "type" thus does not mean it is typical, but simply that it is a **reference specimen for the name**. This use of the word "type" is a very understandable cause for confusion that may well repay attention by the taxonomists of the future. For this reason, the *Code* discourages the use of terms like serotype and recommends instead terms formed from -var, e.g., serovar.

In recent years other type concepts have been suggested. Numerical taxonomists have proposed the hypothetical median organism (Liston et al., 1963), or the centroid; these are mathematical abstractions, not actual organisms. The most typical strain in a collection is commonly taken to be the **centrotype** (Silvestri et al., 1962), which is broadly equivalent to the strain closest to the center (centroid) of a species cluster. Some workers have suggested that several type strains should be designated. Gordon (1967) refers to this as the "population concept". One strain, however, must be the official nomenclatural type in case the species must later be divided. Gibbons (1974b) proposed that the official type strain should be supplemented by reference strains that indicated the range of variation in the species, and that these strains could be termed the "type constellation". It may be noted that some of these concepts are intended to define not merely the center but, in some fashion, the limits of a species. Since these limits may well vary in different ways for different characters, or classes of characters, it will be appreciated that there may be difficulties in extending the type concept in this way. The centrotype, being a very typical strain, has often been chosen as the type strain, but otherwise these new ideas have not had much application to bacterial nomenclature.

Type strains are of the greatest importance for work on both classification and identification. These strains are preserved (by methods to minimize change to their properties) in culture collections from which they are available for study. They are obviously required for new classificatory work, so that the worker can determine if he has new species among his material. They are also needed in diagnostic microbiology, because one of the most important principles in attempting to identify a microorganism that presents difficulties is to compare it with authentic strains of known species. The drawback that the type strain may not be entirely typical is outweighed by the fact that the type strain is by definition authentic.

Not all microorganisms can be cultured, and for some the function of a type can be served by a preserved specimen, a photograph, or some other device. In such instances, these are the nomenclatural types, though it is commonly considered wise to replace them by type strains when this becomes possible. Molecular sequences are increasingly being used as important aspects of organisms, and sometimes they assume the functions of nomenclatural types, although they are not yet explicitly mentioned in the *Code*. Authors should, however, bear in mind the limitations of sequences for distinguishing very closely related organisms.

Sometimes types become lost, and new ones (**neotypes**) have to be set up to replace them; the procedure for this is described in the *Code*. In the past it was necessary to define certain special classes of types, but most of these are now not needed.

Types of species and subspecies are type specimens or type strains. For categories above the species, the function of the type—to serve as a point of reference—is assumed by a *name*, e.g., that of a species or subspecies. The species or subspecies is tied to its type specimen or type strain.

Types of genera are **type species** (one of the included species) and types of higher names are usually **type genera** (one of the included genera). This principle applies up to and including the category, order. This can be illustrated by the types of an example of a taxonomic hierarchy shown in Table 2.

The type specimen or type strain must be considered a member of the species whatever other specimens or strains are excluded. Similarly, the **type species of a genus must be retained in the genus even if all other species are removed from it**. A type, therefore, is sometimes called a **nominifer** or **name bearer**; it is the reference point for the name in question.

Descriptions

The publication of a name, with a designated type, does in a technical sense create a new taxon, insofar as it indicates that the author believes he has observations to support the recognition of a new taxonomic group. But this does not afford evidence that can be readily assessed from the bald facts of a name and designation of a type. From the earliest days of systematic biology, it was thought important to describe the new taxon for two reasons: (a) to show the evidence in support of a new taxon, and (b) to permit others to identify their own material with it—indeed this antedated the type concept (which was introduced later to resolve difficulties with descriptions alone).

It is, therefore, a requirement for valid publication that a description of a new taxon is needed. However, just how full the description should be, and what properties must be listed, is difficult to prescribe.

The codes of nomenclature recognize that the most important aspect of a description is to provide a list of properties that distinguish the new taxon from others that are very similar to it, and that consequently fulfill the two purposes of adducing evidence for a new group and allowing another worker to recognize it. Such a brief differential description is called a **diagnosis**, by analogy with the characteristics of diseases that are associated with the same word. Although it is difficult to legislate for adequate diagnoses, it is usually easy to provide an acceptable one; inability to do so is often because insufficient evidence has been obtained to support the establishment of the new taxon. It is generally unwise to propose a new taxon unless one can provide at least a few properties that distinguish it with good reliability from closely similar taxa.

The *Code* provides guidance on descriptions, in the form of recommendations. Failure to follow the recommendations does not of itself invalidate a name, though it may well lead later workers to dismiss the taxon as unrecognizable or trivial. The code for bacteria recommends that as soon as minimum stan-

TABLE 2. An example of taxonomic types

Category	Taxon	Type
Family	*Pseudomonadaceae*	*Pseudomonas*
Genus	*Pseudomonas*	*Pseudomonas aeruginosa*
Species	*Pseudomonas aeruginosa*	ATTC 10145

4. Generic homonyms will be prohibited across all organisms. At present generic names of animals can be the same as those of plants (thus, *Pieris* is a genus of butterflies and a genus of ericaceous plants). Whether this is practicable remains to be seen. It will be easier to achieve when lists of genus names of plants and animals are more complete and are available in electronic form. The two serial publications, *Index Zoologicus* and *Index Nomina Genericorum Plantarum,* are widely available to check animal and plant genus names. The *Bacteriological Code* already prohibits homonyms among procaryotes, fungi, algae, protozoa, and viruses, as noted earlier.
5. There will be some complex rules on the use of synonyms extending above the genus to the rank of family. These are unfamiliar to bacteriologists, and it is not clear how readily they will be accepted.
6. There will be changes in the formal usage of certain terms. Thus, *effective publication* in bacteriology will become simply *publication* and *valid publication* will become *establishment by registration. Legitimate names* will become *acceptable names.* Synonyms will be *homotypic* and *heterotypic* instead of *objective* and *subjective,* respectively. *Priority* will become *precedence,* and *senior* and *junior* names will become *earlier* and *later* names.
7. Prohibition of genus names ending in -myces, -phyces, -phyta, and -virus has been mentioned earlier.

It is evident that revision of the *Bacteriological Code* will be required to achieve the aims of the BioCode, although it will often be possible to make exceptions for bacteriological work. It is to be hoped that such revision will ultimately lead to a version expressed in language familiar to bacteriologists and illustrated by examples from this discipline.

Etymology in Nomenclature of Procaryotes

Hans G. Trüper

I. Introduction

A. Introductory remark When I was invited to write this chapter I felt flattered. I have always been interested in names, in etymology and semantics. The invitation was probably due to more than 25 years of active membership in the International Committee for Systematic Bacteriology (ICSB) and in the Editorial Board of the *International Journal of Systematic Bacteriology*, and there especially my self-adopted task of watching the correctness of new Latin names by offering advice in etymology and questions of procaryote nomenclature. What I write hereafter is an outflow of the experiences I have gathered in these tasks including correspondence in etymological (often intertwined with nomenclatural) matters with hundreds of colleagues. Therefore, I shall try to write this chapter from the viewpoint of the microbiologist—as a user; for the user—rather than writing it *ex cathedra* as a classicist might want to do. Further, what I write here are my own opinions on these matters and they are not meant to offend anyone who has other or better insights.

B. The Latin/Greek thesaurus of words and word elements Scientific terminology, both in technical terminology and in nomenclature, has to fulfill requirements other than those of everyday language. These requirements have been excellently described by the late Fritz C. Werner (1972), a German zoologist.

The first requirement is that every term must unambiguously circumscribe a clearly conceivable idea and that every name stands for a special object or a special group of objects characterized by determined features.

The second requirement is that the total number of different words and word combinations must exceed the large number of discernible objects and abstract concepts, thereby ensuring that names are unambiguous. This is a real challenge as the number of objects, processes and concepts is continuously growing both in depth and breadth because of new scientific and social developments, and changes in nature due to human activities.

As more scientists from a wider range of nationalities participate in these developments, it is important that scientific terms and names fulfill a third requirement, namely universal comprehensibility.

These three requirements—unambiguousness, a large number of possible combinations, and universal usage—are met, to a high degree, by the fact that the terminology of natural sciences and medicine is largely based on the lexicon of classical Greek and Latin. The fact that these so called "dead" languages no longer undergo natural and living changes makes their word material a thesaurus that has been used and may be used further for contemporary needs. Consequently one has more or less arbitrarily given these classical words and word elements certain new meanings. Using a living and constantly changing language in this way would promptly lead to problems and misunderstandings.

Firstly, the use of ancient word material allows the naming of the many new and—in their numbers—permanently increasing objects and concepts for which there are no respective words in contemporary spoken languages; even circumscriptions and combinations of words would hardly suffice. Latin and Greek offer a wealth of word elements and ways to form words that remain inexhausted thus far and are likely to serve our needs for a long time in the future, although scientists have not always been careful or reasonable in their "creations". By mixing Greek and Latin elements, by dropping syllables, repositioning letters, contracting words and creating arbitrary formations, the antique wealth of words has been changed, at times rather significantly. Furthermore, many other languages have contributed, and the names of scientists and other persons have been latinized.

What Werner (1972) did not emphasize was the fact that Latin remained the international language (*lingua franca*) of philosophy, religion, law, sciences, and politics throughout the European Middle Ages and the Renaissance and for philosophical and scientific publications up into the nineteenth century. Its usage, although limited to these circles, led to an enormous increase in vocabulary, usually adopted from other European or oriental languages (e.g., Arabic). It also needs to be mentioned here that Latin has remained the spoken language in the center of the Catholic Church, the Vatican, and is likely to be so into the future. This is particularly well documented by the fact that the *Libraria Editoria Vaticana* takes all efforts to integrate new Latin words coined for modern objects and concepts into the written and spoken Latin of the Vatican. The *Lexicon Recentis Latinitatis*, that appeared 1992 in Italian and 1998 in German, contains about 15,000 new Latin words, "from astronaut to zabaione", word combinations and circumscriptions of the fields of sciences, technology, religion, medicine, politics, sports, and even common idiomatic terms.

The thesaurus of words, enlarged this way, is thus no longer identical with that of either classical language but represents "something new" that has developed along historical lines and follows special contemporary laws of language.

All of the statements made by Werner (1972) apply to general scientific and medical terminology as well as to biological nomenclature. And they apply especially to the scientific nomenclature of procaryotes (eubacteria and archaebacteria) and viruses because these—in contrast to most animals, plants, and larger fungi—do not have popular or vernacular names in any living language because of their usual invisibility.

Nomenclature ("the system of names used in a branch of learning or activity") is an indispensable tool for correct information in our fast growing scientific world with its rapidly developing information networks. The binomial nomenclature used in biosystematics goes back to 1735 when the Swedish botanist Carolus Linnaeus (Karl von Linné, 1707–1778, ennobled 1757) published his famous "Systema Naturae" in Latin according to the scholarly habits of his times.

By introducing the species concept and the use of Latin and Greek for the names of living beings, Linnaeus laid down the principles of modern biological systematics as well as nomenclature. In our "age of informatics" one could certainly think of other ways to name the vast number of plants, fungi, animals, protists and procaryotes, perhaps by a number and/or letter code. For the human brain, however, names are still easier to memorize and work with as part of a system, as long as they are readable and pronounceable.

For the scientific names of procaryotes the *International Code of Nomenclature of Bacteria* (*ICNB, Bacteriological Code*), issue of 1992, is the compulsory compendium of governing Rules. It is the task of the accompanying chapter on nomenclature by P.H. A. Sneath to explain the *Bacteriological Code* (ICNB), whereas this chapter is intended to deal with etymology. Etymology means "origin and historical development of a word, as evidenced by study of its basic elements, earliest known use, and changes in form and meaning" or "the semantic derivation and evolution of a word". "Etymology" is derived from Greek etymon, "the truth" and thus aims at the true, the literal sense of a word.

Etymology is a necessary element in biological nomenclature as it explains the existing (i.e., so far given) names and helps to form new names. For the average microbiologist, "etymology" is that part of a species or genus description that stands first, describes the accentuation, origin and meaning of the name, contains a lot of strange abbreviations and is often considered as superfluous or nasty. I shall come to appropriate examples at the end of the chapter.

In 1993, the late professor of classical languages, Thomas Ozro MacAdoo of Blacksburg, VA, U.S.A., wrote a marvelous chapter on "Nomenclatural literacy" (MacAdoo, 1993) with the intention of helping bacteriologists form correct names. MacAdoo carefully described and examplified the five Latin declensions, the Greek alphabet and its Latin equivalents, the Greek declensions and their Latin equivalents, adjectives and participles, compounding in Latin and Greek, and the latinization of modern proper names. It cannot and will not be my task to equal this excellent and scholarly piece of work, as it contains an introduction to the two classical languages and requires a basic knowledge of, at least, Latin grammar. I highly recommend reading, or better studying, MacAdoo's paper. But I am afraid that I cannot agree with him on the way personal names should be latinized nowadays. (Additional literature recommended as etymological help for the formation of new bacterial names is marked by an asterisk in the further reading list.)

C. Pronunciation and accentuation For many bacterial names the current common pronunciation differs from the pronunciation that is correct according to Latin rules (cf. common text books for Latin). It is unfortunately strongly influenced by the speaker's mother tongue, a clear indication that Latin is no longer the *lingua franca* of the scientific world. Whereas native speakers of languages that are written close to phonetics, such as Italian, Spanish, Portuguese, Dutch, or German, usually pronounce Latin close to its spelling, native speakers of French and especially of English (languages pronounced rather differently from their spelling) often pronounce Latin according to the pronunciation rules of their languages, i.e., further away from the written form. These differences in pronunciation are not generally that important as differences in spelling, because the name in question is often understood despite differences in pronunciation. Substantially helpful here, however, could be to pronounce at least the vowels as they are pronounced in Spanish and Italian, languages whose pronunciations stayed close to their Latin origin. International science will have to live with this problem until the day when all languages are written according to phonetic rules.

In many Central European high schools Latin pronunciation has gone back to the times of Caesar and Augustus when the Romans always pronounced the letter c as the sound k. As a consequence students pronounce, e.g., Caesar "Kaesar" (origin of the German word Kaiser which means emperor) or Cicero "Kikero". In bacteriology this leads to alternate pronunciations of *Acinetobacter*, *Acetobacter*, etc. (as akinetobakter, aketobakter, etc.) by some younger European microbiologists.

I consider it a pity that, for scientific terms used mainly in chemistry and physics, the writing of Greek k remained (keratin, kinetics) whereas in biological nomenclature it has usually, but not always, been latinized to c (*Triceratops*, *Acinetobacter*). Fortunately, classical Latin already introduced the Greek z for transliterated Greek words, and Medieval Latin introduced the letter j for the consonantic i. Meanwhile several names of bacteria starting with J have been proposed (e.g., *Janthinobacter* and the specific epithet *jejuni*). It makes sense to use the j in Latin names as the first letter of a word or word element when it is followed by a vowel.

One significant problem with pronunciation is that of some personal or geographical names used in generic names or specific epithets, e.g., the bacterial generic name *Buttiauxella*, named after the French microbiologist Buttiaux (pronounced: "buttio"). This generic name and specific epithets like *"bordeauxensis"*, *"leicesterensis"*, or *"worcesterensis"* may be pronounced fully (as Latin would require) or pronounced as though they were spelled "buttioella", "bordoensis", "lesterensis", "woosterensis". I am afraid that we will have to leave the decision of pronunciation in such cases to the single scientist, as a rule for such "problems" seems rather difficult to conceive.

Frequently accentuation of Latin names appears to pose problems, especially when Greek word elements are involved. Here, the correct classical accentuation is often not used in bacterial names, e.g., the accepted accentuation of the name *Pseudomonas* is pseu-do-mo′-nas, whereas the classical Greeks would have accentuated the word pseu-do′-mo-nas. An almost universal guideline for accentuation of generic names is, that the syllable next to the last bears the accent. Although this holds for most specific epithets as well, we do tend to encounter other accentuations more often. The practical sense of natural scientists should prevail and the present common usage of accentuation in bacterial names should be the guideline.

II. Formation of Generic Names and Specific Epithets

Since Linnaeus, biological species bear binomial names, consisting of a *genus* (kind) and a *species* (appearance) name. The latter, if taken by itself, is called "specific epithet". A complete species name thus consists of the genus name and the specific epithet. In principle the language of biological nomenclatural names is Latin. In nomenclature, words of Greek origin as well as those of any other origin are handled as Latin, i.e., they have to be "latinized".

Only those bacterial names contained in the *Approved Lists of Names* (Skerman et al., 1980) and the *Validation Lists* that regularly appear in the *International Journal of Systematic Bacteriology* have standing in nomenclature. Regularly updated non-official lists of legitimate bacterial names (except for cyanobacteria described under the Botanical Code) are published by the German Culture Collection DSMZ, Braunschweig, Germany, twice a year. Dr.J.P. Euzéby, Toulouse, France, provides an even more detailed non-official list electronically on the Web site www- sv.cict.fr/ bacterio/.

A. Compound names Compound names are formed by combining two or more words or word elements of Latin and/or Greek origin into one generic name or specific epithet. In most cases two word elements are used (e.g., *Thio/bacillus, thio/parus*), but up to four elements may be found (e.g., *Ecto/thio/rhodo/spira*).

In principle the formation of such combined or compound names is not at all difficult. There are four basic rules to be followed:

1. Except for the last word element, only the stems are to be used.
2. The connecting vowel is -o- when the preceding element is of Greek, it is -i- when the preceding element is of Latin origin.
3. A connecting vowel is dropped when the following element starts with a vowel.
4. Hyphens are not allowed.

In order to avoid later changes, these recommendations (cf. *Bacteriological Code*, Appendix 9 [Lapage et al., 1992]; Trüper, 1996) should be strictly followed, i.e., they should be considered as rules without exceptions.

The reader may protest here and mention, e.g., *Lactobacillus* as being against this ruling. *Lactibacillus* would indeed be the correct name, however, the name *Lactobacillus* is much older than the *Bacteriological Code* and has become a well established name. The ending *-phile* (or *-philic*) in English is often added to words of Latin origin connected by -o- (e.g., acidophile, francophile, anglophile, nucleophile, lactophile etc.). This is due to the meaning of -phile, "friendly to", which commands the dative case. In the most common Latin declension, the second, the dative is formed by adding an -o to the stem (acidophile, friendly to whom/what? friendly to acid). Therefore in bacteriology we have a number of older compound names of Latin origin with the connecting vowel -o-. By unknowingly taking over such originally dative-derived word elements ending on -o, names like *Lactobacillus* came into existence. Such cases prove that Appendix 9 of the *Bacteriological Code* (Lapage et al., 1992) does not have the power of a Rule yet. In the future new name formations of that kind should be avoided.

There are numerous mistakes with respect to compound names. Sometimes authors want to express that their new organism was isolated from a certain part of an animal' s body, e.g., from the throat of a lion; throat is *pharynx* (Greek word stem: *pharyng-*), lion is *leo* (Latin word stem *leon-*). These stems may be correctly combined in two ways: "*pharyngoleonis*" or "*leonipharyngis*". Unfortunately the authors chose *leopharyngis*, which may be corrected to the latter. This example demonstrates the different connecting vowels as well. Two more examples may emphasize the importance of word stems: so *Obesumbacterium* should be corrected to *Obesibacterium*, as the Latin stem of the first component is *obes-*, and the connecting vowel must be -i-. The generic name *Carbophilus* was formed the wrong way, because the stem of the first component is *carbon-*; the correct name would be *Carboniphilus*. For those scientists without training in Latin, a good Latin dictionary indicates the genitive of a noun thereby allowing them to identify the stem of a Latin noun. Typically, the genitive usually shows the stem (e.g., *carbo, carbonis*, the coal) well. MacAdoo (1993) gives a very useful overview on word stems and declensions for non-classicists. An excellent pocket book on word elements (stems) of Latin and Greek origin for usage in scientific terms and names was published by Werner (1972). However, it has only appeared in German to date. An English translation would be of great value for biologists world wide.

Other typical, yet well established misnomers whose connecting vowels were not dropped include *Acetoanaerobium, Cupriavidus, Haloanaerobacter, Haloanaerobium, Haloarcula, Pseudoalteromonas, Streptoalloteichus, Thermoactinomyces, Thermoanaerobacter, Thermoanaerobacterium*, not to speak of numerous equally malformed specific epithets.

B. Generic names The name of a genus (or subgenus) is a Latin noun (substantive) in the nominative case. If adjectives or participles are chosen to form generic names they have to be transformed into substantives (nouns) and handled as such.

Both Latin and Greek recognize three genders of nouns: masculine, feminine, and neuter. Adjectives associated with nouns follow these in gender. For the correct formation of specific epithets (as adjectives) it is therefore necessary to know the gender of the genus name or of its *last* component, respectively.

The more frequent last components in compound generic names of masculine gender are: *-arcus, -bacillus, -bacter, -coccus, -ferax, -fex, -ger, -globus, -myces, -oides, -philus, -planes, -sinus, -sipho, -vibrio*, and *-vorax*; of feminine gender: *-arcula, -bacca, -cystis, -ella, -ia, -illa, -ina, -musa, -monas, -opsis, -phaga, -pila, rhabdus* (*sic*), *-sarcina, -sphaera, -spira, -spina, -spora, -thrix* , and *-toga*; of neuter gender: *-bacterium, -bactrum, -baculum, -bium, - filamentum, -filum, -genium, -microbium, -nema, -plasma, -spirillum, -sporangium*, and *-tomaculum*.

C. Specific epithets As demanded by Rule 12c of the *Bacteriological Code*, the specific (or subspecific) epithet must be treated in one of the three following ways:

1. as an adjective that must agree in gender with the generic name.
2. as a substantive (noun) in apposition in the nominative case.
3. as a substantive (noun) in the genitive case.

Correct examples of these three ways are *Staphylococcus aureus* (adjective: "golden"), *Desulfovibrio gigas* (nominative noun: "the giant"), and *Escherichia coli* (genitive noun: "of the *colum*/colon"), respectively.

1. Adjectives and participles as specific epithets Latin adjectives belong to the first, second, and third declension. Those of the first and second declension have different endings in the three genders, whereas in the third declension the situation is much more complicated, as there are adjectives that don' t change with gender, others that do, and those that are identical in the masculine and feminine gender and different in the neu-

ter. Table 1 gives some representative examples. Note also that comparative adjectives are listed. I recommend always checking an adjective in the dictionary before using it in the formation of a name.

Participles are treated as if they were adjectives, i.e., they fall under Rule 12c, (2), of the *Bacteriological Code.* Infinitive (also named "present") participles in the singular do not change with gender. According to the four conjugations of Latin they end on *-ans* (e.g., *vorans* devouring, from *vorare* to devour), *-ens* (e.g., *delens* destroying, from *delere* to destroy, *deleo* I destroy), *-ens* (e.g., *legens* reading, from *legere* to read, *lego* I read), *-iens* (e.g., *capiens,* from *capere* to seize, *capio* I seize), *-iens* (e.g., *audiens,* from *audire* to listen, *audio* I listen). Note that the knowledge of the ending of the first person singular in the present is decisive!

Perfect participles change their endings with gender and are handled like adjectives of the first and second declension, e.g., *voratus, vorata, voratum* devoured, *deletus, deleta, deletum* destroyed, *lectus, lecta, lectum* (irregular) read, *captus, capta, captum* (irregular) seized, *auditus, audita, auditum,* listened/heard.

2. Nominative nouns in apposition as specific epithets While the above mentioned first and third ways to form specific epithets are generally well understood and usually do not pose problems, the formation of epithets as substantives in apposition has obviously been misunderstood in several cases. So, for instance, when the name *Mycoplasma leocaptivus* was proposed for an isolate from a lion held in captivity, the authors, probably unintentionally, called their bacterium "the captive lion", whereas they wanted rather to explain the origin of their isolate "from a captive lion". Thus "*captivileonis*" would have been the correct epithet.

A nominative noun in apposition does not just mean that any nominative noun may be added to the generic name to automatically become its acceptable epithet. In grammar, apposition means "the placing of a word or expression beside another so that the second explains and has the same grammatical construction as the first"; i.e., the added nominative noun has an explanatory or specifying function for the generic name, like in general English usage "the Conqueror" has for "William" in "William, (called) the Conqueror". Thus *Desulfovibrio gigas* may be understood as *Desulfovibrio dictus gigas* and translated as "*Desulfovibrio,* called the giant", which, with reference to the unusual cell size of this species, makes sense.

Because all specific epithets ending with the Latin suffixes -*cola* (derived from *incola,* "the inhabitant, dweller") and *-cida* ("the killer") fulfill the above-mentioned requirement, they are to be considered correct.

Most legitimate specific epithets formed in bacteriology as nominative nouns in apposition so far have been mentioned and, where necessary, corrected recently (Trüper and De' Clari, 1997, 1998).

Although they are not explicitly ruled out by the *Bacteriological Code,* I have not yet encountered tautonyms, i.e., specific epithets identical with and repeating the genus name, in bacterial nomenclature (such as in zoology *Canis canis,* the dog). In order to avoid confusion, it would be wise to abstain from proposing such names.

3. Genitive nouns as specific epithets The formation of specific epithets as genitive nouns rarely poses problems, as the singular genitive of substantives (nouns) is usually given in the dictionaries.

If the plural genitive is preferred, as, e.g., in *Rhizobium leguminosarum* ("of legumes"), one has to find out the declension of the noun, as plural genitives are different in different declensions. This question will be addressed below.

D. Formation of bacterial names from personal names Persons may be honored by using their name in forming a generic name or a specific epithet. This is an old custom in the whole area of biology. The *Bacteriological Code,* however, strongly recommends to refrain from naming genera (including subgenera) after persons quite unconnected with bacteriology or at least with natural science (Recommendation 10a) and in the case of specific epithets to ensure that, if taken from the name of a person, it recalls the name of one who discovered or described it, or was in some way connected with it (Recommendation 12c).

It is good style to ask the person to be honored by a scientific name for permission (as long as she/he is alive). Authors should refrain from naming bacteria after themselves or co-authors after each other in the same publication, as this is considered immodest by the majority of the scientific community.

The *Bacteriological Code* provides only two ways to form a ge-

TABLE 1. Examples of Latin adjectives

	Masculine	Feminine	Neuter	English translation
first and second declension:	bonus[a]	bona	bonum	good
	aureus[a]	aurea	aureum	golden
	miser	misera	miserum	wretched
	piger	pigra	pigrum	fat, lazy
	ruber	rubra	rubrum	red
	pulcher	pulchra	pulchrum	beautiful
third declension:	puter	putris	putre	rotten
	celer	celeris	celere	rapid
	facilis[a]	facilis	facile	easy
	facilior	facilior	facilius	easier
	maior	maior	maius	more
	minor	minor	minus	less
	simplex	simplex	simplex	simple
	egens	egens	egens	needy

[a]Most common types.

neric name from a personal name, either directly or as a diminutive: Both are always in feminine gender.

Appendix 9 of the ICBN recommends how such names should be formed. Appendix 9 has, however, not the power of the Rules.

The application of the classical Roman rules for name-giving, as was done by MacAdoo (1993), does not make sense as modern names worldwide follow different and various rules and regulations. A differentiation in *prenomina, nomina,* and *cognomina* is therefore no longer applicable and should not be used as a basis for latinization of names nowadays. Principally, modern family names are either *nomina* or *cognomina* in the classical sense. Continuing latinization of names as practiced in ancient Rome would have the advantage that the practice would not change over time. Rather, it would remain fixed. Therefore MacAdoo (1993) would have preferred to establish a uniform rule for latinization of names. But attention must be paid to the fact that since classical times throughout the Middle Ages up into the nineteenth century, (usually learned) people of others than the Roman nation have latinized their names, and thus several varieties of latinization have developed and must be considered as historically evolved. Thus, if such names are not incorrect, they cannot be denied or refused under the *Bacteriological Code* (Appendix 9). I have therefore tried to give the recommended rulings of Appendix 9 (adopted as editorial policy by the *Bergey's Manual* Trust) a simpler and clearer wording and have given examples according to those latinizations that have historically precedence (Trüper, 1996). The results were revised and are compiled in Table 2.

Some personal names in Europe were already latinized before 1800 and kept since then. If they end on *-us,* replace the ending by *-a* or *-ella* (diminutive) respectively (e.g., the name Bucerius would result in "*Buceria*" or "*Buceriella*"). Beware, however, of Lithuanian names like Didlaukus, Zeikus etc.! These are not latinized but genuine forms and would receive the ending *-ia* according to Table 2.

No more than one person can be honored in a given generic name or epithet. In the case of the Brazilian microbiologist Henrique da Rocha Lima, the generic name *Rochalimaea* was formed by dropping the particle *da* and combining his two family names. Combinations of the names of two or more persons cannot be constructed under this aspect. Here the only possibility would be the provision of the *Bacteriological Code* for forming "arbitrary names". These are treated below.

If an organism is named after a person, the name cannot be shortened, e.g., "*Wigglesia*" after Wigglesworth, "*Stackia*" after Stackebrandt or "*Goodfellia*" after Goodfellow etc., but must fully appear. Certainly titles (*Sir, Lord, Duke, Baron, Graf, Conte,* etc.) and particles (*de, da, af, van, von,* etc.) indicating nobility or local origin of the family should not be included in bacterial names, although they may belong to the name according to the laws of the respective country.

Rarely, generic names or specific epithets have been formed from forenames (first names, given names, Christian names), i.e., not from the family name, so the genus *Erwinia* was named after the American microbiologist Erwin F. Smith. The first name Elizabeth appears in *Bartonella* (formerly *Rochalimaea*) *elizabethae.* One could imagine that, in avoiding the usually long Thai family names first names should be chosen in respective cases. Also unusually long double (hyphenated) names like the (hypothetical) Basingstoke-Thistlethwaite or Saporoshnikov-Shindlefrink hopefully do not occur so often among microbiologists as to be honored by a bacterial name (hyphens are not allowed, anyhow!).

One could think of a simplified standard procedure to ease formation of generic names from personal names:

1. All names ending on consonants or *-a* receive the ending *ia,* all others the ending *-a.*
2. Diminutive formation: All names ending on consonants receive the ending *-ella,* all names ending on vowels receive the ending *-nella.*

This simplified scheme should perhaps be recommended by the *Bacteriological Code* as an optional alternative to Appendix 9. Such a ruling should, however, not be introduced with retroactive power as Principle 1 of the *Bacteriological Code* aims at constancy of names.

TABLE 2. Ways to form generic names from personal names (names in quotation marks are hypothetical)

Personal name ending on	Add ending	Person	Example (direct formation)	Diminuitive ending	Example (diminutive formation)
-a	-ea	da Rocha Lima	Rochalimaea	drop a, add -ella	"Rochalimella"
-e	-a	Benecke	Beneckea	-lla	"Beneckella"
	-ia	Burke	Burkeia	-lla	"Burkella"
-i	-a	Nevski	Nevskia	-ella	"Nevskiella"
-o	-a	Beggiato	Beggiatoa	-nella	"Beggiatonella"
	-nia	Cato	"Catonia"	-nella	Catonella
-u	-ia	Manescu	"Manescuia"	-ella	"Manescuella"
-y	-a	Deley	Deleya	-ella	"Deleyella"
-er	-a	Buchner	Buchnera	-ella	"Buchnerella"
	-ia	Lister	Listeria	-iella	"Listeriella"
any consonant	-ia	Cabot	"Cabotia"	-(i)ella	"Cabot(i)ella"
		Wang	"Wangia"	-(i)ella	"Wang(i)ella"
		Salmon	"Salmonia"	-ella	Salmonella
		Escherich	Escherichia	-(i)ella	"Escherich(i)ella"
		Zeikus[a]	"Zeikusia"	-(i)ella	"Zeikus(i)ella"

[a]This name of Lithuanian origin is not a genuine latinized name. If it were so, the genus names "Zeikia" or "Zeik(i)ella" might have been possible.

To form specific epithets from personal names there are, in principle, two possibilities: the adjective form and the genitive noun form. The adjective form has no means of recognizing the sex of the honored person, which, in principle is not necessary for nomenclatural purposes. The personal names receive appropriate endings according to the gender of the generic name as indicated in Table 3. Thus an adjective epithet is formed that has the meaning of "pertaining/belonging to . . . (the person)".

When the genitive of a latinized personal name is formed for a specific epithet, the sex of the person to be honored may be taken into consideration as indicated in Table 4.

On the basis of classical, medieval, and modern usage any of the forms of latinization listed in Table 4 may be chosen. As evident from Table 4 the formation of specific epithets from personal names as genitive nouns poses certain problems only with names ending on *-a* and *-o.*

Classical Roman names of male persons like Agrippa, Caligula, Caracalla, Galba, Seneca, etc. (predominantly *cognomina*) were used in the first declension like the masculine nouns *poeta* (the poet), *nauta* (the sailor), or *agricola* (the land dweller, farmer), regardless of the fact that most of the nouns in this declension are of feminine gender. If bacteria would have been named after these gentlemen, their specific epithets were *agrippae, caligulae, caracallae, galbae,* and *senecae,* respectively. I think that Volta, Migula, and Komagata are dignified successors in this row.

If authors consider it necessary to indicate the sex of the person to be honored, there are several choices, in the following exemplified by the Japanese name Nakamura:

1. Mr. Nakamura is latinized to Nakamuraus, resulting in a specific epithet "*nakamurai*".
2. Mr. Nakamura is latinized to Nakamuraeus (like Linnaeus or my ancestors Nissaeus and Molinaeus), resulting in a specific epithet "*nakamuraei*".
3. Respectively, Ms. Nakamura may be latinized to Nakamuraea resulting in a specific epithet "*nakamuraeae*".
4. Mr. Nakamura is latinized to Nakamuraius, as in MacAdoo's opinion it should be normative (MacAdoo, 1993), resulting an a specific epithet "*nakamuraii*".
5. Respectively, Ms. Nakamura is latinized to Nakamuraia, resulting in a specific epithet "*nakamuraiae*".

By now the reader will understand that possibilities 2–5, although permissible or even recommended by MacAdoo (1993), look and sound rather awkward and are likely to produce numerous misspellings. Therefore I strongly suggest to use the classical version and version 1 only.

Roman names ending on *-o* usually followed the third declension, i.e., the genitive is formed by adding the ending *-nis,* which also reveals that such words have stems ending on n, e.g., Nero/Neronis, Cicero/Ciceronis, or the noun *leo/leonis* (the lion). Medieval Latin followed this custom. So, for the medieval German emperors named Otto the genitive Ottonis was used in writing, which was all in Latin at that time. Therefore it makes

TABLE 3. Formation of specific epithets from personal names in the adjectival form (examples given are hypothetical)

		Add the endings for gender		
Ending of name	Example: family name	masculine	feminine	neuter
consonant	Grant	-ianus	-iana	-ianum
-a	Kondratieva	-nus	-na	-num
-e	Lee	-anus	-ana	-anum
-i	Bianchi	-anus	-ana	-anum
-o	Guerrero	-anus	-ana	-anum
-u	Manescu	-anus	-ana	-anum
-y	Bergey	-anus	-ana	-anum

TABLE 4. Formation of specific epithets from personal names as genitive nouns (hypothetical epithets in quotation marks)

Ending of name	Add for female	Example (female person)	Add for male	Example (male person)
-a	-e (first declension)	Catarina, "catarinae"	-e (classic)	Komagata, komagatae Volta, voltae
			-i	Thomalla, "thomallai"
	-ea	Julia, "juliaeae"	-ei	Poralla, "porallaei"
	-iae	Mateka, "matekaiae"	-ii	Ventosa, "ventosaii"
-e	-ae	Hesse, "hesseae"	-i	Stille, "stillei"
-i	-ae	Kinski, "kinskiae"	-i	Suzuki, "suzukii"
-o	-niae	Cleo, "cleoniae"	-nii	Guerrero, "guerreronii"
			-nis	Otto, "ottonis"
-u	-iae	Feresu, "feresuiae"	-ii	Manescu, "manescuii"
-y	-ae	Macy, "macyae"	-i	Deley, deleyi
-er	-ae	Miller, "millerae"	-i	Stutzer, stutzeri Stanier, stanieri
any other letter	-iae	Gordon, "gordoniae"	-ii	Pfennig, pfennigii Zeikus,"zeikusii"

sense to treat Spanish, Italian, Portuguese, Japanese, Chinese, Ukrainian, Indonesian, as well as all other names that end on -*o* the same way.

Several European names are derived from classical Greek and end on *-as,* such as Thomas, Andreas, Aeneas, Cosmas, etc. In their genitive form, they receive the ending *-ae*: Thomae, Andreae, Aeneae, Cosmae, etc. Although one could argue for a Latinization to Thomasius, Andreasius, etc., to form the specific epithets *thomasii, andreasii,* etc., I would recommend the use of the classical ending *-ae.*

E. Formation of bacterial names from geographical names Authors often consider it necessary to indicate the geographical origin, provenance, or occurrence of their isolates in the respective specific epithets.

Such epithets are simply constructed by adding the ending -*ensis* (masculine or feminine gender) or *-ense* (neuter gender) to the geographical name in agreement with the latter's gender. If the name of the locality ends on *-a* or *-e* or *-en* these letters are dropped before adding *-ensis/-ense* (e.g., *jenensis* from Jena, *hallensis* from Halle, *bremensis* from Bremen). Sometimes authors make the mistake of adding *iensis/-iense.* This is only correct if the locality's name ends on *-ia* (e.g., California leads to *californiensis*). The advice given above guarantees that such mistakes will not happen.

Specific local landscape names such as tundra, taiga, puszta, prairie, jungle (from Sanskrit *jangala*), steppe and savanna may be dealt with in the same way (*tundrensis, taigensis, pusztensis, prairiensis, jangalensis, steppensis* and *savannensis,* respectively).

Epithets on the basis of geographical names may not be formed as substantives in the genitive case, as if they were derived from personal names (e.g., the city of Austin, Texas, cannot lead to "*austinii*" but must lead to "*austinensis*").

Quite a number of localities in the Old World (Europe, Asia, Africa) have classical Greek, Latin, and medieval Latin names and adjectives derived from these: *europaeus, aegyptius, africanus, asiaticus, ibericus, italicus, romanus* (Rome), *germanicus, britannicus, hibernicus* (Ireland), *indicus* (India), *arabicus* (Arabia), *gallicus* (France), *polonicus, hungaricus, graecus* (Greece), *hellenicus* (Hellas, classical Greece), *hispanicus* (Spain), *rhenanus* (Rhineland), *frisius* (Friesland), *saxonicus* (Saxony), *bavaricus* (Bavaria), *bretonicus* (Brittany), *balticus* (Baltic Sea), *mediterraneus* (Mediterranean Sea), etc.

Since the discovery of the other parts of the world by European sailors and travelers, European geographers have continued to give Latin names to "new" continents and countries, so adjectives like *americanus, cubanus, mexicanus,* etc. were introduced. Wherever older adjectives exist they may be used as specific epithets to indicate geographical origins.

European and Mediterranean cities and places of classical times may have had very different names than those in current useage: e.g., *Lucentum* (Alicante, Spain), *Argentoratum* (Strasbourg, France), *Lutetium* (Paris, France), *Traiectum* (Utrecht, Netherlands), *Ratisbona* (Regensburg, Germany), *Eboracum* (York, U.K.), *Londinium* (London, U.K.), *Hafnia* (Copenhagen, Denmark). Microbiologists are free to demonstrate their knowledge of these ancient names but may use epithets derived from the present names of such places, e.g., *alicantensis, strasburgensis, parisensis, utrechtensis, yorkensis, regensburgensis* (MacAdoo, 1993).

Many localities (mostly lakes, rivers, seas, valleys, islands, capes, rocks or mountains, but also some towns or cities) have names that consist of two words, usually an adjective and a substantive (noun), e.g., Deep Lake, Black Sea, Dead Sea, Red River, Rio Grande, Rio Tinto, Long Island, Blue Mountain, Baton Rouge etc., or of two substantives, e.g., Death Valley, Lake Windermere, Loch Ness, Martha's Vineyard, Ayers Rock, Woods Hole, Cape Cod etc. Although such epithets would be correct in the sense of the *Bacteriological Code,* formation of specific epithets from such localities' names may pose a problem, because the use of the adjectival suffix *-ensis, -ense* may lead to rather strange looking or awkward constructions, such as "deeplakensis" or "bluemountainense". If the name of a locality lends itself to translation into Latin, specific epithets may alternatively well be formed as genitive substantives by forming the genitives of the two components and concatenating them without hyphenation, e.g., like the existing ones *lacusprofundi* (of Deep Lake), *marisnigri* (of the Black Sea), *marismortui* (of the Dead Sea), or (of two nouns) *vallismortis* (of Death Valley). Note that in Latin the basic noun comes first, the determining word (adjective or noun) second. If possible one should avoid the inclusion of articles such as the, *el, il, le, la, de, den, het, der, die, das,* or their plurals *los, les, las, ils, gli, le, de, die,* etc. as they are used for locations in several languages, e.g., La Jolla, La Paz, El Ferrol, El Alamein, Le Havre, The Netherlands, Die Schweiz, Den Haag, Los Angeles, etc. Articles would unnecessarily elongate names without adding information.

F. Formation of names for bacteria living in association or symbiosis with other biota An enormous reservoir of bacteria for future research is the microflora that is more or less tightly associated with other biota. I predict that at least two million new species (Trüper, 1992) will be described for the gut flora of various animal species.

Also the plant microfloras have so far been mainly investigated with respect to nitrogen fixation and diseases of economically important plants. To date, little has been done to investigate the phytopathogens that attack economically unimportant plants or weeds.

It is to be expected that microbiologists working in these fields will want to give new isolates names that relate to their hosts or associates.i.e., Latin nomenclatural names of animals, fungi, plants, and protists have been, and to a much larger extent, will be used.

This area of bacterial name-giving is unfortunately full of traps. Clearly, naming a bacterium after a host animal bearing a tautonym (such as *Picus picus,* the woodpecker) is easier than having to choose between generic name and a different specific epithet of the host. It is therefore important to know what these mean and how they were formed (adjective, substantive in genitive, etc.), in order to avoid nasty, ridiculous, or embarrassing mistakes.

The following example may demonstrate this situation: Certainly a bacterium isolated from the common house fly *Musca domestica* should not receive the epithet *domesticus, -a, -um* ("pertaining to the house"); its epithet should rather be *muscae* (of the fly) or *muscicola* (dwelling in/on the fly) the latter being a nominative noun in apposition.

The *domestica* associated with *Musca* is an adjective. If we theoretically consider it an independent noun meaning "the one pertaining to the house" one could, of course, form the genitive from it and thus produce a bacterial epithet *domesticae.* In this example, however, that would not make much sense as too many things "pertain to a house". But formally it would not violate the Rules of the *Bacteriological Code.*

The easiest way of forming such specific epithets is the use of the genitive case of the generic name of the eucaryote in question, e.g., *suis, equi, bovis, muscae, muris, aquilae, falconis, gypis, elephantis* (of the pig, horse, cow, fly, mouse, eagle, falcon, vulture, elephant), or: *fagi, quercus* (fourth declension genitive, spoken with long u), *castaneae, aesculi, rosae, liliae* (of the beech, the oak, chestnut, horse chestnut, rose, lily).

Alternatively the genitive of the plural is recommended, especially if several species of the eucaryotic genus house the bacterial species in question. The formation of the plural genitive needs the knowledge of the stem and declension of the word. The following examples may be of some principal assistance:

1. First declension: *-arum* (*muscarum*, of flies; *rosarum*, of roses)
2. Second declension: *-orum* (*equorum*, of horses; *pinorum*, of pines)
3. Third declension: *-um* (*leonum*, of lions; *canum*, of dogs)
4. Fourth declension: *-um* (*quercuum*, of oaks)
5. Fifth declension: *-rum* (*scabierum*, of different forms of scabies, a skin disease)

Be aware of irregular forms such as *bos* (the cow), genitive: *bovis*, plural genitive *boum*! Use dictionaries and look up the declension in MacAdoo (1993)!

G. Names taken from languages other than Latin or Greek Besides names of persons or localities, many words from languages other than Latin or Greek have been used in bacterial names and certainly will be in the future. Here a few examples may suffice to demonstrate the width and variety of such cases:

During late medieval and renaissance times alchemy became rather fashionable among European scientists and many Arabic words entered into the terminology that would eventually be used in chemistry. One of these, which is often used in bacterial names, is "alkali" (Arabic *al-qaliy*, the ashes of saltwort) from which the element kalium (K, English: potassium) received its name. As the *-i* at the end of the word belongs to the stem it is wrong to speak and write of alcalophilic instead of alkaliphilic microbes. Latinized names of bacteria containing this stem should therefore be corrected to, e.g., *Alkaligenes, alkaliphilus*, etc., and new ones should be formed correctly!

A rather common mistake occurs with the English suffix -philic (e.g., hydrophilic—friendly to water, water-loving). This is clearly an English transformation of the Latin *-philus, -a, -um* (originating from Greek *philos*, friendly). All names formed thus far ending on *-philicus, -a, -um* are wrong and should, in my opinion, be changed to *-philus, -a, -um* as soon as possible. Here, however, Rule 57a (accordance with the rules of Latin) would have to be weighed against Rule 61 (retaining the original spelling) of the *Bacteriological Code*.

National foods or fermentation products often do not have equivalent Latin names and if typical microorganisms found in them or causing their fermentations are described, they have been (and may be) named after them, e.g., sake, tofu, miso, yogurt, kvas, kefir, pombe, pulque, aiva, etc. However, these names cannot be used unaltered as specific epithets in the form of nominative substantives in apposition (Trüper and De' Clari, 1997). They must be properly Latinized. The best way to do so is to form a neuter substantive from them by adding *-um* (e.g., *sakeum, tofuum, kefirum, pombeum*, etc.) and use the genitive of that (ending: *-i*) in the specific epithet (e.g., *sakei, tofui, kefiri, pombei*, etc.)

Another point worth mentioning is the "unnecessary" usage of words from languages other than Greek or Latin. For instance, the formation of the epithet *simbae* from the East African Swahili word *simba*, lion, for a *Mycoplasma* species was not necessary because in this genus the corresponding Latin epithet *leonis* (of the lion) had not been used before.

H. Formation of bacterial names from names of elements and compounds used in chemistry and pharmacy The almost unlimited biochemical capacities of bacteria is another rather inexhaustible source for new names. Many generic names, as well as specific epithets, have been formed from names of chemical elements, compounds and even pharmaceutical and chemical products or their registered or unregistered trade names.

The late Robert E. Buchanan (1960, reprinted 1994) listed numerous examples of such generic names and specific epithets. Based on the classical Latin/Greek thesaurus and enriched by numerous Arabic words, the pharmaceutical sciences have, since the Middle Ages, developed a Neo-Latin terminology for chemicals of all categories.

The vast majority of names of chemicals are latinized as neuter nouns of the second declension with nominatives ending *-um*, genitives in *-i*. The following groups belong in this category:

1. Most of the chemical elements with the exception of carbon (L. *carbo, carbonis*), phosphorus (L. *phosphorus, phosphori*), and sulfur (L. *sulfur, sulfuris*) have the ending -(*i*)*um*); nitrogen may also be called *azotum* besides *nitrogenium*, calcium may also be called *calx* (genitive: *calcis*).
2. Chemical and biochemical compounds ending on *-ide* (anions), *-in, -ane, -ene, -one, -ol* (only non-alcoholic compounds), *-ose* (sugars), *-an* (polysaccharides), *-ase* (enzymes) (*-um* is added, or the *-e* at the end is replaced by *-um*, respectively).
3. Acids are named by *acidum* (L. neuter noun, acid), followed by a descriptive neuter adjective, e.g., sulfurous acid *acidum sulfurosum*, sulfuric acid *acidum sulfuricum*, acetic acid *acidum aceticum*.

The second largest category of chemicals are treated as neuter nouns of the third declension: these end on *-ol* (the alcohols), *-al* (aldehydes), *-er* (ethers, esters), and *-yl* (organic radicals); latinization does not change their names at the end, whereas the genitive is formed by adding *-is*.

Anions ending in *-ite* and *-ate* are treated as masculine nouns of the third declension. The English ending *-ite* is latinized to *-is*, with the genitive *-itis*, e.g., nitrite becomes *nitris, nitritis*. The English ending *-ate* is latinized to *-as*, with the genitive *-atis*, e.g., nitrate becomes *nitras, nitratis*.

Only few chemicals have names that are latinized in the first declension as feminine nouns, ending on *-a* with a genitive on *-ae*. Besides chemicals that always had names ending on *-a* (like urea), these are drugs found in classical and medieval Latin, such as gentian (*gentiana*) and camphor (*camphora*), further modern drugs, whose Latin names were formed by adding *-a*, like the French ergot becoming *ergota* in Latin.

The most important group of this category are alkaloids and other organic bases, such as nucleic acid bases and amino acids with English names on *-ine*. In Neo-Latin this ending is *-ina*, with the genitive *-inae*, e.g., *betaina, -ae*; *atropina, -ae*; *adenina, -ae*; *alanina, -ae*; etc.

For their use in bacterial generic names and specific epithets word stems and genitives of latinized chemical names are the basis. In principle they are then treated like any other word elements.

I. Arbitrary names Either genus names or specific epithets "may be taken from any source and may even be composed in an arbitrary manner" (*Bacteriological Code*, Rule 10a and Rule 12c). They must, however, be treated as Latin. These "rubber" paragraphs open up a box of unlimited possibilities for people whose Latin is exhausted. But in view of the million names that will have to be formed in the future they are a simple necessity, whether Latin purists like them or not.

Examples for arbitrary generic names are *Cedecea, Afipia*, and in the near future "*Vipia*" and "*Desemzia*", that were derived from the abbreviations CDC (Center for Disease Control), AFIP, VPI (Virginia Polytechnical Institute), and DSMZ (Deutsche Sammlung von Mikroorganismen und Zellkulturen), respectively. Examples for arbitrary specific epithets are, e.g., (*Salmonella*) *etousae*, derived from the abbreviation ETOUSA (European Theater of Operations of the U.S. Army), and (*Bacteroides*) *thetaiotaomicron*, formed from the three Greek letter names *theta, iota*, and *omicron*.

More recently, the new genus *Simkania* was described. The name is a latinized contraction of the first and the family name of the microbiologist Simona Kahane. Certainly an arbitrary name, short, elegant and easy to pronounce, points to future possibilities of bacterial name-giving. Authors should aim at such easily spelled and pronounced short names, when they take advantage of arbitrary name-giving.

III. Some case histories of malformed names

From the viewpoint of classical Latin many of the existing bacterial names are, plainly said, lousy in their grammar and etymology. However, under the Rules of the *Bacteriological Code* they are acceptable. A few case histories of wrong bacterial names are worth mentioning in a chapter on etymology because of their scurrility.

Acetobacter xylinus: This specific epithet goes back to Brown 1886, who described a *Bacterium xylinum*. Several subsequent changes of the genus (Trevisan 1889, *Bacillus xylinus*; Ludwig 1898, *Acetobacterium xylinum*; Pribram 1933, *Ulvina xylina*) prove by the change in gender that the epithet is an adjective. Because before 1951 (*Bacteriological Code*, Opinion 3), the gender of names ending in *-bacter* was not fixed as masculine, *Acetobacter xylinum* (Holland 1920 and Bergey et al. 1925) (all names and dates before 1950 cited were taken from *Index Bergeyana*, Buchanan et al., 1966) was not wrong either. As a consequence of Opinion 3 the species should be named *Acetobacter xylinus*. The *Approved Lists* of names (Skerman et al., 1980), however, listed the organism as *Acetobacter aceti* subspecies *xylinum*! Yamada (1983) revived the species status and correctly called it *A. xylinus*. The compiler of Validation List 14 (*International Journal of Systematic Bacteriology*, 1984) incorrectly put a *sic* after *xylinus* and changed it to the neuter form *xylinum*! (The Latin expression *sic* is used to point out a mistake or other peculiarity.) Unexpectedly the previous authors obeyed this falsifying change and even tried to give the neuter epithet justification by explaining it as a nominative noun in apposition (*xylum*, M.L. neut.n. cotton). "*Acetobacter*, called the cotton" makes little sense and certainly does not meet the requirements of a nominative noun in apposition (cf. Trüper and De' Clari, 1997), Finally, Euzéby (1997) corrected the name to *A. xylinus*.

Methanobrevibacter arboriphilus: In 1975 the new species *Methanobacterium arbophilicum* was described. The organism was isolated from rotting trees and the authors wanted to express "friendly to trees" by the epithet. In Latin, tree is *arbor*, genitive *arboris*, ie., the stem is clearly *arbor-*, not *arbo-*. The second error was that the English ending *-philic* was latinized to *-philicum* instead of correctly to *-philum*. Although this was first pointed out to the authors in 1976, they did not correct the epithet themselves. Then, in a review paper, Balch et al. (1979) rearranged the methanogenic procaryotes and transferred the species to the genus *Methanobrevibacter* as *M. arboriphilus* (the correct form of the epithet). It was again the compiler of the Validation List No. 6 (*International Journal of Systematic Bacteriology*, 1981), who created a new wrong form of the epithet, *arboriphilicus*! Although immediately informed of his error, he did not correct it. And so this wrong epithet still occurred in *Bergey's Manual of Systematic Bacteriology*, Vol. 3 (1989). To my knowledge it has not been corrected!*

Some time ago an author wanted to create the specific epithet "*nakupumuans*" and explained this word as derived from the Maori word *nakupumua*, breaking protein down to fragments.

Becoming informed that there was neither need to use another language than Latin, nor any specific connection between the Maori and protein degradation the author decided to call the isolate *proteoclasticum*. Accepting such name formations in procaryote nomenclature would mean giving up Latin as the basic language of biological nomenclature. As long as names can be formed from the Latin/Greek thesaurus at our hands, names from other languages should be avoided.

In another instance, an author wanted to propose a specific epithet in honor of a colleague and formed an epithet ending in *-icus*. As this is not within the Rules, I advised him to choose either an epithet ending on *-ii* (genitive noun) or on *-ianus* (adjective). His answer was that he did not like the former and felt that the latter sounded like an insult to the colleague to be honored!

Another colleague correctly formed the generic name *Acidianus* (accentuation: a.cid.ia'nus) from the Latin neuter noun *acidum*, acid and the Latin masculine noun *Ianus*, the Roman god with the two faces, by which he wanted to indicate the ability of the organism to both oxidize and reduce elemental sulfur. With this spelling the epithet promptly became mispronounced (a.ci.di.a'nus) suggesting a different meaning and causing suggestive jokes. Here the use of the consonantic i, (i.e., j) would have sufficed to suppress the misinterpretation: *Acidijanus* would be the choice.

These examples also show that nobody is free from making mistakes. During my work in this field I have made several, and sometimes even given wrong advice, quite to my embarrassment afterwards.

IV. Practical etymology in descriptions of genera and species

As mentioned before, for the average microbiologist "etymology" is a kind of nasty linguistic exercise necessary for the description of a new genus or species. In reality he/she has to "create" a new name; the organism has been isolated and determined by the author, not "created"! The better and more modest wording would be, to "propose" a new name.

On the basis of six examples of such "etymologies" I shall try to explain how these are composed.

1. *Escherichia coli*: Esch.er.i'chi.a (better: E.sche.ri'chi.a) M.L. fem.n. *Escherichia*, named after Theodor Escherich, who iso-

---

**Editorial Note:* As of January 2000, this name still appears on the Approved List. No action to correct the name has been taken.

lated the type species of the genus. co'li Gr.n. *colon* large intestine, colon; M.L. gen.n. *coli* of the colon.

2. *Rhodospirillum rubrum*: Rho.do.spi.ril'lum Gr.n. *rhodon*, the rose; M.L. dim neut.n. *Spirillum*, a bacterial genus; M.L. neut n. *Rhodospirillum*, a red *Spirillum*. (Etymology of the latter: Gr. n. *spira*, spiral, M.L. dim. neut n. *Spirillum*, a small spiral.) rub'rum.L. neut. adj. *rubrum*, red.
3. *Azotobacter paspali*: A.zo.to.bac'ter French n. *azote*, nitrogen; M.L. masc.N. *bacter*, the equivalent of Gr. neut.n. *bactrum*, a rod or staff.M.L. masc n. *Azotobacter*, nitrogen rod. pas.pal'i (better: pas.pa'li). M.L. gen n. *paspali*, named for *Paspalum*, generic name of a grass.
4. *Pseudomonas fluorescens*: Pseu.do.mo'nas (seldom: Pseu.do'mo.nas). Gr. adj. *pseudos*, false; Gr.n. *monas*, a unit; M.L. fem.n. *Pseudomonas*, false monad. flu.o.res'cens.M.L. v. *fluorescere* (*fluoresco*), fluoresce; M.L. part adj. *fluorescens*, fluorescing.
5. *Desulfovibrio gigas*: De.sul.fo.vi'brio (or: De.sul.fo.vib'rio). L. pref. *de*, from; L.n. *sulfur*, sulfur; L.v. *vibrare*, vibrate; M.L. masc.n. *Vibrio*, that which vibrates, a bacterial generic name; M.L. masc.n. *Desulfovibrio*, a vibrio that reduces sulfur compounds. (Note: If we were meticulous, the name should either be "*Desulfativibrio*" referring to sulfate, or "*Desulfurivibrio*" referring to sulfur. As *Desulfo-* may cover both, in this case it is certainly the best name for the genus!) gi'gas.L. nom.n. *gigas*, the giant.
6. *Thermoanaerobium aotearoense*: Ther.mo.an.ae.ro'bi.um. Gr. adj. *thermos*, hot; Gr. pref. *an-*, without; Gr.n. *aer*, air; Gr.n. *bios*, life; M.L. neut.n. *Thermoanaerobium*, life in heat without air. a.o.te.a.ro.en'se. Maori n. *Aotearoa*, New Zealand; L. neut. suffix *-ense*, indicating provenance; M.L. neut. adj. *aotearoense*, from or pertaining to Aotearoa (New Zealand).

From these examples several regularities can be deduced:

1. After the name or epithet the "etymology" starts with an indication of accentuation. The word is broken into a row of syllables interrupted by periods. The accent-bearing syllable is indicated by an accent sign behind it (note: never before it!) instead of a period. The classical Latin language did not develop explicit rules about breaking up words into syllables; the Romans broke written words the way they were spoken, and logically split compound words between compounds. As the rules for breaking words into syllables are different for different modern languages, in my opinion, one should continue to follow the Roman custom rather than the rules for any modern language.
2. The accentuation is followed by the etymology proper of the name. The abbreviations commonly in use indicate the language of origin (Gr. classical Greek, L. classical Latin, M.L. modern Latin), the type of word or word element (adj. adjective, n. noun/substantive, v. verb, part. adj. participle used as adjective, dim. diminutive, pref. prefix, suff. suffix), the case (gen. genitive, nom. nominative, the latter being seldom indicated) and the gender of nouns or adjectives (fem. feminine, masc. masculine, neut. neuter).
3. The word elements are explained in the sequence they occur in the name. Then, like a summary, the language, gender, and the word type of the complete name or epithet is given, followed by the Latin name and its translation.

The abbreviation M.L. is very often misunderstood as medieval Latin. I personally would therefore prefer a ruling that M.L. would really mean medieval Latin and that modern Latin, better Neo-Latin, would be abbreviated N.L.

V. Recommendations (from the Viewpoint of Language) for Future Emendations of the *Bacteriological Code*

We should not aim for pure classical Latin in biological nomenclature but rather develop the current Latin/Greek thesaurus further by following the Rules of the ICNB or the respective codes of nomenclature applicable to other fields of biology. This is in reality what has happened since Linnaeus' time. In my opinion the ICNB has excellent provisions to do so. This is already documented by the low number of Opinions that had to be issued by the Judicial Commission of the ICSB during the last ten years.

For several years the development of a uniform code of nomenclature for all biological taxa has been underway, enlisting the participation of well-known taxonomists from bacteriology, botany, mycology, phycology, protozoology, virology, and zoology. This effort has received the support of the International Unions of Biological and Microbiological societies, IUBS and IUMS (Hawksworth and McNeill, 1998). These activities reflect the general scientific need to assess the total extent of biodiversity on Earth, in order to facilitate conservation and, perhaps, prevent further extinction of the biota. For this purpose a unified system of biological names has been considered indispensable. Drafts of the future universal "BioCode" have been published, the latest (fourth) draft by Greuter et al. (1998). As soon as the BioCode is accepted by the taxonomic committees of the different biological disciplines involved, the *Bacteriological Code* will have to be revised to conform with any new recommendation. Changes in etymological rulings should be expected. Unfortunately the recommendations for latinization (Articles 37–39) are not yet formalized, therefore comments and recommendations cannot be offered at this time.

Besides the cases mentioned in the text above, where certain changes or simplifications have been recommended, there are a few other points where, in my opinion, the Rules need further development with respect to etymology:

1. Stronger emphasis should be put on short and easily pronounceable names.
2. Words from languages other than Latin or Greek should be banned as long as an equivalent exists in Greek or Latin or can be constructed by combining word elements from these two languages, and as far as they are not derived from names of geographical localities or local foods or drinks (e.g., sake, kefir, kvas, pombe, tofu, miso, yogurt, etc.), for which no Latin/Greek names exist.
3. Formation of bacterial names on the basis of latinized names of chemical compounds should be regulated under the Code. Here the recommendations of Buchanan (1994), as explained above, should be the basis.
4. The principal ban on ordinal numbers (adjectives) for the formation of bacterial names (ICNB, Rule 52, -2-) only makes sense for those numbers above ten because of their length. Therefore, this part of Rule 52 should be abandoned.
5. In the transliteration of the Greek letter k to the Latin letter c the k sound is lost when the vowels e, i, or y follow. Instead the c is pronounced as a sharp s as in English. Therefore, to preserve the k sound before e, i, and y, the letter K should be kept even in the Latin transliteration (example: *A<u>k</u>inetobacter* as in kinetics instead of *Acinetobacter*).

6. Authors should refrain from naming bacteria after themselves or coauthors after each other in the same publication, as this is considered immodest by the majority of the scientific community.
7. Generic names and specific epithets formed from personal names can only contain the name of one person, not a combination or contraction of the names of two or more persons.
8. In the future, bacteriologists (including those that work on archaebacteria and cyanobacteria) should avoid names that end on *-myces* or *-phyces* in order to avoid confusion with mycology and phycology, i.e., with eucaryote nomenclature. Articles 25–28 of the future BioCode (Greuter et al., 1998) will forbid procaryote names ending in *-myces, -phyta, -phyces,* etc. or in *-virus.*
9. In the etymology given with the description of a taxon, there should be an indication whether a Latin name is from classical Latin ("L.") or Greek ("G."), from a medieval Latin ("M.L.") source or formed as Neo-Latin ("N.L."). This will save time for those who want to look up such names and words in dictionaries, and it will end ambiguous interpretation of M.L. as either "modern" Latin or medieval Latin. Already Buchanan (1960, reprinted 1994) prefered "Neo-Latin" over "modern" Latin.

Acknowledgments

I wish to thank Eckhard Bast (Bonn), Jean P. Euzéby (Toulouse), Lanfranco de' Clari (Lugano), Roy Moore (Ulster), and Bernhard Schink (Konstanz), for their extremely helpful correspondence and discussions on etymology of bacterial names, and the Fonds der Chemischen Industrie for financial support.

Further Reading

*Bailly, A. (Editor).1950. Dictionnaire Grec-Francais, Hachette, Paris.

Balch, W.E., G.E. Fox, L.J. Magrum, C.R. Woese and R.S. Wolfe. 1979. Methanogens: reevaluation of a unique biological group. Microbiol. Rev. *43*: 260–296.

*Brown, W.B. 1956. Composition of Scientific Words - A Manual of Methods and a Lexicon of Materials for the Practice of Logotechnics, Brown, pp. 882

*Buchanan, R.E. 1994. Chemical terminology and microbiological nomenclature. Int. J. Syst. Bacteriol. *44*: 588–590.

Buchanan, R.E., J.G. Holt and E.F. Lessel (Editors). 1966. Index Bergeyana, The Williams & Wilkins Co., Baltimore.

*Calonghi-Badellino, G. (Editor).1966. Dizionario della lingua latina, Rosenberg and Sallier, Torino.

*Diefenback, L. 1857. Glossarium Latino-Germanicum Mediae et Infirmae Aetatis, Frankfurt.

*Egger, C. (Editor).1992. Lexicon Recentis Latinitatis (Italian/Latin), Libraria Editoria Vaticana, Rome.

Euzéby, J.P. 1997. Revised nomenclature of specific or subspecific epithets that do not agree in gender with generic names that end in -bacter. Int. J. Syst. Bacteriol. *47*: 585.

*Farr, E.R., J.A. Leussink and F.A. Stafleu. 1979. Index Nominum Genericorum (Plantarum), Sheltema and Holkema, Utrecht.

*Feihl, S., C. Grau, H. Offen and A. Panella (Editors). 1998. Neues Latein Lexikon, translated from Italian, Libraria Editoria Vaticana, Bonn.

Greuter, W., D.L. Hawksworth, J. McNeill, M.A. Mayo, A. Minelli, P.H.A. Sneath, B.J. Tindall, P. Trehane and P. Tubbs. 1998. Draft Biocode (1997): the prospective international rules for the scientific naming of organisms. Taxon. *47*: 129–150.

*Habel, E. and F. Gröbel. 1989. Mittellateinisches Glossar, Schöningh Verlag, Paderborn.

Hawksworth, D.L. and J. McNeill. 1998. The International Committee on Bionomenclature (ICB), the draft BioCode (1997), and the IUBS resolution on bionomenclature. Taxon. *47*: 123–136.

*Lapage, S.P., P.H.A. Sneath, E.F. Lessel, V.B.D. Skerman, H.P.R. Seeliger and W.A. Clark (Editors). 1992. International Code of Nomenclature of Bacteria (1990) Revision. Bacteriological Code, American Society for Microbiology, Washington, D.C.

*Lewis, C.T. and C. Short (Editors). 1907. A New Latin Dictionary, American Book Company, New York.

*Liddell, H.G., R. Scott, H.S. Jones and R. McKenzie (Editors). 1968. A Greek- English Lexicon, Oxford University Press, Oxford.

*MacAdoo, T.O. 1993. Nomenclatural literacy, *In* Goodfellow, M. and A.G. O' (Editors), Handbook of New Bacterial Systematics, Academic Press Ltd., London. pp. 339–360.

*Noel, F. (Editor).1833. Dictionarium Latino-Gallicum, Le Normant, Paris.

*Simpson, D.P. (Editor).1959. Cassell' s New Latin Dictionary, Funk and Wagnalls, New York.

Skerman, V.B.D., V. McGowan and P.H.A. Sneath. 1980. Approved lists of bacterial names. Int. J. Syst. Bacteriol. *30*: 225–420.

*Stearn, W.T. (Editor).1983. Botanical Latin, David and Charles, Devon.

Trüper, H.G. 1992. Prokaryotes: an overview with respect to biodiversity and environmental importance. Biodivers. Conserv. *1*: 227–236.

*Trüper, H.G. 1996. Help! Latin! How to avoid the most common mistakes while giving Latin names to newly discovered prokaryotes. Microbiologia *12*: 473–475.

Trüper, H.G. and L. de'. 1997. Taxonomic note: Necessary correction of specific epithets formed as substantives (nouns) "in apposition". Int. J. Syst. Bacteriol. *47*: 908–909.

Trüper, H.G. and L. de'. 1998. Taxonomic note: erratum and correction of further specific epithets formed as substantives (nouns) "in apposition". Int. J. Syst. Bacteriol. *48*: 615.

*Werner, F.C. (Editor).1972. Wortelemente Lateinisch-Griechischer Fachausdrücke in den Biologischen Wissenschaften, 3rd ed., Suhrkamp Taschenbuch Verlag, Berlin.

*Woodhouse, S.C. (Editor).1979. English-Greek Dictionary: A Vocabulary of the Attic Language, Routledge and Kegan Paul, London.

Yamada, Y. 1983. *Acetobacter xylinus* sp. nov., nom. rev., for the cellulose-forming and cellulose-less acetate-oxidising acetic acid bacteria with the Q-10 system. J. Gen. Appl. Microbiol. *29*: 417–420.

*Yancey, P.H. 1945. Origin from mythology of biological names and terms. Bios. *16*: 7–19.

Microbial Ecology—New Directions, New Importance

Stephen H. Zinder and Abigail A. Salyers

Introduction: Microbial Ecology—The Core That Links All Branches of Microbiology

Microbial ecology is the study of microorganisms in their natural habitats. In these habitats, they are rarely in pure culture and are usually interacting with other microorganisms, are sometimes interacting with host organisms, and are always interacting with their physicochemical environment. These conditions are usually very different from those used to grow microorganisms in pure culture in the laboratory. Since *Bergey's Manual* is a compendium of properties of pure cultures of procaryotes, it might appear that a discussion of microbial ecology is inappropriate. However, ecological studies have a profound effect on our understanding of pure cultures, and this impact will become more important in the twenty-first century. This chapter will not give a comprehensive overview of microbial ecology, but will, instead, discuss the relevance of microbial ecology to pure culture studies and vice versa.

Microbial ecology has a long history that reaches back to Antony van Leeuwenhoek's microscopic observations of microbial populations in various habitats including rainwater, dental plaque, and feces. Until the late nineteenth century, when the techniques developed by Louis Pasteur and Robert Koch allowed new approaches to be taken, the microscope was essentially the only tool available to study microorganisms, and only natural populations of microorganisms could be studied. Many of Pasteur's early studies on fermentations, spontaneous generation, and the distribution of microorganisms in air had an ecophysiological bent, describing phenomena such as the effect of oxygen on species composition and metabolism. Pasteur eventually used the techniques and concepts he developed in these studies to investigate pathogenesis.

Pure culture microbiology began in the late nineteenth century with the development of isolation techniques, particularly the use of semisolid agar media by Robert Koch. Koch's postulates demanded isolation as an essential step in proving microbial causation of a disease. Koch and his followers, the "microbe hunters", took center stage in microbiology in the first half of the twentieth-century. They isolated and characterized nearly all of the important pathogens, leading to the almost complete elimination of infectious diseases from the Western world through better sanitation and use of vaccines and antibiotics, an achievement that is certainly one of the great triumphs of twentieth-century science.

In the latter half of the twentieth century, molecular biologists took center stage in microbiology, working mainly with *Escherichia coli.* They defined genes and operons, mapped their positions on the chromosome, and studied the regulation of their expression. These studies culminated in the recent determination of the entire genome sequence of *E. coli* K12, as well as those for over 25 other microorganisms at the time this volume went to press, with many more microbial genome sequences in the offing. The molecular characterization of *E. coli* and other bacteria is also a major landmark of twentieth-century science.

Thus, the twentieth century could be considered the "Age of the Pure Culture". Working in a reductionist style with pure cultures was extremely successful and therefore very seductive. Many microbiologists came to believe that only work with pure cultures or with macromolecules could be good science, and forgot the communities from which their microbe had been taken.

During the twentieth century, most microbial ecologists worked mainly at agricultural and technical schools. Their work was often directed towards applied areas such as soil microbiology related to agriculture or the environment. The dearth of researchers, the prevalence of applied rather than fundamental research, relatively low levels of funding, and, as will be described presently, formidable technical difficulties in studying microbial ecology, all contributed to this field lagging behind pure culture microbiology.

Starting in the 1980s, a series of unpleasant surprises on the disease front brought clinical microbiologists face to face with the fact that microbial ecology was indeed central to their interests. The emergence of new diseases such as Lyme disease, AIDS, and ehrlichosis, and the reemergence of old diseases such as cholera in South America and tuberculosis in the United States, demonstrated that changing human practices (human ecology) could create new windows of opportunity for microbes and that understanding the way that they moved into new niches (microbial ecology) was critical for controlling further spread of diseases. Large outbreaks of salmonellosis and *E. coli* O157:H7 raised anew questions concerning the factors that control colonization of animals by these pathogens, and whether the normal microbiota of animals and plants could be manipulated to decrease colonization opportunities. Increased concern about antibiotic resistance, the study of which had long been dominated by molecular biologists, led to the realization that all of the practical questions about how to control the spread of resistance were centered instead on the ecology of resistance—how genes were spreading in various microbial communities. In general, the realization that it was better to prevent disease than intervene after the disease had established a foothold gave new impetus to understanding where disease-causing organisms are normally found, how they fit into their normal ecological niches, and how they adapt to new niches that in some cases were quite different from their usual ones.

Another humbling finding was that, despite decades of intensive research on *E. coli* K12, a function could not be ascribed to 38% of the genes in its genome (Blattner et al., 1997). The unknown genes probably encode functions that help *E. coli* to live in habitats as diverse as the human intestinal tract and freshwater creeks and to make a living under conditions more demanding than those experienced growing in Luria broth or even in a chemostat. Moreover, evidence has been obtained that a considerable fraction (at least 18%) of the genes in the *E. coli* K12 genome were transferred from other organisms (Lawrence and Ochman, 1998). Finally, several *E. coli* strains have genomes several hundred million bases larger than that of strain K12 (Bergthorsson and Ochman, 1995), indicating that we have much to learn about this species which we thought we knew so well. Thus, functional genomics also leads us into microbial ecology.

Meanwhile, environmental microbiologists, who had identified themselves all along as microbial ecologists, began to turn away from characterizing steps in pathways using biochemical analysis of pure cultures, and returned to asking questions about how such pathways operated in nature. At one time such questions would have seemed futile because of the complexity of microbial communities and the suspicion that there remained many uncultivated microbes. Microbial ecologists were also beginning to look at familiar environments in new ways. Bacteriophages began to be recognized as important predators of bacteria in some settings. Horizontal gene transfer assumed new prominence as ecologists began to realize that bacteria in a complex community could interact sexually as well as metabolically, and that gene transfer can even occur between eubacteria and archaea (Doolittle and Logsdon, 1998). The discovery of syntrophic interactions, in which two microbes work together to carry out a reaction that is thermodynamically impossible for one organism (Schink, 1997), opened up a new dimension in metabolic interactions. Indeed, the paper originally describing the resolution of *"Methanobacillus omelianskii"* into two syntrophic organisms (Bryant et al., 1967) was considered to be one of the 100 most important in twentieth-century microbiology (Joklik, 1999). Report after report appeared of microbes that could carry out reactions previously thought to be improbable, if not impossible: anaerobic breakdown of aromatic (Evans et al., 1991) and aliphatic (Aeckersberg et al., 1998) hydrocarbons, "fermentation" of inorganic sulfur compounds (Bak and Pfennig, 1987), utilization of chlorinated organics as respiratory electron acceptors (Mohn and Tiedje, 1992), and methanogenesis in aerobic methane-oxidizing bacteria, using enzymes from the methanogenesis pathway of archaea in the reverse direction (Chistoserdova et al., 1998).

Microbiologists from many different areas have begun to rediscover microbial communities and to recall that the conditions under which microbes normally live are very different from those used to grow them in the laboratory. The pendulum began to swing back to a position where pure culture studies were declared by some to be unscientific and inappropriate, and community analysis became the imperative (Caldwell, 1994). While this represents an extreme position that few would advocate, microbial communities have been neglected too long, and the time is ripe for their study. Moreover, as new technologies have been introduced, another need has become evident—the need for more sophisticated models and theories about community structure and interactions among members of the community. Just as the availability of genome sequences has challenged scientists working on individual microbes to find creative new ways to use this information, the availability of molecular tools for analyzing microbial communities calls for conceptual advances that will make maximal use of new technologies.

Classical Microbial Ecology

Microbial ecology, as practiced through most of the twentieth century, employed nonmolecular biological tools to study natural microbial populations. These consist mainly of activity measurements, biomass measurements, microscopy, and cultivation techniques (Atlas and Bartha, 1993). When applied to procaryotes, all of these techniques suffer from limitations that are mainly due to the small size of these organisms and the complexity of their environments. For example, a single 1-mm crumb of soil contains microhabitats that are aerobic, anaerobic, wet, dry, organic-rich, organic-poor, acidic, and basic. Thus methodological problems arising from the microenvironment are particularly formidable.

As an example of this, consider some commonly used ways of measuring various activities in microbial populations. A compound, sometimes isotopically labeled, is added to the environment. If the compound is at its natural concentration, a chemical transformation can be measured to estimate the rate of that process. Alternatively, the metabolic potential for that process can be determined if a higher concentration is used. A problem with these methods when applied to microbial populations is that they are essentially bulk measurements. For example, one can measure the rate of sulfate reduction in a sample, but several different populations of sulfate reducers may be contributing to that rate. Kinetic measurements and analysis or inhibition studies may provide more fine structure information about the process, and clever use of microelectrodes (Fossing et al., 1995) or microautoradiography (Krumholz et al., 1997) can give spatial information on an appropriate scale for that process. Still, more often than not, the information we obtain from these studies is of low resolution. Moreover, natural microbial populations are often perturbed during sampling, by such processes as mixing or simply placing them in a vial, so that delicate spatial relationships are destroyed.

Microbial biomass can be estimated by a variety of bulk techniques in which some cell constituent such as organic matter, protein, or chlorophyll is extracted and quantified from microbial populations. Measurement of amount of ATP and other nucleotides can give an estimation of the active biomass (Karl, 1980). More specific methods based on quantifying lipids, including those considered "signatures" of various microbial groups such as archaea and eubacterial methylotrophs, have been developed (Hedrick et al., 1991), but their suitability for application to complex natural microbial habitats is uncertain.

Microscopy remains an extremely important technique in microbial ecology, especially since it brings the researcher down to the scale of the microenvironment. Particularly useful are fluorescent microscopical methods, such as staining with nucleic acid specific stains such as acridine orange or DAPI (Amann et al., 1995). A problem with microscopic observation of procaryotic cells is that they are, with certain exceptions, morphologically nondescript to our eyes, so that we cannot simply identify them by looking at them the way we can plants and animals. Even using electron microscopy, many procaryotes are indistinguishable. In some cases, fluorescent antibodies have been useful in identifying microbes *in situ*, but the antibody specificity must be carefully assessed to correlate serotype with taxonomic group (Macario et al., 1991).

The culture of microorganisms is a cornerstone of microbial ecology. Enrichment culture techniques were developed by Beijerinck and Winogradsky at the beginning of the twentieth century and used by them and their followers to cultivate a variety of metabolically diverse organisms from natural habitats. In a manner similar to application of Koch's postulates, organisms that carried out in pure culture processes detected in natural habitats such as nitrogen fixation, pesticide degradation, or pyrite oxidation were isolated.

Despite success in applying cultural techniques to the study of microbial ecosystems, it has long been known that the number of organisms obtained from most natural habitats using cultural techniques is usually one to several orders of magnitude lower than that seen under the microscope, a phenomenon termed the "great plate count anomaly" (Staley and Konopka, 1985). The potential causes of this anomaly will be discussed below, but it should be mentioned here that it was not clear at the time whether the relatively low viable counts were mainly a matter of poor recovery of known organisms, or whether there were entire microbial groups which were not being cultured.

Two sets of classical microbial ecological studies in the twentieth century are particularly notable: those of Robert Hungate on gastrointestinal habitats and those of Thomas Brock on hot springs. Hungate studied the microbiota of the termite gut and the animal rumen from the 1940s until the 1980s (Hungate, 1979). He developed a novel set of anaerobic culture techniques for growing fastidious anaerobes, and enunciated the concept that growth media must simulate the microbial habitat to promote growth of the organisms; for example, by adding sterilized rumen fluid as a nutrient supplement and assuring that the medium, as closely as possible, matched the physicochemical characteristics of the rumen. These studies led to the isolation and characterization of a large variety of previously unknown fastidious anaerobes, most of which had complex nutritional requirements, indicating nutritional interdependence of rumen populations. Once isolated, organisms in the rumen fluid, such as cellulose degraders were enumerated, to determine whether their numbers were sufficiently high to account for a significant fraction of the activity measured directly in the rumen. Hungate and his disciples made the rumen an example of how a microbial habitat can be studied quantitatively. Indeed, in some studies of rumen and colon populations, nearly half of the directly counted organisms were cultured.

Brock studied microbial populations in hot springs, mainly in Yellowstone National Park, in the 1960s to 1970s (Brock, 1995). In the initial isolation studies, he also applied principles of habitat simulation by using a medium containing relatively low concentrations of organic nutrients, a mineral composition similar to that of the hot spring, and an incubation temperature of 70–75°C. Whereas many of his predecessors used rich media, incubated at temperatures near 60°C, and invariably obtained *Bacillus stearothermophilus* and its relatives, Brock and colleagues obtained *Thermus aquaticus*, which grew at temperatures up to 78°C, and produced a thermostable DNA polymerase that made possible automation of the polymerase chain reaction. *T. aquaticus* was the beginning of a flood of thermophiles isolated by Brock and later by Karl Stetter (Stetter, 1995) and others. In other studies, Brock and colleagues applied several physiological and microscopy techniques to study microbial populations in the hot springs. These included studies in which the effects of temperature on processes such as CO_2 fixation were examined in natural populations, and studies in which microautoradiography was employed to examine growth and metabolism of individual cells. Growth of organisms in boiling water was demonstrated by examining colonization of microscope cover slips *in situ*, a technique used by Arthur Henrici in the 1930s to culture (but not isolate) organisms such as *Caulobacter* from aquatic habitats. Cover slips which were repeatedly treated with germicidal ultraviolet radiation had considerably fewer organisms present, demonstrating that the organisms were mainly growing on the slides rather than passively attaching to the cover slips. Electron micrographs of these populations showed unusual ultrastructures that today are easily recognized as archaeal. These boiling water organisms eluded culture until it was realized that most were anaerobes or microaerophiles (Stetter, 1995).

The Woesean Revolution and its Impact on Microbial Ecology

As is amply described elsewhere in this volume, Carl Woese's studies on molecular phylogeny had a revolutionary effect on microbial systematics. In terms of understanding microbial diversity and evolution, Woese's phylogeny exposed the fallacy of assuming that the number of named species is any indication of diversity, a notion put forth by Mayr (1998); for example, that there are millions of animal species and only thousands of bacterial ones. This fallacy should have already been evident considering that if macrobiologists created species by the same criteria of genetic relatedness as microbiologists (two members of the same species have at least 70% DNA–DNA hybridization and 5°C difference in melting temperature of heteroduplexes, equivalent to 4–5% sequence divergence [Stackebrandt and Goebel, 1994]), humans and chimpanzees (~1.6% sequence divergence) would be members of the same species (Sibley et al., 1990; Staley, 1997, 1999). Early microbiologists were well aware of the metabolic diversity of the microbial world, but this sense of microbial diversity had been lost during the era when molecular biology first came to dominate microbiology. Woese's phylogeny made the extent of diversity in the microbial world more apparent by displaying genetic diversity in a way that had high visual impact and showed clearly how much diversity exists in the microbial world.

About ten years after the Woesean revolution in microbial systematics began, it started to have an equally profound effect on microbial ecology. The development, mainly by Norman Pace and his colleagues (Pace, 1997; Hugenholtz, et al., 1998a), of techniques to retrieve rRNA gene sequences from nature has enabled researchers to identify organisms in natural habitats without the need for culturing them.

Many of the features of 16S rRNA that make it a good taxonomic tool, especially its universal distribution and the fact that it contains regions with various degrees of sequence conservation, also make it a powerful tool for ecological analysis. In one of the most straightforward methods using 16S rRNA (Fig. 1), the DNA is extracted from a mixed microbial population, and primers directed at universally conserved regions of the 16S rRNA gene (rDNA) are used to amplify these genes using the polymerase chain reaction (PCR). The resulting population of rDNAs are then cloned and sequenced (Fig. 1). The different 16S rDNA clones can be analyzed phylogenetically by comparison to the databases of known 16S rRNA genes. Thus, a semiquantitative census or community analysis of the organisms present in a habitat can be obtained without culturing them. It should

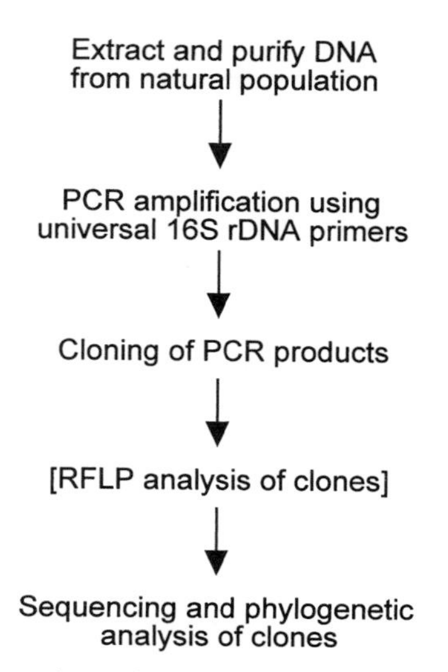

FIGURE 1. Community analysis of 16S rRNA from a natural microbial community. Restriction fragment length polymorphism (RFLP) analysis of clones can be performed to identify potentially identical clones and minimize the amount of sequencing done. However, there can be subtle sequence differences between clones with identical RFLP patterns.

be mentioned that there are biases at each step of these procedures, so that care must be taken in applying and interpreting the results of such analyses (Wilson, 1997; von Wintzingerode et al., 1997; Polz and Cavanaugh, 1998; Suzuki et al., 1998). For example, any organism from which DNA is not extracted by the procedure will not be included in the census.

The nearly universal conclusion obtained from applying this technique and its many variants to microbial habitats is that the diversity of uncultured organisms far exceeds, both in number and in kind, the diversity of those cultured. The studies by Pace and colleagues of a single hot spring, Obsidian Pool, in Yellowstone National Park (Fig. 2) are illustrative. One study (Barns et al., 1996) focusing on *Archaea* demonstrated several new branches of the *Crenarchaeota*, a group that on the basis of the small number of representatives cultured, was not considered to be very diverse. Moreover, two archaeal sequences did not cluster in either the *Crenarchaeota* or the *Euryarchaeota*, and were considered a new archaeal group tentatively named the *"Korarchaeota"*. Other studies have demonstrated that *Crenarchaeota*, all presently cultured members of which are thermophilic, can be found in moderate temperature habitats such as soils, the surface of a sponge (Preston et al., 1996), and even in Antarctic waters (DeLong et al., 1994). A study of the *Bacteria* in Obsidian Pool sediments (Hugenholtz et al., 1998b) revealed several novel phylum-level branches (called division-level branches in the original publication). This and other studies, including those using culturing approaches, greatly increased the diversity of the eubacteria from 11 divisions in 1987 (Woese, 1987) to over 30 (Fig. 3). Finally, PCR amplification of rDNA is being used to re-examine certain chronic diseases which may be caused by an uncultured microorganism, as was the case for ulcer causation by *Helicobacter pylori* (Relman, 1999).

Soil seems to harbor a particularly great diversity of organisms. In one study (Borneman and Triplett, 1997), samples were taken from two Amazonian soils, 16S rDNA clone libraries were generated, and 50 clones from each soil were sequenced. No 2 sequences of the resulting 100 were identical with each other, nor were there any exact matches between the soil clones and sequences of cultured organisms in the database (Table 1). While certain biases in the PCR procedure may have overemphasized diversity (Polz and Cavanaugh, 1998; Suzuki et al., 1998), it is still the equivalent of pulling out 100 jelly beans from a bag, and finding each to be a novel color and a different color from all of the others. Moreover, many of the classical cultural studies of soil have led to the impression that Gram-positive bacteria and *Proteobacteria* are the dominant procaryote groups. However, molecular studies have shown that less well characterized groups, such as the phyla *Verrucomicrobia* and *Acidobacteria*, are apparently of equal quantitative importance.

Molecular ecological studies using 16S rDNA have led to the recognition of many novel phylum-level branches of the procaryotic phylogenetic tree from which only a few or even no organisms have been cultured (Fig. 4). On the basis of number and diversity of sequences, some of these phylum-level branches have phylogenetic depth comparable to that of some of the better characterized phyla such as the *Proteobacteria*, yet our knowledge of these organisms at best is scant. From the ecological perspective, such studies are useful but represent only a promising beginning. The real challenge is to determine how these microbes interact with their environment and each other.

A suite of other molecular techniques has been developed to further characterize microbial populations in natural habitats. Because 16S rRNA has regions that change at different rates, one can design oligonucleotide probes and PCR primers of various specificities such that some are species specific, whereas others cover broader phylogenetic groupings such as a genus, the *Proteobacteria* (a phylum), or all *Bacteria* (a domain). One can use these probes to measure the amount of rRNA or rDNA from various microbial groups in a natural habitat, usually by filter hybridization. Whereas PCR amplification studies are semiquantitative at best, quantitative information can be derived from hybridization reactions. For example, in one study of an anaerobic bioreactor (Raskin et al., 1994b) it was demonstrated that the sum of the hybridizations to probes specific for various methanogenic groups was roughly equal to that of a probe for all *Archaea*, indicating that all significant archaeal groups had been accounted for.

One can also obtain an index of various phylogenetic groups in a population using techniques such as denaturing gradient gel electrophoresis (DGGE) and thermal gradient gel electrophoresis techniques which separate 16S rDNA PCR products on the basis of their mol% G + C (Muyzer and Smalla, 1998). Another indexing method is terminal restriction length polymorphism (T-RFLP) analysis in which one of the PCR primers is end-labeled with a fluorochrome and the resulting PCR products are subjected to restriction enzyme digestion and electrophoretic product analysis (Liu et al., 1997). These and similar methods give a characteristic pattern of either bands or peaks for a given microbial population, and this pattern can then be compared with patterns of other populations. The effects of a change in environmental conditions, such as a temperature shift, on microbial populations can also be examined. The mobility of the bands and peaks can be compared with those of known standards. They can be identified directly by sequencing, so that specific populations can be studied using these techniques. However, since different sequences can exhibit similar migration and behavior using any of these methods, identification is not always conclusive.

In the future, as DNA sequencing technology continues to

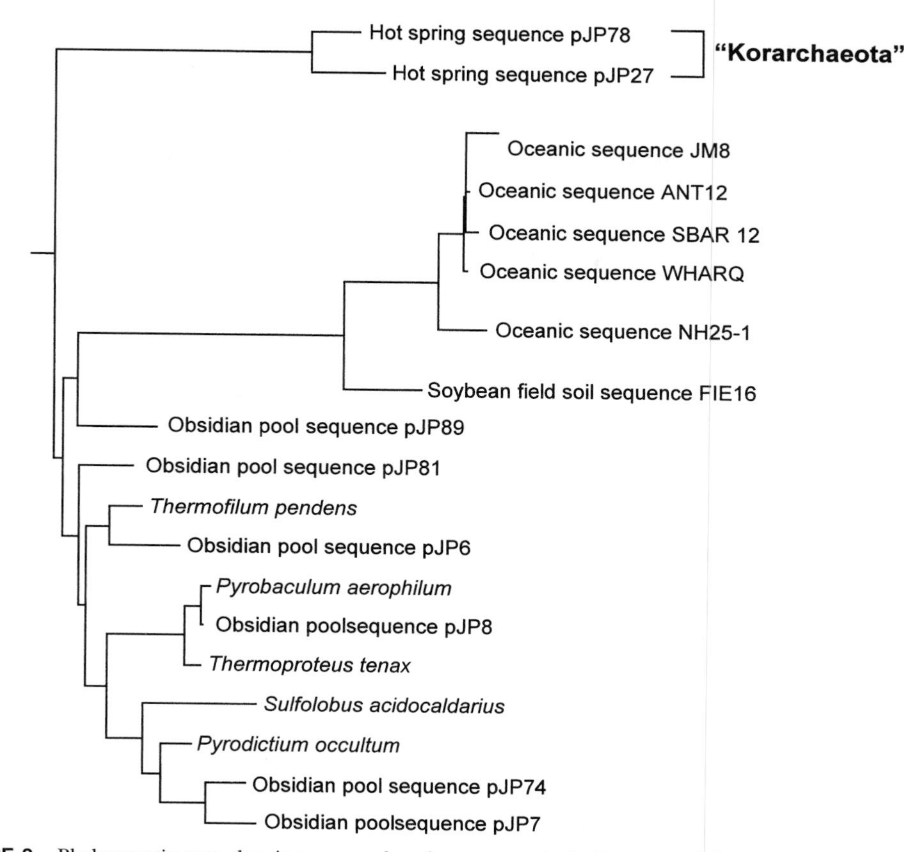

FIGURE 2. Phylogenetic tree showing crenarchaeal sequences including several derived from natural habitats. Also included are the two sequences considered to be members of the new archaeal subdomain, the *"Korarchaeota"*.

improve, it may become feasible to sequence the "genome" of an entire microbial community (Rondon et al., 1999). Information from genome sequencing would suggest hypotheses about the activities of different members of the community, hypotheses which could then be tested biochemically. It is important to note that this exciting possibility requires not just rapid and cheap DNA sequencing technology, but an increased knowledge of bacterial physiology and gene sequences obtained from the study of pure cultures.

An interesting development in modern microbial ecology has been the return to favor of an old ally, the microscope. Microbiologists should not have abandoned their microscopes in the first place, but many of them did. It has taken some fancy new technology to awaken microbiology to the tremendous amount of information they can obtain from microscopic examination of an environmental sample. Fluorescent *in situ* hybridization (FISH), described below, may have stimulated the move back to microscopy, but scientists are rediscovering that a lot can be learned from microscopic examination without resorting to molecular stains. Some bacteria such as cyanobacteria have very distinctive morphologies and naturally fluoresce red when illuminated with light of the appropriate wavelength. The blue-green fluorescence of factor F_{420} is characteristic for many methanogens. A novel use of molecular technology is to insert a gene for green fluorescent protein in an appropriate place in an organism's genome and then use fluorescence microscopy to follow the fate of introduced labeled cells in a habitat. Confocal microscopy provides a three dimensional view of a community. Finally, types of motility and potentially interesting associations can be identified. For example, observation of a cyanobacterium gliding along surrounded by a layer of motile bacteria that contain sulfur granules suggests possible interactions that might be missed by taking a less dynamic view of the population.

FISH is a particularly powerful molecular technique that allows visual identification of phylogenetic groups in natural microbial populations. In its common usage in microbial ecology, a fluorescently labeled oligonucleotide is added to a sample of permeabilized cells, so that it can hybridize to rRNA in the cells and make them fluorescent. Those cells are then viewed using a fluorescence microscope. The specificity of the probe can be adjusted as described above to include all organisms (universal probes), all organisms in a particular phylum, or a single species. Variations of this technique use multiple probes, each with a different fluorescent label (Amann et al., 1995), so that different populations can be visualized in a single sample. FISH is a good way to determine how well a 16S rDNA community analysis reflects the actual composition of the microbial community.

Figure 5 demonstrates the use of FISH to visualize the iron and manganese oxidizing filamentous procaryote *Leptothrix discophora* in a natural aquatic sample. Another study demonstrated that the dominant components of a microbial population in an aerobic sewage digestor were members of the β-*Proteobacteria*, whereas cultural studies indicated that members of the α-*Proteobacteria* were more abundant (Amann et al., 1995). This technique has its own biases and artifacts. It depends upon reliable permeabilization of the target cell populations and on the sample

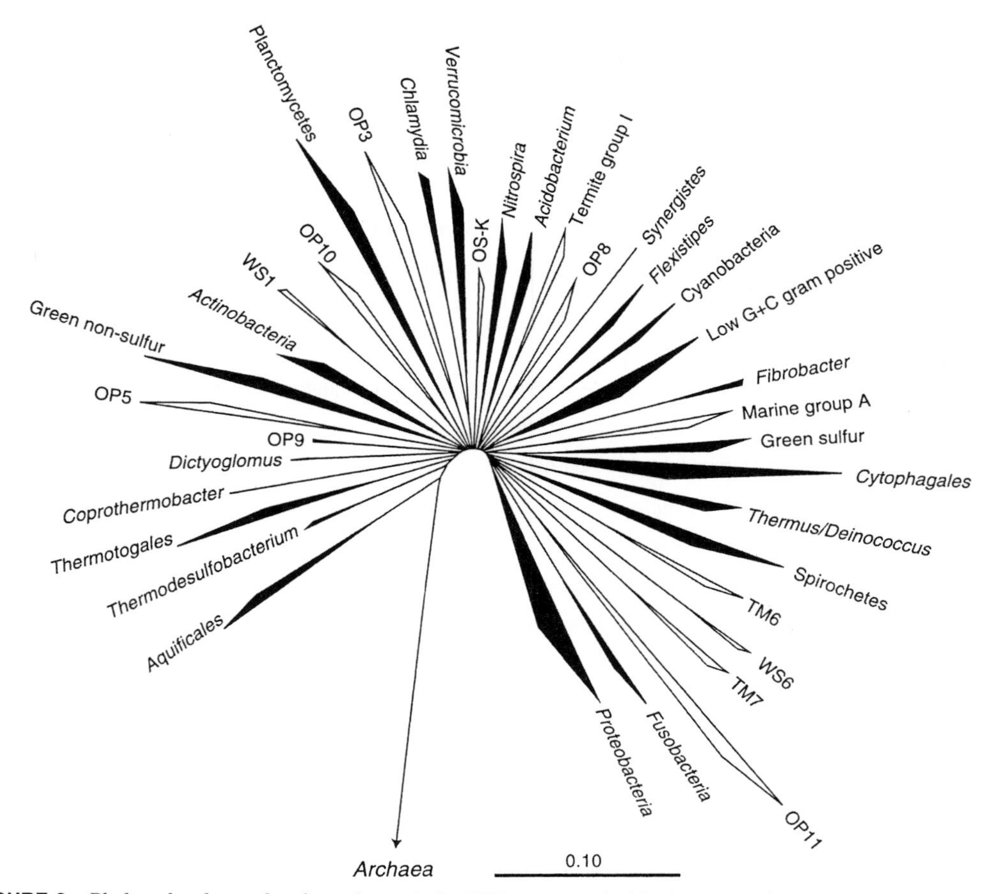

FIGURE 3. Phylum level tree for the eubacteria in 1999 compared with that of 1987. The width of the phylum lines is proportional to the phylogenetic depth within them, and unfilled lines represent uncultured phyla. From Hugenholtz et al. (1998a).

TABLE 1. Phylogenetic assignments of clones derived from PCR amplification of SSU rDNA from two soil samples taken from the Amazon river region[a]

Taxon (Phylum)	Mature forest	Pasture
Crenarchaeota	1	1
Chloroplasts	0	1
Verrucomicrobia	7	6
Planctomycetes	1	0
Low mol% G + C Gram-positive (*Firmicutes*)	11	8
Bacillus	2	8
Clostridium	9	0
High mol% G + C Gram-positive (*Actinobacteria*)	0	3
Cytophaga-Flavobacterium-Bacteroides	3	4
Acidobacteria	9	10
Proteobacteria	6	10
Alpha subdivision	2	2
Beta subdivision	1	0
Delta subdivision	1	4
Epsilon subdivision	2	4
Unclassified	12	7

[a]Although the primers used were universal, no eucaryotes were detected. Data from Borneman and Triplett (1997).

having a low background fluorescence. Moreover, the target cells must contain at least 10,000 ribosomes, a condition that may not be met by slow growing populations often encountered in natural habitats. Finally, hybridization conditions must be optimized to give the desired specificity.

A part of the molecular revolution that remains relatively underdeveloped is the use of molecules other than nucleic acids to characterize microbial communities. For example, clinical microbiologists have long used antibodies to identify pathogens. This approach might be useful for detecting enzyme expression. If the enzyme is encoded by a regulated gene, its expression indicates that the microbes are experiencing a certain set of conditions.

Some microbial ecologists are beginning to exploit another type of technology for monitoring gene expression: tagging a gene with a reporter group such as the gene encoding green fluorescent protein, and monitoring expression of the gene by measuring fluorescence emitted by the microbe in its natural environment. A drawback to this approach is that the organism whose gene expression is to be monitored must be genetically manipulable in order for the reporter gene to be introduced into the bacterial genome. Moreover, addition of the reporter gene to the organism' s natural complement of genes may reduce its fitness in the environment. Nonetheless, the reporter gene approach looks very promising for the future. There are also natural "reporter" traits. For example if an organism growing in pure culture changes shape during starvation, its shape in an environmental sample can give some indication of the degree of starvation it is experiencing. As previously mentioned, the

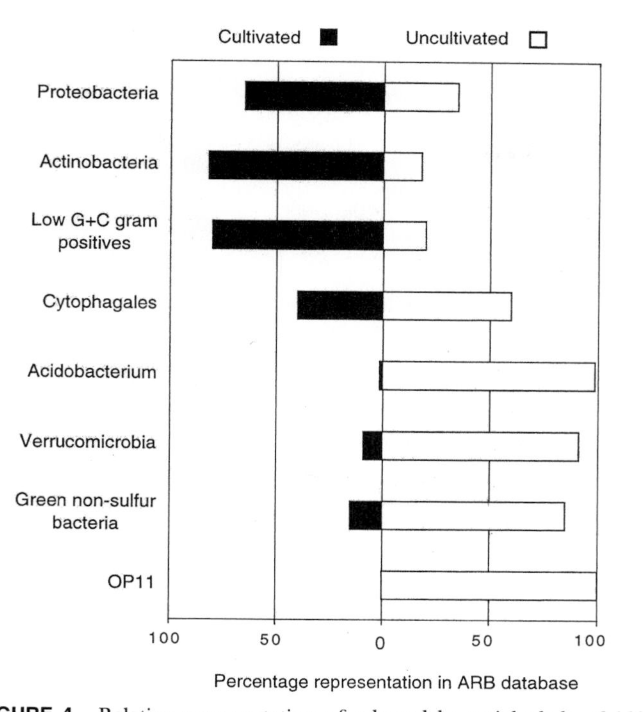

FIGURE 4. Relative representation of selected bacterial phyla of 16S rDNA sequences from cultivated and uncultivated organisms. From Hugenholtz et al. (1998a).

strength of fluorescence from *in situ* hybridization probes may reflect the number of ribosomes and thus the nutritional state of the microbe (Amann et al., 1994). Methods for direct PCR amplification of genes from bacteria that have been fixed on a slide and permeabilized may soon be sufficiently sensitive to allow single copy genes to be detected, providing important clues about the microbe's metabolic potential. For example, if a bacterium tested positive with primers designed to amplify ribulose bis-phosphate carboxylase, the microbe probably is capable of autotrophic growth, fixing carbon dioxide by the Calvin cycle. As more and more genome sequences appear, it should become easier to identify highly conserved regions of genes associated with a particular microbial activity and to design primers to amplify these genes on slides. By incorporating fluorescent nucleotides in the reaction mixture, the organism containing the gene of interest becomes visible under a fluorescence microscope.

The Challenge of Uncultured Species

It is clear that we have cultured only a very small proportion of the procaryotic species in nature. These studies have left us with thousands of what Howard Gest (1999) has termed "virtual bacteria", sequences without a corresponding isolated organism. The increase in number of these sequences shows no signs of abating. Clearly, we would know much more from isolates of these organisms than from their 16S rDNA sequences alone. It has been suggested by some that many of these "uncultured" organisms are "unculturable". There are reports that in natural habitats some microorganisms enter a state, often upon starvation, called "viable, but non-culturable". That is, they cannot be grown using standard culture media but are still viable in their environment or, in the case of pathogens, are still able to cause disease (Colwell, 1993). This concept is controversial (Bogosian et al., 1998), but that controversy is not relevant to this discussion. Even assuming that the concept is completely valid, there is still no reason to equate uncultured with unculturable. We believe that a free-living uncultured organism is culturable until proven otherwise. Our concern with this semantic point is that equating the terms uncultured and unculturable legitimizes not attempting to culture organisms represented by novel rDNA sequences, since if they are nonculturable anyway, why bother trying?

The idea of unculturability is bolstered by the previously mentioned "great plate count anomaly", since direct microscopic counts are so often much higher than viable counts. There are several other reasonable explanations for this phenomenon, however. No single growth medium, even if well designed, can necessarily culture from a sample all of the organisms present, that could include aerobes, anaerobes, heterotrophs, chemolithotrophs, and phototrophs. Moreover, microorganisms in natural habitats are often associated with particles, microcolonies, or biofilms, which must be dispersed to provide accurate counts.

In all too many studies, inappropriate growth media have been used to culture organisms from natural habitats. In the most common examples, overly rich media, often designed for culturing microorganisms from the human body, are used in attempts to culture organisms from soil and water. In addition, most microbiologists are accustomed to growing pure cultures on high concentrations of soluble substrates in batch culture to obtain high organism densities. In many natural habitats, soluble organic compounds are found at vanishingly small steady-state concentrations, and exposure to high nutrient concentrations can be toxic to these starved organisms. Better results can be obtained using appropriately habitat-simulating media. For example, one study (Button et al., 1993) showed that viable counts representing up to 60% of direct microscopic counts could be obtained from seawater if a growth medium consisting of sterilized seawater without organic amendment was used for a most-probable-number enumeration technique in which a fluorescence-activated cell sorter with a detection limit of 10^4 cells/ml was used for detection of tubes positive for growth. Addition of as little as 5 mg/l Casamino acids led to complete inhibition of growth of the natural population. Eventually a hydrocarbon oxidizing organism, less susceptible to this inhibition once in culture, was isolated from these and similar samples (Dyksterhouse et al., 1995; Button et al., 1998). Organic contaminants in agar, especially in "bacteriological grade agars" which are brown from contaminants, are inhibitory to some organisms. Better results can be obtained with more purified agar preparations, other gelling agents such as bacterial polysaccharides, or silica gel, which was used by Winogradski to isolate nitrifiers.

Many organisms require organic nutrients in their growth media, and supplying and identifying those nutrients can pose a challenge. Typical nutrients added to media include vitamins, amino acids, and nucleic acid precursors, or complex nutrient sources such as yeast extract (which, incidentally, lacks vitamin B_{12}). Examples of unconventional nutrients required by microorganisms include polyamines such as putrescine (Cote and Gherna, 1994), the peptidoglycan precursor diaminopimelic acid (Cote and Gherna, 1994), cresol (Stupperich and Eisinger, 1989), and the methanogenic cofactors coenzyme M and coenzyme B (Kuhner et al., 1991). Some organisms require lipids or lipid precursors such as mevalonic acid, an isoprenoid precursor. Certain rumen anaerobes require branched-chain volatile fatty acids as precursors of branched-chain amino acids, but corresponding amino acids sometimes do not support growth because

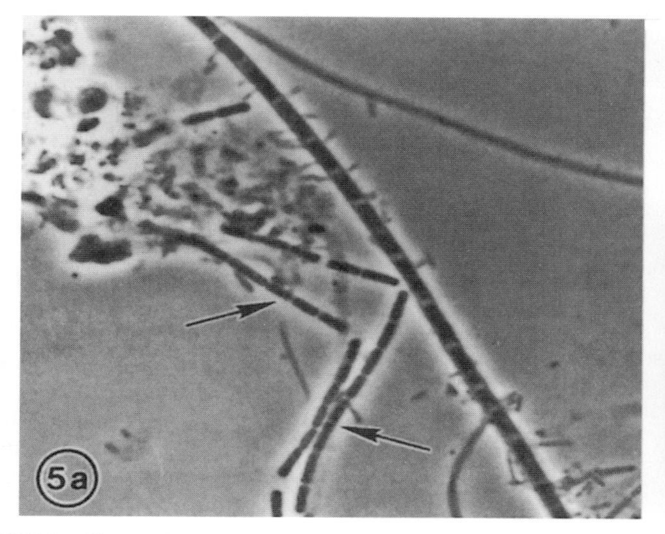

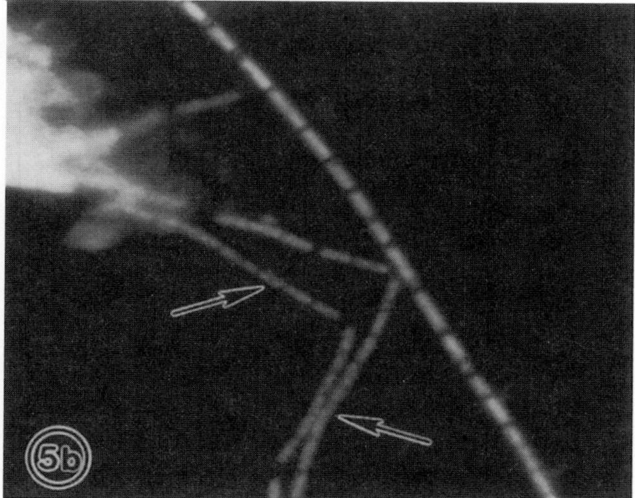

FIGURE 5. Phase contrast (left) and fluorescence *in situ* hybridization (FISH) (right) micrographs of natural populations from an iron/manganese layer in swamp water. The probe used was specific for *Leptothrix discophora*, a sheathed rod with a morphology similar to the cells indicated with arrows. Note that a much thicker filamentous rod is also stained while many other organisms present are not. When the hybridization was done at higher temperature (greater stringency), the thicker rods were not stained. From Ghiorse et al. (1996).

they are not transported into the cell. Often, good sources of unknown nutrients are extracts or supernatants of natural populations or mixed laboratory cultures containing that organism. For example, rumen fluid, a supernatant of rumen contents, is used to cultivate rumen bacteria, and an extract of a mixed culture was used to isolate an organism which utilizes chlorinated ethenes (Maymó-Gatell et al., 1997). These extracts and supernatants help simulate the habitat from which the organism was derived, although the steady state concentrations of essential nutrients in those preparations may not support significant growth in batch culture. Care must also be taken to simulate the physicochemical environment of an organism (Breznak and Costilow, 1994), which includes factors such as pH, oxidation-reduction potential, and concentrations of various solutes. Finally, more sophisticated factors, such as cytokines (Mukamolova et al., 1998), may be needed for the culture of a given microorganism.

Enrichment culture is a powerful tool for isolation of novel metabolic types of organisms, especially ones that carry out a given process rapidly. However, enrichment may not always yield the most numerous organism carrying out that process. As an example, a thermophilic cyanobacterium named *Synechococcus lividus* was routinely enriched from undiluted samples of hot spring photosynthetic mats in Yellowstone National Park. *S. lividus* was morphologically identical to the cyanobacteria seen in microscopic examinations of hot spring mat material. However, molecular community analysis showed that several other cyanobacteria with 16S rRNA sequences highly divergent from that of *S. lividus* were present in the mat material (Ward et al., 1998). Moreover, according to hybridization and DGGE analyses (Ferris et al., 1996), *S. lividus* was only a minor component of the microbial community. Dilutions of Octopus Spring mat material, containing $\sim 10^{10}$ cyanobacterial cells per ml, into liquid growth medium showed that *S. lividus* dominated in low dilutions, whereas in dilutions which received 25–1000 cells, much slower growing strains predominated. Two strains were isolated from these high dilutions: one of these had a 16S rDNA sequence which matched one of the major sequences from the mat, while the other had a sequence not found in the mat material, an instance of culture techniques finding an organism not found using the molecular approach. Thus, one conclusion from these studies was that *S. lividus* was a "weed", able to grow rapidly in low dilutions in enrichments, and thereby overtake the slower growing strains which predominated in the mat material. Simply diluting the material prior to enrichment yielded some more ecologically important strains.

Molecular techniques can provide information that aids in the culturing of target organisms. One can sometimes infer the physiology of the target organism from that of its relatives. For example, early in Brock's studies of Yellowstone National Park hot springs, large pink tufts of a filamentous bacterium were noted in the outflow of Octopus Spring, but the organism eluded culture. Subsequent molecular ecological studies (Reysenbach et al., 1994), including FISH analysis, showed that the 16S rDNA sequence of the dominant organism fell within the *Aquifex–Hydrogenobacter* group, members of which can grow microaerophilically on hydrogen. This information was used to isolate *Thermocrinis ruber* (Huber et al., 1998). One can also use molecular techniques to determine optimal culture conditions for growth of an organism. Do its numbers increase under aerobic or anaerobic conditions, with one nutrient present or with another? FISH can allow visualization of an organism, for example, one member of the *"Korarchaeota"* was shown to be a long crooked rod (Burggraf et al., 1997). In another case, FISH analysis showed that a large coccoid organism in an anaerobic enrichment culture was associated with a particular archaeal rDNA sequence. Although the organism was not numerically dominant, laser tweezers were used to move single cells of this organism into sterile microcapillary tubes from which they were then transferred to culture medium and isolated (Huber et al., 1995).

It should be mentioned that knowing the 16S rDNA sequence of a microorganism may not always allow one to predict whether it possesses a particular phenotype. Thus, the fact that an uncultivated microbe is found to be closely related to a cultivated photosynthetic organism at the rDNA sequence level does not mean that the uncultivated microbe is necessarily photosynthetic. Yet, as demonstrated in the case of *Thermocrinis*, sequence information about genetic relatedness helps to narrow the possibilities and to suggest cultivation strategies that might not otherwise have been tried.

We conclude this section with a plea for greater future efforts to culture uncultured organisms. Whereas much intellectual energy has been expended on developing and optimizing molecular techniques to study microbial populations, considerably less work has gone into culturing members of those populations. Novel approaches and techniques need to be developed. Ideally, the molecular revolution will stimulate new efforts to cultivate the currently uncultivated microorganisms in two ways. First, as already mentioned, a knowledge of what types of microorganisms might be present may suggest the type of media that would be most successful. Second, once scientists know that a particular organism is present in the site, they are less likely to give up after a few halfhearted attempts at cultivation. If a significant fraction of the procaryotic species detected using molecular techniques can be cultured, the next edition of *Bergey's Manual* promises to be greatly expanded and much more indicative of the true diversity of microorganisms in nature.

Relevance of Pure Culture Studies to Microbial Ecology

There are those who have claimed that the study of pure cultures of microorganisms is irrelevant to natural habitats in which microorganisms are interacting with each other and experiencing conditions quite different from those encountered in laboratory media. Indeed, studying a microorganism in pure culture is akin to studying animal behavior at the zoo, and one must be careful about extrapolating results from pure cultures to natural settings. However, to completely reject pure culture results as irrelevant to nature is throwing out the baby with the bathwater. For example, *E. coli* geneticists are baffled by the *lac* operon, especially in view of the fact that lactose is mostly absorbed before it reaches the locations where *E. coli* tends to flourish. If they were more aware of the intestinal milieu, however, they would realize that the ability of *E. coli* to cleave β-galactoside linkages and numerous other sugar linkages would be of great value to a bacterium that is grazing on the mucin layer covering the mucosa of the small and large intestine. The important point, whether the above is the case or not, is that a thorough knowledge of an organism's physiology and the characteristics of its normal location can put a scientist at an enormous advantage in interpreting the results of conventional genetic analysis. Even more important, a microbial ecologist with an eye firmly fixed on the environment understands that the genetic analysis is not an end in itself, but is merely a hypothesis-generating step that generates the hypothesis and the reagents to test this hypothesis in a real environmental sample.

We will provide a single example of the usefulness of pure culture results in the study of natural populations. In the 1970s, it was found that the marine luminescent organism *Vibrio fischeri* was bioluminescent only in late stages of culture (for a recent perspective, see Hastings and Greenberg, 1999). It was eventually determined that the concentration in those cultures of a soluble acylhomoserine lactone (AHSL) autoinducer built up to a critical level, thereby turning on expression of the luminescence genes. Because of this autoinduction phenomenon, the bacteria were luminescent in the light organ of a marine organism, where they were present in high density, but not when in seawater. Eventually, these AHSLs were found in a variety of other *Proteobacteria*, including pathogens, usually regulating some function which is optimal at high organism densities, such as polymer hydrolysis or conjugation. This form of microbial cell-to-cell communication was termed quorum sensing (Fuqua et al., 1994). A similar phenomenon is mediated by small peptides in Gram- positive bacteria and cyanobacteria. It has been demonstrated that *Pseudomonas aeruginosa* production of an AHSL is essential for proper formation of biofilms (Davies et al., 1998), one of the quintessential forms of microbial communities in natural habitats. Moreover, there are now signs of AHSL-mediated communication among different species (Fuqua and Greenberg, 1998) and evidence of their presence in natural aquatic microbial biofilms attached to submerged rocks (McLean et al., 1997). Microbial cell-to-cell communication will be a fertile and fascinating area of study in microbial ecology, and our present level of understanding of this phenomenon is the result of pure culture studies.

Thus, pure culture work can suggest hypotheses, and in some cases give answers, that provide insights into how a natural community operates. In the future, molecules identified in pure culture studies may prove to be valuable indicators of microbial activities in a particular setting. Furthermore, pure culture studies can lead to the development of more sophisticated hypotheses about metabolic interactions between microorganisms in a natural environment.

Finale

The prospects for microbial ecology in the twenty-first century are promising indeed. The application of molecular techniques in conjunction with more classical ones will provide a wealth of information about the natural habits of the most fundamental and arguably the most dominant form of life on Earth (Gould, 1996; Whitman et al., 1998). Most importantly, future editions of *Bergey's Manual* are guaranteed to have a much more complete description of the diversity of procaryotes. We conclude with a quote from the closing chapter of naturalist E.O. Wilson's autobiography (Wilson, 1994) in which he describes what he would do if he were beginning his scientific career again.

"If I could do it all over again, and relive my vision in the twenty-first century, I would be a microbial ecologist. Ten billion bacteria live in a gram of ordinary soil, a mere pinch held between thumb and forefinger. They represent thousands of species, almost none of which are known to science. Into that world I would go with the aid of modern microscopy and molecular analysis. I would cut my way through clonal forests sprawled across grains of sand, travel in an imagined submarine through drops of water proportionately the size of lakes, and track predators and prey in order to discover new life ways and alien food webs."

Culture Collections: An Essential Resource for Microbiology

David P. Labeda

Collections of microbial strains have existed since bacteriologists were first able to isolate and cultivate pure cultures of microorganisms, and have always been an important aspect of microbiology, whether used as a source of strains for teaching purposes or as an archive of reference material for research, taxonomic, or patent purposes. Although the field of microbiology is assuming an increasingly molecular emphasis, the need for culture collections has not diminished. Culture collections still provide a significant degree of continuity with the past through the preservation and distribution of microbial strains described or cited in publications, and collections often maintain novel microorganisms awaiting future exploitation by biotechnology.

Historical Perspective

The history of microbial culture collections has been reviewed in detail by numerous authors over the years (Porter, 1976; Malik and Claus, 1987), but there is consensus that the first culture collection established specifically to preserve and distribute strains to other researchers was that of Professor František Král in Prague during the 1880s (Martinec et al., 1966). Upon the death of Professor Král in 1911, this collection was acquired by Professor Ernst Pribham, who transferred it to the University of Vienna and issued several catalogs listing the holdings of the collection. Part of this collection was brought to Loyola University in Chicago by Pribham in the 1930s. Many of these strains were subsequently transferred to the American Type Culture Collection upon Pribham's death, but others remained in the collection at Loyola University (Porter, 1976). The Vienna portion of the Pribham Collection was apparently lost during World War II (Martinec et al., 1966). The next oldest culture collection, the Centraalbureau voor Schimmelcultures (CBS), was founded in 1906 and is still active in Baarn and Delft, The Netherlands (Malik and Claus, 1987). It has served to rescue the collections of many European laboratories. The large number of filamentous fungi collected by Charles Thom at USDA in Washington, DC, in the early part of this century formed the nucleus of the collections of these microorganisms at both the American Type Culture Collection and the Agricultural Research Service (NRRL) Culture Collection in Peoria, Illinois, when the strains were transferred in the 1940s (Kurtzman, 1986). Both of these collections have also served as repositories for many other orphaned culture collections from individuals and institutions, such as the N.R. Smith *Bacillus* Collection or the U.S. Army Quartermaster Collection.

Role of Culture Collections

The major culture collections throughout the world, some of which are listed in Table 1, have as their primary commission the preservation and distribution of germplasm that has been demonstrated to have significance to the microbiology community. The importance of a particular strain may be as a reference for medical or taxonomic research, as an assay organism for testing or screening, or as an essential component of a patent application for a product or process in which it is involved. Alternatively, the strain may be placed in a collection with reference to the publication in which it was cited as part of the investigation. This latter form deposition is essential on account of the inherent transience of researchers and their research programs, making it possible for later investigators to repeat or advance published research that would be impossible in the absence of the strains involved. As mentioned above, the many national reference and service collections have succeeded in preserving, for later generations of microbiologists, many of the private and specialized collections of microorganisms that may represent an entire career of one microbiologist. In other cases, however, the acquisition of a collection of strains may well result from a change in the direction of the research program in a scientist's laboratory.

Active culture collections represent centers of expertise in the methods of preservation of microbial germplasm and collection management practices, by virtue of their day-to-day activities in these areas. As such, they are an invaluable resource for training others in these important activities.

The major culture collections of the world also serve as centers for excellence in research in systematics and taxonomy. In large part the identification and characterization of strains is an integral function of collections, and the availability of a large collection of strains is essential for this type of research. Culture collections that have contract identification services are also continually searching for faster and more reliable methods to characterize unknown strains for their clients. In many cases, the strains maintained in any collection will directly reflect the taxonomic interests of the curators, in terms of the depth and breadth of particular taxonomic groups.

Role in Preservation of Microbial Biodiversity

The Convention on Biodiversity, also known as the Rio Treaty (Convention on Biological Diversity Secretariat, 1992), has resulted from the recent global emphasis on conservation of biodiversity and, although dealing more specifically with higher or-

TABLE 1. Some of the world's major bacterial culture collections

Collection	Address
ATCC	American Type Culture Collection 10801 University Boulevard, Manassas, VA 20110–2209 USA Telephone: 703–365–2700; Fax: 703–365–2701 Web site: www.atcc.org/
BCCM™/LMG	Belgian Coordinated Collections of Microorganisms Laboratorium voor Microbiologie, Universiteit Gent (RUG) K.L. Ledeganckstraat 35, B-9000 Gent BELGIUM Telephone: 32–9-264 51 08 ; Fax: 32–9-264 53 46 E-mail: bccm.lmg@rug.ac.be Web site: www.belspo.be/bccm/
DSMZ	Deutsche Sammlung von Mikroorganismen und Zellkulturen GmbH (German Collection of Microorganisms and Cell Cultures) Mascheroder Weg 1b, D-38124 Braunschweig GERMANY Telephone: 49–531–2616 Ext. 0; Fax: 49 531–2616 Ext. 418 E-mail: help@dsmz.de Web site: www.dsmz.de
IFO	Institute for Fermentation, Osaka 17–85 Juso-Honmachi 2-chome, Yodogawa-ku, Osaka, 532 JAPAN Telephone: 81–6-300–6555; Fax: 81–6-300–6814 Web site: wwwsoc.nacsis.ac.jp./ifo/index.html
NRRL	Agricultural Research Service Culture Collection National Center for Agricultural Utilization Research 1815 North University Street, Peoria, IL 61604 USA Telephone: 309–681–6560; Fax: 309–681–6672 E-mail: nrrl@mail.ncaur.usda.gov Web site: nrrl.ncaur.usda.gov
JCM	Japan Collection of Microorganisms RIKEN Hirosawa, Wako-shi, Saitama, 351–01 JAPAN Telephone: 81–48–462–1111; Fax: 81–48–462–4617 Web site: www.jcm.riken.go.jp
NCIMB	National Collections of Industrial and Marine Bacteria, Ltd. 23 St. Machar Drive, Aberdeen, AB24 3RY Scotland, UNITED KINGDOM Telephone: 44–0-1224 273332; Fax: 44–0-1224 487658 E-mail: ncimb@abdn.ac.uk Web site: www.ncimb.co.uk
NCTC	National Collection of Type Cultures PHLS Central public Health Laboratory 61 Colindale Avenue, London, NW9 5HT UNITED KINGDOM Telephone: 44–181–2004400; Fax: 44–181–2007874 Web site: www.ukncc.co.uk

ganisms such as plants and animals, suggests *in situ* conservation of genetic resources through the establishment of protected habitats. The very nature of microorganisms makes this concept somewhat untenable, and thus culture collections should play a major role in the cataloging and *ex situ* preservation of microbial germplasm. Moreover, although the Rio Treaty encouraged the establishment of means of conserving genetic resources in the country of origin, this may not be economically or technologically feasible because of the costs and training involved in the *de novo* establishment of a culture collection (Kirsop, 1996). The established national culture collections are staffed with experienced personnel well versed in the preservation of microorganisms. The relative shortage worldwide of trained microbial taxonomists magnifies this problem, and since the large established collections are centers of excellence in systematics, this is additionally supportive of their potential role in conservation of microbial biodiversity.

Type Strains

A mission critical function of the culture collections is the preservation and distribution of type strains as a primary reference for taxonomic research. The importance of type strains to microbial systematics has been reiterated in virtually every edition of *Bergey's Manual*.

Type strains represent the primary reference for taxonomic characterization, whether it is identification of unknown strains, re-characterization of known taxa, or description of new taxa. The advent of molecular phylogenetic characterization and analysis based on sequence determination of the 16S ribosomal RNA gene, or other conserved genes, does not diminish the importance of culture collections; for after all, type strains represent a significant part of the "foliage" on the procaryotic tree of life. Sequence databases have largely been constructed using the type strains of microorganisms held in the international reference and service culture collections. Phylogenetic trees from 16S rRNA gene sequences serve as an indication of the evolutionary relationship among strains, but may underestimate the actual differences between strains. Type strains are thus still necessary for evaluation of subtle phenotypic differences between strains and are essential if other gene sequences are to be determined.

The deposition of type strains of new taxa in one or more of the internationally recognized permanent culture collections, in conjunction with description and valid publication, was a recommendation under Rule 30 of the *Bacteriological Code* through the 1992 Revision (Sneath, 1992). The International Committee on Systematic Bacteriology, upon the recommendation of the Judicial Commission, emended Rule 30 of the code to change this recommendation to an absolute requirement (International Committee of Systematic Bacteriology, 1997). Under this revised rule, a taxon cannot be considered validly published and hence a valid name unless it has been deposited and is available for distribution from a recognized culture collection. Moreover, descriptions of new taxa are not accepted for publication in the *International Journal of Systematic and Evolutionary Microbiology*, the official organ of the International Committee on Systematic Bacteriology, unless type strains have been deposited. Thus, it is the role and responsibility of the permanent culture collections to preserve and distribute to the scientific community type material for all of the validly published taxa. The skill in strain preservation in the major culture collections is such that, barring a major disaster, it is unlikely that type strains held there will be lost, and the frequent distribution and replication of type material among permanent collections is another form of protection against such a loss. Should the type strain of a taxon be lost for any reason, however, the procedure for defining a neotype strain is outlined in the *Bacteriological Code*, and a culture of this strain must be deposited in one or several of the permanent collections.

Networking and Databases

With the advent of the Internet, networking of culture collection information has become commonplace. The first compilation of information regarding the culture collections of the world, based on an international survey, was provided in the first edition of the *World Directory of Culture Collections* (Martin and Skerman, 1972). This was replaced ten years later by the updated second edition (McGowan and Skerman, 1982). A computerized database of this information was maintained as the *World Data Centre for Culture Collections of Microorganisms* (WDCM) at the University of Queensland until 1986, when WDCM was moved to RIKEN, Japan, the site of the Japan Collection for Microorganisms. The WDCM was subsequently moved again in 1996 to the National Institute of Genetics in Japan. Currently 498 culture collections are registered in the WDCM database. The culture collection and strain information compiled and held at the WDCM have been available to microbiologists throughout the world via the World Wide Web since 1994, with approximately 30,000 average accesses per month (H. Sugawara, personal communication). There also has been an explosion in the number of on-line catalogs for culture collections now available on the Internet (Canhos et al., 1996). The WDCM website (wdcm.nig.ac.jp) provides a useful starting point on the Internet to begin a search for this on-line culture collection information. The collection database at WDCM is useful for deciphering the siglas (e.g., ATCC, DSMZ, JCM, NRRL, etc.) used by collections and identifying the location of the collection and contact information. The STRAINS database allows searching for taxonomic names and provides an indication of which culture collections throughout the world have a strain or strains available for a particular species.

The global interest in the study of microbial diversity and biotechnological utilization of microorganisms has greatly accelerated the placement and interrelating of the collection data with databases related to genomics and physiological properties of microorganisms. The efforts toward the total integration of microbial data using the Internet has been well reviewed by Canhos et al. (1996), Larsen et al. (1997), and Sugawara et al. (1996), and so will not be discussed here.

Intellectual Property of Procaryotes

Roy D. Meredith

Procaryotes and their macromolecular components are protectable as intellectual property, which is a composite legal field of mostly federal laws on patents, trademarks, and copyrights. Patents cover scientific inventions evidencing practical application, and provide exclusive rights for a limited period. Trademarks, as well as tradenames and trade dress, are labels designed to identify to the public particular goods or services, and function to preserve the reputation of a business and to prevent confusing similarity. Copyrights protect original works fixed in any tangible medium of expression, and may be applicable to nucleotide or amino acid sequences. All three of these kinds of intellectual property possess the common characteristic of enabling the owner to obtain an injunction against unlicensed use, and to seek monetary damages. Except where specifically noted, the present essay covers only federal laws of the United States.

Patents

An invention is patentable if it is new, useful, and not an obvious variation of what is known. What is held to be new under the law is roughly any invention without its anticipation existing in the public domain, i.e., there is no closely similar invention by another, whether published or publicly known. An invention must be useful to be patented, and this requirement of utility includes some practical application with at least some initial evidence that the invention will work as stipulated, e.g., a DNA sequence capable of expressing a structural protein of medical value with an experiment showing such expression in one host cell. Applications for perpetual motion machines are deemed incredible and lack such utility. Other features of the requirement of utility relate to statutory subject matter, and prevent patenting of mathematical equations, methods of doing business, evolutionary trees, and the like. Finally, a patentable discovery must not be an obvious variant of what is known, the standard of obviousness being defined with reference to a person of ordinary skill in the art. What is an obvious discovery and therefore not patentable under the law is similar in scope to a balanced expert's view of what is obvious in his or her field of expertise.

The patent law on procaryotes in the United States has undergone rapid development ever since a well publicized decision of the U.S. Supreme Court in 1980. This decision was partly responsible for substantial increases in investment and business development in the commercial application of recombinant DNA methods. In *Diamond* v. *Chakrabarty*, a patent claim to a microorganism *per se* was held to be patentable as appropriate statutory subject matter (No. 79–1464, 1980). The U.S. Patent and Trademark Office (USPTO) had rejected a patent claim to a recombinant *Pseudomonas* species capable of metabolizing camphor and octane, two components of oil. The practical application of the microorganism for the clearance of oil spills was not an issue, but the USPTO held the patent claim to be inappropriate statutory subject matter. On appeal after intermediate appellate review, the U.S. Supreme Court held that the patent law permits patenting of "anything under the sun that is made by man." The decision was split 5–4, a hint of potential weakness as binding precedent for future legal decisions of the U.S. Supreme Court. However, no challenges have since been made to overturn or substantially modify *Diamond* v. *Chakrabarty*.

Patenting of the macromolecular components of a procaryotic cell, or of any other cell, largely follow classic guidelines and case law of chemical entities. Since *Diamond* v. *Chakrabarty*, the patent practitioner has available an increasing body of case law relating to polynucleotides or proteins of defined amino acid sequence, vectors, plasmids, and so on. This development is largely consistent with older case law on defined synthetic molecules of an organic or inorganic nature. Patent practitioners can now provide advice of a more certain nature to inventors, providing more opportunities for business development.

Patent law in the United States and Europe differs in various respects. First, a patent claim filed in the United States on a living organism *per se* cannot be properly rejected on moral or ethical grounds, unless the subject matter is repugnant, e.g., claims to a virulent strain of *Yersinia pestis* intended for biological warfare. In contrast, the laws in Europe may prevent patenting of recombinantly altered living organisms wholly confined to the laboratory (e.g., transgenic mice having exclusive uses related to research and drug development). At the time of this writing, the European Union has not resolved the issues, so such ethical concerns may continue as long standing impediments to obtaining coverage in Europe for patent claims to certain kinds of organisms *per se*. The European Patent Convention now permits patent claims to microorganisms alone.

A second important difference is the effect of an inventor's own publication. In the United States, there is a one-year grace period for filing a patent application after the publication date of the invention in a scientific paper, or abstract. In contrast, Europe, Japan, and many other countries have the rule of absolute novelty, which requires the filing of a patent application before publishing. For valuable inventions, an inventor is well advised to follow the absolute novelty rule to obtain non-U.S. patent protection.

A third important difference relates to priority of invention. In the United States, priority depends on the first to invent, not the first to file. In Europe, Japan, Canada, Australia, and many other countries, priority depends on the first to file, prompting

purple sulfur bacteria (*Thiorhodaceae*). Molisch removed the purple sulfur bacteria (*Rhodobacteriaceae* Migula 1900) from the *Thiobacteria* Migula 1900, where these organisms had been combined with the colorless sulfur bacteria (*Beggiatoaceae* Migula 1900). Since that time, pigmentation and ability to perform anoxygenic photosynthesis were considered of primary importance for assignment of bacteria to the *Rhodobacterales*, later called *Rhodospirillales* Pfennig and Trüper 1971. Because the *Rhodospirillaceae* Pfennig and Trüper 1971 do not represent a phylogenetically distinct group of bacteria, it was proposed to abandon the use of the family name. The term purple nonsulfur bacteria (PNSB) has been proposed for the physiological groups of anaerobic phototrophic *Alphaproteobacteria* and *Betaproteobacteria* that contain photosynthetic pigments and are able to perform anoxygenic photosynthesis (Imhoff et al., 1984). Historical aspects of the taxonomy of anoxygenic phototrophic bacteria have been discussed in more detail elsewhere (Imhoff, 1992, 1995, 1999) and in volume 1 of this edition of the *Manual* (Imhoff, 2001a).

The purple nonsulfur bacteria are a highly diverse and heterogeneous group. Furthermore, based on 16S rDNA sequence similarities and chemotaxonomic properties, representatives of this group are closely related to non-phototrophic, strictly chemoheterotrophic bacteria (Woese et al., 1984a, b; 1985; Stackebrandt et al., 1988, Woese, 1987). These similarities are taken as evidence for the development of some non-phototrophic bacteria from phototrophic ancestors. With the recognition of their genetic relationships and with the support from chemotaxonomic data and ecophysiological properties, purple nonsulfur bacteria of the *Alphaproteobacteria* and *Betaproteobacteria* were taxonomically separated and rearranged according to the proposed phylogeny. Despite the fact that many of the phototrophic purple nonsulfur bacteria are closely related to strictly chemotrophic relatives, the phototrophic capability and/or content of photosynthetic pigments are nevertheless included in the genus definitions of these bacteria.

The striking physiological similarities that unify the PNSB are inconsistent with the great variation in the organization of the internal membrane systems, 16S rDNA sequence similarities, cytochrome c_2 amino acid sequences, lipid, quinone and fatty acid compositions, and lipid A structures. As a consequence, it is not appropriate to assign new species to the genera only based on physiological and morphological properties. Chemotaxonomic characteristics and sequence information also have to be taken into consideration. In addition, environmental aspects and ecological distribution should be considered.

***Phototrophic* Alphaproteobacteria** (RHODOSPIRILLALES) Cells are vibrioid to spiral-shaped or spherical to ovoid, motile by polar flagella, and divide by binary fission. Internal photosynthetic membranes consist of vesicles, lamellae, or membrane stacks. Color of cell suspensions is beige, brown, brown-red, red or pink. Photosynthetic pigments are bacteriochlorophyll *a* or *b* (esterified with phytol or geranylgeraniol) and various types of carotenoids.

DIFFERENCES FROM OTHER PHOTOTROPHIC *ALPHAPROTEOBACTERIA*. Several chemotaxonomic properties distinguish the phototrophic *Alphaproteobacteria* in the order *Rhodospirillales* from other phototrophic *Alphaproteobacteria*. Ubiquinones, menaquinones, and rhodoquinones may be present, and the length of their side chain may vary from 7 to 10 isoprene units (Table 1). Members have characteristic phospholipid and fatty acid composition with $C_{18:1}$ as the dominant fatty acid and either $C_{16:1}$ and $C_{16:0}$, $C_{16:0}$ and $C_{18:0}$, or just $C_{16:0}$ as additional major components (Table 1). Based on 16S rDNA sequence analysis, the phototrophic *Alphaproteobacteria* in the order *Rhodospirillales* are phylogenetically distinct from other groups of phototrophic *Alphaproteobacteria*, though they are closely related to several purely chemotrophic representatives of this group.

DIFFERENTIATION OF THE PHOTOTROPHIC *ALPHAPROTEOBACTERIA* IN THE ORDER *Rhodospirillales*. A number of chemotaxonomic properties distinguish *Rhodospirillum* species from *Phaeospirillum* species and other phototrophic spiral-shaped *Alphaproteobacteria* in the order *Rhodospirillales*. These bacteria are characterized by different major quinone and cytochrome *c* structures. Large type cytochromes c_2 are present in *Rhodospirillum rubrum* and *Rhodospirillum photometricum*, whereas small type cytochromes c_2 were found in *Phaeospirillum* species (Ambler et al., 1979). Major differentiating properties between phototrophic *Alphaproteobacteria* in the order *Rhodospirillales* are shown in Table 1. Carbon sources used by these species are listed in Table 2. The phylogenetic relationships of these bacteria as derived from16S rDNA sequences are shown in Fig. 1.

TAXONOMIC COMMENTS Most of the species of the phototrophic *Alphaproteobacteria* in the order *Rhodospirillales* have been previously known as *Rhodospirillum* species and are of spiral shape. These organisms include the genera *Rhodospirillum*, *Phaeospirillum*, *Roseospira*, *Roseospirillum*, *Rhodocista*, *Rhodovibrio*, *Rhodothalassium*, and *Rhodospira* (Imhoff et al., 1998). The only non-spiral representative of this group is *Rhodopila globiformis*. Based on 16S rDNA sequence analysis, this acidophilic phototrophic bacterium is phylogenetically closely related to acidophilic chemotrophic bacteria of the genera *Acetobacter* and *Acidiphilium* (Sievers et al., 1994) and has been placed in the family *Acetobacteraceae*. In addition, phototrophic *Alphaproteobacteria* in the order *Rhodospirillales* are closely related to different chemotrophic representatives of the order. *Phaeospirillum* species, for example, demonstrate close sequence similarity to *Magnetospirillum magnetotacticum* (Burgess et al., 1993), and *Rhodocista centenaria* reveals strong relations to *Azospirillum* species (Xia et al., 1994; Fani et al., 1995). Given the present state of our knowledge, *Rhodothalassium salexigens* cannot be placed in the *Rhodospirillaceae* with confidence because its 16S rDNA gene sequence is equidistant from other *Alphaproteobacteria*. It is treated here with the *Rhodospirillaceae*, but it correct taxonomic placement will depend on further studies.

***Phototrophic* Alphaproteobacteria** (RHIZOBIALES) Cells are ovoid to rod-shaped, motile by polar flagella, and show polar growth and budding. Internal photosynthetic membranes consist of lamellae parallel to and underlying the cytoplasmic membrane. Photosynthetic pigments are bacteriochlorophyll *a* or *b* (esterified with phytol or geranylgeraniol) and various types of carotenoids. Color of cell suspensions can be brown, brown-red, red or pink. Species with bacteriochlorophyll *b* are green to olive-green. Characteristic phospholipids are present; $C_{18:1}$ is the dominant fatty acid and additional major components include either $C_{16:1}$ or $C_{16:0}$, $C_{16:0}$ and $C_{18:0}$, or $C_{16:0}$ alone (Imhoff and Bias-Imhoff, 1995; Table 3).

DIFFERENCES FROM OTHER PHOTOTROPHIC *ALPHAPROTEOBACTERIA* A number of chemotaxonomic properties distinguish the phototrophic *Alphaproteobacteria* of the order *Rhizobiales* from other purple nonsulfur bacteria. Most characteristic is a budding mode of growth and cell division, which is associated with lamellar internal membranes that are lying parallel to the cyto-

TABLE 1. Diagnostic properties of the spiral-shaped anoxygenic phototrophic bacteria belonging to the *Alphaproteobacteria* (*Rhodospirillales* and *Rhodothalassium salexigens*)[a]

Characteristic	*Rhodospirillum rubrum*	*Rhodospirillum photometricum*	*Phaeospirillum fulvum*	*Phaeospirillum molischianum*	*Rhodopila globiformis*	*Rhodocista centenaria*
Cell diameter (μm)	0.8–1.0	1.1–1.5	0.5–0.7	0.7–1.0	1.6–1.8	1.0–2.0
Internal membrane system	Vesicles	Stacks	Stacks	Stacks	Vesicles	Lamellae
Motility	+	+	+	+	+	+
Color	Red	Brown	Brown	Brown	Purple-red	Pink
Bacteriochlorophyll	*a*	*a*	*a*	*a*	*a*	*a*
Growth factors	Biotin	Nicotinamide	*p*-Aminobenzoic acid	Amino acids	Biotin, *p*-aminobenzoic acid	Biotin, B_{12}
Aerobic growth	+	−	−	−	(+)	+
Oxidation of sulfide	+	−	−	−	−	nd
Salt requirement[b]	None	None	None	None	None	None
Optimal temperature (°C)	30–35	25–30	25–30	30	30–35	40–45
Optimal pH	6.8–7.0	6.5–7.5	7.3	7.3	4.8–5.0	6.8
Habitat	Fresh water	Fresh water	Fresh water	Fresh water	Fresh water, acid springs	Fresh water, warm springs
Mol% G + C of the DNA	63.8–65.8[c]	64.8–65.8[c]	64.3–65.3[c]	60.5–64.8[c]	66.3[c]	69.9[d]
Cytochrome *c* size	Large	Large	Small	Small	Small	nd
Major quinones	Q-10, RQ-10	Q-8, RQ-8	Q-9, MK-9	Q-9, MK-9	Q-9/10, MK-9/10; RQ-9/RQ10	Q-9
Major fatty acids:						
$C_{14:0}$	2.1	1.0	0.6	0.7	5.8	nd
$C_{16:0}$	14.0	25.2	15.1	18.1	9.3	nd
$C_{16:1}$	27.1	22.2	25.8	36.5	4.7	nd
$C_{18:0}$	1.3	0.4	1.2	0.7	1	nd
$C_{18:1}$	54.8	51.0	54.5	43.5	74.4	nd

[a]Symbols: +, positive in most strains; −, negative in most strains; (+), weak growth or microaerobic growth only; nd, not determined; Q-7, ubiquinone 7; Q-8, ubiquinone 8; Q-9, ubiquinone 9; Q-10, ubiquinone 10; Q-9/10, ubiquinones 9 and 10; MK-7, menaquinone 7; MK-9, menaquinone 9; MK-10, menaquinone 10; MK-9/10, 10, menaquinones 9 and 10; RQ-8, rhodoquinone 8; RQ-10, rhodoquinone 10; RQ-9/10, rhodoquinones 9 and 10.

[b]Optimum salt concentration (range of salt concentrations tolerated).

[c]Mol% G + C determined by thermal denaturation.

[d]Mol % G + C determined by bouyant density centrifugation.

(*continued*)

TABLE 1. *(cont.)*

Characteristic	*Rhodovibrio salinarum*	*Rhodovibrio sodomensis*	*Roseospirillum parvum*	*Rhodospira trueperi*	*Roseospira mediosalina*	*Rhodothalassium salexigens*
Cell diameter (μm)	0.8–0.9	0.6–0.7	0.4–0.6	0.6–0.8	0.8–1.0	0.6–0.7
Internal membrane system	Vesicles	Vesicles	Lamellae	Vesicles	Vesicles	Lamellae
Motility	+	+	+	+	+	+
Color	Red	Pink	Pink	Beige	Pink	Red
Bacteriochlorophyll	*a*	*a*	*a*	*b*	*a*	*a*
Growth factors	Cobalamine	Cobalamine	Cobalamine	Biotin, thiamine, pantothenate	Thiamine, *p*-aminobenzoic acid, nicotimamide	Glutamic acid
Aerobic growth	+	+	(+)	−	(+)	+
Oxidation of sulfide	−	nd	+	+	+	−
Salt requirement[b]	8–12% (3–24)	12% (6–20)	1–2% (to >6.0%)	2% (0.5–5)	4-7% (0.5–15)	6–8% (5–20%)
Optimal temperature (°C)	42	35–40	30	25–30	30–35	40
Optimal pH	7.5–8.0	7	7.9	7.3–7.5	7	6.6–7.4
Habitat	Saltern	Salt lakes	Marine sediments	Marine sediments	Salty springs	Saltern
Mol% G + C of the DNA	67.4[c]	66.2–66.6[d]	71.2[c]	65.7[c]	66.6[d]	64.0[c]
Cytochrome *c* size	None	None	nd	nd	nd	nd
Major quinones	Q-10, MK-10	nd	nd	Q-7, MK-7	nd	Q-10, MK-10
Major fatty acids:						
$C_{14:0}$	1.0	nd	nd	7.5	nd	3.8
$C_{16:0}$	7.4	nd	nd	27.9	nd	16.1
$C_{16:1}$	0.3	nd	nd	1.2	nd	1.5
$C_{18:0}$	23.0	nd	nd	1.2	nd	17.8
$C_{18:1}$	35.2	nd	nd	60.7	nd	59.9

[a]Symbols: +, positive in most strains; −, negative in most strains; (+), weak growth or microaerobic growth only; nd, not determined; Q-7, ubiquinone 7; Q-8, ubiquinone 8; Q-9, ubiquinone 9; Q–10, ubiquinone 10; Q–9/10, ubiquinones 9 and 10; MK-7, menaquinone 7; MK-9, menaquinone 9; MK-10, menaquinone 10; MK-9/10, 10, menaquinones 9 and 10; RQ-8, rhodoquinone 8; RQ-10, rhodoquinone 10; RQ-9/10, rhodoquinones 9 and 10.

[b]Optimum salt concentration (range of salt concentrations tolerated).

[c]Mol% G + C determined by thermal denaturation.

[d]Mol % G + C determined by bouyant density centrifugation.

TABLE 2. Carbon sources and electron donors used by spiral-shaped anoxygenic phototrophic bacteria belonging to the *Alphaproteobacteria* (*Rhodospirillales* and *Rhodothalassium salexigens*)[a]

Source/donor	*Rhodospirillum rubrum*	*Rhodospirillum photometricum*	*Phaeospirillum molischianum*	*Phaeospirillum fulvum*	*Rhodopila globiformis*	*Rhodocista centenaria*	*Roseospirillum parvum*	*Rhodospira trueperi*	*Roseospira mediosalina*	*Rhodothalassium salexigens*
Carbon sources:										
Acetate	+	+	+	+	−	+	+	+	+	+
Arginine	+	−	−	−	−	nd	nd	nd	−	−
Aspartate	+	+/−	+/−	+/−	−	+	nd	nd	+	nd
Benzoate	−	−	−	+	−	nd	nd	nd	−	nd
Butyrate	+	+	+	+	−	+	+	+	+	nd
Caproate	+	−	+	+	−	+	nd	nd	nd	nd
Caprylate	nd	−	+	+	−	+	nd	nd	−	nd
Citrate	−	−	−	−	−	nd	nd	nd	−	+
Ethanol	+	+	+	+	+	nd	nd	-	−	nd
Formate	−	−	nd	nd	−	nd	nd	nd	−	nd
Fructose	+/−	+	−	−	+	−	+	nd	−	−
Fumarate	+	+	+	+	+	−	+	+	+	nd
Glucose	−	+	−	+/−	+	−	nd	nd	−	+
Glutamate	+	−	nd	nd	−	+	+	nd	+	+
Glycerol	−	+	−	−	−	nd	nd	nd	−	+
Glycolate	nd	+	−	nd	−	nd	nd	nd	−	nd
Lactate	+	+	+/−	−	−	+	+	+	+	−
Malate	+	+	+	+	+	−	+	+	+	nd
Mannitol	−	+	−	−	+	nd	nd	nd	−	nd
Methanol	+/−	−	−	+/−	−	nd	nd	nd	−	nd
Pelargonate	nd	+/−	+	+	−	nd	nd	nd	nd	nd
Propionate	+	+/−	+	+	−	nd	+	+	+	nd
Pyruvate	+	+	+	+	+	+	+	+	+	+
Succinate	+	+	+	+	+	−	+	+	+	+
Tartrate	−	−	−	nd	+	nd	nd	nd	−	nd
Valerate	+	+	+	+	−	+	+	+	−	nd
Electron donors:										
Hydrogen	+	+	nd	nd	nd	nd	nd	nd	nd	nd
Sulfide	+	−	−	−	nd	nd	+	+	+	−
Sulfur	−	−	−	−	−	nd	nd	−	−	−
Thiosulfate	−	−	−	−	−	nd	+	−	−	−

[a]Symbols: +, positive in most strains; −, negative in most strains; +/−, variable in different strains; nd, not determined.

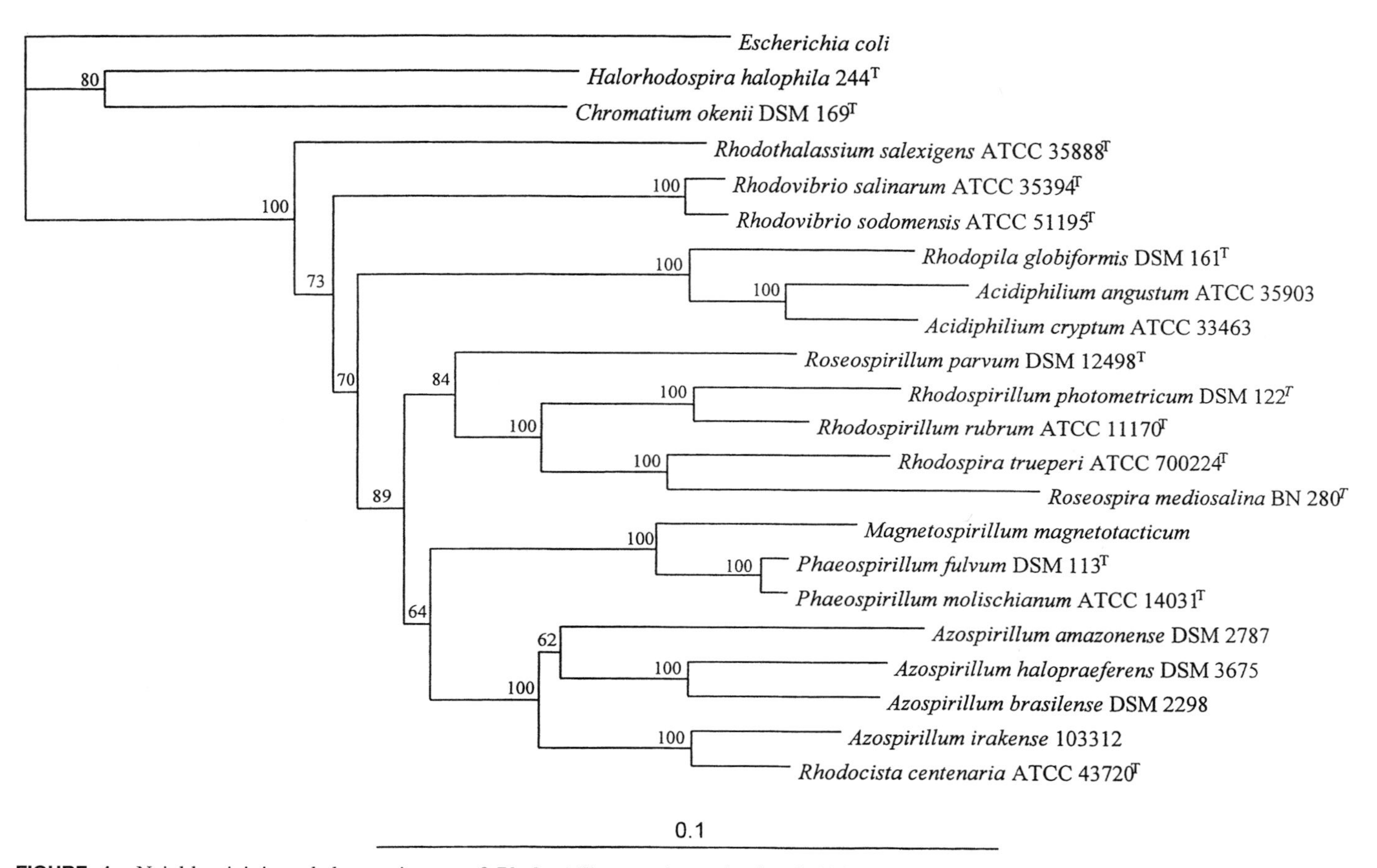

FIGURE. 1. Neighbor-joining phylogenetic tree of *Rhodospirillum* species and related *Alphaproteobacteria* of the order *Rhodospirillales* based on 16S rDNA sequences. The sequence of *Escherichia coli* was used as an outgroup to root the tree. Bar = 10% substitution of nucleotides.

plasmic membrane. There is variation in the presence of either ubiquinone alone, ubiquinone together with either rhodoquinone or menaquinone, or ubiquinone with both menaquinone and rhodoquinone as major components. Most species have side chains with10 isoprenoid units (except *Blastochloris* species). As far as known, either small or large "mitochondrial type" cytochrome c_2 is present.

Differentiation of the Phototrophic *Alphaproteobacteria* in the Order *Rhizobiales* The formation of filamentous stalks and the characteristic growth cycle are the most obvious features that distinguish *Rhodomicrobium* from other phototrophic *Alphaproteobacteria* in the order *Rhizobiales*. According to r-RNA/DNA hybridization studies, *Rhodomicrobium* is clearly distinguished from other purple nonsulfur bacteria (Gillis et al., 1982). *Hyphomicrobium vulgare* is among the closest phylogenetic relatives of *Rhodomicrobium* based on 16S rDNA sequence analysis (Kawasaki et al., 1993). Other outstanding characteristics are the composition of lipid A, the polar lipids and the fatty acids. Major differentiating properties of the genera and species of *Rhodopseudomonas, Rhodobium, Rhodoplanes, Blastochloris, Rhodoblastus,* and *Rhodomicrobium* are shown in Table 3. Carbon sources used by these species are listed in Table 4. The phylogenetic relationships of these genera are shown in Fig. 2.

Taxonomic Comments Most of the species of the phototrophic *Alphaproteobacteria* in the order *Rhizobiales* have been previously known as *Rhodopseudomonas* species and have rod-shaped motile cells. These species are now known to belong to the recently described genera *Rhodopseudomonas, Rhodobium, Rhodoplanes, Rhodoblastus* and *Blastochloris* as well as to the well-known *Rhodomicrobium.* Based on analysis of 16S rDNA sequences, the phototrophic *Alphaproteobacteria* of the order *Rhizobiales* are well separated from other groups of phototrophic *Alphaproteobacteria*; however, these genera are closely related to several purely chemotrophic *Alphaproteobacteria* of the order *Rhizobiales. Rhodopseudomonas palustris*, for example, is closely related to *Nitrobacter* species. In the current Taxonomic Outline (Garrity et al., this volume), the genera *Rhodomicrobium, Blastochloris,* and *Rhodoplanes* are assigned to the family *Hyphomicrobiaceae*, *Rhodobium*, to the family *Rhodobiaceae*, and *Rhodopseudomonas* and *Rhodoblastus*, to the family *Bradyrhizobiaceae*.

***Phototrophic* Alphaproteobacteria** (Rhodobacterales) Cells are ovoid to rod-shaped, motile by polar flagella or non-motile, and multiply by binary fission or show polar growth and budding. Internal photosynthetic membranes consist of vesicles or lamellae. Color of cell suspensions is dependent on the growth conditions and varies between yellowish, beige, brown, brown-red and red. Photosynthetic pigments are bacteriochlorophyll *a* (esterified with phytol) and carotenoids of the spheroidene series. The formation of pigments and the internal membrane systems are repressed under oxic conditions but become derepressed at low oxygen tensions.

Cells grow preferentially photoheterotrophically under anoxic conditions in the light. Photoautotrophic growth with mo-

TABLE 3. Differential characteristics of the anoxygenic phototrophic purple bacteria belonging to the order *Rhizobiales*[a]

Characteristic	*Rhodomicrobium vannielii*	*Rhodobium orientis*	*Rhodobium marinum*	*Rhodoplanes roseus*	*Rhodoplanes elegans*	*Rhodopseudomonas palustris*
Cell diameter (μm)	1.0–1.2	0.7–0.9	0.7–0.9	1.0	0.8–1.0	0.6–0.9
Type of budding	Tube	Sessile	Sessile	Sessile	Tube	Tube
Rosette formation	Complex aggregates	−/+	−	−	+	+
Internal membrane system	Lamellae	Lamellae	Lamellae	Lamellae	Lamellae	Lamellae
Motility	+	+	+	+	+	+
Color of cultures	Orange-brown to red	Pink to red	Pink to red	Pink	Pink	Brown-red to red
Bacteriochlorophyll	*a*	*a*	*a*	*a*	*a*	*a*
Salt requirement	None	4–5%	1–5%	None	None	None
Optimum pH	6.0	7.0–7.5	6.9–7.1	7.0–7.5	7.0	6.9
Optimum temperature	30	30–35	25–30	30	30–35	30–37
Sulfate assimilation[d]	+ (APS)	nd	nd	nd	nd	+ (PAPS)
Aerobic dark growth	+	+	(+)	+	+	+
Denitrification	nd	+	−	+	+	+/−
Fermentation of fructose	nd	−	+	−	−	−
Photoautotrophic growth with	H_2, sulfide	Thiosulfate	Sulfide	Thiosulfate	Thiosulfate	H_2, thiosulfate, sulfide
Growth factors	None	Biotin, *p*-aminobenzoic acid	nd	Niacin	Thiamine, *p*-aminobenzoic acid	*p*-Aminobenzoic acid, (biotin)
Utilization of:						
Benzoate	−	−	−	−	−	+
Citrate	−	−	+/−	+	+	+
Formate	+/−	−	(+)	−	−	−
Glucose	−	+	+	−	−	−
Tartrate	−	−	−	+	+	−
Sulfide	+	−	(+)	−	−	+
Thiosulfate	nd	+	−	+	+	+
Mol % G + C of the DNA	61.8–63.8[e]	65.2–65.7[f]	62.4–64.1[f]	66.8[f]	69.6–69.7[f]	64.8–66.3 [e]
Cytochrome c_2 size	Small	nd	nd	nd	nd	Large
Major quinones	Q-10, RQ-10	Q-10, MK-10	Q-10, MK-10	Q-10, RQ-10	Q-10, RQ-10	Q-10
Major fatty acids						
$C_{14:0}$	2.4	nd	0.4	nd	nd	trace
$C_{16:0}$	3.7	nd	1.9	nd	nd	5.2
$C_{16:1}$	0.6	nd	0.5	nd	nd	3.1
$C_{18:0}$	3.6	nd	14.1	nd	nd	7.3
$C_{18:1}$	85.6	nd	69.0	nd	nd	79.7

[a]Symbols: +, positive in most strains; −, negative in most strains; +/−, variable in different strains; nd, not determined; (+), weak growth or microaerobic growth only; (biotin) biotin is required by some strains; Q-10, ubiquinone 10; MK-10, menaquinone 10; RQ-10, rhodoquinone 10. Bd, buoyant density.

[b]According to 16S rDNA gene sequence analysis, this bacterium belongs to t he genus *Rhodoplanes* rather than to the genus *Rhodopseudomonas* (A. Hirashi, unpublished results).

[c]According to 16S rDNA gene sequence analysis, this bacterium should be considered a species of the genus *Rhodobium* (A. Hirashi, unpublished results).

[d](APS), via adenosine-5′-phosphosulfate; (PAPS), via 3′-phosphoadenosine-5′-phosphosulfate.

[e]Mol% G + C determined by bouyant density centrifugation.

[f]Mol % G + C determined by chemical analysis and HPLC.

(*continued*)

TABLE 3. *(cont.)*

Characteristic	"*Rhodopseudomonas cryptolactis*"[b]	*Rhodopseudomonas julia*[c]	*Rhodopseudomonas rhenobacensis*	*Rhodoblastus acidophilus*	*Blastochloris viridis*	*Blastochloris sulfoviridis*
Cell diameter (μm)	1.0	1.0–1.5	0.4–0.6	1.0–1.3	0.6–0.9	0.5–0.9
Type of budding	Sessile	Sessile	Sessile	Sessile	Tube	Sessile
Rosette formation	+	+	+	−	+	
Internal membrane system	Lamellae	Lamellae	Lamellae	Lamellae	Lamellae	Lamellae
Motility	+	+	+	+	+	+
Color of cultures	Red	Pink	Red	Red to orange-red	Green to olive-green	Olive-green
Bacteriochlorophyll	*a*	*a*	*a*	*a*	*b*	*b*
Salt requirement	None	None	None	None	None	None
Optimum pH	6.8–7.2	6.0–6.5	5.5	5.5–6.0	6.5–7.0	7.0
Optimum temperature	38–40	25–35	(28)	25–30	25–30	28–30
Sulfate assimilation[d]	nd	−	+	+ (APS)	+ (PAPS)	−
Aerobic dark growth	+	+	+	+	(+)	(+)
Denitrification	nd	nd	Reduction of nitrate	−	−	−
Fermentation of fructose	nd	nd	nd	−	nd	nd
Photoautotrophic growth with	−	Sulfide, sulfur	nd	H_2	−	Thiosulfate, sulfide
Growth factors	B_{12}, niacin, *p*-aminobenzoic acid	None	*p*-Aminobenzoic acid	None	Biotin, *p*-aminobenzoic acid	Biotin, *p*-aminobenzoic acid, pyridoxin
Utilization of:						
Benzoate	−	−	−	−	−	−
Citrate	−	−	−	−	−	−
Formate	−	+	+	−	−	−
Glucose	−	+	−	−	(+)	+
Tartrate	−	−	+	−	−	−
Sulfide	−	+	nd	−	−	+
Thiosulfate	−	nd	−	−	−	+
Mol % G + C of the DNA	68.8	63.5	65.4[f]	62.2–66.8[e]	66.3–71.4[e]	67.8–68.4[f]
Cytochrome c_2 size	nd	nd	nd	Small	Small	nd
Major quinones	nd	nd	Q-10	Q-10, MK-10, RQ-10	Q-9, MK-9	Q-8/10, MK-7/8
Major fatty acids						
$C_{14:0}$	nd	nd	nd	0.8	0.5	2.5
$C_{16:0}$	nd	nd	11.7	14.8	8.4	8.6
$C_{16:1}$	nd	nd	9.5	37.2	5.5	9.2
$C_{18:0}$	nd	nd	7.8	0.8	2.2	1.7
$C_{18:1}$	nd	nd	66.1	46.0	74.6	76.5

[a]Symbols: +, positive in most strains; −, negative in most strains; +/−, variable in different strains; nd, not determined; (+), weak growth or microaerobic growth only; (biotin) biotin is required by some strains; Q-10, ubiquinone 10; MK-10, menaquinone 10; RQ-10, rhodoquinone 10. Bd, buoyant density.

[b]According to 16S rDNA gene sequence analysis, this bacterium belongs to t he genus *Rhodoplanes* rather than to the genus *Rhodopseudomonas* (A. Hirashi, unpublished results).

[c]According to 16S rDNA gene sequence analysis, this bacterium should be considered a species of the genus *Rhodobium* (A. Hirashi, unpublished results).

[d](APS), via adenosine-5′-phosphosulfate; (PAPS), via 3′-phosphoadenosine-5′-phosphosulfate.

[e]Mol% G + C determined by bouyant density centrifugation.

[f]Mol % G + C determined by chemical analysis and HPLC.

TABLE 4. Growth substrates of the anoxygenic phototrophic purple bacteria belonging to the order *Rhizobiales*[a]

Source/donor	*Rhodopseudomonas palustris*	*Rhodopseudomonas rhenobacensis*	*Rhodopseudomonas julia*	*"Rhodopseudomonas cryptolactis"*	*Rhodoblastus acidophilus*	*Blastochloris viridis*	*Blastochloris sulfoviridis*	*Rhodomicrobium vannielii*	*Rhodoplanes roseus*	*Rhodoplanes elegans*	*Rhodobium orientis*	*Rhodobium marinum*
Carbon source												
Acetate	+	+	+	+	+	+	+	+	+	+	+	+
Aspartate	+/−	nd	+	−	−	nd	nd	−	−	−	nd	nd
Benzoate	+	−	−	−	−	−	−	−	−	−	−	−
Butyrate	+	+	+	nd	+/−	−	+	+	+	+	+	+
Caproate	+	nd	−	nd	+/−	−	nd	+	−	+/−	+	+
Caprylate	+	nd	−	nd	−	nd	nd	+			−	+/−
Citrate	+/−	−	−	nd	+/−	−	−	−	+	+	−	−
Ethanol	+/−	+	−	nd	+	+	+	+	−	−	−	+/−
Formate	+	+	+	nd	+/−	−	−	+/−	−	−	−	+
D-Fructose	+/−	−	+	−	−	−	+	−	−	−	+	+
Fumarate	+	+	+	nd	+	+	+	+	+	+	+	+
D-Glucose	+/−	−	+	−	+/−	+/−	+	−	−	−	+	+
Glutamate	+	−	−	−	−	+	nd	−	−	−	+/−	nd
Glycerol	+	nd	+	nd	+/−	−	+	+/−	−	−	−	+/−
Glycolate	+	nd	nd	nd	+/−	nd	nd	−	−	−	nd	nd
Lactate	+	+	nd	+	+	+/−	+	+	+	+	+	+/−
Malate	+	+	+	+	+	+	+	+	+	+	+	+
Malonate	+	nd	-	nd	+/−	−	−	+	−	−	nd	nd
Mannitol	+/−	nd	+	nd	−	+/−	−	−	−	−	+/−	+
Methanol	+/−	−	−	nd	+/−	−	−	+/−	−	−	−	−
Propanol	+	nd	−	nd	nd	+/−	+	+	−	−	−	+/−
Propionate	+	−	+	nd	+	−	−	+	+	+	−	+/−
Pyruvate	+	+	+	+	+	+	−	+	+	+	+	+
Sorbitol	+	nd	+	nd	+	+	+	−	−	−	+	+
Succinate	+	+	+	+	+	+	+	+	+	+	+	+
Tartrate	−	+	−	nd	+/−	+/−	−	−	+	+	−	−
Valerate	+	nd	+	nd	+	−	−	+	+	+	+	+
Electron donor												
Sulfide	+	nd	+	nd	−	−	+	+	−	−	−	+/−
Thiosulfate	+	nd	nd	nd	−	−	+	−	+	+	+	−

[a]Symbols: +, positive in most strains; −, negative in most strains; +/− variable in different strains; nd, not determined.

lecular hydrogen, sulfide, thiosulfate and ferrous iron as photosynthetic electron donors may be possible if growth factors are supplied. Most species are capable of chemotrophic growth under microoxic to oxic conditions in the dark. Anaerobic dark growth by fermentation and anaerobic oxidant-dependent growth may also occur.

DIFFERENCES FROM OTHER PHOTOTROPHIC *ALPHAPROTEOBACTERIA*. Characteristic chemotaxonomic features of the phototrophic *Alphaproteobacteria* in the order *Rhodobacterales* are the presence of ubiquinones with 10 isoprenoid units (Q-10) in their side chains (Imhoff, 1984; Hiraishi et al., 1984), large type soluble cytochrome c_2 (Ambler et al., 1979; Dickerson, 1980), C_{18} and C_{16} saturated and monounsaturated fatty acids with $C_{18:1}$ as the predominant component (Imhoff, 1991), and of a phosphate-containing lipid A structure with glucosamine and amide-linked 3-oxo-14:0 and/or 3-OH-14:0 and ester-linked 3-OH-10:0 (Weckesser et al., 1995).

DIFFERENTIATION OF THE PHOTOTROPHIC *ALPHAPROTEOBACTERIA* IN THE ORDER *Rhodobacterales* Characteristic properties of *Rhodobacter* and *Rhodovulum* species are the ovoid to rod-shaped cell morphology, the presence of vesicular internal membranes (except *Rhodobacter blasticus*) and carotenoids of the spheroidene series. *Rhodobacter* species are distinguished from *Rhodovulum* species by the lack of a substantial NaCl requirement for optimal growth, i.e., these species show the typical response of freshwater bacteria. This does not exclude, however, minor requirements for the sodium ion; for example, *Rhodobacter sphaeroides* grows optimally at 4 mM sodium chloride (Sistrom, 1960). The salt requirement for optimal growth of some species does not exclude the possibility that these bacteria may also be able to grow in the absence of salt. The recently described new species *Rhodobaca bogoriensis* is an alkaliphilic slightly halophilic bacterium from African soda lakes and is phylogenetically associated with this group (Milford et al., 2000).

Phototrophic *Alphaproteobacteria* in the order *Rhodobacterales* have a number of characteristic chemotaxonomic properties that enable differentiation. All investigated species have a large type cytochrome c_2 (Ambler et al., 1979) and a single quinone component, Q-10 (Imhoff, 1984). Those species that are able to assimilate sulfate use the pathway via 3′-phosphoadenosine-5′-phosphosulfate (PAPS, Imhoff, 1982). The lipopolysaccharides

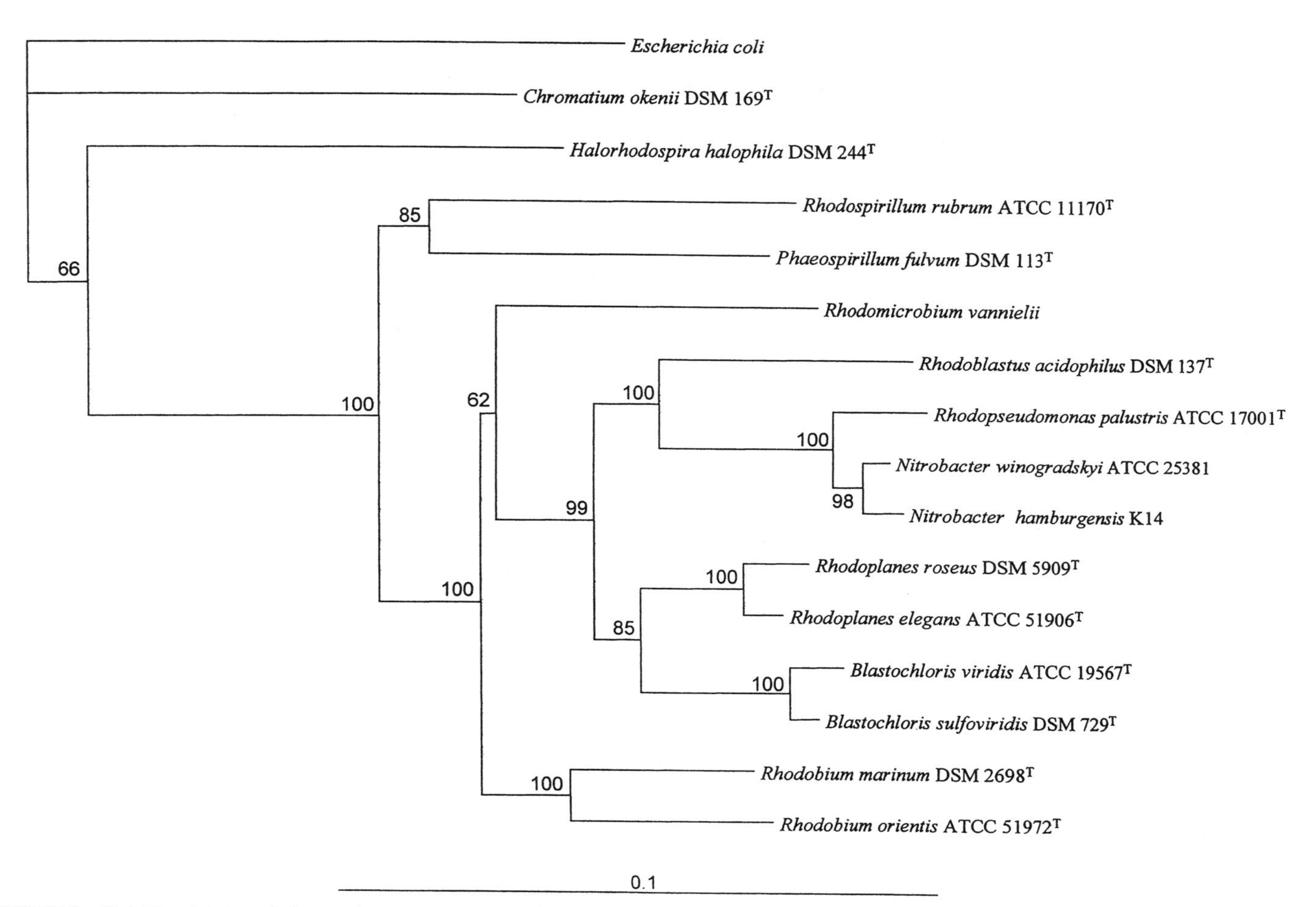

FIGURE 2. Neighbor-joining phylogenetic tree of *Rhodopseudomonas* species and related *Alphaproteobacteria* of the order *Rhizobiales* based on 16S rDNA sequences. The sequence of *Escherichia coli* was used as an outgroup to root the tree. Bar = 10% substitution of nucleotides.

of investigated species contain glucosamine as the sole amino sugar in the lipid A moiety, have phosphate, amide-linked 3-OH-14:0 and/or 3-oxo-14:0, and have ester-linked 3-OH-10:0 (Weckesser et al., 1995). Differentiation of the genera and species of *Rhodobacter*, *Rhodovulum* and *Rhodobaca* is possible based on 16S rDNA sequence analysis (Fig. 3) and DNA–DNA hybridization. Diagnostic properties to distinguish the genera *Rhodobacter*, *Rhodobaca*, and *Rhodovulum* are shown in Table 5.

TAXONOMIC COMMENTS The phototrophic *Alphaproteobacteria* in the order *Rhodobacterales* are phylogenetically well separated from other groups of phototrophic *Alphaproteobacteria*; however, these phototrophic bacteria are closely related to purely chemotrophic *Alphaproteobacteria* in the order *Rhodobacterales*. Phototrophic members of the *Rhodobacterales* have been previously known as *Rhodopseudomonas* species and are currently assigned to the genera *Rhodobacter*, *Rhodovulum* (Pfennig and Trüper, 1974; Imhoff et al., 1984; Hiraishi and Ueda, 1994), and *Rhodobaca* (Milford et al., 2000). *Rhodobacter* species are freshwater bacteria whereas *Rhodovulum* and *Rhodobaca* are true marine bacteria; species of these genera have distinct 16S rDNA sequences (Hiraishi and Ueda, 1994.1995; Hiraishi et al., 1996; Straub et al., 1999).

Intensive DNA–DNA hybridization studies have been performed with both *Rhodobacter* and *Rhodovulum* species. The first detailed study including the 21 strains of the species known at that time gave support for the species recognition of strains of *Rhodobacter veldkampii* and in addition revealed diversity of marine isolates of this group (de Bont et al., 1981). This study also demonstrated the identity on the species level of a denitrifying isolate of *Rhodobacter sphaeroides* and other non-denitrifying strains of this species (Satoh et al., 1976; de Bont et al., 1981). Similarly, several marine and halophilic isolates were shown to be related to *Rhodovulum euryhalinum* by DNA–DNA hybridization but significantly distinct from *Rhodovulum sulfidophilum*, *Rhodobacter sphaeroides* and *Rhodobacter capsulatus* (Ivanova et al., 1988). DNA–DNA hybridization also allowed the genetic distinction of 4 strains of the denitrifying *Rhodobacter azotoformans* from *Rhodobacter sphaeroides* and other *Rhodobacter* species (Hiraishi et al., 1996). Values of 40-50% hybridization between *Rhodobacter azotoformans* and *Rhodobacter sphaeroides* strains correlated with 16S rDNA sequence similarity between their type strains of 98.3% (Hiraishi et al., 1996). Several strains of *Rhodovulum strictum*, which according to 16S rDNA sequence is most similar to *Rhodovulum euryhalinum* (96.8%), were shown to have low DNA–DNA homology (less than 30%) to type strains of all other *Rhodovulum* species, including *Rhodovulum euryhalinum* (Hiraishi and Ueda, 1995). Thus, the species of *Rhodobacter* and *Rhodovulum* not only are well characterized by phenotypic properties, but also are well established based on 16S rDNA sequences and DNA–DNA hybridization studies. The three phototrophic genera of this group are classified in the *Rhodobacteraceae* of the *Rhodobacterales*.

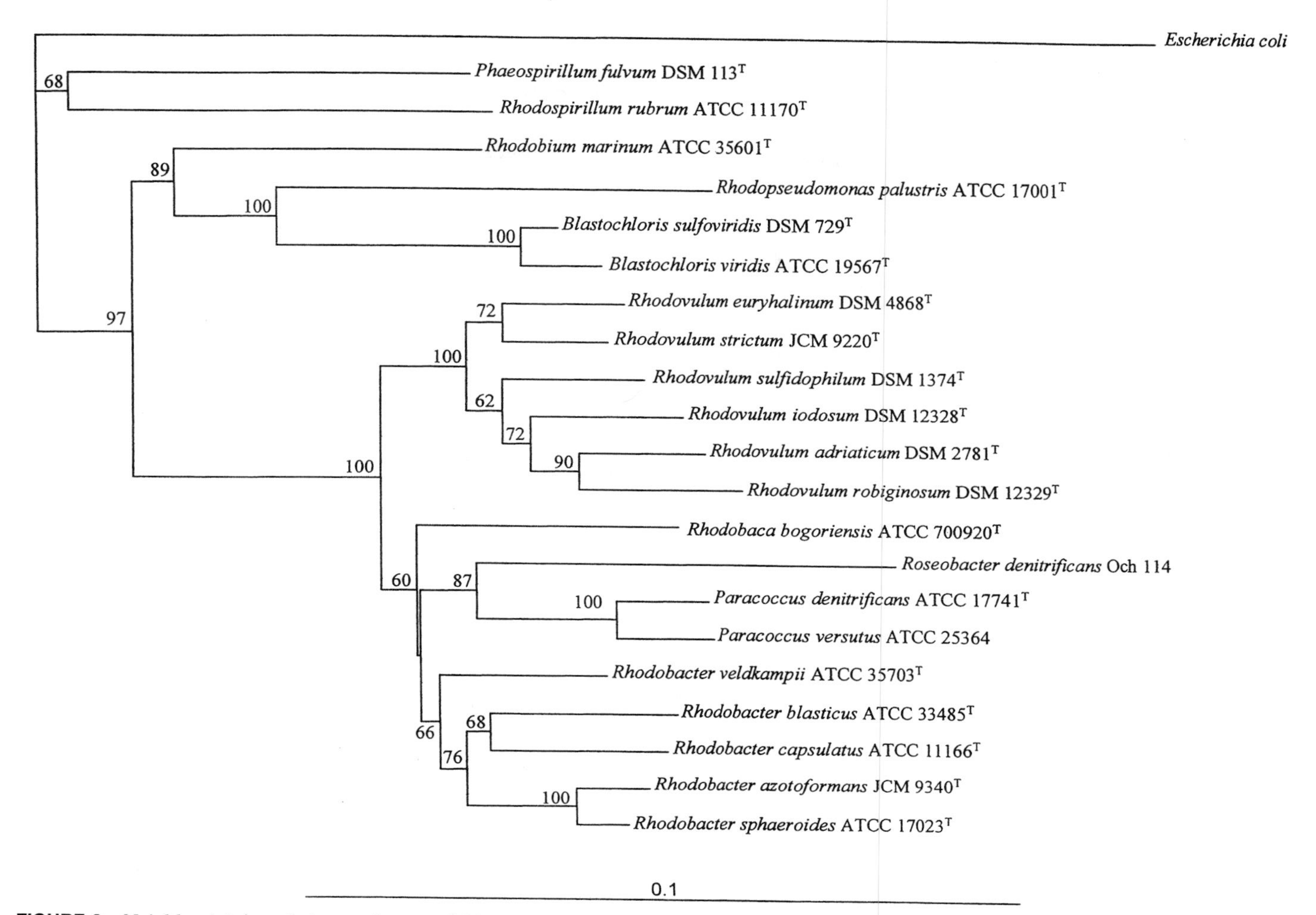

FIGURE 3. Neighbor-joining phylogenetic tree of *Rhodobacter* species and related *Alphaproteobacteria* of the order *Rhodobacterales* based on 16S rDNA sequences. The sequence of *Escherichia coli* was used as an outgroup to root the tree. Bar = 10% substitution of nucleotides.

TABLE 5. Differentiating characteristics of the genera *Rhodobacter*, *Rhodobaca*, and *Rhodovulum*[a]

Characteristic	*Rhodovulum*	*Rhodobacter*	*Rhodobaca*
Salt required for optimal growth	+	−	+
Optimum pH	6.5–7.5	6.5–7.5	9
Final oxidation product of sulfide	SO_4^{2-}	S^0/SO_4^{2-}	S^0
Utilization of:			
Formate	+	+/−	−
Thiosulfate	+	+/−	nd
Polar lipid composition:			
Phosphatidylcholine	−	+/−	nd
Sulfolipid	+	+/−	nd
Mol% G + C of genomic DNA	62–69	64–70	58.8
Light-harvesting complexes	LHI and LHII	LHI and LHII	LHI
Natural habitat	Hypersaline and marine environments	Freshwater and terrestrial environments	Soda lakes
16S rRNA signature(s) at corresponding position(s) in the E. coli sequence:			
359	A	G	G
408	C	C	C
578	G	A	G
1311	C	G	G
1353–1355	CGT	CGT	CGG
1365–1367	ACG	ACG	CCG
1473	A	G	G
1449–1452	TTC/AG	GCAA	CAAT

[a]Symbols: +, positive in most strains; −, negative in most strains; +/−, variable in different strains; nd, not determined.

PHOTOTROPHIC *BETAPROTEOBACTERIA* (*RHODOCYCLALES* AND *BURKHOLDERIALES*)

Bacteria of this group are phototrophic purple nonsulfur bacteria, able to perform anoxygenic photosynthesis with bacteriochlorophylls and carotenoids as photosynthetic pigments. Cells are straight to curved rods, or circles, may be motile by means of polar flagella, divide by binary fission and do not have gas vesicles. Internal photosynthetic membranes are much less developed than in other phototrophic purple bacteria appearing as small finger-like intrusions and are not always evident.

Growth preferentially occurs under photoheterotrophic conditions anaerobically in the light. Reduced sulfur compounds are not used as photosynthetic electron donor and sulfide inhibits growth at low concentrations. Sulfate can be assimilated as the sole sulfur source and is reduced with adenosine-5′-phosphosulfate (APS) as an intermediate (Imhoff, 1982). NADH is used as a cosubstrate in the glutamine synthetase/glutamate synthase reactions, and high potential iron-sulfur protein is present (Ambler et al., 1979).

Phototrophic *Betaproteobacteria* are freshwater bacteria common in stagnant waters that are exposed to light, have an increased load of organic compounds and nutrients, and are deficient in oxygen.

***Differences from phototrophic* Alphaproteobacteria** The phototrophic *Betaproteobacteria* have ubiquinone and menaquinone (or rhodoquinone) derivatives with 8 isoprenoid units in the side chain (Q-8, RQ-8 and MK-8). A "small type" cytochrome c_{551} occurs that is typically found in species of the *Chromatiaceae* and *Ectothiorhodospiraceae*, but not in phototrophic *Alphaproteobacteria* (Ambler et al., 1979; Dickerson, 1980). Characteristic phospholipid and fatty acid compositions occur that have the highest proportions of C_{16} fatty acids ($C_{16:0}$ and $C_{16:1}$) among all phototrophic purple bacteria and correspondingly very low proportions of $C_{18:1}$ (see Hiraishi et al., 1991; Imhoff, 1984; Imhoff and Bias-Imhoff, 1995; Imhoff and Trüper, 1989). Lipopolysaccharides of phototrophic *Betaproteobacteria* characteristically contain significant amounts of phosphate and amide-linked 3-OH-capric acid (3-OH-C-10) in the lipid A moiety (Weckesser et al., 1995). In *Rhodoferax fermentans* 3-OH-C-8:0 was found instead (Hiraishi et al., 1991).

TABLE 6. Differential characteristics of species of the genera *Rhodocyclus*, *Rubrivivax* and *Rhodoferax*[a]

Characteristic	*Rhodocyclus purpureus*	*Rhodocyclus tenuis*	*Rubrivivax gelatinosus*	*Rhodoferax fermentans*	*Rhodoferax antarcticus*
Cell diameter (μm)	0.6–0.7	0.3–0.5	0.4–0.7	0.6–0.9	0.7
Cell shape	Half-circle to circle	Curved rods	Straight to curved rods	Curved rods	Curved rods
Motility	−	+	+	+	+
Slime production	−	+	+	−	−
Color	Purple-violet to violet	Brownish-red or purple-violet	Brown	Peach brown	Peach brown
Major carotenoids	Rhodopin, rhodopinal	Rhodopin, rhodopinal, lycopene[b]	Spheroidene, OH-spheroidene, spirilloxanthin	Spheroidene, OH-spheroidene, spirilloxanthin	Most likely spheroidene and OH-spheroidene
Growth factors	B_{12}, *p*-aminobenzoic acid, biotin	None[d]	thiamine, biotin[c]	thiamine, biotin	biotin
Gelatin liquefaction	−	−	+	+	nd
Fructose fermentation	−	−	−	+	−
Starch hydrolysis	nd	nd	+	−	nd
Tween 80 lysis	nd	nd	+	−	nd
Carbon sources:					
Benzoate	+	−	−	−	−
C_{10} to C_{18} fatty acids	−	+	+	nd	nd
Citrate	−	−	+	+	+
Mannitol	−	−	−	+	−
Sorbitol	−	−	+	+	nd
N_2-fixation	−	+	+	+	+
Fumarate reductase activity:					
With reduced methylviologen	High	High	High	Low	nd
With $FMNH_2$	Low	Low	Low	High	nd
Major fatty acids:					
$C_{16:0}$	33–35	33-36	24–35	33–39	nd
$C_{16:1}$	40–45	43–50	35–45	52–54	nd
$C_{18:0}$	<1	<1	1–3	<1	nd
$C_{18:1}$	18	15–18	16–25	5	nd
3-OH fatty acid	10 : 0	10 : 0	10 : 0	8 : 0	nd
Major quinones	Q-8 + MK-8	Q-8 + MK-8	Q-8 + MK-8	Q-8 + RQ-8	nd
Mol% G + C of DNA					
by HPLC	65.1	64.1–64.8	71.2–72.1	59.8–60.3	nd
by Bd	65.3	64.8	70.5–72.4	nd	nd
by T_m	67.7	64.4 –67.2	70.2–71.9	nd	61.5

[a]Symbols: +, positive in most strains; −, negative in most strains; Q–8, ubiquinone-8; RQ-8, rhodoquinone-8; MK-8, menaquinone-8.

[b]Some strains may contain carotenoids of the spirilloxanthin series and lack rhodopinal (Schmidt, 1978).

[c]Some strains may also require pantothenate.

[d]Some strains may require vitamin B_{12} (Siefert and Koppenhagen, 1982).

***Differentiation of the phototrophic* Betaproteobacteria** At present, five species belonging to three genera of the anoxygenic phototrophic *Betaproteobacteria* are described. Characteristic properties for differentiation of these genera and species are given in Table 6. Carbon sources used by these species are listed in Table 7. The phylogenetic relationships of the phototrophic *Betaproteobacteria* based on 16S rDNA sequences are shown in Fig. 4.

Taxonomic comments Prior to the establishment of the phylogenetic relationship among the phototrophic *Betaproteobacteria,* these species were included in the *Rhodospirillaceae* together with the phototrophic *Alphaproteobacteria* (Pfennig and Trüper, 1974). Three species were known as *Rhodopseudomonas gelatinosa, Rhodospirillum tenue* (Pfennig and Trüper, 1974) and *Rhodocyclus purpureus* (Pfennig, 1978). In addition to a clear phylogenetic separation, both of these groups show significant differences in a number of chemotaxonomic properties. As a consequence, *Rhodospirillum tenue* (Pfennig, 1969) was transferred to *Rhodocyclus tenuis* (Imhoff et al., 1984). In addition, *Rhodopseudomonas gelatinosa* was transferred to this genus as *Rhodocyclus gelatinosus* (Imhoff et al., 1984). Because of its phylogenetic distance from *Rhodocyclus purpureus,* it was assigned later to a new genus as *Rubrivivax gelatinosus* (Willems et al., 1991b). Additional new bacteria have been isolated since then that are also members of the *Betaproteobacteria* and have been described as the new species and genus *Rhodoferax fermentans* (Hiraishi and Kitamura, 1984; Hiraishi et al., 1991) and as an additional species of this genus, *Rhodoferax antarcticus* (Madigan et al., 2000). Based on 16S rDNA sequences, phototrophic *Betaproteobacteria* form different phylogenetic lines within the *Betaproteobacteria* (Hiraishi, 1994). *Rhodocyclus* is classified in the family *Rhodocyclaceae* of the order *Rhodocyclales*; *Rhodoferax* is classified with the *Comamonadaceae* of the order *Burkholderiales,* and *Rubrivivax* is presently classified as *incertae sedis* in the order *Burkholderiales.*

Phototrophic Gammaproteobacteria (Chromatiales)

Phototrophic purple sulfur bacteria that are able to perform photosynthesis under anoxic conditions without oxygen production, that preferentially use reduced sulfur compounds as photosynthetic electron donors, and that grow photolithoautotrophically are *Gammaproteobacteria.* They are anoxygenic phototrophic bacteria and contain bacteriochlorophyll *a* or *b* and various types of carotenoids as photosynthetic pigments located in the cytoplasmic membrane and in internal membrane systems of different fine structure, which originate from and are continuous with the cytoplasmic membrane. These species are classified with the *Chromatiaceae* and *Ectothiorhodospiraceae* of the *Chromatiales.* The *Chromatiaceae* at present represent a family containing 26 genera of phototrophic bacteria and the family Ectothiorhodospiraceae includes three genera of phototrophic bacteria (*Ectothiorhodospira, Thiorhodospira, Halorhodospira*) as well as the phylogenetically distinct chemotrophic genera (*Arhodomonas, Nitrococcus,* and *Alkalispirillum*). The diagnosis and differentiating properties of these bacteria are treated with the description of the *Chromatiales.*

TABLE 7. Carbon sources and electron donors used by species of the genera *Rhodocyclus, Rubrivivax* and *Rhodoferax*[a]

Source/donor	*Rhodocyclus purpureus*	*Rhodocyclus tenuis*	*Rubrivivax gelatinosus*	*Rhodoferax fermentans*	*Rhodoferax antarcticus*
Carbon source					
Acetate	+	+	+	+	+
Arginine	−	−	nd	−	nd
Aspartate	−	−	+	+	+
Benzoate	+	−	−	−	−
Butyrate	+	+	+/−	+	+
Caproate	+	+	nd	−	−
Caprylate	−	+/−	nd	−	−
Citrate	−	−	+	−	+
Ethanol	−	+/−	+	+/−	−
Formate	−	−	+/−	−	−
Fructose	−	−	+	+	+
Fumarate	+	+	+	+	+
Glucose	−	−	+	+	+
Glutamate	−	−	+	+	−
Glycerol	−	−	−	−	−
Glycolate	−	−	nd	−	−
Lactate	−	+	+	+/−	+
Malate	+	+	+	+	+
Malonate	−	−	nd	−	nd
Mannitol	−	−	−	+	−
Mannose	−-	−	+	+	−
Methanol	−	−	+/−	−	−
Pelargonate	−	+	nd	nd	nd
Propionate	−	+/−	+/−	−	−
Pyruvate	+	+	+	+	+
Sorbitol	−	nd	−	+	nd
Succinate	−	+	+	+	+
Tartrate	−	−	+/−	−	nd
Valerate	−	+	+	nd	−
Electron donor:					
Hydrogen	+	+	+	nd	+
Sulfide	−	−	−	−	−
Sulfur	−	−	−	nd	−
Thiosulfate	−	−	−	−	−

[a] Symbols: +, positive in most strains; −, negative in most strains; +/− variable in different strains; nd, not determined.

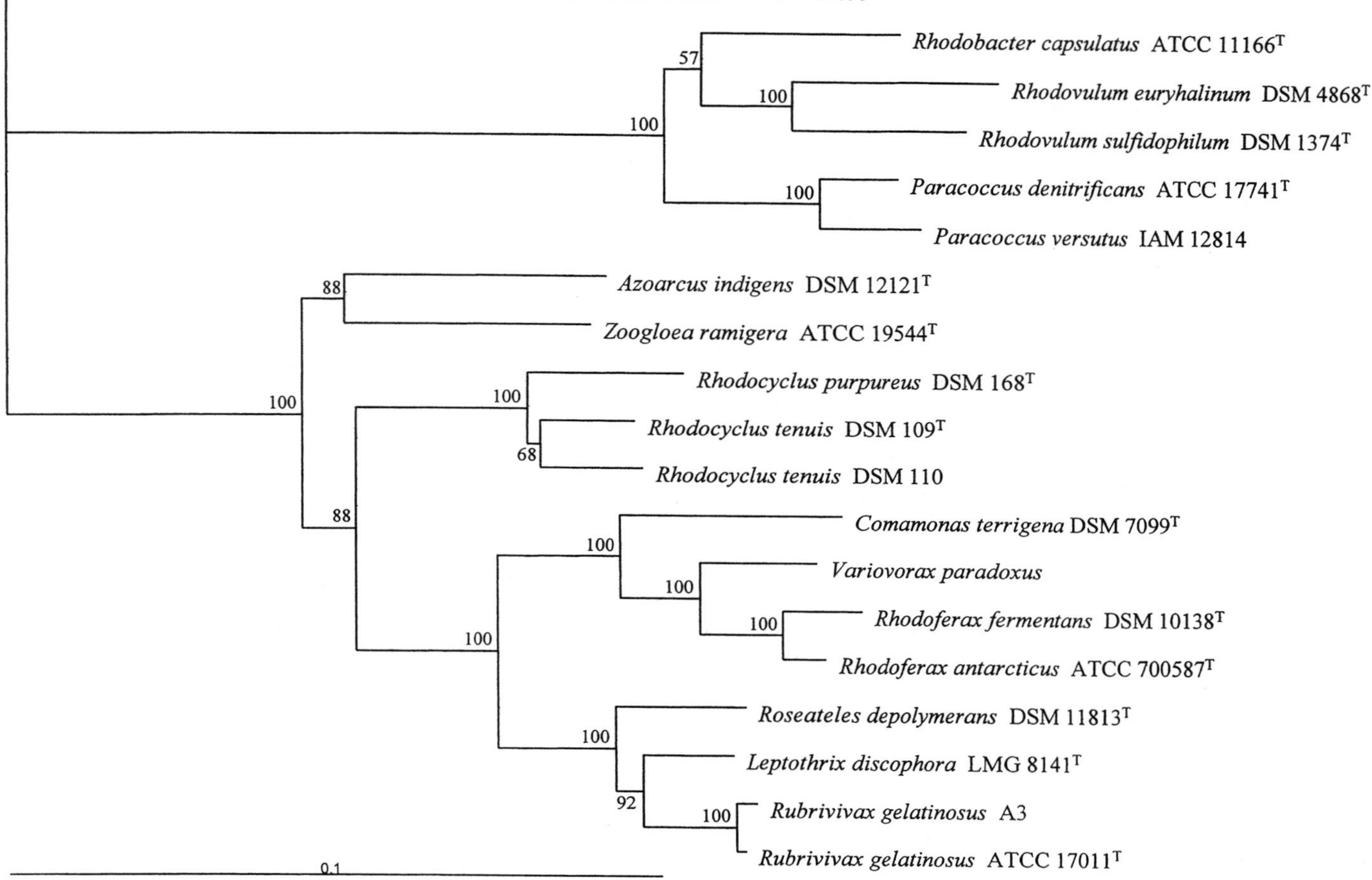

FIGURE 4. Neighbor-joining phylogenetic tree of *Rhodocyclus, Rhodoferax, Rubrivivax* species and purely chemotrophic representatives of the *Betaproteobacteria* based on 16S rDNA sequences. The sequence of *Chromatium vinosum* was used as an outgroup to root the tree. Bar = 10% substitution of nucleotides.

Aerobic Bacteria Containing Bacteriochlorophyll And Belonging To The *Alphaproteobacteria*

Johannes F. Imhoff and Akira Hiraishi

This group includes obligately aerobic chemoheterotrophic bacteria that contain bacteriochlorophyll (BChl) *a* and carotenoids, have a photosynthetic apparatus in which the BChl is integrated, and may develop internal membrane systems as seen in the anoxygenic phototrophic purple bacteria. The BChl content is significantly lower than in the phototrophic purple nonsulfur bacteria.

The aerobic bacteriochlorophyll-containing bacteria ("ABC bacteria") are strictly dependent on energy generation by respiratory electron transport processes with O_2 or alternative electron acceptors (*Roseobacter denitrificans*). They synthesize BChl, as well as the photosynthetic apparatus, only under oxic conditions or in the presence of alternative acceptors, as in the case of *Roseobacter denitrificans* (Takamiya et al., 1992). In fact, under anoxic conditions and in the light, they will neither grow nor produce photosynthetic pigments.

The mol% G + C content of the DNA ranges from 57 to 60 (Td) in *Erythrobacter* species to 74.0–74.8 (Td) in *Rubrimonas* species.

Further Descriptive Information

Some methylotrophic bacteria (Sato, 1978) and isolates from oxic marine environments (Harashima et al., 1978; Shiba et al., 1979) were the first obligately aerobic chemoheterotrophic bacteria known to contain BChl *a*. Both are pink to orange in color. Although initially viewed as an extraordinary and unusual kind of photosynthetic bacteria, their wide distribution in nature and occurrence in a number of phylogenetic lineages within the *Alphaproteobacteria* and *Betaproteobacteria* demonstrates that they are not at all unusual and rare, but may indeed represent phylogenetic offspring of anoxygenic purple nonsulfur bacteria that have adapted to the oxic environment. If their wide distribution within the *Alphaproteobacteria* is due to phylogenetic development (and not multiple lateral gene transfer), and if they have anoxygenic phototrophic purple bacteria as their phylogenetic ancestors, they must have evolved several times and from different phototrophic ancestors. During this adaptation, they have lost some of the properties of their truly phototrophic and anaerobic relatives but have gained others that are of importance in the newly conquered oxic environment. They have lost the ability to grow phototrophically under anoxic conditions in the light and do not produce photosynthetic pigments under anoxic conditions. This is in strict contrast to the phototrophic purple bacteria, where the presence of oxygen precludes the formation of photosynthetic pigments, and its absence induces their synthesis.

Many species can synthesize BChl in permanent darkness in the presence of O_2, but continuous light represses the accumulation of BChl (Harashima et al., 1987; Shioi and Doi, 1988; Sato et al., 1989). While the synthesis of BChl is stimulated under intermittent light/dark cycles in some species, in others such as *Bradyrhizobium* (*"Photorhizobium"*) BTAi and *Methylobacterium rhodesianum*, it is dependent on the provision of such conditions (Sato and Shimizu, 1979; Sato et al., 1985; Evans et al., 1990). Thus, the effect of light is apparently quite complex. It does not repress the synthesis of BChl in all ABC bacteria, but only in some of them. Under certain conditions, light has stimulatory effects upon synthesis.

Light-stimulated CO_2 uptake and increases in cellular ATP pools in *Roseobacter denitrificans* (Shiba, 1984), together with the positive effect of light on survival under conditions of starvation, were the first indications of a functional photosynthetic apparatus in ABC bacteria. Cells of *Roseobacter denitrificans* grown aerobically in the dark which had accumulated significant amounts of BChl showed higher growth rates in a subsequent light incubation than did control cells kept in the dark (Harashima et al., 1987). In addition, light-dependent ATP formation in membrane preparations was demonstrated in some species (Takamiya and Okamura, 1984; Okamura et al., 1986). However, light-stimulated activities were observed only in the presence of O_2 (or auxiliary oxidants as in *Roseobacter denitrificans*), i.e., under conditions that would prevent photopigment synthesis and inhibit photosynthesis in phototrophic purple bacteria. Aerobic bacteriochlorophyll-containing bacteria are not able to gain the major part of their energy from photosynthesis and may be able to use photosynthesis only as a supplementary energy source under certain environmental and growth conditions. This extra supplement of energy might be sufficient, however, to give them a selective advantage over purely chemotrophic bacteria and pays back the expenditure for making these sophisticated molecules and structures. Most likely, ABC bacteria are capable of photosynthetic light utilization when organic carbon as energy source is scarce (Kolber et al., 2001).

Aerobic bacteriochlorophyll-containing bacteria have significantly lower amounts of BChl than phototrophic purple nonsulfur bacteria. The BChl found in most of these bacteria is BChl *a* esterified with phytol (Shiba and Abe, 1987; Harashima and Takamiya, 1989). The *in vivo* absorption spectra of these bacteria show red-shifted spectra of BChl similar to those of phototrophic purple nonsulfur bacteria, indicating the incorporation of BChl into similar pigment-protein complexes. Quite characteristically

TABLE 1. Differential characteristics of genera of aerobic bacteriochlorophyll-containing bacteria

Family	*Acetobacteraceae*					*Methylobacteriaceae*	*Rhodobacteraceae*	
Characteristic	*Acidiphilium*	*Acidisphaera*	*Craurococcus*	*Paracraurococcus*	*Roseococcus*	*Methylobacterium*	*Roseobacter*	*Rubrimonas*
Cell shape	Rods	Cocci, short rods	Cocci	Cocci	Cocci	Rods	Ovoid	Rods
Cell diameter (μm)	0.5–0.7	0.7–0.9	0.8–2.0	0.8–1.5	0.9–1.3	0.8–1.0	0.6–0.9	1.0–1.5
Motility	+	–	–	–	+	+	+	+
Color of colonies	Pink, red	Salmon pink	Pink	Red	Pink	Pink	Pink	Red
Zn-BChl	+	–	–	–	–	–	–	–
Near IR peak of BChl *a* (nm)	864	874	872	856	855	870	868–873	871
Anaerobic phototrophy	–	–	–	–	–	–	–	–
Optimum temperature (°C)	25–35	35	28–32	30–34	nd	25–30	20.30	27–30
Optimum pH	3.0–3.5	4.5–5.0	7.5	6.6–6.8	nd	nd	7.0–8.0	7.5–8.0
Salt requirement/optimum	–	–	–	–	–	nd	+	0.5–7.5%
Quinone(s)	Q-10	Q-10	Q-10	Q-10	Q-10	Q-10	Q-10	Q-10
Mol% G+C of the DNA	63–68	69–70	70.5	70.3–71.0	70.4	64–67	56–60	74.0–74.8

(*continued*)

and in contrast to all other ABC bacteria, *Acidiphilium* species produce zinc-chelated BChl *a* (Zn-BChl *a*; for abbreviation see Takaichi et al., 1999) as the major component (Wakao et al., 1996; Hiraishi et al., 1998; Hiraishi and Shimada, 2001). The absorption maximum of Zn-BChl *a* at 763 nm in acetone–methanol extracts is blue-shifted by 7 nm compared to the corresponding absorption maximum of Mg-BChl *a*. Apparently, all strains described so far have a photosynthetic apparatus similar to that of typical phototrophic purple bacteria. In some species, reaction centers and antenna complexes that are similar to those of phototrophic purple nonsulfur bacteria have been isolated. In *Roseobacter denitrificans*, the most intensively studied species, the genes encoding the photosynthetic apparatus were shown to be similar to those of anoxygenic phototrophic purple bacteria.

Most strains of ABC bacteria have been found in a variety of eutrophic aquatic environments and apparently comprise a significant part of the aerobic, chemoheterotrophic, bacterial community. High proportions of aerobic, chemoheterotrophic, BChl *a*-synthesizing bacteria are present in marine environments from the Australian coast (Shiba et al., 1991), the Pacific Ocean, and probably others (Kolber et al., 2001). They have been isolated from freshwater cyanobacterial mats, from marine coastal habitats, even from the deep sea (Yurkov and Beatty, 1998b) and hot springs (Hanada et al., 1997). *Erythrobacter* and *Roseobacter* species were found to be abundant on the surface of seaweeds (on thalli of *Enteromorpha linza* and *Sargassum horneri*), coastal sands, cyanobacterial mats, and water in the high tidal zone. These bacteria could be isolated from these habitats with a medium rich in complex organic substrates (Shiba et al., 1979, 1991). Representatives of the genera *Erythromicrobium* and *Roseococcus* (Yurkov et al., 1991, 1992, 1994; Yurkov and Gorlenko, 1992a, b) were isolated from cyanobacterial mats formed downstream of alkaline hot springs. *Porphyrobacter* species were isolated from eutrophic fresh waters (Fuerst et al., 1993). *Acidiphilium* species inhabit oligotrophic acidic environments (Hiraishi and Shimada, 2001). *Erythromonas* (basonym: *Blastomonas*; Hiraishi et al., 2000a) and *Sandaracinobacter* (Yurkov et al., 1997) are freshwater bacteria. Facultatively methylotrophic *Methylobacterium* species (the former *"Protomonas"*) were isolated from various sources such as foods, soils, and leaf surfaces (Urakami and Komagata, 1984; Bousfield and Green, 1985).

Enrichment and isolation procedure

A wide variety of media rich in organic components such as yeast extract, peptone, Casamino acids, salts of tricarboxylic acids, or sugars have been used to isolate pure cultures of different aerobic bacteriochlorophyll-containing bacteria. No selective medium has been developed for the isolation of these bacteria. As most of these bacteria grow under aerobic, mesophilic, eutrophic conditions, isolation is achieved aerobically at 20–30°C on agar plates containing media rich in complex organic substrates. In the case of marine bacteria, media such as PPES-II, which are rich in peptones and yeast extract, have been used (Shiba et al., 1979; Shioi, 1986).

Taxonomic comments

The great majority of aerobic bacteriochlorophyll-containing bacteria are *Alphaproteobacteria* (Woese et al., 1984a; Komagata, 1989; Fuerst et al., 1993). The number of such bacteria recognized is steadily increasing and the presence of BChl is apparently widely distributed among the *Alphaproteobacteria*, in anoxygenic phototrophic purple bacteria as well as in the obligately chemotrophic ABC bacteria. The first betaproteobacterium reported to be unable to grow phototrophically under anoxic conditions and which synthesizes BChl *a* is *Roseateles depolymerans* (Suyama et al., 1999).

It must be emphasized that ABC bacteria are neither a phylogenetically nor a taxonomically coherent group. Because the great majority belongs to the *Alphaproteobacteria* and because these bacteria have in common the extraordinary property of BChl synthesis under oxic conditions, a short summary on this group is given in this chapter. Major phylogenetic lineages in the *Alphaproteobacteria* are represented by the following groups of species and genera (Fig. 1).

Roseococcus (Yurkov and Gorlenko, 1992a), *Craurococcus, Paracraurococcus* (Saitoh et al., 1998), and *Acidiphilium* species are *Alphaproteobacteria* related to *Acidocella* and *Acetobacter* species. In particular, *Acidisphaera rubrifaciens* is closely related to the an-

TABLE 1. *(cont.)*

Family	*Rhodobacteraceae*		*Sphingomonadaceae*					
Characteristic	*Roseovarius*	*Roseivivax*	*"Citromicrobium"*	*Erythrobacter*	*Erythromicrobium*	*Erythromonas*	*Porphyrobacter*	*Sandaracinobacter*
Cell shape	Rods	Rods	Pleomorphic	Rods	Rods, branched	Ovoid	Pleomorphic	Long thin rods
Cell diameter (μm)	0.7–1.0	0.5–1.0	0.4–0.5	0.2–0.5	0.6–1.0	0.8–1.0	0.4–0.8	0.3–0.5
Motility	+	+	+	+	+	+	D	+
Color of colonies	Red	Pink	Citron-yellow	Orange	Red-orange	Orange-brown	Orange-red	Yellow-orange
Zn-BChl	–	–	–	–	–	–	–	–
Near IR peak of BChl *a* (nm)	877–879	871–873	867	869	832, 868	867	869	867
Anaerobic phototrophy	–	–	–	–	–	–	–	–
Optimum temperature (°C)	8.0–33.5	27–30	20-42	25-30	25–30	25–35	28–48	25–30
Optimum pH	6.2–9.0	7.5–8.0	6.0–8.0	7.0–8.0	7.0–8.5	7.0–8.0	nd	7.5–8.5
Salt requirement/optimum	1–8%	0–20%	1–5%	0.5–9.6%	–	–	–	–
Quinone(s)	Q-10	Q-10	nd	Q-10	Q-10	Q-10	Q-10	Q-9, Q-10
Mol% G+C of the DNA	62–64	59.7–64.4	67.5	60–67	63.3–64.2	65	65–66	68.5

oxygenic purple nonsulfur bacterium *Rhodopila globiformis* (Hiraishi et al., 2000b).

Obligately chemotrophic representatives of the *Alphaproteobacteria* that contain BChl are *Methylobacterium* and *Bradyrhizobium* (*"Photorhizobium"*) species.

According to 16S rDNA sequence analysis, the genera *Roseobacter* (Shiba, 1991), *Rubrimonas* (Suzuki et al., 1999c), *Roseovarius* (Labrenz et al., 1999), and *Roseivivax* (Suzuki et al., 1999b) are closely related to the chemotrophic *Octadecabacter arcticus* and *Sagittula stellata* and, more distantly, to the anoxygenic phototrophic purple nonsulfur bacteria of the genera *Rhodobacter* and *Rhodovulum* in the *Alphaproteobacteria* (Fuerst et al., 1993; Kawasaki et al., 1993; Yurkov et al., 1994, 1997; Gosink et al., 1997). *Roseinatronobacter thiooxidans* (Sorokin et al., 2000a) constitutes a distinct branch located between the genera *Rhodobacter* and *Rhodovulum*. The genus *Roseibium* (Suzuki et al., 2000b) branches off deeply from other genera within the *Alphaproteobacteria*.

Erythrobacter, Erythromicrobium, Erythromonas (basonym *Blastomonas*), *Porphyrobacter, Sandaracinobacter*, and *"Citromicrobium"* (Yurkov et al., 1999) are the genera of *Alphaproteobacteria* closely related to *Sphingomonas* and allied non-BChl-containing genera (Yurkov et al., 1999; Takeuchi et al., 2001).

Terminology of this Bacterial Group

This group of bacteria has been termed the "aerobic photosynthetic bacteria" (Shiba, 1989) and is treated in this chapter as the "aerobic bacteriochlorophyll-containing bacteria" or, in short, the "ABC bacteria". The term "aerobic anoxygenic phototrophs" (Shimada, 1995) or "aerobic anoxygenic phototrophic bacteria" (Yurkov and Beatty, 1998a), is misleading because these bacteria are not phototrophic in a strict sense, that is they do not grow solely at the expense of light energy, although they might be able to gain energy from the existing photosynthetic machinery (i.e., they are photosynthetic).

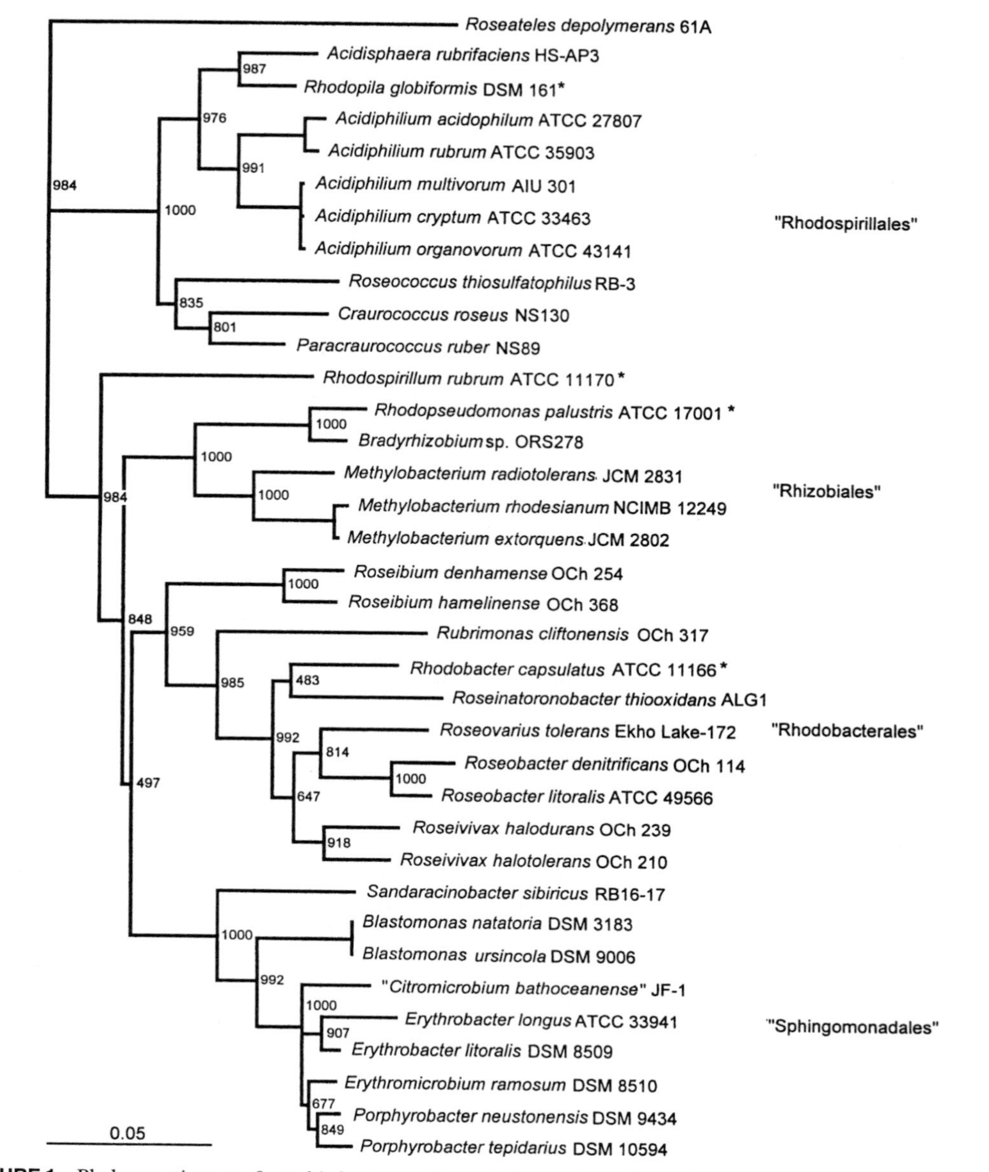

FIGURE 1. Phylogenetic tree of aerobic bacteriochlorophyll-containing bacteria and related phototrophic bacteria based on 16S rDNA sequences.

Nitrifying Bacteria

Eva Spieck and Eberhard Bock

Two groups of highly specialized organisms—the ammonia and the nitrite oxidizers—are called nitrifiers. Although they are not closely related, in the past they have been combined in the family *Nitrobacteraceae* (Buchanan, 1917b; Watson, 1971; Watson et al., 1989). The names of the genera of the ammonia-oxidizing bacteria have the prefix *Nitroso-*, whereas those of the nitrite-oxidizing bacteria start with *Nitro-*. The organisms are characterized by the capacity for lithotrophic growth in which energy is generated by the oxidation of ammonia to nitrite (ammonia oxidizers) or nitrite to nitrate (nitrite oxidizers). Nitrifiers are autotrophic bacteria, fixing CO_2 as the main carbon source via the Calvin cycle. In addition to aerobic lithotrophic growth, the organisms possess a high metabolic diversity. For example, heterotrophic and anaerobic growth by denitrification has been known in *Nitrobacter* for many years (Bock, 1976; Freitag et al., 1987). Anaerobic ammonia oxidation was described recently by Mulder et al. (1995) as well as Schmidt and Bock (1997). In the first case, ammonia and nitrite are combined to form dinitrogen gas (Anammox) by as yet uncultivated planctomycetes (Jetten et al., 1998). Details concerning the lithoautotrophic ammonia-oxidizing bacteria are given in the chapter by Koops and Pommerening-Röser.

Almost twenty years ago, Carl Woese began to analyze the evolutionary relationships of nitrifiers by the use of highly conserved 16S rRNA sequences. In contrast to their common physiology, a high degree of phylogenetic diversity was found among the ammonia and nitrite oxidizers (Woese et al., 1984a, b, 1985). As shown in Fig. 1, these organisms are scattered among the *Alphaproteobacteria, Betaproteobacteria, Gammaproteobacteria,* and *Deltaproteobacteria* (Teske et al., 1994). Nitrite oxidizers of the genus *Nitrospira* occupy a phylogenetically isolated position since they occur in a separate phylum (Ehrich et al., 1995). Most of the nitrifiers were shown to be closely related to photosynthetic bacteria and to resemble them structurally in the arrangement of the intracytoplasmic membranes. According to the conversion hypothesis of Broda (1977), the respiratory membranes of the nitrifiers originated from the photosynthetic apparatus. Ammonia and nitrite oxidizers are peripherally related to methylotrophs that possess similar intracytoplasmic membrane systems.

Ammonia and nitrite oxidizers exist in most aerobic environments where organic matter is mineralized and are widely distributed in soils, fresh water, seawater, sewage, and biofilms. Strains have also been isolated from extreme environments like desert soils, natural stones (Bock and Sand, 1993) or sulfidic ore mines. Nitrifiers have been enriched from permafrost soils up to a depth of 60 m (Wagner et al., 2001) and detected in deep subsurface sediments up to a depth of 260 m (Fredrickson et al., 1989) and in heating systems with temperatures up to 47°C (E. Lebedeva, personal communication). The organisms can be found in suboptimal environments such as acidic soils ($pH < 4$), where they may be protected from adverse conditions by ureolytic activity (De Boer et al., 1989) or by growth on surfaces (Keen and Prosser, 1987) or in aggregates (De Boer et al., 1991). Biofilm matrix may also protect cells from adverse conditions. One acidophilic strain of *Nitrobacter vulgaris* with a pH optimum of 5.5 was isolated from acidic forest soil (Hankinson and Schmidt, 1988). Nitrifiers have also been detected in microaerophilic and anaerobic habitats where reduced nitrogen compounds are formed (Smorczewski and Schmidt, 1991). High concentrations of ammonia and nitrite are rarely found in nature but are often used in laboratory cultures. For example, ammonium ion concentrations in the river Elbe are generally less than 1 ppm, and nitrite is scarcely detectable. K_s values for nitrite oxidation are in the range 15–270 μM for different strains of *Nitrobacter* (reviewed by Prosser, 1989; Both and Laanbroek, 1991). These values are greater than the nitrite concentration in natural environments, where it rarely accumulates (Schmidt, 1982).

The nitrogen of the biosphere exists in the oxidations states of NH_3 (NH_4^+), NO_2^-, and NO_3^-. Lithotrophic nitrification is considered to be the main process by which the transformation of more reduced nitrogen compounds to nitrate occurs. The transfer of ammonia to soil and water is mediated by agricultural fertilization, by microbial activity (e.g., ammonification and N_2-fixation), and by the volatilization of gaseous ammonia. The latter process is important because ammonia and not ammonium ion is the substrate of lithotrophic ammonia oxidizers (Suzuki et al., 1974). At an urban site in Germany (Duisburg), the concentration of volatilized ammonia was high enough to support cell growth as shown by *in situ* experiments over seven years (Mansch and Bock, 1998).

The nitrogen cycle is characterized by mobilization, immobilization, and transformation of the various nitrogen-containing compounds. Nitrification can either prevent nitrogen loss or lead to significant nitrogen loss in local environments by forming the mobile endproducts nitrite and nitrate. At low pH, the oxidation of nitrite is inhibited by the presence of undissociated HNO_2. Nitrite can also be decomposed by chemodenitrification (Chalk and Smith, 1983) especially in acidic environments. Nitrite is chemically unstable and is only formed in quantity when oxygen is limited and at alkaline sites, e.g. on concrete surfaces. Nitrate also can be reduced to N_2 by denitrification.

Nitrifying bacteria are of ecological importance because these organisms convert ammonia, which is often absorbed to soil particles, to nitrate, which is mobile in soil water. The extensive use

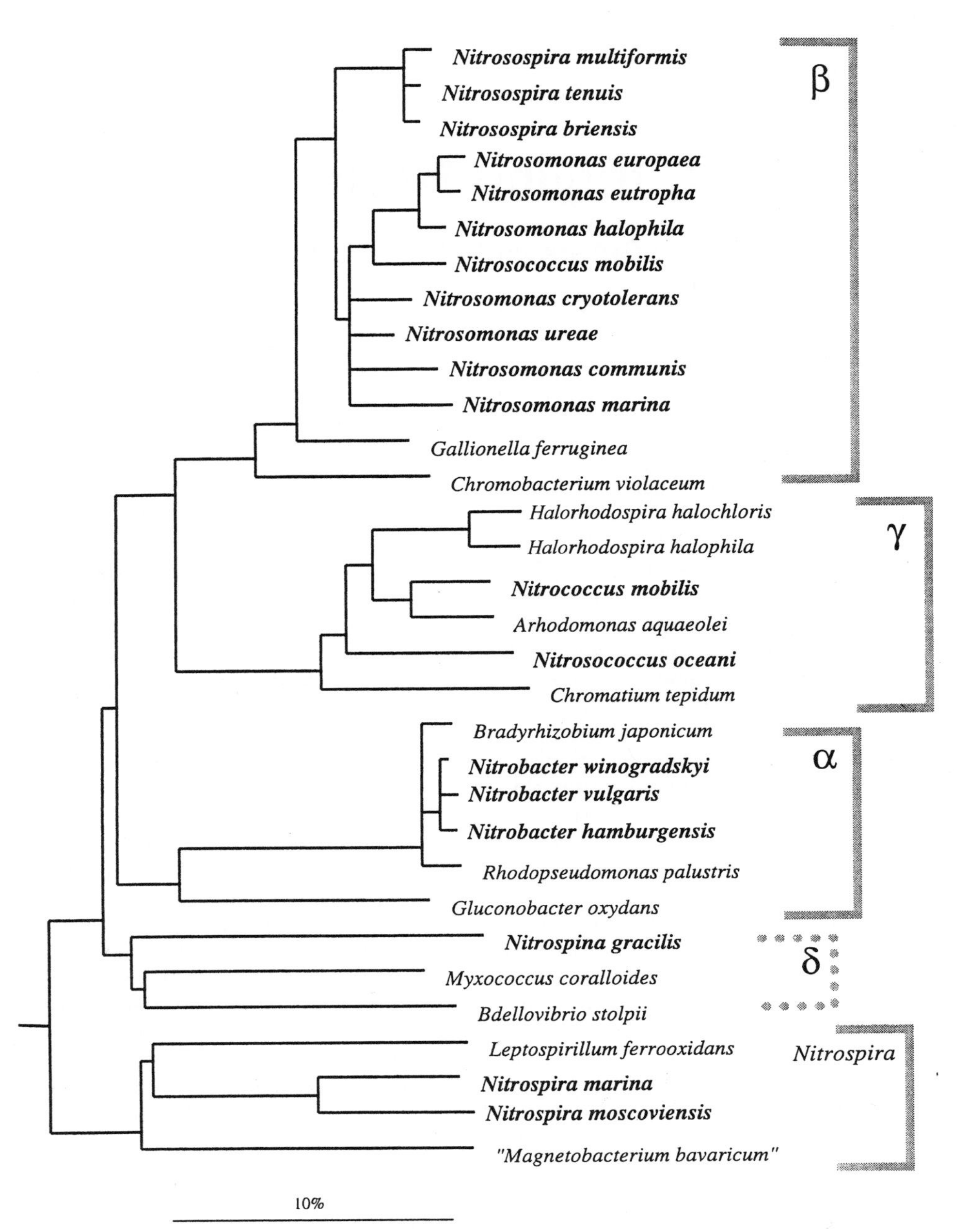

FIGURE 1. 16S rRNA based tree reflecting the phylogenetic relationship of ammonia-and nitrite-oxidizing bacteria. The consensus tree is based on the results of a distance matrix analysis of all available 16S rRNA primary structure data from the members of the phyla shown in the tree and a selection of reference sequences representing the other phyla. Only homologous positions which share identical residues in at least 50% of all available almost complete bacterial 16S rRNA sequences were included for tree reconstruction. The tree topology was corrected according to the results of alternative treeing approaches (maximum parsimony and maximum likelihood) as well as various datasets differing with respect to included sequence positions as well as reference sequences. Multifurcations indicate that a common relative branching order was not significantly supported by the majority of treeing analysis. The classification of *Nitrospina* to the *Deltaproteobacteria* is preliminary. The analysis was performed using the ARB software package (Ludwig and Strunk, 1996). The length bar indicates 10% estimated sequence divergence (Figure courtesy of Wolfgang Ludwig) (Reproduced with permission from E. Spieck and E. Bock, Biospektrum *4:* 25–31, 1998, Biospektrum ©Spektrum Academic Press.)

of nitrogen fertilizers in agriculture in recent decades has influenced the natural balance. One result is increasing nitrate pollution of ground water by the loss of bound nitrogen in agricultural soils. The formation of nitric acid leads to acidification of unbuffered soils followed by the release of positive ions (e.g., Al^{3+}), which are phytotoxic. Nitrifiers also contribute to the biodeterioration of building materials (Bock and Sand, 1993). Nitrogen removal from wastewater by the combined activity of nitrifiers and denitrifiers (Painter, 1986; Eighmy and Bishop, 1989) has become an important area of research with the aim of preventing eutrophication and reducing the toxic effect of ammonia on aquatic organisms (Arthur et al., 1987).

The ecological significance of the nitrogen gases N_2O, NO and NO_2 has become more and more evident in recent years. Nitrifiers are involved in biological transformations of these gases (Goreau et al., 1980; Bock et al., 1995). Nitric oxide (NO) and nitrogen dioxide (NO_2) are trace gases in anthropogenic pollution of the atmosphere; these gases originate mainly from internal combustion engines. Recently it was shown that both gases significantly enhance ammonia oxidation and cell growth of *Nitrosomonas* (Zart and Bock, 1998). Ammonia was oxidized to nitrite in cells of *Nitrosomonas eutropha* grown in the presence of gaseous NO_2 instead of oxygen (Schmidt and Bock, 1997) and in cell-free extracts of *Nitrosomonas eutropha* (Schmidt and Bock, 1998). Nitrifying ammonia oxidizers produce NO (Lipschultz et al., 1981; Kester et al., 1997), and lithotrophic nitrite oxidizers are able to grow with NO to a limited extent (Freitag and Bock, 1990; Mansch and Bock, 1998).

Since lithotrophic nitrifiers are able to grow with inorganic compounds, they can be separated from heterotrophic nitrifying organisms, which are not able to derive energy by the oxidation of nitrogen compounds. Heterotrophic nitrifiers include chemoorganotrophic eubacteria, fungi, and phototrophic algae (Focht and Verstraete, 1977; Killham, 1986). The oxidation of amino groups and free ammonium is mediated by cometabolism. The final product is often nitrite (Castignetti and Gunner, 1980), which is used by lithotrophic nitrite oxidizers in natural environments as an energy source. Only small amounts of nitrite and nitrate are formed by heterotrophic nitrification. Even in unfavorable environmental conditions such as those in acidic forest soils, nitrification by lithotrophic nitrifiers is often the major source of nitrite and nitrate (Stams et al., 1990). Methanotrophic bacteria are also involved in the conversion of nitrogen compounds. These organisms oxidize methane to methanol with the aid of a well-characterized methane monooxygenase (Anthony, 1982); the membrane-bound form of this enzyme is similar to the ammonia monooxygenase of *Nitrosomonas*. Therefore, methane oxidizers are able to oxidize ammonia to nitrite (O'Neill and Wilkinson, 1977), and, conversely, ammonia oxidizers can hydroxylate methane to methanol (Hyman and Wood, 1983). Methane is a competitive inhibitor of the normal substrates of the monooxygenase enzymes and do not support cell growth (Bedard and Knowles, 1989).

Nitrifiers have not been extensively studied because these organisms grow slowly, and cell yield is low. Classical cell counts and isolation of these organisms has been performed by the Most Probable Number (MPN) dilution technique (Matulewich et al., 1975). This method requires incubation periods of several weeks and results have been shown to depend on such factors as the substrate concentration (Both et al., 1990). A fluorescent antibody (FA) technique that employs polyclonal antibodies was developed by Fliermans et al. (1974) for direct microscopic enumeration of nitrifiers. Cell counts obtained by this method yield cell numbers two to three orders of magnitude higher than traditional viable counting techniques (Rennie and Schmidt, 1977). Later the coexistence of several serotypes of *Nitrobacter* and multiple genera of ammonia oxidizers was demonstrated by Belser (1979). However, the antibodies used in these studies were serotype-specific; therefore, unknown and as yet uncultured strains could not be detected.

In order to estimate the activity of natural populations, a potential nitrification assay was developed by Belser and Mays (1982). The authors measured nitrifying activity under optimal laboratory conditions over a short period of time. However, a good correlation between the potential enzyme activity and cell counts was seldom found when the potential nitrification assay method was applied to natural environments (Belser, 1979; Groffman, 1987). In addition, the cell numbers of nitrifying bacteria in stone material (Mansch and Bock, 1998) or in coniferous forest soils (Degrange et al., 1998) were not correlated with the nitrate concentration in these environments. Possible explanations include the loss of nitrate by denitrification, leaching, or the inhibition of nitrification *in situ*. Both counting and activity measurements are dependent on the distribution of the cells in their natural habitat, where ammonia oxidizers as well as nitrite oxidizers aggregate into microcolonies described as zoogloeae or cysts (Watson et al., 1989). Cells in such aggregates produce exopolymeric substances (EPS) that may protect the organisms against toxic substances and changing environmental conditions; however, they hamper separation of individual cells and thereby bias the results obtained from conventional counting techniques. For example, ammonia oxidizers of the genus *Nitrosomonas* in the river Elbe were found attached to flocs (Stehr et al., 1995b).

Molecular techniques are now being used to analyze ammonia and nitrite oxidizers *in situ* without cultivation. For example, the MPN-method was combined with PCR to detect and count *Nitrobacter* in soils (Degrange and Bardin, 1995). Further methods to quantify nitrification include the use of inhibitors (Hall, 1984), ^{15}N dilution techniques (Koike and Hattori, 1978), and isotope pairing (Nielsen, 1992). In the last few years, the introduction of microelectrodes in microbial ecology has enabled scientists to monitor nitrification activity in bacterial communities (Jensen et al., 1993). For example, microprofiles of O_2 and NO_3^- can be measured simultaneously in sediments and biofilms (De Beer et al., 1993). Several sets of PCR primers for amplification of 16S rDNA from ammonia oxidizers were developed to detect these organisms in different habitats (Ward et al., 1997). Oligonucleotide probes have been used to detect nitrite- and ammonia-oxidizing bacteria in biofilms, activated sludge, and composted materials using quantitative 16S RNA whole cell and slot blot hybridization techniques (Mobarry et al., 1996; Wagner et al., 1996). Natural soil populations of ammonia-oxidizing bacteria have been studied using denaturing gradient gel electrophoresis (DGGE) of PCR-amplified 16S rDNA sequences followed by hybridization with oligonucleotide probes designed to detect the 16S sequences of specific groups of ammonia oxidizers (Stephen et al., 1998). The structural gene *amoA* of the ammonia monooxygenase has been used as a functional marker (Rotthauwe et al., 1997). Recently, monoclonal antibodies that recognize the nitrite oxidoreductase were used for the detection and taxonomic classification of nitrite oxidizers at a functional level (Aamand et al., 1996). Immunoblot analysis demonstrated that the similarity of the different enzyme systems reflected the phylogenetic relationships of the four genera of nitrite-oxidizing bacteria (Bartosch et al., 1999). Polyclonal antibodies recognizing the ammonia monooxygenase were developed by Pinck et al. (2001).

Fluorescence *in situ* hybridization (FISH) with 16S rRNA-targeted oligonucleotide probes was used to detect ammonia and nitrite oxidizers in complex environments. The use of such probes allows the diversity and abundance of natural populations to be evaluated on different phylogenetic levels. The distribution of *Nitrosomonas* and *Nitrobacter* in biofilm samples and activated sludge has been visualized by confocal laser scanning microscopy

(Wagner et al., 1995; Mobarry et al., 1996; Schramm et al., 1996). Bacteria that bound probes specific for ammonia oxidizers revealed dense cell clusters, whereas bacteria that bound probes designed to detect *Nitrobacter* revealed less dense aggregates; the two kinds of aggregate were frequently in close contact with each other. Since *Nitrobacter* itself could not regularly be detected (Wagner et al., 1996), the authors suggested that novel nitrifiers were present in the biofilms. Similarly, *Nitrospira* and not *Nitrobacter* was found to dominate in fresh water, biofilm reactors and activated sludge (Burrell et al., 1998; Juretschko et al., 1998). For additional information on *Nitrospira*, see the chapter "Lithoautotrophic Nitrite-Oxidizing Bacteria." Although *Nitrosomonas* and *Nitrobacter* are the two most commonly isolated nitrifiers, they are not necessarily the most abundant ones in natural habitats. Their dominance in enrichment cultures may be explained by the fact that conditions in the commonly used enrichment culture methods promote the growth of these organisms.

The Lithoautotrophic Ammonia-Oxidizing Bacteria

Hans-Peter Koops and Andreas Pommerening-Röser

The physiologically defined group of lithoautotrophic, ammonia-oxidizing bacteria comprises organisms having the ability to utilize ammonia as the major source of energy and carbon dioxide as the main source of carbon. Together with the lithotrophic, nitrite-oxidizing bacteria, the ammonia oxidizers were formerly classified as nitrifying bacteria in the family *Nitrobacteraceae* (Buchanan, 1917b; Starkey, 1948, 1957; Watson, 1971, 1974; Watson et al., 1981, 1989). Phylogenetic investigations, however, have revealed that these two physiologically defined groups of bacteria do not represent a phylogenetically definable unit (Woese et al., 1984a, 1984b, 1985; Teske et al., 1994).

The Organisms

The taxonomic categorization of the lithoautotrophic ammonia oxidizers was based primarily on the early studies of the Winogradskys (Winogradsky 1890a, b, 1891, 1892, 1904, 1930, 1931, 1935a, b, c, 1937; Winogradsky and Winogradsky 1933). This categorization has been difficult because the basic metabolism is identical in all representatives of this group, and only morphological characteristics (shape and ultrastructure of the cells) can be used as discriminating properties (Starkey 1948, 1957; Watson, 1971, 1974; Watson et al., 1981, 1989; Koops and Möller, 1992).

This had led to the definition of five distinct genera, namely *Nitrosomonas* (straight rods with peripherally located flattened vesicles of intracytoplasmic membranes), *Nitrosococcus* (spheres with peripherally or centrally arranged stacks of intracytoplasmic membranes), *Nitrosospira* (tightly wound spirals lacking extensive intracytoplasmic membrane systems), *"Nitrosovibrio"* (curved rods lacking extensive intracytoplasmic membrane systems), and *Nitrosolobus* (pleomorphic lobate cells compartmentalized by intracytoplasmic membranes).

Table 1 lists morphological characteristics useful for the differentiation of the genera of the lithoautotrophic ammonia-oxidizing bacteria.

However, these classical genera represent two phylogenetically distinct groups of ammonia oxidizers (Woese et al., 1984b, 1985; Head et al., 1993; Teske et al., 1994; Pommerening-Röser et al., 1996; Purkhold et al., 2000; Purkhold et al., 2003). The major grouping of ammonia-oxidizers, located within the *Betaproteobacteria*, encompasses two clusters (Fig. 1). The first cluster includes the species of the genus *Nitrosomonas* (six distinct lineages), together with *"Nitrosococcus mobilis"*. Hence, the latter species must be reclassified to the genus *Nitrosomonas* and is therefore listed as *"Nitrosomonas mobilis"* comb. nov. in the chapter "Genus *Nitrosomonas*". The second cluster comprises the species of the classical genera *Nitrosospira*, *"Nitrosovibrio"*, and *Nitrosolobus*. Since the latter three genera reveal a very high level of 16S rDNA similarity to each other and a clear-cut separation on this basis is not practical, Head et al. (1993) have proposed to accommodate them within a single genus, namely *Nitrosospira*. Similar statements have been made by Utåker et al.(1995) and Teske et al. (1994). However, Teske et al. (1994) have pointed out the fact that this group of ammonia oxidizers is currently represented by only a few sufficiently certain 16S rDNA sequences, and that independently reported sequence data often differ in many positions. Another problem is that comparative 16S rRNA gene sequencing does not provide reliable phylogenetic information at levels of 97% similarity and higher among species or groups of species (Ludwig et al., 1998b). Furthermore, the use of 16S rDNA as the sole phylogenetic chronometer of recently evolved diversity within a group of organisms is problematic, as the genome in general evolves more rapidly than does the more conserved rRNA gene (Stackebrandt, 1988). At such high levels of relationship, DNA–DNA reassociation techniques are superior methods for clearing up relationships in detail (Stackebrandt and Goebel, 1994; Ludwig et al., 1998b). Using the S_1 nuclease technique, Pommerening-Röser (1993) has found the striking morphological differences existing among the genera *Nitrosospira*, *"Nitrosovibrio"*, and *Nitrosolobus* to be reflected by a phylogenetic tree constructed on

TABLE 1. Differentiation of the genera of the lithoautotrophic ammonia-oxidizing bacteria

Characteristic	*Nitrosomonas*	*Nitrosospira*	*"Nitrosovibrio"*	*Nitrosolobus*	*Nitrosococcus*
Cell shape	Spherical to rod shaped	Tightly coiled spirals	Slender, curved rods	Lobular	Spherical to ellipsoidal
Intracytoplasmic membranes	Peripherally located flattened vesicles	Occasional tubular invaginations	Occasional tubular invaginations	Cell compartmentalized by cytoplasmic membranes	Centrally located stack of membranes
Flagella	Polar flagella	Peritrichous flagella	Polar to subpolar flagella	Peritrichous flagella	Tuft of flagella

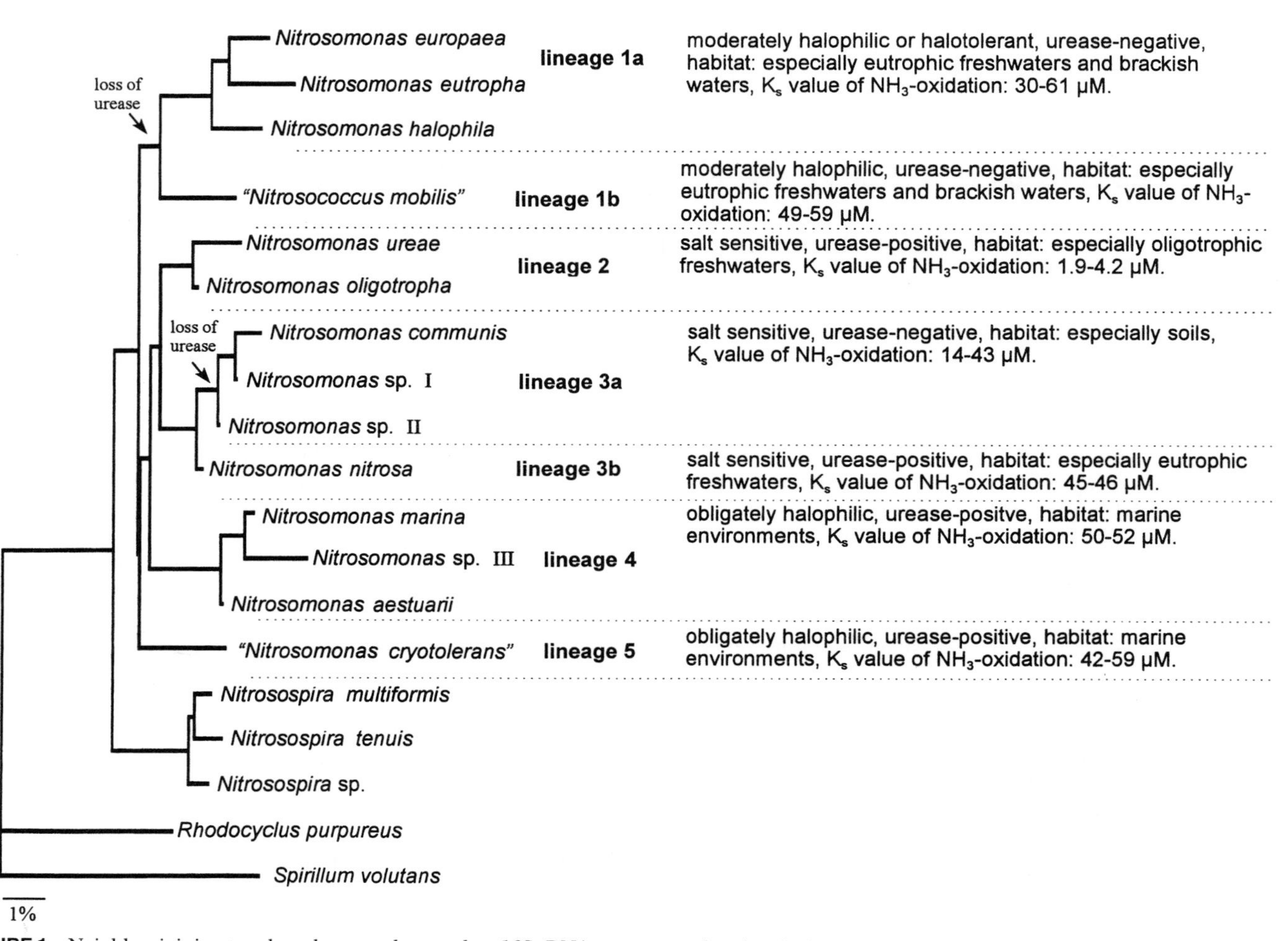

FIGURE 1. Neighbor-joining tree based on nearly complete 16S rDNA sequences, showing the interrelationships among the lithoautotrophic ammonia-oxidizing bacteria of the *Betaproteobacteria* and the most important ecophysiological characteristics of the five distinct lineages of *Nitrosomonas*. *Nitrosospira multiformis* was formerly known as *Nitrosolobus multiformis*; *Nitrosospira tenuis* was formerly known as *"Nitrosovibrio tenuis"*. Bar = 10% sequence divergence.

the basis of DNA–DNA similarity values (Fig. 2), rather than one based on 16S rDNA sequences. Thus, further detailed investigations are needed to ultimately clear up the phylogenetic ultrastructure among members of the three genera *Nitrosospira*, *"Nitrosovibrio"*, and *Nitrosolobus*. Consequently, until that has been done, the original differentiation of the three genera via morphological distinctions should be accepted, although these genera constitute a closely related assemblage of lesser phylogenetic depth than is estimated among the phylogenetic lineages within the genus *Nitrosomonas*.

The second group of ammonia oxidizers, belonging to the *Gammaproteobacteria*, is represented by only two species of the genus *Nitrosococcus*, *Nitrosococcus oceani* and *Nitrosococcus halophilus* (Koops et al., 1990).

Biochemistry of the Organisms

Most of the biochemical investigations have been carried out with *Nitrosomonas europaea* since this species is available from international culture collections. Comparative molecular analyses have indicated the biochemical basis of the ammonia-oxidizing systems of all chemolithotrophic ammonia oxidizers to be relatively uniform (Norton and Klotz, 1991; Rotthauwe et al., 1995; Sinigalliano et al., 1995; Böttcher, 1996).

The oxidation of ammonia to nitrite is generally accepted to be a two-step reaction (Suzuki et al., 1981; Hooper, 1984; Suzuki, 1984; Wood, 1986): $NH_3 + 2[H] + O_2 \rightarrow NH_2OH + H_2O$ and $NH_2OH + H_2O \rightarrow HNO_2 + 4H^+ + 4\,e^-$. The first step of the reaction is catalyzed by the integral membrane enzyme ammonia monooxygenase (AMO). The putative AMO operon is present in two or three copies in the genome of *N. europaea* and *Nitrosospira* spp., respectively (McTavish et al., 1993a; Bergmann and Hooper, 1994b; Norton et al., 1996). The operon reveals two open reading frames coding for the two components of AMO, AmoA (27 kDa) and AmoB (43 kDa) (Bergmann and Hooper, 1994b). However, the existence of a third component, AmoC, has recently been indicated by Klotz et al. (1997). AMO can be inactivated by acetylene; with the use of [^{14}C]-acetylene, a covalently labeled, membrane-bound polypeptide, which is believed to represent the active-site-containing subunit AmoA of AMO, has been detected (Hyman and Wood, 1985; Hyman and Arp, 1992). The AMO reaction incorporates molecular oxygen (Hollocher et al., 1981). The two electrons of the involved reductant presumably stem from the second reaction, the hydroxylamine oxidation. The oxidation of hydroxylamine to nitrite in a dehydrogenation reaction in the periplasm (Andersson and Hooper, 1983; Olson and Hooper, 1983) is catalyzed by hydroxylamine oxidoreductase (HAO), in concert with the tetraheme electron acceptor cytochrome c_{554} (Arciero et al., 1991a; Iverson

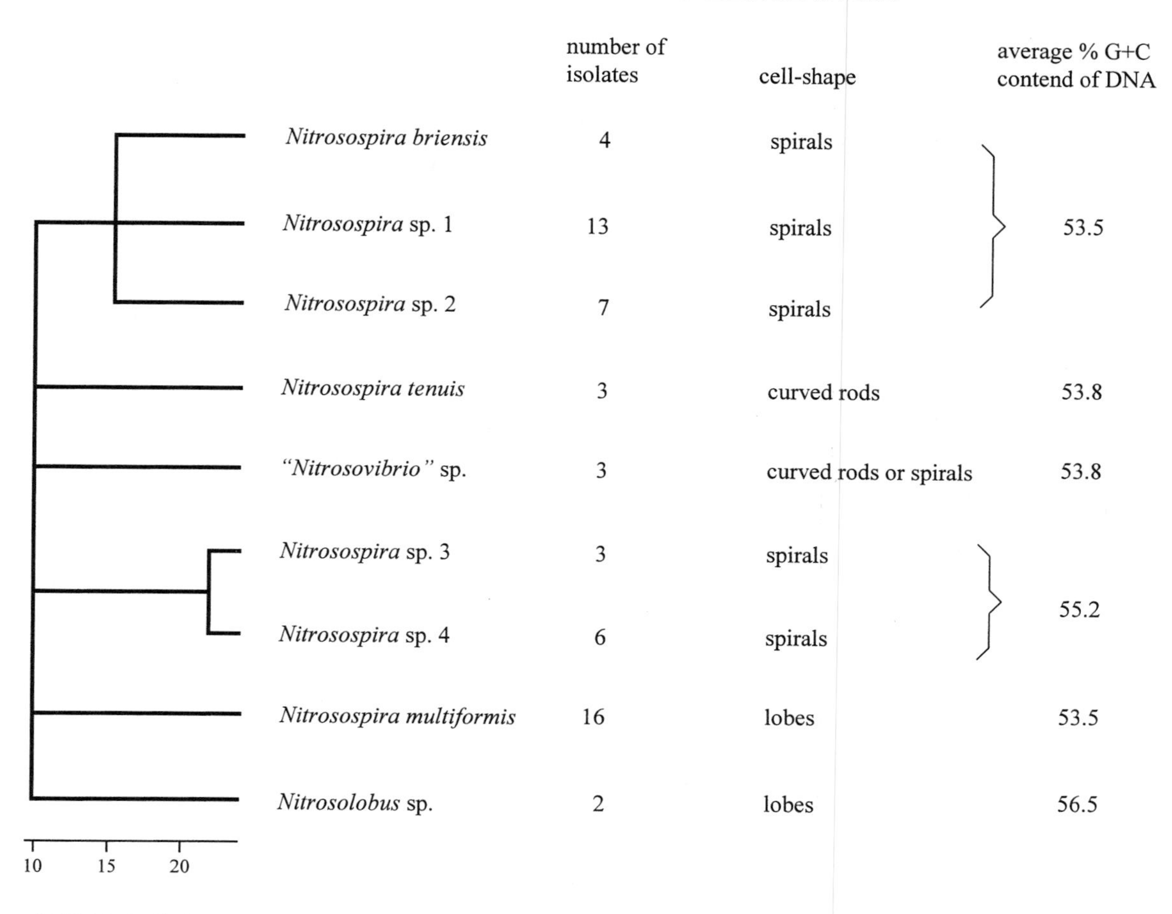

FIGURE 2. DNA-similarity-values-based dendrogram showing the phylogenetic interrelationships among representatives of the genera *Nitrosospira*, *"Nitrosovibrio"*, and *Nitrosolobus*. *Nitrosospira multiformis* was formerly known as *Nitrosolobus multiformis*; *Nitrosospira tenuis* was formerly known as *"Nitrosovibrio tenuis"*.

et al.,1998). This reaction represents the energy-yielding portion of the total process. HAO is a multiheme enzyme that consists of three subunits (63 kDa each), each containing seven or eight *c*-type hemes (Hooper et al., 1997; Igarashi et al., 1997; Arciero et al., 1998) and one P460 center (Arciero and Hooper, 1993; Hooper et al., 1997). The crystal structure of HAO, which indicates pathways by which electron transfer may occur through the precisely arranged hemes, has been described by Igarashi et al. (1997). At least three copies of HAO-coding and of the cytochrome-c_{554}-coding genes are present in a triplicate gene cluster in *N. europaea* (McTavish et al., 1993b; Sayavedra-Soto et al., 1994). A copy of an HAO gene is always located within 2.7 kb of a copy of a cytochrome c_{554} gene (McTavish et al., 1993b). In two of the three clusters, the cytochrome-c_{554}-coding gene is in the same operon as a gene coding for a tetraheme membrane *c* cytochrome, which has an unknown *in vivo* function (Hooper et al., 1997). The *c* type hemes of HAO and the cytochrome c_{554}, respectively, are partially distinguishable by their electron paramagnetic resonance properties (Lipscomb et al., 1982) and oxidation-reduction potentials, which range from −412 to +288 and from −276 to +47 mV, respectively (Arciero et al., 1991b; Collins et al., 1993). The periplasmic cytochrome c_{554} is suggested to be central to a critical electron-transfer branch point, where two of the four electrons derived from the hydroxylamine oxidation to nitrite return to the AMO reaction (see above), while the other two electrons pass to the terminal oxidase reaction and to ATP-dependent reverse electron transfer for the production of reduced pyridine nucleotides, respectively (DiSpirito et al., 1986; Wood, 1988). In accordance with the above concept of energy transduction from the oxidation of ammonia to nitrite, the cell yield obtained per mM nitrite with hydroxylamine as the primary substrate is significantly (about twofold) higher than that obtained with ammonia as the substrate (Böttcher and Koops, 1994; De Bruijn et al., 1995). The electron flow from HAO to the terminal oxidase has been postulated to be: HAO → cytochrome c_{554} → cytochrome c_{552} → terminal oxidase (Yamanaka and Shinra, 1974) or, alternatively, HAO → cytochrome c_{554} → ubiquinone/cytochrome *bc* complex → cytochrome c_{552} → terminal oxidase (Wood, 1986). Beside HAO, cytochrome P460, a trimer of identical 18 kDa subunits, is also able to catalyze the oxidation of hydroxylamine to nitrite (Erickson and Hooper, 1972; Miller et al., 1984; Numata et al., 1990). This cytochrome is suggested to be periplasmic (Bergmann and Hooper, 1994a). The one-copy gene of cytochrome P460 is separate from the three HAO gene copies (McTavish et al., 1993b; Bergmann et al., 1994). The amino acid sequence of cytochrome P460 reveals little sequence homology with other *c*-cytochromes (Bergmann and Hooper, 1994a). The *in vivo* function of cytochrome P460 is not known at this time. The functions of a periplasmic diheme cytochrome c_{553} peroxidase and of a low potential nitrosocyanin are equally unknown (Arciero and Hooper, 1994; Hooper et al., 1997; Whittaker et al., 2000).

As a result of the broad substrate specificity of AMO, the ammonia oxidizers are able to oxidize many alternative, nonphysiological substrates, such as carbon monoxide, alkanes, alkenes, alkynes, and cyclic, aromatic, and halogenated hydrocarbons (Tsang and Suzuki, 1982; Hyman and Wood, 1983, ,1984a b; Jones and Morita, 1983a, b; Hyman and Wood, 1985, 1988; Rasche et al., 1990a, b; Vannelli et al., 1990; Duddleston et al., 2000). Such compounds are competitive inhibitors of the ammonia oxidation, and their oxidation reactions all show the same sensitivity as ammonia oxidation towards different inhibitors (Hooper and Terry, 1973, 1974; Suzuki et al., 1976). Striking similarities have been observed between the hydroxylation of methane and ammonia. However, attempts to grow ammonia oxidizers on methane as the energy source have failed, though carbon from methane was incorporated into cellular components (Jones and Morita, 1983b).

At reduced oxygen tensions, nitrite can serve, via denitrification, as the terminal electron acceptor of the hydroxylamine oxidation, and N_2 and N_2O are the main end products (Blackmer et al., 1980; Goreau et al., 1980; Lipschultz et al., 1981; Poth and Focht, 1985; Poth, 1986). The reaction is catalyzed by a periplasmic, soluble cytochrome oxidase/nitrite reductase (DiSpirito et al., 1985; Miller and Nicholas, 1985) and a recently established NO reductase (Whittaker et al., 2000).

The production of $^{14/15}N_2$ from $^{15}NH_3$ and $^{14}NO_3^-$ has been demonstrated in a denitrifying fluidized bed reactor, indicating the existence of a dinitrogen intermediate in anaerobic ammonia oxidation (Mulder et al., 1995; Van de Graaf et al., 1995). It has been suggested that representatives of the order *Planctomycetales* are involved in such anaerobic nitrification (Strous et al., 1999). However, pure cultures of these species do not yet exist.

Recently, anaerobic growth of pure cultures of *"Nitrosomonas eutropha"* has been described with molecular hydrogen or pyruvate as the energy source and nitrite as the terminal electron acceptor (Bock et al., 1995), as well as with N_2O_4 as the oxygen source for ammonia hydroxylation to hydroxylamine and nitrite as the terminal electron acceptor (Schmidt, 1997; Schmidt and Bock, 1997; Zart, 1997). Details of the underlying biochemical pathways, however, are speculative at the time being.

Urea can be hydrolyzed and used as an ammonia source by strains of many, but not all, species (Koops et al., 1991). Urease activity seems to be common, particularly in oligotrophic environments (Koops et al., 1991). The capability to use urea as an ammonia source might also be important for ammonia oxidizers living in acid soils (Sarathchandra, 1978; Walker and Wickramasinghe, 1979; Hankinson and Schmidt, 1984; De Boer et al., 1989; 1995; Hayatsu and Kosuge, 1993), where the concentration of free ammonia as a result of low pH ($NH_4 \Leftrightarrow NH_3 + H^+$; pK = 9.25 at 25°C) is too low to allow growth of the organisms. At such conditions, sufficient amounts of ammonia can be derived from uptake and subsequent hydrolysis of urea by a urease that is located in the cytoplasm of the cell (Sowitzki, 1994; Dittberner, 1996). The urease of *Nitrosospira* spp. is a 295 kDa polypeptide with a K_m value of 610 μM (Dittberner, 1996). Increasing concentrations of NH_4Cl in the growth medium cause increasing inhibition of urease activity, and no urease activity is measured at concentrations above 80 mM NH_4Cl at pH 7.8 ($\cong$ 3.8 mM NH_3) (Dittberner, 1996).

Some, but not all, species possess polyhedral inclusion bodies called carboxysomes (Wullenweber et al., 1977). The major protein component of these carboxysomes is ribulose-1-5-bisphosphate carboxylase/oxygenase (RuBISCO) (Harms et al., 1981), which is responsible for the carbon dioxide fixation in the RPP cycle.

Various organic compounds, such as formate, acetate, pyruvate, and complex compounds, can affect growth of ammonia oxidizers (Clark and Schmidt, 1966; Krümmel and Harms, 1982). Some of these organic compounds are assimilated, but at a level too low to act as major carbon and energy sources. Characteristic restrictions on the qualitative distribution of assimilated organic carbon among cellular constituents, generally observed with ammonia oxidizers (Smith and Hoare, 1977; Martiny and Koops, 1982), indicate that the lack of some key enzymes is the basic reason why these organisms cannot grow heterotrophically (Williams and Watson, 1968; Hooper, 1969; Wallace et al., 1970; Kelly, 1971; Matin, 1978).

Because the redox potential of the ammonia–nitrite redox couple is near the practical limit of an aerobic energy source, it is problematic to explain the production of reductants that are needed for carbon dioxide assimilation via the RPP-cycle (Wood, 1986). NADH synthesis via reversed electron flow, driven by a proton-motive force, has been suggested as an explanation. However, Aleem (1966) has reported NADH production by cell extracts of *Nitrosomonas europaea* in the presence of hydroxylamine and ATP, though osmotically sealed membranes are missing at such conditions. Another reductant-dependent reaction, the assimilation of nitrogen from ammonia, may be catalyzed by a NADPH-specific reversible glutamate dehydrogenase (Hooper et al., 1967; Wallace and Nicholas, 1969).

The above-described overall conception of the biochemistry of the lithotrophic ammonia-oxidizing bacteria has been obtained from investigations carried out with representatives of the species within the *Betaproteobacteria*. Detailed investigations of the biochemistry of the ammonia-oxidizing system of the second group of ammonia-oxidizers, located within the *Gammaproteobacteria*, are relatively rare. However, many observations indicate that it is generally consistent with the above-described, well-examined corresponding system. Reduced minus oxidized difference spectrum analyses of both phylogenetically distinct groups of ammonia-oxidizing bacteria reveal an identical cytochrome pattern (Watson, 1965; Böttcher, 1996). Beside the cytochromes *a*, *b*, and *c*, the presence of P460 can be shown for both groups. The stoichiometry of the two-step reaction for ammonia-oxidation ($NH_3 + 2[H] + O_2 \rightarrow NH_2OH + H_2O$ and $NH_2OH + H_2O \rightarrow HNO_2 + 4H^+ + 4e^-$) is the same in both groups (Böttcher and Koops, 1994). Furthermore, the nucleotide sequences of the genes coding for the ammonia monooxygenase and the cytochrome P460, respectively, exhibit relatively high levels of homology among representatives of both groups of ammonia-oxidizing bacteria (Böttcher, 1996; Purkhold et al., 2000).

From detailed analyses, Campbell et al. (1966) have concluded that the gamma-group of ammonia-oxidizers also fixes carbon dioxide via the reductive pentose phosphate (RPP) cycle. In the case of *N. oceani*, activities of the enzymes of the RPP-cycle have been estimated to be high enough to account for the observed growth of this bacterium. The presence of a phosphoenolpyruvate–CO_2-fixing system has also been demonstrated (Williams and Watson, 1968). These authors have also found the Embden–Meyerhof enzymes (except phosphofructokinase) and all the tricarboxylic-acid-cycle enzymes, as well as reduced nicotinamide adenine dinucleotide oxidase, to be active in cell-free extracts of *N. oceani.*

Habitats

Species of the beta-group of the lithotrophic ammonia oxidizers are widely distributed. Strains have been isolated from diverse environments. However, several species or groups of species have been observed to occur predominantly or exclusively in special sites, such as acid soils, rivers, freshwater lakes, salt lakes, oceans, brackish waters, sewage disposal plants, and rocks or natural stone buildings (Watson and Mandel,1971; Belser and Schmidt, 1978a; Walker 1978; Walker and Wickramasinghe, 1979; Koops and Harms, 1985; De Boer and Laanbroek, 1989; Meincke et al., 1989; De Boer et al., 1995; Suwa et al., 1997; Stehr et al., 1995b; Speksnijder et al., 1998). Some species have an obligate salt requirement, some prefer eutrophic or oligotrophic environments, and others tolerate either high or low temperatures (Golovacheva, 1976; Jones and Morita, 1985; Koops et al., 1991). Ammonia-oxidizing bacteria are often observed to occur flock- or biofilm-attached, especially in aquatic environments (Stehr et al., 1995b; Wagner et al., 1995; Phillips et al., 1999). This is interpreted to be an important survival strategy of these organisms. Recently, a fast recovery time after ammonium starvation, with no lag phase, has been observed, in biofilm populations. This is in contrast to cell suspensions, which exhibit a significant lag phase prior to exponential nitrite production. Biochemical communication (quorum sensing) within the denser biofilm populations has been discussed as the most probable reason for the observed high speed of recovery (Batchelor et al., 1997).

The environmental distribution differs in detail among the defined, cultured species or groups of these species. Most fundamental information stems from enrichment and isolation of the organisms, but increasing new insights on distribution patterns are obtained from molecular ecological investigations.

The ten described species of the genus *Nitrosomonas,* together with other cultured but undescribed species of this genus and *"Nitrosomonas mobilis"* (*"Nitrosococcus mobilis"*), represent six distinct lines of descent (Fig. 1). As far as is known, these lineages reflect distinct ecophysiological groupings (Pommerening-Röser et al., 1996; Koops and Pommerening-Röser, 2001). Lineage 1a allies the three moderately halophilic and halotolerant species *N. europaea, "N. eutropha"*, and *N. halophila.* The closely related, reclassified *"Nitrosomonas mobilis"* (*"Nitrosococcus mobilis"*), representing lineage 1b, is equally moderately halophilic. All four species are urease negative and eutrophic. With the exception of *N. halophila,* the members of this lineage are common in sewage disposal plants. Occasionally, strains of these species have also been isolated from strongly eutrophicated soils or brackish and freshwater environments. Furthermore, in accordance with their relatively strong salt tolerance, members of the lineage can be detected via PCR-assisted methods in salt lakes (Ward et al., 2000). Recently strains of *N. halophila* have been isolated from soda lakes (Sorokin et al., 2001d). Lineage 2 includes the salt-sensitive and oligotrophic species *"N. ureae"* and *"N. oligotropha"*, together with some undescribed species (Stehr et al., 1995a), which all reveal outstandingly low K_s values for ammonia oxidation and are urease positive. Most isolates of these species originate from oligotrophic freshwater environments, such as rivers or lakes. Lineage 3 comprises two subgroups. One subgroup (lineage 3a) contains three predominantly terrestrial (neutral soils) species, *"N. communis"*, *Nitrosomonas* sp. I, and *Nitrosomonas* sp. II, all of which are urease negative. The second subgroup, lineage 3b, is represented by the species *"N. nitrosa"*, which is urease positive and mainly distributed in eutrophicated freshwater environments. The members of lineage 4, *N. marina, "N. aestuarii"*, and *Nitrosomonas* sp. III, are isolated exclusively from marine environments. All have an obligate salt requirement and all are urease positive. Lineage 5 is represented by a marine species, *"N. cryotolerans"*. Only one isolate, which is obligately halophilic and urease positive, is under culture. Recently, a sixth *Nitrosomonas* lineage was identified (Purkhold et al., 2003). All isolates of this species available in culture were obtained from marine environments (Ward, 1982; Ward and Carlucci, 1985; Purkhold et al., 2003). The only strain investigated in the laboratory was obligately halophilic and urease positive.

Members of the second cluster within the *Betaproteobacteria,* which comprises the three genera *Nitrosospira, "Nitrosovibrio"*, and *Nitrosolobus,* seem to be most common in untreated oligotrophic soils, in mountainous areas and in freshwater environments. Strains of *Nitrosospira* and *"Nitrosovibrio"*, but not of *Nitrosolobus,* have also been isolated from rocks, stone buildings, and even, uniquely among the ammonia oxidizers from acid soils. Strains of *Nitrosolobus* seem to be common in agricultural soils.

Representatives of the gamma-group of ammonia oxidizers have been isolated exclusively from marine environments and salt lakes (Watson, 1965, 1971; Watson et al., 1981, 1989; Koops et al., 1990; Koops and Möller, 1992). Their restricted distribution in nature is in full agreement with their ecophysiological properties, as all isolates investigated are obligately halophilic.

One of the two species of this genus, *Nitrosococcus oceani,* has repeatedly been detected *in situ.* Using specific antisera, the organism has been observed via immunofluorescence in some marine environments (Ward and Perry, 1980; Ward and Carlucci, 1985; Voytek et al., 1998). Using specific 16S rDNA primers, this bacterium has also been detected via PCR and subsequent analyses of the obtained amplicons in samples of seawater from the Southern California Bight (Voytek, 1996; Voytek et al., 1998).

In situ Identification

Description of the nitrification process in special environments requires a detailed knowledge of the responsible bacterial population. However, in the past, *in situ* analyses of natural nitrifying populations have been limited by the methods at disposal. The most-probable-number technique and selective plating are both time-consuming methods. Furthermore, neither method allows discriminations among ammonia oxidizers at the species level.

In the 1970s, serotyping of ammonia-oxidizing bacteria using polyclonal antibodies was introduced, making it possible to detect members of this group of bacteria *in situ.* However, since different serological groups (so-called serovars) exist within some species of the ammonia-oxidizing bacteria, pure cultures of strains from the environment under investigation are required to produce antibodies (Belser and Schmidt, 1978b; Smorczewski and Schmidt, 1991). Recently, polyclonal antibodies against the β-subunit of the ammonia monooxygenase (AmoB) of *Nitrosomonas eutropha* were shown to detect all betaproteobacterial AOB (Pinck et al., 2001).

In recent years, several molecular ecological techniques have been developed, in general using 16S rRNA sequence information for *in situ* analyses. Sets of specific or semispecific PCR primers for amplification of 16S rDNA and direct or cloning-assisted sequence analysis of the obtained fragments have been used for the detection of ammonia oxidizers in environmental samples (McCaig et al., 1994; Stephen et al., 1996; Speksnijder et al.,

1998). Hybridizations with oligonucleotide probes of extracted rRNA from environmental samples or 16S rDNA fragments recovered by PCR have also been applied to analyze natural populations of ammonia oxidizers (Hiorns et al., 1995; Voytek and Ward, 1995; Hovanec and DeLong, 1996; Kowalchuk et al., 1999; Hastings et al., 1997). Denaturing gradient gel electrophoresis of PCR-amplified 16S rDNA fragments, with subsequent sequencing of the obtained distinct amplicons, has proven to be another tool for analyzing the sequence diversity of complex nitrifying bacterial populations (Kowalchuk et al., 1997; Stephen et al., 1998). By employing sets of appropriately specific 16S rRNA targeted oligonucleotide DNA probes, the presence of ammonia oxidizers can successfully be detected via direct cell hybridization (*in situ* fluorescence hybridization) on the level of species or groups of species (Wagner et al., 1995, 1996, 1998a; Mobarry et al., 1996; Schramm et al., 1996; Juretschko et al., 1998). A qualitative evaluation of published oligonucleotides specific for 16S rRNA gene sequences of ammonia oxidizers has been carried out by Utåker and Nes (1998) and, more recently, by Purkhold et al. (2000). Alternatively, *amoA* gene sequences can be used with most of the above described methods.

Attempts to detect *in situ* ammonia monooxygenase gene expression via PCR-based assays targeting partial stretches of the genes which encode the active site polypeptides of amoA and the HAO have been effective (Rotthauwe et al., 1997; Holben et al., 1998; Kloos et al., 1998). This hybridization technique allows analysis of bacterial community structures. Methodical problems that sometimes arise when quantifying the signals obtained by these methods are discussed by Chandler et al. (1998).

However, there are critical aspects to *in situ* analysis of natural bacterial populations via direct DNA extraction and subsequent PCR amplification and sequencing of the amplicon that can make this technique problematic. One of the most important points of uncertainty is that within some groups of ammonia oxidizers comprising closely related species, the degree of 16S rDNA sequence divergence is very low. In such cases, it cannot be stated with certainty how many distinct species are represented by the obtained sequences, nor whether the presence of an as yet uncultured species is indicated. Furthermore, some microorganisms tend to resist classical techniques of cell lysis (Picard et al., 1992), and this can lead to a selective identification of special groups of a targeted population (Speksnijder et al., 1998) and their gene expressions (Juretschko et al., 1998). Combined application of conventional (isolation and physiological characterization of the most abundant species) and molecular (hierarchical sets of specific probes) analyses of nitrifying populations could overcome most of these problems.

Enrichment and isolation procedures

Isolation of lithotrophic ammonia oxidizers is an easy, but time consuming, process. Enrichments are grown in basal salts media (Table 2), and pure cultures are obtained by employing serial dilution or plating techniques. To minimize the organics present in the inoculum or produced by the growing nitrifiers, the enrichments should be serially diluted through several orders of magnitude. Since ammonia oxidizers have prolonged generation times, it takes several months to obtain pure cultures. Purity of isolates can be checked by inoculation of an organic culture medium (containing 0.5 g yeast extract, 0.5 g peptone, and 0.5 g beef extract per liter H_2O; pH 7.4; NaCl must be added for testing isolates originating from haline habitats) and by examination of uniformity of the cells by phase contrast microscopy.

Growth characteristics

Media usable for lithotrophic growth of ammonia-oxidizing bacteria are listed in Table 2. Growth rates are primarily controlled by temperature, by the pH of the medium, and, especially, by the ammonia concentration of the medium. In general, optimum

TABLE 2. Growth media for lithoautotrophic ammonia oxidizers[a]

Ingredients	Terrestrial[b]	Terrestrial[c]	Marine[d]	Brackish[e]
Distilled water (ml)	1000	1000		600
Seawater (ml)			1000	400
$(NH_4)_2SO_4$ (mg/l)	2000		1320	
$(NH_4)Cl$ (mg/l)		535		535
NaCl (mg/l)		584		
$MgSO_4·7H_2O$ (mg/l)	200	49.3	200	
$CaCl_2·2H_2O$ (mg/l)	20	147	20	
KH_2PO_4 (mg/l)		54.4		54.4
K_2HPO_4 (mg/l)	15.9		114	
KCl (mg/l)		74.4		
Chelated iron (13% Geigy Chemical) (mg/l)	1		1	
$FeSO_4·7H_2O$ (μg/l)		973.1		
$Na_2MoO_4·2H_2O$ (μg/l)	100		1	
$(NH_4)_6Mo_7O_{24}·4H_2O$ (μg/l)		37.1		
$MnCl_2·4H_2O$ (μg/l)	200		2	
$MnSO_4·4H_2O$ (μg/l)		44.6		
$CoCl_2·6H_2O$ (μg/l)	2		2	
$CuSO_4·5H_2O$ (μg/l)	20	25	20	
$ZnSO_4·7H_2O$ (μg/l)	100	43.1	100	
H_3BO_3 (μg/l)		49.4		
Phenol red (0.5%) (ml/l)	1		1	
Cresol red (0.05%) (ml/l)		1		1

[a]For stock cultures, 5 g/l $CaCO_3$ must be added.

[b]Watson et al. (1971)

[c]Krümmel and Harms (1982).

[d]Watson (1965).

[e]Koops et al. (1976).

growth is observed at 25–30°C, at pH of 7.5–8.0, and at concentrations between 2 and 10 mM of the respective ammonium compound. Since ammonia is the true energy substrate of the organisms, the optimum concentration of the ammonium compound depends on the pH of the medium and vice versa. The K_s values of the ammonia-oxidizing systems are different among the subgroups of the ammonia oxidizers, ranging from 0.6 to 158 μM NH_3 (Suzuki et al., 1974; Ward, 1986; Hunik et al., 1992; Suwa et al., 1994; Stehr et al., 1995a; Stehr, 1996).

As ammonia is oxidized to nitrite, the pH of the medium drops. Adjustment (addition of 10% $NaHCO_3$) around 7.8 can be made manually or using an automatic pH controller, with the aid of pH indicators such as phenol red or cresol red. Alternatively, the medium can be buffered with sodium carbonate (5 g/l) or HEPES (0.01 M).

Some species, generally originating from marine environments, have an obligate sodium requirement (Koops et al., 1976, 1990, 1991). In general, the culture conditions must be optimized for the respective species under study.

Maintenance procedures of stock cultures

Pure cultures should be maintained in liquid cultures in basal salts medium enriched with ammonia and buffered with $CaCO_3$. Addition of cresol red is useful for indication of successful growth of the cultures. Stock cultures should be stored at room temperature in the dark and must be transferred to fresh medium every 3–5 months.

Alternatively, stock cultures can be stored using liquid nitrogen. Storage by freeze drying of the cells is not recommended.

The Lithoautotrophic Nitrite-Oxidizing Bacteria

Eva Spieck and Eberhard Bock

The lithotrophic nitrite oxidizers are Gram-negative eubacteria that are able to use nitrite as a sole source of energy and CO_2 as the main source of carbon. Some strains are able to grow mixotrophically. These organisms are obligate lithoautotrophs with the exception of *Nitrobacter*, which can grow heterotrophically. *Nitrobacter* has been shown to grow anaerobically by dissimilatory nitrate reduction (Freitag et al., 1987).

Morphological Characteristics

The nitrite oxidizers are a diverse group of rods, cocci, and spirilla. Historically, the classification of genera was founded primarily on cell shape and arrangement of intracytoplasmic membranes, and taxonomic categorization was based on the work of Sergei and Helene Winogradsky (Winogradsky, 1892). Four morphologically distinct genera (*Nitrobacter*, *Nitrococcus*, *Nitrospina*, and *Nitrospira*) have been described (Watson et al., 1989; Bock and Koops, 1992). Cells of *Nitrobacter* are pleomorphic short rods containing a polar cap of intracytoplasmic membranes. *Nitrococcus* occurs in form of coccoid cells with tubular intracytoplasmic membranes. Cells of *Nitrospina* appear as long rods, intracytoplasmic membranes in the form of flattened vesicles or tubes are missing. The genus *Nitrospira* is characterized by a spiral shape and the absence of intracytoplasmic membranes. Some strains are motile by means of a single polar or subpolar flagellum.

Phylogeny

Unlike the ammonia oxidizers, which are restricted to two lineages within the *Proteobacteria*, the nitrite oxidizers are more scattered phylogenetically (Fig. 1 of the chapter "Nitrifying Bacteria"). The genus *Nitrobacter* belongs to the *Alphaproteobacteria* (Woese et al., 1984a; Stackebrandt et al., 1988), whereas *Nitrococcus* is affiliated with the *Gammaproteobacteria* (Woese et al., 1985). The genus *Nitrospina* seems to be a member of the *Deltaproteobacteria* (Teske et al., 1994), although this assignment remains preliminary. *Nitrospira* was thought to be related to *Nitrospina*, but *Nitrospira* was later shown not to be a member of the *Proteobacteria*. Ehrich et al. (1995) demonstrated that this nitrite-oxidizing bacterium occupies a phylogenetically isolated position and represents a new phylum, *Nitrospirae*, of the domain *Bacteria*. *Nitrospira* was described in Volume I of the *Manual*; some data are also presented here to facilitate comparison.

The genus *Nitrobacter* is comprised of four described species (Bock and Koops, 1992; Sorokin et al., 1998), whereas one species is known for each of the genera *Nitrococcus* and *Nitrospina* (Watson and Waterbury, 1971). Two species of *Nitrospira* have been described in the literature (Watson et al., 1986; Ehrich et al., 1995),

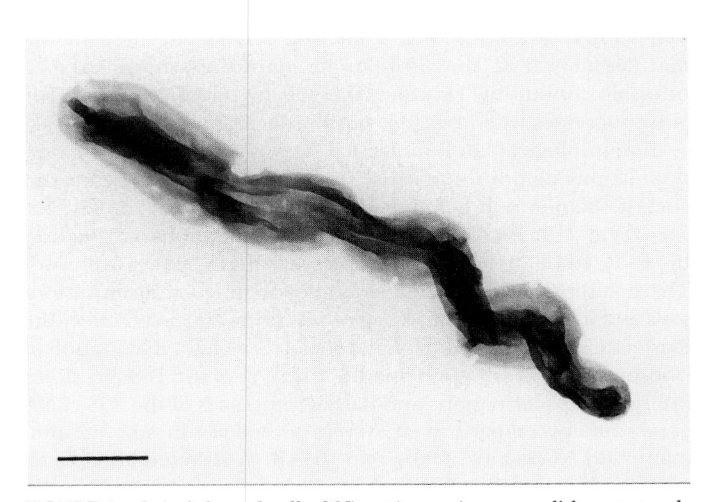

FIGURE 1. Spiral-shaped cell of *Nitrospira marina* grown lithoautotrophically. Negative staining with uranylacetate. Bar = 500 nm.

but much higher phylogenetical diversity may be present in this genus (Schramm et al., 1999).

Ecology and Distribution

The best-investigated nitrite-oxidizing bacterium is *Nitrobacter*, which was believed to dominate in most natural environments except marine ones. This picture has changed significantly over the last few years, and current investigations, especially molecular ones, are focused on the occurrence of *Nitrospira*. Members of the genera *Nitrococcus* and *Nitrospina* have only been found in marine habitats to date.

Nitrobacter is a soil and freshwater organism that is tolerant of changing environmental conditions. Members of this genus also occur in sewage and marine environments. Other isolates were originally obtained from extreme environments such as concrete and natural stones, desert soils and sulfidic ore mines. One acidophilic strain with a pH optimum of 5.5 was isolated from an acidic forest soil (Hankinson and Schmidt, 1988). Facultatively alkalophilic strains of *Nitrobacter* were recently be isolated from soda lakes in Siberia and Kenya and described as a new species, *N. alkalicus* (Sorokin et al., 1998).

Although nitrite oxidizers do not form endospores, they can survive long periods of starvation and dryness. One survival strategy used by these organisms may be the formation and accumulation of extracellular compatible solutes. *Nitrobacter* was found to produce trehalose and was able to accumulate glycine betaine and sucrose from the medium. An increase in the amounts of compatible solutes was reproducibly found in cultures exposed to salt stress and dryness. *Nitrobacter vulgaris* can survive a period of 24 months without water (L. Lin, personal communication). Diab and Shilo (1988) found that adhesion to particles had a positive effect on both the activity and the survival of *Nitrobacter* cells.

When the habitat-specific distribution of three species of *Nitrobacter* was examined using automated pattern matching of proteins, it was found that *Nitrobacter vulgaris* was the dominant species in building stone (T. Krause-Kupsch, personal communication). *Nitrobacter hamburgensis* was only found in soil, whereas *Nitrobacter winogradskyi* occurred in various habitats such as soils, fresh water, sewage and concrete. According to Both et al. (1992) *Nitrobacter winogradskyi* out-competes *Nitrobacter hamburgensis* in well-aerated soils under nitrite-limiting conditions, since the former has a lower K_m for nitrite under autotrophic as well as mixotrophic conditions. However, the activity of *Nitrobacter hamburgensis* increases when oxygen tension decreases.

Immunological and molecular investigations of *Nitrobacter* populations demonstrated that several strains of this genus can coexist (Stanley and Schmidt, 1981; Degrange et al., 1997). Navarro et al. (1992) characterized natural populations of *Nitrobacter* by PCR/RFLP (restriction fragment length polymorphism). These authors differentiated several coexisting strains in various soils and a lake; the coexistence of several strains may reflect the existence of local niches. Genetic distances obtained by amplified ribosomal DNA restriction analysis (ADRA) of the 16S-23S rRNA intergenic spacer regions and partial sequences of the 23S rRNA gene enable comparison of *Nitrobacter* species in soil (Grundmann and Normand, 2000). Two 16S rRNA-targeted oligonucleotide probes specific for *Nitrobacter* have been developed for the *in situ* analysis of nitrite oxidizers (Wagner et al., 1996). Although this genus has been regarded as the most abundant nitrite oxidizer in various environments, it could not be detected in activated sludge samples and reactor biofilms. The authors suggested that still unknown organisms might be responsible for nitrification in these habitats. This hypothesis was confirmed recently when several groups reported that *Nitrospira*-like bacteria seem to be the dominant nitrite oxidizers in freshwater aquaria, biofilms and activated sludge (Burrell et al., 1998; Hovanec et al., 1998; Juretschko et al., 1998). Schramm et al. (1998) found in a nitrifying reactor organisms that formed two phylogenetically distinct groups affiliated with *Nitrospira moscoviensis*. The novel genus *Nitrospira marina* was first isolated by Watson et al. (1986) from the Gulf of Maine (Fig. 1). *Nitrospira moscoviensis* was first isolated from a heating system in Moscow (Ehrich et al., 1995). Similar nitrite-oxidizing organisms have been enriched from soil samples, sediments, beach sands, and salt marshes (Watson et al., 1989). It seems that although the genus *Nitrospira* is ubiquitous, it is outcompeted by *Nitrobacter* when standard isolation procedures are used (Johnson and Sieburth, 1976). In studies using monoclonal antibodies that recognize the nitrite oxidoreductase (NOR) enzyme of *Nitrobacter*, Bartosch et al. (1999) demonstrated that different genera of nitrite oxidizers were enriched from activated sludge depending on the substrate concentration of the media. When enrichments were made in accordance with the instructions of Watson et al. (1989), *Nitrospira* was the most abundant nitrite oxidizer in enrichment cultures grown in mixotrophic medium containing 0.2 g $NaNO_2$ per liter. In contrast, cells of *Nitrobacter* dominated when the medium contained 2 g $NaNO_2$ per liter. Although *Nitrospira* from wastewater treatment plants was postulated to be "unculturable," microcolonies of *Nitrospira* from wastewater samples from Dradenau in Hamburg were highly enriched in laboratory cultures (Fig. 2). *Nitrospira* can be regularly enriched using adapted cultivation techniques that include the avoidance of turbulence, and cultures originating from a wide range of habitats such as permafrost soil (Bartosch et al., 2002), caves, and hot springs are being investigated.

So far, phylogenetic analysis of *Nitrospira* has revealed four sublineages based on environmental sequences from various aquatic environments (summarized by Daims et al., 2001). Two of the sublineages include the described species *N. moscoviensis* and *N. marina*. A third species, originating from a Moscow heating system, will be described in the future (Lebedeva, personal communication). A thermophilic culture derived from a hot spring at Lake Baikal differed from known *Nitrospira* isolates based on DGGE (Alawi and Lebedeva, personal communication). It is likely that the phylogenetic tree of *Nitrospira* will become more complex as new representatives are isolated and sequence analysis of environmental samples is carried out.

Growth Characteristics

Lithotrophic growth of nitrite oxidizers is slow. The generation time varies from 8 h to several days. Growth rates are controlled by substrate concentration, temperature, pH, light, and oxygen concentration. Most nitrite oxidizers grow best at nitrite concentrations of 2–30 mM at a pH of 7.5–8.0 and at temperatures of 25–30°C. Some strains are able to grow mixotrophically; the cell yield from mixotrophically grown cultures can be ten-fold greater than that from lithotrophically grown cultures.

Biochemistry

Most biochemical investigations have been performed on the genus *Nitrobacter*. Initial biochemical studies of *Nitrospira* revealed several significant differences between *Nitrospira* and *Nitrobacter* (Watson et al., 1986). Little is known about the biochemistry of *Nitrococcus* and *Nitrospina*. The genera *Nitrobacter* and *Nitrococcus* are similar in cytochrome content and in the location and molecular masses of the nitrite-oxidizing enzymes, whereas the genera *Nitrospina* and *Nitrospira* differ from *Nitrobacter* and *Nitrococcus* but are similar to each other with respect to these characteristics.

Nitrite, the substrate for aerobic nitrification, is thought to be transported into the bacteria by a nitrite/nitrate antiport system (Wood, 1986). Nitrate, the electron acceptor for the reverse reaction, is assumed to be transferred by the same transporter.

The key enzyme of nitrite oxidation has been studied in *Nitrobacter* and *Nitrospira*; this enzyme is called the nitrite oxidoreductase (NOR) in the genus *Nitrobacter* and nitrite-oxidizing system (NOS) in the genera *Nitrococcus*, *Nitrospina* and *Nitrospira*. The occurrence of membrane-bound particles containing the enzyme is a general characteristic of all members of nitrite-oxidizing bacteria; these particles are densely packed on the surface of the cytoplasmic and intracytoplasmic membranes. The location of the particles is coincident with immunolabeling of the NOR and NOS enzymes (Spieck et al., 1996a). In *Nitrobacter* and *Nitrococcus* the key enzyme is located on the inner side of the cytoplasmic and intracytoplasmic membranes (Watson and Waterbury, 1971; Sundermeyer and Bock, 1981a). In cells of *Nitrospina* and *Nitrospira*, which do not possess intracytoplasmic membranes, the nitrite-oxidizing system is found in the periplasmic space and is associated with the outer surface of the cell membrane in *Nitrospira* (Spieck et al., 1998). This location of the

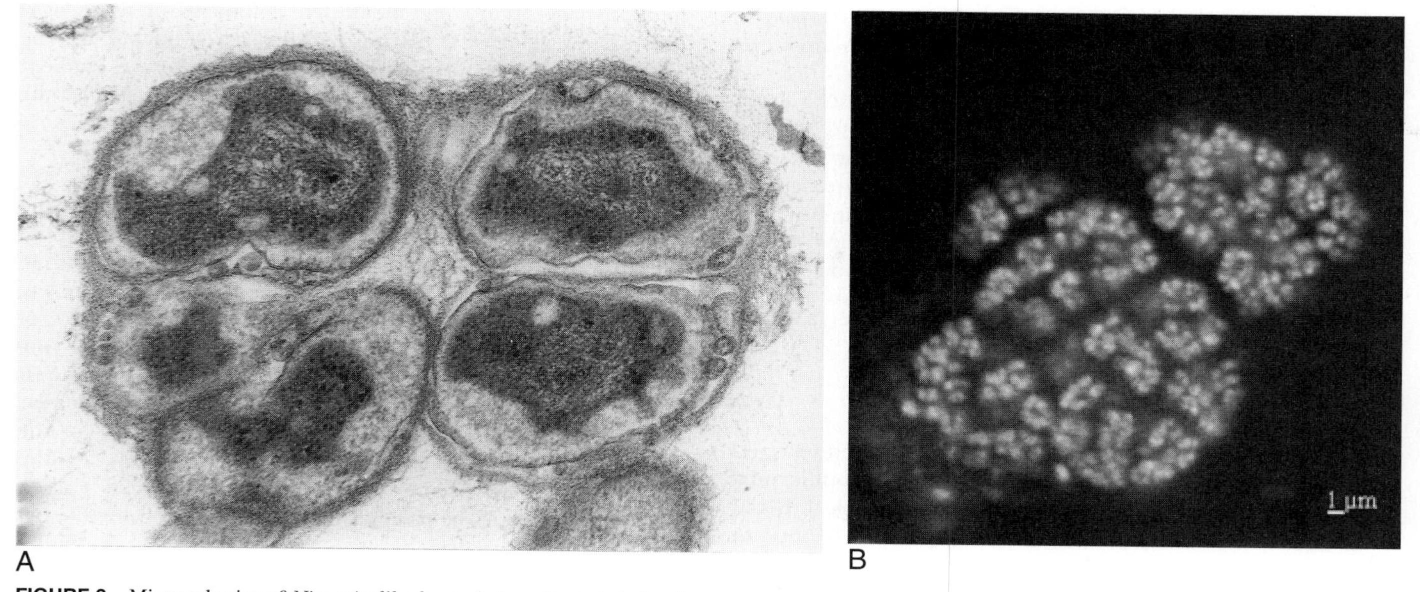

FIGURE 2. Microcolonies of *Nitrospira*-like bacteria in activated sludge from waste water treatment plant in Dradenau, Hamburg. *A*) Ultrathin section of activated sludge. Cells are similar in ultrastructure to those of the genus *Nitrospira* with respect to the extended perimplasmic space and lack of intracytoplasmic membranes. Bar = 250 nm. *B*) Fluorescence *in situ* hybridization (FISH) of a nitrite-oxidizing enrichment culture with oligonucleotide probe S-*Ntspa-1026-a-A-18 specific for *N. moscoviensis* (Juretschko et al., 1998). Cells were grown in mixotrophic medium containing 0.2 g $NaNO_2$/l. Picture courtesy of S. Bartosch

enzyme may explain the higher sensitivity of *Nitrospina* and *Nitrospira* to nitrite in comparison to *Nitrobacter* and *Nitrococcus*. The molecular masses of the β-NOR of *Nitrobacter* and the β-NOS of *Nitrococcus* are identical (65 KDa), whereas the β-NOSs of *Nitrospina* (48 KDa) and *Nitrospira* (46 KDa) differ (Bartosch et al., 1999). Images showing the location and arrangement of the NOR can be found in the chapters describing the genera *Nitrobacter*, *Nitrococcus*, and *Nitrospina*. For *Nitrospira*, see Volume 1.

The NOR of *Nitrobacter* forms a periodic arrangement in paired rows. Tsien and Laudelout (1968) provided the first evidence that a minimum of four particles had to remain associated in order to retain enzymatic activity. The integrity of a structure extending between neighboring particles was assumed to be necessary for conservation of activity. The molecular weight of a single particle was 186 KDa; this result suggests that each particle is an αβ-heterodimer (Spieck et al., 1996b).

The biochemistry of *Nitrobacter* has been reviewed by several authors (Wood, 1986; Yamanaka and Fukumori, 1988; Hooper, 1989; Bock et al. 1991, 1992; Bock and Wagner, 2001.). Two electrons are released during the oxidation of nitrite to nitrate as shown in the following equation. The third oxygen atom in the nitrate molecule is derived from water (Aleem et al., 1965).

$NO_2^- + H_2O \rightarrow NO_3^- + 2H^+ + 2e^-$

$2H^+ + 2e^- + 0.5\ O_2 \rightarrow H_2O$

$NO_2^- + 0.5\ O_2 \rightarrow NO_3^-$

The electron flux from nitrite to oxygen is thought to flow through the following electron carriers.

nitrite→molybdopterin→iron-sulfur-clusters→cytochrome a_1→cytochrome c→cytochrome aa_3→O_2

The electron transfer from nitrite to cytochrome a_1 is catalyzed by the enzyme nitrite oxidoreductase (NOR) which contains molybdopterin and iron-sulfur clusters. Cytochrome a_1 is necessary to channel electrons from nitrite to cytochrome c (Yamanaka and Fukumori, 1988), where the electrons enter the respiratory chain (Cobley, 1976; Aleem and Sewell, 1981). The reduction of cytochrome c is a thermodynamically unfavorable step because the NO_2^-/NO_3^- couple has a redox potential of $E_0' = +420$ mV. Nitrite-oxidizing cells of *Nitrobacter winogradskyi* have a very low energy charge of 0.37 during the logarithmic growth phase (Eigener, 1975). The inefficiency of energy generation in *Nitrobacter* may be compensated for by high levels of NOR, which may comprise 10–30% of total protein (Bock et al., 1991). The primary energy product is NADH (Sundermeyer and Bock, 1981b), which is used for ATP synthesis (Freitag and Bock, 1990). It is not clear how energy conservation occurs because the postulated reverse electron flow for the generation of NADH has not yet been demonstrated. The nitrite oxidase system of *Nitrobacter winogradskyi* was reconstituted in proteoliposomes with isolated nitrite oxidoreductase, cytochrome c oxidase and the subtrate nitrite. In this system oxygen was consumed in the presence of membrane-bound cytochrome c_{550} (Nomoto et al., 1993). A purified ATPase from *N. winogradskyi* has been characterized by Hara et al. (1991).

Cells of *Nitrobacter hamburgensis* seem to utilize different terminal oxidases in response to different growth conditions. During nitrite oxidation, cytochrome aa_3 is active, whereas a *b*-type cytochrome is used as a terminal oxidase for heterotrophic growth (Kirstein et al., 1986). *Nitrobacter* and *Nitrococcus* are rich in cytochromes c and a; dense cell suspensions exhibit a typical red to brownish color. Characteristic peaks occur at 420, 440, 550, 587 and 600 nm in oxidized/dithionite-reduced difference spectra. The other two genera of nitrite-oxidizing bacteria, *Nitrospina* and *Nitrospira*, apparently lack type a cytochromes (Watson et al., 1989).

Lithoautotrophic nitrite oxidizers fix carbon dioxide via the Calvin cycle. About 80% of the energy generated by nitrite oxidation is used for CO_2 fixation. In *Nitrobacter* ribulose-1,5-bisphosphate carboxylase/oxygenase (RubisCO) is responsible for this reaction. In *Nitrobacter* (Shively et al., 1977) and *Nitrococcus*, the enzyme may be soluble as well as carboxysome-bound; carboxysomes are found in most but not all species of *Nitrobacter*. In *Nitrobacter winogradskyi* the soluble form of RubisCO has a molecular mass of 480 KDa. In *Nitrobacter hamburgensis* X14 the enzyme occurs in two forms with different molecular masses—480 KDa and 520 KDa; both forms have an L8S8 quaternary structure. In this species two different genes and gene products for the large subunit of RubisCO have been identified; one is located on the chromosome and the other on a plasmid (Harris et al., 1988). The Calvin cycle genes are located in two separate clusters on the chromosome in *Nitrobacter vulgaris* (Strecker et al., 1994).

Enrichment and Isolation Procedures

Nitrite oxidizers can be isolated using a mineral medium containing nitrite; the compositions of media for lithotrophic, mixotrophic, and heterotrophic growth are given in Table 1. Serial dilutions of enrichment cultures must be incubated for one to several months in the dark. Since nitrite oxidizers are sensitive to high partial pressures of oxygen, cell growth on agar surfaces is limited. Pure cultures of *Nitrobacter alkalicus* were obtained by multiple passages in liquid medium of colonies from nitrite agar (Sorokin et al., 1998). Nitrite oxidizers like *Nitrospira* can be separated from heterotrophic contaminants by Percoll gradient centrifugation and subsequent serial dilution (Ehrich et al., 1995).

TABLE 1. Three different media for lithoautotrophic (medium A for terrestrial strains; medium B for marine strains), mixotrophic (medium C), and heterotrophic (medium C without $NaNO_2$) growth of nitrite oxidizers

	Culture medium		
Ingredient	A[a]	B[b]	C[c, d]
Distilled water (ml)	1000	300	1000
Seawater (ml)		700	
$NaNO_2$ (mg)	200–2000	69	200–2000
$MgSO_4 \cdot 7H_2O$ (mg)	50	100	50
$CaCl_2 \cdot 2H_2O$ (mg)		6	
$CaCO_3$ (mg)	3		3
KH_2PO_4 (mg)	150	1.7	150
$FeSO_4 \cdot 7H_2O$ (mg)	0.15		0.15
Chelated iron (13%, Geigy) (mg)		1	
$Na_2MoO_4 \cdot 2H_2O$ (μg)		30	
$(NH_4)_2Mo_7O_{24} \cdot 4H_2O$ (μg)	50		50
$MnCl_2 \cdot 6H_2O$ (μg)		66	
$CoCl_2 \cdot 6H_2O$ (μg)		0.6	
$CuSO_4 \cdot 5H_2O$ (μg)		6	
$ZnSO_4 \cdot 7H_2O$ (μg)		30	
NaCl (mg)	500		500
Sodium pyruvate (mg)			550
Yeast extract (Difco) (mg)			1,500
Peptone (Difco) (mg)			1,500
pH adjusted to[e]	8.6	6	7.4

[a]For terrestrial strains from Bock et al. (1983).

[b]For marine strains modified from Watson and Waterbury (1971).

[c]For terrestrial strains from Bock et al. (1983).

[d]For heterotrophic growth medium C without $NaNO_2$ is used.

[e]After sterilization pH should be 7.4–7.8.

Maintenance Procedures for Stock Cultures

Nitrifying organisms can survive starvation for more than one year when kept at 17°C in liquid media. Nevertheless, cells should be transferred to fresh media every four months. In Table 1 three different growth media for nitrite oxidizers are listed. Freezing in liquid nitrogen is a suitable technique for maintenance of stock cultures that are suspended in a cryoprotective buffer containing sucrose and histidine. When freeze-dried on lavalite or polyurethane, about 0.5% of *Nitrobacter* cells survive for one year (L. Lin, personal communication). Another possibility for the storage of *Nitrobacter* for several years is cultivation in 1l-bottles filled to the top with complex medium and closed by a screw top. Glycerol should be used instead of pyruvate to keep the pH stable for a long period. Since the bacteria are able to oxidize nitrite to nitrate aerobically and subsequently able to reduce the nitrate anaerobically, a high cell yield can be obtained using this method (Freitag et al., 1987).

Differentiation of the Four Genera of Nitrite-Oxidizing Bacteria

Morphological, genotypic, and chemotaxonomic characteristics that can be used to differentiate the four genera of nitrite-oxidizing bacteria are given in Tables 2, 3, and 4.

Specific reaction patterns of a set of three monoclonal antibodies (MAbs) that recognize the nitrite-oxidizing system (Aamand et al., 1996) were shown to be useful for taxonomic investigations of pure and enrichment cultures by western blot analysis and immunofluorescent labelling (Bartosch et al., 1999). The three different MAbs have different degrees of specificity that permit classification to the genus level. MAb Hyb 153-2 recognizes the α-NOR of the described species of *Nitrobacter*. MAb Hyb 153-1 recognizes the β-NOS of *Nitrobacter* and *Nitrococcus*, whereas MAb Hyb 153-3 reacts with the β-NOS of all known nitrite-oxidizing bacteria. The differing molecular masses of the β-NOSs enable differentiation of the four genera.

The results summarized in Table 2 indicate that the epitope of the β-subunit recognized by MAb Hyb 153-3 is highly conserved. The finding of such conserved regions in the key enzyme of nitrite oxidation does not support the hypothesis of Teske et al. (1994) that the nitrifiers arose independently multiple times, possibly from different photosynthetic ancestors. The specific reactions of the MAbs suggest a close correlation between phylogeny and function and underscore the utility of investigation of the comparative biochemistry of proteins involved in energy metabolism as an approach to the study of bacterial evolution (Brock, 1989).

TABLE 2. Differentiation of the four genera of nitrite-oxidizing bacteria

Characteristic	*Nitrobacter*	*Nitrococcus*	*Nitrospina*	*Nitrospira*
Phylogenetic position	*Alphaproteobacteria*	*Gammaproteobacteria*	*Deltaproteobacteria* (preliminary)	Phylum *Nitrospirae*
Morphology	Pleomorphic short rods	Coccoid cells	Straight rods	Curved rods to spirals
Intracytoplasmic membranes	Polar cap	Tubular	Lacking	Lacking
Size (μm)	0.5–0.9 × 1.0–2.0	1.5–1.8	0.3–0.5 × 1.7–6.6	0.2–0.4 × 0.9–2.2
Motility	+	+	−	−
Reproduction:	Budding or binary fission	Binary fission	Binary fission	Binary fission
Main cytochrome types[a]	*a, c*	*a, c*	*c*	*b, c*
Location of the nitrite oxidizing system on membranes	Cytoplasmic	Cytoplasmic	Periplasmic	Periplasmic
MAb-labeled subunits (KDa)[b]	130 and 65	65	48	46
Crystalline structure of membrane-bound particles	Rows of particle dimers	Particles in rows	Hexagonal pattern	Hexagonal pattern

[a]Lithoautotrophic growth.

[b]MAbs, monoclonal antibodies.

TABLE 3. Properties of the nitrite-oxidizing bacteria

Characteristic	*Nitrobacter winogradskyi*	*Nitrobacter alkalicus*	*Nitrobacter hamburgensis*	*Nitrobacter vulgaris*	*Nitrococcus mobilis*	*Nitrospina gracilis*	*Nitrospira marina*	*Nitrospira moscoviensis*
Mol% G + C of the DNA	61.7	62	61.6	59.4	61.2	57.7	50	56.9
Carboxysomes	+	−	+	+	+	−	−	−
Habitat:								
Fresh water	+			+				
Waste water	+			+				
Brackish water				+				
Oceans	+				+	+	+	
Soda lakes		+						
Soil	+		+	+				
Soda soil		+						
Stones	+			+				
Heating system								+

TABLE 4. Primary fatty acids of the described species of nitrite-oxidizing bacteria[a,b]

Fatty acid	*Nitrobacter winogradskyi* Engel	*Nitrobacter alkalicus* AN4	*Nitrobacter hamburgensis* X14	*Nitrobacter vulgaris* Z	*Nitrococcus mobilis* 231	*Nitrospina gracilis* 3	*Nitrospira marina* 295	*Nitrospira moscoviensis* M1
$C_{14:1 cis9}$						+		
$C_{14:0}$	+		+		+	+ + +	+	+
$C_{16:1 cis7}$							+ + +	+ +
$C_{16:1 cis9}$	+	+	+	+	+ + +	+ + +		
$C_{16:1 cis11}$							+ + +	+ + +
$C_{16:0}$3OH						+		+
$C_{16:0}$	+ +	+ +	+ +	+ +	+ + +	+ +	+ + +	+ + +
$C_{16:0\ 11methyl}$							+	+ + +
$C_{18:1 cis9}$	+		+			+	+	
$C_{18:1 cis11}$	+ + + +	+ + + +	+ + + +	+ + + +	+ + +	+	+	+
$C_{18:0}$	+	+	+		+	+	+ +	+
$C_{19:0 cyclo 11\text{-}12}$	+	+	+	+	+			

[a]Symbols: +, <5%; + +, 6–15%; + + +, 16–60%; + + + +, >60%.

[b]Stirred cultures were grown autotrophically at 28°C (*Nitrospira moscoviensis* at 37°C) and collected at the end of exponential growth. Modified from Lipski et al., (2001).

Bacteria that Respire Oxyanions of Chlorine

John D. Coates

Microbial respiration of oxyanions of chlorine such as chlorate (ClO_3^-) and perchlorate (ClO_4^-) [together referred to as "(per)chlorate" below] under anaerobic conditions has been known for more than half a century (Aslander, 1928). In general, chlorine oxyanions in the environment result from anthropogenic sources including disinfectants, bleaching agents, herbicides (Germgard et al., 1981; Agaev et al., 1986; Rosemarin et al., 1990), and munitions (Urbanski, 1984a, b). No natural source of chlorate exists and the only known natural source of perchlorate is associated with mineral deposits found in Chile, where the perchlorate may represent as much as 6–7% of the total mass (Ericksen, 1983). Although these Chilean deposits have been extensively mined as a mineral and nitrate source for fertilizer manufacture, this is not thought to represent a significant source of perchlorate in the environment (Urbansky et al., 2000). The high reduction potential of (per)chlorate makes them ideal electron acceptors for microbial metabolism (Coates et al., 2000b). Early studies indicated that microorganisms rapidly reduced chlorate that was applied as a herbicide for thistle control (Aslander, 1928), and the application of this reductive metabolism was later proposed for the measurement of sewage and wastewater biological oxygen demand (Bryan and Rohlich, 1954; Bryan, 1966). Initial investigation of the microbiology of chlorate reduction suggested that it was mediated by nitrate-respiring organisms in the environment, and chlorate uptake and reduction was simply a competitive reaction for the nitrate reductase system of these bacteria (Hackenthal et al., 1964; Hackenthal, 1965; de Groot and Stouthamer, 1969). In support of this, many organisms were shown to be capable of the reduction of (per)chlorate including *Escherichia coli, Proteus mirabilis, Rhodobacter capsulatus,* and *Rhodobacter sphaeroides* (de Groot and Stouthamer, 1969; Roldan et al., 1994). Chlorite (ClO_2^-) was generally produced as a toxic end product of this reduction, and there was no evidence that these organisms could couple growth to this metabolism. Furthermore, early studies demonstrated that membrane-bound respiratory nitrate reductases and assimilatory nitrate reductases could alternatively reduce chlorate (Stewart, 1988), and selection for chlorate resistance has been used to obtain mutants that are unable to synthesize the molybdenum cofactor required for nitrate reduction (Neidhardt et al., 1996).

However, this could not explain the presence of specialized enzymes such as the chlorate reductase C purified from *Proteus mirabilis,* which could only use chlorate as a substrate (Oltmann et al., 1976). Now it is known that specialized organisms have evolved that can grow by the anaerobic reductive dissimilation of (per)chlorate into innocuous chloride, and many dissimilatory (per)chlorate-reducing bacteria are now in pure culture (Romanenko et al., 1976; Stepanyuk et al., 1992; Malmqvist et al., 1994; Rikken et al., 1996; Wallace et al., 1996; Bruce et al., 1999; Coates et al., 1999, 2001b; Michaelidou et al., 2000). These organisms have been isolated from a broad diversity of environments including both pristine and contaminated soils and sediments (Romanenko et al., 1976; Stepanyuk et al., 1992; Malmqvist et al., 1994; Rikken et al., 1996; Wallace et al., 1996; Bruce et al., 1999; Coates et al., 1999; Michaelidou et al., 2000). This was unexpected due to the limited natural abundance of (per)chlorate. However, the diverse metabolic capabilities of these organisms may explain their presence in environments where (per)chlorate is not found. Phenotypic characterization revealed that the known dissimilatory (per)chlorate-reducing bacteria exhibit a broad range of metabolic capabilities including the oxidation of hydrogen (Wallace et al., 1996), simple organic acids and alcohols (Malmqvist et al., 1994; Rikken et al., 1996; Bruce et al., 1999; Coates et al., 1999; Michaelidou et al., 2000), aromatic hydrocarbons (Coates et al., 2001b), hexoses (Malmqvist et al., 1994), reduced humic substances (Bruce et al., 1999; Coates et al., 2001b, 2002), both soluble and insoluble ferrous iron (Bruce et al., 1999; Coates et al., 1999; Michaelidou et al., 2000; Chaudhuri et al., 2001; Lack et al., 2002a, b), and hydrogen sulfide (Bruce et al., 1999; Coates et al., 1999). All of the known dissimilatory (per)chlorate-reducing bacteria are facultatively anaerobic or microaerophilic (Rikken et al., 1996; Wallace et al., 1996; Bruce et al., 1999; Coates et al., 1999; Michaelidou et al., 2000), and some, but not all, alternatively respire nitrate, which supports the suggestion that (per)chlorate reduction is unrelated to nitrate reduction (Bruce et al., 1999; Coates et al., 1999). Generally, these organisms are assumed to use either chlorate or perchlorate as terminal electron acceptors (Logan, 1998), although this has been demonstrated only in a few isolated cases (Stepanyuk et al., 1992; Wallace et al., 1996; Bruce et al., 1999). Recent studies (Wu et al., 2001; J.D. Coates, unpublished data) have demonstrated that this assumption was incorrect and there are now several chlorate-reducing bacteria in pure culture, including a novel marine isolate, *"Dechloromarinus chlorophilus"* strain NSS, demonstrated to be incapable of the reductive respiration of perchlorate (J.D. Coates, unpublished data).

(Per)chlorate reducing bacteria are phylogenetically diverse (Wallace et al., 1996; Coates et al., 1999; Michaelidou et al., 2000) with members in the *Alphaproteobacteria, Betaproteobacteria, Gammaproteobacteria,* and *Epsilonproteobacteria* classes of the *Proteobacteria* (Wallace et al., 1996; Coates et al., 1999; Michaelidou et al., 2000; Achenbach et al., 2001). As such, the metabolic capability of (per)chlorate reduction is widespread throughout the *Proteobacteria,* which has some interesting evolutionary implications due to the relatively short time in which (per)chlorate reduction could have evolved. Several of the known (per)chlorate-reducing isolates are representatives of previously defined genera (*Pseudomonas, Magnetospirillum, Wolinella*) (Wallace et al., 1996; Coates et al., 1999) not recognized for the capability of (per)chlorate respiration. However, the majority of the known (per)chlorate-reducing bacteria are closely related to each other and to the bacterial species *Rhodocyclus tenuis* and *Ferribacterium limneticum* in the class *Betaproteobacteria*. In general, the known close relatives to the (per)chlorate-reducing isolates do not grow by (per)chlorate respiration regardless of the similarity of their 16S rDNA sequence, thus making predictions of metabolic functionality based on 16S rDNA sequence analysis futile (Achenbach and Coates, 2000). For example, *R. tenuis* is a phototrophic nonsulfur purple bacterium that contains bacteriochlorophyll and is found on soil surfaces and in shallow waters exposed to sunlight, whereas *F. limneticum* is a strict anaerobic, nonfermenting, dissimilatory Fe(III)-reducer (Cummings et al., 1999). Although the (per)chlorate-reducing bacteria are closely related to these organisms oftentimes with a 16S rDNA sequence divergence of less than 1% (Coates et al., 1999, 2001b; Achenbach and Coates, 2000; Achenbach et al., 2001), they exhibit distinct physiologies. None of the (per)chlorate-reducing isolates can grow by phototrophy or Fe(III)-reduction. By the same token, *F. limneticum* does not grow by phototrophy or by the reduction of (per)chlorate, and *R. tenuis* cannot grow by anaerobic respiration with a broad range of electron acceptors including perchlorate or Fe(III).

The (per)chlorate reducers of the class *Betaproteobacteria* represent two novel genera, the *Dechloromonas* species and the *Dechlorosoma* species (Achenbach et al., 2001). Members of these two groups are ubiquitous (Coates et al., 1999) and have been identified and isolated from nearly all environments screened including pristine and contaminated field samples, *ex situ* bioreactors treating perchlorate-contaminated wastes (Coates et al., 1999; Logan et al., 2001), and even in soil and lake samples collected from Antarctica (JD Coates and LA Achenbach, unpublished). As such, these two groups are considered to represent the dominant (per)chlorate-reducing bacteria in the environment (Coates et al., 1999). Pure culture studies have demonstrated that members of these genera can grow over a broad range of environmental conditions; however, they generally grow optimally at pH values near neutrality in freshwater environments (Bruce et al., 1999; Coates et al., 1999; Michaelidou et al., 2000).

Although there is still relatively little known about the biochemistry of (per)chlorate reduction, some recent studies have yielded important information. Initial investigations have demonstrated the presence of *c*-type cytochrome(s) in perchlorate-reducing bacteria and their involvement in the reduction of (per)chlorate (Bruce et al., 1999; Coates et al., 1999). Difference spectra studies revealed that the H_2-reduced *c*-type cytochrome content of (per)chlorate-reducers was readily reoxidized in the presence of chlorate or perchlorate but was unaffected by nonphysiological electron acceptors for these organisms such as sulfate, fumarate, or Fe(III) (Coates et al., 1999).

More recently, a single oxygen-sensitive perchlorate reductase enzyme of the (per)chlorate-reducing strain GR-1 has been purified and partially characterized (Kengen et al., 1999). This enzyme was located in the periplasm of the organism and was a heterodimer in an $\alpha_3\beta_3$ configuration (Kengen et al., 1999). The perchlorate reductase had a total molecular mass of 420 kDa and contained iron, molybdenum, and selenium (Kengen et al., 1999). In addition to perchlorate, the perchlorate reductase from strain GR-1 also catalyzed the reduction of chlorate, nitrate, iodate, and bromate (Kengen et al., 1999). Perchlorate and chlorate were reduced to chlorite. Subsequent phenotypic studies demonstrated that although selenium can be replaced with alternative cations by perchlorate-reducing bacteria (JD Coates, unpublished), the molybdenum plays a functional role in the reduction of perchlorate (Chaudhuri et al., 2002). Furthermore, molecular studies of the genetic systems associated with perchlorate reduction indicated the presence of a molybdenum-dependent chaperone gene similar to that found in nitrate reductase systems in association with the gene encoding perchlorate reductase in the (per)chlorate-reducers, *Dechloromonas* strain RCB and *Pseudomonas* strain PK (L A Achenbach, unpublished).

The quantitative dismutation of chlorite into chloride and O_2 is now known to be a central step in the reductive pathway of (per)chlorate that is common to all (per)chlorate-reducing bacteria (Coates et al., 1999). Chlorite dismutation by (per)chlorate-reducing bacteria is mediated by a highly conserved single enzyme, chlorite dismutase (CD) (van Ginkel et al., 1996; Coates et al., 1999; Stenklo et al., 2001; O'Connor and Coates, 2002). Studies with washed whole cell suspensions demonstrated that the CD was highly specific for chlorite and none of a broad range of alternative analogous anions tested served as substrates for dismutation (Bruce et al., 1999). The purified CD was a homotetramer with a molecular mass of 120 kDa and a specific activity of 1,928 μmol chlorite dismutated per mg of protein per minute (Coates et al., 1999). This is similar to the molecular mass and specific activity observed for the CD previously purified from the (per)chlorate-reducer strain GR-1 (van Ginkel et al., 1996) and subsequently from *Ideonella dechloratans* (Stenklo et al., 2001). Phenotypic studies with the (per)chlorate-reducers *Dechloromonas agitata* and *Dechlorosoma suillum* indicated that CD activity is present only when the organisms are grown anaerobically on perchlorate or chlorate and expression of the CD is negatively regulated by oxygen and nitrate (Chaudhuri et al., 2002). Furthermore, studies with an immunoprobe specific for purified CD from *Dechloromonas agitata* strain CKB indicated that the CD is present on the outer membrane of all (per)chlorate-reducing bacteria and is highly conserved among these organisms, regardless of their phylogenetic affiliation (O'Connor and Coates, 2002). More recently, the chlorite dismutase gene *cld* was isolated and sequenced from *Dechloromonas agitata* strain CKB (Bender et al., 2002). Sequence analysis identified an open reading frame of 834 bp that encodes a mature protein with an N-terminal sequence identical to that of the previously purified *D. agitata* chlorite dismutase enzyme (Bender et al., 2002). The predicted translation product of the *D. agitata cld* gene is a protein of 277 amino acids including a leader peptide of 26 amino acids. Primer extension analysis identified a single transcription start site directly downstream of an AT–rich region that represented the −10 promoter region of the *D. agitata cld* gene. In support of the previous observations, Northern blot analysis indicated that the *cld* gene is transcriptionally upregulated when *D. agitata* cells are grown under perchlorate-reducing versus aerobic conditions, and slot blot hybridizations with a *D. agitata cld* probe demonstrated the high degree of conservation of the *cld* gene among (per)chlorate-reducing bacteria (Bender et al., 2002).

The role of (per)chlorate-reducing bacteria in environments that have had no previous exposure to chlorine oxyanions has yet to be determined. Environmental contamination with (per)chlorate is predominantly the result of anthropogenic activity over the last hundred years, whereas these organisms have been found in several pristine environments not known to have had any prior contact with perchlorate or chlorate (Coates et al., 1999). As such, the evolution of such a phylogenetically diverse group of organisms with the ability to couple growth to the reduction of (per)chlorate is unexpected and may be the result of horizontal gene transfer events. This possibility is supported by the fact that the reductive pathway is centered on a unique and highly conserved enzyme, chlorite dismutase. Because (per)chlorate-reducing bacteria are found in several pristine environments, the ubiquity of these organisms is unlikely to be related to their ability to grow by dissimilatory (per)chlorate reduction (Coates et al., 1999). Previous studies have demonstrated that these organisms are, in general, very versatile and can use a broad range of alternative electrons donors. As such, the selective pressures for (per)chlorate reducing bacteria in the environment may be based on the diversity of their metabolic capabilities rather than any individual metabolism.

The Revised Road Map to the *Manual*

George M. Garrity, Julia A. Bell and Timothy Lilburn

Introduction

The Second Edition of *Bergey's Manual of Systematic Bacteriology* (the "*Systematics*") represents a major departure from the First Edition, as well as from the Eighth and Ninth Editions of the *Bergey's Manual of Determinative Bacteriology* (the "*Determinative*"), in that the organization of the content follows a phylogenetic framework based on analyses of the nucleotide sequence of the ribosomal small-subunit RNA, rather than one based on phenotypic characters. The Eighth and Ninth Editions of the *Determinative* and the First Edition of the *Systematics* were organized in a non-hierarchical scheme because information about higher taxa was insufficient for construction of a formal hierarchical classification such as those used in all the previous editions. Instead, the genera were organized into phenotypic groupings (e.g., the Gram-positive cocci), and these groupings were called Parts, Sections, or Groups. By the Ninth Edition of the *Determinative*, which was based on the information included in the *Systematics*, there were 35 separate phenotypic groupings.

As early as the 1970s, molecular comparisons suggested a natural phylogenetic classification eventually would be possible. Some sections of both manuals were organized into coherent phylogenetic groups, but the editors were not able to place the bulk of the described genera into a scientifically sound hierarchy based on phylogenetic relationships. By the end of the 1990s, the number of 16S rDNA sequences became large enough, and the taxonomic coverage sufficiently broad, to justify the organization of this edition of the *Systematics* along phylogenetic lines. In adopting 16S rDNA sequences as a means of ordering the taxonomic hierarchy, bacteriologists engaged in a deliberate effort to make an explicit connection between systematics and evolution, a connection that has informed botanical and zoological systematics, which were developed with the aid of the fossil record, for more than a century and a half.

The practice of systematics and taxonomy has been connected to ideas in evolutionary biology since Darwin. For more than half a century, following the emergence of the neo-Darwinian synthesis, biological species of higher eucaryotes have been defined in terms of real or potential gene pools (Mayr, 1942). However, the gene pools of bacteria are difficult to define because of the widespread occurrence of lateral gene transfer. Bacteriologists are therefore caught between the pragmatic definition of a bacterial species, based on phenetic assessments of phenotype and genotype, and a more or less controversial definition of a bacterial species based on evolutionary concepts. (A full treatment of the various views of the nature of bacterial species is outside the scope of this discussion, as are full discussions of the implications of lateral gene transfer for bacterial species concepts and phylogenetic analyses.) Nevertheless, the increasing use of molecular methods to study bacterial evolution and systematics has led to the expectation that a taxonomy reflecting evolutionary lineages can be constructed for bacteria.

This connection between systematics and evolution is pervasive but not logically necessary: purely phenetic taxonomies have been widely used in bacteriology; for many bacteriologists, the current "best practice" in bacterial systematics is polyphasic taxonomy, in which both phenotypic and genotypic information are used. In keeping with the increased emphasis on the use of genetic information in taxonomy, bacteriologists have added the generally accepted requirement of >70% DNA–DNA similarity, the requirement of near-identity in 16S rDNA sequences for species, and a less stringent (but unspecified) level of identity in 16S rDNA sequences for genera. There is a reasonably good correlation between very nearly identical 16S sequences, high levels of DNA–DNA similarity, and similar phenotypes among the species of most bacterial genera. The history of 16S rDNA sequences has thus come to be viewed as a proxy for the history of bacterial evolution. Whether this view is justified is not clear in all cases. For instance, the process of "filling in" and completing the taxonomic hierarchy based on 16S rDNA relationships has indeed produced some discrepancies between genotypic and phenotypic groupings, particularly at higher taxonomic levels. Examples in this volume include both organisms that are phenotypically similar but dissimilar in 16S rDNA sequences (see the introductory essays on nitrifiers, nitrite oxidizers, ammonia oxidizers, and photosynthesizers) and organisms that have nearly identical 16S rDNA sequences but are phenotypically dissimilar (compare *Enhydrobacter* to the genera of the *Moraxellaceae*). Readers may refer to the chapter by Ludwig and Klenk (2001, reprinted in this volume) for a detailed discussion of the methods used and problems encountered in phylogenetic analyses of 16S rDNA sequences.

It was predicted as early as the 1960s that a phylogenetic arrangement of taxa might not be fully compatible with the known phenotypes of the subject taxa (Sokal and Sneath, 1963). While there are a number of clear-cut examples of coincidence between phenotype and phylogeny (e.g., the spirochetes, green-sulfur bacteria [*Chloroflexi*], and the actinomycetes), the picture is much less clear for the many Gram-negative and low G + C Gram-positive genera, which represent a significant proportion of the known, cultivable taxa. Many seemingly contradictory groupings exist, and since there has been relatively little effort to systematically correlate phenotype with phylogenetic grouping, we are left with a dilemma in attempting to present what many bacteriologists require: an organizational scheme that also contains determinative information other than the 16S rDNA sequence.

We have also learned that an incomplete phylogenetic scheme presents some unique challenges when used as the basis for organizing printed material, especially when applied to a multivolume work in which multiple years elapse between the appearance of the first and last volumes. The phylogenetic framework presents fewer mnemonic devices than the earlier scheme and no unique order of appearance. In fact, in many regions within the large-scale phylogenetic trees, the precise location of a given species, genus, or even higher taxon may be uncertain (see Ludwig and Klenk (2001, and this volume for a detailed discussion of this issue)). Consequently, readers might have difficulty in intuitively determining the precise location of a given taxon, especially when searching for less familiar ones. We believe this will be a transient problem for most readers, and, with some guidance, they should be able to rapidly assimilate the "new" taxonomy presented in this edition.

To expedite this process, the Editorial Board decided that this edition required a "road map" to help readers find their way through the *Systematics*. Readers may need to know where a certain genus fits into the overall classification and where it will be found in a given volume; they may need to know the identity of an isolate or the phenotypic characteristics of the closest relatives of an isolate that has been identified by molecular means. Thus, it is our responsibility to provide a guide for the readership to ensure that their needs are met. Initially, this task fell to the past and present Editors-in-Chief, as they were principally responsible for creating and maintaining the taxonomic hierarchy in addition to determining the overall organization of the content of this edition. In this volume, the task falls to us.

Each volume of this edition will have an updated "road map" chapter (at the current rate of description of new taxa, the number of new genera validly published between the first volume and the fifth and last could number well over 500). Also, there will continue to be refinements in the phylogenetic classification and, we hope, the discovery of new phenotypic and genotypic features other than the 16S rDNA sequence to further improve the description of taxa. As more fully sequenced bacterial genomes have become available in the last several years, some steps have been taken to explore the ways in which genomic data can be used to illuminate bacterial phylogeny. However, it is not yet clear what the impact of genomic data on bacterial taxonomy will ultimately be. The status of this endeavor is discussed in the next section.

Genomics and systematics

As this volume goes to press, there are still relatively few phylogenetic studies based on whole genome sequences. Furthermore, the studies available to date are not readily comparable because most are based on different sets of genes, different sets of organisms, different computational methods, and different methods of assessing the statistical significance of the results. Not surprisingly, there are conflicting conclusions on a variety of issues as well as on the details of the derived phylogenies. However, enough progress has been made that it is possible to define some of the problems that must be resolved before genomic data can be used with confidence in phylogenetic analyses. The results of studies using genomic data to explore procaryotic evolution have indicated that the use of a single gene, even the 16S rDNA gene, to establish evolutionary histories may be problematic. However, the extensive 16S rDNA sequence dataset can still be used to provide initial hypotheses regarding the taxonomic placement of a new organism, and the sequences themselves can be used as unique identifiers of bacterial species, serving as a biological index of the taxonomic space. Without systematic studies—at all taxonomic levels—of the efficacy of the various genomic approaches described below, it is unclear whether genomic data will clarify or obscure the picture.

The most common strategy employed in genomic studies is to make alignments of individual or concatenated DNA or protein sequences, calculate distance metrics between all pairs of sequences, and produce distance matrices of the pairwise distances. These distance matrices are then used to produce trees or dendrograms that are interpreted as reflecting the evolutionary relationships among the organisms providing the DNA or protein sequences. If trees are produced for individual genes or proteins rather than for concatenated sequences, the topologies of all of the trees are further examined to identify evolutionary lineages, and so-called "supertrees" are constructed. Examples of studies employing this general strategy include those of Brown et al. (2001d), Nesbø et al. (2001), Wolf et al. (2001), Brochier et al. (2002), Daubin et al. (2002), Raymond et al. (2002), Coenye and Vandamme (2003), Lerat et al. (2003), and Wertz et al. (2003). Many of the studies and methods have been reviewed by Wolf et al. (2002b). In addition to difficulties in aligning distantly related sequences and in editing sequences during the alignment process, problems associated with this strategy include the identification of gene or protein sequences appropriate for analysis at various depths of evolutionary relationship, methods of identification of orthologous sequences, definition of the number and degree of relatedness of taxa that should be employed in a given study (Zwickl and Hillis, 2002), and definition of the degree of conservation of sites appropriate for the taxonomic breadth and evolutionary depth being examined (Hansmann and Martin, 2000). An additional difficulty in the use of genomic sequences for evaluating phylogenetic relationships lies in the fact that genome sequences are only available for one or a few strains of each species; for most non-pathogenic organisms; it is not known to what extent the genome of the sequenced strain is representative of the genetic variation in the higher taxa to which that species belongs. Finally, methods are still being developed for the computation and statistical testing of phylogenetic relationships derived from genomic data.

A related strategy using genomic information for evolutionary studies has been to produce trees based on the presence or absence of orthologous genes in the genomes being analyzed (Wolf et al., 2002b). This strategy employs a series of steps similar to that employed in the first strategy and is subject to many of the same difficulties. For closely related strains, the presence or absence of all the genes in the genome can be measured using microarray technology (Joyce et al., 2002). This approach is useful for assessing genetic variation within species. Another type of presence/absence approach scores insertions and deletions (indels) in the amino acid sequences of particular proteins, usually deduced from gene sequences (Gupta et al., 1999; Gupta, 2000, 2003). The appearance of these "signatures" is thought to be so rare that the likelihood of their chance appearance in more than one genome is negligible.

Another strategy might be termed the genome informational approach. The method uses the entire genome sequence, whereas the aforementioned methods actually use only a subset of those data. The informational approach has been adopted by researchers hoping to avoid the pitfalls and uncertainties of sequence alignments (Vinga and Almeida, 2003). These approaches include measurement of oligonucleotide frequencies (Deschavanne et al., 1999; Stuart et al., 2002) and of complexity (Li et al., 2001; Yu et al., 2001).

Perhaps the most difficult question for genomic assessments of phylogenetic relationships is whether there exists a set of "core" genes common to all organisms for which orthologs can be unambiguously identified and—for the first strategy—properly aligned. Workers have identified "core" genes based either on *a priori* determinations or on the results of computer searches of the genomes to be analyzed. *A priori* identification of "core" genes is usually focused on essential functions in nucleic acid metabolism, protein synthesis, and central metabolic pathways ("housekeeping genes"). The sequences of genes encoding RNAs and proteins that interact with many other molecules are thought to be more constrained evolutionarily than those encoding RNAs and proteins that have fewer such interactions (Jain et al., 1999) and thus are thought to be resistant or immune to lateral gene transfer. (Strictly speaking, the hypothesis is not that "core" genes cannot be transferred but that recipient cells are at a selective disadvantage because chimeric sequences resulting from lateral gene transfer either do not function or do not function as well as the wild type, coevolved forms. Lateral gene transfer events would thus not be preserved over evolutionary time.) The "core" genes include those encoding RNAs and proteins involved in mRNA and protein synthesis. The composition of the "core" set of genes can vary with the taxonomic depth being examined; the genes are chosen so that the sequences exhibit both sufficient conservation that orthologs can be identified and aligned and sufficient variability that a phylogenetic signal can be detected (for examples, see Nesbø et al., 2001; Coenye and Vandamme, 2003; Wertz et al., 2003). Computer searches have usually been performed using BLAST to search genome data in public databases (for examples, see Wolf et al., 2001; Raymond et al., 2002; Lerat et al., 2003).

A related and controversial question concerns the degree to which the core genes, including the 16S rDNA gene, are subject to lateral gene transfer. That lateral gene transfer has been important in procaryotic evolution is not in doubt; lateral gene transfer has been detected in many procaryotes either by the demonstration that particular sequences are mosaic (for examples, see Sneath, 1993; Ueda et al., 1999; Yap et al., 1999) or by the construction of noncongruent phylogenies for different genes in the same organism (for example, Nesbø et al., 2001; Raymond et al., 2002). Even 16S rDNA and 23S rDNA sequences can yield different phylogenies (Raymond et al., 2002). Wertz et al. (2003) point out that there are other possible explanations for noncongruent trees in addition to lateral gene transfer, including differing selective forces acting on different genes and an insufficient number of phylogenetically informative sites in the sequences being examined. Gogarten et al. (2002) point out that the very evolutionary conservation of a "core" gene such as the 16S rDNA gene makes it "vulnerable" to change by lateral gene transfer events. Conclusions have differed on the extent of the difficulties posed by lateral gene transfer for phylogenetic reconstruction studies, particularly since lateral gene transfer may have a more severe confounding effect in some evolutionary lineages and at some taxonomic and evolutionary depths than others. Some authors hold that the extent of lateral gene transfer may be too great to allow construction of anything other than individual gene histories. It may or may not be possible to extract the "deep" evolutionary histories of procaryotes from combined data on the histories of many genes. Brown et al. (2001d) found that removing genes thought likely to have undergone lateral transfer from their analysis substantially altered the branching patterns and phylogenetic affiliations of several lineages, including *"Deinococcus–Thermus"*, *Cyanobacteria*, *"Actinobacteria"*, and *Thermotoga/Aquificales*.

As noted above, conclusions derived from different studies concerning the phylogenetic affiliations of various bacterial lineages can differ; however, some results are supported by more than one study and may provide better clues to evolutionary histories. Because of the distribution of taxa for which genomic data are currently available, the studies attempted so far have been centered either on the elucidation of deep lineages and higher taxa or on variation within genera and families. The reader should be aware that levels of statistical support for these groupings, estimated by bootstrapping or by other methods, vary among the papers cited; because different authors use different methods and adopt different conventions in reporting levels of support, it is often impossible to compare results directly. Furthermore, for groupings that branch at very deep levels, it is expected that the ability to discern relationships will be more limited than it is among more closely related organisms. Three cases involving what are thought to be ancient lineages are discussed below: *Thermotoga/Aquifex*, chlamydia/spirochetes, and actinomycetes/deinococci/cyanobacteria (Wolf et al., 2001). Finally, two studies of more closely related taxa—*Gammaproteobacteria* and lactic acid bacteria—are discussed. As might be expected, the studies of higher taxa and deeper-branching lineages have produced more conflicting results than the studies of more closely related taxa.

In addition to the use of 16S rDNA sequence analysis and the genomically based analyses discussed above, Cavalier-Smith (2002) has employed another strategy. In these analyses, the procedure has been to identify key phenetic characters of the taxa in question and then to use the gain or loss of these characters to define a sequence of evolutionary events. The characters chosen are complex enough that gain and loss should be infrequent evolutionary events. These characters include a suite of complex biochemical traits, together with insertions and deletions in sequences of particular macromolecules and gains, losses, or modifications of particular molecules or macromolecular structures. The results of this analysis with respect to two of the three deeply branching lineages are discussed below.

Thermotoga and *Aquifex* are more closely affiliated with each other than with other deeply branching taxa in most sequence-based studies, although the depth of branching between the two is dependent both on the dataset used and the tree-building method used (Hansmann and Martin, 2000; Brown et al., 2001d; Wolf et al., 2001; Brochier et al., 2002; Daubin et al., 2002; Wolf et al., 2002b). However, in gene content analysis based on the presence or the absence of genes in genomes, *Aquifex* was associated with the *Epsilonproteobacteria*, whereas in analyses based on either the presence or the absence of taxa in lists of Clusters of Orthologous Groups (COGs) of proteins or on degree of sequence identity between probable orthologs, *Thermotoga* and *Aquifex* appeared as independent, deeply branching lineages (Wolf et al., 2002b). In a linear scenario of bacterial evolution derived from indel analysis, Gupta (2003) placed *Thermotoga* between actinobacteria and *"Deinococcus–Thermus"* and placed *Aquifex* between the grouping chlamydiae/cytophagas/flexibacteria/flavobacteria/green sulfur bacteria and the *Deltaproteobacteria* and *Epsilonproteobacteria*.

In the large-scale classification of procaryotes we have applied to the *Systematics* (methods are discussed in more detail below) we find that the *Thermotogae* and *Aquificae* occupy a sparsely populated region within global principal-components analysis (PCA) plots in close proximity to one another and other groups commonly referred to as deeply branching. Graphical inspection of the underlying evolutionary distance matrices used in our models show that *Thermotoga* is more closely related to the low G + C Gram-positives (*"Firmicutes"*), high G + C Gram-positives (*"Actinobacteria"*), and the *Proteobacteria* than is *Aquifex*.

"Chlamydiae" and spirochetes are affiliated with each other in several sequence-based studies (Brown et al., 2001d; Wolf et al., 2001; Brochier et al., 2002). Daubin et al. (2002) recovered this relationship in a set of super tree analyses after restricting the dataset to genes that gave similar individual topologies, i.e., genes whose histories were presumably not obscured by lateral gene transfer or other factors. In the linear evolutionary scenario of Gupta (2003), the emergence of chlamydiae occurs immediately after that of the spirochetes; the same branching order is found in the analysis of Cavalier-Smith (2002). Brochier et al. (2002) found a relationship between chlamydiae, spirochetes, and mycoplasmas based on concatenated 16S and 23S rDNA sequences; this relationship also appeared in gene-content trees derived by Wolf et al. (2001; 2002b). Hansmann and Martin (2000) performed a set of analyses on a concatenation of 35 protein sequences from which variable sites were sequentially removed and each new version of the sequence was reanalyzed; the relationships of *Chlamydia trachomatis* to other taxa changed dramatically as sites were removed. This result illustrates the point that editing of protein sequence data should be considered carefully, even when the intention is to reduce noise and amplify phylogenetic signals.

In our models, we find that the *"Chlamydiae"*, *"Spirochaetes"*, and *Mollicutes* are in close proximity to one another in global PCA plots; however, these organisms form three distinct groups that are clearly separable from one another when the analysis is restricted to those three lineages.

Assessments of the relationships among actinobacteria (mycobacteria and *Streptomyces*), *"Deinococcus–Thermus"*, and cyanobacteria (*Synechocystis* sp. PCC6803) and between these taxa and other lineages vary among different studies. Protein sequence-based studies supporting this grouping of lineages include those of Wolf et al. (2001) and of Brochier et al. (2002), as does the gene content analysis of Wolf et al. (2002b). However, the analysis of Brown et al. (2001d) placed *"Deinococcus–Thermus"* and cyanobacteria together but at a distance from the actinobacteria, as did the 16S–23S rDNA analysis of Brochier et al. (2002) and the analysis of Cavalier-Smith (2002). Daubin et al. (2002) placed the actinobacteria and *"Deinococcus–Thermus"* together and distant from the cyanobacteria. The three groups do not emerge one after the other in the indel-based linear evolutionary scenario of Gupta (2003), in which the order of emergence of the major lineages is: *"Firmicutes"*, *"Actinobacteria"*, *Thermotoga*, *"Deinococcus–Thermus"*, green nonsulfur bacteria, *Cyanobacteria*, *"Spirochaetes"*, *"Chlamydiae"*, *Aquifex*, and *Proteobacteria*.

In our large-scale analyses, we find that the *"Actinobacteria"* are clearly removed from the *"Firmicutes"*, *Cyanobacteria*, and *"Deinococcus–Thermus"*.

Fewer genomically based phylogenetic studies have been performed at the class, family, genus, species, and strain levels because until recently the number of genome sequences available have been insufficient to support such studies. In a study of the *Gammaproteobacteria*, Lerat et al. (2003) used BLAST searches to identify 203 "core" genes in published genome sequences of 12 species of *Gammaproteobacteria* representing five orders and families. These authors found no evidence of lateral transfer of these core genes among these taxa. The same topology was obtained in trees derived from the concatentated protein sequences by the neighbor-joining method with the γ-distance metric, which includes a correction for rate heterogeneity, and in trees derived from 16S rDNA sequences by the neighbor-joining method with the γ-based correction for rate heterogeneity and the Kimura two-parameter distance metric. This topology was also the same as that of a consensus tree calculated from individual trees for each protein. Both 16S rDNA sequences and the concatenated protein sequences yielded different tree topologies when different calculation methods were employed. Of the 12 species, the most unstable was *Vibrio cholerae* (order *"Vibrionales"*) which clustered with five species of the order *"Enterobacteriales"* in three of four trees based on 16S rDNA gene sequences, with two species of the order *Pasteurellales* in one of two trees based on concatenated protein sequences, and independently in the other protein sequence-based tree.

Coenye and Vandamme (2003), in a study of lactic acid bacteria, examined a number of "phenotypes" that can be derived from genomic data (percent of shared orthologs, order of genes on chromosomes, codon bias, G + C content at the third position of synonymous codons, frequency of dinucleotides) and compared results derived from these data to results derived from 16S rDNA sequences and from nine "housekeeping" protein sequences. The study included three strains of *Streptococcus pyogenes*, two strains of *S. agalactiae*, two strains of *S. pneumoniae*, and one strain each of *S. mutans*, *Lactobacillus plantarum*, *Lactococcus lactis*, and with *Bifidobacterium longum* as an outgroup. Only the order of genes on chromosomes appeared not to be taxonomically useful above the species level. There was general agreement between the results obtained with all of the other methods. The results of the Lerat et al. (2003) and the Coenye and Vandamme (2003) studies reflected previous ideas about the phylogenetic relationships of the taxa examined. Since these taxa have been extensively studied, this agreement between genomic and previous analyses is not surprising. It does, however, suggest that the kinds of phenotypic and genotypic data gathered by microbiologists over the years are sufficient to construct accurate taxonomies, at least at the family, genus, species, and strain levels.

In light of results obtained with our large-scale models, the findings of Lerat et al. (2003) are not surprising. Examination of the underlying evolutionary distance matrixes show that members of the *"Enterobacteriales"*, *"Vibrionales"*, and *Pasteurellales* all exhibit a rather high level of sequence similarity to one another ($> 90\%$) and would be expected to have many characters in common, especially those ascribed to "housekeeping" genes. Likewise, one would anticipate the same results when comparing species ascribed to two closely related families within the *"Firmicutes"*. With regard to the study of Coenye and Vandamme (2003), our finding was that members of the *Bifidobacteriales* are distantly removed from the *"Firmicutes"* and appear as a separate group in global PCA analysis. Thus, high levels of genomic similarity would not be expected.

THE TAXONOMIC HIERARCHY

The outline classification presented here is a work in progress. It was started in the early 1990s in the Bergey's Manual Trust Editorial Office. The principal objective was to devise a classification that reflected the phylogeny of procaryotes based upon 16S rDNA sequence analysis and to place all validly named taxa into the classification at a single point, based on the sequence data derived from the type strain, type species, or type genus. We acknowledge that some workers may raise objections to such an approach, as there are a number of existing genera (e.g., *Clostridium*) that are considered to be paraphyletic. It is our view that such instances indicate the need for taxonomic revision, as the species appearing in clades apart from the type strain are clearly misclassified. Authors of individual treatments have been requested to provide readers with detailed discussions of the relevant taxonomic and phylogenetic issues, resolutions of such issues, or proposals as to how such matters are best resolved, where they arise.

Initially, the RDP (Ribosomal Database Project) tree was used to guide the placement of genera within the Taxonomic Outline. However, 16S rDNA sequences for the majority of type strains were not available until recently. As a result, provisional placement, based on phenotypic similarity to sequenced strains, was the only option. As new sequence data became available for existing species, placement in the Taxonomic Outline was changed accordingly. In addition, as new taxa were described they were added to the classification. P.H.A. Sneath and R.G.E. Murray were originally charged with the task of devising a hierarchical classification based upon the phylogenetic trees. However, this responsibility was passed first to J.G. Holt and subsequently to G.M. Garrity and colleagues. All these efforts were combined in the Taxonomic Outline presented in this edition.

In October, 1997, in collaboration with the Center for Microbial Ecology at Michigan State University (East Lansing, MI, USA), the Trust hosted a two-day meeting to discuss progress on updating the RDP tree and to compare the evolving classification with the ARB tree maintained by Ludwig and Strunk at the Technical University of Munich (Germany). A panel of 16 internationally recognized experts in procaryotic phylogeny and taxonomy was assembled to discuss known problems within these two large-scale phylogenetic trees and the ramifications of those problems for the further development of a natural classification of *Bacteria* and *Archaea*. Placement of taxa within the two phylogenetic trees was thoroughly reviewed and areas of uncertainty and discordance were highlighted. In addition, a number of other technical issues, having direct or indirect bearing on the development of a workable taxonomy were raised. These included a lack of control over the quality and authenticity of some sequences, the actual identity of the organisms from which the sequences were obtained, a lack of published documentation on the calculations and algorithms used in the construction of the RDP tree, and the impact of sequence-alignment methods on the resulting phylogenies. This effort led to a significant, albeit slow, improvement in the number and quality of sequences in the trees. More recently, it has also led to experimentation in alternative methods of visualizing very large sets of sequence data.

There were, of course, nomenclatural problems that needed to be addressed. First, the Bacteriological Code (1990 Revision; Lapage et al., 1992) does not cover taxa above the rank of Class, so we have had to follow other Codes of Nomenclature for naming these higher ranks. Second, the trees are of little help in determining the limits of ranks above Order or Class. Third, there is no recognition of rank above Kingdom, yet most phylogeneticists state that the living world is contained in three groups in a rank above Kingdom (variously called Domain or Empire), namely the *Archaea*, *Bacteria*, and *Eucarya*. After considerable deliberation, the Trust has concluded that the rank of Domain should be incorporated into the hierarchy. Furthermore, the rank of Kingdom would not be used to avoid possible conflicts with other Codes of Nomenclature. Within the classification, the *Archaea* and *Bacteria* are divided into Phyla. The phyla are, in turn, successively divided into classes, orders (except for the *Cyanobacteria*, which use the rank of subdivision), families, and genera. In the "*Actinobacteria*", subclasses and suborders are also recognized. As of October 2003, 6466 validly named procaryotic species had appeared either in the Approved Lists of Bacterial Names (Skerman et al., 1980) or in Validation Lists 1–93 published in the *International Journal of Systematic Bacteriology* or the *International Journal of Systematic and Evolutionary Microbiology* (Table 1). In addition, 1007 synonymies had been recorded as a result of taxonomic emendments. A list of new taxa that have been validly published since the appearance of Volume One of the Second Edition of the *Systematics* and the revised genus-level Taxonomic Outline are given in appendices to the roadmap.

Adoption of a hierarchical classification presents several additional challenges. By definition, each species must be a member of successively higher ranks (six of which will be recognized for the majority of taxa in this edition of the *Systematics*). Yet there is considerable reluctance among many workers to place new species and genera into higher taxa, especially at the intermediate levels (family, order, and class). In compiling the Taxonomic Outline we have had to deal with situations where new species were variously assigned to a class or domain without being ascribed membership in any of the intermediate taxa. This may be attributed to a lack of clear rules for delineating higher taxa. Indeed, it may be that taxon delineation requires a more sophisticated approach based on uncovering naturally occurring boundaries that vary from group to group. It may also reflect the inherent limitations of the 16S rRNA gene for defining higher taxonomic structure, especially when contemporary phylogenetic techniques that rely solely on tree graphs are used to analyze small and inherently biased data sets. We have also observed a general lack of consistency in defining the boundaries of genera based on 16S rDNA sequence analysis. This is particularly problematic in "bushy" areas of the phylogenetic trees where uncertainty of branching order is high and clear demarcation of taxonomic groups is impossible in the absence of other supporting data.

In dealing with such problems, we have "filled-in" the missing taxa so that the hierarchy is complete. In such cases, names of higher taxa are based largely on priority, except in instances where such a strategy might lead to unnecessary confusion (e.g.,

TABLE 1. Summary of taxonomic scheme employed in the Second Edition of *Bergey's Manual of Systematic Bacteriology*: increase in numbers of taxa between 2000 and 2003.

	Total		*Archaea*		*Bacteria*	
Taxonomic rank	2000	2003	2000	2003	2000	2003
Domain	2	2	1	1	1	1
Phylum	25	26	2	2	23	24
Class	40	41	8	9	32	32
Subclass	5	5	0	0	5	5
Order/subsection	89	88	12	13	77	75
Suborder	14	17	0	0	14	17
Family	203	240	21	23	182	217
Genus	941	1194	69	79	871	1115
Species	5224	6466	217	281	5007	6185

Helicobacteraceae rather than *Wolinellaceae*). Each of these higher taxa has also been scrutinized for phylogenetic coherence so as to avoid paraphyletic or polyphyletic taxa, wherever possible. However, high-quality 16S rDNA sequences are still not available for all validly named species. As a result, some placement errors are likely to remain. We also acknowledge that some existing taxa are "problematic" and contain misidentified species. There are also areas within the phylogenetic trees that are ambiguous. While corrections have been made to taxa appearing in the first two volumes of this edition, further corrections will have to await publication of subsequent volumes, when authors address these issues in detail.

Despite the above limitations, the use of the well-established phylogeny based on the 16S rRNA gene provides a marked improvement over the earlier artificial classifications. The technique (16S rDNA sequencing) is universal in applicability and will soon provide a single type of data that will be available for all validly named species. Given the rapid advancements in sequencing technology, we expect that other gene sequences (e.g., 23S rRNA gene) will follow in the near future and help in the placement of "problem taxa". Therefore, readers must recognize that the current classification is fluid and as each new "road map" is published there will be changes in the placement of some taxa. If it is true that we have described only about 10% of the extant procaryotes, then it is inevitable that this current classification will expand and change. We are in a period marked by rapid isolation and description of new procaryotic taxa that rivals the expansion of the field in the late 1800s. The current "natural history" approach should begin to provide the basis for a more meaningful and predictive classification of procaryotes for the future.

Mapping the taxonomic space

One of the greatest difficulties we experienced while constructing and updating the Taxonomic Outline was to easily visualize the higher taxonomic structure of the procaryotes based on 16S rDNA sequence analysis. While it was clear that such a structure should exist, based on similarity in the topology of many regions of the ARB and RDP trees (e.g., the separation of *Archaea* and *Bacteria* and the consistent presence of deeply branching taxa in the *Bacteria*), the validity of many of the intermediate taxa appearing in the Taxonomic Outline was less obvious. Summary trees, drawn in various ways (Barns et al., 1996; Hugenholtz et al., 1998; Ludwig and Klenk, 2001 and this volume), suggest the presence of 25–40 major lineages within the procaryotes. However, such trees yield relatively little information about either the number of member taxa or their relatedness. Thus, while these relationships can be examined in larger trees, such trees obscure the spatial relationships among the taxa, especially when the groups of interest may be separated by tens or even hundreds of pages required to print such trees. To that end, we sought alternative methods of exploring the sequence data for evidence of taxonomic structure and of confirming independently the placement of genera within the Taxonomic Outline.

Following the 1997 meeting on phylogenetic trees, P.H.A. Sneath used principal-coordinate analysis to prepare a two-dimensional projection of the major procaryotic groups. His analysis was based on branch lengths (evolutionary distances) of type strains appearing in Version 6.01 of the RDP tree (See Fig. 1 in Krieg and Garrity's chapter, On Using the *Manual*, in this volume). Sneath's analysis supported the clear separation of the procaryotic domains. It also supported separation of the deeply branching bacteria and oxygenic phototrophic bacteria from the Gram-negative and Gram-positive bacteria. There was, however, proportionately less separation among the many phyla of the Gram-positive bacteria than was observed among the Gram-negative bacteria. Although it is unclear how one can use such a plot to expose evolutionary relationships, it seems quite reasonable to infer that points that reside close together in these plots are more likely to have a recent common ancestry than those that do not plot closely together, assuming that the plot of those two principal coordinates accounts for a significant portion of the variance within the data. To show the relationship between the planar projection of evolutionary data and phylogenetic trees, Sneath drew imaginary branches below the plane, along a third axis, time.

Sneath's analysis suggested that ordination techniques might provide a useful alternative to phylogenetic trees, especially for uncovering higher order taxonomic structure within very large sets of sequences (>1000). To that end, a series of experiments was conducted by Garrity and Lilburn using principal components analysis (PCA) to explore aligned 16S rDNA sequence data. While their approach differed from Sneath's in several ways, the results showed remarkable similarity.

After the initial experiments, Garrity and Lilburn settled on the following strategy: first a database of aligned 16S rRNA sequences was built up on Release 8.0 of the RDP-II database, which was updated constantly from GenBank and the RDP. Next, they calculated evolutionary distances for each of the sequences to 223 reference sequences representing, where possible, type strains on which families were based in the initial release of *Bergey's Taxonomic Outline of the Procaryotes* (Garrity et al., 2001). The rationale for this approach was to use these reference sequences as benchmarks, much like those employed in the production and validation of topographical maps. For the analyses that are presented here, we used a set of 6380 sequences with a minimum length of 1399 nucleotides (nts), no more than 3% ambiguities, and with fewer than 11 consecutive "no information" positions. Furthermore, annotation of each sequence record was checked and updated to ensure that the identity of the source strain was correct and the nomenclature associated with the data was current. The sequences were masked to exclude positions that were not conserved and matrices of evolutionary distances were calculated using the Jukes and Cantor correction, as that approach was found to yield the best separation of taxa without significantly distorting either close or distant relationships (Nei and Kumar, 2000). The resulting matrix of distance vectors was then subjected to PCA in S-Plus 6.2 (Insightful, Seattle, WA, USA). Heatmaps, or colorized similarity matrices, comparing each member of a set of taxa to all other members of that set are referred to as "symmetric" heatmaps, while those that make comparisons of a set of taxa to a set of reference points (e.g., benchmarks) are referred to as asymmetric heatmaps.

Like principal-coordinate analysis, PCA provides a means of visualizing high-dimensional data in a lower-dimensional space by finding the uncorrelated single linear combinations of the original variables that explain most of the underlying variability within the data (Mardia et al., 1979; Venables and Ripley, 1994). PCA has been widely used in taxonomic and ecological studies in the past (Sneath and Sokal, 1973; Dunn and Everitt, 1982), it is a well-understood method, and it is suited to answering questions about higher-order structure within data and uncovering outliers (e.g., misclassified taxa). PCA also offers several advantages. Unlike many "treeing" algorithms used in phylogenetic analysis, PCA is a computationally efficient method, allowing the rapid analysis of datasets with thousands of taxa. The reliability of PCA scatter plots (created by plotting scores for each principal component) can easily be tested by estimating the cumulative residual variance explained by the principal components ranked in descending order. If the first two or three account for >85% of the total variance (Mardia et al., 1979; Venables and Ripley, 1994), the plot is generally considered a good depiction of the relationships among the taxa. Furthermore, if the underlying principal component scores are available for further analysis, one can readily create plots identifying the location of subsets of the original taxa projected back into the original coordinate system. This allows one to work in a fixed space of constant dimension and orientation, overcoming one of the more common problems of PCA: recomputation of principal components for subsets, leading to different views of the data. Despite the obvious utility of PCA and other ordination techniques, it does not appear that these methods have been applied previously to the exploration of evolutionary data.

A PCA plot of the full data set (n = 6380) is presented in Fig. 1A. The overall topology of the taxonomic space is quite similar to that observed in our earlier analyses, with relatively little change in the location of major taxonomic groups within the global base-maps. The locations of the domains *Archaea* and *Bacteria* are shown in regions a and b of Fig. 1A. A screeplot (Mardia et al., 1979; Venables and Ripley, 1994) of the cumulative variance reveals that the first component accounts for >74.2% of the total variance, with second and third components accounting for 11.2% and 5.01%, respectively (Fig. 1B). These results are quite consistent with our original findings, despite the addition of 1674 new sequences and 49 new benchmarks to the model. This further confirms our initial observation that the dimensionality of the evolutionary distance matrix, derived from the 16S rDNA sequences, can be significantly reduced with little loss of information. As before, we have augmented the underlying data by adding the names of the higher taxa to which each species (sequence) is currently assigned. This allows us to create different views, in which subsets of taxa of different ranks can be visualized against the background of all taxa and compared to known placements within other classifications. Since we first described this approach, we have used it to validate the Taxonomic Outline and detect misidentified species (sequences).

Since we first described this approach, we have also further explored the potential utility of PCA on subsets of taxa (e.g., phyla) to address issues of overlap and occlusion that occur in global analyses. Such plots have proven quite effective in uncovering sequence annotation errors and instances of unresolved and previously unrecognized synonymies that confound large-scale phylogenetic analyses. These plots, in combination with heatmaps have been used in guiding placement of taxa and exploring alternative classifications (Lilburn and Garrity, 2004).

By using these two complementary visualization techniques, we have been able to assemble a set of "vetted" sequences that are linked to strains of known provenance. We believe that these vetted sequences will provide a more solid foundation on which to extend the taxonomy that is being used in this edition of the *Systematics*. Furthermore, these sequences will be useful for modeling alternative taxonomies, in phylogenetic studies, and as training sets for naive Bayesian aligners and classifiers.

Phenotypic groups within the procaryotes

In the Ninth Edition of the *Determinative*, approximately 590 genera were subdivided into 35 major phenotypic groups. These phenotypic groups were based upon readily recognizable characters that could be used for the presumptive identification of species that are routinely encountered in a wide variety of ecological niches. As stated above, the objective of the *Determinative* was utilitarian, and readers were advised that no attempt at creating a natural classification had been made. A summary of the relationships between the 35 phenotypic groups and the 25 phyla defined in 2001 was presented in the first version of the Road Map, which appeared in Volume One of the Sescond Edition of the *Systematics*. This summary was supplemented with a listing of the genera and their phenotypic groups.

Since publication of the last edition of the *Determinative*, use of molecular methods of identification has become increasingly common, often to the exclusion of traditional methods of phenotypic profiling. Molecular probes, based largely on conserved regions of the genome (principally 16S and 23S rRNA), have come into routine use and provide a universally applicable technique to aid in the detection and identification of bacteria in virtually any sample, without the need to culture them. Despite these advances, there remains a need for incorporation of both phenotypic and genotypic data in formal descriptions and in identification protocols for procaryotes. This problem is discussed in considerable detail by Gillis et al. in their introductory essay, Polyphasic Taxonomy, in this volume. Phenotypic information can also play a role in the separation of closely related taxa, especially in ambiguous regions of phylogenetic trees where 16S rDNA sequence data prove inadequate for resolution of such taxa (Ludwig and Klenk, 2001, and this volume). A prime example is provided by the *"Acidaminococcaceae"*, a well-defined group of cocci possessing a Gram-negative cell wall that is found within the low G + C Gram-positive phylum *"Firmicutes"*.

The procaryotic phyla

As indicated in Table 1, the two procaryotic domains were initially subdivided into 25 phyla, two of which occur within the *Archaea*. The number of phyla was increased to 26 on publication of the proposal of Zhang et al. (2003b) of the phylum *Gemmatimonadetes* in the Domain *Bacteria*. The following are brief, working descriptions that are intended to provide readers with some understanding of the relationship between the Taxonomic Outline and phylogenies proposed by the RDP and ARB trees, along with known problems. It also provides the readers with an indication of where each taxonomic group will appear in this or subsequent volumes of the *Systematics*. Although the contents will for the most part be presented in a phylogenetic context, some practical considerations were necessary in the final layout of the individual volumes. Twelve phyla were presented in Volume One and 11 will be presented in Volume Five, while Volumes Two, Three, and Four will each cover a single phylum (the *Proteobacteria*, the *"Firmicutes"*, and the *"Actinobacteria"*, respectively). Furthermore, Volume One deviated slightly from the phylogenetic model as all of the phototrophic species were presented together as a phenotypically coherent group. Phototrophic species in the phyla *Proteobacteria* and *"Firmicutes"* will also appear in Volumes Two and Three, respectively, in their proper phylogenetic context.

In addition, readers are cautioned that the numbering scheme used in the Taxonomic Outline is, to some extent, arbitrary, especially at the lower levels. As the branching order of species within genera is often ambiguous and the data set is known to be incomplete, the use of phylogenetic trees as a guide to the appearance of taxa in the *Systematics* proved to be untenable. Thus, we have had to adopt a more workable and all-in-

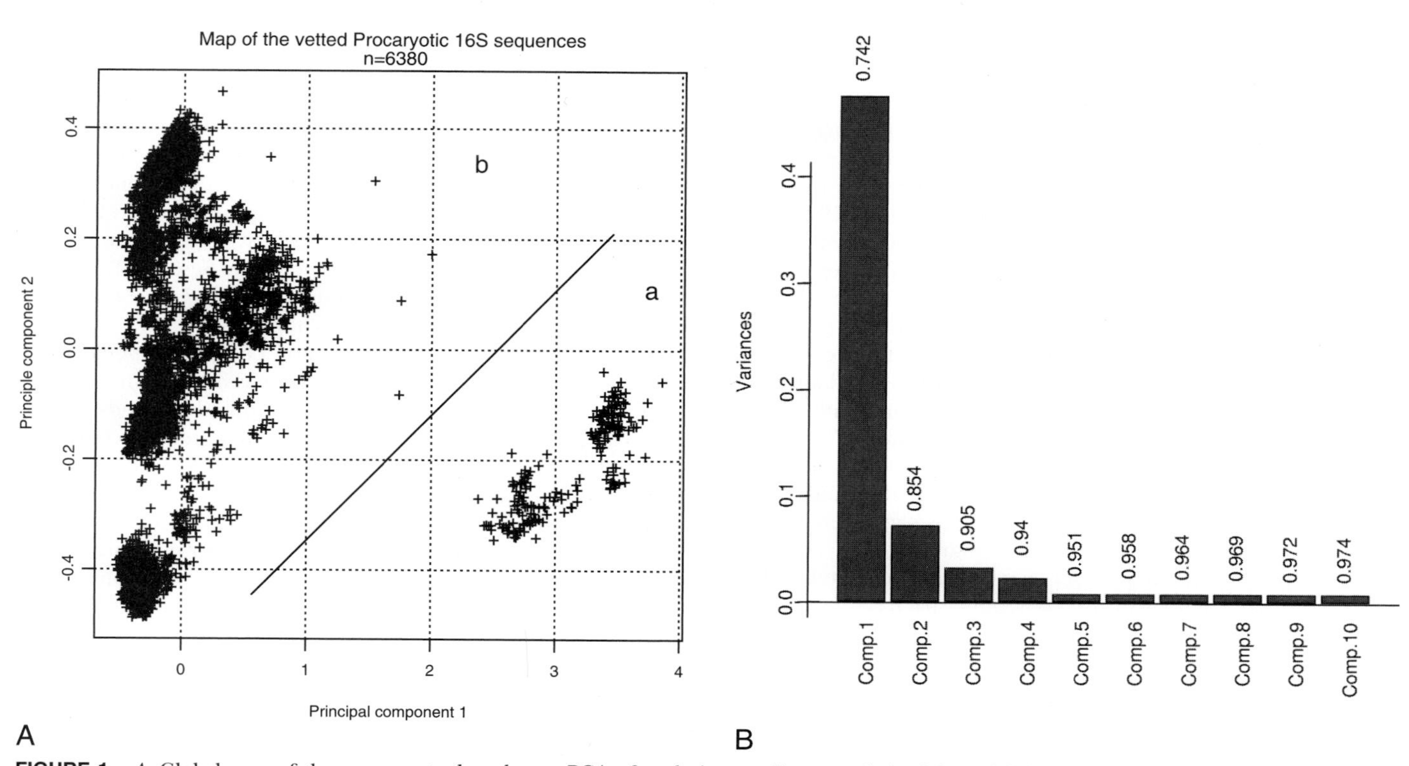

FIGURE 1. *A*, Global map of the procaryotes based on a PCA of evolutionary distances derived from full-length 16S rDNA sequences in the RDP II (Release 8.0). Evolutionary distances (Jukes and Cantor correction) for aligned, full-length 16S rDNA sequences (length >1399 nts, ambiguities ≤3%) were estimated to 223 reference strains on which families were based in the *Bergey's Taxonomic Outline*. The resulting matrix was then subjected to a PCA and the principal component scores were plotted for principal component 1 vs. principal component 2. Region a, *Archaea*; region b, *Bacteria*. *B*, Screeplot of PCA data reveals that the first two principal components account for 85.4% of the total (cumulative) variance within the evolutionary data matrix used to compute the principal components.

clusive strategy. The type genus will always appear first within a family, and all other genera within each family (if more than one genus is included) will appear in alphabetical order. Readers are reminded that the numbering of subordinate taxa is subject to change as new taxa are described and existing taxa are split apart. Likewise, the fact that the *Archaea* and deeply branching *Bacteria* are presented first is based largely on the earlier versions of the RDP tree. There will be some deviation from the original RDP ordering of taxa in subsequent volumes for a variety of practical reasons.

Phylum A1 **Crenarchaeota** Garrity and Holt 2002, 685^VP^ (Effective publication: Garrity and Holt 2001j, 169)

Cren.arch.ae.o'ta. M.L. fem. pl. n. *Crenarchaeota* from the Kingdom *Crenarchaeota* (Woese, Kandler and Wheelis 1990, 4579).

The phylum consists of a single class, the *Thermoprotei*, which is well supported by 16S rDNA sequence data. It is currently subdivided into four orders: the *Thermoproteales*, *Caldisphaerales* (proposed below), *Desulfurococcales*, and *Sulfolobales*. At present, Ludwig and Klenk (2001, and this volume) indicate good support for the *Sulfolobales* and *Desulfurococcales*. In a domain-level PCA plot in which the dataset and benchmarks were restricted to members of the *Archaea* (Fig. 2), the phylum *Crenarchaeota* (Class *Thermoprotei*, region i) is well separated from the *Euryarchaeota* (regions a-h).

Members of the phylum *Crenarchaeota* are morphologically diverse, including rods, cocci, filamentous forms, and disk-shaped cells which stain Gram-negative. Motility is observed in some genera. The organisms are obligately thermophilic, with growth occurring at temperatures ranging from 70–113°C. The organisms are acidophilic and are aerobic, facultatively anaerobic, or strictly anaerobic chemolithoautotrophs or chemoheterotrophs. Most metabolize S^0. Chemoheterotrophs may grow by sulfur respiration. RNA polymerase is of the BAC type.

The phylum *Crenarchaeota* Garrity and Holt 2002, 685^{VP} and class *Thermoprotei* Reysenbach 2002, 687^{VP} are emended to include the order *Caldisphaerales* and the family *Caldisphaeraceae* based on 16S rDNA sequence analysis of the genus *Caldisphaera* (Itoh et al., 2003).

Caldisphaerales ord. nov.

Cal.di.sphae.ra'les. M.L. fem. n. *Caldisphaera* type genus of the order; suff. *-ales* to denote order; M.L. fem. pl. n. *Caldisphaerales* the order *Caldisphaera*.

The order *Caldisphaerales* was circumscribed for this volume based on phylogenetic analysis of 16S rRNA sequences; the order contains the family *Caldisphaeraceae*. The description is the same as that of the genus *Caldisphaera*.

Type genus: *Caldisphaera* Itoh, Suzuki, Sanchez and Nakase 2003, 1153.

Caldisphaeraceae fam. nov.

Cal.di.sphae.ra'ce.ae. M.L. fem. n. *Caldisphaera* type genus of the family; *-aceae* ending to denote family; M.L. fem. pl. n. *Caldisphaeraceae* the family *Caldisphaera*.

The family *Caldisphaeraceae* was circumscribed for this volume based on phylogenetic analysis of 16S rRNA sequences; the family contains the genus *Caldisphaera*. The description is the same as that of the genus *Caldisphaera*.

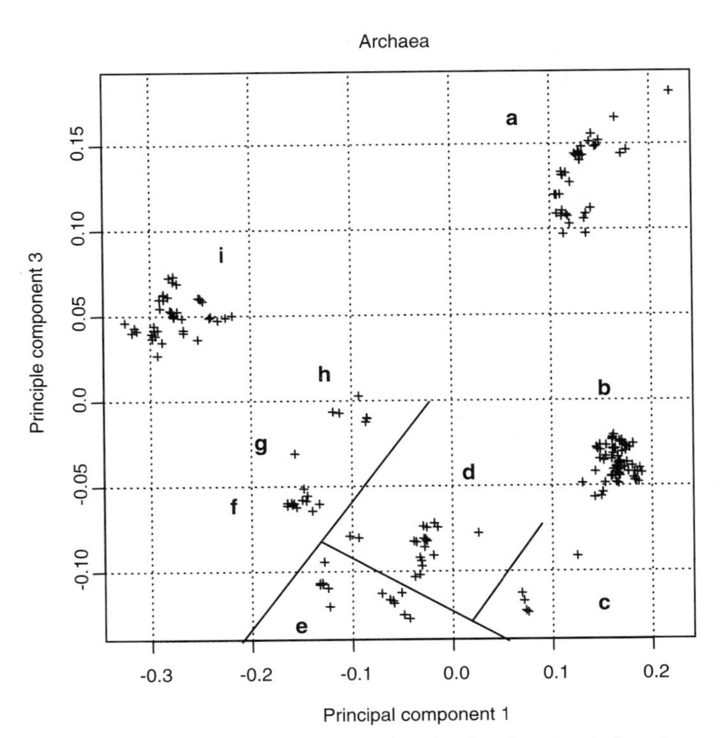

FIGURE 2. Locations of the classes within the domain *Archaea* in a domain-level PCA plot. The dataset and benchmark sequences were restricted to procaryotes belonging to the *Archaea*. Regions a, "*Methanomicrobia*"; b, *Halobacteria*, c, *Thermoplasmata*; d, *Methanobacteria*; e, *Methanococci*; f, *Thermococci*; g, *Methanopyri*, h, *Archaeoglobi* (all in phylum *Euryarchaeota*); i, *Thermoprotei* (phylum *Crenarchaeota*).

Type genus: *Caldisphaera* Itoh, Suzuki, Sanchez and Nakase 2003, 1153.

The order *Desulfurococcales* Huber and Stetter 2002, 685^{VP} and the family *Desulfurococcaceae* Zillig and Stetter 1983, 438^{VP} are emended to include the genus *Acidilobus* Prokofeva et al., 2000, 2007^{VP}.

Phylum A2 **Euryarchaeota** Garrity and Holt 2002, 685^{VP} (Effective publication: Garrity and Holt 2001k, 211)
Eur.y.arch.ae.o'ta. M.L. fem. pl. n. *Euryarchaeota* from the Kingdom *Euryarchaeota* (Woese, Kandler and Wheelis 1990, 4579).

The phylum currently consists of eight classes: the *Methanobacteria, Methanococci*, the *Methanomicrobia* (proposed below), the *Halobacteria*, the *Thermoplasmata*, the *Thermococci*, the *Archaeoglobi*, and the *Methanopyri*. In phylogenetic analyses, Ludwig and Klenk (2001, and this volume) indicate that the *Methanobacteria, Methanomicrobiales, Halobacteria*, and *Thermoplasmata* share a common root. The relationships among the remaining classes are ambiguous. In 2003, we rearranged the higher taxa within the *Euryarchaeota* in response to comments from the community and a careful examination of symmetrical heatmaps of the *Archaea* (Garrity et al., 2003). The *Methanococci*, as we had previously defined the class, were polyphyletic. Following our analysis, we moved the *Methanomicrobiales* and *Methanosarcinales* into a separate class, the *Methanomicrobia*. In the domain-level PCA plot shown in Fig. 2, the eight classes of the *Euryarchaeota* (regions a-h) occupy non-overlapping regions of the plot. The class *Methanomicrobia* (region a) is clearly distant from the remaining classes, which occupy a more diffuse region at the lower right of the plot.

The *Euryarchaeota* are morphologically diverse and occur as rods, cocci, irregular cocci, lancet-shaped, spiral-shaped, disk-shaped, triangular, or square cells. Cells stain Gram-positive or Gram-negative based on the presence or absence of pseudomurein in cell walls. In some classes, cell walls consist entirely of protein or may be completely absent (*Thermoplasmata*). Five major physiological groups have been described: methanogenic archaea, extremely halophilic archaea, archaea lacking a cell wall, sulfate-reducing archaea, and extremely thermophilic S^0 metabolizers.

The phylum *Euryarchaeota* Garrity and Holt 2002, 685^{VP} is emended to include the class *Methanomicrobia*.

Methanomicrobia class. nov.
Me.tha.no.mi.cro.bi'a. M.L. fem. pl. n. *Methanomicrobiales* type order of the class, dropping the ending to denote a class; M.L. fem. pl. n. *Methanomicrobia* the class *Methanomicrobiales*.

The class *Methanomicrobia* is circumscribed based on a phylogenetic analysis of 16S rDNA sequences; the class contains the orders *Methanomicrobiales* Balch and Wolfe 1981 and *Methanosarcinales* Boone et al. 2002.

Type order: *Methanomicrobiales* Balch and Wolfe 1981, 261.

The class *Methanococci* Boone 2002, 686^{VP} is emended to include only the order *Methanococcales* Balch and Wolfe 1981, 216^{VP} (families *Methanococcaceae* Balch and Wolfe 1981, 216^{VP} and *Methanocaldococcaceae* Whitman et al. 2002, 686^{VP}).

The order *Methanosarcinales* Boone et al. 2002, 686^{VP} and family *Methanosarcinaceae* Balch and Wolfe 1981, 216^{VP} are emended to include the genus *Methanimicrococcus* Sprenger et al. 2000, 1998^{VP}.

The class *Halobacteria* Grant et al. 2002, 685^{VP}, order *Halobacteriales* Grant and Larsen 1989, 495^{VP}, and family *Halobacteriaceae* Gibbons 1974a, 269^{AL} are emended to include the genera *Halobiforma* Hezayen et al. 2002, 2278^{VP}, *Halomicrobium* Oren et al. 2002, 1834^{VP}, *Halorhabdus* Wainø et al. 2000, 188^{VP}, and *Halosimplex* Vreeland et al. 2003, 936^{VP}.

The class *Thermoplasmata* Reysenbach 2002, 687^{VP} and order *Thermoplasmatales* Reysenbach 2002, 687^{VP} are emended to include the family *Ferroplasmaceae* Golyshina et al. 2000, 1004^{VP} and the genus *Ferroplasma* Golyshina et al. 2000, 1004^{VP}

The class *Thermococci* Zillig and Reysenbach 2002, 687^{VP}, order *Thermococcales* Zillig 1988, 136^{VP} and family *Thermococcaceae* Zillig 1988, 136^{VP} are emended to include the genus *Palaeococcus* Takai et al. 2000, 498^{VP}.

The class *Archaeoglobi* Garrity and Holt 2002, 685^{VP}, order *Archaeoglobales* Huber and Stetter 2002, 685^{VP}, and family *Archaeoglobaceae* Huber and Stetter 2002, 685^{VP} are emended to include the genus *Geoglobus* Kashefi et al., 2002, 727^{VP}.

Phylum B1 **Aquificae** Reysenbach 2002, 685^{VP} (Effective publication: Reysenbach 2001k, 359)

The phylum *Aquificae* consists of a single class and order. In phylogenetic analysis of 16S and 23S rDNA sequence data, the *Aquificae* are generally considered one of the deepest and earliest branching groups within the *Bacteria*. However, phylogenetic analyses of other protein and gene sequences show the placement with respect to other members of the *Bacteria* to be variable (Huber and Stetter, 2001i). At the time Volume One went to press, high-quality sequence data was available only for the type strain of *Aquifex pyrophilus*. *Desulfurobacterium* was provisionally placed within the phylum; however, it was thought to possibly represent another undefined phylum (Ludwig and Klenk (2001, and this volume; Reysenbach, 2001k). Since that time, additional sequence data have become available and several other genera have been provisionally placed in the phylum, class, and order by us and by others. In Fig. 3 we show the location of member species in a global PCA plot. Region a shows *Hydrogenobaculum*. In region b we show *Sulfurihydrogenibium* and *Hydrogenothermus*. In region c are *Desulfurobacterium*, *Thermovibrio*, and *Persephonella*. In asymmetric heatmaps, *Thermovibrio* and *Persephonella* exhibit a pattern more consistent with that of true *Thermotogae* and may be misplaced in the *Aquificae*. *Aquifex*, *Hydrogenobacter*, and *Thermocrinis* are found in the cluster in region d. In PCA plots, *Thermovibrio* and *Thermotoga* are much closer together.

All members of the *Aquificae* are Gram-negative nonsporulating rods or filaments, thermophilic, with optimum growth in the range 65–85°C. Growth is chemolithoautotrophic or chemoorganotrophic. Many species grow anaerobically and are capable of nitrate reduction. However, both microaerophilic and aerobic species also occur.

***Phylum B2* Thermotogae** Reysenbach 2002, 687[VP] (Effective publication: Reysenbach 2001*l*, 369)

The phylum *Thermotogae* consists of a single class and order. In phylogenetic analyses, this phylum consistently branches deeply, along with the *Aquificae*. In global PCA plots (Fig. 4), the type strains for member species are widely scattered in the same region as many of those listed as *Incertae Sedis* in the *Aquificae*, suggesting both a high level of variability and some uncertainty about their affiliation. The species of *Thermotoga* (region a) fall into three subclusters along with those of *Fervidobacterium* (except for *Fervidobacterium gondwanense*) and *Thermosipho*. *Geotoga*, *Marinitoga*, *Petrotoga* and *Fervidobacterium gondwanense* map in region b and show a distinctive pattern in asymmetric heatmaps based on their low relatedness to most of the bacterial and archaeal lineages.

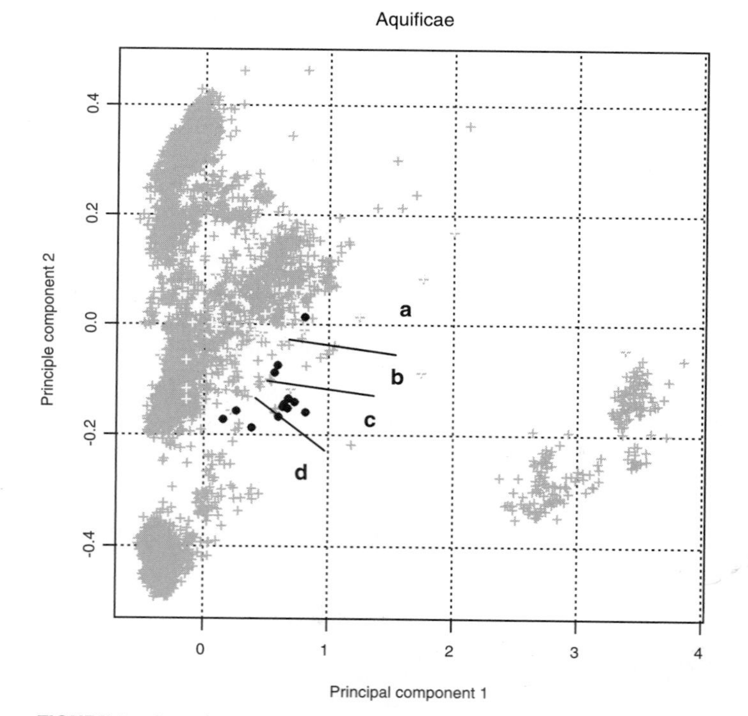

FIGURE 3. Location of phylum *Aquificae* within the global map of the procaryotes. Region a contains *Hydrogenobaculum*. Region b contains *Sulfurihydrogenibium* and *Hydrogenothermus*. Region c contains *Desulfurobacterium*, *Thermovibrio*, and *Persephonella*; region d contains *Aquifex*, *Hydrogenobacter*, and *Thermocrinis*. The affiliation of those genera falling outside the main cluster are regarded as uncertain. In asymmetric heatmaps, *Thermovibrio* and *Persephonella* exhibit a pattern more consistent with that of true *Thermotogae* and may be misplaced within the *Aquificae*.

All *Thermotogae* are Gram-negative, nonsporulating, rod-shaped bacteria that possess a characteristic sheath-like outer layer or "toga". *meso*-Diaminopimelic acid is not present in peptidoglycan. *Thermotogae* are strictly anaerobic heterotrophs, utilizing a broad range of organic compounds for growth. Thiosulfate and/or S^0 are reduced. Growth is inhibited by H_2.

The phylum *Thermotogae* Reysenbach 2002, 687[VP], class *Thermotogae* Reysenbach 2002, 687[VP], order *Thermotogales* Reysenbach 2002, 687[VP], and family *Thermotogaceae* Reysenbach 2002, 687[VP] are emended to include the genus *Marinitoga* Wery et al. 2001a, 502[VP].

***Phylum B3* Thermodesulfobacteria** Garrity and Holt 2002, 687[VP] (Effective publication: Garrity and Holt 2001*l*, 389)

Ther′ mo.de′ sul.fo.bac′ te.ria. M.L. fem. pl. n. *Thermodesulfobacteriales* type order of the phylum, dropping the ending to denote a phylum; M.L. fem. pl. n. *Thermodesulfobacteria* the phylum of *Thermodesulfobacteriales*.

The phylum *Thermodesulfobacteria* is currently represented by a single genus that branches deeply in the ARB tree. In global PCA plots the sequences of three of the four validly named species plot in close proximity to each other (Fig. 5, points labeled d).

Members of the phylum are Gram-negative, rod-shaped cells possessing an outer-membrane layer that forms protrusions; the organisms are thermophilic, strictly anaerobic, and chemoheterotrophic with a dissimilatory sulfate-reducing metabolism.

***Phylum B4* "Deinococcus–Thermus"**

The *"Deinococcus–Thermus"* phylum represents a deep-branching line of descent that is defined largely on the basis of 16S rDNA signature nucleotides (Battista and Rainey, 2001); (da Costa et al., 2001). The phylum is subdivided into two orders, each of which contains a single family. In global PCA plots (Fig. 6), member species group together into two discrete clusters, with *Deinococcus* and *Meiothermus* species (region b) overlapping slightly.

Two major phenotypes, which are consistent with the phylogenetic branching, are observed. Members of the *Deinococcales* are Gram positive and resistant to radiation. Members of the *Thermales* are Gram-negative thermophiles.

The class *Deinococci* Garrity and Holt 2002, 685[VP], order *Thermales* Rainey and da Costa 2002, 687[VP], and family *Thermaceae* da Costa and Rainey 2002, 687[VP] are emended to include the genera *Marinithermus* Sako et al. 2003, 63[VP], *Oceanithermus* Miroshnichenko et al. 2003b, 751[VP], and *Vulcanithermus* Miroshnichenko et al. 2003c, 1147[VP].

***Phylum B5* Chrysiogenetes** Garrity and Holt 2002, 685[VP] (Effective publication: Garrity and Holt 2001a, 421)

Chry.si.o′ ge.netes. M.L. fem. pl. n. *Chrysiogenales* type order of the phylum, dropping the ending to denote a phylum; M.L. fem. pl. n. *Chrysiogenetes* the phylum of *Chrysiogenales*.

The phylum *Chrysiogenetes* is currently represented by a single species, which was reportedly distinct from members of other phyla (Macy et al., 1996). Exact placement is currently uncertain, but a distant relationship to *Deferribacteres* is likely. In global PCA

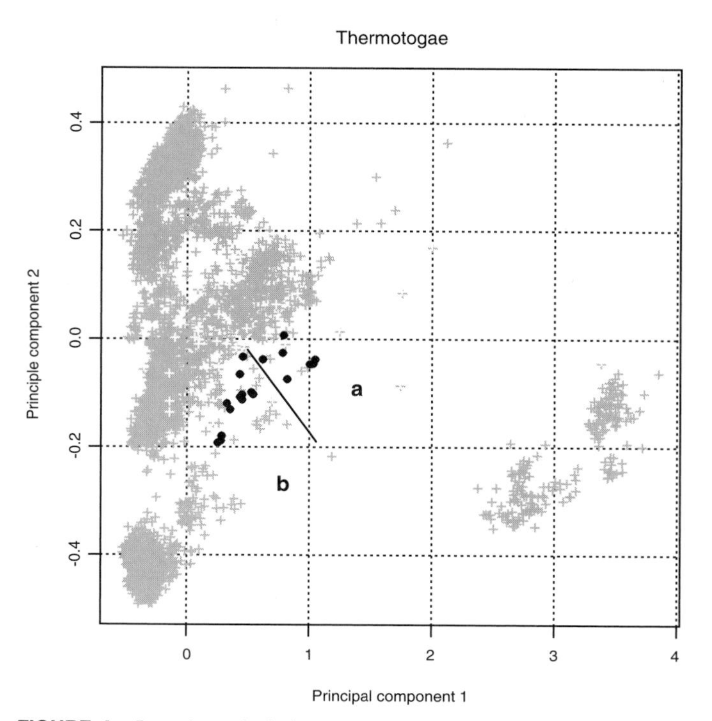

FIGURE 4. Location of phylum *Thermotogae* within the global map of the procaryotes. Region a contains *Thermotoga*, *Fervidobacterium* (except for *Fervidobacterium gondwanense*) and *Thermosipho*; region b contains *Geotoga*, *Marinitoga*, *Petrotoga*, and *Fervidobacterium gondwanense*.

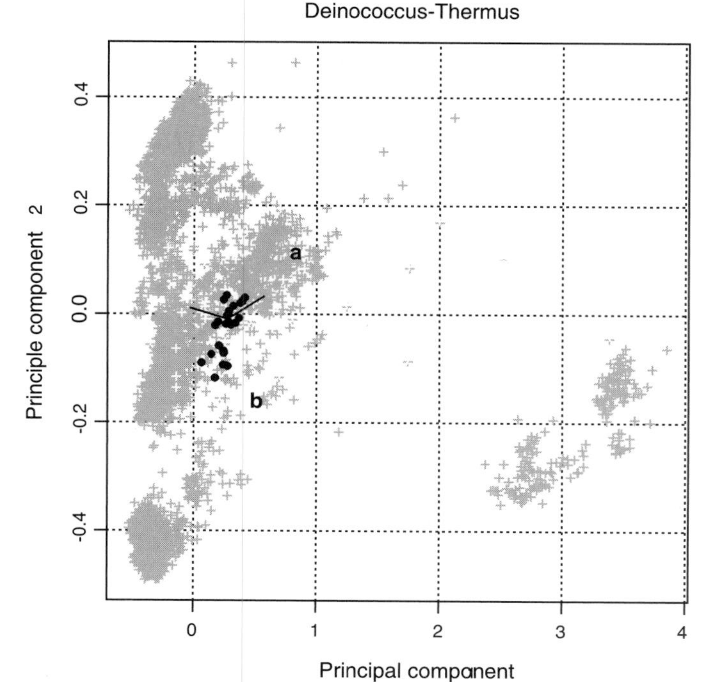

FIGURE 6. Location of phylum *"Deinococcus–Thermus"* within the global map of the procaryotes. *Deinococcus* (region a) is found in close proximity to *Meiothermus* along the boundary line; region b contains other members of the *Thermales*.

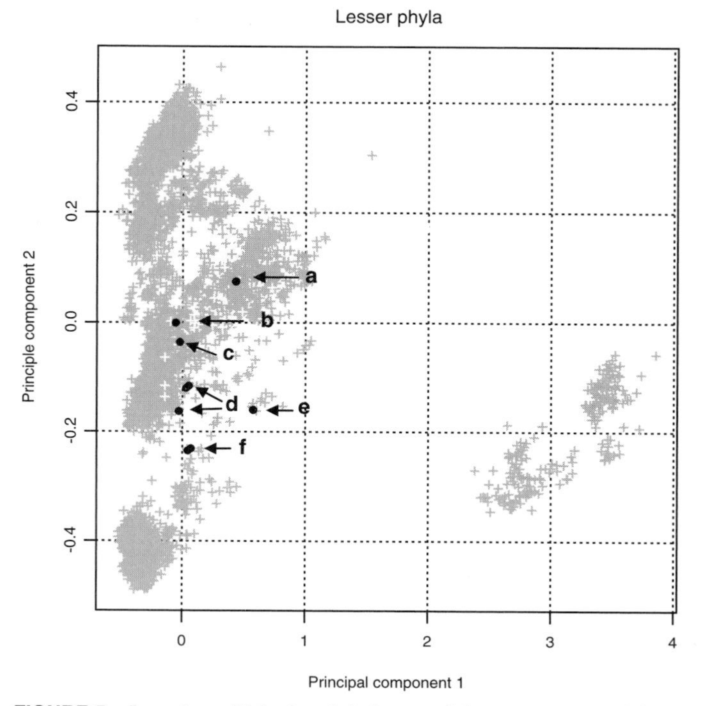

FIGURE 5. Location within the global map of the procaryotes of those phyla each having a single class, order, family, and genus. Points a, *"Fibrobacteres"*; b, *Chrysiogenetes*; c, *Gemmatimonadetes*; d, *Thermodesulfobacteria*; e, *Thermomicrobia*; points f, *"Dictyoglomi"*.

plots (Fig. 5, point b), *Chrysiogenes* is located in a densely populated region, in close proximity to *Geovibrio* as well as the *"Clostridia"*, *"Bacilli"*, and the *Deltaproteobacteria*.

Members of the *Chrysiogenetes* are mesophilic, Gram-negative, motile, curved, rod-shaped cells that exhibit anaerobic respiration in which arsenate serves as the electron acceptor.

Phylum B6 **Chloroflexi** Garrity and Holt 2002, 685^VP^ (Effective publication: Garrity and Holt 2001o, 427)

Chlo.ro.flex′i. M.L. masc. n. *Chloroflexus* genus of the phylum, dropping the ending to denote a phylum; M.L. fem. pl. n. *Chloroflexi* the phylum of *Chloroflexus*.

The phylum *Chloroflexi* is a deep-branching lineage of the *Bacteria*; the position of the phylum in a global PCA plot is shown in Fig. 7 (region a). The single class within *Chloroflexi* is subdivided into two orders: the *"Chloroflexales"* and the *"Herpetosiphonales"*. In phylogenetic trees, the *Chloroflexi* tend to group with the *Thermomicrobia*, which is consistent with the pattern observed in asymmetric heatmaps; however, PCA plots show a clear separation, in keeping with the marked differences in phenotype.

Chloroflexi are Gram negative, filamentous bacteria exhibiting gliding motility. The peptidoglycan contains L-ornithine as the diamino acid. A lipopolysaccharide-containing outer membrane is not present. Members of the *"Chloroflexales"* contain bacteriochlorophyll and are obligate or facultative anoxygenic phototrophs; members of the *"Herpetosiphonales"* do not contain bacteriochlorophyll and are chemoheterotrophs.

The phylum *Chloroflexi* Garrity and Holt 2002, 685^VP^ is emended to include the family *Oscillochloridaceae* Keppen et al. 2000, 1534^VP^. The family *Oscillochloridaceae* is placed in the class *"Chloroflexi"* and the order *"Chloroflexales"*.

Phylum B7 **Thermomicrobia** Garrity and Holt 2002, 687^VP^ (Effective publication: Garrity and Holt 2001p, 447)

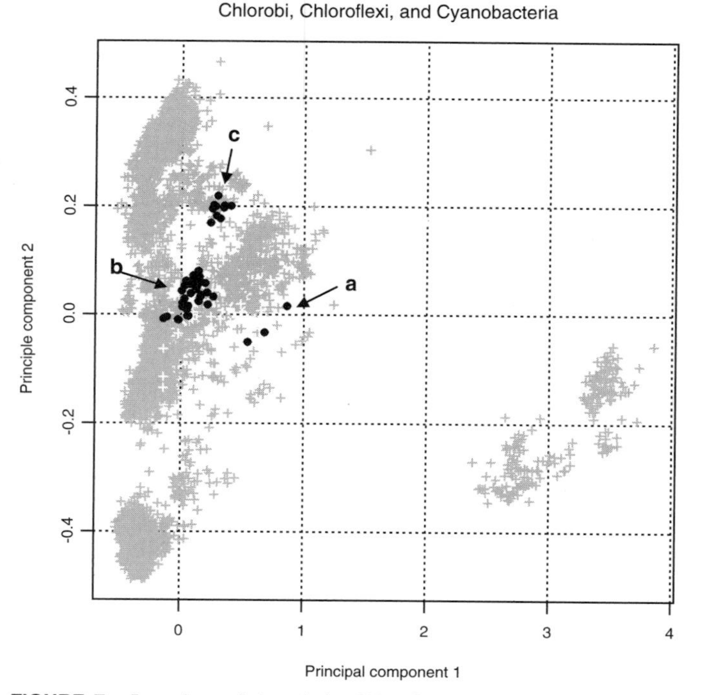

FIGURE 7. Location of the phyla *Chloroflexi* (region a), *Cyanobacteria* (region b), and *"Chlorobi"* (region c) within the global map of the procaryotes.

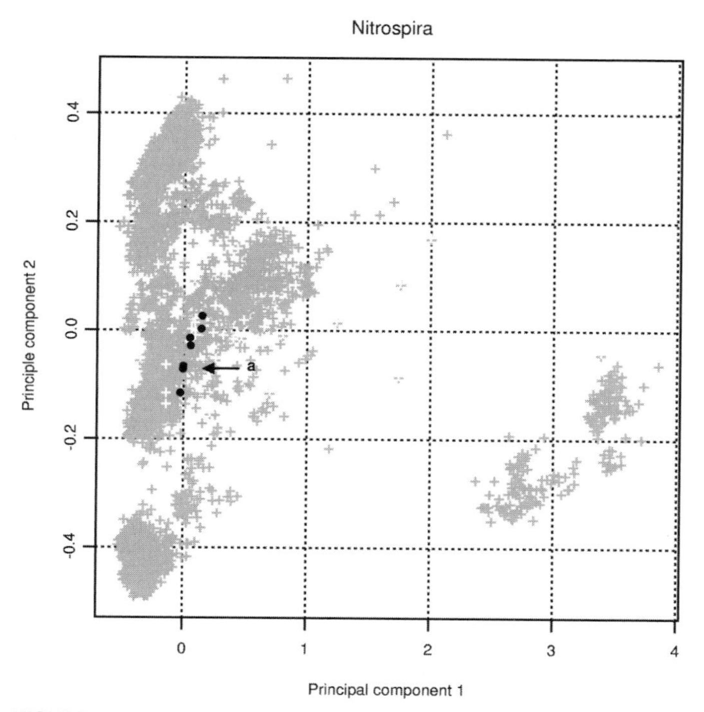

FIGURE 8. Location of *"Nitrospirae"* (point a shows the location of the type genus) within the global map of the procaryotes.

Ther.mo.mi.cro' bia. M.L. fem. pl. n. *Thermomicrobiales* type order of the phylum, dropping the ending to denote a phylum; M.L. fem. pl. n. *Thermomicrobia* the phylum of *Thermomicrobiales.*

The phylum *Thermomicrobia* consists of a single known representative that branches deeply in the RDP and ARB trees and is distantly related to the *Chloroflexi.* Although Ludwig and Klenk (2001, and this volume) argue that the *Thermomicrobia* are monophyletic with the *Chloroflexi,* we have provisionally placed them into a separate phylum. Relatively little work has been done with this organism since the first edition of the *Systematics* was published, and no new relatives have been reported. In global PCA plots (Fig. 5, point e), the type strain maps to the same region as the *Aquificae.*

Thermomicrobia are nonsporulating, Gram-negative, short, irregularly shaped nonmotile rods. No diamino acids are present in significant amounts in the peptidoglycan. The organisms are hyperthermophilic, with optimum growth temperature 70–75°C; obligately aerobic; and chemoorganotrophic.

Phylum B8 **"Nitrospirae"** Garrity and Holt 2001q, 451

Ni.tro. spi' rae. M.L. fem. n. *Nitrospira* genus of the phylum, dropping the ending to denote a phylum; M.L. fem. pl. n. *Nitrospirae* the phylum of *Nitrospira.*

The phylum *"Nitrospirae"* is based mainly on phylogenetic grounds. At present, it consists of a single class, order, and family of the *Bacteria* and includes environmental isolates that branch deeply in the ARB and RDP trees; member taxa consistently group together. In global PCA plots (Fig. 8) the *"Nitrospirae"* map to positions in a densely populated region in close proximity to the *"Clostridia"*, *"Bacilli"*, *Chrysiogenetes* and the *Deltaproteobacteria* (see below). In asymmetric heatmaps, member species appear equidistant from benchmarks representing many of the bacterial lineages.

"Nitrospirae" are Gram-negative, curved, vibrioid or spiral-shaped cells. Metabolically diverse, most genera are aerobic chemolithotrophs, including nitrifiers, dissimilatory sulfate reducers, and magnetotactic forms. One genus (*Thermodesulfovibrio*) is thermophilic and is obligately acidophilic and anaerobic.

Phylum B9 **Deferribacteres** Garrity and Holt 2002, 685[VP] (Effective publication: Garrity and Holt 2001m, 465)

De.fer.ri.bac' teres. M.L. fem. pl. n. *Deferribacterales* type order of the phylum, dropping the ending to denote a phylum; M.L. fem. pl. n. *Deferribacteres* the phylum of *Deferribacterales.*

The phylum *Deferribacteres* is a distinct lineage within the *Bacteria* based on phylogenetic analysis of 16S rDNA sequences. At present the members of this phylum are organized into a single class, order, and family. The relationships within the phylum may, however, be more distant and warrant further subdivision in the future. In global PCA plots (Fig. 9), the member species are widely separated along the Y axis and are in close proximity to the *Deltaproteobacteria* and *"Nitrospirae"*. While such separations are consistent for members of a phylum, it is probable that the individual strains belong to different classes and/or orders. Further subdivision will await the inclusion of additional sequence data in the analyses. As is the case with *Chrysiogenetes* and *"Nitrospirae"*, asymmetric heatmaps show that member taxa are equally distant from representatives of the major bacterial lineages, and show the closest relationship to the benchmark representing *Deferribacteres.*

Deferribacteres are chemoorganotrophic heterotrophs that respire anaerobically with terminal electron acceptors including Fe(II), Mn(IV), S^0, Co(III), and nitrate. Placement of one genus, *Synergistes,* in this phylum is provisional.

The phylum *Deferribacteres* Garrity and Holt 2002, 685[VP], class *Deferribacteres* Huber and Stetter 2002, 685[VP], order *Deferribacterales* Huber and Stetter 2002, 685[VP], and family *Deferribacteraceae* Huber

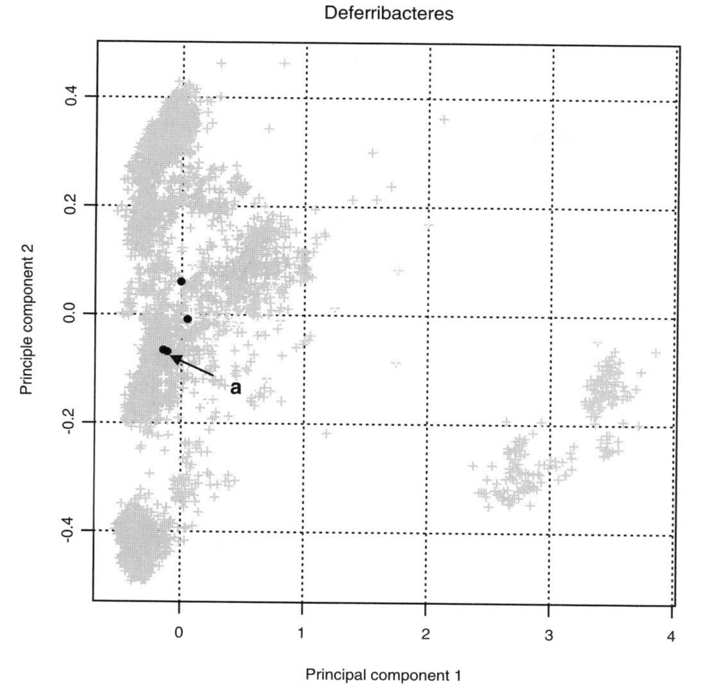

FIGURE 9. Location of *Deferribacteres* (point a shows the location of the type genus) within the global map of the procaryotes.

and Stetter 2002, 685[VP] are emended to include the genera *Denitrovibrio* Myhr and Torsvik 2000, 1617[VP], *Flexistipes* Fiala et al. 2000, 1415[VP], and *Geovibrio* Caccavo et al. 2000, 1415[VP].

Phylum B10 **Cyanobacteria**

Although the *Cyanobacteria* represent a major lineage within the *Bacteria* and include the chloroplasts as a distinct and highly diverse subgroup, the nomenclature of taxa ascribed to this phylum is governed by the International Code of Botanical Nomenclature rather than the Bacteriological Code. The phylum consistently branches close to the low G + C Gram-positive *Bacteria* (the *"Firmicutes"*). In global PCA plots (Fig. 7, region b), the *Cyanobacteria* are located in a region densely populated by the *Mollicutes* (see below). At present, the taxonomy of this group is in a state of flux. In Volume One of this edition, Wilmotte and Herdman (2001) discuss the manifold discrepancies between the current 16S rDNA-based phylogeny and the distinctive phenotypes of these bacteria.

Cyanobacteria are Gram-negative unicellular, colonial, or filamentous oxygenic and photosynthetic bacteria exhibiting complex morphology and life cycles. The principal characters that define all members of this phylum are the presence of two photosystems (PSII and PSI) and the use of H_2O as the photoreductant in photosynthesis. Although facultative photoheterotrophy or chemoheterotrophy may occur in some species or strains, all known members are capable of photoautotrophy (using CO_2 as the primary source of cell carbon). A lipopolysaccharide outer membrane is present along with a thick peptidoglycan layer (2–200 nm). *Cyanobacteria* contain chlorophyll *a*. Phycobiliproteins (allophycocyanin, phycocyanin, and sometimes phycoerythrin) may or may not be present. The *Cyanobacteria* are subdivided into five subsections that are equivalent to orders. The current classification scheme is structured according to phenotypic characteristics rather than the 16S rDNA phylogeny, as such a phylogenetically based scheme is not yet possible.

The form-genera *Prochlorococcus* Chisholm et al. 2001, 264 and *Halospirulina* Nübel et al. 2000, 1265 are placed in the phylum *Cyanobacteria* and class *Cyanobacteria*. The form-genus *Prochlorococcus* is placed in Subsection I, Family 1.1; the form-genus *Halospirulina* is placed in Subsection III, Family 3.1.

Phylum B11 **"Chlorobi"** Garrity and Holt 2001r, 601

Chlo.ro'bi. M.L. neut. n. *Chlorobium* genus of the phylum, dropping the ending to denote a phylum; M.L. fem. pl. n. *Chlorobi* the phylum of *Chlorobium*.

The *"Chlorobi"* share a common root with *"Bacteroidetes"* in both the RDP and the ARB trees. In global PCA plots (Fig. 7, region c), the *"Chlorobi"* are found in closest proximity to the *Epsilonproteobacteria* and the *Flavobacteria* within the *"Bacteroidetes"* (see below). At present, the phylum contains a single class, order, and family.

Chlorobi are Gram-negative, spherical, ovoid, straight, or curved rod-shaped bacteria that are strictly anaerobic and obligately phototrophic. Cells grow preferentially by photoassimilation of simple organic compounds. Some species may utilize sulfide or thiosulfate as an electron donor for CO_2 accumulation. Sulfur globules accumulate on the outside of the cells when grown in the presence of sulfide and light, and sulfur is rarely oxidized further to sulfate. Most genera are able to grow as chemoheterotrophs under microaerobic or aerobic conditions. Ammonia and dinitrogen are used as the nitrogen source. Most genera require one or more growth factors, the most common being biotin, thiamine, niacin, and *p*-aminobenzoic acid.

The order *Chlorobiales* Gibbons and Murray 1978, 4[AL] and family *Chlorobiaceae* Copeland 1956, 31[AL] are emended to include the genus *Chlorobaculum* Imhoff 2003, 949[VP]. The genus *Chlorobaculum* is placed in the phylum *"Chlorobi"* and the class *"Chlorobia"*.

Phylum B12 **Proteobacteria** *phylum nov.* Garrity and Holt 2001s, 130

Pro.te.o.bac.te'ria . M.L. fem. pl. n. *Proteobacteria* class *Proteobacteria* elevated to phylum.

The *Proteobacteria* currently represent the largest phylogenetically coherent group within the *Bacteria*. In terms of taxon definition, it is the region within our classification that has undergone the most significant expansion and reorganization since Volume One was published in 2001. At the end of October 2003, the *Proteobacteria* contained 2279 validly named species that were assigned to 521 genera, increases of approximately 75% and 39%, respectively. This group of organisms was originally proposed as the Class *Proteobacteria* by Stackebrandt et al. (1988) and contained four informally named subclasses; Garrity and Holt (2001s) proposed elevating the class to the rank of phylum. At the time Stackebrandt et al. proposed the class *Proteobacteria*, it was generally agreed that the group had undergone rapid evolution and generated numerous branches in which physiologically and morphologically diverse forms grouped together. Despite this incongruity, the *Proteobacteria* were thought to be monophyletic. More recently, the consensus has changed somewhat, largely because additional sequences have been added to the phylogenetic analyses. At present, five lineages of descent are generally recognized within the phylum. However, it now appears that the *Proteobacteria* may not be monophyletic (Ludwig and Klenk, 2001 and this volume), and the separation between the *Betaproteobacteria* and *Gammaproteobacteria* is much less clear in large-scale trees, with the former appearing as a subgroup of the latter. The *Alphaproteobacteria*, the *Deltaproteobacteria*, and the *Ep-*

silonproteobacteria remain distinct lineages. In global PCA plots, we find the *Proteobacteria* occupying a contiguous region in the upper left-hand quadrant (Fig. 10A). Regions within this taxonomic space that are occupied by the five proposed classes of *Proteobacteria* are shown in Figs. 10B–F. While there is reasonably good separation of the *Alphaproteobacteria*, the *Deltaproteobacteria*, and the *Epsilonproteobacteria* into discrete regions within these maps, we note a significant overlap between the *Betaproteobacteria* and *Gammaproteobacteria*, which is consistent with the observations of Ludwig and Klenk (2001, and this volume). However, these two classes can be readily separated from one another by separate analyses in which principal components are derived from sequence data restricted to *Proteobacteria* (see below).

Phenotypic groups within the *Proteobacteria* include Gram-negative aerobic or microaerophilic rods and cocci; anaerobic straight, curved, and helical Gram-negative rods; anoxygenic phototrophic bacteria; non-photosynthetic, non-fruiting, gliding bacteria; aerobic chemolithotrophic bacteria and associated genera; facultatively anaerobic Gram-negative rods; budding and/or appendaged nonphototrophic bacteria; aerobic/microaerophilic, motile, helical/vibrioid Gram-negative bacteria; symbiotic and parasitic bacteria of vertebrate and invertebrate species; fruiting, gliding bacteria; sheathed bacteria; and nonmotile or rarely motile curved Gram-negative bacteria.

THE ALPHAPROTEOBACTERIA Within our current classification, the *Alphaproteobacteria* are subdivided into seven orders: *Rhodospirillales, Caulobacterales, "Parvularculales", Rhizobiales, Rhodobacterales, Rickettsiales,* and *Sphingomonadales.* In a phylum-level PCA, we find that the *Alphaproteobacteria* form an elongated cluster in the central region of a plot projected along principal components 2 and 3 (Fig. 11A), which yields the optimum view of the five classes. The outlier at point a is *Candidatus* Xenohaliotis californiensis (Friedman et al., 2000), which has been provisionally placed in the *Rickettsiales.* Locations of member species within the seven orders are presented in a series of plots based on a class-level PCA restricted to the *Alphaproteobacteria* (Fig 11B). As reported by Ludwig and Klenk (2001, and this volume), the *Acetobacteraceae* and *Rhodospirillaceae* form two distinct clades. This is consistent with the separation we see in Fig 11B. With the exception of a few outliers, members of the *Caulobacterales, "Parvularculales", Rhizobiales,* and *Sphingomonadales* map to a central location in these plots, overlapping in some cases but in unique locations in others. The positions of *Rhodobacterales* and *Rickettsiales* show that members of these orders are more distantly related to the core members of the *Alphaproteobacteria.* In the case of the *Rickettsiales,* seven discrete groups are formed. Interestingly, we find that *Parvularcula,* which was recently reported to be a deep-branching member of the *Alphaproteobacteria* and the sole member of the order *"Parvularculales"*, maps to a position that is virtually centered within the class. We believe this occurs because the *Parvularcula* 16S rDNA sequence is equally distant from the benchmarks representing other members of the class. At present, this type of relationship may be better viewed using asymmetric heatmaps, where the differences are more clearly revealed.

Since it was first described in Volume One of this edition, the description of the class *Alphaproteobacteria* has been revised to include the order *"Parvularculales"*, family *"Parvularculaceae"*, and genus *Parvularcula* Cho and Giovannoni 2003c, 1035^{VP}. The order *Rhizobiales* has been updated to include the family *"Aurantimonadaceae"* and genus *Aurantimonas* Denner et al. 2003, 1120^{VP}. Dumler et al. (2001) have proposed that the genus *Ehrlichia* be transferred to the family *Anaplasmataceae*; hence the family *Ehrlichiaceae* Moshkovski 1945, 18^{AL} in the order *Rickettsiales* Gieszczykiewicz 1939, 25^{AL}, class *Alphaproteobacteria,* does not appear in the current Taxonomic Outline. Thirty-four newly described genera have been placed in the class *Alphaproteobacteria.*

THE BETAPROTEOBACTERIA The *Betaproteobacteria* have also been subdivided into seven orders: *Burkholderiales, Hydrogenophilales, "Methylophilales", "Neisseriales", Nitrosomonadales, Procabacteriales,* and *Rhodocyclales.* As noted above, we find a clear separation of the *Betaproteobacteria* from the *Gammaproteobacteria* in a phylum-level PCA restricted to the *Proteobacteria* (Fig. 12A). Member species could be further separated in a class-level analysis restricted to the *Betaproteobacteria* (Fig 12B), in which we find that the *Neisseriaceae*, the *Methylophilaceae*, the *Comamonadaceae*, and the *genera incertae sedis* in the family *Comamonadaceae* (as recommended by Willems and Gillis; see Table 2) appear in specific regions of the plot, away from the core taxa of the class. The *Hydrogenophilales* appear to be somewhat problematic, as the member species are widely spaced in the PCA plots. Ludwig and Klenk (2001, and this volume) indicate that this group branches more deeply within the ARB tree.

The class *Betaproteobacteria* has been updated to include the order *Procabacteriales,* family *Procabacteriaceae,* and *Candidatus* Procabacter acanthamoebae (Horn et al. 2002). The genus *Ralstonia,* which appeared in a separate family in Releases 1 and 2 of the Taxonomic Outline (Garrity et al., 2001, 2002) has been combined with the family *"Burkholderiaceae"*. Twenty-three newly described genera have been placed in the class *Betaproteobacteria.*

THE GAMMAPROTEOBACTERIA The *Gammaproteobacteria* are currently subdivided into 14 orders in our classification: *Chromatiales, Acidithiobacillales, Aeromonadales, Alteromonadales, Cardiobacteriales, "Enterobacteriales", Legionellales, Methylococcales, Oceanospirillales, Pasteurellales, Pseudomonadales, Thiotrichales, "Vibrionales"*, and *Xanthomonadales.* In the phylum-level (Fig 13A) and class-level (Fig 13B) PCA plots, we note that most of these orders tend to cluster together, which is consistent with the overall high degree of 16S rDNA sequence similarity among member taxa. These plots also provide some new insights into the overall taxonomic structure of the group that were not evident from the earlier analyses. In both plots, we find that the *Xanthomonadales* form a discrete and well-defined group that is separate from the main lineage of the *Gammaproteobacteria.* In the phylum-level plot, we find that the order *Pasteurellales* appears as a discrete group, separate from the core, and is joined at this location by *Buchnera* and *Alterococcus* species. We also note that the *Francisellaceae* (*Thiotrichales*) plot distal to the main cloud of points.

Examination of asymmetric heatmaps reveals that both of these groups have generally lower levels of sequence similarity to benchmarks than the others within the respective classes. The class-specific analysis reveals additional details about the taxonomic structure. Here too, we find that the *Francisellaceae* and *Legionellaceae* form a distinct cluster, away from the center, as do the *Coxiellaceae* and *Piscirickettsiaceae* and the genera *Halorhodospira* and *Teredinibacter.* The main lineages within the class are found in the central portion of the plot. The *Enterobacteriaceae* form a dense cloud that is localized predominantly above and to the right of the origin and overlaps significantly with the *Aeromonadaceae* and *Vibrionaceae.* We find the *Alteromonadales* split into two subgroups with *Alteromonas, Glaciecola, Pseudoalteromonas,* and *Shewanella* projecting below and predominantly to the right of the origin, while the remaining genera project above and to

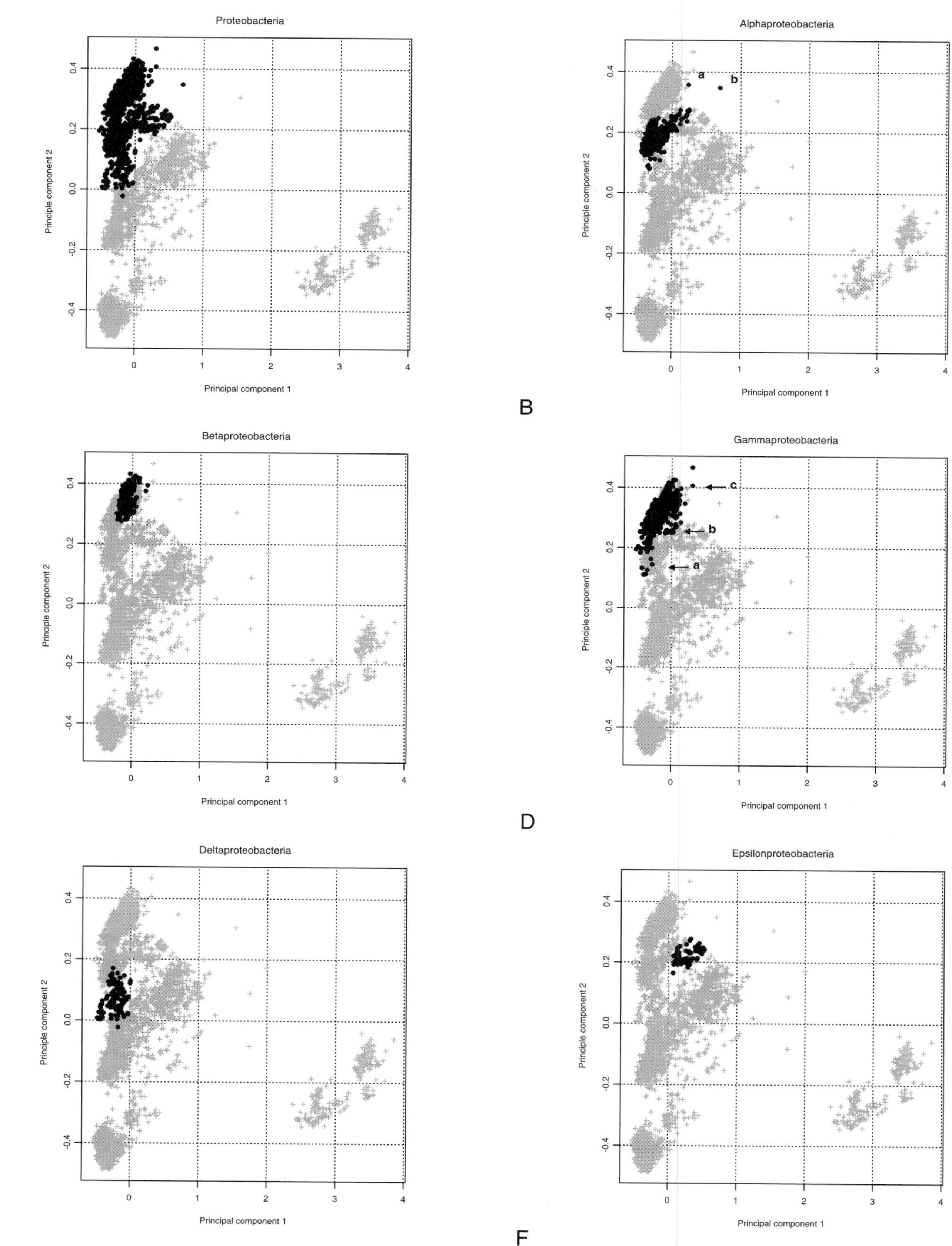

FIGURE 10. Location of Proteobacteria within the global map of the procaryotes. *A*, View of the entire phylum; *B*, The *Alphaproteobacteria* (outliers: a, *Wolbachia pipientis*; b, *Candidatus* Xenohaliotis californiensis); *C*, The *Betaproteobacteria*; *D*, The *Gammaproteobacteria* (outliers: a, *Halorhodospira*; b, *Buchnera*; c, *Alterococcus agarolyticus* and *Moritella japonica*); *E*, The *Deltaproteobacteria*; and *F*, The *Epsilonproteobacteria*.

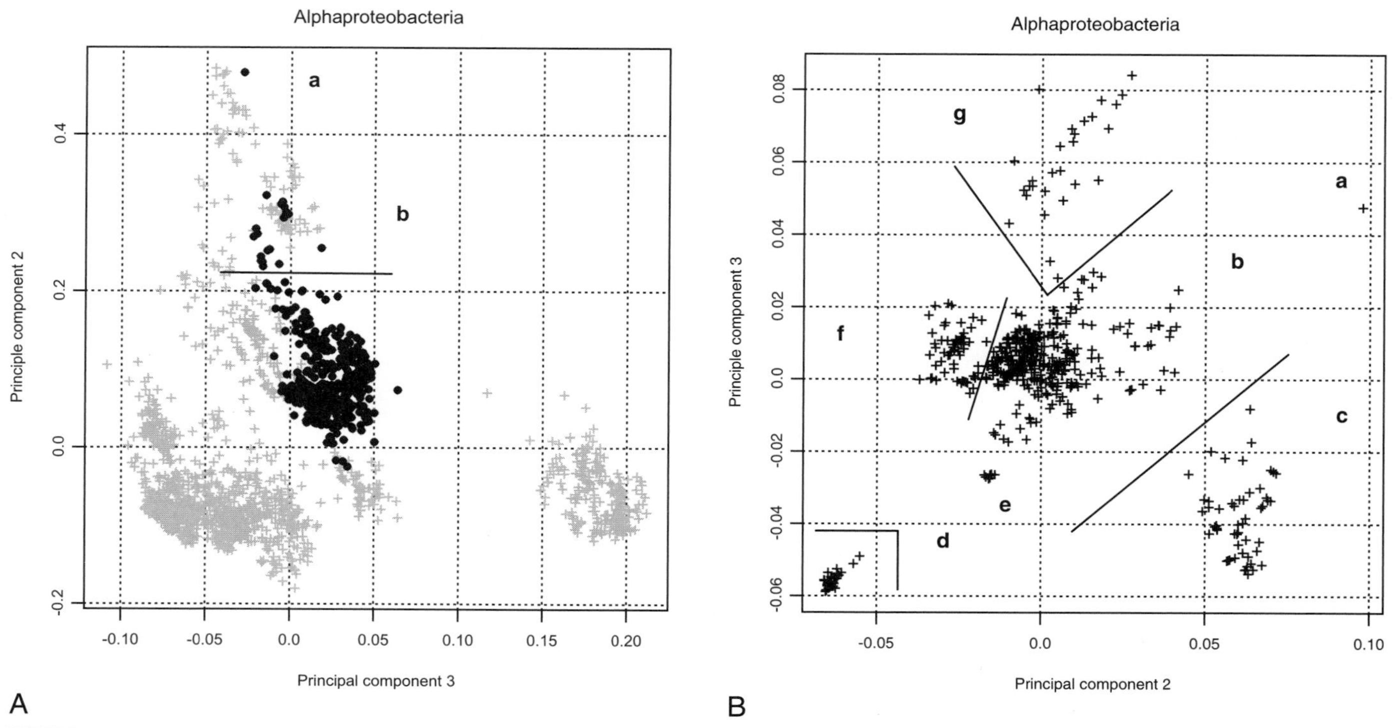

FIGURE 11. PCA of the *Alphaproteobacteria*. *A*, Phylum-level plot in which the dataset and benchmark sequences were restricted to the *Proteobacteria*. The first three dimensions accounted for 87.2% of the total variance. Outliers: a, *Candidatus* Xenohaliotis californiensis; b, the *Rickettsiales*; *B*, Class-level plot of the *Alphaproteobacteria*: point a, *Candidatus* Xenohaliotis; region b, *Rhodospirillales*, *Caulobacterales*, *"Parvularculales"*, *Rhizobiales*, *Caedibacter*, *Neorickettsia*, *Candidatus* Odyssella, *Anaplasma*, *Wolbachia*, and *Ehrlichia*; region c, *Acetobacteraceae*; region d, *Rickettsiaceae*; region e, *Orientia*; region f, *Sphingomonadales*; region g, *Rhodobacterales*.

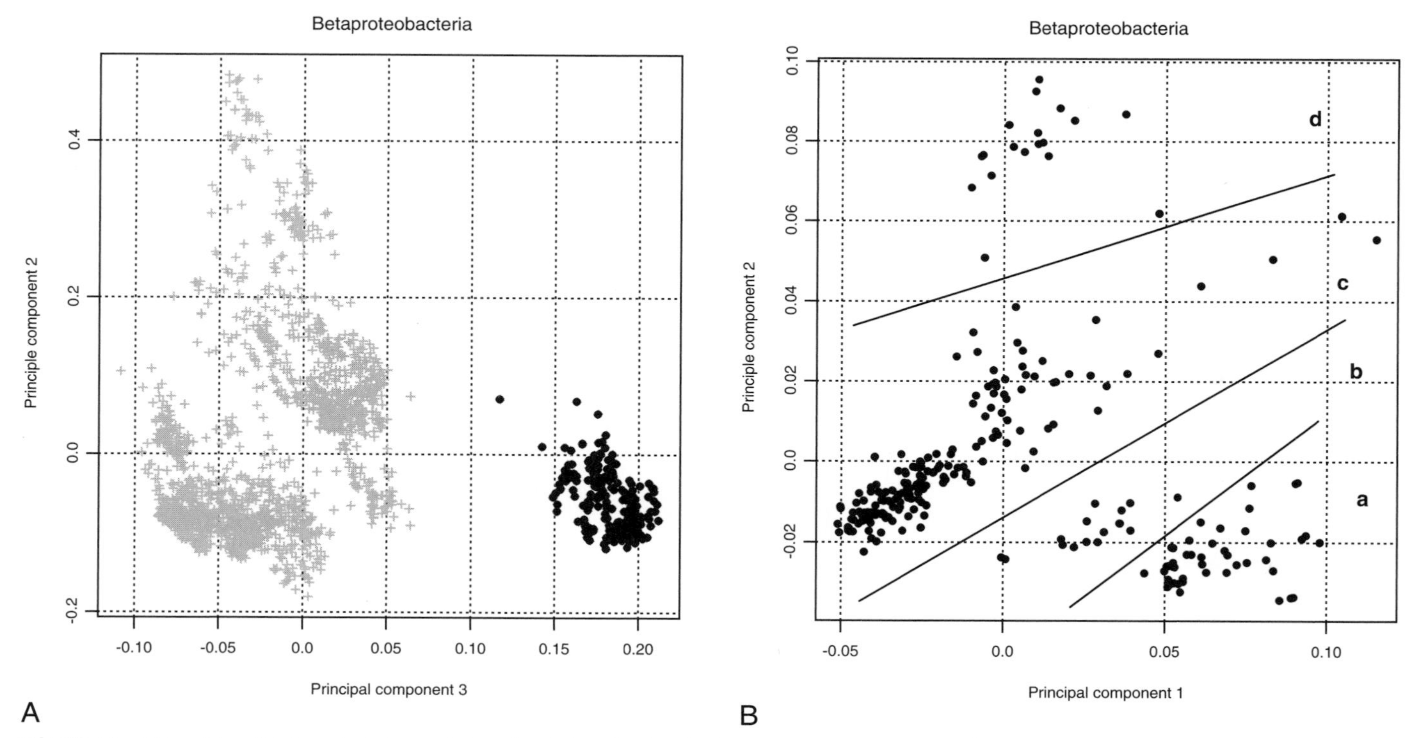

FIGURE 12. PCA of the *Betaproteobacteria*. *A*, Phylum-level plot. *B*, Class-level plot of the *Betaproteobacteria*. Regions a, *Comamonadaceae*; b, *genera incertae sedis*; and d, *"Neisseriales"* and *"Methylophilales"*. Remaining orders plot into the central region of the plot (region c).

TABLE 2. Higher taxa that have been moved within the Taxonomic Outline since the publication of Volume One of the Second Edition of the *Systematics*.[a]

	Taxon	Moved from	Moved to	Comment or authority
Order	*Methanomicrobiales*[b]	*Methanococci*	*Methanomicrobia*	Recommendation of S. Turner and analysis of 16S rDNA sequences
	Methanosarcinales[b]	*Methanococci*	*Methanomicrobia*	Recommendation of S. Turner and analysis of 16S rDNA sequences
Genus	*Alcanivorax*	*Halomonadaceae*	*Alcanivoraceae fam. nov.*	Proposed in this volume by Kelly and Wood
	Allorhizobium[c]	*Phyllobacteriaceae*	*Rhizobiaceae*	Young et al. (2001)
	Anaeroarcus	*Peptococcaceae*	*Acidominococcaceae*	Recommendation of C. Strompl and B.J. Tindall
	Anaerosinus	*Peptococcaceae*	*Acidominococcaceae*	Recommendation of C. Strompl and B.J. Tindall
	Anaerovibrio	*Peptococcaceae*	*Acidominococcaceae*	Recommendation of C. Strompl and B.J. Tindall
	Aquabacterium	*Comamonadaceae*	*Incertae sedis* (*Comamonadaceae*)	Recommendation of A. Willems
	Archangium	*Archangiaceae*	*Cystobacteraceae*	Recommendation of H. Reichenbach
	Bogoriella	*Micrococcaceae*	*Bogoriellaceae*	Stackebrandt and Schumann (2000)
	Caedibacter	*Holosporaceae*	*Incertae sedis* (*Holosporaceae*)	Recommendation of H.-D. Görtz and H. Schmidt
	Catenococcus	*Neisseriaceae*	*Vibrionaceae*	Based on short 16S rDNA sequence
	Centipeda	*Peptococcaceae*	*Acidominococcaceae*	Recommendation of C. Strompl and B.J. Tindall
	Demetria	*Micrococcaceae*	*Dermacoccaceae*	Stackebrandt and Schumann (2000)
	Dendrosporobacter	*Peptococcaceae*	*Acidominococcaceae*	Recommendation of C. Strompl and B.J. Tindall
	Dermacoccus	*Dermatophilaceae*	*Dermacoccaceae*	Stackebrandt and Schumann (2000)
	Derxia	*Beijerinckiaceae*	*Alcaligenaceae*	Recommendation of Hui and Akira
	Desulfobacca	*Nitrospinaceae*	*Syntrophaceae*	Recommendation of J. Kuever
	Desulfomonile	*Nitrospinaceae*	*Syntrophaceae*	Recommendation of J. Kuever
	Desulfonatronum	*Desulfovibrionaceae*	*Desulfonatronumaceae*	Recommendation of J. Kuever
	Desulfotalea	*Desulfobacteraceae*	*Desulfobulbaceae*	Recommendation of J. Kuever
	Enhydrobacter	*Vibrionaceae*	*Incertae sedis* (*Pseudomonadales*)	Recommendation of E. Juni; 16S rDNA sequence closely related to *Moraxella* spp.
	Halothiobacillus	*Chromatiaceae*	*Halothiobacillaceae fam. nov.*	Proposed in this volume by Kelly and Wood
	Ideonella	*Comamonadaceae*	*Incertae sedis* (*Comamonadaceae*)	Recommendation of A. Willems
	Kytococcus	*Dermatophilaceae*	*Dermacoccaceae*	Stackebrandt and Schumann (2000)
	Lampropedia	*Pseudomonadaceae*	*Comamonadaceae*	Recommendation of Hui and Akira
	Leptothrix	*Comamonadaceae*	*Incertae sedis* (*Comamonadaceae*)	Recommendation of A. Willems
	Leucobacter	*Micrococcaceae*	*Microbacteriaceae*	Analysis of 16S rDNA sequences
	Lyticum	*Holosporaceae*	*Incertae sedis* (*Holosporaceae*)	Recommendation of H.-D. Görtz and H. Schmidt
	Macromonas	*Thiotrichaceae*	*Comamonadaceae*	Recommendation of G. A. Dubinina
	Mitsuokella	*Peptococcaceae*	*Acidominococcaceae*	Recommendation of C. Strompl and B.J. Tindall
	Morococcus	*Pseudomonadaceae*	*Neisseriaceae*	Recommenation of L. Sly
	Myroides	*Myroidaceae*	*Flavobacteriaceae*	Bernardet et al., 2002
	Natroniella	*Haloanaerobiaceae*	*Halobacteroidaceae*	Recommendation of Switzer Blum et al. and Zhalina et al.
	Oceanobacillus	*Oceanospirillaceae*	*Bacillaceae*	Correction of data entry error in the original Taxonomic Outline
	Oscillochloris	*Chloroflexaceae*	*Oscillochloridaceae*	Keppen et al., 2000
	Oligella	*Pseudomonadaceae*	*Alcaligenaceae*	Recommendation of K. Kersters
	Odyssella	*Holosporaceae*	*Incertae sedis* (*Holosporaceae*)	Recommendation of H.-D. Görtz and H. Schmidt
	Polynucleobacter	*Holosporaceae*	*Burkholderiaceae*	Based on analysis of Springer et al., 1996
	Propionispora	*Peptococcaceae*	*Acidominococcaceae*	Recommendation of C. Strompl and B.J. Tindall
	Pseudocaedibacter	*Holosporaceae*	*Incertae sedis* (*Holosporaceae*)	Recommendation of H.-D. Görtz and H. Schmidt
	Psychromonas	*Myroidaceae*	*Alteromonadaceae*	Correction of data entry error in the original Taxonomic Outline
	Ralstonia	*"Ralstoniaceae"*	*Burkholderiaceae*	Recommendation of Kuzuko
	Rarobacter	*Cellulomonadaceae*	*Rarobacteraceae*	Stackebrandt and Schumann (2000)
	Rhodothalassium	*Rhodospirillaceae*	*Rhodobacteraceae*	Recommendation of J. Imhoff
	Roseateles	*Comamonadaceae*	*Incertae sedis* (*Comamonadaceae*)	Recommendation of A. Willems
	Rubrivivax	*Comamonadaceae*	*Incertae sedis* (*Comamonadaceae*)	Recommendation of A. Willems
	Sanguibacter	*Intrasporangiaceae*	*Sanguibacteraceae*	Stackebrandt and Schumann (2000)
	Sphaerotilus	*Comamonadaceae*	*Incertae sedis* (*Comamonadaceae*)	Recommendation of A. Willems
	Succinispira	*Peptococcaceae*	*Acidominococcaceae*	Recommendation of C. Strompl and B.J. Tindall
	Symbiotes	*Holosporaceae*	*Incertae sedis* (*Holosporaceae*)	Recommendation of H.-D. Görtz and H. Schmidt
	Tectibacter	*Holosporaceae*	*Incertae sedis* (*Holosporaceae*)	Recommendation of H.-D. Görtz and H. Schmidt
	Terasakiella	*Oceanospirillaceae*	*Methylocystaceae*	Based on 16S rDNA analysis; Satomi et al. (2002) place in *Alphaproteobacteria*.
	Thermoleophilum	*Pseudomonadaceae*	*Rubrobacteraceae*	Yakimov et al. (2003b)
	Thiomonas	*Comamonadaceae*	*Incertae sedis* (*Comamonadaceae*)	Recommendation of A. Willems
	Thiovulum	*Campylobacteraceae*	*Helicobacteraceae*	Recommendation of L.A. Robertson
	Xylophilus	*Pseudomonadaceae*	*Incertae sedis* (*Comamonadaceae*)	Recommendation of A. Willems

[a]The order or family in which *genera incertae sedis* appear in the Taxonomic Outline is given in parentheses.

[b]All families, genera, and species in the order as previously constituted were moved.

[c]The type strain of *Allorhizobium* has been transferred to the genus *Rhizobium* by Young et al. (2001).

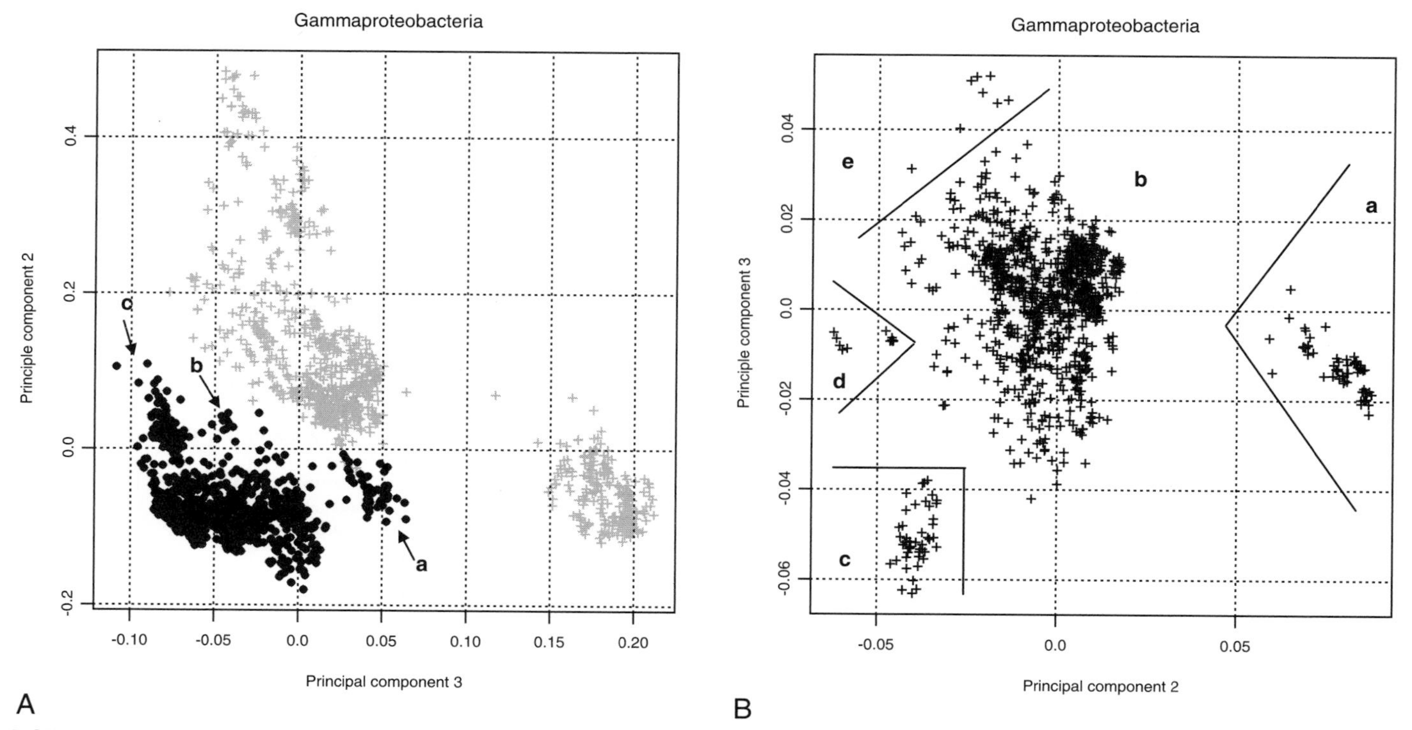

FIGURE 13. PCA of the *Gammaproteobacteria*. A, Phylum-level plot. Regions a, *Xanthomonadales*; b, *Francisellaceae*; c, *Pasteurellaceae*, *Buchnera* and *Alterococcus*. *B*, Class-level plot. Regions a, *Xanthomonadales*; b, core species from the *Aeromonadales*, *Alteromonadales*, *Cardiobacteriales*, *Chromatiales*, *"Enterobacteriales"*, *Methylococcales*, *Oceanospirillaceae*, *Pasteurellaceae*, and *Vibrionales*; c, *Francisellaceae* and *Legionellaceae*; d, *Coxiellaceae* and *Piscirickettsiaceae*; e, *Teredinibacter* and *Halorhodospira*.

the left of the origin. The *Pasteurellales* tend to be localized in the region to the left of the origin, between y = − 0.02 and 0.02 and to partially overlap with the *Oceanospirillales*. The two families of the *Pseudomonadales* are well separated with the *Moraxellaceae* below the origin (y = − 0.01 to − 0.04) and the *Pseudomonadaceae* to the left of the origin in the region of − 0.01 to 0.3), overlapping partially with *Oceanospirillales*, parts of the *Thiotrichales*, *Chromatiales* and *Methylococcales*. These results are quite comparable with those of Ludwig and Klenk (2001, and this volume). We believe that these plots, along with asymmetric heatmaps, provide some insight as to why the relative branching order within the *Gammaproteobacteria* cannot be unambiguously determined. As compared to our earlier classification, the current scheme shows some improvement. Some of the outliers noted in 2001 have been the subject of taxonomic revisions, and many of the synonymies occurring among member species have now been resolved, based on the location of the taxa within these maps. Some problems do, however, remain, such as the composition of the *Cardiobacteriales*, which are probably polyphyletic but cannot yet be resolved because of the small number of member taxa.

The class *Gammaproteobacteria* has been updated to include the order *Acidithiobacillales* and the families *Acidithiobacillaceae* and *Thermithiobacillaceae* (both in the new order *Acidithiobacillales*), *Halothiobacillaceae* (in the order *Chromatiales*), and four new families in the order *Oceanospirillales*: the *Alcanivoraceae*, the *Hahellaceae*, the *Oleiphilaceae*, and the *Saccharospirillaceae*. Thirty-nine newly described genera have been placed in the class *Gammaproteobacteria*.

THE DELTAPROTEOBACTERIA The taxonomy of the *Deltaproteobacteria* has undergone significant change since the publication of Volume One. At present, the class is subdivided into eight orders: the *"Desulfurellales"*, the *"Desulfovibrionales"*, the *"Desulfobacterales"*, the *"Desulfarcales"*, the *"Desulfuromonales"*, the *"Syntrophobacterales"*, the *"Bdellovibrionales"*, and the *Myxococcales*. In the phylum-level PCA plot (Fig. 14A), with the exception of the *"Desulfurellales"*, we find the members of the class plot to a region between the *Alphaproteobacteria*, *Epsilonproteobacteria*, and *Gammaproteobacteria*. While these plots demonstrate that the range of sequence similarity within the phylum is comparable to the others, localization of the separate orders within this plot is not possible. In the class-level analysis (Fig 14B), the *Desulfovibrionales* and *Desulfurellales* clearly separate from the core members of the class. While clustering occurs along order and family lines within the core taxa of the *Deltaproteobacteria*, the groups tend to overlap. To some extent, we believe that this result occurs because the relative number of species within these higher taxa is currently small as compared to the *Alphaproteobacteria*, *Betaproteobacteria* and *Gammaproteobacteria* and because the extent of sequence divergence is relatively high in some of the groups.

The class *Deltaproteobacteria* has been updated by the addition of the order *Desulfarcales*, family *Desulfarculaceae*, and genus *Desulfarculus*. A number of genera have been transferred among families and orders; as a result of these transfers, the family *"Pelobacteraceae"* has been eliminated. The order *Myxococcales* Tchan et al. 1948, 398^{AL} has been updated to include three suborders: *"Cystobacterinea"*, *"Soranginea"*, and *"Nannocystineae"*. The order *Nannocystineae* contains three new families: *"Nannocystaceae"*, *"Hali-*

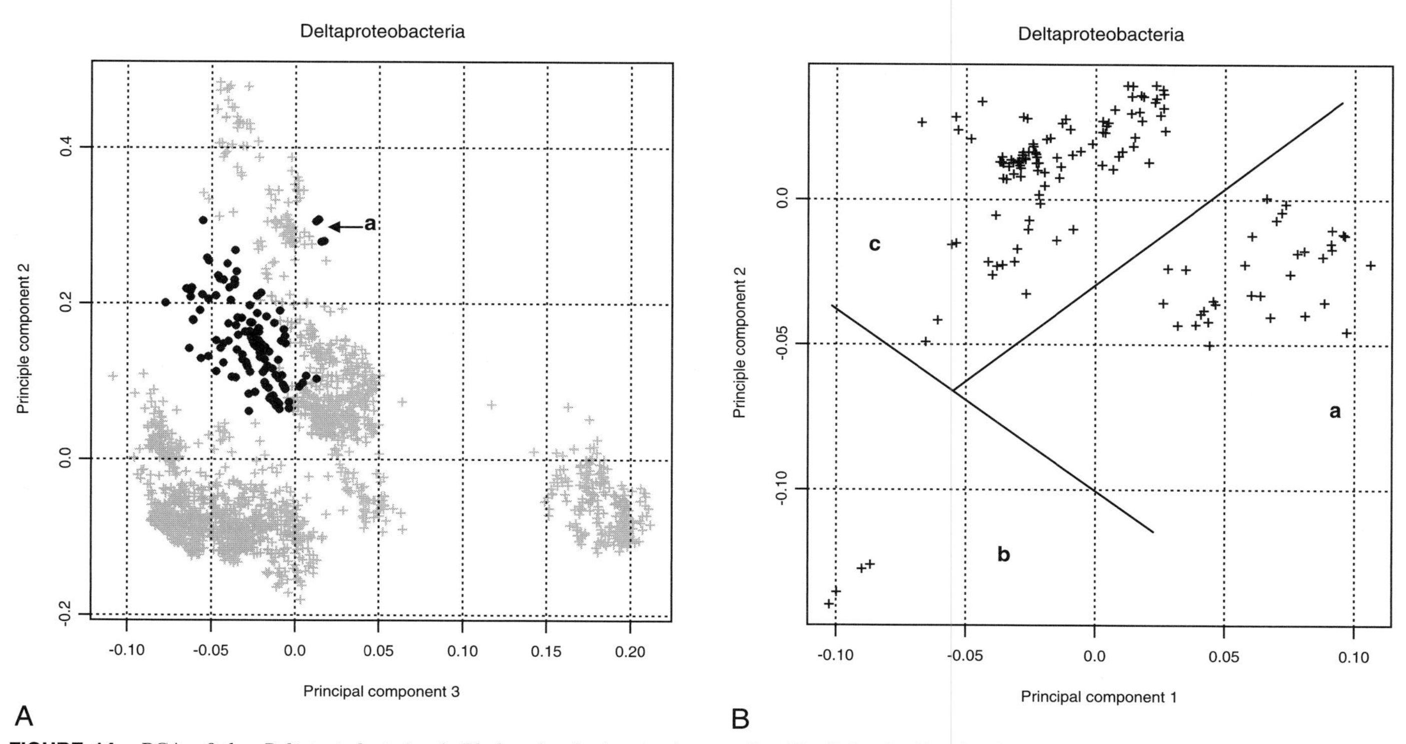

FIGURE 14. PCA of the *Deltaproteobacteria*. *A*, Phylum-level plot; regions a, *Desulfurellales*. *B*, Class-level PCA of the *Deltaproteobacteria*; regions a: *Desulfovibrionales*; b, *Desulfurellales*; c, core members of the class.

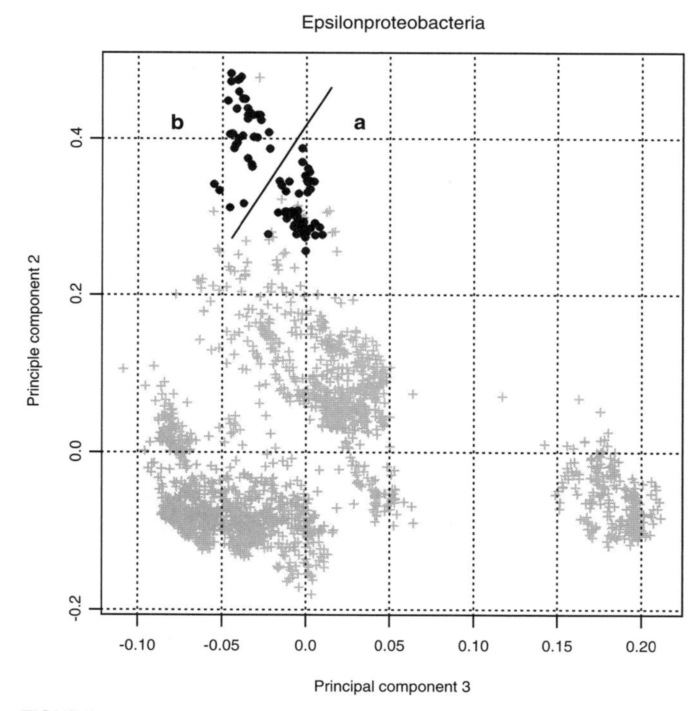

FIGURE 15. Phylum-level PCA plot of the *Epsilonproteobacteria*. Regions a, *Helicobacteraceae* and *Wolinellaceae*; b, *Campylobacteraceae*.

angiaceae", and *"Kofleriaceae"*. The genus *Archangium* Jahn 1924, 66^{AL} has been placed in the family *Cystobacteraceae*; hence, the family *Archangiaceae* Jahn 1924, 66^{AL} does not appear in the current Taxonomic Outline. Twenty newly described genera have been added to the class *Deltaproteobacteria*.

THE EPSILONPROTEOBACTERIA The *Epsilonproteobacteria* represent a more recently recognized line of descent within the *Proteobacteria* and encompass three families within a single order, *"Campylobacterales"*. The group is well supported in phylogenetic trees and appears as a well-separated and tightly clustered group in PCA plots. In the phylum-level analysis (Fig. 15), the *Helicobacteraceae* and *Campylobacteraceae* form two discrete groups. In our earlier study, we noted that there was one potentially misidentified species ascribed to the genus *Helicobacter*, while the other two outliers represent strains of *Wolinella*. These taxonomic errors have been subsequently corrected.

The class *Epsilonproteobacteria* and order *Campylobacterales* have been updated by the addition of the family *"Nautiliaceae"* and genus *Nautilia* (Miroshnichenko et al. 2002). Three newly described genera have been added to the class *Epsilonproteobacteria*.

Phylum B13 **"Firmicutes"** Garrity and Holt 2001s, 133
Fir.mi.cu'tes. M.L. fem. pl. n. *Firmicutes* named for the Division *Firmicutes.*

Originally described by Gibbons and Murray (1978) as the division *"Firmicutes"* and encompassing all of the Gram-positive bacteria; Garrity and Holt (2001s) proposed conservation of the name for the phylum containing the Gram-positive bacteria with a low DNA G + C content.

At the end of October 2003, the phylum contained 1503 species, ascribed to 223 genera belonging to three classes: the *"Clostridia"*, the *Mollicutes*, and the *"Bacilli"*. Ludwig and Klenk (2001, and this volume) note that while the *Mollicutes* appear to be a monophyletic group in reference trees, a common origin of the classical "low G + C Gram positives" is not significantly supported by either the ARB or the RDP trees. Within the global

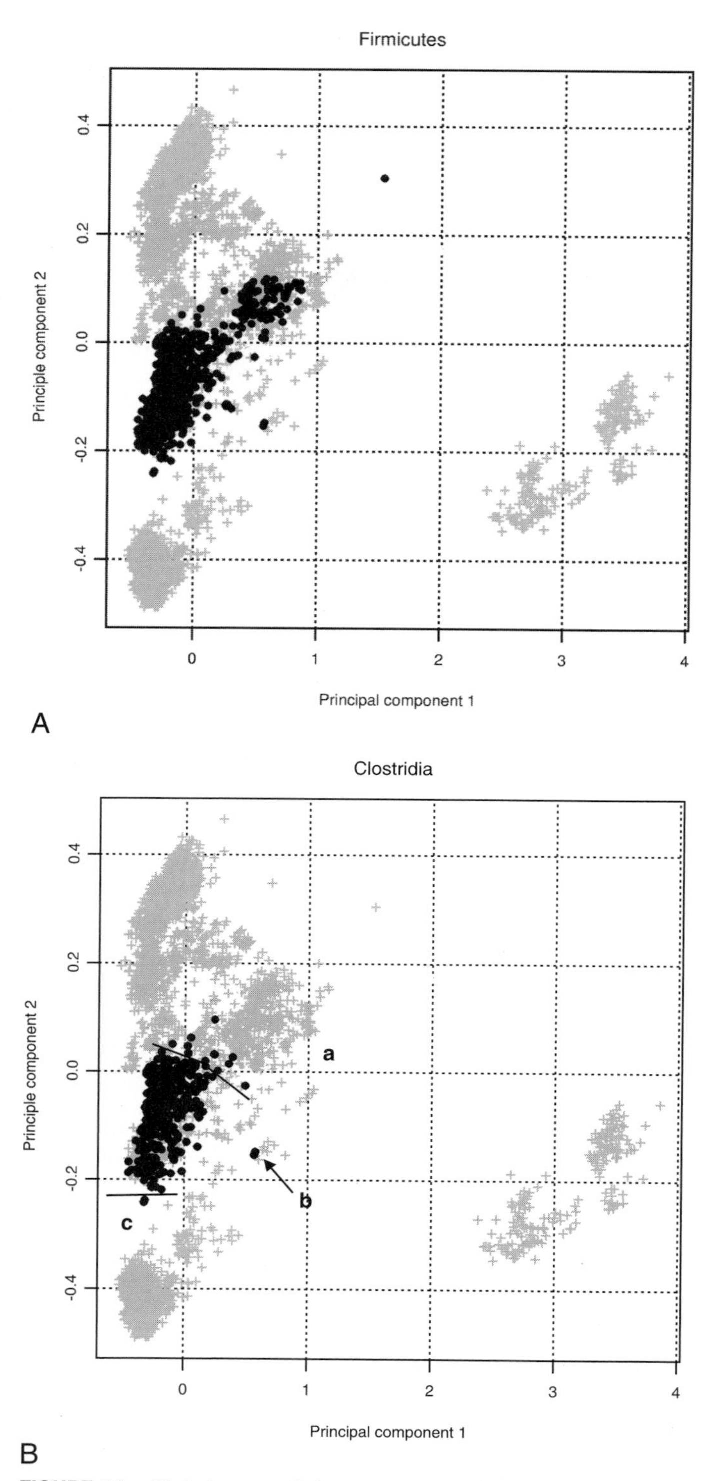

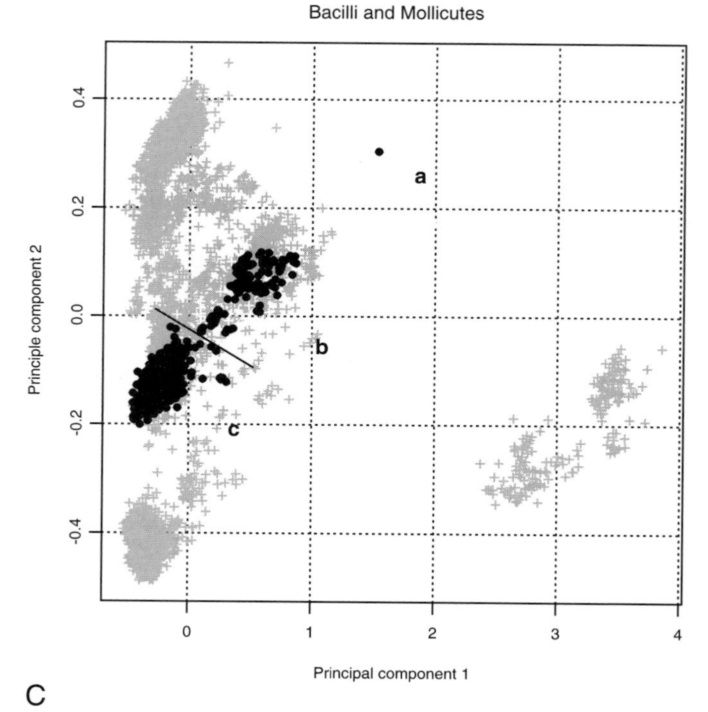

FIGURE 16. Global maps of the procaryotes. *A*, Location of phylum *"Firmicutes"*. *B*, Location of the *"Clostridia"* (Regions a, *Eubacterium yurii, Eubacterium hallii, Halanaerobacter lacunarum* and *Acetohalobium*; Regions b and c, *Coprothermobacter* and *Thermoanaerobacter*, respectively. *C*, Location of the *"Bacilli"* and *Mollicutes* (Point a, *Mycoplasma haemomuris*; region b, *Mollicutes*; region c, *"Bacilli"*.

PCA plot (Fig. 16A), we find that the phylum *"Firmicutes"* occupies a large area bounded by the *Proteobacteria* and the *"Actinobacteria"* along the second principal component, and between the *Proteobacteria* and *Archaea* along the first principal component. The *"Clostridia"* (Fig. 16B) and *"Bacilli"* (Fig. 16C) fall into a densely populated region of the global plot and overlap significantly. (In our earlier model, we noted that these classes overlapped slightly with the *Deltaproteobacteria* in the global plots. The extent of that overlap appears to have diminished somewhat as the size of our dataset has increased. As noted before, in a two-phylum analysis, the *Deltaproteobacteria* could be clearly separated from the *"Clostridia"* and *"Bacilli"*. The *Mollicutes* fall into a separate region of the plots, overlapping slightly with the *Cyanobacteria*. In the global plots, we find a number of outliers that fall into regions occupied by many of the deeply branching phyla; this finding may indicate potential taxonomic misplacements.

The phylum is phenotypically diverse and has been the subject of a number of recent rearrangements. It is expected that further rearrangements and refinements will occur prior to publication of the third volume of this series, which will deal with the *"Firmicutes"* exclusively. The three classes of *"Firmicutes"* can also be separated in a phylum-level analysis (Figs. 17A–C). Examination of the phylum-level plot provides some additional insight into the taxonomic structure of the three classes, as well as revealing some likely misplacements or yet unresolved synonymies.

The *"Clostridia"* (Fig. 17A) form an elongated cluster in the phylum-level PCA plots, covering an area perhaps three times that of the *"Bacilli"*, which is indicative of the proportionately higher level of sequence divergence within the class. In these plots, only a few outliers (notably *Coprothermobacter*) are identifiable. The *"Clostridia"* are subdivided into three orders, which fall into three non-overlapping regions: the *Clostridiales*, the *"Thermoanaerobacteriales"*, and the *Haloanaerobiales*.

The *Mollicutes* (Fig. 17B) represent a well-formed class that is currently subdivided into four orders (the *Mycoplasmatales*, the

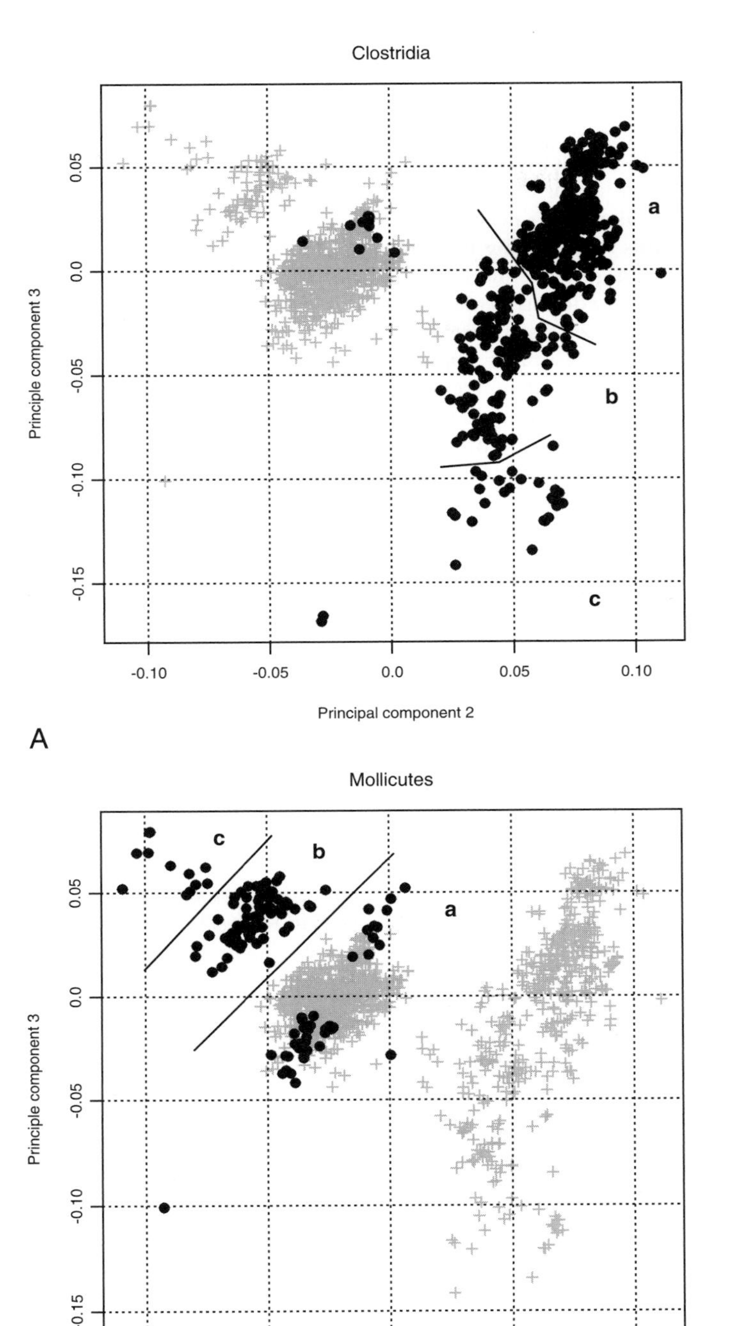

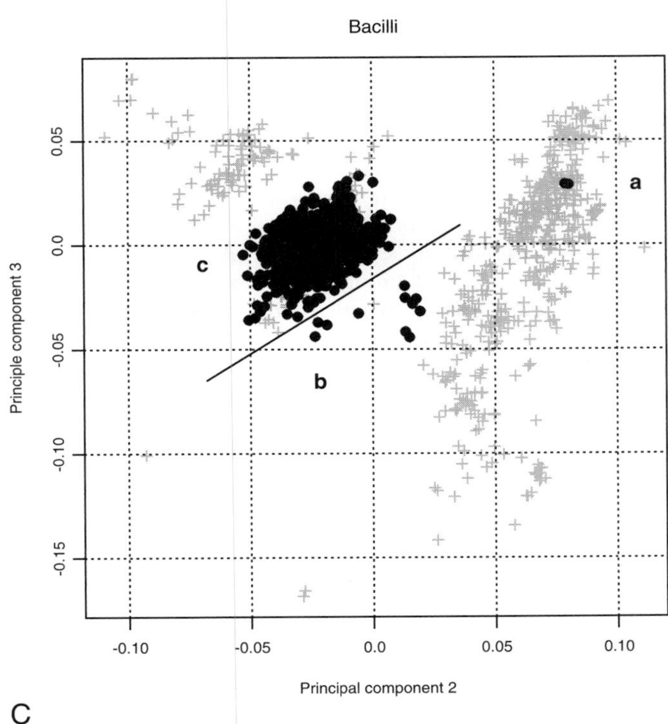

FIGURE 17. Location of *"Firmicutes"* classes within phylum-level PCA plots. *A*, Regions a, *Clostridiales*; b, *"Thermoanaerobacteriales"*; c, *Haloanaerobiales*; and outliers, *Coprothermobacter*. B, *Mollicutes* Regions a, *Acholeplasmatales*, *Anaeroplasmatales*, *Mycoplasmatales*, and *Mycoplasma haemomuris*; b, *Mycoplasmatales* and *Entomoplasmatales*; c, *Mycoplasmatales*. C, *"Bacilli"* Regions a, *Syntrophococcus sucromutans* (listed as *incertae sedis* within the class); b, *Alicyclobacillus*, *Marinococcus*, *Bacillus selenitireducens*, *Thermicanus aegyptius*, and *Sulfobacillus disulfidooxidans*; c, core members of the class.

Entomoplasmatales, the *Acholeplasmatales*, and the *Anaeroplasmatales*), in which the current taxonomy and phylogeny are largely in good agreement. In the global plot (Fig. 16C), the *Mollicutes* form a broad and essentially coherent cluster with only two outliers. One is *Mycoplasma haemomuris* (basonym *Haemobartonella muris*) and the second is *Erysipelothrix rhusiopathiae*, which is currently deemed *incertae sedis* in our taxonomy. In phylum-level plots (Fig. 17B), the *Mollicutes* break apart into four separate clusters, with two outliers. Two clusters (region a) overlap with the *"Bacilli"*. In the cluster above the origin in region a are found members of the *Acholeplasmatales* and the *Anaeroplasmatales* (except for *Asteroleplasma anaerobium*, which appears as an outlier between the *"Bacilli"* and *"Clostridia"*). Below the origin are found *Ureaplasma* and a number of *Mycoplasma* species along with the distant outlier *Mycoplasma haemomuris*. Regions b and c are dominated by *Mycoplasma* species. The position of *Mycoplasma haemomuris* is particularly noteworthy as it falls into a sparsely populated region of the map that is occupied by a few sequences derived from environmental isolates and atypical chloroplasts. *Mycoplasma haemomuris* is an obligately parasitic species, and the unusual position it occupies in PCA plots may be indicative of a highly unusual primary and/or secondary 16S rRNA structure resulting from reductive evolution. Several other parasitic species fall along the periphery of the major clusters and behave like deeply branching taxa.

The *"Bacilli"* also form a coherent class that is currently subdivided into two orders: *Bacillales* and *"Lactobacillales"*. Ludwig and Klenk (2001, and this volume) indicate that the five families of lactic acid bacteria that form the *"Lactobacillales"* constitute a phylogenetically coherent group, whereas four of the nine families in the *Bacillales* branch more deeply than the remaining five. In the phylum-level PCA plot (Fig. 17C), the two orders form a dense cloud that appears to have relatively little discernable structure. We also find 12 outliers. Six of these are centered around *Alicyclobacillus* species in region a and are joined by two species of *Marinococcus* and *Bacillus selenitireducens* in region b as well as by *Thermicanus aegyptius* and *Sulfobacillus disulfidooxidans* in the region between the *"Bacilli"* and *"Clostridia"*. *Syntrophococcus sucromutans* is found in the region occupied by the *"Clostridia"*. Although currently listed as *incertae sedis* in the *"Lactobacillales"*, it is probable that the species is affiliated with the class *"Clostridia"*.

Phenotypic groups of the *"Firmicutes"* include thermophilic and hyperthermophilic bacteria; anaerobic straight, curved, and helical Gram-negative rods; anoxygenic phototrophic bacteria; nonphotosynthetic, nonfruiting, gliding bacteria; aerobic, nonphototrophic, chemolithotrophic bacteria; dissimilatory sulfate- or sulfite- reducing bacteria; symbiotic and parasitic bacteria of vertebrate and invertebrate species; anaerobic Gram-negative cocci; Gram-positive cocci; endospore-forming Gram-positive rods and cocci; regular, nonsporulating Gram-positive rods; irregular, nonsporulating Gram-positive rods; mycoplasmas; and thermoactinomyces.

The phylum *"Firmicutes"* has been updated by the addition of 60 newly described genera.

Phylum B14 **"Actinobacteria"** Garrity and Holt 2001s, 135
Ac.ti.no.bac.te'ria. M.L. fem. pl. n. *Actinobacteria* class of the phylum; M.L. fem. pl. n. *Actinobacteria* the phylum of *"Actinobacteria"*.

In Volume One of this edition, Garrity and Holt proposed elevation of the class *Actinobacteria* (Stackebrandt et al., 1997) to the rank of phylum, recognizing that the phylogenetic depth represented in this lineage is equivalent to that of existing phyla and that the group shows clear separation from the *"Firmicutes"*. Within the phylum, we recognize a single class, *Actinobacteria*, and preserve the complete hierarchical structure of Stackebrandt et al. (1997), including the five subclasses (*Acidimicrobidae, Rubrobacteridae, Coriobacteridae, Sphaerobacteridae,* and *Actinobacteridae*), six orders (the *Acidimicrobiales*, the *Rubrobacterales*, the *Coriobacteriales*, the *Sphaerobacterales*, the *Actinomycetales*, and the *Bifidobacteriales*), and 14 suborders.

In their analysis of the reference trees, Ludwig and Klenk (2001, and this volume) indicate that the phylum is clearly defined and delimited, with the *Rubrobacterales* and *Coriobacteriales* representing the deepest lineages, and the *Acidimicrobiales* occupying a position of intermediate depth. This is consistent with the current classification of *"Actinobacteria"*. No mention was made of *Sphaerobacterales*, which is included in our classification. The sequence of *Sphaerobacter thermophilus* that was in the RDP database was known to be problematic and the sole member of the class is misplaced in the RDP tree. However, a replacement for that sequence has since been published and has been incorporated into our models. Ludwig and Klenk (2001, and this volume) also indicate that neither a significant nor a stable branching order could be established for the families within the order *Actinomycetales* using 16S rDNA, 23S rDNA, or the β subunit of F_1F_0 ATPase.

In the global PCA plot (Fig. 18A), we find that the *"Actinobacteria"* map to a location completely removed from the *"Firmicutes"*, further confirming the likelihood that this group represents a separate line of evolutionary descent. Consistent with the reference trees, we find that *Rubrobacterales* sequences (region f) map into the region of *"Clostridia"*, while the *Sphaerobacterales* (region a) and *Acidimicrobiales* (region e) map into the sparsely populated region between the *"Firmicutes"* and the main cluster of *"Actinobacteria"*. The region occupied by the *Coriobacteriales* has increased significantly since 2001, as additional sequences have been added to the model. As reported before, the *Bifidobacteriales* (region c) comprise a separate group lying adjacent to, but removed from, the major lineages within the *Actinomycetales*, consistent with the published phylogenetic model of Stackebrandt et al. (1997). This differs slightly from the Ludwig and Klenk (2001, and this volume) subtree, in which the *Bifidobacteriales* could not be resolved.

The major cluster of *"Actinobacteria"* in our model has more than doubled since 2001 and currently contains over 1000 data points representing type strains of more than 90% of the validly named genera. The cluster continues to remain quite compact and provides an explanation as to why it is impossible to determine either a stable or a significant branching order within the *Actinomycetales*. It is quite likely that the level of sequence variability, using the current alignment, is simply too low to yield a degree of separation comparable to that found for other, less densely populated phyla.

While the 16S rDNA sequence diversity might appear somewhat lower than that found with some other phyla, the *"Actinobacteria"* have long been recognized for a very high level of morphological, physiological, and genomic diversity. During the past 35 years, considerable effort has been spent in developing a polyphasic approach to the classification and identification of the *"Actinobacteria"*, and most of the characteristics (especially molecular and chemotaxonomic) correspond with the current phylogenetic classification. The level of congruence of the phylogenetic classification with morphology and conventional biochemical approaches is lower. Despite this potential shortcoming and despite the need for specialized microscopy techniques, morphological characteristics are still of value, especially in the preliminary classification and identification of many genera of arthrospore-forming actinobacteria.

Many of the relationships among the *"Actinobacteria"* are evident in the phylum-level PCA plot (Fig 18B). The deeply branching *Acidimicrobidae* (region a), *Coriobacteridae* (region c), *Rubrobacteridae* (region d), and *Sphaerobacteridae* (region b) are located in the lower-right quadrant of the plot, in well-defined regions distal to the main lineages of *"Actinobacteria"*. Region e is dominated by the species-rich *Streptomycetaceae*, along with the species-poor families *Frankiaceae, Sporichthyaceae*, and *Acidothermaceae*. The adjacent, elongated cluster (region g) contains the *Geodermatophilaceae, Nocardiopsaceae, Pseudonocardiaceae* and *Thermomonosporaceae*. *Thermobispora* species fall in region f, far removed from the core of the *Pseudonocardiaceae*, which may be polyphyletic. The *Glycomycetaceae* form a small, distinct cluster in region h and the *Bifidobacteriaceae* are located in the diffuse cluster in region i. The remaining families map into overlapping regions of the large and elongated clusters in region j. While some of the families appear to be localized within specific regions of the cluster, others tend to be spread out in the region. We believe that this may be, in part, an indication of a still suboptimum classification.

The *"Actinobacteria"* can be broadly divided into two major phenotypic groups: unicellular, nonsporulating actinobacteria and the filamentous, sporulating sporoactinomycetes. The unicellular *"Actinobacteria"* include Gram-negative aerobic rods and cocci; aerobic sulfur oxidizers, budding and/or appendaged bacteria; Gram-positive cocci; regular, nonsporulating Gram-positive rods; irregular, nonsporulating Gram-positive rods; and mycobacteria. The sporoactinomycetes include nocardioform actinomycetes, actinomycetes with multilocular sporangia, actinoplanetes, *Streptomyces* and related genera, maduromycetes, *Thermomonospora* and related genera, and other sporoactinomycete genera.

The phylum *"Actinobacteria"* has been updated to include four

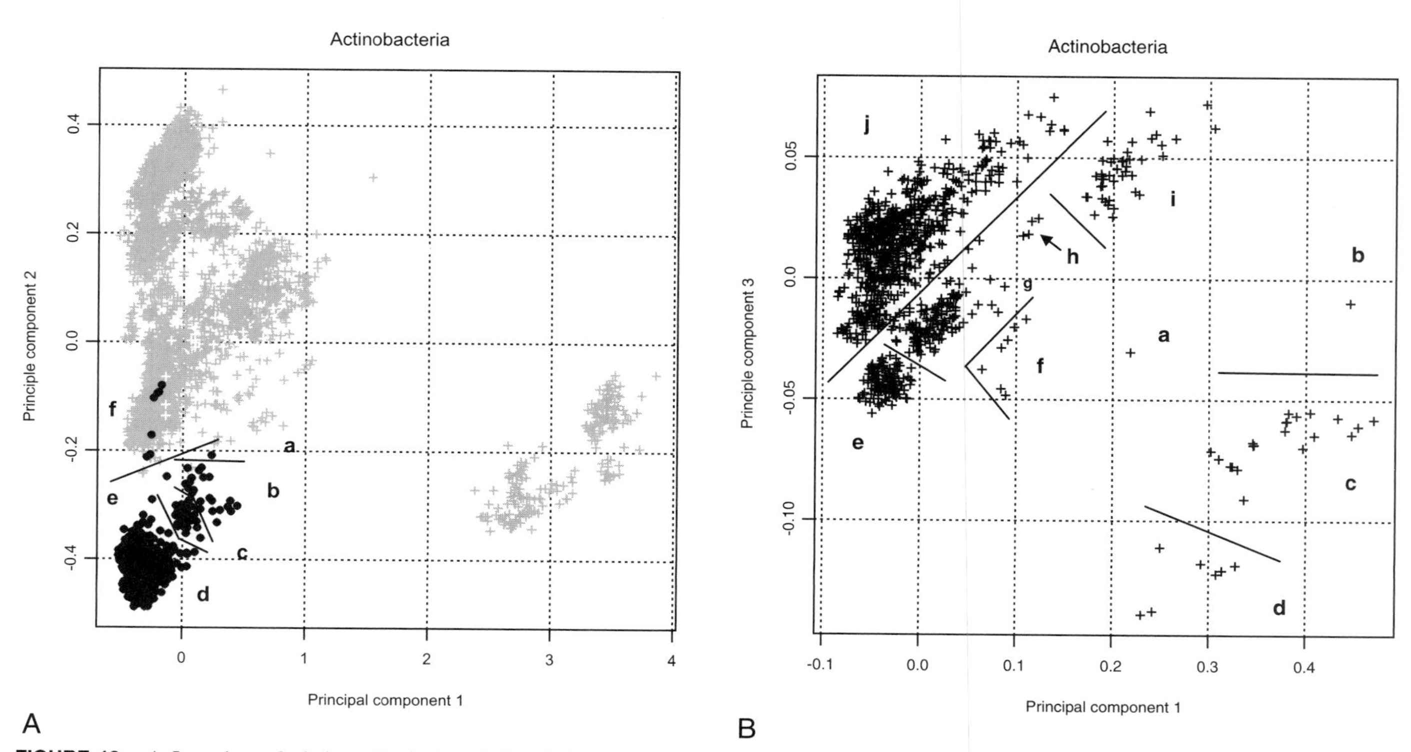

FIGURE 18. *A*, Location of phylum *"Actinobacteria"* and the six orders of the single class *Actinobacteria* within the global map of the procaryotes. Regions a, *Sphaerobacterales*; b, *Coriobacteriales*; c, *Bifidobacteriales*; d, *Actinomycetales*; e, *Acidimicrobiales*; f, *Rubrobacterales*. *B*, Phylum-level PCA of the *"Actinobacteria"*; Regions a, *Acidimicrobiaceae*; b, *Sphaerobacteraceae*; c, *Coriobacteriaceae*; d, *Rubrobacteraceae*; e, *Streptomycetaceae*, *Acidothermaceae*, *Sporichthyaceae*, and *Frankiaceae*; f, *Thermobispora*; g, *Geodermatophilaceae*, *Nocardiopsaceae*, *Pseudonocardiaceae*, and *Thermomonosporaceae*; h, *Glycomycetaceae*; i, *Bifidobacteriaceae*. Region j contains the remaining families of sporoactinomycetes which tend to cluster into well-defined but overlapping regions.

new families in the suborder *Micrococcineae*, order *Actinomycetales*, subclass *Actinobacteridae*, and class *Actinobacteria*: *"Bogoriellaceae"*, *"Rarobacteraceae"*, *"Sanguibacteraceae"*, and *"Dermococcaceae"*. Forty-one newly described genera have been added to the phylum *"Actinobacteria"*.

Phylum B15 **"Planctomycetes"** Garrity and Holt 2001s, 137
Planc.to.my.ce'tes. M.L. fem. pl. n. *Planctomycetales* type order of the phylum, dropping the ending to denote a phylum; M.L. fem. pl. n. *Planctomycetes* the phylum of *Planctomycetales.*

The phylum *"Planctomycetes"* branches deeply within the bacterial radiation in the ARB and RDP trees and has consistently shown a distant relationship to the *"Chlamydiae"*. While the precise location within both the ARB and the RDP trees remains uncertain, Ludwig and Klenk (2001, and this volume) note that the phylum consistently splits into two sister groups, one consisting of *Planctomyces* and *Pirellula*, and the second containing *Gemmata* and *Isosphaera*. Both groups are currently ascribed to the family *Planctomycetaceae* in the order *Planctomycetales* (Schlesner and Stackebrandt, 1986). In global PCA plots (Fig. 19), *Planctomyces* and *Pirellula* cluster closely together and both split into two subgroups that each overlap. The *Gemmata* species (region b) map very close to *Planctomyces/Pirellula* whereas *Isosphaera* (region a) maps to a sparsely populated region some distance from the other genera.

The *"Planctomycetes"* are Gram-negative bacteria that reproduce by budding. Cells are spherical to ovoid or bulbiform. Cells may produce one or more multifibrillar appendages that may terminate in holdfasts. Cell envelope lacks peptidoglycan. Some members of the *"Planctomycetes"* exhibit a membrane-enclosed

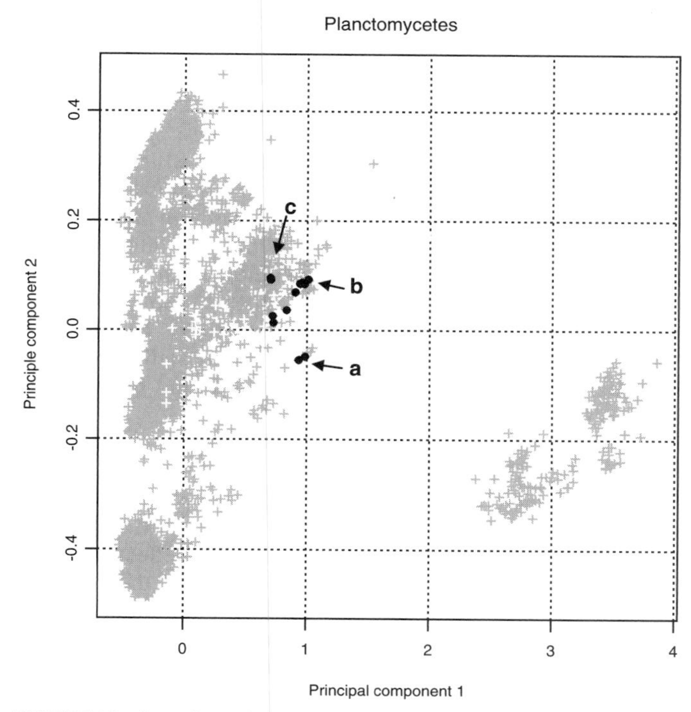

FIGURE 19. Location of phylum *"Planctomycetes"* within the global map of the procaryotes. Outliers are members of the genera *Isosphaera* (region a), *Gemmata* (region b), and *Pirellula* (region c).

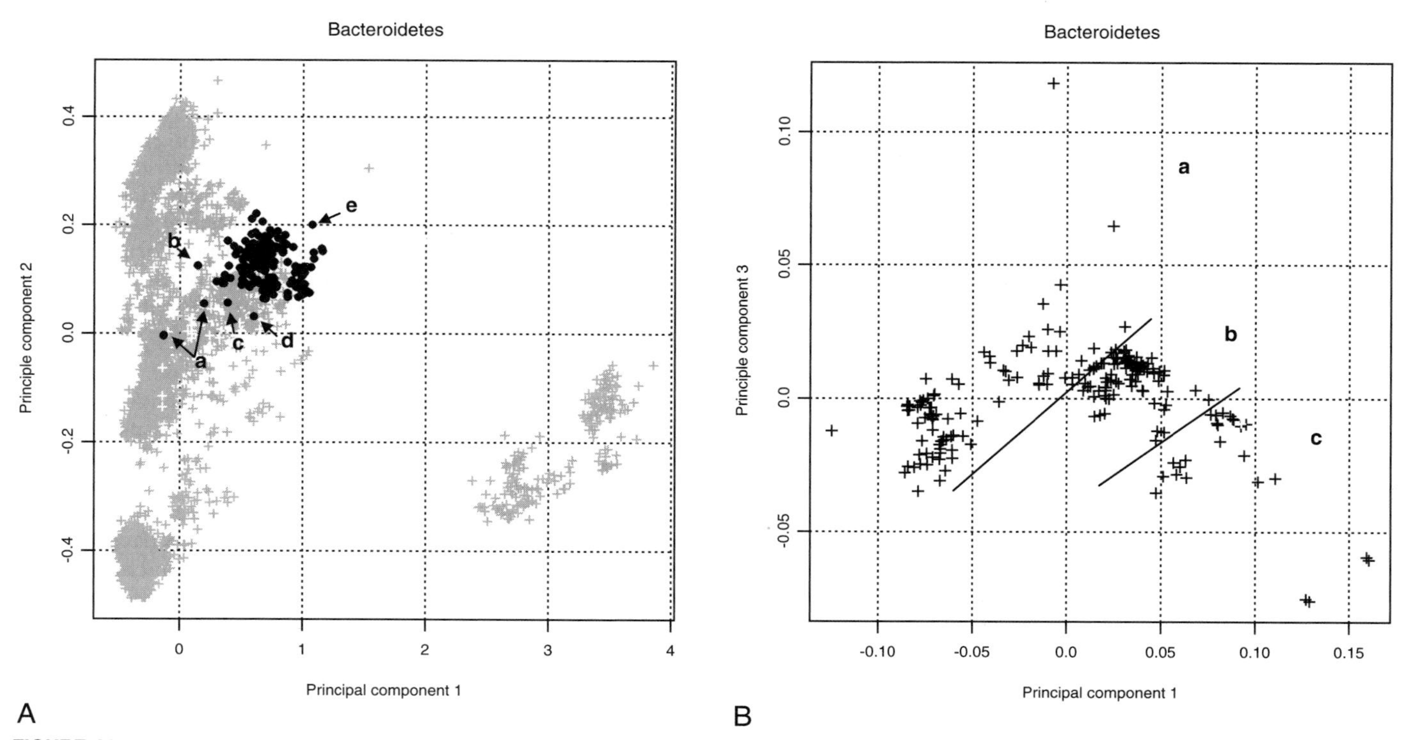

FIGURE 23. *A*, Location of phylum *"Bacteroidetes"* in the global map of the procaryotes. Outliers: a, *Rhodothermus marinus*; b, *Salinospora ruber*; c, *Hymenobacter roseosalivarius*; d, *Muricauda ruestringensis*; and e, *Porphyromonas asaccharolytica*. *B*, Phylum-level PCA of *"Bacteroidetes"*. Regions a, *"Bacteroidales"*; b, *"Flavobacteriales"*; and c, *"Sphingobacteriales"*.

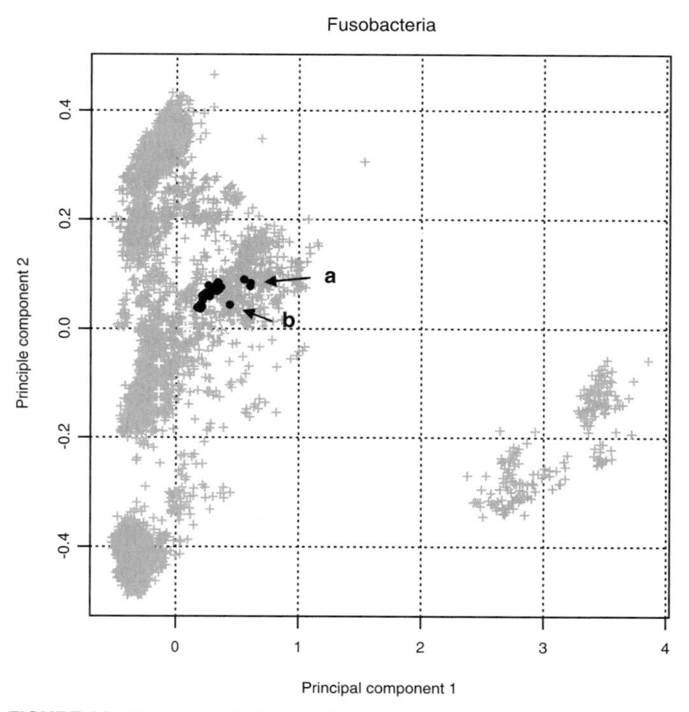

FIGURE 24. Location of phylum *"Fusobacteria"* within the global map of the procaryotes. Outliers: a, *Streptobacillus* and *Leptotrichia*; b, *Sebaldella*.

(2001, and this volume), contains three subclusters. Positions of the member species within *"Fusobacteria"* are presented in the global plot shown in Fig. 24. The quality and size of our dataset has improved since our initial analysis in 2001 and some of the problems alluded to in our earlier discussion have now been resolved (Garrity and Holt, 2001s). For instance, *Fusobacterium prausnitzii*, which was previously noted as falling in the region of the *"Clostridia"*, was the subject of a proposal by Duncan et al. (2002b) to reclassify it as a new combination in the new genus *Faecalibacterium*, in the *"Clostridia"*. As before, the core members of the *"Fusobacteria"* cluster together tightly. However, several outliers remain. The genera *Sebaldella*, *Streptobacillus*, and *Leptotrichia* appear as outliers to the core species. This is consistent with a recent report by Conrads et al. (2002). What remains unclear at this point is whether these genera belong within the radiation of the *"Fusobacteria"* or to another phylum. At present, we recognize a single class, order, and family.

Phenotypically, *"Fusobacteria"* are homogeneous and are characterized as anaerobic, Gram-negative rods with a chemoorganotrophic heterotrophic metabolism.

The phylum *"Fusobacteria"* has been updated by the addition of one newly described genus.

Phylum B22 **"Verrucomicrobia"** (Hedlund, Gosink and Staley 1997) Garrity and Holt 2001s, 140 (Division *"Verrucomicrobia"*Hedlund, Gosink and Staley 1997, 35)

Ver.ru.co.mi.cro'bia. M.L. fem. pl. n. *Verrucomicrobiales* type order of the phylum, dropping the ending to denote a phylum; M.L. fem. pl. n. *Verrucomicrobia* the phylum of *Verrucomicrobiales*.

The phylum *"Verrucomicrobia"* was originally proposed by Hedlund et al. (1997) as a new division within the bacterial domain and re-proposed as a phylum for the sake of consistency. The

"Verrucomicrobia" represent another distinct lineage within the phylogenetic reference trees and contain a number of environ-species as well as a small number of cultured species assigned to four cultivated genera: *Verrucomicrobium, Opitutus, Prosthecobacter,* and *Victivallis,* and one uncultivated genus, *Candidatus* Xiphinematobacter. There is one order, *Verrucomicrobiales.* The phylum has been updated by the addition of two new families, *"Opitutaceae"* and *"Xiphinematobacteriaceae"* (both in the order *Verrucomicrobiales* and the class *Verrucomicrobiae*), based on two newly described genera, *Opitutus* (Chin et al. 2001) and *Candidatus* Xiphinematobacter (Vandekerckhove et al. 2000). In most instances, *"Verrucomicrobia"* have shown a moderate relationship to the *"Planctomycetes"* and *"Chlamydiae"*; however, significance of the common branching is generally low and the relationships among these three phyla are likely to change as additional species are included. Within global PCA plots (Fig. 25), the *"Verrucomicrobia"* map to a region adjacent to the boundaries of the *"Planctomycetes"*. In our models, neither *Opitutus* nor *Victivallis* exhibits a close relationship to the core members of the phylum, and we regard their placement as provisional. This low level of sequence similarity is borne out in asymmetric heatmaps.

Phenotypically, members of the *"Verrucomicrobia"* are Gram-negative bacteria with peptidoglycan containing diaminopimelic acid. Some species are capable of producing prosthecae and fimbriae. They are aerobic or facultatively aerobic; they are chemoheterotrophic, and mesophilic. They multiply by binary fission or asymmetrically by budding. Buds may be produced at the tip of a prostheca or on the cell surface.

Phylum B23 **"Dictyoglomi"** Garrity and Holt 2001s, 140
Dic.ty.o.glo' mi. L. n. *Dictyoglomus* genus of the phylum, dropping the ending to denote a phylum; M.L. fem. pl. n. *Fusobacteria* the phylum of *Dictyoglomus.*

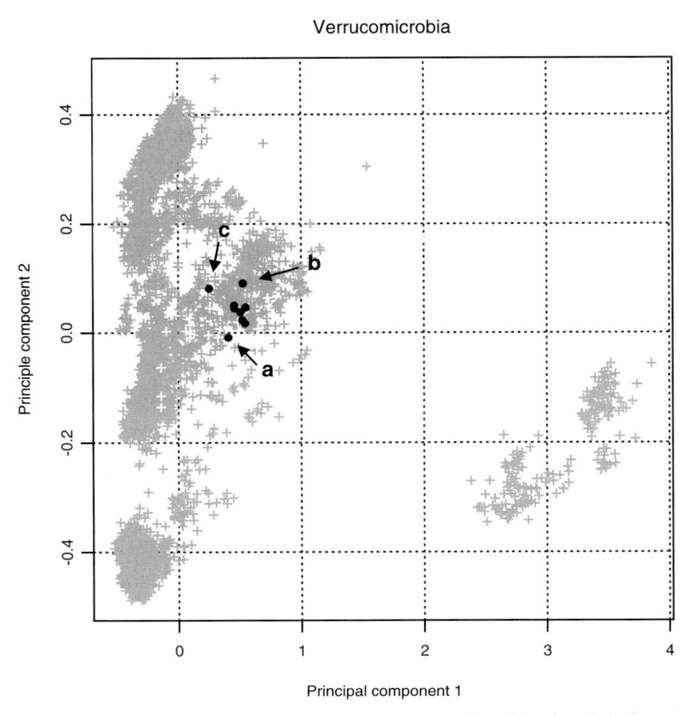

FIGURE 25. Location of phylum *"Verrucomicrobia"* within the global map of the procaryotes. Outliers: a, *Candidatus* Xiphinematobacter, b, *Opitutus;* c, *Victivallis.*

The phylum *"Dictyoglomi"* is currently represented by a single class, order, family, and genus, as well as two species. In the ARB and RDP reference trees, *Dictyoglomus* behaves as a deeply branching group and is found in the vicinity of the *Thermomicrobia* and *"Deinococcus–Thermus"*. In global PCA plots (Fig. 5, point f), *Dictyoglomus* maps to the region occupied by *Coriobacteriales* (*"Actinobacteria"*), nearest to *Atopobium rimae* and *Eggerthella lenta.* However, in asymmetric heatmaps, it is closest to the *Rubrobacter* and *Thermotoga* benchmarks.

"Dictyoglomi" are Gram-negative, nonsporulating, rod-shaped, extremely thermophilic bacteria that are obligately anaerobic and possess a fermentative, chemoorganoheterotrophic metabolism. Single cells may aggregate into spherical, membrane-bound structures of up to several hundred cells.

Phylum B24 **Gemmatimonadetes** Zhang, Sekiguchi, Hanada, Hugenholtz, Kim, Kamagata and Nakamura 2003b, 1161[VP]
Gem.ma' ti.mo.na.det' es. N.L. fem. pl. n. *Gemmatimonas* type genus of the type order of the phylum; N.L. fem. pl. n. *Gemmatimonadetes* the phylum of the genus *Gemmatimonas.*

There is one class (*Gemmatimonadetes*), one order (*Gemmatimonadales*), one family (*Gemmatimonadaceae*), and one genus (*Gemmatimonas*) (Zhang et al., 2003b). Position in global PCA plots is shown in Fig. 5 (point c).

Gemmatimonadetes are Gram-negative bacteria lacking diaminopimelic acid in the cell wall peptidoglycan.

Taxonomic and nomenclatural disagreements of note Readers are advised that there are areas of taxonomic disagreement within the community of microbiologists, some of which are reflected within the pages of this volume. Such disagreement is both natural and healthy, given that we are dealing with a very large taxonomic space approached from a multitude of perspectives. The datasets employed by each author differ in size, taxonomic scope, degree of completeness, and degree of overlap. Their methods of analyzing and interpreting the data have also differed. Moreover, the boundaries between different taxa are rarely clear-cut and tend to become more blurred as datasets grow in size. Our challenge has been to assemble the content of this volume into a more unified view, with full knowledge that the field of systematic microbiology has been moving forward at an accelerating pace. In assembling the content into this comprehensive view, we have attempted to abide by the Bacterial Code (1990 Revision) as well as recent modifications. However, there are equally intense debates within the community about some provisions of the Code, including instances where a strict application of the rules of nomenclature conflicts with generally accepted practice in the field, obfuscates rather than improves communication in the electronic age, or imposes conditions that may ultimately inhibit the taxonomic enterprise.

Brucella Six species of *Brucella* appeared in the Approved Lists and in the first edition of the *Systematics*: *B. abortus, B. canis, B. melitensis, B. neotomae, B. ovis,* and *B. suis.* Differentiation of these species was based on preferred hosts, susceptibility to a panel of lytic phage, metabolic characteristics, and fatty acid composition. However, the results of a DNA–DNA hybridization study by Verger et al. (1985) showed that all the species of *Brucella* were very closely related and led to the proposal that *Brucella* be regarded as a monospecific genus composed of biovars corresponding to the former species. Thus, there is a validly published classification in which these organisms are known as *B. melitensis* biovar abortus, biovar canis, biovar melitensis, biovar neotomae, biovar ovis, and biovar suis. In 1988, the Subcommittee on the taxonomy of *Brucella* indicated that, to avoid confusion, the former species names (nomenspecies) could continue to be used in non-taxonomic contexts. The monospecific system of naming is complicated by the fact that biovars have been designated for several of the nomenspecies: *B. melitensis* biovars 1, 2, and 3; *B. abortus* biovars 1, 2, 3, 4, 5, 6, and 9; and *B. suis* biovars 1, 2, 3, 4, and 5. Subsequent studies (e.g., Michaux-Charachon et al., 1997) have supported the high degree of similarity among the six *Brucella* species/biovars and at the same time have shown that the species/biovars can be distinguished from each other by molecular genetic methods. Finally, in recent years, a heterogeneous group of new *Brucella* isolates has been obtained from marine mammals; it is not yet clear how these isolates are related to *B. abortus, B. canis, B. melitensis, B. neotomae, B. ovis,* and *B. suis* or how they should be classified. In this edition of the *Systematics*, the species of *Brucella* will be presented as *B. abortus, B. canis, B. melitensis, B. neotomae, B. ovis,* and *B. suis.*

RHIZOBIUM In the first edition of the *Systematics*, the family *Rhizobiaceae* contained the genera *Rhizobium, Bradyrhizobium, Agrobacterium,* and *Phyllobacterium.* Many changes in the taxonomy of these organisms have occurred since the first edition. New genera have been described centered around former members of the genus *Rhizobium*: *Sinorhizobium* (Chen et al., 1988), *Mesorhizobium* (Jarvis et al., 1992), and *Allorhizobium* (de Lajudie et al., 1998). Finally, a new family, *Phyllobacteriaceae*, which includes *Mesorhizobium* and *Phyllobacterium* as the type genus, is proposed in this volume. Phylogenetic and taxonomic analyses of rhizobia and their relatives are complicated by lateral gene transfer and coevolution with host plants (Provorov, 1998; Wernegreen and Riley, 1999; Broughton, 2003).

Several of the recent taxonomic proposals regarding the rhizobia and agrobacteria have not been universally accepted by workers in the field. For example, Jarvis et al. (1992) considered *Sinorhizobium* a synonym of *Rhizobium*; the description of *Sinorhizobium* was later emended by de Lajudie et al. (1998), who described additional species of *Sinorhizobium.* In addition, as this volume went to press, two Requests for Opinions regarding a proposed transfer of *Sinorhizobium* to *Ensifer* were pending before the Judicial Commission.

In 2001, Young et al. proposed combining the genera *Rhizobium, Allorhizobium,* and *Agrobacterium* into a single emended genus, *Rhizobium.* This proposal was based primarily on analyses of 16S rDNA sequences in which species of *Agrobacterium* and *Allorhizobium* appeared interspersed among *Rhizobium* species in the resulting dendrograms. Young et al. (2001) also cited a number of published analyses of phenotypic traits that failed to differentiate two or more of these genera. An analysis by Tighe et al. (2000) based on fatty acid content of the genera *Agrobacterium, Bradyrhizobium, Mesorhizobium, Rhizobium,* and *Sinorhizobium* showed that the five genera formed cohesive groups; it also revealed similarities between individual species in different genera, including species of *Agrobacterium* and *Rhizobium.* Farrand et al. (2003) disagreed with Young et al. (2001) and pointed out that other analyses of 16S rDNA sequences and other studies of phenotypic traits can be interpreted to support the retention of *Agrobacterium* as a genus.

It is an open question whether 16S rDNA sequences can properly be used to develop a phylogenetically based taxonomy for this group of bacteria (Broughton, 2003). van Berkum et al. (2003) have provided direct evidence for the possible lateral transfer of segments of 16S rDNA sequence from *Mesorhizobium* to *Bradyrhizobium elkanii* and between *Mesorhizobium* species and *Sinorhizobium* species. Furthermore, these authors showed that analyses of 16S rDNA, 23S rDNA, and ITS region sequences of *Agrobacterium, Rhizobium, Mesorhizobium,* and *Sinorhizobium* produce incongruent phylogenies, a result that is also consistent with the possible occurrence of lateral transfer of these genes. In addition, Turner and Young (2000) found evidence of possible intergeneric transfer of sequences encoding glutamine synthetase II from *Mesorhizobium* to *Bradyrhizobium* and from *Rhizobium* to *Mesorhizobium.* As the matter remains unresolved, we have opted to present both *Agrobacterium* and *Rhizobium* in this edition.

SPHINGOMONAS The genus *Sphingomonas* was proposed by Yabuuchi et al. (1990) to accommodate *Pseudomonas paucimobilis* (which became the type species of the genus), *Flavobacterium capsulatum*, and three newly described species, *Sphingomonas parapaucimobilis, S. yanoikuyae,* and *S. adhaesiva.* Since 1990, the genus description has been emended by Yabuuchi et al. (1999), Takeuchi et al. (2001), Yabuuchi et al. (2002), and Busse et al. (2003a); 30 new species have been described; and *Pseudomonas echinoides, Blastomonas natatoria, Rhizomonas suberifaciens,* and *Erythromonas ursincola* have been transferred into the genus. The main area of disagreement is whether or not there are sufficient grounds to subdivide the genus. In 2001, Takeuchi et al. proposed the division of *Sphingomonas* into four genera: *Sphingomonas sensu stricto, Sphingopyxis, Sphingobium,* and *Novosphingobium,* based on analyses of 16S rDNA sequences, fatty acid and polyamine profiles, and the ability to reduce nitrate. In 2002, Yabuuchi et al. analyzed additional phenotypic characteristics and concluded first, that the phenotypic data did not support the new genera proposed by Takeuchi et al. (2001) and second, that *Sphingopyxis, Sphingobium,* and *Novosphingobium* should be regarded as junior objective synonyms of *Sphingomonas.* In this edition of the *Systematics*, we will present the latter viewpoint.

SALMONELLA NOMENCLATURE Nomenclature in the genus *Salmonella* was in a state of confusion at the time of publication of the first edition of the *Systematics* (Le Minor, 1984) and has remained so ever since. Five species appeared on the Approved Lists: *Salmonella arizonae, S. choleraesuis, S. enteritidis, S. typhi,* and *S. typhimurium.* However, it was recognized that these species were so closely related that combination into a single species was advisable, and in the first edition of the *Systematics*, the members of the genus were presented as a list of "selected serovars" (Le Minor, 1984). In 1999, Euzéby provided a history of several attempts to stabilize the nomenclature. In this volume, Popoff and Le Minor have followed the nomenclature that they proposed in 1987, with the exception that *Salmonella enterica* subsp. *bongori* has since been accorded species status (Reeves et al., 1989). There are two species, *Salmonella enterica* (the type species, formerly *Salmonella choleraesuis* subsp. *choleraesuis*) and *Salmonella bongori*; *Salmonella enterica* contains six subspecies: subsp. *enterica*, subsp. *arizonae*, subsp. *diarizonae*, subsp. *houtenae*, subsp. *indica*, and subsp. *salamae*). *Salmonella enterica* subsp. *enterica* is further subdivided into serovar Choleraesuis, serovar Enteritidis, serovar Gallinarum, serovar Paratyphi A, serovar Paratyphi B, serovar Paratyphi C, serovar Typhi, and serovar Typhimurium. A request for an opinion by Le Minor and Popoff (1987) to adopt this system of nomenclature for *Salmonella* species was initially rejected by the Judicial Commission on the grounds that the Request should have been centered on the nomenclature of the organisms rather than on issues of taxonomy. Despite this apparent rejection, the naming system proposed by Popoff and Le Minor has been widely adopted by many scientists in the field.

As this volume went to press, four Requests for Opinions regarding the nomenclature of *Salmonella* were pending before the Judicial Commission. Euzéby's (1999) request includes the following actions: (1) that the name *Salmonella choleraesuis* be rejected because a species, a subspecies and a serovar have all been named *choleraesuis*; (2) that *Salmonella enterica* be recognized as the type species of the genus, with six subspecies: subsp. *enterica* (subdivided into serovar Enteriditis and serovar Typimurium), subsp. *arizonae*, subsp. *diarizonae*, subsp. *houtenae*, subsp. *indica*, and subsp. *salamae*; and (3) that the species name *Salmonella typhi* be conserved because of the importance of this name in medical communications. Yabuuchi and Ezaki (2000) have recommended against changing the name of the type species of *Salmonella* from *Salmonella choleraesuis* to *Salmonella enterica*. Ezaki et al. (2000a) have requested that the name *Salmonella choleraesuis* subsp. *choleraesuis* serovar Paratyphi A be changed to the conserved name *Salmonella paratyphi* because of the importance of this name in medical communications. Ezaki et al. (2000b) have requested that the names *Salmonella choleraesuis* subsp. *choleraesuis* serovar Enteriditis, serovar Typhi, and serovar Typhimurium be changed to the conserved names *Salmonella enteritidis*, *S. typhi*, and *S. typhimurium*, respectively, again because of the medical importance of these organisms.

The taxonomic outline As noted above, the Taxonomic Outline, now in its fourth revision (http://dx.doi.org/10.1007/bergeysoutline), was intended to serve as a focal point for discussion by the community about our emerging views of the global taxonomy of the *Bacteria* and the *Archaea*. To a large extent, we believe that this has occurred, and over 480 comments have been added to the document. In addition, since the Taxonomic Outline was first released, we have added over 1600 validly published names of new taxa and new combinations. At the recommendation of our authors and of members of the broader taxonomic community, we have also introduced a number of rearrangements that address misplacements and highlight areas of disagreement. One issue that is particularly noteworthy is that the placement of some taxa remains problematic, even when high-quality 16S rDNA sequence data is available. Thus, there are still a number of taxa accorded the status of *incertae sedis* at the family or order level. As new data become available, more precise placement is likely to occur. Higher taxa and genera that have been moved are listed in Table 2, along with the reasons for these moves. While we believe that the current taxonomy is a better reflection of reality than previous versions, we expect that further changes will be made as we work our way through the subsequent volumes and plan for future editions of the *Systematics*.

Acknowledgments

We extend our thanks to Dr. Peter Sneath of the University of Leicester, who provided invaluable assistance in building and testing evolutionary models and statistical concepts that were vital in compilation of this chapter. We would also like to acknowledge Drs. Wolfgang Ludwig and Karl-Heinz Schleifer of the Technical University of Munich for their numerous and helpful discussions on resolving discrepancies between the early versions of *Bergey's Taxonomic Outline* and the phylogenetic reference trees. We would also like to extend our thanks to Drs. Jean Euzéby, Philip Hugenholtz, and Brian J. Tindall for their efforts in reviewing the Taxonomic Outline and for their helpful discussions and constructive comments. We are also indebted to those authors and colleagues who have provided their input in helping us to correct various errors in various releases of the Outline. This work was supported in part by a grant from the U.S. Department of Energy Office of Biological and Environmental Research (DOE # DE-FG02-02ER63315).

We are also indebted to J. Euzéby, B.J. Tindall, and J.-F. Bernardet for bringing various errors in the Taxonomic Outline to our attention and to Conrad Istock for a critical reading of this manuscript.

Appendix 1. New and emended taxa described since publication of Volume One, Second Edition of the Systematics

Basonyms and synonyms[1]

Bacillus thermodenitrificans (ex Klaushofer and Hollaus 1970) Manachini et al. 2000, 1336^{VP}

Blastomonas ursincola (Yurkov et al. 1997) Hiraishi et al. 2000a, 1117^{VP}

Cellulophaga uliginosa (ZoBell and Upham 1944) Bowman 2000, 1867^{VP}

Dehalospirillum Scholz-Muramatsu et al. 2002, 1915^{VP} (Effective publication: Scholz-Muramatsu et al., 1995)

Dehalospirillum multivorans Scholz-Muramatsu et al. 2002, 1915^{VP} (Effective publication: Scholz-Muramatsu et al., 1995)

Desulfotomaculum auripigmentum Newman et al. 2000, 1415^{VP} (Effective publication: Newman et al., 1997)

Enterococcus porcinus^{VP} Teixeira et al. 2001 pro synon. *Enterococcus villorum* Vancanneyt et al. 2001b, 1742^{VP} De Graef et al., 2003

Hongia koreensis Lee et al. 2000d, 197^{VP}

Mycobacterium bovis subsp. *caprae* (Aranaz et al. 1999) Niemann et al. 2002, 435^{VP}

Natronobacterium nitratireducens Xin et al. 2001, 1828^{VP}

Novosphingobium Takeuchi et al. 2001, 1415^{VP}

Saccharococcus caldoxylosilyticus Ahmad et al. 2000, 522^{VP}

Saccharothrix violacea Lee et al. 2000e, 1320^{VP}

Salibacillus marismortui (Arahal et al. 1999) Arahal et al. 2000, 1503^{VP}

Sinorhizobium adhaerens (Casida 1982) Willems et al. 2003, 1215 (Request for an Opinion)^{VP}

Sinorhizobium kummerowiae Wei et al. 2002, 2237^{VP}

Sinorhizobium morelense Wang et al. 2002, 1691^{VP}

Sphingobium Takeuchi et al. 2001, 1415^{VP}

Sphingomonas alaskensis Vancanneyt et al. 2001a, 78^{VP}

Sphingopyxis Takeuchi et al. 2001, 1415^{VP}

Streptococcus pasteurianus Poyart et al. 2002, 1253^{VP}

Subtercola pratensis Behrendt et al. 2002, 1452^{VP}

Vibrio viscosus Lunder et al. 2000, 447^{VP}

New combinations

Acetobacter estunensis (Carr 1958) Lisdiyanti et al. 2001b, 263^{VP} (Effective publication: Lisdiyanti et al., 2000)

Acetobacter lovaniensis (Frateur 1950) Lisdiyanti et al. 2001b, 263 (Effective publication: Lisdiyanti et al., 2000)^{VP}

Acetobacter orleanensis (Henneberg 1906) Lisdiyanti et al. 2001b, 263^{VP} (Effective publication: Lisdiyanti et al., 2000)

Achromobacter denitrificans (Rüger and Tan 1983) Coenye et al. 2003b, 1829^{VP}

Acidithiobacillus albertensis (Bryant et al. 1988) Kelly and Wood 2000, 514^{VP}

Acidithiobacillus caldus (Hallberg and Lindström 1995) Kelly and Wood 2000, 514^{VP})

Acidithiobacillus ferrooxidans (Temple and Colmer 1951) Kelly and Wood 2000, 513^{VP}

Acidithiobacillus thiooxidans (Waksman and Joffe 1922) Kelly and Wood 2000, 513^{VP}

Acrocarpospora corrugata (Williams and Sharples 1976) Tamura et al. 2000a, 1170^{VP}

Actinocorallia aurantiaca (Lavrova and Preobrazhenskaya 1975) Zhang et al. 2001, 381^{VP}

Actinocorallia glomerata (Itoh et al. 1996) Zhang et al. 2001, 381^{VP}

Actinocorallia libanotica (Meyer 1981) Zhang et al. 2001, 381^{VP}

Actinocorallia longicatena (Itoh et al. 1996) Zhang et al. 2001, 381^{VP}

Actinomadura viridilutea (Agre and Guzeva 1975) Zhang et al. 2001, 381^{VP}

Agreia pratensis (Behrendt et al. 2002) Schumann et al. 2003, 2043^{VP}

Alcanivorax jadensis (Bruns and Berthe-Corti 1999) Fernández-Martínez et al. 2003, 337^{VP}

Alistipes putredinis (Weinberg et al. 1937) Rautio et al. 2003b, 1701^{VP} (Effective publication: Rautio et al., 2003a)

Anaerococcus hydrogenalis (Ezaki et al. 1990) Ezaki et al. 2001, 1526^{VP}

Anaerococcus lactolyticus (Li et al. 1992) Ezaki et al. 2001, 1527^{VP}

Anaerococcus octavius (Murdoch et al. 1997) Ezaki et al. 2001, 1527^{VP}

Anaerococcus prevotii (Foubert and Douglas 1948) Ezaki et al. 2001, 1526^{VP}

Anaerococcus tetradius (Ezaki et al. 1983) Ezaki et al. 2001, 1526^{VP}

Anaerococcus vaginalis (Li et al. 1992) Ezaki et al. 2001, 1527^{VP}

Anaplasma phagocytophilum (Foggie 1949) Dumler et al. 2001, 2158^{VP}

Asanoa ferruginea (Kawamoto 1986) Lee and Hah 2002, 970^{VP}

Bacteriovorax starrii (Seidler et al. 1972) Baer et al. 2000, 223^{VP}

Bacteriovorax stolpii (Seidler et al. 1972) Baer et al. 2000, 223^{VP}

Carnobacterium maltaromaticum (Miller et al. 1974) Mora et al. 2003, 677^{VP}

Cellulomonas humilata (Gledhill and Casida 1969) Collins and Pascual 2000, 662^{VP}

Cellulosimicrobium cellulans (Metcalf and Brown 1957) Schumann et al. 2001, 1009^{VP}

Chlorobaculum tepidum (Wahlund et al. 1996) Imhoff 2003, 950^{VP}

Chlorobium clathratiforme (Szafer 1911) Imhoff 2003, 948^{VP}

Chlorobium luteolum (Schmidle 1901) Imhoff 2003, 948^{VP}

Chromohalobacter canadensis (Huval et al. 1996) Arahal et al. 2001a, 1447^{VP}

Chromohalobacter israelensis (Huval et al. 1996) Arahal et al. 2001a, 1447^{VP}

Clostridium estertheticum subsp. *laramiense* (Kalchayanand et al. 1993) Spring et al. 2003, 1028^{VP}

Clostridium stercorarium subsp. *leptospartum* (Toda et al. 1989) Fardeau et al. 2001, 1130^{VP}

Clostridium stercorarium subsp. *thermolacticum* (Le Ruyet et al. 1988) Fardeau et al. 2001, 1130^{VP}

Cobetia marina (Cobet et al. 1970) Arahal et al. 2002b, 1915^{VP} (Effective publication: Arahal et al., 2002a)

Comamonas aquatica (Hylemon et al. 1973) Wauters et al. 2003b, 861^{VP}

Crossiella cryophila (Labeda and Lechevalier 1989) Labeda 2001, 1579^{VP}

Cryptosporangium minutisporangium (Ruan et al. 1986) Tamura and Hatano 2001, 2123^{VP}

---

1. *Citations for emendations and for the original authorities for basonyms, synonyms, and new combinations do not appear in the bibliography unless cited elsewhere in this book.

Dendrosporobacter quercicolus (Stankewich et al. 1971) Strömpl et al. 2000, 105^{VP}

Desulfobacula phenolica (Bak and Widdel 1988) Kuever et al. 2001, 175^{VP}

Desulfomicrobium macestii (Gogotova and Vainstein 1989) Hippe et al. 2003, 1129^{VP}

Desulfosporosinus auripigmenti (Newman et al. 2000) Stackebrandt et al. 2003, 1442^{VP}

Desulfovibrio piger (Moore et al. 1976) Loubinoux et al. 2002, 1307^{VP}

Dorea formicigenerans (Holdeman and Moore 1974) Taras et al. 2002, 426^{VP}

Ehrlichia ruminantium (Cowdry 1925) Dumler et al. 2001, 2158^{VP}

Ensifer arboris (Nick et al. 1999) Young 2003, 2109^{VP}

Ensifer fredii (Scholla and Elkan 1984) Young 2003, 2109^{VP}

Ensifer kostiensis (Nick et al. 1999) Young 2003, 2109^{VP}

Ensifer kummerowiae (Wei et al. 2002) Young 2003, 2109^{VP}

Ensifer medicae (Rome et al. 1996) Young 2003, 2109^{VP}

Ensifer meliloti (Dangeard 1926) Young 2003, 2109^{VP}

Ensifer saheli (de Lajudie et al. 1994) Young 2003, 2109^{VP}

Ensifer terangae (de Lajudie et al. 1994) Young 2003, 2109^{VP}

Ensifer xinjiangensis (Chen et al. 1988) Young 2003, 2109^{VP}

Faecalibacterium prausnitzii (Hauduroy et al. 1937) Duncan et al. 2002b, 2145^{VP}

Finegoldia magna (Prevot 1933) Murdoch and Shah 2000, 1415^{VP} (Effective publication: Murdoch and Shah, 1999)

Gallibacterium anatis (Mutters et al. 1985) Christensen et al. 2003, 285^{VP}

Gallicola barnesae (Schiefer-Ullrich and Andreesen 1986) Ezaki et al. 2001, 1527^{VP}

Geobacillus caldoxylosilyticus (Ahmad et al. 2000) Fortina et al. 2001a, 2069^{VP}

Geobacillus kaustophilus (Priest et al. 1989) Nazina et al. 2001, 444^{VP}

Geobacillus stearothermophilus (Donk 1920) Nazina et al. 2001, 443^{VP}

Geobacillus thermocatenulatus (Golovacheva 1991) Nazina et al. 2001, 444^{VP}

Geobacillus thermodenitrificans (Manachini et al. 2000) Nazina et al. 2001, 444^{VP}

Geobacillus thermoglucosidasius (Suzuki 1984) Nazina et al. 2001, 444^{VP}

Geobacillus thermoleovorans (Zarilla and Perry 1988) Nazina et al. 2001, 444^{VP}

Gluconacetobacter intermedius (Boesch et al. 1998) Yamada 2000, 2226^{VP}

Gluconacetobacter oboediens (Sokollek et al. 1998) Yamada 2000, 2226^{VP}

Granulicatella adiacens (Bouvet et al. 1989) Collins and Lawson 2000, 367^{VP}

Granulicatella balaenopterae (Lawson et al. 1999) Collins and Lawson 2000, 368^{VP}

Granulicatella elegans (Roggenkamp et al. 1999) Collins and Lawson 2000, 367^{VP}

Grimontia hollisae (Hickman et al. 1982) Thompson et al. 2003a, 1617^{VP}

Halobiforma nitratireducens (Xin et al. 2001) Hezayen et al. 2002, 2278^{VP}

Halomicrobium mukohataei (Ihara et al. 1997) Oren et al. 2002, 1834^{VP}

Halothiobacillus halophilus (Wood and Kelly 1995) Kelly and Wood 2000, 515^{VP}

Halothiobacillus hydrothermalis (Durand et al. 1997) Kelly and Wood 2000, 515^{VP}

Halothiobacillus neapolitanus (Parker 1957) Kelly and Wood 2000, 515^{VP}

Hydrogenobacter hydrogenophilus (Kryukov et al. 1984) Stöhr et al. 2001b, 1860^{VP}

Hydrogenobaculum acidophilum (Shima and Suzuki 1993) Stöhr et al. 2001b, 1860^{VP}

Kitasatospora kifunensis (Nakagaito et al. 1993) Groth et al. 2003, 2038^{VP}

Kribbella koreensis (Lee et al. 2000) Sohn et al. 2003, 1007^{VP}

Lamprocystis purpurea (Eichler and Pfennig 1989) Imhoff 2001b, 1700^{VP}

Lechevalieria aerocolonigenes (Labeda 1986) Labeda et al. 2001, 1050^{VP}

Lechevalieria flava (Gauze et al. 1974) Labeda et al. 2001, 1050^{VP}

Leifsonia aquatica (ex Leifson 1962) Evtushenko et al. 2000a, 377^{VP}

Leifsonia cynodontis (Davis et al. 1984) Suzuki et al. 2000a, 1415^{VP} (Effective publication: Suzuki et al., 1999a)

Leifsonia xyli subsp. *cynodontis* (Davis et al. 1984) Evtushenko et al. 2000a, 378^{VP}

Leifsonia xyli subsp. *xyli* (Davis et al. 1984) Evtushenko et al. 2000a, 378^{VP}

Lentzea violacea (Lee et al. 2000) Labeda et al. 2001, 1049^{VP}

Lentzea waywayandensis (Labeda and Lyons 1989) Labeda et al. 2001, 1049^{VP}

Leuconostoc fructosum (Kodama 1956) Antunes et al. 2002, 654^{VP}

Marinibacillus marinus (Rüger and Richter 1979) Yoon et al. 2001e, 2092^{VP}

Marinobacterium jannaschii (Bowditch et al. 1984) Satomi et al. 2002, 745^{VP}

Marinobacterium stanieri (Baumann et al. 1983) Satomi et al. 2002, 746^{VP}

Metallosphaera hakonensis (Takayanagi et al. 1996) Kurosawa et al. 2003, 1608^{VP}

Methanocaldococcus fervens (Jeanthon et al. 1999) Whitman 2002, 686^{VP} (Effective publication: Whitman, 2001a)

Methanocaldococcus infernus (Jeanthon et al. 1998) Whitman 2002, 686^{VP} (Effective publication: Whitman, 2001a)

Methanocaldococcus jannaschii (Jones et al. 1984) Whitman 2002, 686^{VP} (Effective publication: Whitman, 2001a)

Methanocaldococcus vulcanius (Jeanthon et al. 1999) Whitman 2002, 686^{VP} (Effective publication: Whitman, 2001a)

Methanolobus oregonensis (Liu et al. 1990) Boone 2002, 686^{VP} (Effective publication: Boone, 2001c)

Methanosalsum zhilinae (Mathrani et al. 1988) Boone and Baker 2002, 686^{VP} (Effective publication: Boone and Baker, 2001)

Methanothermobacter defluvii (Kotelnikova et al. 1994) Boone 2002, 686^{VP} (Effective publication: Boone, 2001d)

Methanothermobacter thermautotrophicus (Zeikus and Wolfe 1972) Wasserfallen et al. 2000, 51^{VP}

Methanothermobacter thermoflexus (Kotelnikova et al. 1994) Boone 2002, 686^{VP} (Effective publication: Boone, 2001d)

Methanothermobacter thermophilus (Laurinavichus et al. 1990) Boone 2002, 686^{VP} (Effective publication: Boone, 2001d)

Methanothermobacter wolfei (Winter et al. 1985) Wasserfallen et al. 2000, 51^{VP}

Methanothermococcus thermolithotrophicus (Huber et al. 1984) Whitman 2002, 687^{VP} (Effective publication: Whitman, 2001b)

Methanotorris igneus (Burggraf et al. 1990) Whitman 2002, 687^{VP} (Effective publication: Whitman, 2001b)

Microbacterium resistens (Funke et al. 1998) Behrendt et al. 2001, 1275[VP]

Microbulbifer elongatus (Humm 1946) Yoon et al. 2003e, 1360[VP]

Micromonas micros (Prevot 1933) Murdoch and Shah 2000, 1415[VP] (Effective publication: Murdoch and Shah, 1999)

Micromonospora matsumotoense (Asano et al. 1989) Lee et al. 2000b, 3[VP] (Effective publication: Lee et al., 1999)

Micromonospora nigra (Weinstein et al. 1968) Kasai et al. 2000, 131[VP]

Micromonospora pallida (Luedemann and Brodsky 1964) Kasai et al. 2000, 131[VP]

Mogibacterium timidum (Holdeman et al. 1980) Nakazawa et al. 2000, 686[VP]

Moritella viscosa (Lunder et al. 2000) Benediktsdóttir et al. 2000, 487[VP]

Mycobacterium caprae (Aranaz et al. 1999) Aranaz et al. 2003, 1788[VP]

Mycoplasma haemocanis (Kreier and Ristic 1984) Messick et al. 2002, 697[VP]

Mycoplasma haemofelis (Kreier and Ristic 1984) Neimark et al. 2002, 683[VP]

Mycoplasma haemomuris (Mayer 1921) Neimark et al. 2002, 683[VP]

Mycoplasma suis (Splitter 1950) Neimark et al. 2002, 683[VP]

Mycoplasma wenyonii (Adler and Ellenbogen 1934) Neimark et al. 2002, 683[VP]

Neorickettsia risticii (Holland et al. 1985) Dumler et al. 2001, 2159[VP]

Neorickettsia sennetsu (Misao and Kobayashi 1956) Dumler et al. 2001, 2159[VP]

Nonomuraea roseoviolacea subsp. *carminata* (Gauze et al. 1973) Gyobu and Miyadoh 2001, 887[VP]

Novosphingobium aromaticivorans (Balkwill et al. 1997) Takeuchi et al. 2001, 1415[VP]

Novosphingobium capsulatum (Leifson 1962) Takeuchi et al. 2001, 1415[VP]

Novosphingobium rosa (Takeuchi et al. 1995) Takeuchi et al. 2001, 1415[VP]

Novosphingobium stygium (Balkwill et al. 1997) Takeuchi et al. 2001, 1415[VP]

Novosphingobium subarcticum (Nohynek et al. 1996) Takeuchi et al. 2001, 1415[VP]

Novosphingobium subterraneum (Balkwill et al. 1997) Takeuchi et al. 2001, 1415[VP]

Oceanimonas doudoroffii (Baumann et al. 1972) Brown et al. 2001c, 71[VP]

Oceanobacter kriegii (Bowditch et al. 1984) Satomi et al. 2002, 745[VP]

Oerskovia enterophila (Jáger et al. 1983) Stackebrandt et al. 2002a, 1110[VP]

Olsenella uli (Olsen et al. 1991) Dewhirst et al. 2001, 1803[VP]

Pandoraea norimbergensis (Wittke et al. 1998) Coenye et al. 2000, 896[VP]

Parascardovia denticolens (Crociani et al. 1996) Jian and Dong 2002, 811[VP]

Paucimonas lemoignei (Delafield et al. 1965) Jendrossek 2001, 907[VP]

Pectobacterium atrosepticum (van Hall 1902) Gardan et al. 2003a, 390[VP]

Pectobacterium betavasulorum (Thomson et al. 1984) Gardan et al. 2003a, 390[VP]

Pectobacterium wasabiae (Goto and Mazumoto 1987) Gardan et al. 2003a, 390[VP]

Peptoniphilus asaccharolyticus (Distaso 1912) Ezaki et al. 2001, 1525[VP]

Peptoniphilus harei (Murdoch et al. 1997) Ezaki et al. 2001, 1526[VP]

Peptoniphilus indolicus (Christiansen 1934) Ezaki et al. 2001, 1525[VP]

Peptoniphilus ivorii (Murdoch et al. 1997) Ezaki et al. 2001, 1526[VP]

Peptoniphilus lacrimalis (Li et al. 1992) Ezaki et al. 2001, 1526[VP]

Planomicrobium mcmeekinii (Junge et al. 1998) Yoon et al. 2001c, 1519[VP]

Planomicrobium okeanokoites (ZoBell and Upham 1944) Yoon et al. 2001c, 1518[VP]

Propionimicrobium lymphophilum (Torrey 1916) Stackebrandt et al. 2002c, 1926[VP]

Propionivibrio pelophilus (Meijer et al. 1999) Brune et al. 2002b, 444[VP]

Prosthecochloris vibrioformis (Pelsh 1936) Imhoff 2003, 949[VP]

Pseudoalteromonas distincta (Romanenko et al. 1995) Ivanova et al. 2000a, 143[VP]

Pseudoalteromonas elyakovii (Ivanova et al. 1997) Sawabe et al. 2000, 270[VP]

Pseudoalteromonas tetraodonis (Simidu et al. 1990) Ivanova et al. 2001b, 1077[VP]

Pseudonocardia alaniniphila (Xu et al. 1999) Huang et al. 2002, 981[VP]

Pseudonocardia aurantiaca (Xu et al. 1999) Huang et al. 2002, 981[VP]

Pseudonocardia xinjiangensis (Xu et al. 1999) Huang et al. 2002, 981[VP]

Pseudonocardia yunnanensis (Jiang et al. 1991) Huang et al. 2002, 981[VP]

Pseudorhodobacter ferrugineus (Rüger and Höfle 1992) Uchino et al. 2003, 936[VP] (Effective publication: Uchino et al., 2002b)

Pseudospirillum japonicum (Watanabe 1959) Satomi et al. 2002, 745[VP]

Raoultella ornithinolytica (Sakazaki et al. 1989) Drancourt et al. 2001, 931[VP]

Raoultella planticola (Bagley et al. 1982) Drancourt et al. 2001, 931[VP]

Raoultella terrigena (Izard et al. 1981) Drancourt et al. 2001, 931[VP]

Rhizobium radiobacter (Beijerinck and van Delden 1902) Young et al. 2001, 99[VP]

Rhizobium rhizogenes (Riker et al. 1930) Young et al. 2001, 99[VP]

Rhizobium rubi (Hildebrand 1940) Young et al. 2001, 99[VP]

Rhizobium undicola (de Lajudie et al. 1998) Young et al. 2001, 99[VP]

Rhizobium vitis (Ophel and Kerr 1990) Young et al. 2001, 99[VP]

Rhodoblastus acidophilus (Pfennig 1969) Imhoff 2001c, 1865[VP]

Rhodococcus wratislaviensis (Goodfellow et al. 1995) Goodfellow et al. 2002, 752[VP]

Rothia mucilaginosa (Bergan and Kocur 1982) Collins et al. 2000c, 1250[VP]

Saccharothrix albidocapillata (Yassin et al. 1995) Lee et al. 2000e, 1322[VP]

Salegentibacter salegens (Dobson et al. 1993) McCammon and Bowman 2000, 1062[VP]

Scardovia inopinata (Crociani et al. 1996) Jian and Dong 2002, 811[VP]

Sedimentibacter hydroxybenzoicus (Zhang et al. 1994) Breitenstein et al. 2002, 806[VP]

Serratia quinivorans (Grimont et al. 1983) Ashelford et al. 2002, 2288[VP]

Sphingobium chlorophenolicum (Nohynek et al. 1996) Takeuchi et al. 2001, 1415^{VP}
Sphingobium herbicidovorans (Zipper et al. 1997) Takeuchi et al. 2001, 1415^{VP}
Sphingobium yanoikuyae (Yabuuchi et al. 1990) Takeuchi et al. 2001, 1415^{VP}
Sphingopyxis alaskensis (Vancanneyt et al. 2001) Godoy et al. 2003, 476^{VP}
Sphingopyxis macrogoltabida (Takeuchi et al. 1993) Takeuchi et al. 2001, 1416^{VP}
Sphingopyxis terrae (Takeuchi et al. 1993) Takeuchi et al. 2001, 1416^{VP}
Sporosarcina globispora (Larkin and Stokes 1967) Yoon et al. 2001d, 1085^{VP}
Sporosarcina pasteurii (Miquel 1889) Yoon et al. 2001d, 1085^{VP}
Sporosarcina psychrophila (Nakamura 1984) Yoon et al. 2001d, 1085^{VP}
Starkeya novella (Starkey 1934) Kelly et al. 2000, 1800^{VP}
Streptococcus gallolyticus subsp. *macedonicus* (Tsakalidou et al. 1998) Schlegel et al. 2003, 643^{VP}
Streptococcus gallolyticus subsp. *pasteurianus* (Poyart et al. 2002) Schlegel et al. 2003, 643^{VP}
Sulfurospirillum multivorans (Scholz-Muramatsu et al. 2002) Luijten et al. 2003, 791^{VP}
Tannerella forsythensis (Tanner et al. 1986) Sakamoto et al. 2002, 848^{VP}
Tenacibaculum maritimum (Wakabayashi et al. 1986) Suzuki et al. 2001, 1650^{VP}
Tenacibaculum ovolyticum (Hansen et al. 1992) Suzuki et al. 2001, 1650^{VP}
Terasakiella pusilla (Terasaki 1973) Satomi et al. 2002, 745^{VP}
Thermithiobacillus tepidarius (Wood and Kelly 1985) Kelly and Wood 2000, 515^{VP}
Trichococcus palustris (Zhilina et al. 1997) Liu et al. 2002a, 1125^{VP}
Trichococcus pasteurii (Schink 1985) Liu et al. 2002a, 1125^{VP}
Ureibacillus thermosphaericus (Andersson et al. 1996) Fortina et al. 2001b, 453^{VP}
Virgibacillus marismortui (Arahal et al. 1999) Heyrman et al. 2003b, 510^{VP}
Virgibacillus salexigens (Garabito et al. 1997) Heyrman et al. 2003b, 510^{VP}
Zobellia uliginosa (ZoBell and Upham 1944) Barbeyron et al. 2001, 995^{VP}

New Phyla

Aquificae Reysenbach 2002, 685^{VP}(Effective publication: Reysenbach, 2001k)
Chloroflexi Garrity and Holt 2002, 685^{VP} (Effective publication: Garrity and Holt, 2001o)
Chrysiogenetes Garrity and Holt 2002, 685^{VP} (Effective publication: Garrity and Holt, 2001n)
Crenarchaeota Garrity and Holt 2002, 685^{VP} (Effective publication: Garrity and Holt, 2001j)
Deferribacteres Garrity and Holt 2002, 685^{VP} (Effective publication: Garrity and Holt, 2001m)
Euryarchaeota Garrity and Holt 2002, 685^{VP} (Effective publication: Garrity and Holt, 2001k)
Gemmatimonadetes Zhang et al. 2003b, 1161^{VP}
Thermodesulfobacteria Garrity and Holt 2002, 687^{VP} (Effective publication: Garrity and Holt, 2001l)
Thermomicrobia Garrity and Holt 2002, 687^{VP} (Effective publication: Garrity and Holt, 2001p)
Thermotogae Reysenbach 2002, 687^{VP} (Effective publication: Reysenbach, 2001l)

New Classes

Aquificae Reysenbach 2002, 685^{VP} (Effective publication: Reysenbach, 2001a)
Archaeoglobi Garrity and Holt 2002, 685^{VP} (Effective publication: Garrity and Holt, 2001d)
Chrysiogenetes Garrity and Holt 2002, 685^{VP} (Effective publication: Garrity and Holt, 2001a)
Deferribacteres Huber and Stetter 2002, 685^{VP} (Effective publication: Huber and Stetter, 2001a)
Deinococci Garrity and Holt 2002, 685^{VP} (Effective publication: Garrity and Holt, 2001b)
Gemmatimonadetes Zhang et al. 2003b, 1161^{VP}
Halobacteria Grant et al. 2002, 685^{VP} (Effective publication: Grant et al., 2001)
Methanobacteria Boone 2002, 686^{VP} (Effective publication: Boone, 2001a)
Methanococci Boone 2002, 686^{VP} (Effective publication: Boone, 2001b)
Methanopyri Garrity and Holt 2002, 686^{VP} (Effective publication: Garrity and Holt, 2001e)
Thermococci Zillig and Reysenbach 2002, 687^{VP} (Effective publication: Zillig and Reysenbach, 2001)
Thermodesulfobacteria Hatchikian et al. 2002, 687^{VP} (Effective publication: Hatchikian et al., 2001a)
Thermomicrobia Garrity and Holt 2002, 687^{VP} (Effective publication: Garrity and Holt, 2001c)
Thermoplasmata Reysenbach 2002, 687^{VP} (Effective publication: Reysenbach, 2001d)
Thermoprotei Reysenbach 2002, 687^{VP} (Effective publication: Reysenbach, 2001b)
Thermotogae Reysenbach 2002, 687^{VP} (Effective publication: Reysenbach, 2001c)

New Orders

Aquificales Reysenbach 2002, 685^{VP} (Effective publication: Reysenbach, 2001h)
Archaeoglobales Huber and Stetter 2002, 685^{VP} (Effective publication: Huber and Stetter, 2001e)
Chrysiogenales Garrity and Holt 2002, 685^{VP} (Effective publication: Garrity and Holt, 2001h)
Deferribacterales Huber and Stetter 2002, 685^{VP} (Effective publication: Huber and Stetter, 2001f)
Desulfurococcales Huber and Stetter 2002, 685^{VP} (Effective publication: Huber and Stetter, 2001h)
Gemmatimonadales Zhang et al. 2003b, 1161^{VP}
Methanopyrales Huber and Stetter 2002, 686^{VP} (Effective publication: Huber and Stetter, 2001g)
Methanosarcinales Boone et al. 2002, 686^{VP} (Effective publication: Boone et al., 2001c)
Thermales Rainey and da Costa 2002, 687^{VP} (Effective publication: Rainey and da Costa, 2001)
Thermodesulfobacteriales Hatchikian et al. 2002, 687^{VP} (Effective publication: Hatchikian et al., 2001c)
Thermomicrobiales Garrity and Holt 2002, 687^{VP} (Effective publication: Garrity and Holt, 2001i)

Thermoplasmatales Reysenbach 2002, 687^{VP} (Effective publication: Reysenbach, 2001i)

Thermotogales Reysenbach 2002, 687^{VP} (Effective publication: Reysenbach, 2001j)

New Families

Actinosynnemataceae Labeda and Kroppenstedt 2000, 335^{VP}

Alteromonadaceae Ivanova and Mikhailov 2001b, 1229^{VP} (Effective publication: Ivanova and Mikhailov, 2001a)

Aquificaceae Reysenbach 2002, 685^{VP} (Effective publication: Reysenbach, 2001e)

Archaeoglobaceae Huber and Stetter 2002, 685^{VP} (Effective publication: Huber and Stetter, 2001b)

Bogoriellaceae Stackebrandt and Schumann 2000, 1283^{VP}

Chrysiogenaceae Garrity and Holt 2002, 685^{VP} (Effective publication: Garrity and Holt, 2001f)

Cryomorphaceae Bowman et al. 2003, 1353^{VP}

Deferribacteraceae Huber and Stetter 2002, 685^{VP} (Effective publication: Huber and Stetter, 2001c)

Dermacoccaceae Stackebrandt and Schumann 2000, 1283^{VP}

Ferroplasmaceae Golyshina et al., 2000, 1004^{VP}

Gemmatimonadaceae Zhang et al. 2003b, 1161^{VP}

Methanocaldococcaceae Whitman et al., 2002, 686^{VP} (Effective publication: Whitman et al., 2001)

Methanopyraceae Huber and Stetter 2002, 686^{VP} (Effective publication: Huber and Stetter, 2001d)

Methanosaetaceae Boone et al. 2002, 686^{VP} (Effective publication: Boone et al., 2001a)

Methanospirillaceae Boone et al. 2002, 686^{VP} (Effective publication: Boone et al., 2001b)

Oleiphilaceae Golyshin et al. 2002, 909^{VP}

Oscillochloridaceae Keppen et al., 2000, 1534^{VP}

Rarobacteraceae Stackebrandt and Schumann 2000, 1284^{VP}

Sanguibacteraceae Stackebrandt and Schumann 2000, 1284^{VP}

Sphingomonadaceae Kosako et al. 2000b, 1953^{VP} (Effective publication: Kosako et al., 2000a)

Thermaceae da Costa and Rainey 2002, 687^{VP} (Effective publication: da Costa and Rainey, 2001)

Thermodesulfobacteriaceae Hatchikian et al. 2002, 687^{VP} (Effective publication: Hatchikian et al., 2001b)

Thermomicrobiaceae Garrity and Holt 2002, 687^{VP} (Effective publication: Garrity and Holt, 2001g)

Thermoplasmataceae Reysenbach 2002, 687^{VP} (Effective publication: Reysenbach, 2001f)

Thermotogaceae Reysenbach 2002, 687^{VP} (Effective publication: Reysenbach, 2001g)

New Genera

Acidilobus Prokofeva et al. 2000, 2007^{VP}

Acidisphaera Hiraishi et al. 2000b, 1545^{VP}

Acidithiobacillus Kelly and Wood 2000, 513^{VP}

Acrocarpospora Tamura et al. 2000a, 1170^{VP}

Actinoalloteichus Tamura et al. 2000b, 1439^{VP}

Actinopolymorpha Wang et al. 2001b, 471^{VP}

Aequorivita Bowman and Nichols 2002, 1538^{VP}

Agreia Evtushenko et al. 2001, 2077^{VP}

Albibacter Doronina et al. 2001b, 1056^{VP}

Albidovulum Albuquerque et al. 2003, 1^{VP} (Effective publication: Albuquerque et al., 2002)

Algoriphagus Bowman et al. 2003, 1351^{VP}

Alicycliphilus Mechichi et al. 2003, 149^{VP}

Alishewanella Fonnesbech Vogel et al. 2000, 1140^{VP}

Alistipes Rautio et al. 2003b, 1701^{VP} (Effective publication: Rautio et al., 2003a)

Alkalibacterium Ntougias and Russell 2001, 1169^{VP}

Alkalilimnicola Yakimov et al. 2001, 2142^{VP}

Alkaliphilus Takai et al. 2001b^{VP} emend. Cao et al., 2003

Alkalispirillum Rijkenberg et al. 2002, 1075^{VP} (Effective publication: Rijkenberg et al., 2001)

Alkanindiges Bogan et al. 2003, 1394^{VP}

Allisonella Garner et al. 2003, 373^{VP} (Effective publication: Garner et al., 2002)

Allofustis Collins et al. 2003a, 813^{VP}

Anaerococcus Ezaki et al. 2001, 1526^{VP}

Anaeroglobus Carlier et al. 2002, 986^{VP}

Anaerolinea Sekiguchi et al. 2003, 1848^{VP}

Anaeromyxobacter Sanford et al. 2002b, 1075^{VP} (Effective publication: Sanford et al., 2002a)

Anaerophaga Denger et al. 2002, 177^{VP}

Anaerostipes Schwiertz et al. 2002b, 1437^{VP} (Effective publication: Schwiertz et al., 2002a)

Anaerovorax Matthies et al. 2000a, 1593^{VP}

Anoxybacillus Pikuta et al. 2000a^{VP} emend. Pikuta et al., 2003

Anoxynatronum Garnova et al. 2003b, 1219^{VP} (Effective publication: Garnova et al., 2003a)

Arenibacter Ivanova et al. 2001a, 1992^{VP}

Asaia Yamada et al. 2000, 828^{VP}

Asanoa Lee and Hah 2002, 970^{VP}

Atopobacter Lawson et al. 2000c, 1758^{VP}

Aurantimonas Denner et al. 2003, 1120^{VP}

Azonexus Reinhold-Hurek and Hurek 2000, 658^{VP}

Azospira Reinhold-Hurek and Hurek 2000, 658^{VP}

Azovibrio Reinhold-Hurek and Hurek 2000, 657^{VP}

Bacteriovorax Baer et al. 2000, 222^{VP}

Balnearium Takai et al. 2003c, 1952^{VP}

Brackiella Willems et al. 2002, 184^{VP}

Brumimicrobium Bowman et al. 2003, 1352^{VP}

Bulleidia Downes et al. 2000, 982^{VP}

Caenibacterium Manaia et al. 2003b, 1380^{VP}

Caldilinea Sekiguchi et al. 2003, 1850^{VP}

Caldimonas Takeda et al. 2002a, 899^{VP}

Caldisphaera Itoh et al. 2003, 1153^{VP}

Caldithrix Miroshnichenko et al. 2003a, 327^{VP}

Caloranaerobacter Wery et al. 2001b, 1795^{VP}

Caminibacter Alain et al. 2002c, 1322^{VP}

Caminicella Alain et al. 2002b, 1627^{VP}

Carboxydibrachium Sokolova et al. 2001, 146^{VP}

Carboxydocella Sokolova et al. 2002, 1965^{VP}

Catenibacterium Kageyama and Benno 2000a, 1598^{VP}

Cellulosimicrobium Schumann et al. 2001, 1009^{VP}

Chlorobaculum Imhoff 2003, 950^{VP}

Citricoccus Altenburger et al. 2002a, 2098^{VP}

Cobetia Arahal et al. 2002b, 1915^{VP} (Effective publication: Arahal et al., 2002a)

Conexibacter Monciardini et al. 2003, 574^{VP}

Coprobacillus Kageyama and Benno 2000e, 949^{VP} (Effective publication: Kageyama and Benno, 2000b)

Croceibacter Cho and Giovannoni 2003d, 935^{VP} (Effective publication: Cho and Giovannoni, 2003a)

Crocinitomix Bowman et al. 2003, 1353^{VP}

Crossiella Labeda 2001, 1578^{VP}

Cryomorpha Bowman et al. 2003, 1352^{VP}

Dechloromonas Achenbach et al. 2001, 531^{VP}

Dechlorosoma Achenbach et al. 2001, 531^{VP}
Dendrosporobacter Strömpl et al. 2000, 105^{VP}
Denitrobacterium Anderson et al. 2000, 636^{VP}
Denitrovibrio Myhr and Torsvik 2000, 1618^{VP}
Desulfobacula Rabus et al. 2000^{VP} emend. Kuever et al., 2001 (Effective publication: Rabus et al., 1993)
Desulfomusa Finster et al. 2001, 2060^{VP}
Desulfonauticus Audiffrin et al. 2003, 1589^{VP}
Desulforegula Rees and Patel 2001, 1915^{VP}
Desulfotignum Kuever et al. 2001, 176^{VP}
Desulfovirga Tanaka et al. 2000, 643^{VP}
Diaphorobacter Khan and Hiraishi 2003, 936^{VP} (Effective publication: Khan and Hiraishi, 2002)
Dorea Taras et al. 2002, 426^{VP}
Dyadobacter Chelius and Triplett 2000, 755^{VP}
Dysgonomonas Hofstad et al. 2000, 2194^{VP}
Enhygromyxa Iizuka et al. 2003c, 1219^{VP} (Effective publication: Iizuka et al., 2003b)
Enterovibrio Thompson et al. 2002, 2019^{VP}
Faecalibacterium Duncan et al. 2002b, 2145^{VP}
Ferribacterium Cummings et al. 2000, 1953^{VP} (Effective publication: Cummings et al., 1999)
Ferroplasma Golyshina et al. 2000, 1004^{VP}
Filobacillus Schlesner et al. 2001, 430^{VP}
Finegoldia Murdoch and Shah 2000, 1415^{VP} (Effective publication: Murdoch and Shah, 1999)
Flexistipes Fiala et al. 2000, 1415^{VP} (Effective publication: Fiala et al., 1990)
Frigoribacterium Kämpfer et al. 2000, 362^{VP}
Fulvimarina Cho and Giovannoni 2003b, 1857^{VP}
Fulvimonas Mergaert et al. 2002, 1288^{VP}
Gallibacterium Christensen et al. 2003, 284^{VP}
Gallicola Ezaki et al. 2001, 1527^{VP}
Garciella Miranda-Tello et al. 2003, 1512^{VP}
Gelria Plugge et al. 2002b, 406^{VP}
Gemmatimonas Zhang et al. 2003b, 1161^{VP}
Geobacillus Nazina et al. 2001, 442^{VP}
Geoglobus Kashefi et al. 2002, 727^{VP}
Georgenia Altenburger et al. 2002b, 880^{VP}
Geovibrio Caccavo et al. 2000, 1415^{VP} (Effective publication: Caccavo et al., 1996)
Granulicatella Collins and Lawson 2000, 367^{VP}
Grimontia Thompson et al., 2003a, 1617^{VP}
Hahella Lee et al. 2001a, 664^{VP}
Haliangium Fudou et al. 2002b, 1437^{VP} (Effective publication: Fudou et al., 2002a)
Halobiforma Hezayen et al. 2002, 2278^{VP}
Halomicrobium Oren et al. 2002, 1834^{VP}
Halonatronum Zhilina et al. 2001c, 263^{VP} (Effective publication: Zhilina et al., 2001b)
Halorhabdus Wainø et al. 2000, 188^{VP}
Halosimplex Vreeland et al. 2003, 936^{VP} (Effective publication: Vreeland et al., 2002)
Halospirulina Nübel et al. 2000, 1275^{VP}
Halothiobacillus Kelly and Wood 2000^{VP} emend. Sievert et al., 2000
Heliorestis Bryantseva et al. 2000a, 949^{VP} (Effective publication: Bryantseva et al., 1999)
Histophilus Angen et al. 2003, 1454^{VP}
Hongia Lee et al. 2000d, 197^{VP}
Hydrogenobaculum Stöhr et al. 2001b, 1860^{VP}
Hydrogenothermus Stöhr et al. 2001b, 1860^{VP}
Idiomarina Ivanova et al. 2000b, 906^{VP}
Ignicoccus Huber et al. 2000a, 2098^{VP}
Inquilinus Coenye et al. 2002b, 1437^{VP} (Effective publication: Coenye et al., 2002a)
Isobaculum Collins et al. 2002d, 209^{VP}
Jannaschia Wagner-Döbler et al. 2003, 735^{VP}
Jeotgalibacillus Yoon et al. 2001e, 2092^{VP}
Jeotgalicoccus Yoon et al. 2003h, 600^{VP}
Kerstersia Coenye et al. 2003b, 1830^{VP}
Ketogulonicigenium Urbance et al. 2001, 1068^{VP}
Kineosphaera Liu et al. 2002b, 1847^{VP}
Knoellia Groth et al. 2002, 81^{VP}
Kozakia Lisdiyanti et al. 2002b, 816^{VP}
Lachnobacterium Whitford et al. 2001, 1980^{VP}
Laribacter Yuen et al. 2002, 1437^{VP} (Effective publication: Yuen et al., 2001)
Lechevalieria Labeda et al. 2001, 1049^{VP}
Leifsonia Evtushenko et al. 2000a, 377^{VP}
Leisingera Schaefer et al. 2002, 857^{VP}
Lentibacillus Yoon et al. 2002a, 2047^{VP}
Leptospirillum (ex Markosyan 1972) Hippe 2000, 502^{VP}
Limnobacter Spring et al. 2001, 1469^{VP}
Longispora Matsumoto et al. 2003, 1558^{VP}
Luteimonas Finkmann et al. 2000, 280^{VP}
Marinibacillus Yoon et al. 2001e, 2092^{VP}
Marinilactibacillus Ishikawa et al. 2003b, 719^{VP}
Marinithermus Sako et al. 2003, 63^{VP}
Marinitoga Wery et al. 2001a, 502^{VP}
Marmoricola Urcì et al. 2000, 534^{VP}
Massilia La Scola et al. 2000, 423^{VP} (Effective publication: La Scola et al., 1998)
Mesonia Nedashkovskaya et al. 2003a, 1970^{VP}
Methanocaldococcus Whitman 2002, 686^{VP} (Effective publication: Whitman, 2001a)
Methanomicrococcus Sprenger et al. 2000, 1998^{VP}
Methanosalsum Boone and Baker 2002, 686^{VP} (Effective publication: Boone and Baker, 2001)
Methanothermobacter Wasserfallen et al. 2000, 51^{VP}
Methanothermococcus Whitman 2002, 687^{VP} (Effective publication: Whitman, 2001b)
Methanotorris Whitman 2002, 687^{VP} (Effective publication: Whitman, 2001c)
Methylarcula Doronina et al. 2000b, 1857^{VP}
Methylocapsa Dedysh et al. 2002, 259^{VP}
Methylocella Dedysh et al. 2000, 967^{VP}
Methylosarcina Wise et al. 2001, 620^{VP}
Micromonas Murdoch and Shah 2000, 1415^{VP} (Effective publication: Murdoch and Shah, 1999)
Micropruina Shintani et al. 2000, 205^{VP}
Microvirga Kanso and Patel 2003, 404^{VP}
Modestobacter Mevs et al. 2000, 344^{VP}
Mogibacterium Nakazawa et al. 2000, 686^{VP}
Muricauda Bruns et al. 2001, 2005^{VP}
Muricoccus Kämpfer et al. 2003b, 936^{VP} (Effective publication: Kämpfer et al., 2003a)
Mycetocola Tsukamoto et al. 2001, 943^{VP}
Nautilia Miroshnichenko et al. 2002, 1302^{VP}
Neochlamydia Horn et al. 2001, 1229^{VP} (Effective publication: Horn et al., 2000)
Oceanicaulis Strömpl et al. 2003, 1905^{VP}
Oceanimonas Brown et al. 2001c, 71^{VP}
Oceanisphaera Romanenko et al. 2003a, 1887^{VP}
Oceanithermus Miroshnichenko et al. 2003b, 751^{VP}

Oceanobacillus Lu et al. 2002, 687^{VP} (Effective publication: Lu et al., 2001a)
Oceanobacter Satomi et al. 2002, 745^{VP}
Okibacterium Evtushenko et al. 2002, 991^{VP}
Oleiphilus Golyshin et al. 2002, 909^{VP}
Oleispira Yakimov et al. 2003a, 784^{VP}
Olsenella Dewhirst et al. 2001, 1802^{VP}
Opitutus Chin et al. 2001, 1967^{VP}
Ornithinimicrobium Groth et al. 2001, 85^{VP}
Oxalicibacterium Tamer et al. 2003, 627^{VP} (Effective publication: Tamer et al., 2002)
Palaeococcus Takai et al. 2000, 498^{VP}
Pandoraea Coenye et al. 2000, 895^{VP}
Pannonibacter Borsodi et al. 2003, 559^{VP}
Papillibacter Defnoun et al. 2000, 1227^{VP}
Paralactobacillus Leisner et al. 2000, 22^{VP}
Paraliobacillus Ishikawa et al. 2003a, 627^{VP} (Effective publication: Ishikawa et al., 2002)
Parascardovia Jian and Dong 2002, 811^{VP}
Parvularcula Cho and Giovannoni 2003c, 1035^{VP}
Paucimonas Jendrossek 2001, 906^{VP}
Pelospora Matthies et al. 2000b, 647^{VP}
Pelotomaculum Imachi et al. 2002, 1734^{VP}
Peptoniphilus Ezaki et al. 2001, 1524^{VP}
Persephonella Götz et al. 2002, 1357^{VP}
Phocoenobacter Foster et al. 2000, 138^{VP}
Pigmentiphaga Blümel et al. 2001b, 1870^{VP}
Planomicrobium Yoon et al. 2001c, 1518^{VP}
Plantibacter Behrendt et al. 2002, 1451^{VP}
Plesiocystis Iizuka et al. 2003a, 194^{VP}
Prochlorococcus Chisholm et al. 2001, 264^{VP} (Effective publication: Chisholm et al., 1992)
Propionicimonas Akasaka et al. 2003, 1996^{VP}
Propionimicrobium (Torrey 1916) Stackebrandt et al. 2002c, 1926^{VP}
Propionispora Biebl et al. 2001, 793^{VP} (Effective publication: Biebl et al., 2000)
Pseudorhodobacter Uchino et al. 2003, 936^{VP} (Effective publication: Uchino et al., 2002b)
Pseudospirillum Satomi et al. 2002, 745^{VP}
Pseudoxanthomonas Finkmann et al. 2000, 280^{VP}
Quadricoccus Maszenan et al. 2002, 227^{VP}
Ramlibacter Heulin et al. 2003, 593^{VP}
Raoultella Drancourt et al. 2001, 930^{VP}
Reichenbachia Nedashkovskaya et al. 2003b, 82^{VP}
Rheinheimera Brettar et al. 2002a, 1856^{VP}
Rhodobaca Milford et al. 2001, 793^{VP} (Effective publication: Milford et al., 2000)
Rhodoblastus Imhoff 2001c, 1865^{VP}
Rhodoglobus Sheridan et al. 2003, 992^{VP}
Roseibium Suzuki et al. 2000b, 2155^{VP}
Roseiflexus Hanada et al. 2002b, 192^{VP}
Roseinatronobacter Sorokin et al. 2000b, 1415^{VP} (Effective publication: Sorokin et al., 2000a)
Roseospirillum Glaeser and Overmann 2001, 793^{VP} (Effective publication: Glaeser and Overmann, 1999)
Rubritepida Alarico et al. 2002b, 1915^{VP} (Effective publication: Alarico et al., 2002a)
Saccharospirillum Labrenz et al. 2003, 659^{VP}
Salana von Wintzingerode et al. 2001a, 1659^{VP}
Salegentibacter McCammon and Bowman 2000, 1062^{VP}
Salinibacter Antón et al. 2002, 490^{VP}
Salinibacterium Han et al. 2003a, 2065^{VP}
Salinisphaera Antunes et al. 2003b, 1219^{VP} (Effective publication: Antunes et al., 2003a)
Samsonia Sutra et al. 2001, 1301^{VP}
Scardovia Jian and Dong 2002, 811^{VP}
Schineria Tóth et al. 2001, 406^{VP}
Schlegelella Elbanna et al. 2003, 1167^{VP}
Sedimentibacter Breitenstein et al. 2002, 806^{VP}
Selenihalanaerobacter Switzer Blum et al. 2001c, 1229^{VP} (Effective publication: Switzer Blum et al., 2001b)
Shuttleworthia Downes et al. 2002, 1473^{VP}
Sneathia Collins et al. 2002b, 687^{VP} (Effective publication: Collins et al., 2001d)
Soehngenia Parshina et al. 2003, 1797^{VP}
Solirubrobacter Singleton et al. 2003, 489^{VP}
Solobacterium Kageyama and Benno 2000f, 1415^{VP} (Effective publication: Kageyama and Benno, 2000d)
Sporanaerobacter Hernandez-Eugenio et al. 2002b, 1221^{VP}
Staleya Labrenz et al. 2000, 310^{VP}
Starkeya Kelly et al. 2000, 1800^{VP}
Sterolibacterium Tarlera and Denner 2003, 1089^{VP}
Streptacidiphilus Kim et al. 2003d, 1219^{VP} (Effective publication: Kim et al., 2003c)
Streptomonospora Cui et al. 2001^{VP} emend. Li et al., 2003
Subtercola Männistö et al. 2000, 1737^{VP}
Sulfurihydrogenibium Takai et al. 2003b, 826^{VP}
Sulfurimonas Inagaki et al. 2003, 1805^{VP}
Symbiobacterium Ohno et al. 2000, 1832^{VP}
Syntrophothermus Sekiguchi et al. 2000, 778^{VP}
Tannerella Sakamoto et al. 2002, 848^{VP}
Teichococcus Kämpfer et al. 2003b, 936^{VP} (Effective publication: Kämpfer et al., 2003a)
Tenacibaculum Suzuki et al. 2001, 1650^{VP}
Tepidibacter Slobodkin et al. 2003, 1133^{VP}
Tepidimonas Moreira et al. 2000, 741^{VP}
Tepidiphilus Manaia et al. 2003a, 1409^{VP}
Terasakiella Satomi et al. 2002, 745^{VP}
Teredinibacter Distel et al. 2002, 2267^{VP}
Tetrasphaera Maszenan et al. 2000, 601^{VP}
Thalassomonas Macián et al. 2001b, 1287^{VP}
Thalassospira López-López et al. 2002, 1282^{VP}
Thermacetogenium Hattori et al. 2000, 1608^{VP}
Thermanaeromonas Mori et al. 2002, 1679^{VP}
Thermicanus Gößner et al. 2000, 423^{VP} (Effective publication: Gößner et al., 1999)
Thermithiobacillus Kelly and Wood 2000, 515^{VP}
Thermobacillus Touzel et al. 2000, 318^{VP}
Thermodiscus Stetter 2003, 1^{VP} (Effective publication: Stetter, 2001)
Thermohalobacter Cayol et al. 2000, 562^{VP}
Thermomonas Busse et al. 2002 emend. Mergaert et al., 2003^{VP}
Thermovenabulum Zavarzina et al. 2002, 1741^{VP}
Thermovibrio Huber et al. 2002, 1864^{VP}
Thioalkalicoccus Bryantseva et al. 2000b, 2161^{VP}
Thioalkalimicrobium Sorokin et al. 2001a emend. Sorokin et al., 2002^{VP}
Thioalkalivibrio Sorokin et al. 2001a, 578^{VP}
Thioalkalispira Sorokin et al. 2002b, 2181^{VP}
Thiobaca Rees et al. 2002, 677^{VP}
Thioflavicoccus Imhoff and Pfennig 2001, 109^{VP}
Tistrella Shi et al. 2003, 936^{VP} (Effective publication: Shi et al., 2002)

Trichlorobacter De Wever et al. 2001, 1^{VP} (Effective publication: De Wever et al., 2000)
Tropheryma La Scola et al. 2001, 1478^{VP}
Turicibacter Bosshard et al. 2002b, 1266^{VP}
Ureibacillus Fortina et al. 2001b, 453^{VP}
Varibaculum Hall et al. 2003h, 627^{VP} (Effective publication: Hall et al., 2003g)
Victivallis Zoetendal et al. 2003, 214^{VP}
Virgisporangium Tamura et al. 2001, 1814^{VP}
Vitellibacter Nedashkovskaya et al. 2003c, 1285^{VP}
Vulcanisaeta Itoh et al. 2002, 1103^{VP}
Vulcanithermus Miroshnichenko et al. 2003c, 1147^{VP}
Xenophilus Blümel et al. 2001a, 1835^{VP}
Xylanimonas Rivas et al. 2003a, 102^{VP}
Zobellia Barbeyron et al. 2001, 993^{VP}
Zooshikella Yi et al. 2003, 1016^{VP}

New species

Acetobacter cerevisiae Cleenwerck et al. 2002, 1557^{VP}
Acetobacter cibinongensis Lisdiyanti et al. 2002a, 3^{VP} (Effective publication: Lisdiyanti et al., 2001a)
Acetobacter indonesiensis Lisdiyanti et al. 2001b, 263^{VP} (Effective publication: Lisdiyanti et al., 2000)
Acetobacter malorum Cleenwerck et al. 2002, 1557^{VP}
Acetobacter orientalis Lisdiyanti et al. 2002a, 3^{VP} (Effective publication: Lisdiyanti et al., 2001a)
Acetobacter syzygii Lisdiyanti et al. 2002a, 3^{VP} (Effective publication: Lisdiyanti et al., 2001a)
Acetobacter tropicalis Lisdiyanti et al. 2001b, 263^{VP} (Effective publication: Lisdiyanti et al., 2000)
Acetobacterium tundrae Simankova et al. 2001a, 793^{VP} (Effective publication: Simankova et al., 2000)
Acholeplasma vituli Angulo et al. 2000, 1130^{VP}
Achromobacter insolitus Coenye et al. 2003c, 1823^{VP}
Achromobacter spanius Coenye et al. 2003c, 1823^{VP}
Acidilobus aceticus Prokofeva et al. 2000, 2007^{VP}
Acidisphaera rubrifaciens Hiraishi et al. 2000b, 1545^{VP}
Acidovorax anthurii Gardan et al. 2000, 245^{VP}
Acidovorax valerianellae Gardan et al. 2003b, 799^{VP}
Acinetobacter baylyi Carr et al. 2003, 960^{VP}
Acinetobacter bouvetii Carr et al. 2003, 961^{VP}
Acinetobacter gerneri Carr et al. 2003, 961^{VP}
Acinetobacter grimontii Carr et al. 2003, 961^{VP}
Acinetobacter parvus Nemec et al. 2003, 1566^{VP}
Acinetobacter schindleri Nemec et al. 2001, 1898^{VP}
Acinetobacter tandoii Carr et al. 2003, 962^{VP}
Acinetobacter tjernbergiae Carr et al. 2003, 961^{VP}
Acinetobacter towneri Carr et al. 2003, 961^{VP}
Acinetobacter ursingii Nemec et al. 2001, 1898^{VP}
Acrocarpospora macrocephala Tamura et al. 2000a, 1170^{VP}
Acrocarpospora pleiomorpha Tamura et al. 2000a, 1170^{VP}
Actinoalloteichus cyanogriseus Tamura et al. 2000b, 1439^{VP}
Actinobacillus arthritidis Christensen et al. 2002a, 1244^{VP}
Actinobaculum urinale Hall et al. 2003b, 682^{VP}
Actinokineospora auranticolor Otoguro et al. 2003, 1^{VP} (Effective publication: Otoguro et al., 2001)
Actinokineospora enzanensis Otoguro et al. 2003, 1^{VP} (Effective publication: Otoguro et al., 2001)
Actinomadura catellatispora Lu et al. 2003, 140^{VP}
Actinomadura glauciflava Lu et al. 2003, 141^{VP}
Actinomadura namibiensis Wink et al. 2003c, 724^{VP}
Actinomyces canis Hoyles et al. 2000a, 1549^{VP}
Actinomyces cardiffensis Hall et al. 2003a, 1^{VP} (Effective publication: Hall et al., 2002)
Actinomyces catuli Hoyles et al. 2001b, 681^{VP}
Actinomyces coleocanis Hoyles et al. 2002a, 1203^{VP}
Actinomyces funkei Lawson et al. 2001e, 855^{VP}
Actinomyces marimammalium Hoyles et al. 2001d, 154^{VP}
Actinomyces nasicola Hall et al. 2003f, 1448^{VP}
Actinomyces oricola Hall et al. 2003c, 1518^{VP}
Actinomyces radicidentis Collins et al. 2001c, 1^{VP} (Effective publication: Collins et al., 2000a)
Actinomyces suimastitidis Hoyles et al. 2001a, 1326^{VP}
Actinomyces urogenitalis Nikolaitchouk et al. 2000, 1653^{VP}
Actinomyces vaccimaxillae Hall et al. 2003d, 605^{VP}
Actinoplanes capillaceus Matsumoto et al. 2001, 793^{VP} (Effective publication: Matsumoto et al., 2000)
Actinoplanes friuliensis Aretz et al. 2001, 793^{VP} (Effective publication: Aretz et al., 2000)
Actinopolymorpha singaporensis Wang et al. 2001b, 472^{VP}
Aequorivita antarctica Bowman and Nichols 2002, 1539^{VP}
Aequorivita crocea Bowman and Nichols 2002, 1540^{VP}
Aequorivita lipolytica Bowman and Nichols 2002, 1539^{VP}
Aequorivita sublithincola Bowman and Nichols 2002, 1540^{VP}
Aerococcus sanguinicola Lawson et al. 2001d, 478^{VP}
Aerococcus urinaehominis Lawson et al. 2001c, 685^{VP}
Aeromicrobium marinum Bruns et al. 2003, 1922^{VP}
Aeromonas culicicola Pidiyar et al. 2002, 1727^{VP}
Afipia birgiae La Scola et al. 2002, 1779^{VP}
Afipia massiliensis La Scola et al. 2002, 1780^{VP}
Agreia bicolorata Evtushenko et al. 2001, 2077^{VP}
Agrobacterium larrymoorei Bouzar and Jones 2001, 1025^{VP}
Agrococcus baldri Zlamala et al. 2002a, 1215^{VP}
Agromyces albus Dorofeeva et al. 2003, 1438^{VP}
Agromyces aurantiacus Li et al. 2003e, 306^{VP}
Agromyces bracchium Takeuchi and Hatano 2001, 1536^{VP}
Agromyces luteolus Takeuchi and Hatano 2001, 1535^{VP}
Agromyces rhizosphaerae Takeuchi and Hatano 2001, 1536^{VP}
Albibacter methylovorans Doronina et al. 2001b, 1056^{VP}
Albidovulum inexpectatum Albuquerque et al. 2003, 1^{VP} (Effective publication: Albuquerque et al., 2002)
Alcanivorax venustensis Fernández-Martínez et al. 2003, 337^{VP}
Algoriphagus ratkowskyi Bowman et al. 2003, 1352^{VP}
Alicycliphilus denitrificans Mechichi et al. 2003, 151^{VP}
Alicyclobacillus acidiphilus Matsubara et al. 2002, 1684^{VP}
Alicyclobacillus herbarius Goto et al. 2002, 112^{VP}
Alicyclobacillus hesperidum Albuquerque et al. 2000, 454^{VP}
Alicyclobacillus pomorum Goto et al. 2003, 1542^{VP}
Alicyclobacillus sendaiensis Tsuruoka et al. 2003, 1084^{VP}
Alishewanella fetalis Fonnesbech Vogel et al. 2000, 1141^{VP}
Alistipes finegoldii Rautio et al. 2003b, 1701^{VP} (Effective publication: Rautio et al., 2003a)
Alkalibacterium olivapovliticus Ntougias and Russell 2001, 1169^{VP}
Alkalilimnicola halodurans Yakimov et al. 2001, 2142^{VP}
Alkaliphilus crotonatoxidans Cao et al. 2003, 973^{VP}
Alkaliphilus transvaalensis Takai et al. 2001b, 1254^{VP}
Alkalispirillum mobile Rijkenberg et al. 2002, 1075^{VP} (Effective publication: Rijkenberg et al., 2001)
Alkanindiges illinoisensis Bogan et al. 2003, 1394^{VP}
Allisonella histaminiformans Garner et al. 2003, 373^{VP} (Effective publication: Garner et al., 2002)
Allofustis seminis Collins et al. 2003a, 813^{VP}
Alteromonas marina Yoon et al. 2003c, 1629^{VP}

Aminobacterium mobile Baena et al. 2000, 263^{VP}
Amphibacillus fermentum Zhilina et al. 2002, 685^{VP} (Effective publication: Zhilina et al., 2001a)
Amphibacillus tropicus Zhilina et al. 2002, 685^{VP} (Effective publication: Zhilina et al., 2001a)
Amycolatopsis albidoflavus Lee and Hah 2001, 649^{VP}
Amycolatopsis balhimycina Wink et al. 2003b, 935^{VP} (Effective publication: Wink et al., 2003a)
Amycolatopsis eurytherma Kim et al. 2002a, 893^{VP}
Amycolatopsis kentuckyensis Labeda et al. 2003, 1603^{VP}
Amycolatopsis lexingtonensis Labeda et al. 2003, 1603^{VP}
Amycolatopsis pretoriensis Labeda et al. 2003, 1605^{VP}
Amycolatopsis rubida Huang et al. 2001, 1096^{VP}
Amycolatopsis sacchari Goodfellow et al. 2001, 191^{VP}
Amycolatopsis tolypomycina Wink et al. 2003b, 935^{VP} (Effective publication: Wink et al., 2003a)
Amycolatopsis vancoresmycina Wink et al. 2003b, 935^{VP} (Effective publication: Wink et al., 2003a)
Anaerobaculum mobile Menes and Muxí 2002, 163^{VP}
Anaerobranca gottschalkii Prowe and Antranikian 2001, 464^{VP}
Anaeroglobus geminatus Carlier et al. 2002, 986^{VP}
Anaerolinea thermophila Sekiguchi et al. 2003, 1850^{VP}
Anaeromyxobacter dehalogenans Sanford et al. 2002b, 1075^{VP} (Effective publication: Sanford et al., 2002a)
Anaerophaga thermohalophila Denger et al. 2002, 177^{VP}
Anaerostipes caccae Schwiertz et al. 2002b, 1437^{VP} (Effective publication: Schwiertz et al., 2002a)
Anaerovorax odorimutans Matthies et al. 2000a, 1593^{VP}
Anaplasma bovis Dumler et al. 2001, 2158^{VP}
Anaplasma platys Dumler et al. 2001, 2159^{VP}
Anoxybacillus flavithermus Pikuta et al. 2000a, 2116^{VP}
Anoxybacillus gonensis Belduz et al. 2003, 1319^{VP}
Anoxybacillus pushchinoensis Pikuta et al. 2000a^{VP} emend. Pikuta et al., 2003
Anoxynatronum sibiricum Garnova et al. 2003b, 1219^{VP} (Effective publication: Garnova et al., 2003a)
Arcanobacterium hippocoleae Hoyles et al. 2002b, 619^{VP}
Arcanobacterium pluranimalium Lawson et al. 2001b, 58^{VP}
Arenibacter latericius Ivanova et al. 2001a, 1994^{VP}
Arenibacter troitsensis Nedashkovskaya et al. 2003c, 1289^{VP}
Arthrobacter albus Wauters et al. 2000b, 1699^{VP} (Effective publication: Wauters et al., 2000a)
Arthrobacter chlorophenolicus Westerberg et al. 2000, 2090^{VP}
Arthrobacter flavus Reddy et al. 2000, 1559^{VP}
Arthrobacter gandavensis Storms et al. 2003, 1883^{VP}
Arthrobacter koreensis Lee et al. 2003, 1280^{VP}
Arthrobacter luteolus Wauters et al. 2000b, 1699^{VP} (Effective publication: Wauters et al., 2000a)
Arthrobacter methylotrophus Borodina et al. 2002b, 685^{VP} (Effective publication: Borodina et al., 2002a)
Arthrobacter nasiphocae Collins et al. 2002a, 571^{VP}
Arthrobacter psychrolactophilus Loveland-Curtze et al. 2000, 3^{VP} (Effective publication: Loveland-Curtze et al., 1999)
Arthrobacter roseus Reddy et al. 2002a, 1020^{VP}
Arthrobacter sulfonivorans Borodina et al. 2002b, 685^{VP} (Effective publication:Borodina et al., 2002a)
Asaia bogorensis Yamada et al. 2000, 828^{VP}
Asaia siamensis Katsura et al. 2001, 562^{VP}
Asanoa ishikariensis Lee and Hah 2002, 971^{VP}
Atopobacter phocae Lawson et al. 2000c, 1759^{VP}
Aurantimonas coralicida Denner et al. 2003, 1120^{VP}
Azoarcus buckelii Mechichi et al. 2002b, 1437^{VP} (Effective publication: Mechichi et al., 2002a)
Azonexus fungiphilus Reinhold-Hurek and Hurek 2000, 658^{VP}
Azospira oryzae Reinhold-Hurek and Hurek 2000^{VP} emend. Tan and Reinhold-Hurek, 2003
Azospirillum doebereinerae Eckert et al. 2001, 24^{VP}
Azovibrio restrictus Reinhold-Hurek and Hurek 2000, 657^{VP}
Bacillus aeolius Gugliandolo et al. 2003b, 1701^{VP} (Effective publication: Gugliandolo et al., 2003a)
Bacillus aquimaris Yoon et al. 2003d, 1302^{VP}
Bacillus arseniciselenatis Switzer Blum et al. 2001a, 793^{VP} (Effective publication: Switzer Blum et al., 1998)
Bacillus barbaricus Täubel et al. 2003, 729^{VP}
Bacillus decolorationis Heyrman et al. 2003a, 462^{VP}
Bacillus endophyticus Reva et al. 2002, 106^{VP}
Bacillus fumarioli Logan et al. 2000b, 1751^{VP}
Bacillus funiculus Ajithkumar et al. 2002, 1143^{VP}
Bacillus jeotgali Yoon et al. 2001b, 1091^{VP}
Bacillus krulwichiae Yumoto et al. 2003b, 1536^{VP}
Bacillus luciferensis Logan et al. 2002b, 1988^{VP}
Bacillus marisflavi Yoon et al. 2003d, 1302^{VP}
Bacillus nealsonii Venkateswaran et al. 2003, 171^{VP}
Bacillus neidei Nakamura et al. 2002, 504^{VP}
Bacillus okuhidensis Li et al. 2002c, 1208^{VP}
Bacillus psychrodurans Abd El-Rahman et al. 2002, 2132^{VP}
Bacillus psychrotolerans Abd El-Rahman et al. 2002, 2131^{VP}
Bacillus pycnus Nakamura et al. 2002, 504^{VP}
Bacillus selenitireducens Switzer Blum et al. 2001c, 793^{VP}
Bacillus siralis Pettersson et al. 2000, 2186^{VP}
Bacillus sonorensis Palmisano et al. 2001, 1678^{VP}
Bacillus subterraneus Kanso et al. 2002, 873^{VP}
Bacillus thermantarcticus Nicolaus et al. 2002, 3^{VP} (Effective publication: Nicolaus et al., 1996)
Bacillus vulcani Caccamo et al. 2000, 2011^{VP}
Bacteroides acidifaciens Miyamoto and Itoh 2000, 148^{VP}
Balnearium lithotrophicum Takai et al. 2003c, 1953^{VP}
Bartonella birtlesii Bermond et al. 2000, 1978^{VP}
Bartonella bovis Bermond et al. 2002, 388^{VP}
Bartonella capreoli Bermond et al. 2002, 388^{VP}
Bartonella koehlerae Droz et al. 2000, 423 (Effective publication: Droz et al., 1999)^{VP}
Bartonella schoenbuchensis Dehio et al. 2001, 1563^{VP}
Bifidobacterium scardovii Hoyles et al. 2002c, 998^{VP}
Bordetella petrii von Wintzingerode et al. 2001b, 1263^{VP}
Borrelia sinica Masuzawa et al. 2001, 1823^{VP}
Bosea eneae La Scola et al. 2003, 19^{VP}
Bosea massiliensis La Scola et al. 2003, 19^{VP}
Bosea minatitlanensis Ouattara et al. 2003, 1250^{VP}
Bosea vestrisii La Scola et al. 2003, 20^{VP}
Brachybacterium fresconis Heyrman et al. 2002a, 1644^{VP}
Brachybacterium muris Buczolits et al. 2003, 1959^{VP}
Brachybacterium sacelli Heyrman et al. 2002a, 1644^{VP}
Brackiella oedipodis Willems et al. 2002, 184^{VP}
Bradyrhizobium yuanmingense Yao et al. 2002, 2228^{VP}
Brevibacillus invocatus Logan et al. 2002a, 964^{VP}
Brevibacterium lutescens Wauters et al. 2003a, 1324^{VP}
Brevibacterium paucivorans Wauters et al. 2001, 1706^{VP}
Brumimicrobium glaciale Bowman et al. 2003, 1352^{VP}
Bulleidia extructa Downes et al. 2000, 981^{VP}
Burkholderia ambifaria Coenye et al. 2001b, 1488^{VP}
Burkholderia anthina Vandamme et al. 2002c, 1437^{VP} (Effective publication: Vandamme et al., 2002b)

Burkholderia caledonica Coenye et al. 2001a, 1106[VP]
Burkholderia cenocepacia Vandamme et al. 2003c, 935[VP] (Effective publication: Vandamme et al., 2003b)
Burkholderia fungorum Coenye et al. 2001a, 1105[VP]
Burkholderia hospita Goris et al. 2003, 1[VP] (Effective publication: Goris et al., 2002)
Burkholderia kururiensis Zhang et al. 2000, 747[VP]
Burkholderia phymatum Vandamme et al. 2003a, 627[VP] (Effective publication: Vandamme et al., 2002a)
Burkholderia sacchari Brämer et al. 2001, 1711[VP]
Burkholderia sordidicola Lim et al. 2003, 1635[VP]
Burkholderia stabilis Vandamme et al. 2000b, 1415[VP] (Effective publication: Vandamme et al., 2000a)
Burkholderia terricola Goris et al. 2003, 1[VP] (Effective publication: Goris et al., 2002)
Burkholderia tuberum Vandamme et al. 2003a, 627[VP] (Effective publication: Vandamme et al., 2002a)
Burkholderia ubonensis Yabuuchi et al. 2000b, 1415[VP] (Effective publication: Yabuuchi et al., 2000a)
Butyrivibrio hungatei Kopecny et al. 2003, 208[VP]
Caenibacterium thermophilum Manaia et al. 2003b, 1380[VP]
Caldilinea aerophila Sekiguchi et al. 2003, 1850[VP]
Caldimonas manganoxidans Takeda et al. 2002a, 899[VP]
Caldisphaera lagunensis Itoh et al. 2003, 1153[VP]
Caldithrix abyssi Miroshnichenko et al. 2003a, 327[VP]
Caloramator coolhaasii Plugge et al. 2000, 1161[VP]
Caloramator viterbiensis Seyfried et al. 2002, 1183[VP]
Caloranaerobacter azorensis Wery et al. 2001b, 1795[VP]
Caminibacter hydrogeniphilus Alain et al. 2002c, 1322[VP]
Caminicella sporogenes Alain et al. 2002b, 1627[VP]
Campylobacter hominis Lawson et al. 2001a, 658[VP]
Campylobacter lanienae Logan et al. 2000a, 870[VP]
Carboxydibrachium pacificum Sokolova et al. 2001, 147[VP]
Carboxydocella thermautotrophica Sokolova et al. 2002, 1965[VP]
Carnobacterium viridans Holley et al. 2002, 1884[VP]
Catellatospora koreensis Lee et al. 2000c, 1110[VP]
Catenibacterium mitsuokai Kageyama and Benno 2000a, 1598[VP]
Cellulomonas iranensis Elberson et al. 2000, 996[VP]
Cellulomonas persica Elberson et al. 2000, 995[VP]
Cellulophaga algicola Bowman 2000, 1866[VP]
Cellulosimicrobium variabile Bakalidou et al. 2002, 1189[VP]
Cellvibrio fibrivorans Mergaert et al. 2003b, 471[VP]
Cellvibrio fulvus (ex Stapp and Bortels 1934) Humphry et al. 2003, 399[VP]
Cellvibrio gandavensis Mergaert et al. 2003b, 471[VP]
Cellvibrio japonicus Humphry et al. 2003, 398[VP]
Cellvibrio ostraviensis Mergaert et al. 2003b, 470[VP]
Cellvibrio vulgaris (ex Stapp and Bortels 1934) Humphry et al. 2003, 399[VP]
Cetobacterium somerae Finegold et al. 2003b, 1219[VP] (Effective publication: Finegold et al., 2003a)
Chlorobaculum limnaeum Imhoff 2003, 950[VP]
Chlorobaculum parvum Imhoff 2003, 950[VP]
Chlorobaculum thiosulfatiphilum Imhoff 2003, 950[VP]
Chromohalobacter salexigens Arahal et al. 2001b, 1460[VP]
Chryseobacterium defluvii Kämpfer et al. 2003d, 96[VP]
Chryseobacterium joostei Hugo et al. 2003, 776[VP]
Citricoccus muralis Altenburger et al. 2002a, 2099[VP]
Citrobacter gillenii Brenner et al. 2000, 423[VP] (Effective publication: Brenner et al., 1999)
Citrobacter murliniae Brenner et al. 2000, 423[VP] (Effective publication: Brenner et al., 1999)
Clostridium acidisoli Kuhner et al. 2000, 880[VP]
Clostridium akagii Kuhner et al. 2000, 879[VP]
Clostridium algidixylanolyticum Broda et al. 2000a, 629[VP]
Clostridium amygdalinum Parshina et al. 2003, 1797[VP]
Clostridium bolteae Song et al. 2003b, 935[VP] (Effective publication: Song et al., 2003a)
Clostridium bowmanii Spring et al. 2003, 1027[VP]
Clostridium caminithermale Brisbarre et al. 2003, 1047[VP]
Clostridium colicanis Greetham et al. 2003, 261[VP]
Clostridium diolis Biebl and Spröer 2003, 627[VP] (Effective publication: Biebl and Spröer, 2002)
Clostridium frigoris Spring et al. 2003, 1027[VP]
Clostridium gasigenes Broda et al. 2000b, 116[VP]
Clostridium hathewayi Steer et al. 2002, 685[VP] (Effective publication: Steer et al., 2001)
Clostridium hiranonis Kitahara et al. 2001, 43[VP]
Clostridium hungatei Monserrate et al. 2001, 130[VP]
Clostridium hylemonae Kitahara et al. 2000, 977[VP]
Clostridium lactatifermentans van der Wielen et al. 2002, 925[VP]
Clostridium lacusfryxellense Spring et al. 2003, 1027[VP]
Clostridium peptidivorans Mechichi et al. 2000a, 1263[VP]
Clostridium phytofermentans Warnick et al. 2002, 1158[VP]
Clostridium psychrophilum Spring et al. 2003, 1028[VP]
Clostridium saccharobutylicum Keis et al. 2001, 2101[VP]
Clostridium saccharoperbutylacetonicu Keis et al. 2001, 2101[VP]
Clostridium thiosulfatireducens Hernandez-Eugenio et al. 2002a, 1466[VP]
Clostridium uliginosum Matthies et al. 2001, 1124[VP]
Clostridium xylanovorans Mechichi et al. 2000b, 3[VP] (Effective publication: Mechichi et al., 1999)
Collinsella intestinalis Kageyama and Benno 2000c, 1773[VP]
Collinsella stercoris Kageyama and Benno 2000c, 1773[VP]
Comamonas denitrificans Gumaelius et al. 2001, 1005[VP]
Comamonas kerstersii Wauters et al. 2003b, 862[VP]
Comamonas koreensis Chang et al. 2002, 380[VP]
Comamonas nitrativorans Etchebehere et al. 2001, 982[VP]
Conexibacter woesei Monciardini et al. 2003, 575[VP]
Coprobacillus cateniformis Kageyama and Benno 2000e, 949[VP] (Effective publication: Kageyama and Benno, 2000b)
Corynebacterium appendicis Yassin et al. 2002a, 1168[VP]
Corynebacterium aquilae Fernández-Garayzábal et al. 2003, 1138[VP]
Corynebacterium atypicum Hall et al. 2003e, 1067[VP]
Corynebacterium aurimucosum Yassin et al. 2002b, 1004[VP]
Corynebacterium auriscanis Collins et al. 2000b, 423[VP] (Effective publication: Collins et al., 1999)
Corynebacterium capitovis Collins et al. 2001a, 858[VP]
Corynebacterium casei Brennan et al. 2001a, 850[VP]
Corynebacterium efficiens Fudou et al. 2002c, 1130[VP]
Corynebacterium felinum Collins et al. 2001b, 1351[VP]
Corynebacterium freneyi Renaud et al. 2001, 1728[VP]
Corynebacterium glaucum Yassin et al. 2003a, 708[VP]
Corynebacterium mooreparkense Brennan et al. 2001a, 848[VP]
Corynebacterium simulans Wattiau et al. 2000, 351[VP]
Corynebacterium sphenisci Goyache et al. 2003a, 1012[VP]
Corynebacterium spheniscorum Goyache et al. 2003b, 46[VP]
Corynebacterium suicordis Vela et al. 2003b, 2030[VP]
Corynebacterium testudinoris Collins et al. 2001b, 1351[VP]
Croceibacter atlanticus Cho and Giovannoni 2003d, 935[VP] (Effective publication: Cho and Giovannoni, 2003a)
Crocinitomix catalasitica Bowman et al. 2003, 1353[VP]
Crossiella equi Donahue et al. 2002, 2172[VP]
Cryomorpha ignava Bowman et al. 2003, 1353[VP]

Cryptosporangium aurantiacum Tamura and Hatano 2001, 2124^{VP}
Curtobacterium herbarum Behrendt et al. 2002, 1452^{VP}
Dechloromonas agitata Achenbach et al. 2001, 531^{VP}
Dechlorosoma suillum Achenbach et al. 2001 pro synon. *Azospira oryzae*, 532^{VP}
Deferribacter abyssi Miroshnichenko et al. 2003d, 1640^{VP}
Deferribacter desulfuricans Takai et al. 2003a, 845^{VP}
Delftia tsuruhatensis Shigematsu et al. 2003, 1482^{VP}
Denitrobacterium detoxificans Anderson et al. 2000, 637^{VP}
Denitrovibrio acetiphilus Myhr and Torsvik 2000, 1618^{VP}
Desulfacinum hydrothermale Sievert and Kuever 2000, 1244^{VP}
Desulfitobacterium chlororespirans Sanford et al. 2001, 793^{VP} (Effective publication: Sanford et al., 1996)
Desulfitobacterium metallireducens Finneran et al. 2002, 1934^{VP}
Desulfobacula toluolica Rabus et al. 2000, 1415^{VP} (Effective publication: Rabus et al., 1993)
Desulfobulbus mediterraneus Sass et al. 2002b, 1437^{VP} (Effective publication: Sass et al., 2002a)
Desulfocapsa sulfexigens Finster et al. 2000, 1699^{VP} (Effective publication: Finster et al., 1998)
Desulfomicrobium orale Langendijk et al. 2001, 1042^{VP}
Desulfomonile limimaris Sun et al. 2001b, 370^{VP}
Desulfomusa hansenii Finster et al. 2001, 2060^{VP}
Desulfonatronum thiodismutans Pikuta et al. 2003c, 1331^{VP}
Desulfonauticus submarinus Audiffrin et al. 2003, 1589^{VP}
Desulfonema ishimotonii Fukui et al. 2000, 1415^{VP} (Effective publication: Fukui et al., 1999)
Desulforegula conservatrix Rees and Patel 2001, 1915^{VP}
Desulforhopalus singaporensis Lie et al. 2000, 1699^{VP} (Effective publication: Lie et al., 1999)
Desulfosporosinus meridiei Robertson et al. 2001b, 139^{VP}
Desulfotignum balticum Kuever et al. 2001, 176^{VP}
Desulfotignum phosphitoxidans Schink et al. 2002b, 1437^{VP} (Effective publication: Schink et al., 2002a)
Desulfotomaculum alkaliphilum Pikuta et al. 2000b, 32^{VP}
Desulfotomaculum solfataricum Goorissen et al. 2003, 1228^{VP}
Desulfovibrio dechloracetivorans Sun et al. 2001a, 1^{VP} (Effective publication: Sun et al., 2000)
Desulfovibrio hydrothermalis Alazard et al. 2003, 173^{VP}
Desulfovibrio indonesiensis Feio et al. 2000, 1415^{VP} (Effective publication: Feio et al., 1998)
Desulfovibrio magneticus Sakaguchi et al. 2002, 219^{VP}
Desulfovibrio mexicanus Hernandez-Eugenio et al. 2001, 263^{VP} (Effective publication: Hernandez-Eugenio et al., 2000)
Desulfovibrio oxyclinae Krekeler et al. 2000, 1699^{VP} (Effective publication: Krekeler et al., 1997)
Desulfovibrio vietnamensis Dang et al. 2002, 1075^{VP} (Effective publication: Dang et al., 1996)
Desulfovirga adipica Tanaka et al. 2000, 643^{VP}
Desulfurococcus amylolyticus Bonch-Osmolovskaya et al. 2001, 1619^{VP} (Effective publication: Bonch-Osmolovskaya et al., 1988)
Desulfuromonas palmitatis Coates et al. 2000a, 1699^{VP} (Effective publication: Coates et al., 1995)
Dethiosulfovibrio acidaminovorans Surkov et al. 2001, 335^{VP}
Dethiosulfovibrio marinus Surkov et al. 2001, 335^{VP}
Dethiosulfovibrio russensis Surkov et al. 2001, 335^{VP}
Devosia neptuniae Rivas et al. 2003b, 935^{VP} (Effective publication: Rivas et al., 2003c)
Dialister invisus Downes et al. 2003, 1940^{VP}
Diaphorobacter nitroreducens Khan and Hiraishi 2003, 936^{VP} (Effective publication: Khan and Hiraishi, 2002)
Dietzia psychralcaliphila Yumoto et al. 2002b, 89^{VP}
Dorea longicatena Taras et al. 2002, 427^{VP}
Dyadobacter fermentans Chelius and Triplett 2000, 756^{VP}
Dysgonomonas capnocytophagoides Hofstad et al. 2000, 2194^{VP}
Dysgonomonas gadei Hofstad et al. 2000, 2194^{VP}
Dysgonomonas mossii Lawson et al. 2002b, 1915^{VP} (Effective publication: Lawson et al., 2002a)
Enhygromyxa salina Iizuka et al. 2003c, 1219^{VP} (Effective publication: Iizuka et al., 2003b)
Enterobacter cowanii Inoue et al. 2001, 1619^{VP} (Effective publication: Inoue et al., 2000)
Enterococcus canis De Graef et al. 2003, 1073^{VP}
Enterococcus gilvus Tyrrell et al. 2002b, 1075^{VP} (Effective publication: Tyrrell et al., 2002a)
Enterococcus haemoperoxidus Svec et al. 2001, 1571^{VP}
Enterococcus moraviensis Svec et al. 2001, 1572^{VP}
Enterococcus pallens Tyrrell et al. 2002b, 1075^{VP} (Effective publication: Tyrrell et al., 2002a)
Enterococcus phoeniculicola Law-Brown and Meyers 2003, 685^{VP}
Enterococcus ratti Teixeira et al. 2001, 1742^{VP}
Enterococcus villorum Vancanneyt et al. 2001b, 398^{VP}
Enterovibrio norvegicus Thompson et al. 2002, 2019^{VP}
Erythrobacter citreus Denner et al. 2002b, 1659^{VP}
Erythrobacter flavus Yoon et al. 2003f, 1173^{VP}
Escherichia albertii Huys et al. 2003a, 810^{VP}
Eubacterium aggregans Mechichi et al. 2000c, 1699^{VP} (Effective publication: Mechichi et al., 1998)
Eubacterium pyruvativorans Wallace et al. 2003, 969^{VP}
Exiguobacterium antarcticum Frühling et al. 2002, 1175^{VP}
Exiguobacterium undae Frühling et al. 2002, 1173^{VP}
Facklamia miroungae Hoyles et al. 2001c, 1403^{VP}
Ferribacterium limneticum Cummings et al. 2000, 1953^{VP} (Effective publication: Cummings et al., 1999)
Ferroplasma acidiphilum Golyshina et al. 2000, 1005^{VP}
Filobacillus milosensis Schlesner et al. 2001, 430^{VP}
Flavobacterium frigidarium Humphry et al. 2001, 1242^{VP}
Flavobacterium gelidilacus Van Trappen et al. 2003, 1244^{VP}
Flavobacterium gillisiae McCammon and Bowman 2000, 1059^{VP}
Flavobacterium limicola Tamaki et al. 2003, 525^{VP}
Flavobacterium omnivorum Zhu et al. 2003a, 857^{VP}
Flavobacterium tegetincola McCammon and Bowman 2000, 1060^{VP}
Flavobacterium xanthum McCammon and Bowman 2000, 1060^{VP}
Flavobacterium xinjiangense Zhu et al. 2003a, 857^{VP}
Flexistipes sinusarabici Fiala et al. 2000, 1415^{VP} (Effective publication: Fiala et al., 1990)
Friedmanniella lacustris Lawson et al. 2000b, 1953^{VP} (Effective publication: Lawson et al., 2000a)
Frigoribacterium faeni Kämpfer et al. 2000, 362^{VP}
Fulvimarina pelagi Cho and Giovannoni 2003b, 1858^{VP}
Fulvimonas soli Mergaert et al. 2002, 1289^{VP}
Fusobacterium equinum Dorsch et al. 2001, 1962^{VP}
Garciella nitratireducens Miranda-Tello et al. 2003, 1513^{VP}
Gelidibacter mesophilus Macián et al. 2002, 1328^{VP}
Gelria glutamica Plugge et al. 2002b, 406^{VP}
Gemella cuniculi Hoyles et al. 2000b, 2039^{VP}
Gemmatimonas aurantiaca Zhang et al. 2003b, 1161^{VP}
Geobacillus subterraneus Nazina et al. 2001, 443^{VP}
Geobacillus toebii Sung et al. 2002, 2254^{VP}
Geobacillus uzenensis Nazina et al. 2001, 443^{VP}
Geobacter bremensis Straub and Buchholz-Cleven 2001, 1807^{VP}
Geobacter chapellei Coates et al. 2001a, 586^{VP}
Geobacter grbiciae Coates et al. 2001a, 587^{VP}

Geobacter hydrogenophilus Coates et al. 2001a, 586^{VP}
Geobacter pelophilus Straub and Buchholz-Cleven 2001, 1807^{VP}
Geoglobus ahangari Kashefi et al. 2002, 727^{VP}
Georgenia muralis Altenburger et al. 2002b, 880^{VP}
Geovibrio ferrireducens Caccavo et al. 2000, 1415^{VP} (Effective publication: Caccavo et al., 1996)
Geovibrio thiophilus Janssen et al. 2002, 1346^{VP}
Glaciecola mesophila Romanenko et al. 2003e, 647^{VP}
Globicatella sulfidifaciens Vandamme et al. 2001, 1748^{VP}
Gluconacetobacter azotocaptans Fuentes-Ramírez et al. 2001, 1312^{VP}
Gluconacetobacter entanii Schüller et al. 2000, 2019^{VP}
Gluconacetobacter johannae Fuentes-Ramírez et al. 2001, 1312^{VP}
Gordonia amicalis Kim et al. 2000c, 2033^{VP}
Gordonia namibiensis Brandão et al. 2002, 685^{VP} (Effective publication: Brandão et al., 2001)
Gordonia nitida Yoon et al. 2000e, 1208^{VP}
Gordonia paraffinivorans Xue et al. 2003, 1645^{VP}
Gordonia sihwensis Kim et al. 2003b, 1432^{VP}
Gordonia sinesedis Maldonado et al. 2003a, 1219^{VP} (Effective publication: Maldonado et al., 2003b)
Gordonia westfalica Linos et al. 2002, 1137^{VP}
Hahella chejuensis Lee et al. 2001a, 665^{VP}
Halanaerobium fermentans Kobayashi et al. 2000, 1626^{VP}
Haliangium ochraceum Fudou et al. 2002b, 1437^{VP} (Effective publication: Fudou et al., 2002a)
Haliangium tepidum Fudou et al. 2002b, 1437^{VP} (Effective publication: Fudou et al., 2002a)
Halobacillus karajensis Amoozegar et al. 2003, 1062^{VP}
Halobacillus salinus Yoon et al. 2003a, 691^{VP}
Halobiforma haloterrestris Hezayen et al. 2002, 2278^{VP}
Halococcus dombrowskii Stan-Lotter et al. 2002, 1813^{VP}
Haloferax alexandrinus Asker and Ohta 2002, 736^{VP}
Halomonas alimentaria Yoon et al. 2002b, 128^{VP}
Halomonas campisalis Mormile et al. 2000, 949^{VP} (Effective publication: Mormile et al., 1999)
Halomonas halocynthiae Romanenko et al. 2002b, 1771^{VP}
Halomonas magadiensis Duckworth et al. 2000b, 1415^{VP} (Effective publication: Duckworth et al., 2000a)
Halomonas marisflavi Yoon et al. 2001a, 1176^{VP}
Halomonas maura Bouchotroch et al. 2001, 1630^{VP}
Halomonas muralis Heyrman et al. 2002b, 2053^{VP}
Halonatronum saccharophilum Zhilina et al. 2001c, 263^{VP} (Effective publication: Zhilina et al., 2001b)
Halorhabdus utahensis Wainø et al. 2000, 189^{VP}
Halorhodospira neutriphila Hirschler-Réa et al. 2003, 162^{VP}
Halorubrum tebenquichense Lizama et al. 2002, 154^{VP}
Halosimplex carlsbadense Vreeland et al. 2003, 936^{VP} (Effective publication: Vreeland et al., 2002)
Halospirulina tapeticola Nübel et al. 2000, 1275^{VP}
Haloterrigena thermotolerans Montalvo-Rodríguez et al. 2000, 1070^{VP}
Halothiobacillus kellyi Sievert et al. 2000, 1235^{VP}
Helicobacter aurati Patterson et al. 2002, 3^{VP} (Effective publication: Patterson et al., 2000)
Helicobacter canadensis Fox et al. 2002, 3^{VP} (Effective publication: Fox et al., 2000)
Helicobacter ganmani Robertson et al. 2001a, 1888^{VP}
Helicobacter mesocricetorum Simmons et al. 2000b, 1699^{VP} (Effective publication: Simmons et al., 2000a)
Helicobacter typhlonius Franklin et al. 2002, 686^{VP} (Effective publication: Franklin et al., 2001)
Heliobacterium sulfidophilum Bryantseva et al. 2001b, 1^{VP} (Effective publication: Bryantseva et al., 2000d)
Heliobacterium undosum Bryantseva et al. 2001b, 1^{VP} (Effective publication: Bryantseva et al., 2000d)
Heliorestis baculata Bryantseva et al. 2001a, 264^{VP} (Effective publication: Bryantseva et al., 2000c)
Heliorestis daurensis Bryantseva et al. 2000a, 949^{VP} (Effective publication: Bryantseva et al., 1999)
Herbaspirillum frisingense Kirchhof et al. 2001, 166^{VP}
Herbaspirillum lusitanum Valverde et al. 2003, 1982^{VP}
Histophilus somni Angen et al. 2003, 1455^{VP}
Hydrogenobacter subterraneus Takai et al. 2001a, 1433^{VP}
Hydrogenophaga intermedia Contzen et al. 2001, 793^{VP} (Effective publication: Contzen et al., 2000)
Hydrogenophilus hirschii Stöhr et al. 2001a, 488^{VP}
Hydrogenothermus marinus Stöhr et al. 2001b, 1861^{VP}
Hymenobacter actinosclerus Collins et al. 2000e, 733^{VP}
Hymenobacter aerophilus Buczolits et al. 2002, 454^{VP}
Hyphomicrobium chloromethanicum McDonald et al. 2001, 121^{VP}
Hyphomicrobium sulfonivorans Borodina et al. 2002b, 686^{VP} (Effective publication: Borodina et al., 2002a)
Hyphomonas adhaerens Weiner et al. 2000, 467^{VP}
Hyphomonas johnsonii Weiner et al. 2000, 467^{VP}
Hyphomonas rosenbergii Weiner et al. 2000, 467^{VP}
Idiomarina abyssalis Ivanova et al. 2000b, 906^{VP}
Idiomarina baltica Brettar et al. 2003, 412^{VP}
Idiomarina loihiensis Donachie et al. 2003, 1878^{VP}
Idiomarina zobellii Ivanova et al. 2000b, 906^{VP}
Ignicoccus islandicus Huber et al. 2000a, 2099^{VP}
Ignicoccus pacificus Huber et al. 2000a, 2099^{VP}
Ilyobacter insuetus Brune et al. 2002a, 431^{VP}
Inquilinus limosus Coenye et al. 2002b, 1437^{VP} (Effective publication: Coenye et al., 2002a)
Isobaculum melis Collins et al. 2002d, 209^{VP}
Janibacter brevis Imamura et al. 2000, 1902^{VP}
Janibacter terrae Yoon et al. 2000d^{VP} emend. Lang et al., 2003
Jannaschia helgolandensis Wagner-Döbler et al. 2003, 736^{VP}
Jeotgalibacillus alimentarius Yoon et al. 2001e, 2092^{VP}
Jeotgalicoccus halotolerans Yoon et al. 2003h, 601^{VP}
Jeotgalicoccus psychrophilus Yoon et al. 2003h, 601^{VP}
Kerstersia gyiorum Coenye et al. 2003b, 1830^{VP}
Ketogulonicigenium robustum Urbance et al. 2001, 1069^{VP}
Ketogulonicigenium vulgare Urbance et al. 2001, 1069^{VP}
Kineococcus radiotolerans Phillips et al. 2002, 937^{VP}
Kineosphaera limosa Liu et al. 2002b, 1848^{VP}
Kitasatospora cineracea Tajima et al. 2001, 1770^{VP}
Kitasatospora niigatensis Tajima et al. 2001, 1770^{VP}
Kitasatospora putterlickiae Groth et al. 2003, 2037^{VP}
Knoellia sinensis Groth et al. 2002, 82^{VP}
Knoellia subterranea Groth et al. 2002, 82^{VP}
Kocuria polaris Reddy et al. 2003b, 187^{VP}
Kozakia baliensis Lisdiyanti et al. 2002b, 817^{VP}
Kytococcus schroeteri Becker et al. 2002, 1613^{VP}
Lachnobacterium bovis Whitford et al. 2001, 1980^{VP}
Lactobacillus acidipiscis Tanasupawat et al. 2000, 1481^{VP}
Lactobacillus algidus Kato et al. 2000, 1148^{VP}
Lactobacillus arizonensis Swezey et al. 2000, 1808^{VP}
Lactobacillus coleohominis Nikolaitchouk et al. 2001, 2084^{VP}
Lactobacillus cypricasei Lawson et al. 2001f, 48^{VP}
Lactobacillus diolivorans Krooneman et al. 2002, 645^{VP}
Lactobacillus durianis Leisner et al. 2002, 929^{VP}
Lactobacillus equi Morotomi et al. 2002, 214^{VP}

Lactobacillus ferintoshensis Simpson et al. 2002, 1075^{VP} (Effective publication: Simpson et al., 2001)
Lactobacillus fornicalis Dicks et al. 2000, 1258^{VP}
Lactobacillus frumenti Müller et al. 2000, 2132^{VP}
Lactobacillus fuchuensis Sakala et al. 2002, 1153^{VP}
Lactobacillus ingluviei Baele et al. 2003, 136^{VP}
Lactobacillus kimchii Yoon et al. 2000c, 1794^{VP}
Lactobacillus kitasatonis Mukai et al. 2003, 2058^{VP}
Lactobacillus mindensis Ehrmann et al. 2003, 12^{VP}
Lactobacillus mucosae Roos et al. 2000, 256^{VP}
Lactobacillus nagelii Edwards et al. 2000, 700^{VP}
Lactobacillus pantheris Liu and Dong 2002, 1747^{VP}
Lactobacillus perolens Back et al. 2000, 3^{VP} (Effective publication: Back et al., 1999)
Lactobacillus psittaci Lawson et al. 2001g, 969^{VP}
Lactobacillus thermotolerans Niamsup et al. 2003, 267^{VP}
Lactobacillus versmoldensis Kröckel et al. 2003, 516^{VP}
Laribacter hongkongensis Yuen et al. 2002, 1437^{VP} (Effective publication: Yuen et al., 2001)
Legionella beliardensis Lo Presti et al. 2001, 1956^{VP}
Legionella busanensis Park et al. 2003b, 79^{VP}
Legionella drozanskii Adeleke et al. 2001, 1158^{VP}
Legionella fallonii Adeleke et al. 2001, 1158^{VP}
Legionella gresilensis Lo Presti et al. 2001, 1956^{VP}
Legionella rowbothamii Adeleke et al. 2001, 1158^{VP}
Leifsonia aurea Reddy et al. 2003c, 983^{VP}
Leifsonia naganoensis Suzuki et al. 2000a, 1415^{VP} (Effective publication: Suzuki et al., 1999a)
Leifsonia poae Evtushenko et al. 2000a, 378^{VP}
Leifsonia rubra Reddy et al. 2003c, 782^{VP}
Leifsonia shinshuensis Suzuki et al. 2000a, 1415^{VP} (Effective publication: Suzuki et al., 1999a)
Leisingera methylohalidivorans Schaefer et al. 2002, 857^{VP}
Lentibacillus salicampi Yoon et al. 2002a, 2047^{VP}
Lentzea albida Labeda et al. 2001, 1049^{VP}
Lentzea californiensis Labeda et al. 2001, 1049^{VP}
Lentzea flaviverrucosa Xie et al. 2002, 1818^{VP}
Leptospirillum ferriphilum Coram and Rawlings 2002b, 1075^{VP} (Effective publication: Coram and Rawlings, 2002a)
Leptospirillum ferrooxidans (ex Markosyan 1972) Hippe 2000, 502^{VP}
Leptospirillum thermoferrooxidans (ex Golovacheva et al. 1992) Hippe 2000, 502^{VP}
Leptotrichia trevisanii Tee et al. 2002, 686^{VP} (Effective publication: Tee et al., 2001)
Leuconostoc ficulneum Antunes et al. 2002, 653^{VP}
Leuconostoc gasicomitatum Björkroth et al. 2001, 264^{VP} (Effective publication: Björkroth et al., 2000)
Leuconostoc inhae Kim et al. 2003a, 1126^{VP}
Leuconostoc kimchii Kim et al. 2000b, 1918^{VP}
Limnobacter thiooxidans Spring et al. 2001, 1469^{VP}
Longispora albida Matsumoto et al. 2003, 1558^{VP}
Luteimonas mephitis Finkmann et al. 2000, 280^{VP}
Luteococcus peritonei Collins et al. 2000f, 181^{VP}
Luteococcus sanguinis Collins et al. 2003b, 1891^{VP}
Macrococcus brunensis Mannerova et al. 2003, 1653^{VP}
Macrococcus hajekii Mannerova et al. 2003, 1653^{VP}
Macrococcus lamae Mannerova et al. 2003, 1653^{VP}
Maricaulis parjimensis Abraham et al. 2002, 2199^{VP}
Maricaulis salignorans Abraham et al. 2002, 2199^{VP}
Maricaulis virginensis Abraham et al. 2002, 2200^{VP}
Maricaulis washingtonensis Abraham et al. 2002, 2200^{VP}
Marinilactibacillus psychrotolerans Ishikawa et al. 2003b, 719^{VP}
Marinithermus hydrothermalis Sako et al. 2003, 64^{VP}
Marinitoga camini Wery et al. 2001a, 502^{VP}
Marinitoga piezophila Alain et al. 2002a, 1337^{VP}
Marinobacter excellens Gorshkova et al. 2003, 2077^{VP}
Marinobacter lipolyticus Martin et al. 2003, 1386^{VP}
Marinobacter litoralis Yoon et al. 2003j, 567^{VP}
Marinobacter lutaoensis Shieh et al. 2003c, 1701^{VP} (Effective publication: Shieh et al., 2003b)
Marinomonas primoryensis Romanenko et al. 2003c, 831^{VP}
Marinospirillum alkaliphilum Zhang et al. 2002e, 1437^{VP} (Effective publication: Zhang et al., 2002d)
Marmoricola aurantiacus Urcì et al. 2000, 534^{VP}
Massilia timonae La Scola et al. 2000, 423^{VP} (Effective publication: La Scola et al., 1998)
Megasphaera micronuciformis Marchandin et al. 2003, 552^{VP}
Meiothermus taiwanensis Chen et al. 2002a, 1653^{VP}
Mesonia algae Nedashkovskaya et al. 2003a, 1970^{VP}
Mesorhizobium chacoense Velázquez et al. 2001, 1019^{VP}
Methanobacterium congolense Cuzin et al. 2001, 492^{VP}
Methanobacterium oryzae Joulian et al. 2000, 527^{VP}
Methanobrevibacter acididurans Savant et al. 2002, 1086^{VP}
Methanobrevibacter gottschalkii Miller and Lin 2002, 820^{VP}
Methanobrevibacter thaueri Miller and Lin 2002, 820^{VP}
Methanobrevibacter woesei Miller and Lin 2002, 821^{VP}
Methanobrevibacter wolinii Miller and Lin 2002, 821^{VP}
Methanocalculus pumilus Mori et al. 2000, 1728^{VP}
Methanocalculus taiwanensis Lai et al. 2002, 1805^{VP}
Methanocaldococcus indicus L'Haridon et al. 2003, 1934^{VP}
Methanoculleus chikugoensis Dianou et al. 2001, 1667^{VP}
Methanoculleus submarinus Mikucki et al. 2003b, 1701^{VP} (Effective publication: Mikucki et al., 2003a)
Methanofollis aquaemaris Lai and Chen 2001, 1878^{VP}
Methanogenium marinum Chong et al. 2003, 1701^{VP} (Effective publication: Chong et al., 2002)
Methanomicrococcus blatticola Sprenger et al. 2000, 1998^{VP}
Methanosarcina baltica von Klein et al. 2002b, 686^{VP} (Effective publication: von Klein et al., 2002a)
Methanosarcina lacustris Simankova et al. 2002, 686^{VP} (Effective publication: Simankova et al., 2001b)
Methanosarcina semesiae Lyimo et al. 2000, 177^{VP}
Methanothermobacter marburgensis Wasserfallen et al. 2000, 52^{VP}
Methanothermococcus okinawensis Takai et al. 2002, 1094^{VP}
Methylarcula marina Doronina et al. 2000b, 1858^{VP}
Methylarcula terricola Doronina et al. 2000b, 1858^{VP}
Methylobacter psychrophilus Omel'chenko et al. 2000, 423^{VP} (Effective publication: Omel'chenko et al., 1996)
Methylobacterium chloromethanicum McDonald et al. 2001, 121^{VP}
Methylobacterium dichloromethanicum Doronina et al. 2000d, 1953^{VP} (Effective publication: Doronina et al., 2000c)
Methylobacterium lusitanum Doronina et al. 2002, 775^{VP}
Methylobacterium suomiense Doronina et al. 2002, 775^{VP}
Methylocapsa acidiphila Dedysh et al. 2002, 260^{VP}
Methylocella palustris Dedysh et al. 2000, 967^{VP}
Methylocella silvestris Dunfield et al. 2003, 1238^{VP}
Methylomicrobium buryatense Kalyuzhnaya et al. 2001, 1945^{VP} (Effective publication: Kaluzhnaya et al., 2001)
Methylomonas scandinavica Kalyuzhnaya et al. 2000, 949^{VP} (Effective publication: Kalyuzhnaya et al., 1999)
Methylophaga alcalica Doronina et al. 2003, 228^{VP}
Methylophilus leisingeri Doronina and Trotsenko 2001b, 1^{VP} (Effective publication: Doronina and Trotsenko 1994)

Methylopila helvetica Doronina et al. 2000d, 1953^{VP} (Effective publication: Doronina et al., 2000c)
Methylosarcina fibrata Wise et al. 2001, 620^{VP}
Methylosarcina quisquiliarum Wise et al. 2001, 620^{VP}
Methylovorus mays Doronina et al. 2001a, 1619^{VP} (Effective publication: Doronina et al., 2000a)
Microbacterium aerolatum Zlamala et al. 2002b, 1233^{VP}
Microbacterium foliorum Behrendt et al. 2001, 1273^{VP}
Microbacterium gubbeenense Brennan et al. 2001b, 1974^{VP}
Microbacterium paraoxydans Laffineur et al. 2003b, 936^{VP} (Effective publication: Laffineur et al., 2003a)
Microbacterium phyllosphaerae Behrendt et al. 2001, 1273^{VP}
Microbulbifer salipaludis Yoon et al. 2003g, 57^{VP}
Micrococcus antarcticus Liu et al. 2000, 718^{VP}
Micropruina glycogenica Shintani et al. 2000, 206^{VP}
Microvirga subterranea Kanso and Patel 2003, 405^{VP}
Mitsuokella jalaludinii Lan et al. 2002, 717^{VP}
Modestobacter multiseptatus Mevs et al. 2000, 344^{VP}
Mogibacterium diversum Nakazawa et al. 2002, 121^{VP}
Mogibacterium neglectum Nakazawa et al. 2002, 121^{VP}
Mogibacterium pumilum Nakazawa et al. 2000, 686^{VP}
Mogibacterium vescum Nakazawa et al. 2000, 686^{VP}
Moritella abyssi Xu et al. 2003, 537^{VP}
Moritella profunda Xu et al. 2003, 536^{VP}
Muricauda ruestringensis Bruns et al. 2001, 2005^{VP}
Muricoccus roseus Kämpfer et al. 2003b, 936^{VP} (Effective publication: Kämpfer et al., 2003a)
Mycetocola lacteus Tsukamoto et al. 2001, 942^{VP}
Mycetocola saprophilus Tsukamoto et al. 2001, 942^{VP}
Mycetocola tolaasinivorans Tsukamoto et al. 2001, 942^{VP}
Mycobacterium botniense Torkko et al. 2000, 288^{VP}
Mycobacterium doricum Tortoli et al. 2001, 2011^{VP}
Mycobacterium elephantis Shojaei et al. 2000, 1819^{VP}
Mycobacterium frederiksbergense Willumsen et al. 2001, 1719^{VP}
Mycobacterium heckeshornense Roth et al. 2001, 264^{VP} (Effective publication: Roth et al., 2000)
Mycobacterium holsaticum Richter et al. 2002, 1995^{VP}
Mycobacterium immunogenum Wilson et al. 2001, 1762^{VP}
Mycobacterium kubicae Floyd et al. 2000, 1814^{VP}
Mycobacterium lacus Turenne et al. 2002, 2138^{VP}
Mycobacterium montefiorense Levi et al. 2003b, 1701^{VP} (Effective publication: Levi et al., 2003a)
Mycobacterium palustre Torkko et al. 2002, 1524^{VP}
Mycobacterium pinnipedii Cousins et al. 2003, 1312^{VP}
Mycobacterium septicum Schinsky et al. 2000, 580^{VP}
Mycobacterium shottsii Rhodes et al. 2003, 424^{VP}
Mycobacterium vanbaalenii Khan et al. 2002, 2001^{VP}
Mycoplasma agassizii Brown et al. 2001e, 417^{VP}
Mycoplasma alligatoris Brown et al. 2001a, 423^{VP}
Mycoplasma microti Brown et al. 2001b, 412^{VP}
Natrialba aegyptia Hezayen et al. 2001, 1140^{VP}
Natrialba chahannaoensis Xu et al. 2001, 1697^{VP}
Natrialba hulunbeirensis Xu et al. 2001, 1696^{VP}
Natrialba taiwanensis Hezayen et al. 2001, 1140^{VP}
Natrinema versiforme Xin et al. 2000, 1302^{VP}
Nautilia lithotrophica Miroshnichenko et al. 2002, 1303^{VP}
Neochlamydia hartmannellae Horn et al. 2001, 1229^{VP} (Effective publication: Horn et al., 2000)
Nesterenkonia lacusekhoensis Collins et al. 2002e, 1149^{VP}
Nitrobacter alkalicus Sorokin et al., 2001b, 1^{VP} (Effective publication: Sorokin et al., 1998)
Nitrobacter hamburgensis Bock et al. 2001b, 1^{VP} (Effective publication: Bock et al., 1983)
Nitrobacter vulgaris Bock et al. 2001a, 1^{VP} (Effective publication: Bock et al., 1990)
Nitrosomonas aestuarii Koops et al. 2001, 1945^{VP} (Effective publication: Koops et al., 1991)
Nitrosomonas communis Koops et al. 2001, 1945^{VP} (Effective publication: Koops et al., 1991)
Nitrosomonas eutropha Koops et al. 2001, 1945^{VP} (Effective publication: Koops et al., 1991)
Nitrosomonas halophila Koops et al. 2001, 1945^{VP} (Effective publication: Koops et al., 1991)
Nitrosomonas marina Koops et al. 2001, 1945^{VP} (Effective publication: Koops et al., 1991)
Nitrosomonas nitrosa Koops et al. 2001, 1945^{VP} (Effective publication: Koops et al., 1991)
Nitrosomonas oligotropha Koops et al. 2001, 1945^{VP} (Effective publication: Koops et al., 1991)
Nitrosomonas ureae Koops et al. 2001, 1945^{VP} (Effective publication: Koops et al., 1991)
Nitrospira moscoviensis Ehrich et al. 2001, 1^{VP} (Effective publication: Ehrich et al., 1995)
Nocardia abscessus Yassin et al. 2000b, 1492^{VP}
Nocardia africana Hamid et al. 2001b, 1229^{VP} (Effective publication: Hamid et al., 2001a)
Nocardia beijingensis Wang et al. 2001a, 1785^{VP}
Nocardia caishijiensis Zhang et al. 2003c, 1003^{VP}
Nocardia cerradoensis Albuquerque de Barros et al. 2003, 32^{VP}
Nocardia cummidelens Maldonado et al. 2001, 1619^{VP} (Effective publication: Maldonado et al., 2000)
Nocardia cyriacigeorgica Yassin et al. 2001a, 1422^{VP}
Nocardia fluminea Maldonado et al. 2001, 1619^{VP} (Effective publication: Maldonado et al., 2000)
Nocardia ignorata Yassin et al. 2001b, 2130^{VP}
Nocardia paucivorans Yassin et al. 2000a, 807^{VP}
Nocardia pseudovaccinii Kim et al. 2002b, 1828^{VP}
Nocardia puris Yassin et al. 2003b, 1598^{VP}
Nocardia soli Maldonado et al. 2001, 1619^{VP} (Effective publication: Maldonado et al., 2000)
Nocardia veterana Gürtler et al. 2001, 935^{VP}
Nocardia vinacea Kinoshita et al. 2002, 3^{VP} (Effective publication: Kinoshita et al., 2001)
Nocardioides aquaticus Lawson et al. 2000b, 1953^{VP} (Effective publication: Lawson et al., 2000a)
Nocardiopsis composta Kämpfer et al. 2002, 627^{VP}
Nocardiopsis exhalans Peltola et al. 2002, 3^{VP} (Effective publication: Peltola et al., 2001)
Nocardiopsis halotolerans Al-Zarban et al. 2002a, 528^{VP}
Nocardiopsis kunsanensis Chun et al. 2000, 1911^{VP}
Nocardiopsis metallicus Schippers et al. 2002, 2294^{VP}
Nocardiopsis trehalosi Evtushenko et al. 2000b, 79^{VP}
Nocardiopsis tropica Evtushenko et al. 2000b, 79^{VP}
Nocardiopsis umidischolae Peltola et al. 2002, 3^{VP} (Effective publication: Peltola et al., 2001)
Nocardiopsis xinjiangensis Li et al. 2003a, 320^{VP}
Nonomuraea dietziae Stackebrandt et al. 2001, 1439^{VP}
Novosphingobium hassiacum Kämpfer et al. 2002e, 1437^{VP} (Effective publication: Kämpfer et al., 2002c)
Novosphingobium tardaugens Fujii et al. 2003, 51^{VP}
Oceanicaulis alexandrii Strömpl et al. 2003, 1905^{VP}
Oceanimonas baumannii Brown et al. 2001c, 71^{VP}
Oceanisphaera litoralis Romanenko et al. 2003a, 1888^{VP}

Oceanithermus profundus Miroshnichenko et al. 2003b, 751^{VP}
Oceanobacillus iheyensis Lu et al. 2002, 687^{VP} (Effective publication: Lu et al., 2001a)
Ochrobactrum gallinifaecis Kämpfer et al. 2003c, 896^{VP}
Ochrobactrum grignonense Lebuhn et al. 2000, 2221^{VP}
Ochrobactrum tritici Lebuhn et al. 2000, 2222^{VP}
Oerskovia jenensis Stackebrandt et al. 2002a, 1110^{VP}
Oerskovia paurometabola Stackebrandt et al. 2002a, 1110^{VP}
Okibacterium fritillariae Evtushenko et al. 2002, 992^{VP}
Oleiphilus messinensis Golyshin et al. 2002, 910^{VP}
Oleispira antarctica Yakimov et al. 2003a, 784^{VP}
Olsenella profusa Dewhirst et al. 2001, 1803^{VP}
Opitutus terrae Chin et al. 2001, 1968^{VP}
Orenia salinaria Mouné et al. 2000, 728^{VP}
Orenia sivashensis Zhilina et al. 2000, 3^{VP} (Effective publication: Zhilina et al., 1999)
Ornithinimicrobium humiphilum Groth et al. 2001, 85^{VP}
Oxalicibacterium flavum Tamer et al. 2003, 627^{VP} (Effective publication: Tamer et al., 2002)
Paenibacillus agarexedens (ex Wieringa 1941) Uetanabaro et al. 2003, 1056^{VP}
Paenibacillus agaridevorans Uetanabaro et al. 2003, 1056^{VP}
Paenibacillus azoreducens Meehan et al. 2001, 1684^{VP}
Paenibacillus borealis Elo et al. 2001, 542^{VP}
Paenibacillus brasilensis von der Weid et al. 2002, 2152^{VP}
Paenibacillus chinjuensis Yoon et al. 2002c, 419^{VP}
Paenibacillus daejeonensis Lee et al. 2002b, 2110^{VP}
Paenibacillus glycanilyticus Dasman et al. 2002, 1671^{VP}
Paenibacillus graminis Berge et al. 2002, 613^{VP}
Paenibacillus granivorans van der Maarel et al. 2001, 264^{VP} (Effective publication: van der Maarel et al., 2000)
Paenibacillus jamilae Aguilera et al. 2001, 1691^{VP}
Paenibacillus koleovorans Takeda et al. 2002b, 1600^{VP}
Paenibacillus koreensis Chung et al. 2000b, 1499^{VP}
Paenibacillus kribbensis Yoon et al. 2003i, 300^{VP}
Paenibacillus naphthalenovorans Daane et al. 2002, 137^{VP}
Paenibacillus nematophilus Enright et al. 2003, 440^{VP}
Paenibacillus odorifer Berge et al. 2002, 614^{VP}
Paenibacillus stellifer Suominen et al. 2003, 1373^{VP}
Paenibacillus terrae Yoon et al. 2003i, 300^{VP}
Paenibacillus turicensis Bosshard et al. 2002a, 2247^{VP}
Palaeococcus ferrophilus Takai et al. 2000, 498^{VP}
Pandoraea apista Coenye et al. 2000, 896^{VP}
Pandoraea pnomenusa Coenye et al. 2000, 896^{VP}
Pandoraea pulmonicola Coenye et al. 2000, 896^{VP}
Pandoraea sputorum Coenye et al. 2000, 897^{VP}
Pannonibacter phragmitetus Borsodi et al. 2003, 560^{VP}
Papillibacter cinnamivorans Defnoun et al. 2000, 1227^{VP}
Paracoccus kondratievae Doronina and Trotsenko 2001a^{VP} emend. Doronina et al., 2002 (Effective publication: Doronina and Trotsenko, 2000)
Paracoccus seriniphilus Pukall et al. 2003, 446^{VP}
Paracoccus yeei Daneshvar et al. 2003b, 936^{VP} (Effective publication: Daneshvar et al., 2003a)
Paracoccus zeaxanthinifaciens Berry et al. 2003, 237^{VP}
Paralactobacillus selangorensis Leisner et al. 2000, 23^{VP}
Paraliobacillus ryukyuensis Ishikawa et al. 2003a, 627^{VP} (Effective publication: Ishikawa et al., 2002)
Parvularcula bermudensis Cho and Giovannoni 2003c, 1035^{VP}
Pasteurella skyensis Birkbeck et al. 2002, 703^{VP}
Pediococcus claussenii Dobson et al. 2002, 2009^{VP}
Pedobacter cryoconitis Margesin et al. 2003, 1295^{VP}
Pelospora glutarica Matthies et al. 2000b, 647^{VP}
Pelotomaculum thermopropionicum Imachi et al. 2002, 1734^{VP}
Persephonella guaymasensis Götz et al. 2002, 1358^{VP}
Persephonella hydrogeniphila Nakagawa et al. 2003, 868^{VP}
Persephonella marina Götz et al. 2002, 1357^{VP}
Petrotoga olearia L'Haridon et al. 2002, 1720^{VP}
Petrotoga sibirica L'Haridon et al. 2002, 1720^{VP}
Phocoenobacter uteri Foster et al. 2000, 139^{VP}
Pigmentiphaga kullae Blümel et al. 2001b, 1870^{VP}
Planococcus alkanoclasticus Engelhardt et al. 2001b, 1229^{VP} (Effective publication: Engelhardt et al., 2001a)
Planococcus antarcticus Reddy et al. 2002c, 1437^{VP} (Effective publication: Reddy et al., 2002b)
Planococcus maritimus Yoon et al. 2003k, 2016^{VP}
Planococcus psychrophilus Reddy et al. 2002c, 1437^{VP} (Effective publication: Reddy et al., 2002b)
Planococcus rifietoensis Romano et al. 2003b, 1701^{VP} (Effective publication: Romano et al., 2003a)
Planomicrobium koreense Yoon et al. 2001c, 1518^{VP}
Plantibacter flavus Behrendt et al. 2002, 1451^{VP}
Plesiocystis pacifica Iizuka et al. 2003a, 195^{VP}
Porphyrobacter cryptus Rainey et al. 2003, 41^{VP}
Porphyrobacter sanguineus Hiraishi et al. 2002b, 1915^{VP} (Effective publication: Hiraishi et al., 2002a)
Porphyromonas gulae Fournier et al. 2001, 1187^{VP}
Prauserella alba Li et al. 2003c, 1548^{VP}
Prauserella halophila Li et al. 2003c, 1548^{VP}
Prochlorococcus marinus Chisholm et al. 2001, 264^{VP} (Effective publication: Chisholm et al., 1992)
Promicromonospora aerolata Busse et al. 2003b, 1506^{VP}
Promicromonospora vindobonensis Busse et al. 2003b, 1505^{VP}
Propionibacterium australiense Bernard et al. 2002b, 1915^{VP} (Effective publication: Bernard et al., 2002a)
Propionibacterium microaerophilum Koussémon et al. 2001, 1380^{VP}
Propionicimonas paludicola Akasaka et al. 2003, 1996^{VP}
Propionispora vibrioides Biebl et al. 2001, 793^{VP} (Effective publication: Biebl et al., 2000)
Propionivibrio limicola Brune et al. 2002b, 443^{VP}
Proteus hauseri O'Hara et al. 2000, 1874^{VP}
Pseudoalteromonas agarivorans Romanenko et al. 2003f, 130^{VP}
Pseudoalteromonas flavipulchra Ivanova et al. 2002d, 269^{VP}
Pseudoalteromonas issachenkonii Ivanova et al. 2002b, 233^{VP}
Pseudoalteromonas maricaloris Ivanova et al. 2002d, 269^{VP}
Pseudoalteromonas mariniglutinosa (ex Berland et al. 1969) Romanenko et al., 2003d, 1108^{VP}
Pseudoalteromonas paragorgicola Ivanova et al. 2002c, 1765^{VP}
Pseudoalteromonas peptidolytica Venkateswaran and Dohmoto 2000, 572^{VP}
Pseudoalteromonas phenolica Isnansetyo and Kamei 2003, 586^{VP}
Pseudoalteromonas ruthenica Ivanova et al. 2002b, 239^{VP}
Pseudoalteromonas sagamiensis Kobayashi et al., 2003, 1810^{VP}
Pseudoalteromonas translucida Ivanova et al. 2002c, 1765^{VP}
Pseudoalteromonas ulvae Egan et al. 2001, 1503^{VP}
Pseudobutyrivibrio xylanivorans Kopecny et al. 2003, 208^{VP}
Pseudomonas alcaliphila Yumoto et al. 2001b, 354^{VP}
Pseudomonas brassicacearum Achouak et al. 2000, 16^{VP}
Pseudomonas brenneri Baïda et al. 2002b, 1437^{VP} (Effective publication: Baïda et al., 2001)
Pseudomonas cedrina Dabboussi et al. 2002a, 1437^{VP} (Effective publication: Dabboussi et al., 1999)
Pseudomonas chloritidismutans Wolterink et al. 2002, 2188^{VP}
Pseudomonas congelans Behrendt et al. 2003, 1467^{VP}

Pseudomonas costantinii Munsch et al. 2002, 1981^{VP}

Pseudomonas cremoricolorata Uchino et al. 2002a, 687^{VP} (Effective publication: Uchino et al., 2001)

Pseudomonas extremorientalis Ivanova et al. 2002a, 2118^{VP}

Pseudomonas frederiksbergensis Andersen et al. 2000, 1962^{VP}

Pseudomonas grimontii Baïda et al. 2002a, 1502^{VP}

Pseudomonas indica Pandey et al. 2002, 1566^{VP}

Pseudomonas jinjuensis Kwon et al. 2003, 27^{VP}

Pseudomonas kilonensis Sikorski et al. 2001, 1554^{VP}

Pseudomonas koreensis Kwon et al. 2003, 26^{VP}

Pseudomonas lini Delorme et al. 2002, 521^{VP}

Pseudomonas mediterranea Catara et al. 2002, 1756^{VP}

Pseudomonas mosselii Dabboussi et al. 2002b, 374^{VP}

Pseudomonas orientalis Dabboussi et al. 2002a, 1438^{VP} (Effective publication: Dabboussi et al., 1999)

Pseudomonas palleroniana Gardan et al. 2002, 2074^{VP}

Pseudomonas parafulva Uchino et al. 2002a, 687^{VP} (Effective publication: Uchino et al., 2001)

Pseudomonas plecoglossicida Nishimori et al. 2000, 87^{VP}

Pseudomonas poae Behrendt et al. 2003, 1467^{VP}

Pseudomonas psychrophila Yumoto et al. 2002a, 687^{VP} (Effective publication: Yumoto et al., 2001a)

Pseudomonas rhizosphaerae Peix et al. 2003, 2070^{VP}

Pseudomonas salomonii Gardan et al. 2002, 2073^{VP}

Pseudomonas thermotolerans Manaia and Moore 2002, 2208^{VP}

Pseudomonas thivervalensis Achouak et al. 2000, 17^{VP}

Pseudomonas trivialis Behrendt et al. 2003, 1467^{VP}

Pseudomonas umsongensis Kwon et al. 2003, 26^{VP}

Pseudonocardia kongjuensis Lee et al. 2001c, 1509^{VP}

Pseudonocardia spinosispora Lee et al. 2002c, 1607^{VP}

Pseudonocardia zijingensis Huang et al. 2002, 971^{VP}

Pseudoxanthomonas broegbernensis Finkmann et al. 2000, 280^{VP}

Pseudoxanthomonas taiwanensis Chen et al. 2002b, 2160^{VP}

Psychrobacter faecalis Kämpfer et al. 2002e, 1438^{VP} (Effective publication: Kämpfer et al., 2002a)

Psychrobacter fozii Bozal et al. 2003, 1099^{VP}

Psychrobacter jeotgali Yoon et al. 2003b, 453^{VP}

Psychrobacter luti Bozal et al. 2003, 1098^{VP}

Psychrobacter marincola Romanenko et al. 2002a, 1296^{VP}

Psychrobacter okhotskensis Yumoto et al. 2003a, 1988^{VP}

Psychrobacter pacificensis Maruyama et al. 2000, 845^{VP}

Psychrobacter proteolyticus Denner et al. 2001b, 1619^{VP} (Effective publication: Denner et al., 2001a)

Psychrobacter pulmonis Vela et al. 2003a, 418^{VP}

Psychrobacter submarinus Romanenko et al. 2002a, 1296^{VP}

Psychromonas arctica Groudieva et al. 2003, 544^{VP}

Psychromonas kaikoae Nogi et al. 2002, 1531^{VP}

Psychromonas marina Kawasaki et al. 2002, 1458^{VP}

Psychromonas profunda Xu et al. 2003, 531^{VP}

Pyrobaculum arsenaticum Huber et al. 2001, 264^{VP} (Effective publication: Huber et al., 2000b)

Pyrobaculum oguniense Sako et al. 2001, 308^{VP}

Quadricoccus australiensis Maszenan et al. 2002, 227^{VP}

Ralstonia campinensis Goris et al. 2001, 1780^{VP}

Ralstonia insidiosa Coenye et al. 2003a, 1079^{VP}

Ralstonia mannitolilytica De Baere et al. 2001, 556^{VP}

Ralstonia metallidurans Goris et al. 2001, 1780^{VP}

Ralstonia respiraculi Coenye et al. 2003d, 1341^{VP}

Ralstonia taiwanensis Chen et al. 2001, 1734^{VP}

Ramlibacter henchirensis Heulin et al. 2003, 594^{VP}

Ramlibacter tataouinensis Heulin et al. 2003, 593^{VP}

Rathayibacter caricis Dorofeeva et al. 2002, 1921^{VP}

Rathayibacter festucae Dorofeeva et al. 2002, 1921^{VP}

Reichenbachia agariperforans Nedashkovskaya et al. 2003b, 84^{VP}

Rheinheimera baltica Brettar et al. 2002a, 1856^{VP}

Rheinheimera pacifica Romanenko et al. 2003b, 1976^{VP}

Rhizobium indigoferae Wei et al. 2002, 2237^{VP}

Rhizobium loessense Wei et al. 2003, 1582^{VP}

Rhizobium sullae Squartini et al. 2002, 1274^{VP}

Rhizobium yanglingense Tan et al. 2001, 913^{VP}

Rhodobaca bogoriensis Milford et al. 2001, 793^{VP} (Effective publication: Milford et al., 2000)

Rhodocista pekingensis Zhang et al. 2003a, 1114^{VP}

Rhodococcus jostii Takeuchi et al. 2002, 413^{VP}

Rhodococcus koreensis Yoon et al. 2000a, 1199^{VP}

Rhodococcus maanshanensis Zhang et al. 2002b, 2124^{VP}

Rhodococcus pyridinivorans Yoon et al. 2000b, 2178^{VP}

Rhodococcus tukisamuensis Matsuyama et al. 2003, 1335^{VP}

Rhodoferax antarcticus Madigan et al. 2001, 793^{VP} (Effective publication: Madigan et al., 2000)

Rhodoferax ferrireducens Finneran et al. 2003, 673^{VP}

Rhodoglobus vestalii Sheridan et al. 2003, 992^{VP}

Rhodopseudomonas faecalis Zhang et al. 2002, 2059^{VP}

Rhodopseudomonas rhenobacensis Hougardy et al. 2000, 991^{VP}

Rickettsia felis Bouyer et al. 2001^{VP} emend. La Scola et al., 2002

Roseburia intestinalis Duncan et al. 2002a, 1619^{VP}

Roseibium denhamense Suzuki et al. 2000b, 2155^{VP}

Roseibium hamelinense Suzuki et al. 2000b, 2155^{VP}

Roseiflexus castenholzii Hanada et al. 2002b, 192^{VP}

Roseinatronobacter thiooxidans Sorokin et al. 2000b, 1415^{VP} (Effective publication: Sorokin et al., 2000a)

Roseomonas mucosa Han et al. 2003c, 1701^{VP} (Effective publication: Han et al., 2003b)

Roseospira marina Guyoneaud et al. 2003, 1701^{VP} (Effective publication: Guyoneaud et al., 2002)

Roseospira navarrensis Guyoneaud et al. 2003, 1701^{VP} (Effective publication: Guyoneaud et al., 2002)

Roseospirillum parvum Glaeser and Overmann 2001, 793^{VP} (Effective publication: Glaeser and Overmann, 1999)

Roseovarius nubinhibens González et al. 2003, 1268^{VP}

Rothia amarae Fan et al. 2002, 2259^{VP}

Rothia nasimurium Collins et al. 2000c, 1250^{VP}

Rubritepida flocculans Alarico et al. 2002b, 1915^{VP} (Effective publication: Alarico et al., 2002a)

Ruminococcus luti Simmering et al. 2002b, 1915^{VP} (Effective publication: Simmering et al., 2002a)

Runella zeae Chelius et al. 2002, 2062^{VP}

Saccharomonospora halophila Al-Zarban et al. 2002b, 557^{VP}

Saccharomonospora paurometabolica Li et al. 2003b, 1593^{VP}

Saccharopolyspora flava Lu et al. 2001b, 322^{VP}

Saccharopolyspora thermophila Lu et al. 2001b, 322^{VP}

Saccharospirillum impatiens Labrenz et al. 2003, 659^{VP}

Saccharothrix tangerinus Kinoshita et al. 2000, 949^{VP} (Effective publication: Kinoshita et al., 1999)

Salana multivorans von Wintzingerode et al. 2001a, 1660^{VP}

Salinibacter ruber Antón et al. 2002, 490^{VP}

Salinibacterium amurskyense Han et al. 2003a, 2065^{VP}

Salinicoccus alkaliphilus Zhang et al. 2002f, 792^{VP}

Salinisphaera shabanensis Antunes et al. 2003b, 1219^{VP} (Effective publication: Antunes et al., 2003a)

Samsonia erythrinae Sutra et al. 2001, 1301^{VP}

Schineria larvae Tóth et al. 2001, 406^{VP}

Schlegelella thermodepolymerans Elbanna et al. 2003, 1167^{VP}

Sedimentibacter saalensis Breitenstein et al. 2002, 806^{VP}

Selenihalanaerobacter shriftii Switzer Blum et al. 2001c, 1229^{VP} (Effective publication: Switzer Blum et al., 2001b)
Shewanella denitrificans Brettar et al. 2002b, 2216^{VP}
Shewanella fidelis Ivanova et al. 2003b, 581^{VP}
Shewanella japonica Ivanova et al. 2001c, 1032^{VP}
Shewanella livingstonensis Bozal et al. 2002, 202^{VP}
Shewanella marinintestina Satomi et al. 2003, 497^{VP}
Shewanella olleyana Skerratt et al. 2002, 2104^{VP}
Shewanella sairae Satomi et al. 2003, 497^{VP}
Shewanella schlegeliana Satomi et al. 2003, 497^{VP}
Shewanella waksmanii Ivanova et al. 2003a, 1476^{VP}
Shuttleworthia satelles Downes et al. 2002, 1474^{VP}
Silicibacter pomeroyi González et al. 2003, 1268^{VP}
Sneathia sanguinegens Collins et al. 2002b, 687^{VP} (Effective publication: Collins et al., 2001d)
Soehngenia saccharolytica Parshina et al. 2003, 1797^{VP}
Solirubrobacter pauli Singleton et al. 2003, 489^{VP}
Solobacterium moorei Kageyama and Benno 2000f, 1415^{VP} (Effective publication: Kageyama and Benno, 2000d)
Sphingobium amiense Ushiba et al. 2003, 2048^{VP}
Sphingomonas aerolata Busse et al. 2003a, 1259^{VP}
Sphingomonas aquatilis Lee et al. 2001b, 1495^{VP}
Sphingomonas aurantiaca Busse et al. 2003a, 1259^{VP}
Sphingomonas chungbukensis Kim et al. 2000d, 1646^{VP}
Sphingomonas cloacae Fujii et al. 2001, 608^{VP}
Sphingomonas faeni Busse et al. 2003a, 1259^{VP}
Sphingomonas koreensis Lee et al. 2001b, 1496^{VP}
Sphingomonas melonis Buonaurio et al. 2002, 2086^{VP}
Sphingomonas pituitosa Denner et al. 2001c, 837^{VP}
Sphingomonas roseiflava Yun et al. 2000b, 1415^{VP} (Effective publication: Yun et al., 2000a)
Sphingomonas taejonensis Lee et al. 2001b, 1497^{VP}
Sphingomonas wittichii Yabuuchi et al. 2001, 289^{VP}
Sphingomonas xenophaga Stolz et al. 2000, 40^{VP}
Sphingopyxis chilensis Godoy et al. 2003, 476^{VP}
Sphingopyxis witflariensis Kämpfer et al. 2002d, 2032^{VP}
Spirochaeta americana Hoover et al. 2003, 820^{VP}
Sporanaerobacter acetigenes Hernandez-Eugenio et al. 2002b, 1221^{VP}
Sporomusa aerivorans Boga et al. 2003, 1403^{VP}
Sporosarcina aquimarina Yoon et al. 2001d, 1084^{VP}
Sporosarcina macmurdoensis Reddy et al. 2003a, 1366^{VP}
Sporotomaculum syntrophicum Qiu et al. 2003b, 936^{VP} (Effective publication: Qiu et al., 2003a)
Staleya guttiformis Labrenz et al. 2000, 311^{VP}
Staphylococcus fleurettii Vernozy-Rozand et al. 2000, 1523^{VP}
Staphylococcus nepalensis Spergser et al. 2003, 2010^{VP}
Staphylothermus hellenicus Arab et al. 2000, 2106^{VP}
Stenotrophomonas acidaminiphila Assih et al. 2002, 567^{VP}
Stenotrophomonas nitritireducens Finkmann et al. 2000, 281^{VP}
Stenotrophomonas rhizophila Wolf et al. 2002a, 1943^{VP}
Sterolibacterium denitrificans Tarlera and Denner 2003, 1090^{VP}
Streptacidiphilus albus Kim et al. 2003d, 1219^{VP} (Effective publication: Kim et al., 2003c)
Streptacidiphilus carbonis Kim et al. 2003d, 1219^{VP} (Effective publication: Kim et al., 2003c)
Streptacidiphilus neutrinimicus Kim et al. 2003d, 1219^{VP} (Effective publication: Kim et al., 2003c)
Streptococcus australis Willcox et al. 2001, 1281^{VP}
Streptococcus didelphis Rurangirwa et al. 2000, 765^{VP}
Streptococcus entericus Vela et al. 2002, 668^{VP}
Streptococcus gallinaceus Collins et al. 2002c, 1163^{VP}
Streptococcus infantarius Schlegel et al. 2000, 1432^{VP}
Streptococcus lutetiensis Poyart et al. 2002, 1253^{VP}
Streptococcus oligofermentans Tong et al. 2003, 1103^{VP}
Streptococcus orisratti Zhu et al. 2000, 60^{VP}
Streptococcus ovis Collins et al. 2001e, 1149^{VP}
Streptococcus sinensis Woo et al. 2002b, 1438^{VP} (Effective publication: Woo et al., 2002a)
Streptococcus urinalis Collins et al. 2000d, 1177^{VP}
Streptomonospora alba Li et al. 2003d, 1424^{VP}
Streptomonospora salina Cui et al. 2001, 362^{VP}
Streptomyces asiaticus Sembiring et al. 2001, 1619^{VP} (Effective publication: Sembiring et al., 2000)
Streptomyces aureus Manfio et al. 2003b, 1219^{VP} (Effective publication: Manfio et al., 2003a)
Streptomyces avermectinius Takahashi et al. 2002, 2167^{VP}
Streptomyces avermitilis Kim and Goodfellow 2002a, 2013^{VP}
Streptomyces beijiangensis Li et al. 2002b, 1698^{VP}
Streptomyces cangkringensis Sembiring et al. 2001, 1619^{VP} (Effective publication: Sembiring et al., 2000)
Streptomyces europaeiscabiei Bouchek-Mechiche et al. 2000, 97^{VP}
Streptomyces indonesiensis Sembiring et al. 2001, 1619^{VP} (Effective publication: Sembiring et al., 2000)
Streptomyces javensis Sembiring et al. 2001, 1619^{VP} (Effective publication: Sembiring et al., 2000)
Streptomyces laceyi Manfio et al. 2003b, 1219^{VP} (Effective publication: Manfio et al., 2003a)
Streptomyces luridiscabiei Park et al. 2003a, 2053^{VP}
Streptomyces mexicanus Petrosyan et al. 2003, 273^{VP}
Streptomyces niveiciscabiei Park et al. 2003a, 2053^{VP}
Streptomyces puniciscabiei Park et al. 2003a, 2053^{VP}
Streptomyces reticuliscabiei Bouchek-Mechiche et al. 2000, 98^{VP}
Streptomyces rhizosphaericus Sembiring et al. 2001, 1619^{VP} (Effective publication: Sembiring et al., 2000)
Streptomyces sanglieri Manfio et al. 2003b, 1219^{VP} (Effective publication: Manfio et al., 2003a)
Streptomyces scopiformis Li et al.2002a, 1632^{VP}
Streptomyces speibonae Meyers et al. 2003, 804^{VP}
Streptomyces stelliscabiei Bouchek-Mechiche et al. 2000, 98^{VP}
Streptomyces thermocoprophilus Kim et al. 2000a, 506^{VP}
Streptomyces thermospinosisporus Kim and Goodfellow 2002b, 1227^{VP}
Streptomyces yatensis Saintpierre et al. 2003b, 1219^{VP} (Effective publication: Saintpierre et al., 2003a)
Streptomyces yogyakartensis Sembiring et al. 2001, 1619^{VP} (Effective publication: Sembiring et al., 2000)
Streptomyces yunnanensis Zhang et al. 2003d, 220^{VP}
Streptosporangium subroseum Zhang et al. 2002c, 1237^{VP}
Subtercola boreus Männistö et al. 2000, 1737^{VP}
Subtercola frigoramans Männistö et al. 2000, 1737^{VP}
Sulfitobacter brevis Labrenz et al. 2000, 311^{VP}
Sulfolobus tokodaii Suzuki et al. 2002b, 1438^{VP} (Effective publication: Suzuki et al., 2002a)
Sulfurihydrogenibium subterraneum Takai et al. 2003b, 826^{VP}
Sulfurimonas autotrophica Inagaki et al. 2003, 1805^{VP}
Sulfurospirillum halorespirans Luijten et al. 2003, 791^{VP}
Symbiobacterium thermophilum Ohno et al. 2000, 1832^{VP}
Syntrophothermus lipocalidus Sekiguchi et al. 2000, 778^{VP}
Syntrophus aciditrophicus Jackson et al. 2001, 793^{VP} (Effective publication: Jackson et al., 1999)
Taylorella asinigenitalis Jang et al. 2001, 975^{VP}
Teichococcus ludipueritiae Kämpfer et al. 2003b, 936^{VP} (Effective publication: Kämpfer et al., 2003a)
Tenacibaculum amylolyticum Suzuki et al. 2001, 1650^{VP}
Tenacibaculum mesophilum Suzuki et al. 2001, 1650^{VP}

Tepidibacter thalassicus Slobodkin et al. 2003, 1133^{VP}
Tepidimonas aquatica Freitas et al. 2003b, 1701^{VP} (Effective publication: Freitas et al., 2003a)
Tepidimonas ignava Moreira et al. 2000, 741^{VP}
Tepidiphilus margaritifer Manaia et al. 2003a, 1409^{VP}
Teredinibacter turnerae Distel et al. 2002, 2268^{VP}
Tetrasphaera australiensis Maszenan et al. 2000, 601^{VP}
Tetrasphaera elongata Hanada et al. 2002a, 886^{VP}
Tetrasphaera japonica Maszenan et al. 2000, 601^{VP}
Thalassomonas viridans Macián et al. 2001b, 1288^{VP}
Thalassospira lucentensis López-López et al. 2002, 1282^{VP}
Thauera aminoaromatica Mechichi et al. 2002b, 1438^{VP} (Effective publication: Mechichi et al., 2002a)
Thauera chlorobenzoica Song et al. 2001, 600^{VP}
Thauera phenylacetica Mechichi et al. 2002b, 1438^{VP} (Effective publication: Mechichi et al., 2002a)
Thermacetogenium phaeum Hattori et al. 2000, 1608^{VP}
Thermaerobacter nagasakiensis Nunoura et al. 2002b, 1075^{VP} (Effective publication: Nunoura et al., 2002a)
Thermaerobacter subterraneus Spanevello et al. 2002, 799^{VP}
Thermanaeromonas toyohensis Mori et al. 2002, 1679^{VP}
Thermanaerovibrio velox Zavarzina et al. 2000, 1293^{VP}
Thermicanus aegyptius Gößner et al. 2000, 423^{VP} (Effective publication: Gößner et al., 1999)
Thermoanaerobacter subterraneus Fardeau et al. 2000, 2145^{VP}
Thermoanaerobacter tengcongensis Xue et al. 2001, 1340^{VP}
Thermoanaerobacter yonseiensis Kim et al. 2001, 1546^{VP}
Thermoanaerobacterium polysaccharolyticum Cann et al. 2001, 299^{VP}
Thermoanaerobacterium zeae Cann et al. 2001, 300^{VP}
Thermobacillus xylanilyticus Touzel et al. 2000, 319^{VP}
Thermobifida cellulosilytica Kukolya et al. 2002, 1198^{VP}
Thermococcus acidaminovorans Dirmeier et al. 2001, 793^{VP} (Effective publication: Dirmeier et al., 1998)
Thermococcus aegaeus Arab et al. 2000, 2106^{VP}
Thermococcus gammatolerans Jolivet et al. 2003, 851^{VP}
Thermococcus litoralis Neuner et al. 2001, 1619^{VP} (Effective publication: Neuner et al., 1990)
Thermococcus sibiricus Miroshnichenko et al. 2001b, 1619^{VP} (Effective publication: Miroshnichenko et al., 2001a)
Thermococcus siculi Grote et al. 2000, 949^{VP} (Effective publication: Grote et al., 1999)
Thermococcus waiotapuensis González et al. 2001, 793^{VP} (Effective publication: González et al., 1999)
Thermocrinis albus Eder and Huber 2002b, 1915^{VP} (Effective publication: Eder and Huber, 2002a)
Thermodesulfobacterium hveragerdense Sonne-Hansen and Ahring 2000, 949^{VP} (Effective publication: Sonne-Hansen and Ahring, 1999)
Thermodesulfobacterium hydrogeniphilum Jeanthon et al. 2002, 770^{VP}
Thermodesulfovibrio islandicus Sonne-Hansen and Ahring 2000, 949^{VP} (Effective publication: Sonne-Hansen and Ahring, 1999)
Thermodiscus maritimus Stetter 2003, 1^{VP} (Effective publication: Stetter, 2001)
Thermohalobacter berrensis Cayol et al. 2000, 562^{VP}
Thermomonas brevis Mergaert et al. 2003a, 1966^{VP}
Thermomonas fusca Mergaert et al. 2003a, 1965^{VP}
Thermomonas haemolytica Busse et al. 2002, 480^{VP}
Thermomonas hydrothermalis Alves et al. 2003b, 936^{VP} (Effective publication: Alves et al., 2003a)
Thermoproteus uzoniensis Bonch-Osmolovskaya et al. 2001, 1619^{VP} (Effective publication: Bonch-Osmolovskaya et al., 1990)
Thermosipho geolei L'Haridon et al. 2001, 1332^{VP}
Thermosipho japonicus Takai and Horikoshi 2000b, 1699^{VP} (Effective publication: Takai and Horikoshi, 2000a)
Thermotoga lettingae Balk et al. 2002, 1367^{VP}
Thermotoga naphthophila Takahata et al. 2001, 1907^{VP}
Thermotoga petrophila Takahata et al. 2001, 1907^{VP}
Thermotoga subterranea Jeanthon et al. 2000, 1699^{VP} (Effective publication: Jeanthon et al., 1995)
Thermovenabulum ferriorganovorum Zavarzina et al. 2002, 1741^{VP}
Thermovibrio ruber Huber et al. 2002, 1864^{VP}
Thermus antranikianii Chung et al. 2000a, 216^{VP}
Thermus igniterrae Chung et al. 2000a, 216^{VP}
Thioalkalicoccus limnaeus Bryantseva et al. 2000b, 2162^{VP}
Thioalkalimicrobium aerophilum Sorokin et al. 2001a, 578^{VP}
Thioalkalimicrobium cyclicum Sorokin et al. 2002a, 919^{VP}
Thioalkalimicrobium sibiricum Sorokin et al. 2001a, 578^{VP}
Thioalkalivibrio denitrificans Sorokin et al. 2001a, 579^{VP}
Thioalkalivibrio jannaschii Sorokin et al. 2002a, 919^{VP}
Thioalkalivibrio nitratireducens Sorokin et al. 2003, 1783^{VP}
Thioalkalivibrio nitratis Sorokin et al. 2001a, 579^{VP}
Thioalkalivibrio paradoxus Sorokin et al. 2002c, 663^{VP}
Thioalkalivibrio thiocyanoxidans Sorokin et al. 2002c, 663^{VP}
Thioalkalivibrio versutus Sorokin et al. 2001a, 579^{VP}
Thioalkalispira microaerophila Sorokin et al. 2002b, 2181^{VP}
Thiobaca trueperi Rees et al. 2002, 677^{VP}
Thiocapsa litoralis Puchkova et al. 2000, 1446^{VP}
Thioflavicoccus mobilis Imhoff and Pfennig 2001, 109^{VP}
Thiothrix disciformis Aruga et al. 2002, 1315^{VP}
Thiothrix flexilis Aruga et al. 2002, 1315^{VP}
Tindallia californiensis Pikuta et al. 2003b, 1701^{VP} (Effective publication: Pikuta et al., 2003a)
Tistrella mobilis Shi et al. 2003, 936^{VP} (Effective publication: Shi et al., 2002)
Treponema parvum Wyss et al. 2001, 960^{VP}
Trichlorobacter thiogenes De Wever et al. 2001, 1^{VP} (Effective publication: De Wever et al., 2000)
Trichococcus collinsii Liu et al. 2002a, 1124^{VP}
Tropheryma whipplei La Scola et al. 2001, 1478^{VP}
Tsukamurella spumae Nam et al. 2003b, 1701^{VP} (Effective publication: Nam et al., 2003a)
Tsukamurella strandjordii Kattar et al. 2002, 1075^{VP} (Effective publication: Kattar et al., 2001)
Turicibacter sanguinis Bosshard et al. 2002b, 1266^{VP}
Ureaplasma parvum Robertson et al. 2002, 593^{VP}
Ureibacillus terrenus Fortina et al. 2001b, 454^{VP}
Vagococcus fessus Hoyles et al. 2000c, 1154^{VP}
Varibaculum cambriense Hall et al. 2003h, 627^{VP} (Effective publication: Hall et al., 2003g)
Vibrio aerogenes Shieh et al. 2000, 327^{VP}
Vibrio agarivorans Macián et al. 2001c, 2035^{VP}
Vibrio brasiliensis Thompson et al. 2003b, 250^{VP}
Vibrio calviensis Denner et al. 2002a, 552^{VP}
Vibrio chagasii Thompson et al. 2003d, 758^{VP}
Vibrio coralliilyticus Ben-Haim et al. 2003, 314^{VP}
Vibrio cyclitrophicus Hedlund and Staley 2001, 65^{VP}
Vibrio fortis Thompson et al. 2003c, 1499^{VP}
Vibrio hepatarius Thompson et al. 2003c, 1500^{VP}
Vibrio kanaloae Thompson et al. 2003d, 757^{VP}
Vibrio lentus Macián et al. 2001a, 1454^{VP}
Vibrio neptunius Thompson et al. 2003b, 249^{VP}
Vibrio pacinii Gomez-Gil et al. 2003a, 1572^{VP}
Vibrio pomeroyi Thompson et al. 2003d, 757^{VP}
Vibrio rotiferianus Gomez-Gil et al. 2003b, 242^{VP}

Vibrio ruber Shieh et al. 2003a, 483[VP]
Vibrio shilonii Kushmaro et al. 2001, 1387[VP]
Vibrio superstes Hayashi et al. 2003, 1816[VP]
Vibrio tasmaniensis Thompson et al. 2003e, 1701[VP] (Effective publication: Thompson et al., 2003f)
Vibrio wodanis Lunder et al. 2000, 447[VP]
Vibrio xuii Thompson et al. 2003b, 251[VP]
Victivallis vadensis Zoetendal et al. 2003, 214[VP]
Virgibacillus carmonensis Heyrman et al. 2003b, 507[VP]
Virgibacillus necropolis Heyrman et al. 2003b, 509[VP]
Virgibacillus picturae Heyrman et al. 2003b, 509[VP]
Virgisporangium aurantiacum Tamura et al. 2001, 1815[VP]
Virgisporangium ochraceum Tamura et al. 2001, 1815[VP]
Vitellibacter vladivostokensis Nedashkovskaya et al. 2003c, 1285[VP]
Vulcanisaeta distributa Itoh et al. 2002, 1103[VP]
Vulcanisaeta souniana Itoh et al. 2002, 1103[VP]
Vulcanithermus mediatlanticus Miroshnichenko et al. 2003c, 1147[VP]
Weissella cibaria Björkroth et al. 2002, 147[VP]
Weissella kimchii Choi et al. 2002, 510[VP]
Weissella koreensis Lee et al. 2002a, 1260[VP]
Weissella soli Magnusson et al. 2002, 833[VP]
Weissella thailandensis Tanasupawat et al. 2000, 1484[VP]
Xanthobacter aminoxidans Doronina and Trotsenko 2003, 181[VP]
Xanthobacter viscosus Doronina and Trotsenko 2003, 181[VP]
Xanthomonas cynarae Trébaol et al. 2000, 1476[VP]
Xenophilus azovorans Blümel et al. 2001b, 1835[VP]
Xylanimonas cellulosilytica Rivas et al. 2003a, 103[VP]
Zobellia galactanivorans Barbeyron et al. 2001, 994[VP]
Zooshikella ganghwensis Yi et al. 2003, 1016[VP]

New subspecies

Actinobacillus equuli subsp. *haemolyticus* Christensen et al. 2002b, 1575[VP]
Aeromonas hydrophila subsp. *dhakensis* Huys et al. 2002, 710[VP]
Aeromonas hydrophila subsp. *ranae* Huys et al. 2003b, 890[VP]
Aeromonas salmonicida subsp. *pectinolytica* Pavan et al. 2000, 1123[VP]
Alcaligenes faecalis subsp. *parafaecalis* Schroll et al. 2001b, 1619[VP] (Effective publication: Schroll et al., 2001a)
Alicyclobacillus acidocaldarius subsp. *rittmannii* Nicolaus et al. 2002, 3[VP] (Effective publication: Nicolaus et al., 1998)
Amycolatopsis keratiniphila subsp. *keratiniphila* Al-Musallam et al. 2003[VP] emend. Wink et al., 2003 (Effective publication: Wink et al., 2003a)
Amycolatopsis keratiniphila subsp. *nogabecina* Wink et al. 2003b, 935[VP] (Effective publication: Wink et al., 2003a)
Bartonella vinsonii subsp. *arupensis* Welch et al. 2000, 3[VP] (Effective publication: Welch et al., 1999)
Bifidobacterium thermacidophilum subsp. *porcinum* Zhu et al. 2003b, 1622[VP]
Bifidobacterium thermacidophilum subsp. *thermacidophilum* Dong et al. 2000, 124; 53:1622[VP]
Desulfotomaculum thermobenzoicum subsp. *thermosyntrophicum* Plugge et al. 2002a, 398[VP]
Nocardiopsis dassonvillei subsp. *albirubida* (Grund and Kroppenstedt 1990) Evtushenko et al. 2000b, 80[VP]
Prochlorococcus marinus subsp. *pastoris* Rippka et al. 2001, 264[VP] (Effective publication: Rippka et al., 2000)
Roseomonas gilardii subsp. *rosea* Han et al. 2003c, 1701[VP] (Effective publication: Han et al., 2003b)
Salinivibrio costicola subsp. *vallismortis* Huang et al. 2000, 621[VP]
Serratia marcescens subsp. *sakuensis* Ajithkumar et al. 2003, 258[VP]
Staphylococcus equorum subsp. *linens* Place et al. 2003c, 1219[VP] (Effective publication: Place et al., 2003b)
Staphylococcus succinus subsp. *casei* Place et al. 2003a, 1[VP] (Effective publication: Place et al., 2002)
Streptococcus infantarius subsp. *coli* Schlegel et al. 2003, 642[VP]
Streptococcus infantarius subsp. *infantarius* Schlegel et al. 2003, 642[VP]
Yersinia enterocolitica subsp. *palearctica* Neubauer et al. 2000a, 1416[VP] (Effective publication: Neubauer et al., 2000b)

Administrative

Chlorobaculum chlorovibrioides (Gorlenko et al. 1974) Imhoff 2003, 951[VP]
Ralstonia oxalatica Sahin et al. 2000b, 1953[VP] (Effective publication: Sahin et al., 2000a)
Sinorhizobium adhaerens (Casida 1982) Willems et al. 2003 (Request for an Opinion), 1215[VP]
Streptomyces luteireticuli (ex Katoh and Arai 1957) Hatano et al. 2003, 1528[VP]

Appendix 2. Taxonomic Outline of the *Archaea* and *Bacteria*

Readers are advised that the taxonomic scheme presented here is a work-in-progress and is based on data available in October 2003. Some rearrangement and emendment is expected to occur as new data become available and subsequent volumes go to press.

- Domain *Archaea*[VP]
  - Phylum AI. *Crenarchaeota*[VP]
    - Class I. *Thermoprotei*[VP]
      - Order I. *Thermoproteales*[VP (T)]
        - Family I. *Thermoproteaceae*[VP]
          - Genus I. *Thermoproteus*[VP (T)]
          - Genus II. *Caldivirga*[VP]
          - Genus III. *Pyrobaculum*[VP]
          - Genus IV. *Thermocladium*[VP]
          - Genus V. *Vulcanisaeta*[VP]
        - Family II. *Thermofilaceae*[VP]
          - Genus I. *Thermofilum*[VP (T)]
      - Order II. *"Caldisphaerales"*
        - Family I. *"Caldisphaeraceae"*
          - Genus I. *Caldisphaera*[VP (T)]
      - Order III. *Desulfurococcales*[VP]
        - Family I. *Desulfurococcaceae*[VP]
          - Genus I. *Desulfurococcus*[VP (T)]
          - Genus II. *Acidilobus*[VP]
          - Genus III. *Aeropyrum*[VP]
          - Genus IV. *Ignicoccus*[VP]
          - Genus V. *Staphylothermus*[VP]
          - Genus VI. *Stetteria*[VP]
          - Genus VII. *Sulfophobococcus*[VP]
          - Genus VIII. *Thermodiscus*[VP]
          - Genus IX. *Thermosphaera*[VP]
        - Family II. *Pyrodictiaceae*[VP]
          - Genus I. *Pyrodictium*[VP (T)]
          - Genus II. *Hyperthermus*[VP]
          - Genus III. *Pyrolobus*[VP]
      - Order IV. *Sulfolobales*[VP]
        - Family I. *Sulfolobaceae*[VP]
          - Genus I. *Sulfolobus*[AL (T)]
          - Genus II. *Acidianus*[VP]
          - Genus III. *Metallosphaera*[VP]
          - Genus IV. *Stygiolobus*[VP]
          - Genus V. *Sulfurisphaera*[VP]
          - Genus VI. *Sulfurococcus*[VP]
  - Phylum AII. *Euryarchaeota*[VP]
    - Class I. *Methanobacteria*[VP]
      - Order I. *Methanobacteriales*[VP (T)]
        - Family I. *Methanobacteriaceae*[AL]
          - Genus I. *Methanobacterium*[AL (T)]
          - Genus II. *Methanobrevibacter*[VP]
          - Genus III. *Methanosphaera*[VP]
          - Genus IV. *Methanothermobacter*[VP]
        - Family II. *Methanothermaceae*[VP]
          - Genus I. *Methanothermus*[VP (T)]
    - Class II. *Methanococci*[VP]
      - Order I. *Methanococcales*[VP (T)]
        - Family I. *Methanococcaceae*[VP]
          - Genus I. *Methanococcus*[AL (T)]
          - Genus II. *Methanothermococcus*[VP]
        - Family II. *Methanocaldococcaceae*[VP]
          - Genus I. *Methanocaldococcus*[VP (T)]
          - Genus II. *Methanotorris*[VP]
    - Class III. *"Methanomicrobia"*
      - Order I. *Methanomicrobiales*[VP (T)]
        - Family I. *Methanomicrobiaceae*[VP]
          - Genus I. *Methanomicrobium*[VP (T)]
          - Genus II. *Methanoculleus*[VP]
          - Genus III. *Methanofollis*[VP]
          - Genus IV. *Methanogenium*[VP]
          - Genus V. *Methanolacinia*[VP]
          - Genus VI. *Methanoplanus*[VP]
        - Family II. *Methanocorpusculaceae*[VP]
          - Genus I. *Methanocorpusculum*[VP (T)]
        - Family III. *Methanospirillaceae*[VP]
          - Genus I. *Methanospirillum*[AL (T)]
        - Genera incertae sedis
          - Genus I. *Methanocalculus*[VP]
      - Order II. *Methanosarcinales*[VP]
        - Family I. *Methanosarcinaceae*[VP]
          - Genus I. *Methanosarcina*[AL (T)]
          - Genus II. *Methanococcoides*[VP]
          - Genus III. *Methanohalobium*[VP]
          - Genus IV. *Methanohalophilus*[VP]
          - Genus V. *Methanolobus*[VP]
          - Genus VI. *Methanimicrococcus*[VP]
          - Genus VII. *Methanosalsum*[VP]
        - Family II. *Methanosaetaceae*[VP]
          - Genus I. *Methanosaeta*[VP (T)]
    - Class IV. *Halobacteria*[VP]
      - Order I. *Halobacteriales*[VP (T)]
        - Family I. *Halobacteriaceae*[AL]
          - Genus I. *Halobacterium*[AL (T)]
          - Genus II. *Haloarcula*[VP]
          - Genus III. *Halobaculum*[VP]
          - Genus IV. *Halobiforma*[VP]
          - Genus V. *Halococcus*[AL]
          - Genus VI. *Haloferax*[VP]
          - Genus VII. *Halogeometricum*[VP]
          - Genus VIII. *Halomicrobium*[VP]
          - Genus IX. *Halorhabdus*[VP]
          - Genus X. *Halorubrum*[VP]
          - Genus XI. *Halosimplex*[VP]
          - Genus XII. *Haloterrigena*[VP]
          - Genus XIII. *Natrialba*[VP]
          - Genus XIV. *Natrinema*[VP]
          - Genus XV. *Natronobacterium*[VP]
          - Genus XVI. *Natronococcus*[VP]
          - Genus XVII. *Natronomonas*[VP]
          - Genus XVIII. *Natronorubrum*[VP]
    - Class V. *Thermoplasmata*[VP]
      - Order I. *Thermoplasmatales*[VP (T)]
        - Family I. *Thermoplasmataceae*[VP]
          - Genus I. *Thermoplasma*[AL (T)]
        - Family II. *Picrophilaceae*[VP]
          - Genus I. *Picrophilus*[VP (T)]
        - Family III. *"Ferroplasmaceae"*
          - Genus I. *Ferroplasma*[VP]

AL - Approved Lists, VP - validly published, NP - new proposal appearing in Volume Two of *Bergey's Manual of Systematic Bacteriology*, 2nd Edition.

Class VI. *Thermococci*[VP]
Order I. *Thermococcales*[VP (T)]
Family I. *Thermococcaceae*[VP]
Genus I. *Thermococcus*[VP (T)]
Genus II. *Palaeococcus*[VP]
Genus III. *Pyrococcus*[VP]
Class VII. *Archaeoglobi*[VP]
Order I. *Archaeoglobales*[VP (T)]
Family I. *Archaeoglobaceae*[VP]
Genus I. *Archaeoglobus*[VP (T)]
Genus II. *Ferroglobus*[VP]
Genus III. *Geoglobus*[VP]
Class VIII. *Methanopyri*[VP]
Order I. *Methanopyrales*[VP (T)]
Family I. *Methanopyraceae*[VP]
Genus I. *Methanopyrus*[VP (T)]
Domain *Bacteria*[VP]
Phylum BI. *Aquificae*[VP]
Class I. *Aquificae*[VP]
Order I. *Aquificales*[VP (T)]
Family I. *Aquificaceae*[VP]
Genus I. *Aquifex*[VP (T)]
Genus II. *Calderobacterium*[VP]
Genus III. *Hydrogenobaculum*[VP]
Genus IV. *Hydrogenobacter*[VP]
Genus V. *Hydrogenothermus*[VP]
Genus VI. *Persephonella*[VP]
Genus VII. *Sulfurihydrogenibium*[VP]
Genus VIII. *Thermocrinis*[VP]
Genera incertae sedis
Genus I. *Desulfurobacterium*[VP]
Genus II. *Thermovibrio*[VP]
Phylum BII. *Thermotogae*[VP]
Class I. *Thermotogae*[VP]
Order I. *Thermotogales*[VP (T)]
Family I. *Thermotogaceae*[VP]
Genus I. *Thermotoga*[VP (T)]
Genus II. *Fervidobacterium*[VP]
Genus III. *Geotoga*[VP]
Genus IV. *Marinitoga*[VP]
Genus V. *Petrotoga*[VP]
Genus VI. *Thermosipho*[VP]
Phylum BIII. *Thermodesulfobacteria*[VP]
Class I. *Thermodesulfobacteria*[VP]
Order I. *Thermodesulfobacteriales*[VP (T)]
Family I. *Thermodesulfobacteriaceae*[VP]
Genus I. *Thermodesulfobacterium*[VP (T)]
Phylum BIV. *Deinococcus-Thermus*[VP]
Class I. *Deinococci*[VP]
Order I. *Deinococcales*[VP (T)]
Family I. *Deinococcaceae*[VP]
Genus I. *Deinococcus*[VP (T)]
Order II. *Thermales*[VP]
Family I. *Thermaceae*[VP]
Genus I. *Thermus*[AL (T)]
Genus II. *Marinithermus*[VP]
Genus III. *Meiothermus*[VP]
Genus IV. *Oceanithermus*[VP]
Genus V. *Vulcanithermus*[VP]
Phylum BV. *Chrysiogenetes*[VP]
Class I. *Chrysiogenetes*[VP]
Order I. *Chrysiogenales*[VP (T)]
Family I. *Chrysiogenaceae*[VP]
Genus I. *Chrysiogenes*[VP (T)]
Phylum BVI. *"Chloroflexi"*
Class I. *"Chloroflexi"*
Order I. *"Chloroflexales"*
Family I. *"Chloroflexaceae"*
Genus I. *Chloroflexus*[AL]
Genus II. *Chloronema*[AL]
Genus III. *Heliothrix*[VP]
Genus IV. *Roseiflexus*[VP]
Family II. *Oscillochloridaceae*[VP]
Genus I. *Oscillochloris*[VP (T)]
Order II. *"Herpetosiphonales"*
Family I. *"Herpetosiphonaceae"*
Genus I. *Herpetosiphon*[AL]
Phylum BVII. *Thermomicrobia*[VP]
Class I. *Thermomicrobia*[VP]
Order I. *Thermomicrobiales*[VP (T)]
Family I. *Thermomicrobiaceae*[VP]
Genus I. *Thermomicrobium*[AL (T)]
Phylum BVIII. *"Nitrospira"*
Class I. *"Nitrospira"*
Order I. *"Nitrospirales"*
Family I. *"Nitrospiraceae"*
Genus I. *Nitrospira*[VP]
Genus II. *Leptospirillum*[VP]
Genus III. *Magnetobacterium*[VP]
Genus IV. *Thermodesulfovibrio*[VP]
Phylum BIX. *Deferribacteres*[VP]
Class I. *Deferribacteres*[VP]
Order I. *Deferribacterales*[VP (T)]
Family I. *Deferribacteraceae*[VP]
Genus I. *Deferribacter*[VP (T)]
Genus II. *Denitrovibrio*[VP]
Genus III. *Flexistipes*[VP]
Genus IV. *Geovibrio*[VP]
Genera incertae sedis
Genus I. *Synergistes*[VP]
Phylum BX. *Cyanobacteria*
Class I. *Cyanobacteria*
Subsection I
Family
Form genus I. *Chamaesiphon*
Form genus II. *Chroococcus*
Form genus III. *Cyanobacterium*
Form genus IV. *Cyanobium*
Form genus V. *Cyanothece*
Form genus VI. *Dactylococcopsis*
Form genus VII. *Gloeobacter*
Form genus VIII. *Gloeocapsa*
Form genus IX. *Gloeothece*
Form genus X. *Microcystis*
Form genus XI. *Prochlorococcus*
Form genus XII. *Prochloron*
Form genus XIII. *Synechococcus*
Form genus XIV. *Synechocystis*
Subsection II.
Family
Form genus I. *Cyanocystis*
Form genus II. *Dermocarpella*
Form genus III. *Stanieria*
Form genus IV. *Xenococcus*

Family
- Form genus I. *Chroococcidiopsis*
- Form genus II. *Myxosarcina*
- Form genus III. *Pleurocapsa*

Subsection III.

Family
- Form genus I. *Arthrospira*
- Form genus II. *Borzia*
- Form genus III. *Crinalium*
- Form genus IV. *Geitlerinema*
- Form genus V. *Halospirulina*
- Form genus VI. *Leptolyngbya*
- Form genus VII. *Limnothrix*
- Form genus VIII. *Lyngbya*
- Form genus IX. *Microcoleus*
- Form genus X. *Oscillatoria*
- Form genus XI. *Planktothrix*
- Form genus XII. *Prochlorothrix*
- Form genus XIII. *Pseudanabaena*
- Form genus XIV. *Spirulina*
- Form genus XV. *Starria*
- Form genus XVI. *Symploca*
- Form genus XVII. *Trichodesmium*
- Form genus XVIII. *Tychonema*

Subsection IV.

Family
- Form genus I. *Anabaena*
- Form genus II. *Anabaenopsis*
- Form genus III. *Aphanizomenon*
- Form genus IV. *Cyanospira*
- Form genus V. *Cylindrospermopsis*
- Form genus VI. *Cylindrospermum*
- Form genus VII. *Nodularia*
- Form genus VIII. *Nostoc*
- Form genus IX. *Scytonema*

Family
- Form genus I. *Calothrix*
- Form genus II. *Rivularia*
- Form genus III. *Tolypothrix*

Subsection V

Family I
- Form genus I. *Chlorogloeopsis*
- Form genus II. *Fischerella*
- Form genus III. *Geitleria*
- Form genus IV. *Iyengariella*
- Form genus V. *Nostochopsis*
- Form genus VI. *Stigonema*

Phylum BXI. *Chlorobi*[VP]

Class I. *"Chlorobia"*

Order I. *Chlorobiales*[AL]

Family I. *Chlorobiaceae*[AL]
- Genus I. *Chlorobium*[AL (T)]
- Genus II. *Ancalochloris*[AL]
- Genus III. *Chlorobaculum*[VP]
- Genus IV. *Chloroherpeton*[VP]
- Genus V. *Pelodictyon*[AL]
- Genus VI. *Prosthecochloris*[AL]

Phylum BXII. *Proteobacteria*[NP]

Class I. *"Alphaproteobacteria"*[NP]

Order I. *Rhodospirillales*[AL (T)]

Family I. *Rhodospirillaceae*[AL]
- Genus I. *Rhodospirillum*[AL (T)]
- Genus II. *Azospirillum*[AL]
- Genus III. *Inquilinus*[VP]
- Genus IV. *Magnetospirillum*[VP]
- Genus V. *Phaeospirillum*[VP]
- Genus VI. *Rhodocista*[VP]
- Genus VII. *Rhodospira*[VP]
- Genus VIII. *Rhodovibrio*[VP]
- Genus IX. *Roseospira*[VP]
- Genus X. *Skermanella*[VP]
- Genus XI. *Thalassospira*[VP]
- Genus XII. *Tistrella*[VP]

Family II. *Acetobacteraceae*[VP]
- Genus I. *Acetobacter*[AL (T)]
- Genus II. *Acidiphilium*[VP]
- Genus III. *Acidisphaera*[VP]
- Genus IV. *Acidocella*[VP]
- Genus V. *Acidomonas*[VP]
- Genus VI. *Asaia*[VP]
- Genus VII. *Craurococcus*[VP]
- Genus VIII. *Gluconacetobacter*[VP]
- Genus IX. *Gluconobacter*[AL]
- Genus X. *Kozakia*[VP]
- Genus XI. *Muricoccus*[VP]
- Genus XII. *Paracraurococcus*[VP]
- Genus XIII. *Rhodopila*[VP]
- Genus XIV. *Roseococcus*[VP]
- Genus XV. *Rubritepida*[VP]
- Genus XVI. *Stella*[VP]
- Genus XVII. *Teichococcus*[VP]
- Genus XVIII. *Zavarzinia*[VP]

Order II. *Rickettsiales*[AL]

Family I. *Rickettsiaceae*[AL]
- Genus I. *Rickettsia*[AL (T)]
- Genus II. *Orientia*[VP]

Family II. *Anaplasmataceae*[AL]
- Genus I. *Anaplasma*[AL (T)]
- Genus II. *Aegyptianella*[AL]
- Genus III. *Cowdria*[AL]
- Genus IV. *Ehrlichia*[AL]
- Genus V. *Neorickettsia*[AL]
- Genus VI. *Wolbachia*[AL]
- Genus VII. *Xenohaliotis*[VP]

Family III. *"Holosporaceae"*[NP]
- Genus I. *Holospora*[VP (T)]

Genera incertae sedis
- Genus I. *Caedibacter*
- Genus II. *Lyticum*[VP]
- Genus III. *Odyssella*[VP]
- Genus IV. *Polynucleobacter*[VP]
- Genus V. *Pseudocaedibacter*[VP]
- Genus VI. *Symbiotes*[AL]
- Genus VII. *Tectibacter*[VP]

Order III. *"Rhodobacterales"*[NP]

Family I. *"Rhodobacteraceae"*[NP]
- Genus I. *Rhodobacter*[VP (T)]
- Genus II. *Ahrensia*[VP]
- Genus III. *Albidovulum*[VP]
- Genus IV. *Amaricoccus*[VP]
- Genus V. *Antarctobacter*[VP]
- Genus VI. *Gemmobacter*[VP]
- Genus VII. *Hirschia*[VP]
- Genus VIII. *Hyphomonas*[VP]

Genus IX. *Jannaschia*^{VP}
Genus X. *Ketogulonicigenium*^{VP}
Genus XI. *Leisingera*^{VP}
Genus XII. *Maricaulis*^{VP}
Genus XIII. *Methylarcula*^{VP}
Genus XIV. *Octadecabacter*^{VP}
Genus XV. *Pannonibacter*^{VP}
Genus XVI. *Paracoccus*^{VP}
Genus XVII. *Pseudorhodobacter*^{VP}
Genus XVIII. *Rhodobaca*^{VP}
Genus XIX. *Rhodothalassium*^{VP}
Genus XX. *Rhodovulum*^{VP}
Genus XXI. *Roseibium*^{VP}
Genus XXII. *Roseinatronobacter*^{VP}
Genus XXIII. *Roseivivax*^{VP}
Genus XXIV. *Roseobacter*^{VP}
Genus XXV. *Roseovarius*^{VP}
Genus XXVI. *Rubrimonas*^{VP}
Genus XXVII. *Ruegeria*^{VP}
Genus XXVIII. *Sagittula*^{VP}
Genus XXIX. *Staleya*^{VP}
Genus XXX. *Stappia*^{VP}
Genus XXXI. *Sulfitobacter*^{VP}
Order IV. *"Sphingomonadales"*^{NP}
Family I. *Sphingomonadaceae*^{VP}
Genus I. *Sphingomonas*$^{VP\ (T)}$
Genus II. *Blastomonas*^{VP}
Genus III. *Erythrobacter*^{VP}
Genus IV. *Erythromicrobium*^{VP}
Genus V. *Erythromonas*^{VP}
Genus VI. *Novosphingobium*^{VP}
Genus VII. *Porphyrobacter*^{VP}
Genus VIII. *Rhizomonas*^{VP}
Genus IX. *Sandaracinobacter*^{VP}
Genus X. *Sphingobium*^{VP}
Genus XI. *Sphingopyxis*^{VP}
Genus XII. *Zymomonas*^{AL}
Order V. *Caulobacterales*^{AL}
Family I. *Caulobacteraceae*^{AL}
Genus I. *Caulobacter*$^{AL\ (T)}$
Genus II. *Asticcacaulis*^{AL}
Genus III. *Brevundimonas*^{VP}
Genus IV. *Phenylobacterium*^{VP}
Order VI. *"Rhizobiales"*^{NP}
Family I. *Rhizobiaceae*^{AL}
Genus I. *Rhizobium*$^{AL\ (T)}$
Genus II. *Agrobacterium*^{AL}
Genus III. *Allorhizobium*^{VP}
Genus IV. *Carbophilus*^{VP}
Genus V. *Chelatobacter*^{VP}
Genus VI. *Ensifer*^{VP}
Genus VII. *Sinorhizobium*^{VP}
Family II. *"Aurantimonadaceae"*$^{NP\ (T)}$
Genus I. *Aurantimonas*$^{VP\ (T)}$
Family III. *Bartonellaceae*^{AL}
Genus I. *Bartonella*$^{AL\ (T)}$
Family IV. *Brucellaceae*^{AL}
Genus I. *Brucella*$^{AL\ (T)}$
Genus II. *Mycoplana*^{AL}
Genus III. *Ochrobactrum*^{VP}
Family V. *"Phyllobacteriaceae"*^{NP}
Genus I. *Phyllobacterium*$^{VP\ (T)}$
Genus II. *Aminobacter*^{VP}
Genus III. *Aquamicrobium*^{VP}
Genus IV. *Defluvibacter*^{VP}
Genus V. *"Candidatus Liberibacter"*
Genus VI. *Mesorhizobium*^{VP}
Genus VII. *Pseudaminobacter*^{VP}
Family VI. *"Methylocystaceae"*^{NP}
Genus I. *Methylocystis*$^{VP\ (T)}$
Genus II. *Albibacter*^{VP}
Genus III. *Methylopila*^{VP}
Genus IV. *Methylosinus*^{VP}
Genus V. *Terasakiella*^{VP}
Family VII. *"Beijerinckiaceae"*^{NP}
Genus I. *Beijerinckia*$^{AL\ (T)}$
Genus II. *Chelatococcus*^{VP}
Genus III. *Methylocapsa*^{VP}
Genus IV. *Methylocella*^{VP}
Family VIII. *"Bradyrhizobiaceae"*^{NP}
Genus I. *Bradyrhizobium*$^{VP\ (T)}$
Genus II. *Afipia*^{VP}
Genus III. *Agromonas*^{VP}
Genus IV. *Blastobacter*^{AL}
Genus V. *Bosea*^{VP}
Genus VI. *Nitrobacter*^{AL}
Genus VII. *Oligotropha*^{VP}
Genus VIII. *Rhodoblastus*^{VP}
Genus IX. *Rhodopseudomonas*^{AL}
Family IX. *Hyphomicrobiaceae*^{AL}
Genus I. *Hyphomicrobium*$^{AL\ (T)}$
Genus II. *Ancalomicrobium*^{AL}
Genus III. *Ancylobacter*^{VP}
Genus IV. *Angulomicrobium*^{VP}
Genus V. *Aquabacter*^{VP}
Genus VI. *Azorhizobium*^{VP}
Genus VII. *Blastochloris*^{VP}
Genus VIII. *Devosia*^{VP}
Genus IX. *Dichotomicrobium*^{VP}
Genus X. *Filomicrobium*^{VP}
Genus XI. *Gemmiger*^{AL}
Genus XII. *Labrys*^{VP}
Genus XIII. *Methylorhabdus*^{VP}
Genus XIV. *Pedomicrobium*^{AL}
Genus XV. *Prosthecomicrobium*^{AL}
Genus XVI. *Rhodomicrobium*^{AL}
Genus XVII. *Rhodoplanes*^{VP}
Genus XVIII. *Seliberia*^{AL}
Genus XIX. *Starkeya*^{VP}
Genus XX. *Xanthobacter*^{AL}
Family X. *"Methylobacteriaceae"*^{NP}
Genus I. *Methylobacterium*$^{AL\ (T)}$
Genus II. *Microvirga*^{VP}
Genus III. *Protomonas*^{VP}
Genus IV. *Roseomonas*^{VP}
Family XI. *"Rhodobiaceae"*^{NP}
Genus I. *Rhodobium*$^{VP\ (T)}$
Genus II. *Roseospirillum*^{VP}
Order VII. *"Parvularculales"*^{NP}
Family I. *"Parvularculaceae"*^{NP}
Genus I. *Parvularcula*$^{VP\ (T)}$
Class II. *"Betaproteobacteria"*^{NP}
Order I. *"Burkholderiales"*$^{NP\ (T)}$
Family I. *"Burkholderiaceae"*^{NP}

Genus I. *Burkholderia*VP (T)
Genus II. *Cupriavidus*VP
Genus III. *Lautropia*VP
Genus IV. *Limnobacter*VP
Genus V. *Pandoraea*VP
Genus VI. *Paucimonas*VP
Genus VII. *Polynucleobacter*VP
Genus VIII. *Ralstonia*VP
Genus IX. *Thermothrix*VP
Family II. *"Oxalobacteraceae"*NP
Genus I. *Oxalobacter*VP (T)
Genus II. *Duganella*VP
Genus III. *Herbaspirillum*VP
Genus IV. *Janthinobacterium*AL
Genus V. *Massilia*VP
Genus VI. *Oxalicibacterium*VP
Genus VII. *Telluria*VP
Family III. *Alcaligenaceae*VP
Genus I. *Alcaligenes*AL (T)
Genus II. *Achromobacter*VP
Genus III. *Bordetella*AL
Genus IV. *Brackiella*VP
Genus V. *Derxia*AL
Genus VI. *Oligella*VP
Genus VII. *Pelistega*VP
Genus VIII. *Pigmentiphaga*VP
Genus IXI. *Sutterella*VP
Genus X. *Taylorella*VP
Family IV. *Comamonadaceae*VP
Genus I. *Comamonas*VP (T)
Genus II. *Acidovorax*VP
Genus III. *Alicycliphilus*VP
Genus IV. *Brachymonas*VP
Genus V. *Caldimonas*VP
Genus VI. *Delftia*VP
Genus VII. *Diaphorobacter*VP
Genus VIII. *Hydrogenophaga*VP
Genus IX. *Lampropedia*AL
Genus X. *Macromonas*AL
Genus XI. *Polaromonas*VP
Genus XII. *Ramlibacter*VP
Genus XIII. *Rhodoferax*VP
Genus XIV. *Variovorax*VP
Genus XV. *Xenophilus*VP
Genera incertae sedis
Genus I. *Aquabacterium*VP
Genus II. *Ideonella*VP
Genus III. *Leptothrix*AL
Genus IV. *Roseateles*VP
Genus V. *Rubrivivax*VP
Genus VI. *Schlegelella*VP
Genus VII. *Sphaerotilus*AL
Genus VIII. *Tepidimonas*VP
Genus IX. *Thiomonas*VP
Genus X. *Xylophilus*VP
Order II. *"Hydrogenophilales"*NP
Family I. *"Hydrogenophilaceae"*NP
Genus I. *Hydrogenophilus*VP (T)
Genus II. *Thiobacillus*AL
Order III. *"Methylophilales"*NP
Family I. *"Methylophilaceae"*NP
Genus I. *Methylophilus*VP (T)
Genus II. *Methylobacillus*AL
Genus III. *Methylovorus*VP
Order IV. *"Neisseriales"*NP
Family I. *Neisseriaceae*AL
Genus I. *Neisseria*AL (T)
Genus II. *Alysiella*AL
Genus III. *Aquaspirillum*AL
Genus IV. *Chromobacterium*AL
Genus V. *Eikenella*AL
Genus VI. *Formivibrio*VP
Genus VII. *Iodobacter*VP
Genus VIII. *Kingella*AL
Genus IX. *Laribacter*VP
Genus X. *Microvirgula*VP
Genus XI. *Morococcus*VP
Genus XII. *Prolinoborus*VP
Genus XIII. *Simonsiella*AL
Genus XIV. *Vitreoscilla*AL
Genus XV. *Vogesella*VP
Order V. *"Nitrosomonadales"*NP
Family I. *"Nitrosomonadaceae"*NP
Genus I. *Nitrosomonas*AL (T)
Genus II. *Nitrosolobus*AL
Genus III. *Nitrosospira*AL
Family II. *Spirillaceae*AL
Genus I. *Spirillum*AL (T)
Family III. *Gallionellaceae*AL
Genus I. *Gallionella*AL (T)
Order VI. *"Rhodocyclales"*NP
Family I. *"Rhodocyclaceae"*NP
Genus I. *Rhodocyclus*AL (T)
Genus II. *Azoarcus*VP
Genus III. *Azonexus*VP
Genus IV. *Azospira*VP
Genus V. *Azovibrio*VP
Genus VI. *Dechloromonas*VP
Genus VII. *Dechlorosoma*VP
Genus VIII. *Ferribacterium*VP
Genus IX. *Propionibacter*VP
Genus X. *Propionivibrio*VP
Genus XI. *Quadricoccus*VP
Genus XII. *Sterolibacterium*VP
Genus XIII. *Thauera*VP
Genus XIV. *Zoogloea*AL
Order VII. *"Procabacteriales"*NP
Family I. *"Procabacteriaceae"*NP
Genus I. *"Procabacter"*
Class III. *"Gammaproteobacteria"*NP
Order I. *"Chromatiales"*NP (T)
Family I. *Chromatiaceae*AL
Genus I. *Chromatium*AL (T)
Genus II. *Allochromatium*VP
Genus III. *Amoebobacter*AL
Genus IV. *Halochromatium*VP
Genus V. *Isochromatium*VP
Genus VI. *Lamprobacter*VP
Genus VII. *Lamprocystis*AL
Genus VIII. *Marichromatium*VP
Genus IX. *Nitrosococcus*AL
Genus X. *Pfennigia*VP
Genus XI. *Rhabdochromatium*VP
Genus XII. *Rheinheimera*VP

- Genus XIII. *Thermochromatium*[VP]
- Genus XIV. *Thioalkalicoccus*[VP]
- Genus XV. *Thiobaca*[VP]
- Genus XVI. *Thiocapsa*[AL]
- Genus XVII. *Thiococcus*[VP]
- Genus XVIII. *Thiocystis*[AL]
- Genus XIX. *Thiodictyon*[AL]
- Genus XX. *Thioflavicoccus*[VP]
- Genus XXI. *Thiohalocapsa*[VP]
- Genus XXII. *Thiolamprovum*[VP]
- Genus XXIII. *Thiopedia*[AL]
- Genus XXIV. *Thiorhodococcus*[VP]
- Genus XXV. *Thiorhodovibrio*[VP]
- Genus XXVI. *Thiospirillum*[AL]

- Family II. *Ectothiorhodospiraceae*[VP]
  - Genus I. *Ectothiorhodospira*[AL]
  - Genus II. *Alcalilimnicola*[VP]
  - Genus III. *Alkalispirillum*[VP]
  - Genus IV. *Arhodomonas*[VP]
  - Genus V. *Halorhodospira*[VP]
  - Genus VI. *Nitrococcus*[AL]
  - Genus VII. *Thioalkalispira*[VP]
  - Genus VIII. *Thioalkalivibrio*[VP]
  - Genus IX. *Thiorhodospira*[VP]
- Family III. *"Halothiobacillaceae"*[NP]
  - Genus I. *Halothiobacillus*[VP (T)]

- Order II. *"Acidithiobacillales"*[NP]
  - Family I. *"Acidithiobacillaceae"*[NP]
    - Genus I. *Acidithiobacillus*[VP (T)]
  - Family II. *"Thermithiobacillaceae"*[NP]
    - Genus I. *Thermithiobacillus*[VP (T)]
- Order III. *"Xanthomonadales"*[NP]
  - Family I. *"Xanthomonadaceae"*[NP]
    - Genus I. *Xanthomonas*[AL (T)]
    - Genus II. *Frateuria*[VP]
    - Genus III. *Fulvimonas*[VP]
    - Genus IV. *Luteimonas*[VP]
    - Genus V. *Lysobacter*[AL]
    - Genus VI. *Nevskia*[AL]
    - Genus VII. *Pseudoxanthomonas*[VP]
    - Genus VIII. *Rhodanobacter*[VP]
    - Genus IX. *Schineria*[VP]
    - Genus X. *Stenotrophomonas*[VP]
    - Genus XI. *Thermomonas*[VP]
    - Genus XII. *Xylella*[VP]
- Order IV. *"Cardiobacteriales"*[NP]
  - Family I. *Cardiobacteriaceae*[VP]
    - Genus I. *Cardiobacterium*[AL (T)]
    - Genus II. *Dichelobacter*[VP]
    - Genus III. *Suttonella*[VP]
- Order V. *"Thiotrichales"*[NP]
  - Family I. *"Thiotrichaceae"*[NP]
    - Genus I. *Thiothrix*[AL (T)]
    - Genus II. *Achromatium*[AL]
    - Genus III. *Beggiatoa*[AL]
    - Genus IV. *Leucothrix*[AL]
    - Genus V. *Thiobacterium*[VP]
    - Genus VII. *Thioploca*[AL]
    - Genus VIII. *Thiospira*[AL]
  - Family II. *"Piscirickettsiaceae"*[NP]
    - Genus I. *Piscirickettsia*[VP (T)]
    - Genus II. *Cycloclasticus*[VP]
    - Genus III. *Hydrogenovibrio*[VP]
    - Genus IV. *Methylophaga*[VP]
    - Genus V. *Thioalkalimicrobium*[VP]
    - Genus VI. *Thiomicrospira*[AL]
  - Family III. *"Francisellaceae"*[NP]
    - Genus I. *Francisella*[AL (T)]
- Order VI. *"Legionellales"*[NP]
  - Family I. *Legionellaceae*[AL]
    - Genus I. *Legionella*[AL (T)]
  - Family II. *"Coxiellaceae"*[NP]
    - Genus I. *Coxiella*[AL (T)]
    - Genus II. *Rickettsiella*[AL]
- Order VII. *"Methylococcales"*[NP]
  - Family I. *Methylococcaceae*[VP]
    - Genus I. *Methylococcus*[AL (T)]
    - Genus II. *Methylobacter*[VP]
    - Genus III. *Methylocaldum*[VP]
    - Genus IV. *Methylomicrobium*[VP]
    - Genus V. *Methylomonas*[VP]
    - Genus VI. *Methylosarcina*[VP]
    - Genus VII. *Methylosphaera*[VP]
- Order VIII. *"Oceanospirillales"*[NP]
  - Family I. *"Oceanospirillaceae"*[NP]
    - Genus I. *Oceanospirillum*[AL (T)]
    - Genus II. *Balneatrix*[VP]
    - Genus III. *Marinomonas*[VP]
    - Genus IV. *Marinospirillum*[VP]
    - Genus V. *Neptunomonas*[VP]
    - Genus VI. *Oceanobacter*[VP]
    - Genus VII. *Oleispira*[VP]
    - Genus VIII. *Pseudospirillum*[VP]
  - Family II. *"Alcanivoraceae"*[NP]
    - Genus I. *Alcanivorax*[VP (T)]
    - Genus II. *Fundibacter*[VP]
  - Family III. *"Hahellaceae"*[NP]
    - Genus I. *Hahella*[VP (T)]
    - Genus II. *Zooshikella*[VP]
  - Family IV. *Halomonadaceae*[VP]
    - Genus I. *Halomonas*[VP (T)]
    - Genus II. *Carnimonas*[VP]
    - Genus III. *Chromohalobacter*[VP]
    - Genus IV. *Cobetia*[VP]
    - Genus V. *Deleya*[VP]
    - Genus VI. *Zymobacter*[VP]
  - Family V. *Oleiphilaceae*[VP]
    - Genus I. *Oleiphilus*[VP (T)]
  - Family VI. *"Saccharospirillaceae"*[NP]
    - Genus I. *Saccharospirillum*[VP (T)]
- Order IX. *Pseudomonadales*[AL]
  - Family I. *Pseudomonadaceae*[AL]
    - Genus I. *Pseudomonas*[AL (T)]
    - Genus II. *Azomonas*[AL]
    - Genus III. *Azotobacter*[AL]
    - Genus IV. *Cellvibrio*[VP]
    - Genus V. *Chryseomonas*[VP]
    - Genus VI. *Flavimonas*[VP]
    - Genus VII. *Mesophilobacter*[VP]
    - Genus VIII. *Rhizobacter*[VP]
    - Genus IX. *Rugamonas*[VP]
    - Genus X. *Serpens*[AL]
  - Family II. *Moraxellaceae*[VP]
    - Genus I. *Moraxella*[AL (T)]

Genus II. *Acinetobacter*^{AL}
Genus III. *Enhydrobacter*^{VP}
Genus IV. *Psychrobacter*^{VP}
Order X. *"Alteromonadales"*
Family I. *Alteromonadaceae*^{VP}
Genus I. *Alteromonas*$^{AL\ (T)}$
Genus II. *Alishewanella*^{VP}
Genus III. *Colwellia*^{VP}
Genus IV. *Ferrimonas*^{VP}
Genus V. *Glaciecola*^{VP}
Genus VI. *Idiomarina*^{VP}
Genus VII. *Marinobacter*^{VP}
Genus VIII. *Marinobacterium*^{VP}
Genus IX. *Microbulbifer*^{VP}
Genus X. *Moritella*^{VP}
Genus XI. *Pseudoalteromonas*^{VP}
Genus XII. *Psychromonas*^{VP}
Genus XIII. *Shewanella*^{VP}
Genus XIV. *Thalassomonas*^{VP}
Family II. *Incertae sedis*
Genus I. *Teredinibacter*^{VP}
Order XI. *"Vibrionales"*^{NP}
Family I. *Vibrionaceae*^{AL}
Genus I. *Vibrio*$^{AL\ (T)}$
Genus II. *Allomonas*^{VP}
Genus III. *Catenococcus*^{VP}
Genus IV. *Enterovibrio*^{VP}
Genus V. *Grimontia*^{VP}
Genus VI. *Listonella*^{VP}
Genus VII. *Photobacterium*^{AL}
Genus VIII. *Salinivibrio*^{VP}
Order XII. *"Aeromonadales"*^{NP}
Family I. *Aeromonadaceae*^{VP}
Genus I. *Aeromonas*$^{AL\ (T)}$
Genus II. *Oceanimonas*^{VP}
Genus III. *Tolumonas*^{VP}
Family II. *Succinivibrionaceae*^{VP}
Genus I. *Succinivibrio*$^{AL\ (T)}$
Genus II. *Anaerobiospirillum*^{AL}
Genus III. *Ruminobacter*^{VP}
Genus IV. *Succinimonas*^{VP}
Order XIII. *"Enterobacteriales"*^{NP}
Family I. *Enterobacteriaceae*^{AL}
Genus I. *Escherichia*$^{AL\ (T)}$
Genus II. *Alterococcus*^{VP}
Genus III. *Arsenophonus*^{VP}
Genus IV. *Brenneria*^{VP}
Genus V. *Buchnera*^{VP}
Genus VI. *Budvicia*^{VP}
Genus VII. *Buttiauxella*^{VP}
Genus VIII. *Calymmatobacterium*^{VP}
Genus IX. *Cedecea*^{VP}
Genus X. *Citrobacter*^{AL}
Genus XI. *Edwardsiella*^{AL}
Genus XII. *Enterobacter*^{AL}
Genus XIII. *Erwinia*^{AL}
Genus XIV. *Ewingella*^{VP}
Genus XV. *Hafnia*^{AL}
Genus XVI. *Klebsiella*^{AL}
Genus XVII. *Kluyvera*^{VP}
Genus XVIII. *Leclercia*^{VP}
Genus XIX. *Leminorella*^{VP}
Genus XX. *Moellerella*^{VP}
Genus XXI. *Morganella*^{AL}
Genus XXII. *Obesumbacterium*^{AL}
Genus XXIII. *Pantoea*^{VP}
Genus XXIV. *Pectobacterium*^{AL}
Genus XXV. *"Phlomobacter"*
Genus XXVI. *Photorhabdus*^{VP}
Genus XXVII. *Plesiomonas*^{AL}
Genus XXVIII. *Pragia*^{VP}
Genus XXIX. *Proteus*^{AL}
Genus XXX. *Providencia*^{AL}
Genus XXXI. *Rahnella*^{VP}
Genus XXXII. *Raoultella*^{VP}
Genus XXXIII. *Saccharobacter*^{VP}
Genus XXXIV. *Salmonella*^{AL}
Genus XXXV. *Samsonia*^{VP}
Genus XXXVI. *Serratia*^{AL}
Genus XXXVII. *Shigella*^{AL}
Genus XXXVIII. *Sodalis*^{VP}
Genus XXXIX. *Tatumella*^{VP}
Genus XL. *Trabulsiella*^{VP}
Genus XLI. *Wigglesworthia*^{VP}
Genus XLII. *Xenorhabdus*^{AL}
Genus XLIII. *Yersinia*^{AL}
Genus XLIV. *Yokenella*^{VP}
Order XIV. *"Pasteurellales"*^{NP}
Family I. *Pasteurellaceae*^{VP}
Genus I. *Pasteurella*$^{AL\ (T)}$
Genus II. *Actinobacillus*^{AL}
Genus III. *Gallibacterium*^{VP}
Genus IV. *Haemophilus*^{AL}
Genus V. *Lonepinella*^{VP}
Genus VI. *Mannheimia*^{VP}
Genus VII. *Phocoenobacter*^{VP}
Class IV. *"Deltaproteobacteria"*^{NP}
Order I. *"Desulfurellales"*$^{NP\ (T)}$
Family I. *"Desulfurellaceae"*^{NP}
Genus I. *Desulfurella*$^{VP\ (T)}$
Genus II. *Hippea*^{VP}
Order II. *"Desulfovibrionales"*^{NP}
Family I. *"Desulfovibrionaceae"*^{NP}
Genus I. *Desulfovibrio*$^{AL\ (T)}$
Genus II. *Bilophila*^{VP}
Genus III. *Lawsonia*^{VP}
Family II. *"Desulfomicrobiaceae"*
Genus I. *Desulfomicrobium*$^{VP\ (T)}$
Family III. *"Desulfohalobiaceae"*^{NP}
Genus I. *Desulfohalobium*$^{VP\ (T)}$
Genus II. *Desulfomonas*^{AL}
Genus III. *Desulfonatronovibrio*^{VP}
Genus IV. *"Desulfothermus"*^{NP}
Family IV. *"Desulfonatronumaceae"*^{NP}
Genus I. *Desulfonatronum*$^{VP\ (T)}$
Order III. *"Desulfobacterales"*^{NP}
Family I. *"Desulfobacteraceae"*^{NP}
Genus I. *Desulfobacter*$^{VP\ (T)}$
Genus II. *Desulfobacterium*^{VP}
Genus III. *Desulfobacula*^{VP}
Genus IV. *"Desulfobotulus"*^{NP}
Genus V. *Desulfocella*^{VP}
Genus VI. *Desulfococcus*^{VP}
Genus VII. *Desulfofaba*^{VP}

Family I. *Acholeplasmataceae*[AL]
Genus I. *Acholeplasma*[AL]
Order IV. *Anaeroplasmatales*[VP]
Family I. *Anaeroplasmataceae*[VP]
Genus I. *Anaeroplasma*[AL (T)]
Genus II. *Asteroleplasma*[VP]
Order V. *Incertae sedis*
Family I. *"Erysipelotrichaceae"*
Genus I. *Erysipelothrix*[AL]
Genus II. *Bulleidia*[VP]
Genus III. *Holdemania*[VP]
Genus IV. *Solobacterium*[VP]
Class III. *"Bacilli"*
Order I. *Bacillales*[AL]
Family I. *Bacillaceae*[AL]
Genus I. *Bacillus*[AL (T)]
Genus II. *Amphibacillus*[VP]
Genus III. *Anoxybacillus*[VP]
Genus IV. *Exiguobacterium*[VP]
Genus V. *Filobacillus*[VP]
Genus VI. *Geobacillus*[VP]
Genus VII. *Gracilibacillus*[VP]
Genus VIII. *Halobacillus*[VP]
Genus IX. *Jeotgalibacillus*[VP]
Genus X. *Lentibacillus*[VP]
Genus XI. *Marinibacillus*[VP]
Genus XII. *Oceanobacillus*[VP]
Genus XIII. *Paraliobacillus*[VP]
Genus XIV. *Saccharococcus*[VP]
Genus XV. *Salibacillus*[VP]
Genus XVI. *Ureibacillus*[VP]
Genus XVII. *Virgibacillus*[VP]
Family II. *"Alicyclobacillaceae"*
Genus I. *Alicyclobacillus*[VP]
Genus II. *Pasteuria*[AL]
Genus III. *Sulfobacillus*[VP]
Family III. *Caryophanaceae*[AL]
Genus I. *Caryophanon*[AL (T)]
Family IV. *"Listeriaceae"*
Genus I. *Listeria*[AL]
Genus II. *Brochothrix*[AL]
Family V. *"Paenibacillaceae"*
Genus I. *Paenibacillus*[VP]
Genus II. *Ammoniphilus*[VP]
Genus III. *Aneurinibacillus*[VP]
Genus IV. *Brevibacillus*[VP]
Genus V. *Oxalophagus*[VP]
Genus VI. *Thermicanus*[VP]
Genus VII. *Thermobacillus*[VP]
Family VI. *Planococcaceae*[AL]
Genus I. *Planococcus*[AL (T)]
Genus II. *Filibacter*[VP]
Genus III. *Kurthia*[AL]
Genus IV. *Planomicrobium*[VP]
Genus V. *Sporosarcina*[AL]
Family VII. *"Sporolactobacillaceae"*
Genus I. *Sporolactobacillus*[AL]
Genus II. *Marinococcus*[VP]
Family VIII. *"Staphylococcaceae"*
Genus I. *Staphylococcus*[AL]
Genus II. *Gemella*[AL]
Genus III. *Jeotgalicoccus*[VP]
Genus IV. *Macrococcus*[VP]
Genus V. *Salinicoccus*[VP]
Family IX. *"Thermoactinomycetaceae"*
Genus I. *Thermoactinomyces*[AL]
Family X. *"Turicibacteraceae"*
Genus I. *Turicibacter*[VP (T)]
Order II. *"Lactobacillales"*
Family I. *Lactobacillaceae*[AL]
Genus I. *Lactobacillus*[AL (T)]
Genus II. *Paralactobacillus*[VP]
Genus III. *Pediococcus*[AL]
Family II. *"Aerococcaceae"*
Genus I. *Aerococcus*[AL]
Genus II. *Abiotrophia*[VP]
Genus III. *Dolosicoccus*[VP]
Genus IV. *Eremococcus*[VP]
Genus V. *Facklamia*[VP]
Genus VI. *Globicatella*[VP]
Genus VII. *Ignavigranum*[VP]
Family III. *"Carnobacteriaceae"*
Genus I. *Carnobacterium*[VP]
Genus II. *Agitococcus*[VP]
Genus III. *Alkalibacterium*[VP]
Genus IV. *Allofustis*[VP]
Genus V. *Alloiococcus*[VP]
Genus VI. *Desemzia*[VP]
Genus VII. *Dolosigranulum*[VP]
Genus VIII. *Granulicatella*[VP]
Genus IX. *Isobaculum*[VP]
Genus X. *Lactosphaera*[VP]
Genus XI. *Marinilactibacillus*[VP]
Genus XII. *Trichococcus*[VP]
Family IV. *"Enterococcaceae"*
Genus I. *Enterococcus*[VP]
Genus II. *Atopobacter*[VP]
Genus III. *Melissococcus*[VP]
Genus IV. *Tetragenococcus*[VP]
Genus V. *Vagococcus*[VP]
Family V. *"Leuconostocaceae"*
Genus I. *Leuconostoc*[AL]
Genus II. *Oenococcus*[VP]
Genus III. *Weissella*[VP]
Family VI. *Streptococcaceae*[AL]
Genus I. *Streptococcus*[AL (T)]
Genus II. *Lactococcus*[VP]
Family VII. *Incertae sedis*
Genus I. *Acetoanaerobium*[VP]
Genus II. *Oscillospira*[AL]
Genus III. *Syntrophococcus*[VP]
Phylum BXIV. *"Actinobacteria"*[NP]
Class I. *Actinobacteria*[VP]
Subclass I. *Acidimicrobidae*[VP]
Order I. *Acidimicrobiales*[VP]
Suborder I. *"Acidimicrobineae"*
Family I. *Acidimicrobiaceae*[VP]
Genus I. *Acidimicrobium*[VP (T)]
Subclass II. *Rubrobacteridae*[VP]
Order I. *Rubrobacterales*[VP]
Suborder I. *"Rubrobacterineae"*
Family I. *Rubrobacteraceae*[VP]
Genus I. *Rubrobacter*[VP (T)]
Genus II. *Conexibacter*[VP]

Genus III. *Solirubrobacter*^{VP}
Genus IV. *Thermoleophilum*^{VP}
Subclass III. *Coriobacteridae*^{VP}
Order I. *Coriobacteriales*^{VP}
Suborder I. *"Coriobacterineae"*
Family I. *Coriobacteriaceae*^{VP}
Genus I. *Coriobacterium*$^{VP (T)}$
Genus II. *Atopobium*^{VP}
Genus III. *Collinsella*^{VP}
Genus IV. *Cryptobacterium*^{VP}
Genus V. *Denitrobacterium*^{VP}
Genus VI. *Eggerthella*^{VP}
Genus VII. *Olsenella*^{VP}
Genus VIII. *Slackia*^{VP}
Subclass IV. *Sphaerobacteridae*^{VP}
Order I. *Sphaerobacterales*^{VP}
Suborder I. *"Sphaerobacterineae"*
Family I. *Sphaerobacteraceae*^{VP}
Genus I. *Sphaerobacter*$^{VP (T)}$
Subclass V. *Actinobacteridae*^{VP}
Order I. *Actinomycetales*^{AL}
Suborder I. *Actinomycineae*^{VP}
Family I. *Actinomycetaceae*^{AL}
Genus I. *Actinomyces*$^{AL (T)}$
Genus II. *Actinobaculum*^{VP}
Genus III. *Arcanobacterium*^{VP}
Genus IV. *Mobiluncus*^{VP}
Genus V. *Varibaculum*^{VP}
Suborder II. *Micrococcineae*^{VP}
Family I. *Micrococcaceae*^{AL}
Genus I. *Micrococcus*$^{AL (T)}$
Genus II. *Arthrobacter*^{AL}
Genus III. *Citricoccus*^{VP}
Genus IV. *Kocuria*^{VP}
Genus V. *Nesterenkonia*^{VP}
Genus VI. *Renibacterium*^{VP}
Genus VII. *Rothia*^{AL}
Genus VIII. *Stomatococcus*^{VP}
Family II. *Bogoriellaceae*^{VP}
Genus I. *Bogoriella*$^{VP (T)}$
Family III. *Rarobacteraceae*^{VP}
Genus I. *Rarobacter*$^{VP (T)}$
Family IV. *Sanguibacteraceae*^{VP}
Genus I. *Sanguibacter*$^{VP (T)}$
Family V. *Brevibacteriaceae*^{AL}
Genus I. *Brevibacterium*$^{AL (T)}$
Family VI. *Cellulomonadaceae*^{VP}
Genus I. *Cellulomonas*$^{AL (T)}$
Genus II. *Oerskovia*^{AL}
Genus III. *Tropheryma*^{VP}
Family VII. *Dermabacteraceae*^{VP}
Genus I. *Dermabacter*$^{VP (T)}$
Genus II. *Brachybacterium*^{VP}
Family VIII. *Dermatophilaceae*^{AL}
Genus I. *Dermatophilus*$^{AL (T)}$
Genus II. *Kineosphaera*^{VP}
Family IX. *Dermacoccaceae*^{VP}
Genus I. *Dermacoccus*$^{VP (T)}$
Genus II. *Demetria*^{VP}
Genus III. *Kytococcus*^{VP}
Family X. *Intrasporangiaceae*^{VP}
Genus I. *Intrasporangium*$^{AL (T)}$
Genus II. *Janibacter*^{VP}
Genus III. *Knoellia*^{VP}
Genus IV. *Ornithinicoccus*^{VP}
Genus V. *Ornithinimicrobium*^{VP}
Genus VI. *Nostocoidia*^{VP}
Genus VII. *Terrabacter*^{VP}
Genus VIII. *Terracoccus*^{VP}
Genus IX. *Tetrasphaera*^{VP}
Family XI. *Jonesiaceae*^{VP}
Genus I. *Jonesia*$^{VP (T)}$
Family XII. *Microbacteriaceae*^{VP}
Genus I. *Microbacterium*$^{AL (T)}$
Genus II. *Agreia*^{VP}
Genus III. *Agrococcus*^{VP}
Genus IV. *Agromyces*^{AL}
Genus V. *Aureobacterium*^{VP}
Genus VI. *Clavibacter*^{VP}
Genus VII. *Cryobacterium*^{VP}
Genus VIII. *Curtobacterium*^{AL}
Genus IX. *Frigoribacterium*^{VP}
Genus X. *Leifsonia*^{VP}
Genus XI. *Leucobacter*^{VP}
Genus XII. *Mycetocola*^{VP}
Genus XIII. *Okibacterium*^{VP}
Genus XIV. *Plantibacter*^{VP}
Genus XV. *Rathayibacter*^{VP}
Genus XVI. *Rhodoglobus*^{VP}
Genus XVII. *Subtercola*^{VP}
Family XIII. *"Beutenbergiaceae"*
Genus I. *Beutenbergia*$^{VP (T)}$
Genus II. *Georgenia*^{VP}
Genus III. *Salana*^{VP}
Family XIV. *Promicromonosporaceae*^{VP}
Genus I. *Promicromonospora*$^{AL (T)}$
Genus II. *Cellulosimicrobium*^{VP}
Genus III. *Xylanimonas*^{VP}
Suborder III. *Corynebacterineae*^{VP}
Family I. *Corynebacteriaceae*^{AL}
Genus I. *Corynebacterium*$^{AL (T)}$
Family II. *Dietziaceae*^{VP}
Genus I. *Dietzia*$^{VP (T)}$
Family III. *Gordoniaceae*^{VP}
Genus I. *Gordonia*$^{VP (T)}$
Genus II. *Skermania*^{VP}
Family IV. *Mycobacteriaceae*^{AL}
Genus I. *Mycobacterium*$^{AL (T)}$
Family V. *Nocardiaceae*^{AL}
Genus I. *Nocardia*$^{AL (T)}$
Genus II. *Rhodococcus*^{AL}
Family VI. *Tsukamurellaceae*^{VP}
Genus I. *Tsukamurella*$^{VP (T)}$
Family VII. *"Williamsiaceae"*
Genus I. *Williamsia*^{VP}
Suborder IV. *Micromonosporineae*^{VP}
Family I. *Micromonosporaceae*^{AL}
Genus I. *Micromonospora*$^{AL (T)}$
Genus II. *Actinoplanes*^{AL}
Genus III. *Asanoa*^{VP}
Genus IV. *Catellatospora*^{VP}
Genus V. *Catenuloplanes*^{VP}
Genus VI. *Couchioplanes*^{VP}
Genus VII. *Dactylosporangium*^{AL}

Genus VIII. *Pilimelia*^AL
Genus IX. *Spirilliplanes*^VP
Genus X. *Verrucosispora*^VP
Genus XI. *Virgisporangium*^VP
Suborder V. *Propionibacterineae*^VP
Family I. *Propionibacteriaceae*^AL
Genus I. *Propionibacterium*^AL (T)
Genus II. *Luteococcus*^VP
Genus III. *Microlunatus*^VP
Genus IV. *Propioniferax*^VP
Genus V. *Propionimicrobium*^VP
Genus VI. *Tessaracoccus*^VP
Family II. *Nocardioidaceae*^VP
Genus I. *Nocardioides*^AL (T)
Genus II. *Aeromicrobium*^VP
Genus III. *Actinopolymorpha*^VP
Genus IV. *Friedmanniella*^VP
Genus V. *Hongia*^VP
Genus VI. *Kribbella*^VP
Genus VII. *Micropruina*^VP
Genus VIII. *Marmoricola*^VP
Suborder VI. *Pseudonocardineae*^VP
Family I. *Pseudonocardiaceae*^VP
Genus I. *Pseudonocardia*^AL (T)
Genus II. *Actinoalloteichus*^VP
Genus III. *Actinopolyspora*^AL
Genus IV. *Amycolatopsis*^VP
Genus V. *Crossiella*^VP
Genus VI. *Kibdelosporangium*^VP
Genus VII. *Kutzneria*^VP
Genus VIII. *Prauserella*^VP
Genus IX. *Saccharomonospora*^AL
Genus X. *Saccharopolyspora*^AL
Genus XI. *Streptoalloteichus*^VP
Genus XII. *Thermobispora*^VP
Genus XIII. *Thermocrispum*^VP
Family II. *Actinosynnemataceae*^VP
Genus I. *Actinosynnema*^AL (T)
Genus II. *Actinokineospora*^VP
Genus III. *Lechevalieria*^VP
Genus IV. *Lentzea*^VP
Genus V. *Saccharothrix*^VP
Suborder VII. *Streptomycineae*^VP
Family I. *Streptomycetaceae*^AL
Genus I. *Streptomyces*^AL (T)
Genus II. *Kitasatospora*^VP
Genus III. *Streptoverticillium*^AL
Suborder VIII. *Streptosporangineae*^VP
Family I. *Streptosporangiaceae*^VP
Genus I. *Streptosporangium*^AL (T)
Genus II. *Acrocarpospora*^VP
Genus III. *Herbidospora*^VP
Genus IV. *Microbispora*^AL
Genus V. *Microtetraspora*^AL
Genus VI. *Nonomuraea*^VP
Genus VII. *Planobispora*^AL
Genus VIII. *Planomonospora*^AL
Genus IX. *Planopolyspora*^VP
Genus X. *Planotetraspora*^VP
Family II. *Nocardiopsaceae*^VP
Genus I. *Nocardiopsis*^AL (T)
Genus II. *Streptomonospora*^VP
Genus III. *Thermobifida*^VP
Family III. *Thermomonosporaceae*^VP
Genus I. *Thermomonospora*^AL (T)
Genus II. *Actinomadura*^AL
Genus III. *Spirillospora*^AL
Suborder IX. *Frankineae*^VP
Family I. *Frankiaceae*^AL
Genus I. *Frankia*^AL (T)
Family II. *"Geodermatophilaceae"*
Genus I. *Geodermatophilus*^AL
Genus II. *Blastococcus*^AL
Genus III. *Modestobacter*^VP
Family III. *Microsphaeraceae*^VP
Genus I. *Microsphaera*^VP (T)
Family IV. *Sporichthyaceae*^VP
Genus I. *Sporichthya*^AL (T)
Family V. *Acidothermaceae*^VP
Genus I. *Acidothermus*^VP (T)
Family VI. *"Kineosporiaceae"*
Genus I. *Kineosporia*^AL
Genus II. *Cryptosporangium*^VP
Genus III. *Kineococcus*^VP
Suborder X. *Glycomycineae*^VP
Family I. *Glycomycetaceae*^VP
Genus I. *Glycomyces*^VP (T)
Order II. *Bifidobacteriales*^VP
Family I. *Bifidobacteriaceae*^VP
Genus I. *Bifidobacterium*^AL (T)
Genus II. *Falcivibrio*^VP
Genus III. *Gardnerella*^VP
Genus IV. *Parascardovia*^VP
Genus V. *Scardovia*^VP
Family II. *Unknown Affiliation*^VP
Genus I. *Actinobispora*^VP
Genus II. *Actinocorallia*^VP
Genus III. *Excellospora*^AL
Genus IV. *Pelczaria*^VP
Genus V. *Turicella*^VP
Phylum BXV. *"Planctomycetes"*
Class I. *"Planctomycetacia"*
Order I. *Planctomycetales*^VP
Family I. *Planctomycetaceae*^VP
Genus I. *Planctomyces*^AL (T)
Genus II. *Gemmata*^VP
Genus III. *Isosphaera*^VP
Genus IV. *Pirellula*^VP
Phylum BXVI. *"Chlamydiae"*
Class I. *Chlamydiae*^VP
Order I. *Chlamydiales*^AL (T)
Family I. *Chlamydiaceae*^AL
Genus I. *Chlamydia*^AL (T)
Genus II. *Chlamydophila*^VP
Family II. *Parachlamydiaceae*^VP
Genus I. *Parachlamydia*^VP (T)
Genus II. *Neochlamydia*^VP
Family III. *Simkaniaceae*^VP
Genus I. *Simkania*^VP (T)
Family IV. *Waddliaceae*^VP
Genus I. *Waddlia*^VP (T)
Phylum BXVII. *"Spirochaetes"*^NP
Class I. *"Spirochaetes"*
Order I. *Spirochaetales*^AL

- Family I. *Spirochaetaceae*[AL]
  - Genus I. *Spirochaeta*[AL (T)]
  - Genus II. *Borrelia*[AL]
  - Genus III. *Brevinema*[VP]
  - Genus IV. *Clevelandina*[VP]
  - Genus V. *Cristispira*[AL]
  - Genus VI. *Diplocalyx*[VP]
  - Genus VII. *Hollandina*[VP]
  - Genus VIII. *Pillotina*[VP]
  - Genus IX. *Treponema*[AL]
- Family II. *"Serpulinaceae"*
  - Genus I. *Serpulina*[VP]
  - Genus II. *Brachyspira*[VP]
- Family III. *Leptospiraceae*[AL]
  - Genus I. *Leptospira*[AL (T)]
  - Genus II. *Leptonema*[VP]

- Phylum BXVIII. *"Fibrobacteres"*
  - Class I. *"Fibrobacteres"*
    - Order I. *"Fibrobacterales"*
      - Family I. *"Fibrobacteraceae"*
        - Genus I. *Fibrobacter*[VP]
- Phylum BXIX. *"Acidobacteria"*
  - Class I. *Acidobacteria*[VP]
    - Order I. *Acidobacteriales*[VP (T)]
      - Family I. *"Acidobacteriaceae"*
        - Genus I. *Acidobacterium*[VP]
        - Genus II. *Geothrix*[VP]
        - Genus III. *Holophaga*[VP]
- Phylum BXX. *"Bacteroidetes"*
  - Class I. *"Bacteroidetes"*
    - Order I. *"Bacteroidales"*
      - Family I. *Bacteroidaceae*[AL]
        - Genus I. *Bacteroides*[AL (T)]
        - Genus II. *Acetofilamentum*[VP]
        - Genus III. *Acetomicrobium*[VP]
        - Genus IV. *Acetothermus*[VP]
        - Genus V. *Anaerophaga*[VP]
        - Genus VI. *Anaerorhabdus*[VP]
        - Genus VII. *Megamonas*[VP]
      - Family II. *"Rikenellaceae"*
        - Genus I. *Rikenella*[VP]
        - Genus II. *Marinilabilia*[VP]
      - Family III. *"Porphyromonadaceae"*
        - Genus I. *Porphyromonas*[VP]
        - Genus II. *Dysgonomonas*[VP]
        - Genus III. *Tannerella*[VP]
      - Family IV. *"Prevotellaceae"*
        - Genus I. *Prevotella*[VP]
  - Class II. *Flavobacteria*[VP]
    - Order I. *"Flavobacteriales"*
      - Family I. *Flavobacteriaceae*[VP]
        - Genus I. *Flavobacterium*[AL (T)]
        - Genus II. *Aequorivita*[VP]
        - Genus III. *Arenibacter*[VP]
        - Genus IV. *Bergeyella*[VP]
        - Genus V. *Capnocytophaga*[VP]
        - Genus VI. *Cellulophaga*[VP]
        - Genus VII. *Chryseobacterium*[VP]
        - Genus VIII. *Coenonia*[VP]
        - Genus IX. *Croceibacter*[VP]
        - Genus X. *Empedobacter*[VP]
        - Genus XI. *Gelidibacter*[VP]
        - Genus XII. *Muricauda*[VP]
        - Genus XIII. *Myroides*[VP]
        - Genus XIV. *Ornithobacterium*[VP]
        - Genus XV. *Polaribacter*[VP]
        - Genus XVI. *Psychroflexus*[VP]
        - Genus XVII. *Psychroserpens*[VP]
        - Genus XVIII. *Riemerella*[VP]
        - Genus XIX. *Saligentibacter*[VP]
        - Genus XX. *Tenacibaculum*[VP]
        - Genus XXI. *Weeksella*[VP]
        - Genus XXII. *Zobellia*[VP]
      - Family II. *"Myroidaceae"*
        - Genus I. *Myroides*[VP]
      - Family III. *"Blattabacteriaceae"*
        - Genus I. *Blattabacterium*[AL]
  - Class III. *"Sphingobacteria"*
    - Order I. *"Sphingobacteriales"*
      - Family I. *Sphingobacteriaceae*[VP]
        - Genus I. *Sphingobacterium*[VP (T)]
        - Genus II. *Pedobacter*[VP]
      - Family II. *"Saprospiraceae"*
        - Genus I. *Saprospira*[AL]
        - Genus II. *Haliscomenobacter*[AL]
        - Genus III. *Lewinella*[VP]
      - Family III. *"Flexibacteraceae"*
        - Genus I. *Flexibacter*[AL]
        - Genus II. *Cyclobacterium*[VP]
        - Genus III. *Cytophaga*[AL]
        - Genus IV. *Dyadobacter*[VP]
        - Genus V. *Flectobacillus*[AL]
        - Genus VI. *Hymenobacter*[VP]
        - Genus VII. *Meniscus*[AL]
        - Genus VIII. *Microscilla*[AL]
        - Genus IX. *Reichenbachia*[VP]
        - Genus X. *Runella*[AL]
        - Genus XI. *Spirosoma*[AL]
        - Genus XII. *Sporocytophaga*[AL]
      - Family IV. *"Flammeovirgaceae"*
        - Genus I. *Flammeovirga*[VP]
        - Genus II. *Flexithrix*[AL]
        - Genus III. *Persicobacter*[VP]
        - Genus IV. *Thermonema*[VP]
      - Family V. *Crenotrichaceae*[AL]
        - Genus I. *Crenothrix*[AL (T)]
        - Genus II. *Chitinophaga*[VP]
        - Genus III. *Rhodothermus*[VP]
        - Genus IV. *Salinibacter*[VP]
        - Genus V. *Toxothrix*[AL]
- Phylum BXXI. *"Fusobacteria"*
  - Class I. *"Fusobacteria"*
    - Order I. *"Fusobacteriales"*
      - Family I. *"Fusobacteriaceae"*
        - Genus I. *Fusobacterium*[AL]
        - Genus II. *Ilyobacter*[VP]
        - Genus III. *Leptotrichia*[AL]
        - Genus IV. *Propionigenium*[VP]
        - Genus V. *Sebaldella*[VP]
        - Genus VI. *Streptobacillus*[AL]
        - Genus VII. *Sneathia*[VP]
      - Family II. *Incertae sedis*[VP]
        - Genus I. *Cetobacterium*[VP]
- Phylum BXXII. *"Verrucomicrobia"*

- Class I. *Verrucomicrobiae* VP
  - Order I. *Verrucomicrobiales* VP (T)
    - Family I. *Verrucomicrobiaceae* VP
      - Genus I. *Verrucomicrobium* VP (T)
      - Genus II. *Prosthecobacter* VP
    - Family II. *"Opitutaceae"*
      - Genus I. *Opitutus* VP (T)
    - Family III. *"Victivallaceae"*
      - Genus I. *Victivallis* VP
    - Family IV. *"Xiphinematobacteriaceae"*
      - Genus I. *Xiphinematobacter* VP

- Phylum BXXIII. *"Dictyoglomi"*
  - Class I. *"Dictyoglomi"*
    - Order I. *"Dictyoglomales"*
      - Family I. *"Dictyoglomaceae"*
        - Genus I. *Dictyoglomus* VP
- Phylum BXXIV. *Gemmatimonadetes* VP
  - Class I. *Gemmatimonadetes* VP
    - Order I. *Gemmatimonadales* VP (T)
      - Family I. *Gemmatimonadaceae* VP
        - Genus I. *Gemmatimonas* VP (T)

Bibliography

Aamand, J., T. Ahl and E. Spieck. 1996. Monoclonal antibodies recognizing nitrite oxidoreductase of *Nitrobacter hamburgensis*, *N. winogradskyi*, and *N. vulgaris*. Appl. Environ. Microbiol. *62*: 2352–2355.

Abd El-Rahman, H.A., D. Fritze, C. Sproer and D. Claus. 2002. Two novel psychrotolerant species, *Bacillus psychrotolerans* sp. nov. and *Bacillus psychrodurans* sp. nov., which contain ornithine in their cell walls. Int. J. Syst. Evol. Microbiol. *52*: 2127–2133.

Abraham, W.R., C. Strompl, A. Bennasar, M. Vancanneyt, C. Snauwaert, J. Swings, J. Smit and E.R. Moore. 2002. Phylogeny of *Maricaulis* Abraham et al. 1999 and proposal of *Maricaulis virginensis* sp. nov., *M. parjimensis* sp. nov., *M. washingtonensis* sp. nov. and *M. salignorans* sp. nov. Int. J. Syst. Evol. Microbiol. *52*: 2191–2201.

Achenbach, L.A. and J.D. Coates. 2000. Disparity between bacterial phylogeny and physiology. ASM News. *66*: 714–716.

Achenbach, L.A., U. Michaelidou, R.A. Bruce, J. Fryman and J.D. Coates. 2001. *Dechloromonas agitata* gen. nov., sp. nov. and *Dechlorosoma suillum* gen. nov., sp. nov., two novel environmentally dominant (per)chlorate-reducing bacteria and their phylogenetic position. Int. J. Syst. Evol. Microbiol. *51*: 527–533.

Achouak, W., L. Sutra, T. Heulin, J.-M. Meyer, N. Fromin, S. Degraeve, R. Christen and L. Gardan. 2000. *Pseudomonas brassicacearum* sp. nov. and *Pseudomonas thivervalensis* sp. nov., two root-associated bacteria isolated from *Brassica napus* and *Arabidopsis thaliana*. Int. J. Syst. Evol. Microbiol. *50*: 9–18.

Adeleke, A.A., B.S. Fields, R.F. Benson, M.I. Daneshvar, J.M. Pruckler, R.M. Ratcliff, T.G. Harrison, R.S. Weyant, R.J. Birtles, D. Raoult and M.A. Halablab. 2001. *Legionella drozanskii* sp. nov., *Legionella rowbothamii* sp. nov. and *Legionella fallonii* sp. nov.: three unusual new *Legionella* species. Int. J. Syst. Evol. Microbiol. *51*: 1151–1160.

Adnan, S., N. Li, H. Miura, Y. Hashimoto, H. Yamamoto and T. Ezaki. 1993. Covalently immobilized DNA plate for luminometric DNA–DNA hybridization to identify viridans streptococci in under 2 hours. FEMS Microbiol. Lett. *106*: 139–142.

Aeckersberg, F., F.A. Rainey and F. Widdel. 1998. Growth, natural relationships, cellular fatty acids and metabolic adaptation of sulfate-reducing bacteria that utilize long-chain alkanes under anoxic conditions. Arch. Microbiol. *170*: 361–369.

Agaev, R., V. Danilov, V. Khachaturov, B. Kasymov and B. Tishabaev. 1986. The toxicity to warm-blooded animals and fish of new defoliants based on sodium and magnesium chlorates. Uzb. Biol. Zh. *1*: 40–43.

Aguilera, M., M. Monteoliva-Sanchez, A. Suarez, V. Guerra, C. Lizama, A. Bennasar and A. Ramos-Cormenzana. 2001. *Paenibacillus jamilae* sp. nov., an exopolysaccharide-producing bacterium able to grow in olive-mill wastewater. Int. J. Syst. Evol. Microbiol. *51*: 1687–1692.

Ahmad, S., R.K. Scopes, G.N. Rees and B.K. Patel. 2000. *Saccharococcus caldoxylosilyticus* sp. nov., an obligately thermophilic, xylose-utilizing, endospore-forming bacterium. Int. J. Syst. Evol. Microbiol. *50*: 517–523.

Ainsworth, G.C. and P.H.A. Sneath. 1962. Microbial Classification: Appendix I. Symp. Soc. Gen. Microbiol. *12*: 456–463.

Ajithkumar, B., V.P. Ajithkumar, R. Iriye, Y. Doi and T. Sakai. 2003. Spore-forming *Serratia marcescens* subsp. *sakuensis* subsp. nov., isolated from a domestic wastewater treatment tank. Int. J. Syst. Evol. Microbiol. *53*: 253–258.

Ajithkumar, V.P., B. Ajithkumar, R. Iriye and T. Sakai. 2002. *Bacillus funiculus* sp. nov., novel filamentous isolates from activated sludge. Int. J. Syst. Evol. Microbiol. *52*: 1141–1144.

Akasaka, H., A. Ueki, S. Hanada, Y. Kamagata and K. Ueki. 2003. *Propionicimonas paludicola* gen. nov., sp. nov., a novel facultatively anaerobic, Gram-positive, propionate-producing bacterium isolated from plant residue in irrigated rice-field soil. Int. J. Syst. Evol. Microbiol. *53*: 1991–1998.

Al-Musallam, A.A., S.S. Al-Zarban, Y.A. Fasasi, R.M. Kroppenstedt and E. Stackebrandt. 2003. *Amycolatopsis keratiniphila* sp. nov., a novel keratinolytic soil actinomycete from Kuwait. Int. J. Syst. Evol. Microbiol. *53*: 871–874.

Al-Zarban, S.S., I. Abbas, A.A. Al-Musallam, U. Steiner, E. Stackebrandt and R.M. Kroppenstedt. 2002a. *Nocardiopsis halotolerans* sp. nov., isolated from salt marsh soil in Kuwait. Int. J. Syst. Evol. Microbiol. *52*: 525–529.

Al-Zarban, S.S., A.A. Al-Musallam, I. Abbas, E. Stackebrandt and R.M. Kroppenstedt. 2002b. *Saccharomonospora halophila* sp. nov., a novel halophilic actinomycete isolated from marsh soil in Kuwait. Int. J. Syst. Evol. Microbiol. *52*: 555–558.

Alain, K., V.T. Marteinsson, M.L. Miroshnichenko, E.A. Bonch-Osmolovskaya, D. Prieur and J.L. Birrien. 2002a. *Marinitoga piezophila* sp. nov., a rod-shaped, thermo-piezophilic bacterium isolated under high hydrostatic pressure from a deep-sea hydrothermal vent. Int. J. Syst. Evol. Microbiol. *52*: 1331–1339.

Alain, K., P. Pignet, M. Zbinden, M. Quillevere, F. Duchiron, J.P. Donval, F. Lesongeur, G. Raguenes, P. Crassous, J. Querellou and M.A. Cambon-Bonavita. 2002b. *Caminicella sporogenes* gen. nov., sp. nov., a novel thermophilic spore-forming bacterium isolated from an East-Pacific Rise hydrothermal vent. Int. J. Syst. Evol. Microbiol. *52*: 1621–1628.

Alain, K., J. Querellou, F. Lesongeur, P. Pignet, P. Crassous, G. Raguenes, V. Cueff and M.A. Cambon-Bonavita. 2002c. *Caminibacter hydrogeniphilus* gen. nov., sp. nov., a novel thermophilic, hydrogen-oxidizing bacterium isolated from an East Pacific Rise hydrothermal vent. Int. J. Syst. Evol. Microbiol. *52*: 1317–1323.

Alarico, S., F.A. Rainey, N. Empadinhas, P. Schumann, M.F. Nobre and M.S. da Costa. 2002a. *Rubritepida flocculans* gen. nov., sp. nov., a new slightly thermophilic member of the α-1 subclass of the *Proteobacteria*. Syst. Appl. Microbiol. *25*: 198–206.

Alarico, S., F.A. Rainey, N. Empadinhas, P. Schumann, M.F. Nobre and M.S. da Costa. 2002b. *In* Validation of publication of new names and new combinations previously effectively published outside the IJSEM. List No. 88. Int. J. Syst. Evol. Microbiol. *52*: 1915–1916.

Alazard, D., S. Dukan, A. Urios, F. Verhe, N. Bouabida, F. Morel, P. Thomas, J.L. Garcia and B. Ollivier. 2003. *Desulfovibrio hydrothermalis* sp. nov., a novel sulfate-reducing bacterium isolated from hydrothermal vents. Int. J. Syst. Evol. Microbiol. *53*: 173–178.

Albuquerque, L., F.A. Rainey, A.P. Chung, A. Sunna, M.F. Nobre, R. Grote,

G. Antranikian and M.S. da Costa. 2000. *Alicyclobacillus hesperidum* sp. nov. and a related genomic species from solfataric soils of Sao Miguel in the Azores. Int. J. Syst. Evol. Microbiol. *50*: 451–457.

Albuquerque, L., J. Santos, P. Travassos, M.F. Nobre, F.A. Rainey, R. Wait, N. Empadinhas, M.T. Silva and M.S. da Costa. 2002. *Albidovulum inexpectatum* gen. nov., sp nov., a nonphotosynthetic and slightly thermophilic bacterium from a marine hot spring that is very closely related to members of the photosynthetic genus Rhodovulum. Appl. Environ. Microbiol. *68*: 4266–4273.

Albuquerque, L., J. Santos, P. Travassos, M.F. Nobre, F.A. Rainey, R. Wait, N. Empadinhas, M.T. Silva and M.S. da Costa. 2003. *In* Validation of the publication of new names and new combinations previously effectively published outside the IJSEM. List no. 89. Int. J. Syst. Evol. Microbiol. *53*: 1–2.

Albuquerque de Barros, E.V., G.P. Manfio, V. Ribiero Maitan, L.A. Mendes Bataus, S.B. Kim, L.A. Maldonado and M. Goodfellow. 2003. *Nocardia cerradoensis* sp. nov., a novel isolate from Cerrado soil in Brazil. Int. J. Syst. Evol. Microbiol. *53*: 29–33.

Aleem, M.I. 1966. Generation of reducing power in chemosynthesis. II. Energy-linked reduction of pyridine nucleotides in the chemoautotroph, *Nitrosomonas europaea*. Biochim Biophys. Acta. *113*: 216–224.

Aleem, M.I., G.E. Hoch and J.E. Varner. 1965. Water as the source of oxidant and reductant in bacterial chemosynthesis. Proc. Natl. Acad. Sci. U.S.A. *54*: 869–873.

Aleem, M.I.H. and D.L. Sewell. 1981. Mechanism of nitrite oxidation and oxidoreductase systems in *Nitrobacter agilis*. Curr. Micriobiol. *5*: 267–272.

Alfreider, A., J. Pernthaler, R.I. Amann, B. Sattler, F.O. Glöckner, A. Wille and R. Psenner. 1996. Community analysis of the bacterial assemblages in the winter cover and pelagic layers of a high Mountain Lake by *in situ* hybridization. Appl. Environ. Microbiol. *62*: 2138–2144.

Alm, E.W., D.B. Oerther, N. Larsen, D.A. Stahl and L. Raskin. 1996. The oligonucleotide probe database. Appl. Environ. Microbiol. *62*: 3557–3559.

Altenburger, P., P. Kampfer, P. Schumann, R. Steiner, W. Lubitz and H.J. Busse. 2002a. *Citricoccus muralis* gen. nov., sp. nov., a novel actinobacterium isolated from a medieval wall painting. Int. J. Syst. Evol. Microbiol. *52*: 2095–2100.

Altenburger, P., P. Kampfer, P. Schumann, D. Vybiral, W. Lubitz and H.J. Busse. 2002b. *Georgenia muralis* gen. nov., sp. nov., a novel actinobacterium isolated from a medieval wall painting. Int. J. Syst. Evol. Microbiol. *52*: 875–881.

Alves, M.P., F.A. Rainey, M.F. Nobre and M.S. da Costa. 2003a. *Thermomonas hydrothermalis* sp. nov., a new slightly thermophilic gamma-proteobacterium isolated from a hot spring in central Portugal. System. Appl. Microbiol. *26*: 70–75.

Alves, M.P., F.A. Rainey, M.F. Nobre and M.S. da Costa. 2003b. *In* Validation of publication of new names and new combinations previously effectively published outside the IJSEM. List No. 92. Int. J. Syst. Evol. Microbiol. *53*: 935–937.

Amann, R.I. 1995a. Fluorescently labeled, rRNA-targeted oligonucleotide probes in the study of microbial ecology. Mol. Ecol. *4*: 543–554.

Amann, R.I. 1995b. *In situ* identification of micro-organisms by whole cell hybridization with rRNA-targeted nucleic acid probes. *In* Akkermans, van Elsas and de Bruijn (Editors), Molecular Microbial Ecology Manual, Vol. 3.3.6, Kluwer Academic Publishers, Dordrecht. pp. 1–15.

Amann, R.I., B.J. Binder, R.J. Olson, S.W. Chisholm, R. Devereux and D.A. Stahl. 1990a. Combination of 16S rRNA-targeted oligonucleotide probes with flow cytometry for analyzing mixed microbial populations. Appl. Environ. Microbiol. *56*: 1919–1925.

Amann, R.I. and M. Kühl. 1998. *In situ* methods for assessment of microorganisms and their activities. Curr. Opin. Microbiol. *1*: 352–358.

Amann, R.I., L. Krumholz and D.A. Stahl. 1990b. Fluorescent-oligonucleotide probing of whole cells for determinative, phylogenetic, and environmental studies in microbiology. J. Bacteriol. *172*: 762–770.

Amann, R.I., W. Ludwig and K.H. Schleifer. 1994. Identification of uncultured bacteria: a challenging task for molecular taxonomists. ASM News. *60*: 360–365.

Amann, R.I., W. Ludwig and K.H. Schleifer. 1995. Phylogenetic identification and *in situ* detection of individual microbial cells without cultivation. Microbiol. Rev. *59*: 143–169.

Amann, R.I., W. Ludwig, R. Schulze, S. Spring, E. Moore and K.H. Schleifer. 1996a. rRNA-targeted oligonucleotide probes for the identification of genuine and former pseudomonads. Syst. Appl. Microbiol. *19*: 501–509.

Amann, R.I., J. Snaidr, M. Wagner, W. Ludwig and K.H. Schleifer. 1996b. *In situ* visualization of high genetic diversity in a natural microbial community. J. Bacteriol. *178*: 3496–3500.

Amann, R.I., N. Springer, W. Ludwig, H.-D. Görtz and K.H. Schleifer. 1991. Identification *in situ* and phylogeny of uncultured bacterial endosymbionts. Nature *351*: 161–164.

Amann, R.I., J. Stromley, R. Devereux, R. Key and D.A. Stahl. 1992a. Molecular and microscopic identification of sulfate-reducing bacteria in multispecies biofilms. Appl. Environ. Microbiol. *58*: 614–623.

Amann, R.I., B. Zarda, D.A. Stahl and K.H. Schleifer. 1992b. Identification of individual prokaryotic cells by using enzyme-labeled, rRNA-targeted oligonucleotide probes. Appl. Environ. Microbiol. *58*: 3007–3011.

Ambler, R.P., M. Daniel, J. Hermoso, T.E. Meyer, T.G. Bartsch and M.D. Kamen. 1979. Cytochrome c_2 sequence variation among the recognized species of purple nonsulfur photosynthetic bacteria. Nature *278*: 659–660.

Amoozegar, M.A., F. Malekzadeh, K.A. Malik, P. Schumann and C. Sproer. 2003. *Halobacillus karajensis* sp. nov., a novel moderate halophile. Int. J. Syst. Evol. Microbiol. *53*: 1059–1063.

Amy, P.S. and H.D. Hiatt. 1989. Survival and detection of bacteria in an aquatic environment. Appl. Environ. Microbiol. *55*: 788–793.

Andersen, S.M., K. Johnsen, J. Sorensen, P. Nielsen and C.S. Jacobsen. 2000. *Pseudomonas frederiksbergensis* sp. nov., isolated from soil at a coal gasification site. Int. J. Syst. Evol. Microbiol. *50*: 1957–1964.

Anderson, R.C., M.A. Rasmussen, N.S. Jensen and M.J. Allison. 2000. *Denitrobacterium detoxificans* gen. nov., sp. nov., a ruminal bacterium that respires on nitrocompounds. Int. J. Syst. Evol. Microbiol. *50*: 633–638.

Andersson, K.K. and A.B. Hooper. 1983. O_2 and H_2O are each the source of one O in NO_2 produced from NH_3 by *Nitrosomonas* ^{15}N-NMR evidence. FEBS Lett. *164*: 236–240.

Angen, O., P. Ahrens, P. Kuhnert, H. Christensen and R. Mutters. 2003. Proposal of *Histophilus somni* gen. nov., sp. nov. for the three species incertae sedis *'Haemophilus somnus'*, *'Haemophilus agni'* and *'Histophilus ovis'*. Int. J. Syst. Evol. Microbiol. *53*: 1449–1456.

Angert, E.R., K.D. Clements and N.R. Pace. 1993. The largest bacterium. Nature *362*: 239–241.

Angulo, A.F., R. Reijgers, J. Brugman, I. Kroesen, F.E. Hekkens, P. Carle, J.M. Bove, J.G. Tully, A.C. Hill, L.M. Schouls, C.S. Schot, P.J. Roholl and A.A. Polak-Vogelzang. 2000. *Acholeplasma vituli* sp. nov., from bovine serum and cell cultures. Int. J. Syst. Evol. Microbiol. *50*: 1125–1131.

Anthony, C. 1982. The Biochemistry of Methylotrophs, Academic Press, Ltd., London.

Antón, J., A. Oren, S. Benlloch, F. Rodriguez-Valera, R. Amann and R. Rosselló-Mora. 2002. *Salinibacter ruber* gen. nov., sp. nov., a novel, extremely halophilic member of the *Bacteria* from saltern crystallizer ponds. Int. J. Syst. Evol. Microbiol. *52*: 485–491.

Antunes, A., W. Eder, P. Fareleira, H. Santos and R. Huber. 2003a. *Salinisphaera shabanensis* gen. nov., sp. nov., a novel, moderately halophilic bacterium from the brine-seawater interface of the Shaban Deep, Red Sea. Extremophiles *7*: 29–34.

Antunes, A., W. Eder, P. Fareleira, H. Santos and R. Huber. 2003b. *In* Validation of publication of new names and new combinations previously effectively published outside the IJSEM. List No. 93. Int. J. Syst. Evol. Microbiol. *53*: 1219–1220.

Antunes, A., F.A. Rainey, M.F. Nobre, P. Schumann, A.M. Ferreira, A. Ramos, H. Santos and M.S. da Costa. 2002. *Leuconostoc ficulneum* sp.

nov., a novel lactic acid bacterium isolated from a ripe fig, and reclassification of *Lactobacillus fructosus* as *Leuconostoc fructosum* comb. nov. Int. J. Syst. Evol. Microbiol. *52*: 647–655.

Arab, H., H. Volker and M. Thomm. 2000. *Thermococcus aegaeicus* sp. nov. and *Staphylothermus hellenicus* sp. nov., two novel hyperthermophilic archaea isolated from geothermally heated vents off Palaeochori Bay, Milos, Greece. Int. J. Syst. Evol. Microbiol. *50*: 2101–2108.

Arahal, D.R., A.M. Castillo, W. Ludwig, K.-H. Schleifer and A. Ventosa. 2002a. Proposal of *Cobetia marina* gen. nov., comb. nov., within the family *Halomonadaceae*, to include the species *Halomonas marina*. Syst. Appl. Microbiol. *25*: 207–211.

Arahal, D.R., A.M. Castillo, W. Ludwig, K.-H. Schleifer and A. Ventosa. 2002b. *In* Validation of the publication of new names and new combinations previously effectively published outside the IJSEM. List no. 88. Int. J. Syst. Evol. Microbiol. *52*: 1915–1916.

Arahal, D.R., M.T. Garcia, W. Ludwig, K.H. Schleifer and A. Ventosa. 2001a. Transfer of *Halomonas canadensis* and *Halomonas israelensis* to the genus *Chromohalobacter* as *Chromohalobacter canadensis* comb. nov. and *Chromohalobacter israelensis* comb. nov. Int. J. Syst. Evol. Microbiol. *51*: 1443–1448.

Arahal, D.R., M.T. Garcia, C. Vargas, D. Cánovas, J.J. Nieto and A. Ventosa. 2001b. *Chromohalobacter salexigens* sp. nov., a moderately halophilic species that includes *Halomonas elongata* DSM 3043 and ATCC 33174. Int. J. Syst. Evol. Microbiol. *51*: 1457–1462.

Arahal, D.R., M.C. Marquez, B.E. Volcani, K.H. Schleifer and A. Ventosa. 2000. Reclassification of *Bacillus marismortui* as *Salibacillus marismortui* comb. nov. Int. J. Syst. Evol. Microbiol. *50*: 1501–1503.

Aranaz, A., D. Cousins, A. Mateos and L. Dominguez. 2003. Elevation of *Mycobacterium tuberculosis* subsp. *caprae* Aranaz et al. 1999 to species rank as *Mycobacterium caprae* comb. nov., sp. nov. Int. J. Syst. Evol. Microbiol. *53*: 1785–1789.

Arciero, D.M., C. Balny and A.B. Hooper. 1991a. Spectroscopic and rapid kinetic studies of reduction of cytochrome c_{554} by hydroxylamine oxidoreductase from *Nitrosomonas europaea*. Biochemistry *30*: 11466–11472.

Arciero, D.M., M.J. Collins, J. Haladjian, P. Bianco and A.B. Hooper. 1991b. Resolution of the four hemes of cytochrome c_{554} from *Nitrosomonas europaea* by redox potentiometry and optical spectroscopy. Biochemistry *30*: 11459–11465.

Arciero, D.M., A. Golombek, M.P. Hendrich and A.B. Hooper. 1998. Correlation of optical and EPR signals with the P460 heme of hydroxylamine oxidoreductase from *Nitrosomonas europaea*. Biochemistry *37*: 523–529.

Arciero, D.M. and A.B. Hooper. 1993. Hydroxylamine oxidoreductase from *Nitrosomonas europaea* is a multimer of an octa-heme subunit. J. Biol. Chem. *268*: 14645–14654.

Arciero, D.M. and A.B. Hooper. 1994. A di-heme cytochrome *c* peroxidase from *Nitrosomonas europaea* catalytically active in both the oxidized and half-reduced states. J. Biol. Chem. *269*: 11878–11886.

Aretz, W., J. Meiwes, G. Seibert, G. Vobis and J. Wink. 2000. Friulimicins: Novel lipopeptide antibiotics with peptidoglycan synthesis inhibiting activity from *Actinoplanes friuliensis* sp. nov. I. Taxonomic studies of the producing microorganism and fermentation. J. Antibiot. *53*: 807–815.

Aretz, W., J. Meiwes, G. Seibert, G. Vobis and J. Wink. 2001. *In* Validation of publication of new names and new combinations previously effectively published outside the IJSEM. List No. 80. Int. J. Syst. Evol. Microbiol. *51*: 793–794.

Arthur, J.W., C.W. West, K.N. Allen and S.F. Hedtke. 1987. Seasonal toxicity of ammonia to 5 fish and 9 invertebrate species. Bull. Environ. Contam. Toxicol. *38*: 324–331.

Aruga, S., Y. Kamagata, T. Kohno, S. Hanada, K. Nakamura and T. Kanagawa. 2002. Characterization of filamentous Eikelboom type 021N bacteria and description of *Thiothrix disciformis* sp. nov. and *Thiothrix flexilis* sp. nov. Int. J. Syst. Evol. Microbiol. *52*: 13009–1316.

Ashelford, K.E., J.C. Fry, M.J. Bailey and M.J. Day. 2002. Characterization of *Serratia* isolates from soil, ecological implications and transfer of *Serratia proteamaculans* subsp. *quinovora* Grimont et al. 1983 to *Serratia quinivorans* corrig., sp. nov. Int. J. Syst. Evol. Microbiol. *52*: 2281–2289.

Asker, D. and Y. Ohta. 2002. *Haloferax alexandrinus* sp. nov., an extremely halophilic canthaxanthin-producing archaeon from a solar saltern in Alexandria (Egypt). Int. J. Syst. Evol. Microbiol. *52*: 729–738.

Aslander, A. 1928. Experiments on the eradication of Canada thistle *Cirsium arvense* with chlorates and other herbicides. J. Agr. Res. *36*: 915–935.

Assih, E.A., A.S. Ouattara, S. Thierry, J.L. Cayol, M. Labat and H. Macarie. 2002. *Stenotrophomonas acidaminiphila* sp. nov., a strictly aerobic bacterium isolated from an upflow anaerobic sludge blanket (UASB) reactor. Int. J. Syst. Evol. Microbiol. *52*: 559–568.

Assmus, B., P. Hutzler, G. Kirchhof, R.I. Amann, J.R. Lawrence and A. Hartmann. 1995. *In situ* localization of *Azospirillum brasilense* in the rhizosphere of wheat with fluorescently labeled, rRNA-targeted oligonucleotide probes and scanning confocal laser microscopy. Appl. Environ. Microbiol. *61*: 1013–1019.

Atlas, R.M. and R. Bartha. 1993. Microbial Ecology—Fundamentals and Applications, Benjamin-Cummings Publishing Co., Redwood City.

Audiffrin, C., J.L. Cayol, C. Joulian, L. Casalot, P. Thomas, J.L. Garcia and B. Ollivier. 2003. *Desulfonauticus submarinus* gen. nov., sp. nov., a novel sulfate-reducing bacterium isolated from a deep-sea hydrothermal vent. Int. J. Syst. Evol. Microbiol. *53*: 1585–1590.

Back, W., I. Bohak, M. Ehrmann, T. Ludwig, B. Pot and K.H. Schleifer. 1999. *Lactobacillus perolens* sp. nov., a soft drink spoilage bacterium. Syst. Appl. Microbiol. *22*: 354–359.

Back, W., I. Bohak, M. Ehrmann, T. Ludwig, B. Pot and K.H. Schleifer. 2000. *In* Validation of publication of new names and new combinations previously effectively published outside the IJSEM. List No. 72. Int. J. Syst. Evol. Microbiol. *50*: 3–4.

Baele, M., M. Vancanneyt, L.A. Devriese, K. Lefebvre, J. Swings and F. Haesebrouck. 2003. *Lactobacillus ingluviei* sp. nov., isolated from the intestinal tract of pigeons. Int. J. Syst. Evol. Microbiol. *53*: 133–136.

Baena, S., M.L. Fardeau, M. Labat, B. Ollivier, J.L. Garcia and B.K. Patel. 2000. *Aminobacterium mobile* sp. nov., a new anaerobic amino-acid-degrading bacterium. Int. J. Syst. Evol. Microbiol. *50*: 259–264.

Baer, M.L., J. Ravel, J. Chun, R.T. Hill and H.N. Williams. 2000. A proposal for the reclassification of *Bdellovibrio stolpii* and *Bdellovibrio starrii* into a new genus, *Bacteriovorax* gen. nov. as *Bacteriovorax stolpii* comb. nov. and *Bacteriovorax starrii* comb. nov., respectively. Int. J. Syst. Evol. Microbiol. *50*: 219–224.

Baïda, N., A. Yazourh, E. Singer and D. Izard. 2001. *Pseudomonas brenneri* sp. nov., a new species isolated from natural mineral waters. Res. Microbiol. *152*: 493–502.

Baïda, N., A. Yazourh, E. Singer and D. Izard. 2002a. *Pseudomonas grimontii* sp. nov. Int. J. Syst. Evol. Microbiol. *52*: 1497–1503.

Baïda, N., A. Yazourh, E. Singer and D. Izard. 2002b. *In* Validation of publication of new names and new combinations previously effectively published outside the IJSEM. List No. 87. Int. J. Syst. Evol. Microbiol. *52*: 1437–1438.

Bak, F. and N. Pfennig. 1987. Chemolithotrophic growth of *Desulfovibrio sulfodismutans*, new species by disproportionation of inorganic sulfur compounds. Arch. Microbiol. *147*: 184–189.

Bakalidou, A., P. Kämpfer, M. Berchtold, T. Kuhnigk, M. Wenzel and H. Konig. 2002. *Cellulosimicrobium variabile* sp. nov., a cellulolytic bacterium from the hindgut of the termite *Mastotermes darwiniensis*. Int. J. Syst. Evol. Microbiol. *52*: 1185–1192.

Balch, W.E., G.E. Fox, L.J. Magrum, C.R. Woese and R.S. Wolfe. 1979. Methanogens: reevaluation of a unique biological group. Microbiol. Rev. *43*: 260–296.

Balch, W.E. and R.S. Wolfe. 1981. *In* Validation of the publication of new names and new combinations previously effectively published outside the IJSB. List No. 6. Int. J. Syst. Bacteriol. *31*: 215–218.

Balk, M., J. Weijma and A.J. Stams. 2002. *Thermotoga lettingae* sp. nov., a novel thermophilic, methanol-degrading bacterium isolated from a thermophilic anaerobic reactor. Int. J. Syst. Evol. Microbiol. *52*: 1361–1368.

Barbeyron, T., S. L'Haridon, E. Corre, B. Kloareg and P. Potin. 2001.

Zobellia galactanovorans gen. nov., sp. nov., a marine species of *Flavobacteriaceae* isolated from a red alga, and classification of *[Cytophaga] uliginosa* (ZoBell and Upham 1944) Reichenbach 1989 as *Zobellia uliginosa* gen. nov., comb. nov. Int. J. Syst. Evol. Microbiol. *51*: 985–997.

Barkay, T., D.L. Fouts and B.H. Olson. 1985. Preparation of a DNA gene probe for detection of mercury resistance genes in Gram-negative bacterial communities. Appl. Environ. Microbiol. *49*: 686–692.

Barns, S.M., C.F. Delwiche, J.D. Palmer and N.R. Pace. 1996. Perspectives on archael diversity, thermophily, and monophyly from environmental rRNA sequences. Proc. Natl. Acad. Sci. U.S.A. *93*: 9188–9193.

Barrett, D.M., D.O. Faigel, D.C. Metz, K. Montone and E.E. Furth. 1997. *In situ* hybridization for *Helicobacter pylori* in gastric mucosal biopsy specimens: quantitative evaluation of test performance in comparison with the CLO test and thiazine stain. J. Clin. Lab. Anal. *11*: 374–379.

Bartosch, S., C. Hartwig, E. Spieck and E. Bock. 2002. Immunological detection of *Nitrospira*-like bacteria in various soils. Microb. Ecol. *43*: 26–33.

Bartosch, S., I. Wolgast, E. Spieck and E. Bock. 1999. Identification of nitrite-oxidizing bacteria with monoclonal antibodies recognizing the nitrite oxidoreductase. Appl. Environ. Microbiol. *65*: 4126–4133.

Batchelor, S.E., M. Cooper, S.R. Chhabra, L.A. Glover, G.S. Stewart, P. Williams and J.I. Prosser. 1997. Cell density-regulated recovery of starved biofilm populations of ammonia-oxidizing bacteria. Appl. Environ. Microbiol. *63*: 2281–2286.

Battista, J.R. and F.A. Rainey. 2001. Genus I. *Deinococcus*. *In* Boone, Castenholz and Garrity (Editors), Bergey's Manual of Systematic Bacteriology, 2nd Edition, Volume 1. The archaea and the deeply branching and phototrophic bacteria. Springer, New York. 396–403.

Becker, K., P. Schumann, J. Wullenweber, M. Schulte, H.P. Weil, E. Stackebrandt, G. Peters and C. von Eiff. 2002. *Kytococcus schroeteri* sp. nov., a novel Gram-positive actinobacterium isolated from a human clinical source. Int. J. Syst. Evol. Microbiol. *52*: 1609–1614.

Bedard, C. and R. Knowles. 1989. Physiology, biochemistry, and specific inhibitors of CH_4, NH_4^+, and co-oxidation by methanotrophs and nitrifiers. Microbiol. Rev. *53*: 68–84.

Behrendt, U., A. Ulrich and P. Schumann. 2001. Description of *Microbacterium foliorum* sp. nov. and *Microbacterium phyllosphaerae* sp. nov., isolated from the phyllosphere of grasses and the surface litter after mulching the sward, and reclassification of *Aureobacterium resistens* (Funke et al. 1998) as *Microbacterium resistens* comb. nov. Int. J. Syst. Evol. Microbiol. *51*: 1267–1276.

Behrendt, U., A. Ulrich and P. Schumann. 2003. Fluorescent pseudomonads associated with the phyllosphere of grasses; *Pseudomonas trivialis* sp. nov., *Pseudomonas poae* sp. nov. and *Pseudomonas congelans* sp. nov. Int. J. Syst. Evol. Microbiol. *53*: 1461–1469.

Behrendt, U., A. Ulrich, P. Schumann, D. Naumann and K. Suzuki. 2002. Diversity of grass-associated *Microbacteriaceae* isolated from the phyllosphere and litter layer after mulching the sward; polyphasic characterization of *Subtercola pratensis* sp. nov., *Curtobacterium herbarum* sp. nov. and *Plantibacter flavus* gen. nov., sp. nov. Int. J. Syst. Evol. Microbiol. *52*: 1441–1454.

Beimfohr, C., W. Ludwig and K.H. Schleifer. 1997. Rapid genotypic differentiation of *Lactococcus lactis* subspecies and biovar. Syst. Appl. Microbiol. *20*: 216–221.

Belduz, A.O., S. Dulger and Z. Demirbag. 2003. *Anoxybacillus gonensis* sp. nov., a moderately thermophilic, xylose-utilizing, endospore-forming bacterium. Int. J. Syst. Evol. Microbiol. *53*: 1315–1320.

Belser, L.W. 1979. Population ecology of nitrifying bacteria. Annu. Rev. Microbiol. *33*: 309–333.

Belser, L.W. and E.L. Mays. 1982. Use of nitrifier activity measurements to estimate the efficiency of viable nitrifier counts in soils and sediments. Appl. Environ. Microbiol. *43*: 945–948.

Belser, L.W. and E.L. Schmidt. 1978a. Nitrification in soils. *In* Schlessinger (Editor), Microbiology 1978, American Society for Microbiology, Washington DC. 348–351.

Belser, L.W. and E.L. Schmidt. 1978b. Serological diversity within a terrestrial ammonia-oxidizing population. Appl. Environ. Microbiol. *36*: 589–593.

Ben-Haim, Y., F.L. Thompson, C.C. Thompson, M.C. Cnockaert, B. Hoste, J. Swings and E. Rosenberg. 2003. *Vibrio coralliilyticus* sp. nov., a temperature-dependent pathogen of the coral *Pocillopora damicornis*. Int. J. Syst. Evol. Microbiol. *53*: 309–315.

Bender, K.S., S.M. O'Connor, R. Chakraborty, J.D. Coates and L.A. Achenbach. 2002. Sequencing and transcriptional analysis of the chlorite dismutase gene of *Dechloromonas agitata* and its use as a metabolic probe. Appl. Environ. Microbiol. *68*: 4820–4826.

Benediktsdøttir, E., L. Verdonck, C. Sproer, S. Helgasön and J. Swings. 2000. Characterization of *Vibrio viscosus* and *Vibrio wodanis* isolated at different geographical locations: a proposal for reclassification of *Vibrio viscosus* as *Moritella viscosa* comb. nov. Int. J. Syst. Evol. Microbiol. *50*: 479–488.

Berge, O., M.H. Guinebretiere, W. Achouak, P. Normand and T. Heulin. 2002. *Paenibacillus graminis* sp. nov. and *Paenibacillus odorifer* sp. nov., isolated from plant roots, soil and food. Int. J. Syst. Evol. Microbiol. *52*: 607–616.

Bergey, D.H., R.S. Breed, B.W. Hammer, F.M. Huntoon, E.G.D. Murray and F.C. Harrison (Editors). 1934. Bergey's Manual of Determinative Bacteriology, 4th Ed., The Williams & Wilkins Co., Baltimore.

Bergey, D.H., R.S. Breed, E.G.D. Murray and A.P. Hitchens (Editors). 1939. Bergey's Manual of Determinative Bacteriology, 5th Ed., The Williams & Wilkins Co., Baltimore.

Bergey, D.H., F.C. Harrison, R.S. Breed, B.W. Hammer and F.M. Huntoon. 1930. Bergey's Manual of Determinative Bacteriology, 3rd edition, The Williams & Wilkins Co, Baltimore. 1–589.

Bergmann, D.J., D.M. Arciero and A.B. Hooper. 1994. Organization of the *hao* gene cluster of *Nitrosomonas europaea*: genes for two tetraheme *c* cytochromes. J. Bacteriol. *176*: 3148–3153.

Bergmann, D.J. and A.B. Hooper. 1994a. The primary structure of cytochrome P_{460} of *Nitrosomonas europaea*: presence of a c-heme binding motif. FEBS Lett. *353*: 324–326.

Bergmann, D.J. and A.B. Hooper. 1994b. Sequence of the gene, *amoB*, for the 43 kDa polypeptide of ammonia monoxygenase of *Nitrosomonas europaea*. Biochem. Biophys. Res. Commun. *204*: 759–762.

Bergthorsson, U. and H. Ochman. 1995. Heterogeneity of genome sizes among natural isolates of *Escherichia coli*. J. Bacteriol. *177*: 5784–5789.

Bermond, D., H.J. Boulouis, R. Heller, G. Van Laere, H. Monteil, B.B. Chomel, A. Sander, C. Dehio and Y. Piemont. 2002. *Bartonella bovis* Bermond et al. sp. nov. and *Bartonella capreoli* sp. nov., isolated from European ruminants. Int. J. Syst. Evol. Microbiol. *52*: 383–390.

Bermond, D., R. Heller, F. Barrat, G. Delacour, C. Dehio, A. Alliot, H. Monteil, B. Chomel, H.J. Boulouis and Y. Piémont. 2000. *Bartonella birtlesii* sp. nov., isolated from small mammals (*Apodemus* spp.). Int. J. Syst. Evol. Microbiol . *50*: 1973–1979.

Bernard, K.A., L. Shuttleworth, C. Munro, J.C. Forbes-Faulkner, D. Pitt, J.H. Norton and A.D. Thomas. 2002a. *Propionibacterium australiense* sp. nov. derived from granulomatous bovine lesions. Anaerobe *8*: 41–47.

Bernard, K.A., L. Shuttleworth, C. Munro, J.C. Forbes-Faulkner, D. Pitt, J.H. Norton and A.D. Thomas. 2002b. *In* Validation of publication of new names and new combinations previously effectively published outside the IJSEM. List No. 88. Int. J. Syst. Evol. Microbiol. *52*: 1915–1916.

Bernardet, J.F., Y. Nakagawa and B. Holmes. 2002. Proposed minimal standards for describing new taxa of the family *Flavobacteriaceae* and emended description of the family. Int. J. Syst. Evol. Microbiol. *52*: 1049–1070.

Berry, A., D. Janssens, M. Hümbelin, J.P.M. Jore, B. Hoste, I. Cleenwerck, M. Vancanneyt, W. Bretzel, A.F. Mayer, R. Lopez-Ulibarri, B. Shanmugam, J. Swings and L. Pasamontes. 2003. *Paracoccus zeaxanthinifaciens* sp. nov., a zeaxanthin-producing bacterium. Int. J. Syst. Evol. Microbiol. *53*: 231–238.

Betzl, D., W. Ludwig and K.H. Schleifer. 1990. Identification of lactococci and enterococci by colony hybridization with 23S rRNA-targeted oligonucleotide probes. Appl. Environ. Microbiol. *56*: 2927–2929.

Bidnenko, E., C. Mercier, J. Tremblay, P. Tailliez and S. Kulakauskas. 1998. Estimation of the state of the bacterial cell wall by fluorescent *in situ* hybridization. Appl. Environ. Microbiol. *64*: 3059–3062.

Biebl, H., H. Schwab-Hanisch, C. Sproer and H. Lunsdorf. 2000. *Propionispora vibrioides*, nov. gen., nov. sp., a new Gram-negative, spore-forming anaerobe that ferments sugar alcohols. Arch. Microbiol. *174*: 239–247.

Biebl, H., H. Schwab-Hanisch, C. Sproer and H. Lunsdorf. 2001. *In* Validation of publication of new names and new combinations previously effectively published outside the IJSEM. List No. 80. Int. J. Syst. Evol. Microbiol. *51*: 793–794.

Biebl, H. and C. Spröer. 2002. Taxonomy of the glycerol fermenting clostridia and description of *Clostridium diolis* sp. nov. Syst. Appl. Microbiol. *25*: 491–497.

Biebl, H. and C. Spröer. 2003. *In* Validation of publication of new names and new combinations previously effectively published outside the IJSEM. List No. 91. Int. J. Syst. Evol. Microbiol. *53*: 627–628.

Birkbeck, T.H., L.A. Laidler, A.N. Grant and D.I. Cox. 2002. *Pasteurella skyensis* sp. nov., isolated from Atlantic salmon (*Salmo salar* L.). Int. J. Syst. Evol. Microbiol. *52*: 699–704.

Bizio, B. 1823. Lettera di Bartolomeo Bizio al chiarissimo canonico Angelo Bellani sopra il fenomeno della polenta porporina. Biblioteca Italiana o sia Giornale di Letteratura Scienze e Arti (Anno VIII). *30*: 275–295.

Björkroth, K.J., R. Geisen, U. Schillinger, N. Weiss, P. De Vos, W.H. Holzapfel, H.J. Korkeala and P. Vandamme. 2000. Characterization of *Leuconostoc gasicomitatum* sp. nov., associated with spoiled raw tomato-marinated broiler meat strips packaged under modified-atmosphere conditions. Appl. Environ. Microbiol. *66*: 3764–3772.

Björkroth, K.J., R. Geisen, U. Schillinger, N. Weiss, B. Hoste, W.H. Holzapfel, H.J. Korkeala and P. Vandamme. 2001. *In* Validation of publication of new names and new combinations previously effectively published outside the IJSEM. List No. 79. Int. J. Syst. Evol. Microbiol. *51*: 263–265.

Björkroth, K.J., U. Schillinger, R. Geisen, N. Weiss, B. Hoste, W.H. Holzapfel, H.J. Korkeala and P. Vandamme. 2002. Taxonomic study of *Weissella confusa* and description of *Weissella cibaria* sp. nov., detected in food and clinical samples. Int. J. Syst. Evol. Microbiol. *52*: 141–148.

Bjourson, A.J., C.E. Stone and J.E. Cooper. 1992. Combined subtraction hybridization and polymerase chain reaction amplification procedure for isolation of strain-specific *Rhizobium* DNA sequences. Appl. Environ. Microbiol. *58*: 2296–2301.

Blackmer, A.M., J.M. Bremner and E.L. Schmidt. 1980. Production of nitrous oxide by ammonia-oxidizing chemoautotrophic microorganisms in soil. Appl. Environ. Microbiol. *40*: 1060–1066.

Blattner, F.R., G.I. Plunkett, III, C.A. Bloch, N.T. Perna, V. Burland, M. Riley, J. Collado-Vides, J.D. Glasner, C.K. Rode, G.F. Mayhew, J. Gregor, N.W. Davis, H.A. Kirkpatrick, M.A. Goeden, D.J. Rose, B. Mau and Y. Shao. 1997. The complete genome sequence of *Escherichia coli* K-12. Science *277*: 1453–1462.

Blümel, S., H.-J. Busse, A. Stolze and P. Kämpfer. 2001a. *Xenophilus azovorans* gen. nov., sp. nov., a soil bacterium that is able to degrade azo dyes of the Orange II type. Int. J. Syst. Evol. Microbiol. *51*: 1831–1837.

Blümel, S., B. Mark, H.J. Busse, P. Kämpfer and A. Stolz. 2001b. *Pigmentiphaga kullae* gen. nov., sp. nov., a novel member of the family *Alcaligenaceae* with the ability to decolorize azo dyes aerobically. Int. J. Syst. Evol. Microbiol. *51*: 1867–1871.

Bock, E. 1976. Growth of *Nitrobacter* in the presence of organic matter. II. Chemoorganotrophic growth of *Nitrobacter agilis*. Arch. Microbiol. *108*: 305–312.

Bock, E., H.-P. Koops, B. Ahlers and H. Harms. 1992. Oxidation of inorganic nitrogen compounds as energy sources. *In* Balows, Trüper, Dworkin, Harder and Schleifer (Editors), The Prokaryotes. A Handbook on the Biology of Bacteria: Ecophysiology, Isolation, Identification, Applications, 2nd ed., Vol. 1, Springer-Verlag, New York. pp. 414–430.

Bock, E., H.-P. Koops, H. Harms and B. Ahlers. 1991. The biochemistry of nitrifying organisms. *In* Shively and Barton (Editors), Variations in Autotrophic Life, Academic Press, Ltd., London. pp. 171–199.

Bock, E. and H.-P., Koops. 1992. The genus *Nitrobacter* and related genera. *In* Balows, Trüper, Dworkin, Harder and Schleifer (Editors), The Prokaryotes. A Handbook on the Biology of Bacteria: Ecophysiology, Isolation, Identification, Applications, 2nd ed., Vol. 3, Springer-Verlag, New York. pp. 2302–2309.

Bock, E., H.P. Koops, U.C. Möller and M. Rudert. 1990. A new facultatively nitrite oxidizing bacterium, *Nitrobacter vulgaris*, new species. Arch. Microbiol. *153*: 105–110.

Bock, E., H.P. Koops, U.C. Möller and M. Rudert. 2001a. *In* Validation of the publication of new names and new combinations previously effectively published outside the IJSEM. List No.78. Int. J.Syst. Bacteriol. *51*: 1–2.

Bock, E. and W. Sand. 1993. The microbiology of masonry biodeterioration. J. Appl. Bacteriol. *74*: 503–514.

Bock, E., I. Schmidt, R. Stüven and D. Zart. 1995. Nitrogen loss caused by denitrifying *Nitrosomonas* cells using ammonium or hydrogen as electron donors and nitrite as electron acceptor. Arch. Microbiol. *163*: 16–20.

Bock, E., H. Sundermeyer-Klinger and E. Stackebrandt. 1983. New facultative lithoautotrophic nitrite-oxidizing bacteria. Arch. Microbiol. *136*: 281–284.

Bock, E., H. Sundermeyer-Klinger and E. Stackebrandt. 2001b. *In* Validation of the publication of new names and new combinations previously effectively published outside the IJSEM. List No. 78. Int. J.Syst. Bacteriol. *51*: 1–2.

Bock, E. and M. Wagner. 2001. Oxidation of inorganic nitrogen compounds as an energy source. *In* Schleifer and Stackbrandt (Editors), The Prokaryotes, 2nd ed. Online, Springer-Verlag, New York.

Boga, H.I., W. Ludwig and A. Brune. 2003. *Sporomusa aerivorans* sp. nov., an oxygen-reducing homoacetogenic bacterium from the gut of a soil-feeding termite. Int. J. Syst. Evol. Microbiol. *53*: 1397–1404.

Bogan, B.W., W.R. Sullivan, K.J. Kayser, K.D. Derr, H.C. Aldrich and J.R. Paterek. 2003. *Alkanindiges illinoisensis* gen. nov., sp. nov., an obligately hydrocarbonoclastic, aerobic squalane-degrading bacterium isolated from oilfield soils. Int. J. Syst. Evol. Microbiol. *53*: 1389–1395.

Bogosian, G., P.J.L. Morris and J.P. O'Neil. 1998. A mixed culture recovery method indicates that enteric bacteria do not enter the viable but nonculturable state. Appl. Environ. Microbiol. *64*: 1736–1742.

Bonch-Osmolovskaya, E.A., M.L. Miroshnichenko, N.A. Kostrikina, N.A. Chernych and G.A. Zavarzin. 1990. *Thermoproteus uzoniensis* sp. nov., a new extremely thermophilic archaebacterium from Kamchatka continental hot springs. Arch. Microbiol. *154*: 556–559.

Bonch-Osmolovskaya, E.A., A.I. Slesarev, M.L. Miroshnichenko, T.P. Svetlichnaya and V.A. Alekseev. 1988. Characteristics of *Desulfurococcus amylolyticus* nov. sp.–a new extremely thermophilic archaebacterium isolated from thermal springs of Kamchatka and Kunashir Island. Microbiology *57*: 78–85.

Bonch-Osmolovskaya, E.A., A.I. Slesarev, M.L. Miroshnichenko, T.P. Svetlichnaya and V.A. Alekseev. 2001. *In* Validation of publication of new names and new combinations previously effectively published outside the IJSEM. List No. 82. Int. J. Syst. Evol. Microbiol. *51*: 1619–1620.

Boone, D.R. 2001a. Class I. *Methanobacteria* class. nov. *In* Boone, Castenholz and Garrity (Editors), Bergey's Manual of Systematic Bacteriology, 2nd Edition, Volume 1. The archaea and the deeply branching and phototrophic bacteria. Springer, New York. 213.

Boone, D.R. 2001b. Class II. *Methanococci* class. nov. *In* Boone, Castenholz and Garrity (Editors), Bergey's Manual of Systematic Bacteriology, 2nd Edition, Volume 1. The archaea and the deeply branching and phototrophic bacteria. Springer, New York. 235–236.

Boone, D.R. 2001c. Genus IV. *Methanolobus*. *In* Boone, Castenholz and Garrity (Editors), Bergey's Manual of Systematic Bacteriology, 2nd Edition, Vol. 1, Springer, New York. 283–287.

Boone, D.R. 2001d. Genus IV. *Methanothermobacter*. *In* Boone, Castenholz and Garrity (Editors), Bergey's Manual of Systematic Bacteriology, 2nd Edition, Vol. 1, Springer, New York. 232–233.

Boone, D.R. 2002. *In* Validation of publication of new names and new combinations previously effectively published outside the IJSEM. List No. 85. Int. J. Syst. Evol. Microbiol. *52*: 685–690.

Boone, D.R. and C.C. Baker. 2001. Genus VI. *Methanosalsum*. *In* Boone, Castenholz and Garrity (Editors), Bergey's Manual of Systematic Bacteriology, 2nd Edition, Vol. 1, Springer, New York. 287–289.

Boone, D. R. and C.C. Baker. 2002. *In* Validation of publication of new names and new combinations previously effectively published outside the IJSEM. List No. 85. Int. J. Syst. Evol. Microbiol. *52*: 685–690.

Boone, D.R., W.B.Whitman and Y. Koga. 2001a. Family II. *Methanosaetaceae* fam. nov. *In* Boone, Castenholz and Garrity (Editors), Bergey's Manual of Systematic Bacteriology, 2nd Edition, Volume 1. The archaea and the deeply branching and phototrophic bacteria. Springer, New York. 289.

Boone, D.R., W.B.Whitman and Y. Koga. 2001b. Family III. *Methanospirillaceae* fam. nov. *In* Boone, Castenholz and Garrity (Editors), Bergey's Manual of Systematic Bacteriology, 2nd Edition, Volume 1. The archaea and the deeply branching and phototrophic bacteria. Springer, New York. 264.

Boone, D.R., W.B.Whitman and Y. Koga. 2001c. Order III. *Methanosarcinales* ord. nov. *In* Boone, Castenholz and Garrity (Editors), Bergey's Manual of Systematic Bacteriology, 2nd Edition, Volume 1. The archaea and the deeply branching and phototrophic bacteria. Springer, New York. 268.

Boone, D.R., W.B.Whitman and Y. Koga. 2002. *In* Validation of publication of new names and new combinations previously effectively published outside the IJSEM. List No. 85. Int. J. Syst. Evol. Microbiol. *52*: 685–690.

Borneman, J. and E.W. Triplett. 1997. Molecular microbial diversity in soils from eastern Amazonia: evidence for unusual microorganisms and microbial population shifts associated with deforestation. Appl. Environ. Microbiol. *63*: 2647–2653.

Borodina, E., D.P. Kelly, P. Schumann, F.A. Rainey, N.L. Ward-Rainey and A.P. Wood. 2002a. Enzymes of dimethylsulfone metabolism and the phylogenetic characterization of the facultative methylotrophs *Arthrobacter sulfonivorans* sp. nov., *Arthrobacter methylotrophus* sp nov., and *Hyphomicrobium sulfonivorans* sp. nov. Arch. Microbiol. *177*: 173–183.

Borodina, E., D.P. Kelly, P. Schumann, F.A. Rainey, N.L. Ward-Rainey and A.P. Wood. 2002b. *In* Validation of publication of new names and new combinations previously effectively published outside the IJSEM. List No. 85. Int. J. Syst. Evol. Microbiol. *52*: 685–690.

Borsodi, A. K. , A. Micsinai, G. Kovacs, E. Toth, P. Schumann, A. L. Kovacs, B. Boddi and K. Marialigeti. 2003. *Pannonibacter phragmitetus* gen. nov., sp nov., a novel alkalitolerant bacterium isolated from decomposing reed rhizomes in a Hungarian soda lake. Int. J. Syst. Evol. Microbiol. *53*: 555–561.

Bosshard, P.P., R. Zbinden and M. Altwegg. 2002a. *Paenibacillus turicensis* sp. nov., a novel bacterium harbouring heterogeneities between 16S rRNA genes. Int. J. Syst. Evol. Microbiol. *52*: 2241–2249.

Bosshard, P.P., R. Zbinden and M. Altwegg. 2002b. *Turicibacter sanguinis* gen. nov., sp. nov., a novel anaerobic, Gram-positive bacterium. Int. J. Syst. Evol. Microbiol. *52*: 1263–1266.

Both, G.J., S. Gerards and H.J. Laanbroek. 1990. Most probable numbers of chemolitho-autotrophic nitrite-oxidizing bacteria in well drained grassland soils: stimulation by high nitrite concentrations. FEMS Microbiol. Lett. *74*: 287–294.

Both, G.J., S. Gerards and H.J. Laanbroek. 1992. Kinetics of nitrite oxidation in two *Nitrobacter* spp. grown in nitrite-limited chemostats. Arch. Microbiol. *157*: 436–441.

Both, G.J. and H.J. Laanbroek. 1991. The effect of the incubation period on the result of MPN enumerations of nitrite-oxidizing bacteria: theoretical considerations. FEMS Microbiol. Lett. *85*: 335–344.

Böttcher, B. 1996. Untersuchungen zur phylogenie des ammoniak oxidierenden systems nitrifizierender bakterien, University of Hamburg

Böttcher, B. and H.P. Koops. 1994. Growth of lithotrophic ammonia oxidizing bacteria on hydroxylamine. FEMS Microbiol. Lett. *122*: 263–266.

Bouchek-Mechiche, K., L. Gardan, P. Normand and B. Jouan. 2000. DNA relatedness among strains of *Streptomyces* pathogenic to potato in France: description of three new species, *S. europaeiscabiei* sp. nov. and *S. stelliscabiei* sp. nov. associated with common scab, and *S. reticuliscabiei* sp. nov. associated with netted scab. Int. J. Syst. Evol. Microbiol. *50*: 91–99.

Bouchotroch, S., E. Quesada, A. del Moral, I. Llamas and V. Bejar. 2001. *Halomonas maura* sp nov., a novel moderately halophilic, exopolysaccharide-producing bacterium. Int. J. Syst. Evol. Microbiol. *51*: 1625–1632.

Bousfield, I.J. and P.N. Green. 1985. Reclassification of bacteria of the genus *Protomonas* Urakami and Komagata 1984 in the genus *Methylobacterium* (Patt, Cole, and Hanson) emend. Green and Bousfield 1983. Int. J. Syst. Bacteriol. *35*: 209.

Bouvet, P.J.M. and S. Jeanjean. 1989. Delineation of new proteolytic genomic species in the genus *Acinetobacter*. Res. Microbiol. *140*: 291–300.

Bouyer, D.H., J. Stenos, P. Crocquet-Valdes, C.G. Moron, V.L. Popov, J.E. Zavala-Velazquez, L.D. Foil, D.R. Stothard, A.F. Azad and D.H. Walker. 2001. *Rickettsia felis*: Molecular characterization of a new member of the spotted fever group. Int. J. Syst. Evol. Microbiol. *51*: 339–347.

Bouzar, H. and J.B. Jones. 2001. *Agrobacterium larrymoorei* sp. nov., a pathogen isolated from aerial tumours of *Ficus benjamina*. Int. J. Syst. Evol. Microbiol. *51*: 1023–1026.

Bowman, J.P. 2000. Description of *Cellulophaga algicola* sp. nov., isolated from the surfaces of Antarctic algae, and reclassification of *Cytophaga uliginosa* (ZoBell and Upham 1944) Reichenbach 1989 as *Cellulophaga uliginosa* comb. nov. Int. J. Syst. Evol. Microbiol. *50*: 1861–1868.

Bowman, J.P. and D.S. Nichols. 2002. *Aequorivita* gen. nov., a member of the family *Flavobacteriaceae* isolated from terrestrial and marine Antarctic habitats. Int. J. Syst. Evol. Microbiol. *52*: 1533–1541.

Bowman, J.P., C.M. Nichols and J.A. Gibson. 2003. *Algoriphagus ratkowskyi* gen. nov., sp. nov., *Brumimicrobium glaciale* gen. nov., sp. nov., *Cryomorpha ignava* gen. nov., sp. nov. and *Crocinitomix catalasitica* gen. nov., sp. nov., novel flavobacteria isolated from various polar habitats. Int. J. Syst. Evol. Microbiol. *53*: 1343–1355.

Bozal, N., M.J. Montes, E. Tudela and J. Guinea. 2003. Characterization of several *Psychrobacter* strains isolated from Antarctic environments and description of *Psychrobacter luti* sp. nov. and *Psychrobacter fozii* sp. nov. Int. J. Syst. Evol. Microbiol. *53*: 1093–1100.

Bozal, N., M.J. Montes, E. Tudela, F. Jimenez and J. Guinea. 2002. *Shewanella frigidimarina* and *Shewanella livingstonensis* sp. nov. isolated from Antarctic coastal areas. Int. J. Syst. Evol. Microbiol. *52*: 195–205.

Brämer, C.O., P. Vandamme, L.F. da Silva, J.G. Gomez and A. Steinbuchel. 2001. Polyhydroxyalkanoate-accumulating bacterium isolated from soil of a sugar-cane plantation in Brazil. Int. J. Syst. Evol. Microbiol. *51*: 1709–1713.

Brandão, P.F.B., L.A. Maldonado, A.C. Ward, A.T. Bull and M. Goodfellow. 2001. *Gordonia namibiensis* sp. nov., a novel nitrile metabolising actinomycete recovered from an African sand. Syst. Appl. Microbiol. *24*: 510–515.

Brandão, P.F.B., L.A. Maldonado, A.C. Ward, A.T. Bull and M. Goodfellow. 2002. *In* Validation of publication of new names and new combinations previously effectively published outside the IJSEM. List No. 85. Int. J. Syst. Evol. Microbiol. *52*: 685–690.

Breed, R.S. , E.G.D. Murray and A.P. Hitchens (Editors). 1948a. Bergey's Manual of Determinative Bacteriology, 6th Ed., The Williams & Wilkins Co., Baltimore.

Breed, R.S., E.G.D. Murray and A.P. Hitchens (Editors). 1948b. Bergey's Manual of Determinative Bacteriology, abridged 6th Ed., Biotech Publications, Geneva.

Breed, R.S., E.G.D. Murray and N.R. Smith (Editors). 1957. Bergey's Manual of Determinative Bacteriology, 7th Ed., The Williams & Wilkins Co., Baltimore.

Breitenstein, A., J. Wiegel, C. Haertig, N. Weiss, J.R. Andreesen and U. Lechner. 2002. Reclassification of *Clostridium hydroxybenzoicum* as *Sedimentibacter hydroxybenzoicus* gen. nov., comb. nov., and description of *Sedimentibacter saalensis* sp. nov. Int. J. Syst. Evol. Microbiol. *52*: 801–807.

Brennan, N.M., R. Brown, M. Goodfellow, A.C. Ward, T.P. Beresford, P.J. Simpson, P.F. Fox and T.M. Cogan. 2001a. *Corynebacterium mooreparkense* sp. nov. and *Corynebacterium casei* sp. nov., isolated from the surface of a smear-ripened cheese. Int. J. Syst. Evol. Microbiol. *51*: 843–852.

Brennan, N.M., R. Brown, M. Goodfellow, A.C. Ward, T.P. Beresford, M. Vancanneyt, T.M. Cogan and P.F. Fox. 2001b. *Microbacterium gubbeenense* sp. nov., from the surface of a smear-ripened cheese. Int. J. Syst. Evol. Microbiol. *51*: 1969–1976.

Brenner, D.J. 1984. Family I. *Enterobacteriaceae*. *In* Krieg and Holt (Editors), Bergey's Manual of Systematic Bacteriology, 1st Ed., Vol. 1, The Williams & Wilkins Co., Baltimore. pp. 408–420.

Brenner, D.J., A.C. McWhorter, J.K.L. Knutson and A.G. Steigerwalt. 1982. *Escherichia vulneris*: a new species of *Enterobacteriaceae* associated with human wounds. J. Clin. Microbiol. *15*: 1133–1140.

Brenner, D.J., C.M. O'Hara, P.A.D. Grimont, J.M. Janda, E. Falsen, E. Aldová, E. Ageron, J. Schindler, S.L. Abbott and A.G. Steigerwalt. 1999. Biochemical identification of *Citrobacter* species defined by DNA hybridization and description of *Citrobacter gillenii* sp. nov. (formerly *Citrobacter* genomospecies 10) and *Citrobacter murliniae* sp. nov. (formerly *Citrobacter* genomospecies 11). J. Clin. Microbiol. *37*: 2619–2624.

Brenner, D.J., C.M. O'Hara, P.A.D. Grimont, J.M. Janda, E. Falsen, E. Aldová, E. Ageron, J. Schindler, S.L. Abbott and A.G. Steigerwalt. 2000. *In* Validation of the publication of new names and new combinations previously effectively published outside the IJSEM. List No. 73. Int. J. Syst. Evol. Microbiol. *50*: 423–424.

Brettar, I., R. Christen and M.G. Hofle. 2002a. *Rheinheimera baltica* gen. nov., sp. nov., a blue-coloured bacterium isolated from the central Baltic Sea. Int. J. Syst. Evol. Microbiol. *52*: 1851–1857.

Brettar, I., R. Christen and M.G. Hofle. 2002b. *Shewanella denitrificans* sp. nov., a vigorously denitrifying bacterium isolated from the oxic-anoxic interface of the Gotland Deep in the central Baltic Sea. Int. J. Syst. Evol. Microbiol. *52*: 2211–2217.

Brettar, I., R. Christen and M.G. Hofle. 2003. *Idiomarina baltica* sp. nov., a marine bacterium with a high optimum growth temperature isolated from surface water of the central Baltic Sea. Int. J. Syst. Evol. Microbiol. *53*: 407–413.

Breznak, J.A. and R.N. Costilow. 1994. Physiochemical factors in growth. *In* Gerhardt, Murray, Wood and Krieg (Editors), Methods for General and Molecular Bacteriology, American Society for Microbiology, Washington, DC. pp. 137–154.

Brinkmann, H., P. Martinez, F. Quigley, W. Martin and R. Cerff. 1987. Endosymbiotic origin and codon bias of the nuclear gene for chloroplast glyceraldehyde-3-phosphate dehydrogenase from maize. J. Mol. Evol. *26*: 320–328.

Brisbarre, N., M.L. Fardeau, V. Cueff, J.L. Cayol, G. Barbier, V. Cilia, G. Ravot, P. Thomas, J.L. Garcia and B. Ollivier. 2003. *Clostridium caminithermale* sp. nov., a slightly halophilic and moderately thermophilic bacterium isolated from an Atlantic deep-sea hydrothermal chimney. Int. J. Syst. Evol. Microbiol. *53*: 1043–1049.

Brochier, C., E. Bapteste, D. Moreira and H. Philippe. 2002. Eubacterial phylogeny based on translational apparatus proteins. Trends Genet. *18*: 1–5.

Brock, T.D. 1989. Evolutionary relationships of the autotrophic bacteria. *In* Bowien (Editor), Autotrophic Bacteria, Science Tech, Madison. pp. 499–512.

Brock, T.D. 1995. The road to Yellowstone—and beyond. Annu. Rev. Microbiol. *49*: 1–28.

Brockmann, E., B.L. Jacobsen, C. Hertel, W. Ludwig and K.H. Schleifer. 1996. Monitoring of genetically modified *Lactococcus lactis* in gnotobiotic and conventional rats by using antibiotic resistance markers and specific probe or primer based methods. Syst. Appl. Microbiol. *19*: 203–212.

Broda, D.M., D.J. Saul, R.G. Bell and D.R. Musgrave. 2000a. *Clostridium algidixylanolyticum* sp. nov., a psychrotolerant, xylan-degrading, spore-forming bacterium. Int. J. Syst. Evol. Microbiol. *50*: 623–631.

Broda, D.M., D.J. Saul, P.A. Lawson, R.G. Bell and D.R. Musgrave. 2000b. *Clostridium gasigenes* sp. nov., a psychrophile causing spoilage of vacuum-packed meat. Int. J. Syst. Evol. Microbiol. *50*: 107–118.

Broda, E. 1977. 2 kinds of lithotrophs missing in nature. Z. Allg. Mikrobiol. *17*: 491–493.

Broughton, W.J. 2003. Roses by other names: Taxonomy of the *Rhizobiaceae*. J. Bacteriol. *185*: 2975–2979.

Brown, D.R., J.M. Farley, L.A. Zacher, J.M. Carlton, T.L. Clippinger, J.G. Tully and M.B. Brown. 2001a. *Mycoplasma alligatoris* sp. nov., from American alligators. Int. J. Syst. Evol. Microbiol. *51*: 419–424.

Brown, D.R., D.F. Talkington, W.L. Thacker, M.B. Brown, D.L. Dillehay and J.G. Tully. 2001b. *Mycoplasma microti* sp. nov., isolated from the respiratory tract of prairie voles (*Microtus ochrogaster*). Int. J. Syst. Evol. Microbiol. *51*: 409–412.

Brown, G.R., I.C. Sutcliffe and S.P. Cummings. 2001c. Reclassification of *[Pseudomonas] doudoroffii* (Baumann et al. 1983) in the genus *Oceanaomonas* gen. nov. as *Oceanomonas doudoroffii* comb. nov., and description of a phenol-degrading bacterium from estuarine water as *Oceanomonas baumannii* sp. nov. Int. J. Syst. Evol. Microbiol. *51*: 67–72.

Brown, J.R. and W.F. Doolittle. 1997. Archaea and the prokaryote-to-eukaryote transition. Microbiol. Mol. Biol. Rev. *61*: 456–502.

Brown, J.R., C.J. Douady, M.J. Italia, W.E. Marshall and M.J. Stanhope. 2001d. Universal trees based on large combined protein sequence data sets. Nature Genet. *28*: 281–285.

Brown, M.B., D.R. Brown, P.A. Klein, G.S. McLaughlin, I.M. Schumacher, E.R. Jacobson, H.P. Adams and J.G. Tully. 2001e. *Mycoplasma agassizii* sp. nov., isolated from the upper respiratory tract of the desert tortoise (*Gopherus agassizii*) and the gopher tortoise (*Gopherus polyphemus*). Int. J. Syst. Evol. Microbiol. *51*: 413–418.

Bruce, R.A., L.A. Achenbach and J.D. Coates. 1999. Reduction of (per)chlorate by a novel organism isolated from paper mill waste. Environ. Microbiol. *1*: 319–329.

Brune, A., S. Evers, G. Kaim, W. Ludwig and B. Schink. 2002a. *Ilyobacter insuetus* sp. nov., a fermentative bacterium specialized in the degradation of hydroaromatic compounds. Int. J. Syst. Evol. Microbiol. *52*: 429–432.

Brune, A., W. Ludwig and B. Schink. 2002b. *Propionivibrio limicola* sp. nov., a fermentative bacterium specialized in the degradation of hydroaromatic compounds, reclassification of *Propionibacter pelophilus* and *Propionivibrio pelophilus* comb. nov. and amended description of the genus *Propionivibrio*. Int. J. Syst. Evol. Microbiol. *52*: 441–444.

Bruns, A., H. Philipp, H. Cypionka and T. Brinkhoff. 2003. *Aeromicrobium marinum* sp. nov., an abundant pelagic bacterium isolated from the German Wadden Sea. Int. J. Syst. Evol. Microbiol. *53*: 1917–1923.

Bruns, A., M. Rohde and L. Berthe-Corti. 2001. *Muricauda ruestringensis* gen. nov., sp. nov., a facultatively anaerobic, appendaged bacterium from German North Sea intertidal sediment. Int. J. Syst. Evol. Microbiol. *51*: 1997–2006.

Bryan, E.H. 1966. Application of the chlorate BOD procedure to routine measurement of wastewater strength. J. Water Pollut. Cont. Fed. *38*: 1350–1362.

Bryan, E.H. and G.A. Rohlich. 1954. Biological reduction of sodium chlorate as applied to measurement of sewage BOD. Sewage Ind. Waste. *26*: 1315–1324.

Bryant, M.P., E.A. Wolin, M.J. Wolin and R.S. Wolfe. 1967. *Methanobacillus omelianskii*, a symbiotic association of two species of bacteria. Arch. Mikrobiol. *59*: 20–31.

Bryantseva, I.A., V.M. Gorlenko, E.I. Kompantseva, L.A. Achenbach and M.T. Madigan. 1999. *Heliorestis daurensis*, gen. nov. sp. nov., an alkaliphilic rod-to-coiled-shaped phototrophic heliobacterium from a Siberian soda lake. Arch. Microbiol. *172*: 167–174.

Bryantseva, I.A., V.M. Gorlenko, E.I. Kompantseva, L.A. Achenbach and M.T. Madigan. 2000a. *In* Validation of publication of new names and new combinations previously effectively published outside the IJSEM. List No. 74. Int. J. Syst. Evol. Microbiol. *50*: 949–950.

Bryantseva, I.A., V.M. Gorlenko, E.I. Kompantseva and J.F. Imhoff. 2000b. *Thioalkalicoccus limnaeus* gen. nov., sp. nov., a new alkaliphilic purple sulfur bacterium with bacteriochlorophyll *b*. Int. J. Syst. Evol. Microbiol. *50*: 2157–2163.

Bryantseva, I.A., V.M. Gorlenko, E.I. Kompantseva, T.P. Tourova, B.B. Kuznetsov and G.A. Osipov. 2000c. Alkaliphilic heliobacterium *Heliorestis baculata* sp. nov. and emended description of the genus *Heliorestis*. Arch. Microbiol. *174*: 283–291.

Bryantseva, I.A., V.M. Gorlenko, E.I. Kompantseva, T.P. Tourova, B.B. Kuznetsov and G.A. Osipov. 2001a. *In* Validation of publication of new names and new combinations previously effectively published outside the IJSEM. List No. 79. Int. J. Syst. Evol. Microbiol. *51*: 263–265.

Bryantseva, I.A., V.M. Gorlenko, T.P. Tourova, B.B. Kuznetsov, A.M. Lysenko, S.A. Bykova, V.F. Gal'chenko, L.L. Mityushina and G.A. Osipov. 2000d. *Heliobacterium sulfidophilum* sp. nov. and *Heliobacterium undosum* sp. nov.: sulfide-oxidizing heliobacteria from thermal sulfidic springs. Microbiology *69*: 325–334.

Bryantseva, I.A., V.M. Gorlenko, T.P. Tourova, B.B. Kuznetsov, A.M. Lysenko, S.A. Bykova, V.F. Gal'chenko, L.L. Mityushina and G.A. Osipov. 2001b. *In* Validation of publication of new names and new combinations previously effectively published outside the IJSEM. List No. 78. Int. J. Syst. Evol. Microbiol. *51*: 1–2.

Bsat, N. and C.A. Batt. 1993. A combined modified reverse dot-blot and nested PCR assay for the specific non-radioactive detection of *Listeria monocytogenes*. Mol. Cell. Probes *7*: 199–207.

Buchanan, R.E. 1916. Studies in the nomenclature and classification of the bacteria. I. The problem of bacterial nomenclature. J. Bacteriol. *1*: 591–596.

Buchanan, R.E. 1917a. Studies in the nomenclature and classification of the bacteria. II. The primary subdivisions of the *Schizomycetes*. J. Bacteriol. *2*: 155–164.

Buchanan, R.E. 1917b. Studies in the nomenclature and classification of the bacteria. III. The families of the *Eubacteriales*. J. Bacteriol. *2*: 347–350.

Buchanan, R.E. 1917c. Studies in the nomenclature and classification of the bacteria. IV. Subgroups and genera of the *Coccaceae*. J. Bacteriol. *2*: 603–617.

Buchanan, R.E. 1918a. Studies in the nomenclature and classification of the bacteria. V. Subgroups and genera of the *Bacteriaceae*. J. Bacteriol. *3*: 27–61.

Buchanan, R.E. 1918b. Studies in the nomenclature and classification of the bacteria. VI. Subdivisions and genera of the *Spirillaceae* and *Nitrobacteriaceae*. J. Bacteriol. *3*: 175–181.

Buchanan, R.E. 1918c. Studies in the nomenclature and classification of the bacteria. VII. The subgroups and genera of the *Chlamydobacteriales*. J. Bacteriol. *3*: 301–306.

Buchanan, R.E. 1918d. Studies in the nomenclature and classification of the bacteria. VIII. The subgroups and genera of the *Actinomycetales*. J. Bacteriol. *3*: 403–406.

Buchanan, R.E. 1918e. Studies in the nomenclature and classification of the bacteria. IX. The subgroups and genera of the *Thiobacteriales*. J. Bacteriol. *3*: 461–474.

Buchanan, R.E. 1918f. Studies in the nomenclature and classification of the bacteria. X. Subgroups and genera of the *Myxobacteriales* and *Spirochaetales*. J. Bacteriol. *3*: 541–545.

Buchanan, R.E. 1925. General Systematic Bacteriology, The Williams & Wilkins Co, Baltimore.

Buchanan, R.E. 1948. How bacteria are named and identified. *In* Breed, Murray and Hitchens (Editors), Bergey's Manual of Determinative Bacteriology, 6th Ed., The Williams & Wilkins Co., Baltimore. pp. 39–48.

Buchanan, R.E. 1994. Chemical terminology and microbiological nomenclature. Int. J. Syst. Bacteriol. *44*: 588–590.

Buchanan, R.E. and W.E. Gibbons (Editors). 1974. Bergey's Manual of Determinative Bacteriology, 8th Ed., The Williams & Wilkins Co., Baltimore.

Buchanan, R.E., J.G. Holt and E.F.J. Lessel (Editors). 1966. Index Bergeyana, The Williams & Wilkins Co., Baltimore.

Buchanan, R.E., R. St. John-Brooks and R.S. Breed. 1948. International Bacteriological Code of Nomenclature. J. Bacteriol. *55*: 287–306.

Buczolits, S., E.B. Denner, D. Vybiral, M. Wieser, P. Kämpfer and H.J. Busse. 2002. Classification of three airborne bacteria and proposal of *Hymenobacter aerophilus* sp. nov. Int. J. Syst. Evol. Microbiol. *52*: 445–456.

Buczolits, S., P. Schumann, G. Weidler, C. Radax and H.J. Busse. 2003. *Brachybacterium muris* sp. nov., isolated from the liver of a laboratory mouse strain. Int. J. Syst. Evol. Microbiol. *53*: 1955–1960.

Buonaurio, R., V.M. Stravato, Y. Kosako, N. Fujiwara, T. Naka, K. Kobayashi, C. Cappelli and E. Yabuuchi. 2002. *Sphingomonas melonis* sp. nov., a novel pathogen that causes brown spots on yellow Spanish melon fruits. Int. J. Syst. Evol. Microbiol. *52*: 2081–2087.

Burgess, J.G., R. Kawaguchi, T. Sakaguchi, R.H. Thornhill and T. Matsunaga. 1993. Evolutionary relationships among *Magnetospirillum* strains inferred from phylogenetic analysis of 16S rDNA sequences. J. Bacteriol. *175*: 6689–6694.

Burggraf, S., P. Heyder and N. Eis. 1997. A pivotal Archaea group. Nature *385*: 780.

Burrell, P.C., J. Keller and L.L. Blackall. 1998. Microbiology of a nitrite-oxidizing bioreactor. Appl. Environ. Microbiol. *64*: 1878–1883.

Busse, H.J., E.B. Denner, S. Buczolits, M. Salkinoja-Salonen, A. Bennasar and P. Kämpfer. 2003a. *Sphingomonas aurantiaca* sp. nov., *Sphingomonas aerolata* sp. nov. and *Sphingomonas faeni* sp. nov., air- and dustborne and Antarctic, orange-pigmented, psychrotolerant bacteria, and emended description of the genus *Sphingomonas*. Int. J. Syst. Evol. Microbiol. *53*: 1253–1260.

Busse, H.J., P. Kämpfer, E.R.B. Moore, J. Nuutinen, I.V. Tsitko, E.B.M. Denner, L. Vauterin, M. Valens, R. Rosselló-Mora and M.S. Salkinoja-Salonen. 2002. *Thermomonas haemolytica* gen. nov., sp. nov., a g -proteobacterium from kaolin slurry. Int. J. Syst. Evol. Microbiol. *52*: 473–483.

Busse, H.J., C. Zlamala, S. Buczolits, W. Lubitz, P. Kämpfer and M. Takeuchi. 2003b. *Promicromonospora vindobonensis* sp. nov. and *Promicromonospora aerolata* sp. nov., isolated from the air in the medieval 'Virgilkapelle' in Vienna. Int. J. Syst. Evol. Microbiol. *53*: 1503–1507.

Button, D.K., B.R. Robertson, P.W. Lepp and T.M. Schmidt. 1998. A small, dilute-cytoplasm, high-affinity, novel bacterium isolated by extinction culture and having kinetic constants compatible with growth at ambient concentrations of dissolved nutrients in seawater. Appl. Environ. Microbiol. *64*: 4467–4476.

Button, D.K., F. Schut, P. Quang, R. Martin and B.R. Robertson. 1993. Viability and isolation of marine bacteria by dilution culture: theory, procedures, and initial results. Appl. Environ. Microbiol. *59*: 881–891.

Caccamo, D., C. Gugliandolo, E. Stackebrandt and T.L. Maugeri. 2000. *Bacillus vulcani* sp. nov., a novel thermophilic species isolated from a shallow marine hydrothermal vent. Int. J. Syst. Evol. Microbiol. *50*: 2009–2012.

Caccavo, F., J.D. Coates, R.A. Rosselló-Mora, W. Ludwig, K.H. Schleifer, D.R. Lovley and M.J. McInerney. 1996. *Geovibrio ferrireducens*, a phylogenetically distinct dissimilatory Fe(III)-reducing bacterium. Arch. Microbiol. *165*: 370–376.

Caccavo, F., J.D. Coates, R.A. Rosselló-Mora, W. Ludwig, K.H. Schleifer, D.R. Lovley and M.J. McInerney. 2000. *In* Validation of publication of new names and new combinations previously effectively published outside the IJSEM. List No. 75. Int. J. Syst. Evol. Microbiol. *50*: 1415–1417.

Caldwell, D.E. 1994. End of the pure culture era? ASM News. *60*: 231–232.

Campbell, A.E., J.A. Hellebust and S.W. Watson. 1966. Reductive pentose phosphate cycle in *Nitrosocystis oceanus*. J. Bacteriol. *91*: 1178–1185.

Canhos, V.P., G.P. Manfio and D.A.L. Canhos. 1996. Networking the microbial diversity information. J. Ind. Microbiol. *17*: 498–504.

Cann, I.K., P.G. Stroot, K.R. Mackie, B.A. White and R.I. Mackie. 2001. Characterization of two novel saccharolytic, anaerobic thermophiles, *Thermoanaerobacterium polysaccharolyticum* sp. nov. and *Thermoanaerobacterium zeae* sp. nov., and emendation of the genus *Thermoanaerobacterium*. Int. J. Syst. Evol. Microbiol. *51*: 293–302.

Cao, X., X. Liu and X. Dong. 2003. *Alkaliphilus crotonatoxidans* sp. nov., a strictly anaerobic, crotonate-dismutating bacterium isolated from a methanogenic environment. Int. J. Syst. Evol. Microbiol. *53*: 971–975.

Carlier, J.P., H. Marchandin, E. Jumas-Bilak, V. Lorin, C. Henry, C. Carriere and H. Jean-Pierre. 2002. *Anaeroglobus geminatus* gen. nov., sp. nov., a novel member of the family *Veillonellaceae*. Int. J. Syst. Evol. Microbiol. *52*: 983–986.

Carr, E.L., P. Kämpfer, B.K. Patel, V. Gurtler and R.J. Seviour. 2003. Seven novel species of *Acinetobacter* isolated from activated sludge. Int. J. Syst. Evol. Microbiol. *53*: 953–963.

Castignetti, D. and H.B. Gunner. 1980. Sequential nitrification by an *Alcaligenes* sp. and *Nitrobacter agilis*. Can. J. Microbiol. *26*: 1114–1119.

Catara, V., L. Sutra, A. Morineau, W. Achouak, R. Christen and L. Gardan. 2002. Phenotypic and genomic evidence for the revision of *Pseudomonas corrugata* and proposal of *Pseudomonas mediterranea* sp. nov. Int. J. Syst. Evol. Microbiol. *52*: 1749–1758.

Cavalier-Smith, T. 2002. The neomuran origin of archaebacteria, the negibacterial root of the universal tree and bacterial megaclassification. Int. J. Syst. Evol. Microbiol. *52*: 7–76.

Cayol, J.L., S. Ducerf, B.K. Patel, J.L. Garcia, P. Thomas and B. Ollivier. 2000. *Thermohalobacter berrensis* gen. nov., sp. nov., a thermophilic, strictly halophilic bacterium from a solar saltern. Int. J. Syst. Evol. Microbiol. *50*: 559–564.

Chalk, P.M. and C.J. Smith. 1983. Chemodenitrification. *In* Freney and Simpson (Editors), Gaseous Loss of Nitrogen, Nijhoff, La Hague, The Netherlands. pp. 65–89.

Chandler, D.P., C.A. Wagnon and H.J. Bolton. 1998. Reverse transcriptase (RT) inhibition of PCR at low concentrations of template and its implications for quantitative RT-PCR. Appl. Environ. Microbiol. *64*: 669–677.

Chang, Y.-H., J. Han, J. Chun, K.C. Lee, M.S. Rhee, Y.B. Kim and K.S. Bae. 2002. *Comamonas koreensis* sp. nov., a non-motile species from wetland in Woopo, Korea. Int. J. Syst. Evol. Microbiol. *52*: 377–381.

Chatzinotas, A., R.A. Sandaa, W. Schönhuber, R.I. Amann, F.L. Daae, V. Torsvik, J. Zeyer and D. Hahn. 1998. Analysis of broad-scale differences in microbial community composition of two pristine forest soils. Syst. Appl. Microbiol. *21*: 579–587.

Chaudhuri, S.K., J.G. Lack and J.D. Coates. 2001. Biogenic magnetite formation through anaerobic biooxidation of Fe(II). Appl. Environ. Microbiol. *67*: 2844–2848.

Chaudhuri, S.K., S.M. O'Connor, R.L. Gustavson, L.A. Achenbach and J.D. Coates. 2002. Environmental factors that control microbial perchlorate reduction. Appl. Environ. Microbiol. *68*: 4425–4430.

Chelius, M.K., J.A. Henn and E.W. Triplett. 2002. *Runella zeae* sp. nov., a novel Gram-negative bacterium from the stems of surface-sterilized *Zea mays*. Int. J. Syst. Evol. Microbiol. *52*: 2061–2063.

Chelius, M.K. and E.W. Triplett. 2000. *Dyadobacter fermentans* gen. nov., sp. nov., a novel Gram-negative bacterium isolated from surface-sterilized *Zea mays* stems. Int. J. Syst. Evol. Microbiol. *50*: 751–758.

Chen, M.Y., G.H. Lin, Y.T. Lin and S.S. Tsay. 2002a. *Meiothermus taiwanensis* sp. nov., a novel filamentous, thermophilic species isolated in Taiwan. Int. J. Syst. Evol. Microbiol. *52*: 1647–1654.

Chen, M.Y., S.S. Tsay, K.Y. Chen, Y.C. Shi, Y.T. Lin and G.H. Lin. 2002b. *Pseudoxanthomonas taiwanensis* sp. nov., a novel thermophilic, N_2O-producing species isolated from hot springs. Int. J. Syst. Evol. Microbiol. *52*: 2155–2161.

Chen, W.M., S. Laevens, T.M. Lee, T. Coenye, P. De Vos, M. Mergeay and P. Vandamme. 2001. *Ralstonia taiwanensis* sp. nov., isolated from root nodules of *Mimosa* species and sputum of a cystic fibrosis patient. Int. J. Syst. Evol. Microbiol. *51*: 1729–1735.

Chen, W.X., G.H. Yan and J.L. Li. 1988. Numerical taxonomic study of fast-growing soybean rhizobia and a proposal that *Rhizobium fredii* be assigned to *Sinorhizbium* gen. nov. Int. J. Syst. Bacteriol. *38*: 392–397.

Chester, F.D. 1897. Report of the mycologist: bacteriological work. Del. Agr. Exp. Stn. Bull. *9*: 38–145.

Chester, F.D. 1898. Report of the mycologist: bacteriological work. Del. Agr. Exp. Stn. Bull. *10*: 47–137.

Chester, F.D. 1901. A manual of Determinative Bacteriology, The Macmillan Co., New York. 1–401.

Chin, K.J., W. Liesack and P.H. Janssen. 2001. *Opitutus terrae* gen. nov., sp. nov., to accommodate novel strains of the division '*Verrucomicrobia*' isolated from rice paddy soil. Int. J. Syst. Evol. Microbiol. *51*: 1965–1968.

Chisholm, S.W., S.L. Frankel, R. Goericke, R.J. Olson, B. Palenik, J.B. Waterbury, L. Westjohnsrud and E.R. Zettler. 1992. *Prochlorococcus marinus* nov. gen. nov. sp.–an oxyphototrophic marine prokaryote containing divinyl chlorophyll *a* and chlorophyll *b*. Arch. Microbiol. *157*: 297–300.

Chisholm, S.W., S.L. Frankel, R. Goericke, R.J. Olson, B. Palenik, J.B. Waterbury, L. Westjohnsrud and E.R. Zettler. 2001. *In* Validation of publication of new names and new combinations previously effectively published outside the IJSEM. List No. 79. Int. J. Syst. Evol. Microbiol. *51*: 263–265.

Chistoserdova, L., J.A. Vorholt, R.K. Thauer and M.E. Lidstrom. 1998. C_1 transfer enzymes and coenzymes linking methylotrophic bacteria and methanogenic archaea. Science *281*: 99–102.

Cho, J.C. and S.J. Giovannoni. 2003a. *Croceibacter atlanticus* gen. nov., sp. nov., a novel marine bacterium in the family *Flavobacteriaceae*. Syst. Appl. Microbiol. *26*: 76–83.

Cho, J.C. and S.J. Giovannoni. 2003b. *Fulvimarina pelagi* gen. nov., sp. nov., a marine bacterium that forms a deep evolutionary lineage of descent in the order "*Rhizobiales*". Int. J. Syst. Evol. Microbiol. *53*: 1853–1859.

Cho, J.C. and S.J. Giovannoni. 2003c. *Parvularcula bermudensis* gen. nov., sp. nov., a marine bacterium that forms a deep branch in the α-*Proteobacteria*. Int. J. Syst. Evol. Microbiol. *53*: 1031–1036.

Cho, J.C. and S.J. Giovannoni. 2003d. *In* Validation of publication of new names and new combinations previously effectively published outside the IJSEM. List No. 92. Int. J. Syst. Evol. Microbiol. *53*: 935–937.

Choi, H.J., C.I. Cheigh, S.B. Kim, J.C. Lee, D.W. Lee, S.W. Choi, J.M. Park and Y.R. Pyun. 2002. *Weissella kimchii* sp. nov., a novel lactic acid bacterium from kimchi. Int. J. Syst. Evol. Microbiol. *52*: 507–511.

Chong, S.C., Y.T. Liu, M. Cummins, D.L. Valentine and D.R. Boone. 2002. *Methanogenium marinum* sp. nov., a H_2-using methanogen from Skan Bay, Alaska, and kinetics of H_2 utilization. Antonie Leeuwenhoek Int. J. of Gen. and Molec. Microbiol. *81*: 263–270.

Chong, S.C., Y.T. Liu, M. Cummins, D.L. Valentine and D.R. Boone. 2003. *In* Validation of publication of new names and new combinations previously effectively published outside the IJSEM. List No. 94. Int. J. Syst. Evol. Microbiol. *53*: 1701–1702.

Christensen, H., M. Bisgaard, O. Angen and J.E. Olsen. 2002a. Final classification of Bisgaard taxon 9 as *Actinobacillus arthritidis* sp. nov. and recognition of a novel genomospecies for the equine strains of *Actinobacillus lignieresii*. Int. J. Syst. Evol. Microbiol. *52*: 1239–1246.

Christensen, H., M. Bisgaard, A.M. Bojesen, R. Mutters and J.E. Olsen. 2003. Genetic relationships among avian isolates classified as *Pasteurella haemolytica*, '*Actinobacillus salpingitidis*' or *Pasteurella anatis* with proposal of *Gallibacterium anatis* gen. nov., comb. nov. and description of additional genomospecies within *Gallibacterium* gen. nov. Int. J. Syst. Evol. Microbiol. *53*: 275–287.

Christensen, H., M. Bisgaard and J.E. Olsen. 2002b. Reclassification of equine isolates previously reported as *Actinobacillus equuli*, variants of *A. equuli*, *Actinobacillus suis* or Bisgaard taxon 11 and proposal of *A. equuli* subsp. *equuli* subsp. nov. and *A. equuli* subsp. *haemolyticus* subsp. nov. Int. J. Syst. Evol. Microbiol. *52*: 1569–1576.

Chun, J., K.S. Bae, E.Y. Moon, S.O. Jung, H.K. Lee and S.J. Kim. 2000. *Nocardiopsis kunsanensis* sp. nov., a moderately halophilic actinomycete isolated from a saltern. Int. J. Syst. Evol. Microbiol. *50*: 1909–1913.

Chung, A.P., F.A. Rainey, M. Valente, M.F. Nobre and M.S. da Costa. 2000a. *Thermus igniterrae* sp. nov. and *Thermus antranikianii* sp. nov., two new species from Iceland. Int. J. Syst. Evol. Microbiol. *50*: 209–217.

Chung, Y.R., C.H. Kim, I. Hwang and J. Chun. 2000b. *Paenibacillus koreensis* sp. nov., a new species that produces an iturin-like antifungal compound. Int. J. Syst. Evol. Microbiol. *50*: 1495–1500.

Clark, C. and E.L. Schmidt. 1966. Effect of mixed culture on *Nitrosomonas europaea* simulated by uptake and utilization of pyruvate. J. Bacteriol. *91*: 367–373.

Clark-Curtiss, J.E. and M.A. Docherty. 1989. A species-specific repetitive sequence in *Mycobacterium leprae* DNA. J. Infect. Dis. *159*: 7–15.

Clayton, R.A., G. Sutton, P.S. Hinkle, Jr, C. Bult and C. Fields. 1995. Intraspecific variation in small-subunit rRNA sequences in GenBank:

why single sequences may not adequately represent prokaryotic taxa. Int. J. Syst. Bacteriol. *45*: 595–599.

Cleenwerck, I., K. Vandemeulebroecke, D. Janssens and J. Swings. 2002. Re-examination of the genus *Acetobacter*, with descriptions of *Acetobacter cerevisiae* sp. nov. and *Acetobacter malorum* sp. nov. Int. J. Syst. Evol. Microbiol. *52*: 1551–1558.

Coates, J.D., V.K. Bhupathiraju, L.A. Achenbach, M.J. McInerney and D.R. Lovley. 2001a. *Geobacter hydrogenophilus, Geobacter chapellei* and *Geobacter grbiciae*—three new, strictly anaerobic, dissimilatory Fe(III)-reducers. Int. J. Syst. Evol. Microbiol. *51*: 581–588.

Coates, J.D., R. Chakraborty, J.G. Lack, S.M. O'Connor, K.A. Cole, K.S. Bender and L.A. Achenbach. 2001b. Anaerobic benzene oxidation coupled to nitrate reduction in pure culture by two strains of *Dechloromonas.* Nature *411*: 1039–1043.

Coates, J.D., K.A. Cole, R. Chakraborty, S.M. O'Connor and L.A. Achenbach. 2002. The diversity and ubiquity of bacteria utilizing humic substances as an electron donor for anaerobic respiration. Appl. Environ. Microbiol. *68*: 2445–2452.

Coates, J.D., D.J. Lonergan, E.J. Philips, H. Jenter and D.R. Lovley. 1995. *Desulfuromonas palmitatis* sp. nov., a marine dissimilatory Fe(III) reducer that can oxidize long-chain fatty acids. Arch. Microbiol. *164*: 406–413.

Coates, J.D., D.J. Lonergan, E.J.P. Philips, H. Jenter and D.R. Lovley. 2000a. *In* Validation of the publication of new names and new combinations previously effectively published outside the IJSB. List No. 76. Int. J. Syst. Evol. Microbiol. *50*: 1699–1700.

Coates, J.D., U. Michaelidou, R.A. Bruce, S.M. O'Connor, J.N. Crespi and L.A. Achenbach. 1999. Ubiquity and diversity of dissimilatory (per)chlorate-reducing bacteria. Appl. Environ. Microbiol. *65*: 5234–5241.

Coates, J.D., U. Michaelidou, S.M. O'Connor, R.A. Bruce and L.A. Achenbach. 2000b. The diverse microbiology of (per)chlorate reduction. *In* Urbansky (Editor), Perchlorate in the Environment, Kluwer Academic/Plenum, New York. pp. 257–270.

Cobley, J.G. 1976. Reduction of cytochromes by nitrite in electron-transport particles from *Nitrobacter winogradskyi*: proposal of a mechanism for H^+ translocation. Biochem. J. *156*: 493–498.

Coenye, T., E. Falsen, B. Hoste, M. Ohlen, J. Goris, J.R.W. Govan, M. Gillis and P. Vandamme. 2000. Description of *Pandoraea* gen. nov. with *Pandoraea apista* sp. nov., *Pandoraea pulmonicola* sp. nov., *Pandoraea pnomenusa* sp. nov., *Pandoraea sputorum* sp. nov. and *Pandoraea norimbergensis* comb. nov. Int. J. Syst. Evol. Microbiol. *50*: 887–899.

Coenye, T., J. Goris, P. De Vos, P. Vandamme and J.J. LiPuma. 2003a. Classification of *Ralstonia pickettii*-like isolates from the environment and clinical samples as *Ralstonia insidiosa* sp. nov. Int. J. Syst. Evol. Microbiol. *53*: 1075–1080.

Coenye, T., J. Goris, T. Spilker, P. Vandamme and J.J. LiPuma. 2002a. Characterization of unusual bacteria isolated from respiratory secretions of cystic fibrosis patients and description of *Inquilinus limosus* gen. nov., sp nov. J. Clin. Microbiol. *40*: 2062–2069.

Coenye, T., J. Goris, T. Spilker, P. Vandamme and J.J. LiPuma. 2002b. *In* Validation of the publication of new names and new combinations previously effectively published outside the IJSEM. List no. 87. Int. J. Syst. Evol. Microbiol. *52*: 1437–1438.

Coenye, T., S. Laevens, A. Willems, M. Ohlen, W. Hannant, J.R.W. Govan, M. Gillis, E. Falsen and P. Vandamme. 2001a. *Burkholderia fungorum* sp. nov., and *Burkholderia caledonica* sp. nov., two new species isolated from the environment, animals and human clinical samples. Int. J. Syst. Evol. Microbiol. *51*: 1099–1107.

Coenye, T., E. Mahenthiralingam, D. Henry, J.J. LiPuma, S. Laevens, M. Gillis, D.P. Speert and P. Vandamme. 2001b. *Burkholderia ambifaria* sp. nov., a novel member of the *Burkholderia cepacia* complex including biocontrol and cystic fibrosis-related isolates. Int. J. of Syst. Evol. Microbiol. *51*: 1481–1490.

Coenye, T., M. Vancanneyt, M.C. Cnockaert, E. Falsen, J. Swings and P. Vandamme. 2003b. *Kerstersia gyiorum* gen. nov., sp. nov., a novel *Alcaligenes faecalis*-like organism isolated from human clinical samples, and reclassification of *Alcaligenes denitrificans* Ruger and Tan 1983 as *Achromobacter denitrificans* comb. nov. Int. J. Syst. Evol. Microbiol. *53*: 1825–1831.

Coenye, T., M. Vancanneyt, E. Falsen, J. Swings and P. Vandamme. 2003c. *Achromobacter insolitus* sp. nov. and *Achromobacter spanius* sp. nov., from human clinical samples. Int. J. Syst. Evol. Microbiol. *53*: 1819–1824.

Coenye, T. and P. Vandamme. 2003. Extracting phylogenetic information from whole-genome sequencing projects: the lactic acid bacteria as a test case. Microbiology (Read.) *149*: 3507–3517.

Coenye, T., P. Vandamme and J.J. LiPuma. 2003d. *Ralstonia respiraculi* sp. nov., isolated from the respiratory tract of cystic fibrosis patients. Int. J. Syst. Evol. Microbiol. *53*: 1339–1342.

Cohn, F. 1872. Untersuchungen über Bakterien. Beitr. Biol. Pflanz. *1875 1 (Heft 2)*: 127–224.

Cohn, F. 1875. Untersuchungen über Bakterien II. Beitr. Biol. Pflanz. *1875 1(Heft 3)*: 141–207.

Cole, S.T., R. Brosch, J. Parkhill, T. Garnier, C. Churcher, D. Harris, S.V. Gordon, K. Eiglmeier, S. Gas, C.E. Barry, III, F. Tekaia, K. Badcock, D. Basham, D. Brown, T. Chillingworth, R. Connor, R. Davies, K. Devlin, T. Feltwell, S. Gentles, N. Hamlin, S. Holroyd, T. Hornsby, K. Jagels, A. Krogh, J. McLeah, S. Moule, L. Murphy, K. Oliver, J. Osborne, M.A. Quail, M.A. Rajandream, J. Rogers, S. Rutter, K. Soeger, J. Skelton, R. Squares, S. Squares, J.E. Sulston, K. Taylor, S. Whitehead and B.G. Barrett. 1998. Deciphering the biology of *Mycobacterium tuberculosis* from the complete genome sequence. Nature *393*: 537–544.

Collins, M.J., D.M. Arciero and A.B. Hooper. 1993. Optical spectropotentiometric resolution of the hemes of hydroxylamine oxidoreductase. Heme quantitation and pH dependence of E_m. J. Biol. Chem. *268*: 14655–14662.

Collins, M.D., R. Higgins, S. Messier, M. Fortin, R.A. Hutson, P.A. Lawson and E. Falsen. 2003a. *Allofustis seminis* gen. nov., sp. nov., a novel Gram-positive, catalase-negative, rod-shaped bacterium from pig semen. Int. J. Syst. Evol. Microbiol. *53*: 811–814.

Collins, M.D., L. Hoyles, G. Foster, E. Falsen and N. Weiss. 2002a. *Arthrobacter nasiphocae* sp. nov., from the common seal (*Phoca vitulina*). Int. J. Syst. Evol. Microbiol. *52*: 569–571.

Collins, M.D., L. Hoyles, G. Foster, B. Sjoden and E. Falsen. 2001a. *Corynebacterium capitovis* sp. nov., from a sheep. Int. J. Syst. Evol. Microbiol. *51*: 857–860.

Collins, M.D., L. Hoyles, R.A. Hutson, G. Foster and E. Falsen. 2001b. *Corynebacterium testudinoris* sp. nov., from a tortoise, and *Corynebacterium felinum* sp. nov., from a Scottish wild cat. Int. J. Syst. Evol. Microbiol. *51*: 1349–1352.

Collins, M.D., L. Hoyles, S. Kalfas, G. Sundquist, T. Monsen, N. Nikolaitchouk and E. Falsen. 2000a. Characterization of *Actinomyces* isolates from infected root canals of teeth: Description of *Actinomyces radicidentis* sp. nov. J. Clin. Microbiol. *38*: 3399–3403.

Collins, M.D., L. Hoyles, S. Kalfas, G. Sundquist, T. Monsen, N. Nikolaitchouk and E. Falsen. 2001c. *In* Validation of publication of new names and new combinations previously effectively published outside the IJSEM. List No. 78. Int. J. Syst. Evol. Microbiol. *51*: 1–2.

Collins, M.D., L. Hoyles, P.A. Lawson, E. Falsen, R.L. Robson and G. Foster. 1999. Phenotypic and phylogenetic characterization of a new *Corynebacterium* species from dogs: Description of *Corynebacterium auriscanis* sp. nov. J. Clin. Microbiol. *37*: 3443–3447.

Collins, M.D., L. Hoyles, P.A. Lawson, E. Falsen, R.L. Robson and G. Foster. 2000b. *In* Validation of publication of new names and new combinations previously effectively published outside the IJSEM. List No. 73. Int. J. Syst. Evol. Microbiol. *50*: 423–424.

Collins, M.D., L. Hoyles, E. Tornqvist, R. von Essen and E. Falsen. 2001d. Characterization of some strains from human clinical sources which resemble *"Leptotrichia sanguinegens"*: Description of *Sneathia sanguinegens* sp. nov., gen. nov. Syst. Appl. Microbiol. *24*: 358–361.

Collins, M.D., L. Hoyles, E. Tornqvist, R. von Essen and E. Falsen. 2002b. *In* Validation of publication of new names and new combinations previously effectively published outside the IJSEM. List No. 85. Int. J. Syst. Evol. Microbiol. *52*: 685–690.

Collins, M.D., R.A. Hutson, V. Baverud and E. Falsen. 2000c. Characterization of a *Rothia*-like organism from a mouse: description of *Rothia nasimurium* sp. nov. and reclassification of *Stomatococcus mucilaginosus* as *Rothia mucilaginosa* comb. nov. Int. J. Syst. Evol. Microbiol. *50*: 1247–1251.

Collins, M.D., R.A. Hutson, E. Falsen, E. Inganas and M. Bisgaard. 2002c. *Streptococcus gallinaceus* sp. nov., from chickens. Int. J. Syst. Evol. Microbiol. *52*: 1161–1164.

Collins, M.D., R.A. Hutson, E. Falsen, N. Nikolaitchouk, L. LaClaire and R.R. Facklam. 2000d. An unusual *Streptococcus* from human urine, *Streptococcus urinalis* sp. nov. Int. J. Syst. Evol. Microbiol. *50*: 1173–1178.

Collins, M.D., R.A. Hutson, G. Foster, E. Falsen and N. Weiss. 2002d. *Isobaculum melis* gen. nov., sp. nov., a *Carnobacterium*-like organism isolated from the intestine of a badger. Int. J. Syst. Evol. Microbiol. *52*: 207–210.

Collins, M.D., R.A. Hutson, I.R. Grant and M.F. Patterson. 2000e. Phylogenetic characterization of a novel radiation-resistant bacterium from irradiated pork: description of *Hymenobacter actinosclerus* sp. nov. Int. J. Syst. Evol. Microbiol. *50*: 731–734.

Collins, M.D., R.A. Hutson, L. Hoyles, E. Falsen, N. Nikolaitchouk and G. Foster. 2001e. *Streptococcus ovis* sp. nov., isolated from sheep. Int. J. Syst. Evol. Microbiol. *51*: 1147–1150.

Collins, M.D., R.A. Hutson, N. Nikolaitchouk, A. Nyberg and E. Falsen. 2003b. *Luteococcus sanguinis* sp. nov., isolated from human blood. Int. J. Syst. Evol. Microbiol. *53*: 1889–1891.

Collins, M.D. and P.A. Lawson. 2000. The genus *Abiotrophia* (Kawamura et al.) is not monophyletic: proposal of *Granulicatella* gen. nov., *Granulicatella adiacens* comb. nov., *Granulicatella elegans* comb. nov. and *Granulicatella balaenopterae* comb. nov. Int. J. Syst. Evol. Microbiol. *50*: 365–369.

Collins, M.D., P.A. Lawson, M. Labrenz, B.J. Tindall, N. Weiss and P. Hirsch. 2002e. *Nesterenkonia lacusekhoensis* sp. nov., isolated from hypersaline Ekho Lake, East Antarctica, and emended description of the genus *Nesterenkonia*. Int. J. Syst. Evol. Microbiol. *52*: 1145–1150.

Collins, M.D., P.A. Lawson, N. Nikolaitchouk and E. Falsen. 2000f. *Luteococcus peritonei* sp. nov., isolated from the human peritoneum. Int. J. Syst. Evol. Microbiol. *50*: 179–181.

Collins, M.D. and C. Pascual. 2000. Reclassification of *Actinomyces humiferus* (Gledhill and Casida) as *Cellulomonas humilata* nom. corrig., comb. nov. Int. J. Syst. Evol. Microbiol. *50*: 661–663.

Colwell, R.R. 1970. Polyphasic taxonomy of the genus *Vibrio*: numerical taxonomy of *Vibrio cholerae*, *Vibrio parahaemolyticus*, and related *Vibrio* species. J. Bacteriol. *104*: 410–433.

Colwell, R.R. 1973. Genetic and phenetic classification of bacteria. Adv. Appl. Microbiol. *16*: 137–175.

Colwell, R.R. 1993. Nonculturable but still viable and potentially pathogenic. Int. J. Med. Microbiol. Virol. Parasitol. Infect. Dis. *279*: 154–156.

Conrads, G., M.C. Claros, D.M. Citron, K.L. Tyrrell, V. Merriam and E.J. Goldstein. 2002. 16S-23S rDNA internal transcribed spacer sequences for analysis of the phylogenetic relationships among species of the genus *Fusobacterium*. Int. J. Syst. Evol. Microbiol. *52*: 493–499.

Contzen, M., E.R.B. Moore, S. Blümel, A. Stolz and P. Kämpfer. 2000. *Hydrogenophaga intermedia* sp. nov., a 4-aminobenzene-sulfonate degrading organism. Syst. Appl. Microbiol. *23*: 487–493.

Contzen, M., E.R.B. Moore, S. Blümel, A. Stolz and P. Kämpfer. 2001. *In* Validation of the publication of new names and new combinations previously effectively published outside the IJSEM. List No. 80. Int. J. Syst. Evol. Microbiol. *51*: 793–794.

Convention on Biological Diversity Secretariat. 1992. Conference of the Parties Documentation on COP I. UNEP, 15, chemin des Anemones, CP 356, CH 1219, Geneva, Switzerland.

Cooper, I.P. 1997. Biotechnology and the Law, Clark Boardman Callaghan Co., New York.

Copeland, H.F. 1956. The Classification of Lower Organisms, Pacific Book, Palo Alto.

Coram, N.J. and D.E. Rawlings. 2002a. Molecular relationship between two groups of the genus *Leptospirillum* and the finding that *Leptospirillum ferriphilum* sp. nov. dominates South African commercial biooxidation tanks that operate at 40°C. Appl. Environ. Microbiol. *68*: 838–845.

Coram, N.J. and D.E. Rawlings. 2002b. *In* Validation of publication of new names and new combinations previously effectively published outside the IJSEM. List No. 86. Int. J. Syst. Evol. Microbiol. *52*: 1075–1076.

Cote, R.J. and R.L. Gherna. 1994. Nutrition and media. *In* Gerhardt, Murray, Wood and Krieg (Editors), Methods for General and Molecular Bacteriology, American Society for Microbiology, Washington, DC. pp. 155–178.

Cousins, D.V., R. Bastida, A. Cataldi, V. Quse, S. Redrobe, S. Dow, P. Duignan, A. Murray, C. Dupont, N. Ahmed, D.M. Collins, W.R. Butler, D. Dawson, D. Rodriguez, J. Loureiro, M.I. Romano, A. Alito, M. Zumarraga and A. Bernardelli. 2003. Tuberculosis in seals caused by a novel member of the *Mycobacterium tuberculosis* complex: *Mycobacterium pinnipedii* sp. nov. Int. J. Syst. Evol. Microbiol. *53*: 1305–1314.

Cowan, S.T. 1965. Principles and practice of bacterial taxonomy—a forward look. J. Gen. Microbiol. *37*: 143–153.

Cowan, S.T. 1971. Sense and nonsense in bacterial taxonomy. J. Gen. Microbiol. *67*: 1–8.

Cowan, S.T. 1974. Cowan and Steel's Manual for the Identification of Medical Bacteria, 2nd Ed., Cambridge University Press, Cambridge.

Cowan, S.T. and Hill, L.R. (Editors). 1978. A Dictionary of Microbial Taxonomy, Cambridge University Press, Cambridge. 285 pp.

Crosa, J., D.J. Brenner and S. Falkow. 1973. Use of a single-strand specific nuclease for the analysis of bacterial and plasmid DNA homo- and heteroduplexes. J. Bacteriol. *115*: 904–911.

Cui, X.L., P.H. Mao, M. Zeng, W.J. Li, L.P. Zhang, L.H. Xu and C.L. Jiang. 2001. *Streptimonospora salina* gen. nov., sp. nov., a new member of the family *Nocardiopsaceae*. Int. J. Syst. Evol. Microbiol. *51*: 357–363.

Cummings, D.E., F. Caccavo, S. Spring and R.F. Rosenzweig. 1999. *Ferribacterium limneticum*, gen. nov., sp. nov., an Fe(III)- reducing microorganism isolated from mining-impacted freshwater lake sediments. Arch. Microbiol. *171*: 183–188.

Cummings, D.E., F. Caccavo, S. Spring and R.F. Rosenzweig. 2000. *In* Validation of publication of new names and new combinations previously effectively published outside the IJSEM. List No.77. Int. J. Syst. Evol. Microbiol. *50*: 1953.

Cuzin, N., A.S. Ouattara, M. Labat and J.L. Garcia. 2001. *Methanobacterium congolense* sp. nov., from a methanogenic fermentation of cassava peel. Int. J. Syst. Evol. Microbiol. *51*: 489–493.

da Costa, M.S., M.F. Nobre and F.A. Rainey. 2001. Genus I. *Thermus*. *In* Boone, Castenholz and Garrity (Editors), Bergey's Manual of Systematic Bacteriology, 2nd Edition, Volume 1. The archaea and the deeply branching and phototrophic bacteria. Springer, New York. 404–414.

da Costa, M.S. and F.A. Rainey. 2001. Family I. *Thermaceae* fam. nov. *In* Boone, Castenholz and Garrity (Editors), Bergey's Manual of Systematic Bacteriology, 2nd Edition, Volume 1. The archaea and the deeply branching and phototrophic bacteria. Springer, New York. 403–404.

da Costa, M.S. and F.A. Rainey. 2002. *In* Validation of publication of new names and new combinations previously effectively published outside the IJSEM. List No. 85. Int. J. Syst. Evol. Microbiol. *52*: 685–690.

Daane, L.L., I. Harjono, S.M. Barns, L.A. Launen, N.J. Palleroni and M.M. Haggblom. 2002. PAH-degradation by *Paenibacillus* spp. and description of *Paenibacillus naphthalenovorans* sp. nov., a naphthalene-degrading bacterium from the rhizosphere of salt marsh plants. Int. J. Syst. Evol. Microbiol. *52*: 131–139.

Dabboussi, F. , M. Hamze, M. Elomari, S. Verhille, N. Baïda, D. Izard and H. Leclerc. 2002a. *In* Validation of publication of new names and new combinations previously effectively published outside the IJSEM. List No. 87. Int. J. Syst. Evol. Microbiol. *52*: 1437–1438.

Dabboussi, F., M. Hamze, M. Elomari, S. Verhille, N. Baida, D. Izard and H. Leclerc. 1999. Taxonomic study of bacteria isolated from Lebanese spring waters: proposal for *Pseudomonas cedrella* sp. nov. and *P. orientalis* sp. nov. Res. Microbiol. *150*: 303–316.

Dabboussi, F., M. Hamze, E. Singer, V. Geoffroy, J.M. Meyer and D. Izard.

2002b. *Pseudomonas mosselii* sp. nov., a novel species isolated from clinical specimens. Int. J. Syst. Evol. Microbiol. *52*: 363–376.

Daims, H., J.L. Nielsen, P.H. Nielsen, K.-H. Schleifer and M. Wagner. 2001. *In situ* characterization of *Nitrospira*-like nitrite oxidizing bacteria active in wastewater treatment plants. Appl. Environ. Microbiol. *67*: 5273–5284.

Daneshvar, M.I., D.G. Hollis, R.S. Weyant, A.G. Steigerwalt, A.M. Whitney, M.P. Douglas, J.P. Macgregor, J.G. Jordan, L.W. Mayer, S.M. Rassouli, W. Barchet, C. Munro, L. Shuttleworth and K. Bernard. 2003a. *Paracoccus yeeii* sp. nov. (formerly CDC group EO-2), a novel bacterial species associated with human infection. J. Clin. Microbiol. *41*: 1289–1294.

Daneshvar, M.I., D.G. Hollis, R.S. Weyant, A.G. Steigerwalt, A.M. Whitney, M.P. Douglas, J.P. Macgregor, J.G. Jordan, L.W. Mayer, S.M. Rassouli, W. Barchet, C. Munro, L. Shuttleworth and K. Bernard. 2003b. *In* Validation of publication of new names and new combinations previously effectively published outside the IJSEM. List No. 92. Int. J. Syst. Evol. Microbiol. *53*: 935–937.

Dang, P.N., T.C.H. Dang, T.H. Lai and H. Stan-Lotter. 1996. *Desulfovibrio vietnamensis* sp. nov., a halophilic sulfate-reducing bacterium from Vietnamese oil fields. Anaerobe *2*: 385–392.

Dang, P.N., T.C.H. Dang, T.H. Lai and H. Stan-Lotter. 2002. *In* Validation of publication of new names and new combinations previously effectively published outside the IJSEM. List No. 86. Int. J. Syst. Evol. Microbiol. *52*: 1075–1076.

Dasman, S. Kajiyama, H. Kawasaki, M. Yagi, T. Seki, E. Fukusaki and A. Kobayashi. 2002. *Paenibacillus glycanilyticus* sp. nov., a novel species that degrades heteropolysaccharide produced by the cyanobacterium *Nostoc commune*. Int. J. Syst. Evol. Microbiol. *52*: 1669–1674.

Daubin, V., M. Gouy and G. Perriere. 2002. A phylogenomic approach to bacterial phylogeny: Evidence of a core of genes sharing a common history. Genome Res. *12*: 1080–1090.

Davies, D.G., M.R. Parsek, J.P. Pearson, B.H. Iglewski, J.W. Costerton and E.P. Greenberg. 1998. The involvement of cell-to-cell signals in the development of a bacterial biofilm. Science *280*: 295–298.

De Baere, T., S. Steyaert, G. Wauters, P. De Vos, J. Goris, T. Coenye, T. Suyama, G. Verschraegen and M. Vaneechoutte. 2001. Classification of *Ralstonia pickettii* biovar 3/'thomasii' strains (Pickett 1994) and of new isolates related to nosocomial recurrent meningitis as *Ralstonia mannitolytica* sp. nov. Int. J. Syst. Evol. Microbiol. *51*: 547–558.

De Beer, D., J.C. Vandenheuvel and S.P.P. Ottengraf. 1993. Microelectrode measurements of the activity distribution in nitrifying bacterial aggregates. Appl. Environ. Microbiol. *59*: 573–579.

De Boer, W., H. Duyts and H.J. Laanbroek. 1989. Urea stimulated autotrophic nitrification in suspensions of fertilized, acid heath soil. Soil Biol. Biochem. *21*: 349–354.

De Boer, W., P.A.K. Gunnewiek and H.J. Laanbroek. 1995. Ammonium oxidation at low pH by a chemolithotrophic bacterium belonging to the genus *Nitrosospira*. Soil Biol. Biochem. *27*: 127–132.

De Boer, W., P.J.A. Gunnewiek, M. Veenhuis, E. Bock and H.J. Laanbroek. 1991. Nitrification at low pH by aggregated chemolithotrophic bacteria. Appl. Environ. Microbiol. *57*: 3600–3604.

De Boer, W. and H.J. Laanbroek. 1989. Ureolytic nitrification at low pH by *Nitrosospira* species. Arch. Microbiol. *152*: 178–181.

de Bont, J.A., A. Scholten and T.A. Hansen. 1981. DNA-DNA hybridization of *Rhodopseudomonas capsulata*, *Rhodopseudomonas sphaeroides* and *Rhodopseudomonas sulfidophila* strains. Arch. Microbiol. *128*: 271–274.

De Bruijn, P., A.A. Vandegraaf, M.S.M. Jetten, L.A. Robertson and J.G. Kuenen. 1995. Growth of *Nitrosomonas europaea* on hydroxylamine. FEMS Microbiol. Lett. *125*: 179–184.

De Graef, E.M., L.A. Devriese, M. Vancanneyt, M. Baele, M.D. Collins, K. Lefebvre, J. Swings and F. Haesebrouck. 2003. Description of *Enterococcus canis* sp. nov. from dogs and reclassification of *Enterococcus porcinus* Teixeira et al. 2001 as a junior synonym of *Enterococcus villorum* Vancanneyt et al. 2001. Int. J. Syst. Evol. Microbiol. *53*: 1069–1074.

de Groot, G.N. and A.H. Stouthamer. 1969. Regulation of reductase formation in *Proteus mirabilis*. I. Formation of reductases and enzymes of the formic hydrogenlyase complex in the wild type and in chlorate resistant mutants. Arch. Microbiol. *66*: 220–233.

De Lajudie, P., A. Laurent-Fulele, U. Willems, R. Torck, R. Coopman, M.D. Collins, K. Kersters, B. Dreyfus and M. Gillis. 1998. Description of *Allorhizobium undicola* gen. nov., sp. nov., for nitrogen-fixing bacteria efficiently nodulating *Neptunia natans* in Senegal. Int. J. Syst. Bacteriol. *48*: 1277–1290.

De Ley, J. 1992. The *Proteobacteria* ribosomal RNA cistron similarities and bacterial taxonomy. *In* Balows, Trüper, Dworkin, Harder and Schleifer (Editors), The Prokaryotes-A Handbook on the Biology of Bacteria: Ecophysiology, Isolation, Identification, Applications., 2nd Ed., Vol. 2, Springer-Verlag, New York. pp. 2111–2140.

De Ley, J. and J. De Smedt. 1975. Improvement of the membrane filter method for DNA–rRNA hybridization. Antonie Leeuwenhoek *41*: 287–307.

de Rijk, P., E. Robbrecht, S. de Hoog, A. Caers, Y. van de Peer and R. de Wachter. 1999. Database on the structure of large subunit ribosomal RNA. Nucleic Acids Res. *27*: 174–178.

de Rijk, P., Y. van de Peer, I. van den Broeck and R. de Wachter. 1995. Evolution according to large ribosomal subunit RNA. J. Mol. Evol. *41*: 366–375.

De Wever, H., J.R. Cole, M.R. Fettig, D.A. Hogan and J.M. Tiedje. 2000. Reductive dehalogenation of trichloroacetic acid by *Trichlorobacter thiogenes* gen. nov., sp. nov. Appl. Environ. Microbiol. *66*: 2297–2301.

De Wever, H., J.R. Cole, M.R. Fettig, D.A. Hogan and J.M. Tiedje. 2001. *In* Validation of the publication of new names and new combinations previously effectively published oustide the IJSEM. List No. 78. Int. J. Syst. Evol. Microbiol. *51*: 1–2.

Dedysh, S.N. , V.N. Khmelenina, N.E. Suzina, Y.A. Trotsenko, J.D. Semrau, W. Liesack and J.M. Tiedje. 2002. *Methylocapsa acidiphila* gen. nov., sp. nov., a novel methane-oxidizing and dinitrogen-fixing acidophilic bacterium from Sphagnum bog. Int. J. Syst. Evol. Microbiol. *52*: 251–261.

Dedysh, S.N., W. Liesack, V.N. Khmelenina, N.E. Suzina, Y.A. Trotsenko, J.D. Semrau, A.M. Bares, N.S. Panikov and J.M. Tiedje. 2000. *Methylocella palustris* gen. nov., sp. nov., a new methane oxidizing acidophilic bacterium from peat bags, representing a novel subtype of serine pathway methanotrophs. Int. J. Syst. Evol. Microbiol. *50*: 955–969.

Defnoun, S., M. Labat, M. Ambrosio, J.L. Garcia and B.K. Patel. 2000. *Papillibacter cinnamivorans* gen. nov., sp. nov., a cinnamate-transforming bacterium from a shea cake digester. Int. J. Syst. Evol. Microbiol. *50*: 1221–1228.

Degrange, V. and R. Bardin. 1995. Detection and counting of *Nitrobacter* populations in soil by PCR. Appl. Environ. Microbiol. *61*: 2093–2098.

Degrange, V., M.M. Couteaux, J.M. Anderson, M.P. Berg and R. Lensi. 1998. Nitrification and occurrence of *Nitrobacter* by MPN-PCR in low and high nitrifying coniferous forest soils. Plant Soil *198*: 201–208.

Degrange, V., R. Lensi and R. Bardin. 1997. Activity, size and structure of a *Nitrobacter* community as affected by organic carbon and nitrite in sterile soil. FEMS Microbiol. Ecol. *24*: 173–180.

Dehio, C., C. Lanz, R. Pohl, P. Behrens, D. Bermond, Y. Piemont, K. Pelz and A. Sander. 2001. *Bartonella schoenbuchii* sp. nov., isolated from the blood of wild roe deer. Int. J. Syst. Evol. Microbiol. *51*: 1557–1565.

DeLong, E.F. 1992. *Archaea* in coastal marine environments. Proc. Natl. Acad. Sci. U.S.A. *89*: 5685–5689.

DeLong, E.F., G.S. Wickham and N.R. Pace. 1989. Phylogenetic stains: ribosomal RNA-based probes for the identification of single cells. Science *243*: 1360–1363.

DeLong, E.F., K.Y. Wu, B.B. Prezelin and R.V.M. Jovine. 1994. High abundance of *Archaea* in Antarctic marine picoplankton. Nature *371*: 695–697.

Delorme, S., P. Lemanceau, R. Christen, T. Corberand, J.M. Meyer and L. Gardan. 2002. *Pseudomonas lini* sp. nov., a novel species from bulk and rhizospheric soils. Int. J. Syst. Evol. Microbiol. *52*: 513–523.

Denger, K., R. Warthmann, W. Ludwig and B. Schink. 2002. *Anaerophaga thermohalophila* gen. nov., sp. nov., a moderately thermohalophilic, strictly anaerobic fermentative bacterium. Int. J. Syst. Evol. Microbiol. *52*: 173–178.

Denhardt, D.T. 1966. A membrane-filter technique for the detection of complementary DNA. Biochem. Biophys. Res. Commun. *23*: 641–646.

Denner, E.B.M., B. Mark, H.J. Busse, M. Turkiewicz and W. Lubitz. 2001a. *Psychrobacter proteolyticus* sp. nov., a psychrotrophic, halotolerant bacterium isolated from the antarctic krill *Euphausia superba* Dana, excreting a cold-adapted metalloprotease. Syst. Appl. Microbiol. *24*: 44–53.

Denner, E.B.M., B. Mark, H.J. Busse, M. Turkiewicz and W. Lubitz. 2001b. *In* Validation of the publication of new names and new combinations previously effectively published outside the IJSEM. List No. 82. Int. J. Syst. Evol. Microbiol. *51*: 1619–1620.

Denner, E.B., S. Paukner, P. Kämpfer, E.R. Moore, W.R. Abraham, H.J. Busse, G. Wanner and W. Lubitz. 2001c. *Sphingomonas pituitosa* sp. nov., an exopolysaccharide-producing bacterium that secretes an unusual type of sphingan. Int. J. Syst. Evol. Microbiol. *51*: 827–841.

Denner, E.B., G.W. Smith, H.J. Busse, P. Schumann, T. Narzt, S.W. Polson, W. Lubitz and L.L. Richardson. 2003. *Aurantimonas coralicida* gen. nov., sp. nov., the causative agent of white plague type II on Caribbean scleractinian corals. Int. J. Syst. Evol. Microbiol. *53*: 1115–1122.

Denner, E.B., D. Vybiral, U.R. Fischer, B. Velimirov and H.J. Busse. 2002a. *Vibrio calviensis* sp. nov., a halophilic, facultatively oligotrophic 0.2 µm-filterable marine bacterium. Int. J. Syst. Evol. Microbiol. *52*: 549–553.

Denner, E.B., D. Vybiral, M. Koblizek, P. Kämpfer, H.J. Busse and B. Velimirov. 2002b. *Erythrobacter citreus* sp. nov., a yellow-pigmented bacterium that lacks bacteriochlorophyll *a*, isolated from the western Mediterranean Sea. Int. J. Syst. Evol. Microbiol. *52*: 1655–1661.

Deschavanne, P.J., A. Giron, J. Vilain, G. Fagot and B. Fertil. 1999. Genomic signature: Characterization and classification of species assessed by chaos game representation of sequences. Mol. Biol. Evol. *16*: 1391–1399.

Devereux, R., M.D. Kane, J. Winfrey and D.A. Stahl. 1992. Genus- and group-specific hybridization probes for determinative and environmental studies of sulfate-reducing bacteria. Syst. Appl. Microbiol. *15*: 601–609.

Dewhirst, F.E., B.J. Paster, N. Tzellas, B. Coleman, J. Downes, D.A. Spratt and W.G. Wade. 2001. Characterization of novel human oral isolates and cloned 16S rDNA sequences that fall in the family *Coriobacteriaceae*: description of *Olsenella* gen. nov., reclassification of *Lactobacillus uli* as *Olsenella uli* comb. nov. and description of *Olsenella profusa* sp. nov. Int. J. Syst. Evol. Microbiol. *51*: 1797–1804.

Diab, S. and M. Shilo. 1988. Effect of adhesion to particles on the survival and activity of *Nitrosomonas* sp. and *Nitrobacter* sp. Arch. Microbiol. *150*: 387–393.

Dianou, D., T. Miyaki, S. Asakawa, H. Morii, K. Nagaoka, H. Oyaizu and S. Matsumoto. 2001. *Methanoculleus chikugoensis* sp. nov., a novel methanogenic archaeon isolated from paddy field soil in Japan, and DNA-DNA hybridization among *Methanoculleus* species. Int. J. Syst. Evol. Microbiol. *51*: 1663–1669.

Dickerson, R.E. 1980. Evolution and gene transfer in purple photosynthetic bacteria. Nature *283*: 210–212.

Dicks, L.M., M. Silvester, P.A. Lawson and M.D. Collins. 2000. *Lactobacillus fornicalis* sp. nov., isolated from the posterior fornix of the human vagina. Int. J. Syst. Evol. Microbiol. *50*: 1253–1258.

Dirmeier, R., M. Keller, D. Hafenbradl, F.J. Braun, R. Rachel, S. Burggraf and K.O. Stetter. 1998. *Thermococcus acidaminovorans* sp. nov., a new hyperthermophilic alkalophilic archaeon growing on amino acids. Extremophiles *2*: 109–114.

Dirmeier, R., M. Keller, D. Hafenbradl, F.J. Braun, R. Rachel, S. Burggraf and K.O. Stetter. 2001. *In* Validation of publication of new names and new combinations previously effectively published outside the IJSEM. List No. 80. Int. J. Syst. Evol. Microbiol. *51*: 793–794.

DiSpirito, A.A., J.D. Lipscomb and A.B. Hooper. 1986. Cytochrome aa_3 from *Nitrosomonas europaea*. J. Biol. Chem. *261*: 17048–17056.

DiSpirito, A.A., L.R. Taaffe and A.B. Hooper. 1985. Localization and concentration of hydroxylamine oxidoreductase and cytochrome c-552, cytochrome c-554, cytochrome c_m-533, cytochrome c_m-552 and cytochrome *a* in *Nitrosomonas europaea*. Biochim. Biophys. Acta *806*: 320–330.

Distel, D.L., W. Morrill, N. MacLaren-Toussaint, D. Franks and J. Waterbury. 2002. *Teredinibacter turnerae* gen. nov., sp. nov., a dinitrogen-fixing, cellulolytic, endosymbiotic γ-proteobacterium isolated from the gills of wood-boring molluscs (*Bivalvia*: *Teredinidae*). Int. J. Syst. Evol. Microbiol. *52*: 2261–2269.

Dittberner, P. 1996. Untersuchungen zur phylogenie und biochemie des Harnstoff-hydrolyse systems bei ammoniak oxidierenden bakterien, University of Hamburg.

Dobson, C.M., H. Deneer, S. Lee, S. Hemmingsen, S. Glaze and B. Ziola. 2002. Phylogenetic analysis of the genus *Pediococcus*, including *Pediococcus claussenii* sp. nov., a novel lactic acid bacterium isolated from beer. Int. J. Syst. Evol. Microbiol. *52*: 2003–2010.

Doi, R.H. and R.T. Igarashi. 1965. Conservation of ribosomal and messenger ribonucleic acid cistrons in *Bacillus* species. J. Bacteriol. *90*: 384–390.

Donachie, S.P., S. Hou, T.S. Gregory, A. Malahoff and M. Alam. 2003. *Idiomarina loihiensis* sp. nov., a halophilic γ-Proteobacterium from the Lo'ihi submarine volcano, Hawai'i. Int. J. Syst. Evol. Microbiol. *53*: 1873–1879.

Donahue, J.M., N.M. Williams, S.F. Sells and D.P. Labeda. 2002. *Crossiella equi* sp. nov., isolated from equine placentas. Int. J. Syst. Evol. Microbiol. *52*: 2169–2173.

Dong, X., Y. Xin, W. Jian, X. Liu and D. Ling. 2000. *Bifidobacterium thermacidophilum* sp. nov., isolated from an anaerobic digester. Int. J. Syst. Evol. Microbiol. *50*: 119–125.

Doolittle, W.F. and J.M. Logsdon, Jr. 1998. Archaeal genomics: do archaea have a mixed heritage? Curr. Biol. *8*: R209–R211.

Dorofeeva, L.V., L.I. Evtushenko, V.I. Krausova, A.V. Karpov, S.A. Subbotin and J.M. Tiedje. 2002. *Rathayibacter caricis* sp. nov. and *Rathayibacter festucae* sp. nov., isolated from the phyllosphere of *Carex* sp. and the leaf gall induced by the nematode *Anguina graminis* on *Festuca rubra* L., respectively. Int. J. Syst. Evol. Microbiol. *52*: 1917–1923.

Dorofeeva, L.V., V.I. Krausova, L.I. Evtushenko and J.M. Tiedje. 2003. *Agromyces albus* sp. nov., isolated from a plant (*Androsace* sp.). Int. J. Syst. Evol. Microbiol. *53*: 1435–1438.

Doronina, N.V., T.D. Darmaeva and Y.A. Trotsenko. 2003. *Methylophaga alcalica* sp. nov., a novel alkaliphilic and moderately halophilic, obligately methylotrophic bacterium from an East Mongolian saline soda lake. Int. J. Syst. Evol. Microbiol. *53*: 223–229.

Doronina, N.V., L.V. Kudinova and Y.A. Trotsenko. 2000a. *Methylovorus mays* sp. nov.: A new species of aerobic, obligately methylotrophic bacteria associated with plants. Microbiology *69*: 599–603.

Doronina, N.V., L.V. Kudinova and Y.A. Trotsenko. 2001a. *In* Validation of publication of new names and new combinations previously effectively published outside the IJSEM. List No. 82. Int. J. Syst. Evol. Microbiol. *51*: 1619–1620.

Doronina, N.V. and Y.A. Trotsenko. 1994. *Methylophilus leisingerii* sp. nov., a new species of restricted facultatively methylotrophic bacteria. Mikrobiologiya *63*: 529–536.

Doronina, N.V. and Y.A. Trotsenko. 2000. A novel plant-associated thermotolerant alkaliphilic methylotroph of the genus *Paracoccus*. Microbiology *69*: 593–598.

Doronina, N.V. and Y.A. Trotsenko. 2001a. *In* Validation of the publication of new names and new combinations previously effectively published outside the IJSB. List No. 82. Int. J. Syst. Evol. Microbiol. *51*: 1619–1620.

Doronina, N.V. and Y.A. Trotsenko. 2001b. *In* Validation of the publication of new names and new combinations previously effectively published outside the IJSEM. List No. 78. Int. J. Syst. Evol. Bacteriol. *51*: 1.

Doronina, N.V. and Y.A. Trotsenko. 2003. Reclassification of *'Blastobacter viscosus'* 7d and *'Blastobacter aminooxidans'* 14a as *Xanthobacter viscosus* sp. nov. and *Xanthobacter aminoxidans* sp. nov. Int. J. Syst. Evol. Microbiol. *53*: 179–182.

Doronina, N.V., Y.A. Trotsenko, B.B. Kuznetsov, T.P. Tourova and M.S. Salkinoja-Salonen. 2002. *Methylobacterium suomiense* sp. nov. and *Methylobacterium lusitanum* sp. nov., aerobic, pink-pigmented, facultatively methylotrophic bacteria. Int. J. Syst. Evol. Microbiol. *52*: 773–776.

Doronina, N.V., Y.A. Trotsenko and T.P. Tourova. 2000b. *Methylarcula marina* gen. nov., sp. nov. and *Methylarcula terricola* sp. nov.: novel aerobic, moderately halophilic, facultatively methylotrophic bacteria from coastal saline environments. Int. J. Syst. Evol. Microbiol. *50*: 1849–1859.

Doronina, N.V., Y.A. Trotsenko, T.P. Tourova, B.B. Kuznetsov and T. Leisinger. 2000c. *Methylopila helvetica* sp. nov. and *Methylobacterium dichloromethanicum* sp. nov.—novel aerobic facultatively methylotrophic bacteria utilizing dichloromethane. Syst. Appl. Microbiol. *23*: 210–218.

Doronina, N.V., Y.A. Trotsenko, T.P. Tourova, B.B. Kuznetsov and T. Leisinger. 2000d. *In* Validation of the publication of new names and new combinations previously effectively published outside the IJSB. List No. 77. Int. J. Syst. Evol. Microbiol. *50*: 1953.

Doronina, N.V., Y.A. Trotsenko, T.P. Tourova, B.B. Kuznetsov and T. Leisinger. 2001b. *Albibacter methylovorans* gen. nov., sp. nov., a novel aerobic, facultatively autotrophic and methylotrophic bacterium that utilizes dichloromethane. Int. J. Syst. Evol. Microbiol. *51*: 1051–1058.

Dorsch, M., D.N. Lovet and G.D. Bailey. 2001. *Fusobacterium equinum* sp. nov., from the oral cavity of horses. Int. J. Syst. Evol. Microbiol. *51*: 1959–1963.

Downes, J., M.A. Munson, D.R. Radford, D.A. Spratt and W.G. Wade. 2002. *Shuttleworthia satelles* gen. nov., sp. nov., isolated from the human oral cavity. Int. J. Syst. Evol. Microbiol. *52*: 1469–1475.

Downes, J., M. Munson and W.G. Wade. 2003. *Dialister invisus* sp. nov., isolated from the human oral cavity. Int. J. Syst. Evol. Microbiol. *53*: 1937–1940.

Downes, J., B. Olsvik, S.J. Hiom, D.A. Spratt, S.L. Cheeseman, I. Olsen, A.J. Weightman and W.G. Wade. 2000. *Bulleidia extructa* gen. nov., sp. nov., isolated from the oral cavity. Int. J. Syst. Evol. Microbiol. *50*: 979–983.

Drancourt, M., C. Bollet, A. Carta and P. Rousselier. 2001. Phylogenetic analyses of *Klebsiella* species delineate *Klebsiella* and *Raoultella* gen. nov., with description of *Raoultella ornithinolytica* comb. nov., *Raoultella terrigena* comb. nov. and *Raoultella planticola* comb. nov. Int. J. Syst. Evol. Microbiol. *51*: 925–932.

Drews, G. and J.F. Imhoff. 1991. Phototrophic purple bacteria. *In* Shively and Barton (Editors), Variations in Autotrophic Life, Academic Press, London. pp. 51–97.

Droz, S., B. Chi, E. Horn, A.G. Steigerwalt, A.M. Whitney and D.J. Brenner. 1999. *Bartonella koehlerae* sp. nov., isolated from cats. J. Clin. Microbiol. *37*: 1117–1122.

Droz, S., B. Chi, E. Horn, A.G. Steigerwalt, A.M. Whitney and D.J. Brenner. 2000. *In* Validation of the publication of new names and new combinations previously effectively published outside the IJSB. List No. 73. Int. J. Syst. Evol. Microbiol. *50*: 423–424.

Dubnau, D., I. Smith, P. Morell and J. Marmur. 1965. Gene conservation in *Bacillus* species. I. Conserved genetic and nucleic acid base sequence homologies. Proc. Natl. Acad. Sci. U.S.A. *54*: 491–498.

Duckworth, A.W., W.D. Grant, B.E. Jones, D. Meijer, M.C. Marquez and A. Ventosa. 2000a. *Halomonas magadii* sp. nov., a new member of the genus *Halomonas*, isolated from a soda lake of the East African Rift Valley. Extremophiles *4*: 53–60.

Duckworth, A.W., W.D. Grant, B.E. Jones, D. Meijer, M.C. Marquez and A. Ventosa. 2000b. *In* Validation of the publication of new names and new combinations previously effectively published outside the IJSEM, List No. 75. Int. J. Syst. Evol. Microbiol. *50*: 1415–1417.

Duddleston, K.N., P.J. Bottomley, A.J. Porter and D.J. Arp. 2000. New insights into methyl bromide cooxidation by *Nitrosomonas europaea* obtained by experimenting with moderately low density cell suspensions. Appl. Environ. Microbiol. *66*: 2726–2731.

Dumler, J.S., A.F. Barbet, C.P.J. Bekker, G.A. Dasch, G.H. Palmer, S.C. Ray, Y. Rikihisa and F.R. Rurangirwa. 2001. Reorganization of genera in the families *Rickettsiaceae* and *Anaplasmataceae* in the order *Rickettsiales*: Unification of some species of *Ehrlichia* with *Anaplasma*, *Cowdria* with *Ehrlichia* and *Ehrlichia* with *Neorickettsia*, descriptions of six new species combinations and designation of *Ehrlichia equi* and 'HGE agent' as subjective synonyms of *Ehrlichia phagocytophila*. Int. J. Syst. Evol. Microbiol. *51*: 2145–2165.

Duncan, S.H., G.L. Hold, A. Barcenilla, C.S. Stewart and H.J. Flint. 2002a. *Roseburia intestinalis* sp. nov., a novel saccharolytic, butyrate-producing bacterium from human faeces. Int. J. Syst. Evol. Microbiol. *52*: 1615–1620.

Duncan, S.H., G.L. Hold, H.J. Harmsen, C.S. Stewart and H.J. Flint. 2002b. Growth requirements and fermentation products of *Fusobacterium prausnitzii*, and a proposal to reclassify it as *Faecalibacterium prausnitzii* gen. nov., comb. nov. Int. J. Syst. Evol. Microbiol. *52*: 2141–2146.

Dunfield, P.F., V.N. Khmelenina, N.E. Suzina, Y.A. Trotsenko and S.N. Dedysh. 2003. *Methylocella silvestris* sp. nov., a novel methanotroph isolated from an acidic forest cambisol. Int. J. Syst. Evol. Microbiol. *53*: 1231–1239.

Dunn, G. and B.S. Everitt. 1982. An Introduction to Mathematical Taxonomy, Cambridge University Press, Cambridge.

Dyksterhouse, S.E., J.P. Gray, R.P. Herwig, J.C. Lara and J.T. Staley. 1995. *Cycloclasticus pugetii* gen. nov., sp. nov., an aromatic hydrocarbon-degrading bacterium from marine sediments. Int. J. Syst. Bacteriol. *45*: 116–123.

Eckert, B., O.B. Weber, G. Kirchhof, A. Halbritter, M. Stoffels and A. Hartmann. 2001. *Azospirillum doebereinerae* sp. nov., a nitrogen-fixing bacterium associated with the C4-grass *Miscanthus*. Int. J. Syst. Evol. Bacteriol. *51*: 17–26.

Eder, W. and R. Huber. 2002a. New isolates and physiological properties of the *Aquificales* and description of *Thermocrinis albus* sp. nov. Extremophiles *6*: 309–318.

Eder, W. and R. Huber. 2002b. *In* Validation of publication of new names and new combinations previously effectively published outside the IJSEM. List No. 88. Int. J. Syst. Evol. Microbiol. *52*: 1915–1916.

Edgell, D.R., N.M. Fast and W.F. Doolittle. 1996. Selfish DNA: the best defense is a good offense. Curr. Biol. *6*: 385–388.

Edwards, C.G., M.D. Collins, P.A. Lawson and A.V. Rodriguez. 2000. *Lactobacillus nagelii* sp. nov., an organism isolated from a partially fermented wine. Int. J. Syst. Evol. Microbiol. *50*: 699–702.

Edwards, P.R. and W.H. Ewing. 1962. Identification of *Enterobacteriaceae*, 2nd Ed., Burgess Publishing Company, Minneapolis.

Egan, S., C. Holmstrom and S. Kjelleberg. 2001. *Pseudoalteromonas ulvae* sp. nov., a bacterium with antifouling activities isolated from the surface of a marine alga. Int. J. Syst. Evol. Microbiol. *51*: 1499–1504.

Ehrenberg, C.G. 1838. Die Infusionthierchen als vollkommene Organismen: ein Blick in das tiefere organische Leben der Natur, L. Voss, Leipzig. pp. i–xvii; 1–547.

Ehrenreich, A. and F. Widdel. 1994. Anaerobic oxidation of ferrous iron by purple bacteria, a new type of phototrophic metabolism. Appl. Environ. Microbiol. *60*: 4517–4526.

Ehrich, S., D. Behrens, E. Lebedeva, W. Ludwig and E. Bock. 1995. A new obligately chemolithoautotrophic, nitrite-oxidizing bacterium, *Nitrospira moscoviensis* sp. nov. and its phylogenetic relationship. Arch. Microbiol. *164*: 16–23.

Ehrich, S., D. Behrens, E. Lebedeva, W. Ludwig and E. Bock. 2001. *In* Validation of publication of new names and new combinations previously effectively published outside the IJSEM. List No. 78. Int. J. Syst. Evol. Microbiol. *51*: 1–2.

Ehrmann, M., W. Ludwig and K.H. Schleifer. 1994. Reverse dot blot hybridization: a useful method for the direct identification of lactic acid bacteria in fermented food. FEMS Microbiol. Lett. *117*: 143–150.

Ehrmann, M.A., M.R. Muller and R.F. Vogel. 2003. Molecular analysis of sourdough reveals *Lactobacillus mindensis* sp. nov. Int. J. Syst. Evol. Microbiol. *53*: 7–13.

Eigener, U. 1975. Adenine nucleotide pool variations in intact *Nitrobacter winogradskyi* cells. Arch. Microbiol. *102*: 233–240.

Eighmy, T.T. and P.L. Bishop. 1989. Distribution and role of bacterial nitrifying populations in nitrogen removal in aquatic treatment systems. Water Res. *23*: 947–955.

Eisen, J.A. 1995. The RecA protein as a model molecule for molecular systematic studies of bacteria: comparison of trees of RecAs and 16S rRNAs from the same species. J. Mol. Evol. *41*: 1105–1023.

Elbanna, K., T. Lutke-Eversloh, S. Van Trappen, J. Mergaert, J. Swings and A. Steinbuchel. 2003. *Schlegelella thermodepolymerans* gen. nov., sp. nov., a novel thermophilic bacterium that degrades poly(3-hydroxybutyrate-*co*-3-mercaptopropionate). Int. J. Syst. Evol. Microbiol. *53*: 1165–1168.

Elberson, M.A., F. Malekzadeh, M.T. Yazdi, N. Kameranpour, M.R. Noori-Daloii, M.H. Matte, M. Shahamat, R.R. Colwell and K.R. Sowers. 2000. *Cellulomonas persica* sp. nov. and *Cellulomonas iranensis* sp. nov., mesophilic cellulose-degrading bacteria isolated from forest soils. Int. J. Syst. Evol. Microbiol. *50*: 993–996.

Elo, S., I. Suominen, P. Kämpfer, J. Juhanoja, M. Salkinoja-Salonen and K. Haahtela. 2001. *Paenibacillus borealis* sp. nov., a nitrogen-fixing species isolated from spruce forest humus in Finland. Int. J. Syst. Evol. Microbiol. *51*: 535–545.

Engel, M. 1999. Untersuchungen zur Sequenzheterogenität multipler rRNS-Operone bei Vertretern verschiedener Entwicklungslinien der Bacteria, Thesis, Technical University, Munich.

Engelhardt, M.A., K. Daly, R.P.J. Swannell and I.M. Head. 2001a. Isolation and characterization of a novel hydrocarbon-degrading, Gram-positive bacterium, isolated from intertidal beach sediment, and description of *Planococcus alkanoclasticus* sp. nov. J. Appl. Microbiol. *90*: 237–247.

Engelhardt, M.A., K. Daly, R.P.J. Swannell and I.M. Head. 2001b. *In* Validation of publication of new names and new combinations previously effectively published outside the IJSEM. List No. 81. Int. J. Syst. Evol. Microbiol. *51*: 1229.

Enright, M.R., J.O. McInerney and C.T. Griffin. 2003. Characterization of endospore-forming bacteria associated with entomopathogenic nematodes, *Heterorhabditis* spp., and description of *Paenibacillus nematophilus* sp. nov. Int. J. Syst. Evol. Microbiol. *53*: 435–441.

Erhart, R., D. Bradford, R.J. Seviour, R.I. Amann and L.L. Blackall. 1997. Development and use of fluorescent *in situ* hybridization probes for the detection and identification of *"Microthrix parvicella"* in activated sludge. Syst. Appl. Microbiol. *20*: 310–318.

Ericksen, G.E. 1983. The Chilean nitrate deposits. Am. Sci. *71*: 366–374.

Erickson, R.H. and A.B. Hooper. 1972. Preliminary characterization of a variant CO-binding heme protein from *Nitrosomonas*. Biochim. Biophys. Acta *275*: 231–244.

Etchebehere, C., M.I. Errazquin, R. Dabert, R. Moletta and L. Muxí. 2001. *Comamonas nitrativorans* sp. nov., a novel denitrifier isolated from a denitifying reactor treating landfill leachate. Int. J. Syst. Evol. Microbiol. *51*: 977–983.

Euzéby, J.P. 1997. Revised nomenclature of specific or subspecific epithets that do not agree in gender with generic names that end in -bacter. Int. J. Syst. Bacteriol. *47*: 585.

Euzéby, J.P. 1999. Revised *Salmonella* nomenclature: designation of *Salmonella enterica* (ex Kauffmann and Edwards 1952) Le Minor and Popoff 1987 sp. nov., nom. rev. as the neotype species of the genus *Salmonella* Lignieres 1900 (Approved Lists 1980), rejection of the name *Salmonella choleraesuis* (Smith 1894) Weldin 1927 (Approved Lists 1980), and conservation of the name *Salmonella typhi* (Schroeter 1886) Warren and Scott 1930 (Approved Lists 1980). Request for an Opinion. Int. J. Syst. Bacteriol. *49*: 927–930.

Evans, P.J., D.T. Mang, K.S. Kim and L.Y. Young. 1991. Anaerobic degradation of toluene by a denitrifying bacterium. Appl. Environ. Microbiol. *57*: 1139–1145.

Evans, W.R., D.E. Fleischman, H.E. Calvert, P.V. Pyati, G.M. Alter and N.S.S. Rao. 1990. Bacteriochlorophyll and photosynthetic reaction centers in *Rhizobium* strain BTAi 1. Appl. Environ. Microbiol. *56*: 3445–3449.

Everett, K.D.E., R.M. Bush and A.A. Andersen. 1999. Emended description of the order *Chlamydiales*, proposal of *Parachlamydiaceae* fam. nov. and *Simkaniaceae* fam. nov., each containing one monotypic genus, revised taxonomy of the family *Chlamydiaceae*, including a new genus and five new species, and standards for the identification of organisms. Int. J. Syst. Bacteriol. *49*: 415–440.

Evtushenko, L.I., L.V. Dorofeeva, T.G. Dobrovolskaya, G.M. Streshinskaya, S.A. Subbotin and J.M. Tiedje. 2001. *Agreia bicolorata* gen. nov., sp. nov., to accommodate actinobacteria isolated from narrow reed grass infected by the nematode *Heteroanguina graminophila*. Int. J. Syst. Evol. Microbiol. *51*: 2073–2079.

Evtushenko, L.I., L.V. Dorofeeva, V.I. Krausova, E.Y. Gavrish, S.G. Yashina and M. Takeuchi. 2002. *Okibacterium fritillariae* gen. nov., sp. nov., a novel genus of the family *Microbacteriaceae*. Int. J. Syst. Evol. Microbiol. *52*: 987–993.

Evtushenko, L.I., L.V. Dorofeeva, S.A. Subbotin, J.R. Cole and J.M. Tiedje. 2000a. *Leifsonia poae* gen. nov., sp. nov., isolated from nematode galls on *Poa annua*, and reclassification of *'Corynebacterium aquaticum'* Leifson 1962 as *Leifsonia aquatica* (ex Leifson 1962) gen. nov., nom. rev., comb. nov. and *Clavibacter xyli* Davis et al. 1984 with two subspecies as *Leifsonia xyli* (Davis et al. 1984) gen. nov., comb. nov. Int. J. Syst. Evol. Microbiol. *50*: 371–380.

Evtushenko, L.I., V.V. Taran, V.N. Akimov, R.M. Kroppenstedt, J.M. Tiedje and E. Stackebrandt. 2000b. *Nocardiopsis tropica* sp. nov., *Nocardiopsis trehalosi* sp. nov., nom. rev. and *Nocardiopsis dassonvillei* subsp. *albirubida* subsp. nov., comb. nov. Int. J. Syst. Evol. Microbiol. *50*: 73–81.

Ezaki, T., M. Amano, Y. Kawamura and E. Yabuuchi. 2000a. Proposal of *Salmonella paratyphi* sp. nov., nom. rev. and request for an opinion to conserve the specific epithet *paratyphi* in the binary combination *Salmonella paratyphi* as *nomen epitheton conservandum*. Int. J. Syst. Evol. Microbiol. *50*: 941–944.

Ezaki, T., Y. Hashimoto and E. Yabuuchi. 1989. Fluorometric deoxyribonucleic acid-deoxyribonucleic acid hybridization in microdilution wells as an alternative to membrane filter hybridization in which radioisotopes are used to determine genetic relatedness among bacterial strains. Int. J. Syst. Bacteriol. *39*: 224–229.

Ezaki, T., Y. Kawamura, N. Li, Z.Y. Li, L. Zhao and S. Shu. 2001. Proposal of the genera *Anaerococcus* gen. nov., *Peptoniphilus* gen. nov. and *Gallicola* gen. nov. for members of the genus *Peptostreptococcus*. Int. J. Syst. Evol. Microbiol. *51*: 1521–1528.

Ezaki, T., Y. Kawamura and E. Yabuuchi. 2000b. Recognition of nomenclatural standing of *Salmonella typhi* (Approved Lists 1980), *Salmonella enteritidis* (Approved Lists 1980) and *Salmonella typhimurium* (Approved Lists 1980), and conservation of the specific epithets *enteritidis* and *typhimurium*. Request for an opinion. Int. J. Syst. Evol. Microbiol. *50*: 945–947.

Fan, Y., Z. Jin, J. Tong, W. Li, M. Pasciak, A. Gamian, Z. Liu and Y. Huang. 2002. *Rothia amarae* sp. nov., from sludge of a foul water sewer. Int. J. Syst. Evol. Microbiol. *52*: 2257–2260.

Fani, R., C. Bandi, M. Bazzicalupo, M.T. Ceccherini, S. Fancelli, E. Gallori, L. Gerace, A. Grifoni, N. Miclaus and G. Damiani. 1995. Phylogeny of the genus *Azospirillum* based on 16S rDNA sequence. FEMS Microbiol. Lett. *129*: 195–200.

Fardeau, M.L., M. Magot, B.K. Patel, P. Thomas, J.L. Garcia and B. Ollivier. 2000. *Thermoanaerobacter subterraneus* sp. nov., a novel thermophile isolated from oilfield water. Int. J. Syst. Evol. Microbiol. *50*: 2141–2149.

Fardeau, M.L., B. Ollivier, J.L. Garcia and B.K. Patel. 2001. Transfer of *Thermobacteroides leptospartum* and *Clostridium thermolacticum* as *Clostridium stercorarium* subsp. *leptospartum* subsp. *thermolacticum* subsp. nov., comb. nov. and *C. stercorarium* subsp. *thermolacticum* subsp. nov., comb. nov. Int. J. Syst. Evol. Microbiol. *51*: 1127–1131.

Farrand, S.K., P.B. Van Berkum and P. Oger. 2003. *Agrobacterium* is a definable genus of the family *Rhizobiaceae*. Int. J. Syst. Evol. Microbiol. *53*: 1681–1687.

Feio, M.J., I.B. Beech, M. Carepo, J.M. Lopes, C.W.S. Cheung, R. Franco, J. Guezennec, J.R. Smith, J.I. Mitchell, J.J.G. Moura and A.R. Lino. 1998. Isolation and characterization of a novel sulphate-reducing bacterium of the *Desulfovibrio* genus. Anaerobe *4*: 117–130.

Feio, M.J., I.B. Beech, M. Carepo, J.M. Lopes, C.W.S. Cheung, R. Franco, J. Guezennec, J.R. Smith, J.I. Mitchell, J.J.G. Moura and A.R. Lino. 2000. *In* Validation of the publication of new names and new combinations previously effectively published outside the IJSB. List no. 75. Int. J. Syst. Evol. Microbiol. *50*: 1415–1417.

Felsenstein, J. 1982. Numerical methods for inferring evolutionary trees. Q. Rev. Biol. *27*: 44–57.

Felske, A., H. Rheims, A. Wolterink, E. Stackebrandt and A.D.L. Akkermans. 1997. Ribosome analysis reveals prominent activity of an uncultured member of the class Actinobacteria in grassland soils. Microbiology (Reading) *143*: 2983–2989.

Feltham, R.K.A., P.A. Wood and P.H.A. Sneath. 1984. A general-purpose system for characterizing medically important bacteria to genus level. J. Appl. Bacteriol. *57*: 279–290.

Fernández-Garayzábal, J.F., R. Egido, A.I. Vela, V. Briones, M.D. Collins, A. Mateos, R.A. Hutson, L. Dominguez and J. Goyache. 2003. Isolation of *Corynebacterium falsenii* and description of *Corynebacterium aquilae* sp. nov., from eagles. Int. J. Syst. Evol. Microbiol. *53*: 1135–1138.

Fernández-Martínez, J., M.J. Pujalte, J. García-Martínez, M. Mata, E. Garay and F. Rodríguez-Valera. 2003. Description of *Alcanivorax venustensis* sp. nov. and reclassification of *Fundibacter jadensis* DSM 12178(T) (Bruns and Berthe-Corti 1999) as *Alcanivorax jadensis* comb. nov., members of the emended genus *Alcanivorax*. Int. J. Syst. Evol. Microbiol. *53*: 331–338.

Ferris, M.J., S.C. Nold, N.P. Revsbech and D.M. Ward. 1997. Population structure and physiological changes within a hot spring microbial mat community following disturbance. Appl. Environ. Microbiol. *63*: 1367–1374.

Ferris, M.J., A.L. Ruff Roberts, E.D. Kopczynski, M.M. Bateson and D.M. Ward. 1996. Enrichment culture and microscopy conceal diverse thermophilic *Synechococcus* populations in a single hot spring microbial mat habitat. Appl. Environ. Microbiol. *62*: 1045–1050.

Ferris, M.J. and D.M. Ward. 1997. Seasonal distributions of dominant 16S rRNA-defined populations in a hot spring microbial mat examined by denaturing gradient gel electrophoresis. Appl. Environ. Microbiol. *63*: 1375–1381.

Festl, H., W. Ludwig and K.H. Schleifer. 1986. DNA hybridization probe for the *Pseudomonas fluorescens* group. Appl. Environ. Microbiol. *52*: 1190–1194.

Fiala, G., C.R. Woese, T.A. Langworthy and K.O. Stetter. 1990. *Flexistipes sinusarabici*, a novel genus and species of eubacteria occurring in the Atlantis II deep brines of the Red Sea. Arch. Microbiol. *154*: 120–126.

Fiala, G., C.R. Woese, T.A. Langworthy and K.O. Stetter. 2000. *In* Validation of publication of new names and new combinations previously effectively published outside the IJSEM. List No. 75. Int. J. Syst. Evol. Microbiol. *50*: 1415–1417.

Finegold, S.M., M.L. Vaisanen, D.R. Molitoris, T.J. Tomzynski, Y. Song, C. Liu, M.D. Collins and P.A. Lawson. 2003a. *Cetobacterium somerae* sp. nov. from human feces and emended description of the genus *Cetobacterium*. Syst. Appl. Microbiol. *26*: 177–181.

Finegold, S.M., M.L. Vaisanen, D.R. Molitoris, T.J. Tomzynski, Y. Song, C. Liu, M.D. Collins and P.A. Lawson. 2003b. *In* Validation of publication of new names and new combinations previously effectively published outside the IJSEM. List No. 93. Int. J. Syst. Evol. Microbiol. *53*: 1219–1220.

Finkmann, W., K. Altendorf, E. Stackebrandt and A. Lipski. 2000. Characterization of N_2O-producing *Xanthomonas*-like isolates from biofilters as *Stenotrophomonas nitritireducens* sp. nov., *Luteimonas mephitis* gen. nov., sp. nov. and *Pseudoxanthomonas broegbernensis* gen. nov., sp. nov. Int. J. Syst. Evol. Microbiol. *50*: 273–282.

Finneran, K.T., H.M. Forbush, C.V. VanPraagh and D.R. Lovley. 2002. *Desulfitobacterium metallireducens* sp. nov., an anaerobic bacterium that couples growth to the reduction of metals and humic acids as well as chlorinated compounds. Int. J. Syst. Evol. Microbiol. *52*: 1929–1935.

Finneran, K.T., C.V. Johnsen and D.R. Lovley. 2003. *Rhodoferax ferrireducens* sp. nov., a psychrotolerant, facultatively anaerobic bacterium that oxidizes acetate with the reduction of Fe(III). Int. J. Syst. Evol. Microbiol. *53*: 669–673.

Finster, K., W. Liesack and B. Thamdrup. 1998. Elemental sulfur and thiosulfate disproportionation by *Desulfocapsa sulfoexigens* sp. nov., a new anaerobic bacterium isolated from marine surface sediment. Appl. Environ. Microbiol. *64*: 119–125.

Finster, K., W. Liesack and B. Thamdrup. 2000. *In* Validation of the publication of new names and new combinations previously published outside the IJSB. List No. 76. Int. J. Syst. Bacteriol. *50*: 1699–1700.

Finster, K., T.R. Thomsen and N.B. Ramsing. 2001. *Desulfomusa hansenii* gen. nov., sp. nov., a novel marine propionate-degrading, sulfate-reducing bacterium isolated from *Zostera* marina roots. Int. J. Syst. Evol. Microbiol. *51*: 2055–2061.

Fischer, A. 1895. Untersuchungen über Bakterien. J. Wiss. Bot. *27*: 1–163.

Fitch, W.M. and E. Margoliash. 1967. Construction of phylogenetic trees: a method based on mutational distances as estimated from cytochrome *c* sequences of general applicability. Science *155*: 279–284.

Flärdh, K., P.S. Cohen and S. Kjelleberg. 1992. Ribosomes exist in large excess over the apparent demand for protein synthesis during carbon starvation in marine *Vibrio* sp. strain CCUG 15956. J. Bacteriol. *174*: 6780–6788.

Fleischmann, R.D., M.D. Adams, O. White, R.A. Clayton, E.F. Kirkness, A.R. Kerlavage, C.J. Bult, J.F. Tomb, B.A. Dougherty, J.M. Merrick, K. McKenney, G. Sutton, W. Fitzhugh, C. Fields, J.D. Gocayne, J. Scott, R. Shirley, L.I. Liu, A. Glodek, J.M. Kelley, J.F. Weidman, C.A. Phillips, T. Spriggs, E. Hedblom, M.D. Cotton, T.R. Utterback, M.C. Hanna, D.T. Nguyen, D.M. Saudek, R.C. Brandon, L.D. Fine, J.L. Fritchman, J.L. Fuhrmann, N.S.M. Geoghagen, C.L. Gnehm, L.A. McDonald, K.V. Small, C.M. Fraser, H.O. Smith and J.C. Venter. 1995. Whole-genome random sequencing and assembly of *Haemophilus influenzae*. Science (Wash. D. C.) *269*: 496–512.

Fliermans, C.B., B.B. Bohlool and E.L. Schmidt. 1974. Autecological study of the chemoautotroph *Nitrobacter* by immunofluorescence. Appl. Microbiol. *27*: 124–129.

Floyd, M.M., W.M. Gross, D.A. Bonato, V.A. Silcox, R.W. Smithwick, B. Metchock, J.T. Crawford and W.R. Butler. 2000. *Mycobacterium kubicae* sp. nov., a slowly growing, scotochromogenic *Mycobacterium*. Int. J. Syst. Evol. Microbiol. *50*: 1811–1816.

Flügge, C. 1886. Die Microorganismen, F. C. W. Vogel, Leipzig.

Focht, D.D. and W. Verstraete. 1977. Biochemical ecology of nitrification and denitrification. Adv. Microb. Ecol. *1*: 135–214.

Fodor, S.P.A., J.L. Read, M.C. Pirrung, L. Stryer, A.T. Lu and D. Solas. 1991. Light-directed, spatially addressable parallel chemical synthesis. Science *251*: 767–773.

Fonnesbech Vogel, B., K. Venkateswaran, H. Christensen, E. Falsen, G. Christiansen and L. Gram. 2000. Polyphasic taxonomic approach in the description of *Alishewanella fetalis* gen. nov., sp. nov., isolated from a human foetus. Int. J. Syst. Evol. Microbiol. *50*: 1133–1142.

Forster, A.C., J.L. McInnes, D.C. Skingle and R.H. Symons. 1985. Non-radioactive hybridization probes prepared by the chemical labeling of DNA and RNA with a novel reagent, photobiotin. Nucleic Acids Res. *13*: 745–761.

Fortina, M.G., D. Mora, P. Schumann, C. Parini, P.L. Manachini and E. Stackebrandt. 2001a. Reclassification of *Saccharococcus caldoxylosilyticus* as *Geobacillus caldoxylosilyticus* (Ahmad et al. 2000) comb. nov. Int. J. Syst. Evol. Microbiol. *51*: 2063–2071.

Fortina, M.G., R. Pukall, P. Schumann, D. Mora, C. Parini, P.L. Manachini and E. Stackebrandt. 2001b. *Ureibacillus* gen. nov., a new genus to accommodate *Bacillus thermosphaericus* (Andersson et al. 1995), emendation of *Ureibacillus thermosphaericus* and description of *Ureibacillus terrenus* sp. nov. Int. J. Syst. Evol. Microbiol. *51*: 447–455.

Fossing, H., V.A. Gallardo, B.B. Jørgensen, M. Hüttel, L.P. Nielsen, H. Schulz, D.E. Canfield, S. Forster, R.N. Glud, J.K. Gundersen, J. Küver, N.B. Ramsing, A. Teske, B. Thamdrup and O. Ulloa. 1995. Concentration and transport of nitrate by the mat-forming sulphur bacterium *Thioploca*. Nature *374*: 713–715.

Foster, G., H.M. Ross, H. Malnick, A. Willems, R.A. Hutson, R.J. Reid and M.D. Collins. 2000. *Phocoenobacter uteri* gen. nov., sp. nov., a new member of the family *Pasteurellaceae* Pohl (1979) 1981 isolated from a harbour porpoise (*Phocoena phocoena*). Int. J. Syst. Evol. Microbiol. *50*: 135–139.

Fournier, D., R. Lemieux and D. Couillard. 1998. Genetic evidence for highly diversified bacterial populations in wastewater sludge during biological leaching of metals. Biotechnol. Lett. *20*: 27–31.

Fournier, D., C. Mouton, P. Lapierre, T. Kato, K. Okuda and C. Menard. 2001. *Porphyromonas gulae* sp. nov., an anaerobic, Gram-negative coccobacillus from the gingival sulcus of various animal hosts. Int. J. Syst. Evol. Microbiol. *51*: 1179–1189.

Fox, G.E., K.R. Pechman and C.R. Woese. 1977. Comparative cataloging of 16S ribosomal ribonucleic acid - molecular approach to procaryotic systematics. Int. J. Syst. Bacteriol. *27*: 44–57.

Fox, G.E., J.D. Wisotzkey and P. Jurtshuk Jr.. 1992. How close is close: 16S rRNA sequence identity may not be sufficient to guarantee species identity. Int. J. Syst. Bacteriol. *41*: 166–170.

Fox, J.G., C.C. Chien, F.E. Dewhirst, B.J. Paster, Z. Shen, P.L. Melito, D.L. Woodward and F.G. Rodgers. 2000. *Helicobacter canadensis* sp. nov. isolated from humans with diarrhea as an example of an emerging pathogen. J. Clin. Microbiol. *38*: 2546–2549.

Fox, J.G., C.C. Chien, F.E. Dewhirst, B.J. Paster, Z. Shen, P.L. Melito, D.L. Woodward and F.G. Rodgers. 2002. *In* Validation of new names and new combinations previously effectively published outside the IJSEM. List No. 84. Int. J. Syst. Evol. Microbiol. *52*: 3–4.

Franklin, C.L., P.L. Gorelick, L.K. Riley, F.E. Dewhirst, R.S. Livingston, J.M. Ward, C.S. Beckwith and J.G. Fox. 2001. *Helicobacter typhlonius* sp. nov., a novel urease-negative *Helicobacter* species. J. Clin. Microbiol. *39*: 3920–3926.

Franklin, C.L., P.L. Gorelick, L.K. Riley, F.E. Dewhirst, R.S. Livingston, J.M. Ward, C.S. Beckwith and J.G. Fox. 2002. *In* Validation of new names and new combinations previously effectively published outside the IJSEM. List No. 85. Int. J. Syst. Evol. Microbiol. *52*: 685–690.

Fredrickson, J.K., T.R. Garland, R.J. Hicks, J.M. Thomas, S.W. Li and K.M. McFadden. 1989. Lithotrophic and heterotrophic bacteria in deep subsurface sediments and their relations to sediment properties. Geomicrobiol. J. *7*: 53–66.

Freitag, A. and E. Bock. 1990. Energy conservation in *Nitrobacter*. FEMS Microbiol. Lett. *66*: 157–162.

Freitag, A., M. Rudert and E. Bock. 1987. Growth of *Nitrobacter* by dissimilatoric nitrate reduction. FEMS Microbiol. Lett. *48*: 105–109.

Freitas, M., F.A. Rainey, M.F. Nobre, A.J. Silvestre and M.S. da Costa. 2003a. *Tepidimonas aquatica* sp. nov., a new slightly thermophilic β-proteobacterium isolated from a hot water tank. Syst. Appl. Microbiol. *26*: 376–381.

Freitas, M., F.A. Rainey, M.F. Nobre, A.J. Silvestre and M.S. da Costa. 2003b. *In* Validation of publication of new names and new combinations previously effectively published outside the IJSEM. List No. 94. Int. J. Syst. Evol. Microbiol. *53*: 1701–1702.

Friedman, C.S., K.B. Andree, K.A. Beauchamp, J.D. Moore, T.T. Robbins, J.D. Shields and R.P. Hedrick. 2000. *Xenohaliotis californiensis*, a newly described pathogen of abalone, *Haliotis* spp., along the west coast of North America. Int. J. Syst. Evol. Microbiol. *50*: 847–855.

Frischer, M.E., P.J. Floriani and S.A. Nierzwicki Bauer. 1996. Differential sensitivity of 16S rRNA targeted oligonucleotide probes used for fluorescence *in situ* hybridization is a result of ribosomal higher order structure. Can. J. Microbiol. *42*: 1061–1071.

Frühling, A., P. Schumann, H. Hippe, B. Straubler and E. Stackebrandt. 2002. *Exiguobacterium undae* sp. nov. and *Exiguobacterium antarcticum* sp. nov. Int. J. Syst. Evol. Microbiol. *52*: 1171–1176.

Fuchs, B.M., G. Wallner, W. Beisker, I. Schwippl, W. Ludwig and R.I. Amann. 1998. Flow cytometric analysis of the *in situ* accessibility of *Escherichia coli* 16S rRNA for fluorescently labeled oligonucleotide probes. Appl. Environ. Microbiol. *64*: 4973–4982.

Fudou, R., Y. Jojima, T. Iizuka and S. Yamanaka. 2002a. *Haliangium ochraceum* gen. nov., sp. nov. and *Haliangium tepidum* sp. nov.: Novel moderately halophilic myxobacteria isolated from coastal saline environments. J. Gen. Appl. Microbiol. *48*: 109–115.

Fudou, R., Y. Jojima, T. Iizuka and S. Yamanaka. 2002b. *In* Validation of publication of new names and new combinations previously effectively published outside the IJSEM. List No. 87. Int. J. Syst. Evol. Microbiol. *52*: 1437–1438.

Fudou, R., Y. Jojima, A. Seto, K. Yamada, E. Kimura, T. Nakamatsu, A. Hiraishi and S. Yamanaka. 2002c. *Corynebacterium efficiens* sp. nov., a glutamic-acid-producing species from soil and vegetables. Int. J. Syst. Evol. Microbiol. *52*: 1127–1131.

Fuentes-Ramírez, L.E., R. Bustillos-Cristales, A. Tapia-Hernández, T. Jiménez-Salgado, E.T. Wang, E. Martínez-Romero and J. Caballero-Mellado. 2001. Novel nitrogen-fixing acetic acid bacteria, *Gluconacetobacter johannae* sp. nov. and *Gluconacetobacter azotocaptans* sp. nov., associated with coffee plants. Int. J. Syst. Evol. Microbiol. *51*: 1305–1314.

Fuerst, J.A., J.A. Hawkins, A. Holmes, L.I. Sly, C.J. Moore and E. Stackebrandt. 1993. *Porphyrobacter neustonensis* gen. nov., sp. nov., an aerobic bacteriochlorophyll-synthesizing budding bacterium from fresh water. Int. J. Syst. Bacteriol. *43*: 125–134.

Fuhrman, J.A., K. McCallum and A.A. Davis. 1992. Novel major archaebacterial group from marine plankton. Nature *356*: 148–149.

Fuhrman, J.A. and C.C. Ouverney. 1998. Marine microbial diversity studied via 16S rRNA sequences: cloning results from coastal waters and counting of native archaea with fluorescent single cell probes. Aquat. Ecol. *32*: 3–15.

Fujii, K., M. Satomi, N. Morita, T. Motomura, T. Tanaka and S. Kikuchi. 2003. *Novosphingobium tardaugens* sp. nov., an oestradiol-degrading bacterium isolated from activated sludge of a sewage treatment plant in Tokyo. Int. J. Syst. Evol. Microbiol. *53*: 47–52.

Fujii, K., N. Urano, H. Ushio, M. Satomi and S. Kimura. 2001. *Sphingomonas cloacae* sp. nov., a nonylphenol-degrading bacterium isolated from wastewater of a sewage-treatment plant in Tokyo. Int. J. Syst. Evol. Microbiol. *51*: 603–610.

Fukui, M., A. Teske, B. Assmus, G. Muyzer and F. Widdel. 1999. Physiology, phylogenetic relationships, and ecology of filamentous sulfate-reducing bacteria (genus *Desulfonema*). Arch. Microbiol. *172*: 193–203.

Fukui, M., A. Teske, B. Assmus, G. Muyzer and F. Widdel. 2000. *In* Validation of the publication of new names and new combinations previously effectively published outside the IJSEM. List No. 75. Int. J. Syst. Evol. Microbiol. *50*: 1415–1417.

Fuqua, C. and E.P. Greenberg. 1998. Cell-to-cell communication in *Escherichia coli* and *Salmonella typhimurium*: they may be talking, but who's listening? Proc. Natl. Acad. Sci. U.S.A. *95*: 6571–6572.

Fuqua, W.C., S.C. Winans and E.P. Greenberg. 1994. Quorum sensing in bacteria: the *LuxR-LuxI* family of cell density-responsive transcriptional regulators. J. Bacteriol. *176*: 269–275.

Gaasterland, T. and M.A. Ragan. 1999. Microbial genescapes: phyletic and functional patterns of ORF distribution among prokaryotes. Microbiol. Comp. Genomics. *3*: 199–217.

Galindo, I., R. Rangel-Aldao and J.L. Ramirez. 1993. A combined polymerase chain reaction-colour development hybridization assay in a microtitre format for the detection of *Clostridium* spp. Appl. Microbiol. Biotechnol. *39*: 553–557.

Gardan, L., P. Bella, J.M. Meyer, R. Christen, P. Rott, W. Achouak and R. Samson. 2002. *Pseudomonas salomonii* sp. nov., pathogenic on garlic, and *Pseudomonas palleroniana* sp. nov., isolated from rice. Int. J. Syst. Evol. Microbiol. *52*: 2065–2074.

Gardan, L., C. Dauga, P. Prior, M. Gillis and G.S. Saddler. 2000. *Acidovoarx anthurii* sp. nov., a new phytopathogenic bacterium which causes bacterial leaf-spot of anthurium. Int. J. Syst. Evol. Microbiol. *50*: 235–246.

Gardan, L., C. Gouy, R. Christen and R. Samson. 2003a. Elevation of three subspecies of *Pectobacterium carotovorum* to species level: *Pectobacterium atrosepticum* sp. nov., *Pectobacterium betavasculorum* sp. nov. and *Pectobacterium wasabiae* sp. nov. Int. J. Syst. Evol. Microbiol. *53*: 381–391.

Gardan, L., D.E. Stead, C. Dauga and M. Gillis. 2003b. *Acidovorax valerianellae* sp. nov., a novel pathogen of lamb's lettuce [*Valerianella locusta* (L.) Laterr.]. Int. J. Syst. Evol. Microbiol. *53*: 795–800.

Garner, M.R., J.F. Flint and J.B. Russell. 2002. *Allisonella histaminiformans* gen. nov., sp. nov. A novel bacterium that produces histamine, utilizes histicline as its sole energy source, and could play a role in bovine and equine laminitis. Syst. Appl. Microbiol. *25*: 498–506.

Garner, M.R., J.F. Flint and J.B. Russell. 2003. *In* Validation of publication of new names and new combinations previously effectively published outside the IJSEM. List No. 90. Int. J. Syst. Evol. Microbiol. *53*: 373.

Garnova, E.S., T.N. Zhilina, T.P. Tourova and A.M. Lysenko. 2003a. *Anoxynatronum sibiricum* gen. nov., sp. nov. alkaliphilic saccharolytic anaerobe from cellulolytic community of Nizhnee Beloe (Transbaikal region). Extremophiles *7*: 213–220.

Garnova, E.S., T.N. Zhilina, T.P. Tourova and A.M. Lysenko. 2003b. *In* Validation of publication of new names and new combinations previously effectively published outside the IJSEM. List No. 93. Int. J. Syst. Evol. Microbiol. *53*: 1219–1220.

Garrity, G.M., J.A. Bell and T.G. Lilburn. 2003. Bergey's Taxonomic Outline: Release 4.0: http://dx.doi.org/10.1007/bergeysoutline200310.

Garrity, G.M. and J.G. Holt. 2001a. Class I. *Chrysiogenetes* class. nov. *In* Boone, Castenholz and Garrity (Editors), Bergey's Manual of Systematic Bacteriology, 2nd Edition, Volume 1. The archaea and the deeply branching and phototrophic bacteria. Springer, New York. 421.

Garrity, G.M. and J.G. Holt. 2001b. Class I. *Deinococci* class. nov. *In* Boone, Castenholz and Garrity (Editors), Bergey's Manual of Systematic Bacteriology, 2nd Edition, Volume 1. The archaea and the deeply branching and phototrophic bacteria. Springer, New York. 395.

Garrity, G.M. and J.G. Holt. 2001c. Class I. *Thermomicrobia* class. nov. *In* Boone, Castenholz and Garrity (Editors), Bergey's Manual of Systematic Bacteriology, 2nd Edition, Volume 1. The archaea and the deeply branching and phototrophic bacteria. Springer, New York. 447.

Garrity, G.M. and J.G. Holt. 2001d. Class IV. *Archaeoglobi* class. nov. *In* Boone, Castenholz and Garrity (Editors), Bergey's Manual of Systematic Bacteriology, 2nd Edition, Volume 1. The archaea and the deeply branching and phototrophic bacteria. Springer, New York. 349.

Garrity, G.M. and J.G. Holt. 2001e. Class VII. *Methanopyri* class. nov. *In* Boone, Castenholz and Garrity (Editors), Bergey's Manual of Systematic Bacteriology, 2nd Edition, Volume 1. The archaea and the deeply branching and phototrophic bacteria. Springer, New York. 535.

Garrity, G.M. and J.G. Holt. 2001f. Family I. *Chrysiogenaceae* fam nov. *In* Boone, Castenholz and Garrity (Editors), Bergey's Manual of Systematic Bacteriology, 2nd Edition, Volume 1. The archaea and the deeply branching and phototrophic bacteria. Springer, New York. 421.

Garrity, G.M. and J.G. Holt. 2001g. Family I. *Thermomicrobiaceae* ord. nov. *In* Boone, Castenholz and Garrity (Editors), Bergey's Manual of Systematic Bacteriology, 2nd Edition, Volume 1. The archaea and the deeply branching and phototrophic bacteria. Springer, New York. 447.

Garrity, G.M. and J.G. Holt. 2001h. Order I. *Chrysiogenales* ord. nov. *In* Boone, Castenholz and Garrity (Editors), Bergey's Manual of Systematic Bacteriology, 2nd Edition, Volume 1. The archaea and the deeply branching and phototrophic bacteria. Springer, New York. 421.

Garrity, G.M. and J.G. Holt. 2001i. Order I. *Thermomicrobiales* ord. nov. *In* Boone, Castenholz and Garrity (Editors), Bergey's Manual of Systematic Bacteriology, 2nd Edition, Volume 1. The archaea and the deeply branching and phototrophic bacteria. Springer, New York. 447.

Garrity, G.M. and J.G. Holt. 2001j. Phylum AI. *Crenarchaeota* phyl. nov. *In* Boone, Castenholz and Garrity (Editors), Bergey's Manual of Systematic Bacteriology, 2nd Edition, Volume 1. The archaea and the deeply branching and phototrophic bacteria. Springer, New York. 169.

Garrity, G.M. and J.G. Holt. 2001k. Phylum AII. *Euryarchaeota* phyl. nov. *In* Boone, Castenholz and Garrity (Editors), Bergey's Manual of Systematic Bacteriology, 2nd Edition, Volume 1. The archaea and the deeply branching and phototrophic bacteria. Springer, New York. 211.

Garrity, G.M. and J.G. Holt. 2001*l*. Phylum BIII. *Thermodesulfobacteria* phyl. nov. *In* Boone, Castenholz and Garrity (Editors), Bergey's Manual of Systematic Bacteriology, 2nd Edition, Volume 1. The archaea and the deeply branching and phototrophic bacteria. Springer, New York. 389.

Garrity, G.M. and J.G. Holt. 2001m. Phylum BIX. *Deferribacteres* phyl. nov. *In* Boone, Castenholz and Garrity (Editors), Bergey's Manual of Systematic Bacteriology, 2nd Edition, Volume 1. The archaea and the deeply branching and phototrophic bacteria. Springer, New York. 465.

Garrity, G.M. and J.G. Holt. 2001n. Phylum BV. *Chrysiogenetes* phyl. nov. *In* Boone, Castenholz and Garrity (Editors), Bergey's Manual of Systematic Bacteriology, 2nd Edition, Volume 1. The archaea and the deeply branching and phototrophic bacteria. Springer, New York. 421.

Garrity, G.M. and J.G. Holt. 2001o. Phylum BVI. *Chloroflexi* phyl. nov. *In* Boone, Castenholz and Garrity (Editors), Bergey's Manual of Systematic Bacteriology, 2nd Edition, Volume 1. The archaea and the deeply branching and phototrophic bacteria. Springer, New York. 427.

Garrity, G.M. and J.G. Holt. 2001p. Phylum BVII. *Thermomicrobia* phyl. nov. *In* Boone, Castenholz and Garrity (Editors), Bergey's Manual of Systematic Bacteriology, 2nd Edition, Volume 1. The archaea and the deeply branching and phototrophic bacteria. Springer, New York. 447.

Garrity, G.M. and J.G. Holt. 2001q. Phylum BVIII. *Nitrospirae* phyl. nov. *In* Boone, Castenholz and Garrity (Editors), Bergey's Manual of Systematic Bacteriology, 2nd Edition, Volume 1. The archaea and the deeply branching and phototrophic bacteria. Springer, New York. 451.

Garrity, G.M. and J.G. Holt. 2001r. Phylum BXI. *Chlorobi* phyl. nov. *In* Boone, Castenholz and Garrity (Editors), Bergey's Manual of Systematic Bacteriology, 2nd Edition, Volume 1. The archaea and the deeply branching and phototrophic bacteria. Springer, New York. 601.

Garrity, G.M. and J.G. Holt. 2001s. The Road Map to the Manual. *In* Boone, Castenholz and Garrity (Editors), Bergey's Manual of Systematic Bacteriology, 2nd Edition, Volume 1. The archaea and the deeply branching and phototrophic bacteria. Springer, New York. 119–166.

Garrity, G.M. and J.G. Holt. 2002. *In* Validation of publication of new names and new combinations previously effectively published outside the IJSEM. List No. 85. Int. J. Syst. Evol. Microbiol. *52*: 685–690.

Garrity, G.M., M. Winters, A.W. Kuo and D.B. Searles. 2002. Bergey's Taxonomic Outline: Release 2.0: http://dx.doi.org/10.1007/bergeysoutline200310.

Garrity, G.M., M. Winters and D.B. Searles. 2001. Bergey's Taxonomic Outline: Release 1.0: http://dx.doi.org/10.1007/bergeysoutline200310.

Germgard, U., A. Teder and D. Tormund. 1981. Chlorate formation during chlorine dioxide bleaching of softwood kraft pulp. Paperi ja Puu. *3*: 127–133.

Gest, H. 1999. Gest's postulates. ASM News. *65*: 123.

Gibbons, N.E. 1974a. Family V. *Halobacteriaceae*. *In* Buchanan and Gibbons (Editors), Bergey's Manual of Determinative Bacteriology, 8th Ed., The Williams & Wilkins Co., Baltimore. pp. 269–273.

Gibbons, N.E. 1974b. Reference collections of bacteria—the need and requirements for type and neotype strains. *In* Buchanan and Gibbons (Editors), Bergey's Manual of Determinative Bacteriology, 8th Ed., The Williams & Wilkins Co., Baltimore. pp. 14–17.

Gibbons, N.E. and R.G.E. Murray. 1978. Proposals concerning the higher taxa of bacteria. Int. J. Syst. Bacteriol. *28*: 1–6.

Gibbons, N.E., K.B. Pattee and J.G. Holt (Editors). 1981. Supplement to Index Bergeyana, The Williams & Wilkins Co., Baltimore.

Gieszczykiewicz, M. 1939. Zagadniene systematihki w bakteriologii — Zür Frage der Bakterien-Systematic. Bull. Acad. Polon. Sci., Ser. Sci. Biol. *1*: 9—27.

Gillis, M., J. Dejonghe, A. Smet, G. Onghenae and J. De Ley. 1982. Intra- and intergeneric similarities of the ribosomal ribonucleic acid cistrons in the *Rhodospirillaceae*. Abstract A-16. IV. Intern. Symp. Photosynthetic Procaryotes. Bombannes-Bordeaux

Giovannoni, S.J. 1991. The polymerase chain reaction. *In* Stackebrandt and Goodfellow (Editors), Nucleic Acid Techniques in Bacterial Systematics, John Wiley and Sons, Chichester. pp. 177–203.

Giovannoni, S.J., T.B. Britschgi, C.L. Moyer and K.G. Field. 1990. Genetic diversity in Sargasso Sea bacterioplankton. Nature *345*: 60–63.

Giovannoni, S.J., M.S. Rappé, K.L. Vergin and N.L. Adair. 1996. 16S rRNA genes reveal stratified open ocean bacterioplankton populations related to the green non-sulfur bacteria. Proc. Natl. Acad. Sci. U.S.A. *93*: 7979–7984.

Glaeser, J. and J. Overmann. 1999. Selective enrichment and characterization of *Roseospirillum parvum*, gen. nov. and sp. nov., a new purple nonsulfur bacterium with unusual light absorption properties. Arch. Microbiol. *171*: 405–416.

Glaeser, J. and J. Overmann. 2001. *In* Validation of the publication of new names and new combinations previously effectively published outside the IJSB. List No. 80. Int. J. Syst. Evol. Bacteriol. *51*: 793–794.

Glöckner, F.O., R.I. Amann, A. Alfreider, J. Pernthaler, R. Psenner, K. Trebesius and K.H. Schleifer. 1996. An *in situ* hybridization protocol for detection and identification of planktonic bacteria. Syst. Appl. Microbiol. *19*: 403–406.

Godoy, F., M. Vancanneyt, M. Martinez, A. Steinbuchel, J. Swings and

B.H. Rehm. 2003. *Sphingopyxis chilensis* sp. nov., a chlorophenol-degrading bacterium that accumulates polyhydroxyalkanoate, and transfer of *Sphingomonas alaskensis* to *Sphingopyxis alaskensis* comb. nov. Int. J. Syst. Evol. Microbiol. *53*: 473–477.

Gogarten, J.P., W.F. Doolittle and J.G. Lawrence. 2002. Prokaryotic evolution in light of gene transfer. Mol. Biol. Evol. *19*: 2226–2238.

Golovacheva, R.S. 1976. Thermophilic nitrifying bacteria from hot springs. Mikrobiologiia *45*: 298–301.

Golyshin, P.N., T.N. Chernikova, W.R. Abraham, H. Lunsdorf, K.N. Timmis and M.M. Yakimov. 2002. *Oleiphilaceae* fam. nov., to include *Oleiphilus messinensis* gen. nov., sp. nov., a novel marine bacterium that obligately utilizes hydrocarbons. Int. J. Syst. Evol. Microbiol. *52*: 901–911.

Golyshina, O.V., T.A. Pivovarova, G.I. Karavaiko, T.F. Kondrateva, E.R. Moore, W.R. Abraham, H. Lunsdorf, K.N. Timmis, M.M. Yakimov and P.N. Golyshin. 2000. *Ferroplasma acidiphilum* gen. nov., sp. nov., an acidophilic, autotrophic, ferrous-iron-oxidizing, cell-wall-lacking, mesophilic member of the *Ferroplasmaceae* fam. nov., comprising a distinct lineage of the *Archaea*. Int. J. Syst. Evol. Microbiol. *50*: 997–1006.

Gomez-Gil, B., F.L. Thompson, C.C. Thompson and J. Swings. 2003a. *Vibrio pacinii* sp. nov., from cultured aquatic organisms. Int. J. Syst. Evol. Microbiol. *53*: 1569–1573.

Gomez-Gil, B., F.L. Thompson, C.C. Thompson and J. Swings. 2003b. *Vibrio rotiferianus* sp. nov., isolated from cultures of the rotifer *Brachionus plicatilis*. Int. J. Syst. Evol. Microbiol. *53*: 239–243.

González, J.M., J.S. Covert, W.B. Whitman, J.R. Henriksen, F. Mayer, B. Scharf, R. Schmitt, A. Buchan, J.A. Fuhrman, R.P. Kiene and M.A. Moran. 2003. *Silicibacter pomeroyi* sp. nov. and *Roseovarius nubinhibens* sp. nov., dimethylsulfoniopropionate-demethylating bacteria from marine environments. Int. J. Syst. Evol. Microbiol. *53*: 1261–1269.

González, J.M., D. Sheckells, M. Viebahn, D. Krupatkina, K.M. Borges and F.T. Robb. 1999. *Thermococcus waiotapuensis* sp. nov., an extremely thermophilic archaeon isolated from a freshwater hot spring. Arch. Microbiol. *172*: 95–101.

González, J.M., D. Sheckells, M. Viebahn, D. Krupatkina, K.M. Borges and F.T. Robb. 2001. *In* Validation of publication of new names and new combinations previously effectively published outside the IJSEM. List No. 80. Int. J. Syst. Evol. Microbiol. *51*: 793–794.

Goodfellow, M., J. Chun, E. Stackebrandt and R.M. Kroppenstedt. 2002. Transfer of *Tsukamurella wratislaviensis* Goodfellow et al. 1995 to the genus *Rhodococcus* as *Rhodococcus wratislaviensis* comb. nov. Int. J. Syst. Evol. Microbiol. *52*: 749–755.

Goodfellow, M., S.B. Kim, D.E. Minnikin, D. Whitehead, Z.H. Zhou and A.D. Mattinson-Rose. 2001. *Amycolatopsis sacchari* sp. nov., a moderately thermophilic actinomycete isolated from vegetable matter. Int. J. Syst. Evol. Microbiol. *51*: 187–193.

Goorissen, H.P., H.T. Boschker, A.J. Stams and T.A. Hansen. 2003. Isolation of thermophilic *Desulfotomaculum* strains with methanol and sulfite from solfataric mud pools, and characterization of *Desulfotomaculum solfataricum* sp. nov. Int. J. Syst. Evol. Microbiol. *53*: 1223–1229.

Gordon, R.E. 1967. The taxonomy of soil bacteria. *In* Gray and Parkinson (Editors), The Ecology of Soil Bacteria. An International Symposium, University of Toronto Press, Toronto. pp. 293–321.

Goreau, T.J., W.A. Kaplan, S.C. Wofsy, M.B. McElroy, F.W. Valois and S.W. Watson. 1980. Production of NO_2^- and N_2O by nitrifying bacteria at reduced concentrations of oxygen. Appl. Environ. Microbiol. *40*: 526–532.

Goris, J., P. De Vos, T. Coenye, B. Hoste, D. Janssens, H. Brim, L. Diels, M. Mergeay, K. Kersters and P. Vandamme. 2001. Classification of metal-resistant bacteria from industrial biotopes as *Ralstonia campinensis* sp. nov., *Ralstonia metallidurans* sp. nov. and *Ralstonia basilensis* Steinle et al. 1998 emend. Int. J. Syst. Evol. Microbiol. *51*: 1773–1782.

Goris, J., W. Dejonghe, E. Falsen, E. De Clerck, B. Geeraerts, A. Willems, E.M. Top, P. Vandamme and P. De Vos. 2002. Diversity of transconjugants that acquired plasmid pJP4 or pEMT1 after inoculation of a donor strain in the A- and B-horizon of an agricultural soil and description of *Burkholderia hospita* sp. nov. and *Burkholderia terricola* sp. nov. Syst. Appl. Microbiol. *25*: 340–352.

Goris, J., W. Dejonghe, E. Falsen, E. De Clerck, B. Geeraerts, A. Willems, E.M. Top, P. Vandamme and P. De Vos. 2003. *In* Validation of publication of new names and new combinations previously effectively published outside the IJSEM. List No. 89. Int. J. Syst. Evol. Microbiol. *53*: 1–2.

Gorshkova, N.M., E.P. Ivanova, A.F. Sergeev, N.V. Zhukova, Y. Alexeeva, J.P. Wright, D.V. Nicolau, V.V. Mikhailov and R. Christen. 2003. *Marinobacter excellens* sp. nov., isolated from sediments of the Sea of Japan. Int. J. Syst. Evol. Microbiol. *53*: 2073–2078.

Gosink, J.J., R.P. Herwig and J.T. Staley. 1997. *Octadecabacter arcticus* gen. nov., sp. nov., nonpigmented, psychrophilic gas vacuolate bacteria from polar sea ice and water. Syst. Appl. Microbiol. *20*: 356–365.

Gosink, J.J., C.R. Woese and J.T. Staley. 1998. *Polaribacter* gen. nov., with three new species, *P. irgensii* sp. nov., *P. franzmannii* sp. nov. and *P. filamentus* sp. nov., gas vacuolate polar marine bacteria of the *Cytophaga-Flavobacterium-Bacteroides* group and reclassification of *'Flectobacillus glomeratus'* as *Polaribacter glomeratus* comb. nov. Int. J. Syst. Bacteriol. *48*: 223–235.

Gossner, A.S., R. Devereux, N. Ohnemuller, G. Acker, E. Stackebrandt and H.L. Drake. 1999. *Thermicanus aegyptius* gen. nov., sp. nov., isolated from oxic soil, a fermentative microaerophile that grows commensally with the thermophilic acetogen *Moorella thermoacetica*. Appl. Environ. Microbiol. *65*: 5124–5133.

Gossner, A.S., R. Devereux, N. Ohnemuller, G. Acker, E. Stackebrandt and H.L. Drake. 2000. *In* Validation of publication of new names and new combinations previously effectively published outside the IJSEM. List No. 73. Int. J. Syst. Evol. Microbiol. *50*: 423–424.

Goto, K., H. Matsubara, K. Mochida, T. Matsumura, Y. Hara, M. Niwa and K. Yamasato. 2002. *Alicyclobacillus herbarius* sp. nov., a novel bacterium containing ω-cycloheptane fatty acids, isolated from herbal tea. Int. J. Syst. Evol. Microbiol. *52*: 109–113.

Goto, K., K. Mochida, M. Asahara, M. Suzuki, H. Kasai and A. Yokota. 2003. *Alicyclobacillus pomorum* sp. nov., a novel thermo-acidophilic, endospore-forming bacterium that does not possess omega-alicyclic fatty acids, and emended description of the genus *Alicyclobacillus*. Int. J. Syst. Evol. Microbiol. *53*: 1537–1544.

Götz , D., A. Banta, T.J. Beveridge, A.I. Rushdi, B.R. Simoneit and A.L. Reysenbach. 2002. *Persephonella marina* gen. nov., sp. nov. and *Persephonella guaymasensis* sp. nov., two novel, thermophilic, hydrogen-oxidizing microaerophiles from deep-sea hydrothermal vents. Int. J. Syst. Evol. Microbiol. *52*: 1349–1359.

Gould, S.J. 1996. Full house: the spread of excellence from Plato to Darwin, Harmony Books, New York.

Goyache, J., C. Ballesteros, A.I. Vela, M.D. Collins, V. Briones, R.A. Hutson, J. Potti, P. Garcia-Borboroglu, L. Dominguez and J.F. Fernandez-Garayzabal. 2003a. *Corynebacterium sphenisci* sp. nov., isolated from wild penguins. Int. J. Syst. Evol. Microbiol. *53*: 1009–1012.

Goyache, J., A.I. Vela, M.D. Collins, C. Ballesteros, V. Briones, J. Moreno, P. Yorio, L. Dominguez, R. Hutson and J.F. Fernandez-Garayzabal. 2003b. *Corynebacterium spheniscorum* sp. nov., isolated from the cloacae of wild penguins. Int. J. Syst. Evol. Microbiol. *53*: 43–46.

Grant, W.D., M. Kamekura, T.J. McGenity and A. Ventosa. 2001. Class III. *Halobacteria*, class. nov. *In* Boone, Castenholz and Garrity (Editors), Bergey's Manual of Systematic Bacteriology, 2nd Edition, Volume 1. The archaea and the deeply branching and phototrophic bacteria. Springer, New York. 294.

Grant, W.D., M. Kamekura, T.J. McGenity and A. Ventosa. 2002. *In* Validation of publication of new names and new combinations previously effectively published outside the IJSEM. List No. 85. Int. J. Syst. Evol. Microbiol. *52*: 685–690.

Grant, W.D. and H. Larsen. 1989. *In* Validation of the publication of new names and new combinations previously effectively published outside the IJSB. List No. 31. Int. J. Syst. Bacteriol. *39*: 495–497.

Greetham, H.L., G.R. Gibson, C. Giffard, H. Hippe, B. Merkhoffer, U. Steiner, E. Falsen and M.D. Collins. 2003. *Clostridium colicanis* sp. nov., from canine faeces. Int. J. Syst. Evol. Microbiol. *53*: 259–262.

Greuter, W., D.L. Hawksworth, J. McNeill, M.A. Mayo, A. Minelli, P.H.A. Sneath, B.J. Tindall, P. Trehane and P. Tubbs. 1998. Draft Biocode (1997): the prospective international rules for the scientific naming of organisms. Taxon. *47*: 129–150.

Grimm, D., H. Merkert, W. Ludwig, K.H. Schleifer, J. Hacker and B.C. Brand. 1998. Specific detection of *Legionella pneumophila*: construction of a new 16S rRNA-targeted oligonucleotide probe. Appl. Environ. Microbiol. *64*: 2686–2690.

Grimont, P.A.D., F. Grimont, N. Desplaces and P. Tchen. 1985. DNA probe specific for *Legionella pneumophila*. J. Clin. Microbiol. *21*: 431–437.

Grimont, P.A.D., M.Y. Popoff, F. Grimont, C. Coynault and M. Lemelin. 1980. Reproducibility and correlation study of three deoxyribonucleic acid hybridization procedures. Curr. Microbiol. *4*: 325–330.

Groffman, P.M. 1987. Nitrification and denitrification in soil: a comparison of enzyme assay, incubation and enumeration methods. Plant Soil *97*: 445–450.

Groot Obbink, D.J., L.J. Ritchie, F.H. Cameron, J.S. Mattick and V.P. Ackerman. 1985. Construction of a gentamicin resistance gene probe for epidemiological studies. Antimicrob. Agents Chemother. *28*: 96–102.

Grote, R., L.N. Li, J. Tamaoka, C. Kato, K. Horikoshi and G. Antranikian. 1999. *Thermococcus siculi* sp. nov., a novel hyperthermophilic archaeon isolated from a deep-sea hydrothermal vent at the Mid-Okinawa Trough. Extremophiles *3*: 55–62.

Grote, R., L.N. Li, J. Tamaoka, C. Kato, K. Horikoshi and G. Antranikian. 2000. *In* Validation of publication of new names and new combinations previously effectively published outside the IJSEM. List No. 74. Int. J. Syst. Evol. Microbiol. *50*: 949–950.

Groth, I., P. Schumann, B. Schutze, K. Augsten and E. Stackebrandt. 2002. *Knoellia sinensis* gen. nov., sp. nov. and *Knoellia subterranea* sp. nov., two novel actinobacteria isolated from a cave. Int. J. Syst. Evol. Microbiol. *52*: 77–84.

Groth, I., P. Schumann, N. Weiss, B. Schuetze, K. Augsten and E. Stackebrandt. 2001. *Ornithinimicrobium humiphilum* gen. nov., sp. nov., a novel soil actinomycete with L-ornithine in the peptidoglycan. Int. J. Syst. Evol. Microbiol. *51*: 81–87.

Groth, I., B. Schutze, T. Boettcher, C.B. Pullen, C. Rodriguez, E. Leistner and M. Goodfellow. 2003. *Kitasatospora putterlickiae* sp. nov., isolated from rhizosphere soil, transfer of *Streptomyces kifunensis* to the genus *Kitasatospora* as *Kitasatospora kifunensis* comb. nov., and emended description of *Streptomyces aureofaciens* Duggar 1948. Int. J. Syst. Evol. Microbiol. *53*: 2033–2040.

Groudieva, T., R. Grote and G. Antranikian. 2003. *Psychromonas arctica* sp. nov., a novel psychrotolerant, biofilm-forming bacterium isolated from Spitzbergen. Int. J. Syst. Evol. Microbiol. *53*: 539–545.

Grundmann, G.L. and P. Normand. 2000. Microscale diversity of the genus *Nitrobacter* in soil on the basis of analysis of genes encoding rRNA. Appl. Environ. Microbiol. *66*: 4543–4546.

Grunstein, M. and D.S. Hogness. 1975. Colony hybridization: a method for the isolation of cloned DNAs that contain a specific gene. Proc. Natl. Acad. Sci. U.S.A. *72*: 3961–3965.

Gugliandolo, C., T.L. Maugeri, D. Caccamo and E. Stackebrandt. 2003a. *Bacillus aeolius* sp. nov. a novel thermophilic, halophilic marine *Bacillus* species from Eolian islands (Italy). Syst. Appl. Microbiol. *26*: 172–176.

Gugliandolo, C., T.L. Maugeri, D. Caccamo and E. Stackebrandt. 2003b. *In* Validation of publication of new names and new combinations previously effectively published outside the IJSEM. List No. 94. Int. J. Syst. Evol. Microbiol. *53*: 1701–1702.

Gumaelius, L., G. Magnusson, B. Pettersson and G. Dalhammar. 2001. *Comamonas denitrificans* sp. nov., an efficient denitrifying bacterium isolated from activated sludge. Int. J. Syst. Evol. Microbiol. *51*: 999–1006.

Gupta, R.S. 1996. Evolutionary relationships of chaperonins. *In* Ellis (Editor), The Chaperonins, Academic Press, New York. pp. 27–64.

Gupta, R.S. 1998. Protein phylogenies and signature sequences: a reappraisal of evolutionary relationships among archaebacteria, eubacteria, and eukaryotes. Microbiol. Mol. Biol. Rev. *62*: 1435–1491.

Gupta, R.S. 2000. The phylogeny of proteobacteria: relationships to other eubacterial phyla and eukaryotes. FEMS Microbiol Rev. *24*: 367–402.

Gupta, R.S. 2003. Evolutionary relationships among photosynthetic bacteria. Photosynth. Res. *76*: 173–183.

Gupta, R.S., T. Mukhtar and B. Singh. 1999. Evolutionary relationships among photosynthetic prokaryotes (*Heliobacterium chlorum*, *Chloroflexus aurantiacus*, cyanobacteria, *Chlorobium tepidum* and proteobacteria): implications regarding the origin of photosynthesis. Mol. Microbiol. *32*: 893–906.

Gürtler, V., R. Smith, B.C. Mayall, G. Potter-Reinemann, E. Stackebrandt and R.M. Kroppenstedt. 2001. *Nocardia veterana* sp. nov., isolated from human bronchial lavage. Int. J. Syst. Evol. Microbiol. *51*: 933–936.

Guschin, D.Y., B.K. Mobarry, D. Proudnikov, D.A. Stahl, B.E. Rittmann and A.D. Mirzabekov. 1997. Oligonucleotide microchips as genosensors for determinative and environmental studies in microbiology. Appl. Environ. Microbiol. *63*: 2397–2402.

Guyoneaud, R., S. Moune, C. Eatock, V. Bothorel, A.S. Hirschler-Rea, J. Willison, R. Duran, W. Liesack, R. Herbert, R. Matheron and P. Caumette. 2002. Characterization of three spiral-shaped purple nonsulfur bacteria isolated from coastal lagoon sediments, saline sulfur springs, and microbial mats: emended description of the genus *Roseospira* and description of *Roseospira marina* sp. nov., *Roseospira navarrensis* sp. nov., and *Roseospira thiosulfatophila* sp. nov. Arch. Microbiol. *178*: 315–324.

Guyoneaud, R., S. Moune, C. Eatock, V. Bothorel, A.S. Hirschler-Rea, J. Willison, R. Duran, W. Liesack, R. Herbert, R. Matheron and P. Caumette. 2003. *In* Validation of publication of new names and new combinations previously effectively published outside the IJSEM. List No. 94. Int. J. Syst. Evol. Microbiol. *53*: 1701–1702.

Gyobu, Y. and S. Miyadoh. 2001. Proposal to transfer *Actinomadura carminata* to a new subspecies of the genus *Nonomuraea* as *Nonomuraea roseoviolacea* subsp. *carminata* comb. nov. Int. J. Syst. Evol. Microbiol. *51*: 881–889.

Hackenthal, E. 1965. Die Reduktion von Pperchlorat durch Bacterien. II. Die Identitat der Nitratreduktase und des Perchlorat reduzierenden Enzyms aus *B. cereus*. Biochem. Pharm. *14*: 1313–1324.

Hackenthal, E., W. Mannheim, R. Hackenthal and R. Becher. 1964. Die Reduktion von Perchlorat durch Bakterien. I. Untersuchungen an Intaken Zellen. Biochem. Pharmacol. *13*: 195–206.

Hahn, D., R.I. Amann, W. Ludwig, A.D.L. Akkermans and K.H. Schleifer. 1992. Detection of microorganisms in soil after *in situ* hybridization with rRNA-targeted, fluorescently labelled oligonucleotides. J. Gen. Microbiol. *138*: 879–887.

Hahn, D., R.I. Amann and J. Zeyer. 1993. Detection of mRNA in *Streptomyces* cells by whole-cell hybridization with digoxigenin-labeled probes. Appl. Environ. Microbiol. *59*: 2753–2757.

Hall, B.D. and S. Spiegelman. 1961. Sequence complementarity of T2-DNA and T2-specific RNA. Proc. Natl. Acad. Sci. U.S.A. *47*: 137–146.

Hall, G.H. 1984. Measurement of nitrification rates in lake sediments—comparison of the nitrification inhibitors nitrapyrin and allylthiourea. Microb. Ecol. *10*: 25–36.

Hall, I.C. 1927. Some fallacious tendencies in bacteriological taxonomy. J. Bacteriol. *13*: 245–253.

Hall, V., M.D. Collins, R. Hutson, E. Falsen and B.I. Duerden. 2002. *Actinomyces cardiffensis* sp. nov. from human clinical sources. J. Clin. Microbiol. *40*: 3427–3431.

Hall, V., M.D. Collins, R. Hutson, E. Falsen and B.I. Duerden. 2003a. *In* Validation of publication of new names and new combinations previously effectively published outside the IJSEM. List No. 89. Int. J. Syst. Evol. Microbiol. *53*: 1–2.

Hall, V., M.D. Collins, R.A. Hutson, E. Falsen, E. Inganas and B.I. Duerden. 2003b. *Actinobaculum urinale* sp. nov., from human urine. Int. J. Syst. Evol. Microbiol. *53*: 679–682.

Hall, V., M.D. Collins, R.A. Hutson, E. Inganas, E. Falsen and B.I. Duerden. 2003c. *Actinomyces oricola* sp. nov., from a human dental abscess. Int. J. Syst. Evol. Microbiol. *53*: 1515–1518.

Hall, V., M.D. Collins, R. Hutson, E. Inganas, E. Falsen and B.I. Duerden. 2003d. *Actinomyces vaccimaxillae* sp. nov., from the jaw of a cow. Int. J. Syst. Evol. Microbiol. *53*: 603–606.

Hall, V., M.D. Collins, R.A. Hutson, P.A. Lawson, E. Falsen and B.I. Duerden. 2003e. *Corynebacterium atypicum* sp. nov., from a human clinical source, does not contain corynomycolic acids. Int. J. Syst. Evol. Microbiol. *53*: 1065–1068.

Hall, V., M.D. Collins, P.A. Lawson, E. Falsen and B.I. Duerden. 2003f. *Actinomyces nasicola* sp. nov., isolated from a human nose. Int. J. Syst. Evol. Microbiol. *53*: 1445–1448.

Hall, V., M.D. Collins, P.A. Lawson, R.A. Hutson, E. Falsen, E. Inganas and B. Duerden. 2003g. Characterization of some *Actinomyces*-like isolates from human clinical sources: Description of *Varibaculum cambriensis* gen. nov., sp. nov. J. Clin. Microbiol. *41*: 640–644.

Hall, V., M.D. Collins, P.A. Lawson, R.A. Hutson, E. Falsen, E. Inganas and B. Duerden. 2003h. *In* Validation of publication of new names and new combinations previously effectively published outside the IJSEM. List No. 91. Int. J. Syst. Evol. Microbiol. *53*: 627–628.

Hamid, M.E., L. Maldonado, G.S.S. Eldin, M.F. Mohamed, N.S. Saeed and M. Goodfellow. 2001a. *Nocardia africana* sp. nov., a new pathogen isolated from patients with pulmonary infections. J. Clin. Microbiol. *39*: 625–630.

Hamid, M.E., L. Maldonado, G.S.S. Eldin, M.F. Mohamed, N.S. Saeed and M. Goodfellow. 2001b. *In* Validation of publication of new names and new combinations previously effectively published outside the IJSEM. List No. 81. Int. J. Syst. Evol. Microbiol. *51*: 1229.

Han, S.K., O.I. Nedashkovskaya, V.V. Mikhailov, S.B. Kim and K.S. Bae. 2003a. *Salinibacterium amurskyense* gen. nov., sp. nov., a novel genus of the family *Microbacteriaceae* from the marine environment. Int. J. Syst. Evol. Microbiol. *53*: 2061–2066.

Han, X.Y., A.S. Pham, J.J. Tarrand, K.V. Rolston, L.O. Helsel and P.N. Levett. 2003b. Bacteriologic characterization of 36 strains of *Roseomonas* species and proposal of *Roseomonas mucosa* sp. nov. and *Roseomonas gilardii* subsp. *rosea* subsp. nov. Am. J. Clin. Pathol. *120*: 256–264.

Han, X.Y., A.S. Pham, J.J. Tarrand, K.V. Rolston, L.O. Helsel and P.N. Levett. 2003c. *In* Validation of publication of new names and new combinations previously effectively published outside the IJSEM. List No. 94. Int. J. Syst. Evol. Microbiol. *53*: 1701–1702.

Hanada, S., Y. Kawase, A. Hiraishi, S. Takaichi, K. Matsuura, K. Shimada and K.V.P. Nagashima. 1997. *Porphyrobacter tepidarius* sp. nov., a moderately thermophilic aerobic photosynthetic bacteium isolated from a hot spring. Int. J. Syst. Bacteriol. *47*: 408–413.

Hanada, S., W.T. Liu, T. Shintani, Y. Kamagata and K. Nakamura. 2002a. *Tetrasphaera elongata* sp. nov., a polyphosphate-accumulating bacterium isolated from activated sludge. Int. J. Syst. Evol. Microbiol. *52*: 883–887.

Hanada, S., S. Takaichi, K. Matsuura and K. Nakamura. 2002b. *Roseiflexus castenholzii* gen. nov., sp. nov., a thermophilic, filamentous, photosynthetic bacterium that lacks chlorosomes. Int. J. Syst. Evol. Microbiol. *52*: 187–193.

Hankinson, T.R. and E.L. Schmidt. 1984. Examination of an acid forest soil for ammonia-oxidizing and nitrite-oxidizing autotrophic bacteria. Can. J. Microbiol. *30*: 1125–1132.

Hankinson, T.R. and E.L. Schmidt. 1988. An acidophilic and a neutrophilic *Nitrobacter* strain isolated from the numerically predominant nitrite-oxidizing population of an acid forest soil. Appl. Environ. Microbiol. *54*: 1536–1540.

Hansmann, S. and W. Martin. 2000. Phylogeny of 33 ribosomal and six other proteins encoded in an ancient gene cluster that is conserved across prokaryotic genomes: influence of excluding poorly alignable sites from analysis. Int. J. Syst. Evol. Microbiol. *50*: 1655–1663.

Hara, T., A.P. Villalobos, Y. Fukumori and T. Yamanaka. 1991. Purification and characterization of ATPase from *Nitrobacter winogradskyi*. FEMS Microbiol. Lett. *82*: 49–54.

Harashima, K., K. Kawazoe, I. Yoshida and H. Kamata. 1987. Light-stimultated aerobic growth of *Erythrobacter* species Och-114. Plant Cell Physiol. *28*: 365–374.

Harashima, K., T. Shiba, T. Totsuka, U. Simidu and N. Taga. 1978. Occurrence of bacteriochlorophyll-*a* in a strain of an aerobic heterotrophic bacterium. Agr. Biol. Chem. Tokyo. *42*: 1627–1628.

Harashima, K. and K. Takamiya. 1989. Photosynthesis and photosynthetic apparatus. *In* Harashima, Shiba and Murata (Editors), Aerobic Photosynthetic Bacteria, Springer-Verlag, Berlin. 39–72.

Harms, H., H.P. Koops, H. Martiny and M. Wullenweber. 1981. D-Ribulose 1,5-bisphosphate carboxylase and polyhedral inclusion bodies in *Nitrosomonas* spec. Arch. Microbiol. *128*: 280–281.

Harmsen, H.J.M., A.D.L. Akkermans, A.J.M. Stams and W.M. de Vos. 1996a. Population dynamics of propionate-oxidizing bacteria under methanogenic and sulfidogenic conditions in anaerobic granular sludge. Appl. Environ. Microbiol. *62*: 2163–2168.

Harmsen, H.J.M., H.M.P. Kengen, A.D.L. Akkermans, A.J.M. Stams and W.M. de Vos. 1996b. Detection and localization of syntrophic propionate-oxidizing bacteria in granular sludge by *in situ* hybridization using 16S rRNA-based oligonucleotide probes. Appl. Environ. Microbiol. *62*: 1656–1663.

Harris, S., A. Ebert, E. Schütze, M. Diercks, E. Bock and J.M. Shively. 1988. Two different genes and gene products for the large subunit of ribulose-1,5-bisphosphate carboxylase/oxygenase (RuBisCOase) in *Nitrobacter hamburgensis*. FEMS Microbiol. Lett. *49*: 267–271.

Hastings, J.W. and E.P. Greenberg. 1999. Quorum sensing: the explanation of a curious phenomenon reveals a common characteristic of bacteria. J. Bacteriol. *181*: 2667–2668.

Hastings, R.C., M.T. Ceccherini, N. Miclaus, J.R. Saunders, M. Bazzicalupo and A.J. McCarthy. 1997. Direct molecular biological analysis of ammonia oxidizing bacteria populations in cultivated soil plots treated with swine manure. FEMS Microbiol. Ecol. *23*: 45–54.

Hatano, K., T. Nishii and H. Kasai. 2003. Taxonomic re-evaluation of whorl-forming *Streptomyces* (formerly *Streptoverticillium*) species by using phenotypes, DNA-DNA hybridization and sequences of *gyrB*, and proposal of *Streptomyces luteireticuli* (ex Katoh and Arai 1957) corrig., sp. nov., nom. rev. Int. J. Syst. Evol. Microbiol. *53*: 1519–1529.

Hatchikian, E.C., B. Ollivier and J-L. Garcia. 2001a. Class I. *Thermodesulfobacteria* class. nov. *In* Boone, Castenholz and Garrity (Editors), Bergey's Manual of Systematic Bacteriology, 2nd Edition, Volume 1. The archaea and the deeply branching and phototrophic bacteria. Springer, New York. 389.

Hatchikian, E.C., B. Ollivier and J-L. Garcia. 2001b. Family I. *Thermodesulfobacteriaceae* fam. nov. *In* Boone, Castenholz and Garrity (Editors), Bergey's Manual of Systematic Bacteriology, 2nd Edition, Volume 1. The archaea and the deeply branching and phototrophic bacteria. Springer, New York. 390.

Hatchikian, E.C., B. Ollivier and J-L. Garcia. 2001c. Order I. *Thermodesulfobacteriales* ord. nov. *In* Boone, Castenholz and Garrity (Editors), Bergey's Manual of Systematic Bacteriology, 2nd Edition, Volume 1. The archaea and the deeply branching and phototrophic bacteria. Springer, New York. 389–390.

Hatchikian, E.C., B. Ollivier and J-L. Garcia. 2002. *In* Validation of publication of new names and new combinations previously effectively published outside the IJSEM. List No. 85. Int. J. Syst. Evol. Microbiol. *52*: 685–690.

Hattori, S., Y. Kamagata, S. Hanada and H. Shoun. 2000. *Thermacetogenium phaeum* gen. nov., sp. nov., a strictly anaerobic, thermophilic, syntrophic acetate-oxidizing bacterium. Int. J. Syst. Evol. Microbiol. *50*: 1601–1609.

Hauser, N.C., M. Vingron, M. Scheidler, B. Krems, K. Hellmuth, K.D. Entian and J.D. Hoheisel. 1998. Transcriptional profiling on all open reading frames of *Saccharomyces cerevisiae*. Yeast. *14*: 1209–1221.

Hawksworth, D.L. and J. McNeill. 1998. The International Committee on Bionomenclature (ICB), the draft BioCode (1997), and the IUBS resolution on bionomenclature. Taxon. *47*: 123–136.

Hayashi, K., J. Moriwaki, T. Sawabe, F.L. Thompson, J. Swings, N. Gudkovs, R. Christen and Y. Ezura. 2003. *Vibrio superstes* sp. nov., isolated from the gut of Australian abalones *Haliotis laevigata* and *Haliotis rubra*. Int. J. Syst. Evol. Microbiol. *53*: 1813–1817.

Hayatsu, M. and N. Kosuge. 1993. Effects of difference in fertilization treatments on nitrification activity in tea soils. Soil Sci. Plant Nutr. *39*: 373–378.

Head, I.M., W.D. Hiorns, T.M. Embley, A.J. McCarthy and J.R. Saunders.

1993. The phylogeny of autotrophic ammonia-oxidizing bacteria as determined by analysis of 16S ribosomal RNA gene sequences. J. Gen. Microbiol. *139*: 1147–1153.

Hedlund, B.P., J.J. Gosink and J.T. Staley. 1996. Phylogeny of *Prosthecobacter*, the fusiform caulobacters: members of a recently discovered division of the Bacteria. Int. J. Syst. Bacteriol. *46*: 960–966.

Hedlund, B.P., J.J. Gosink and J.T. Staley. 1997. *Verrucomicrobia* div. nov., a new division of the *Bacteria* containing three new species of *Prosthecobacter*. Antonie Leeuwenhoek *72*: 29–38.

Hedlund, B.P. and J.T. Staley. 2001. *Vibrio cyclotrophicus* sp. nov., a polycyclic aromatic hydrocarbon (PAH)-degrading marine bacterium. Int. J. Syst. Evol. Microbiol. *51*: 61–66.

Hedrick, D.B., J.B. Guckert and D.C. White. 1991. Archaebacterial ether lipid diversity analyzed by supercritical fluid chromatography: integration with a bacterial lipid protocol. J. Lipid Res. *32*: 659–666.

Henze, K., A. Badr, M. Wettern, R. Cerff and W. Martin. 1995. A nuclear gene of eubacterial origin in *Euglena gracilis* reflects cryptic endosymbioses during protist evolution. Proc. Natl. Acad. Sci. U.S.A. *92*: 9122–9126.

Hernandez-Eugenio, G., M.L. Fardeau, J.L. Cayol, B.K. Patel, P. Thomas, H. Macarie, J.L. Garcia and B. Ollivier. 2002a. *Clostridium thiosulfatireducens* sp. nov., a proteolytic, thiosulfate- and sulfur-reducing bacterium isolated from an upflow anaerobic sludge blanket (UASB) reactor. Int. J. Syst. Evol. Microbiol. *52*: 1461–1468.

Hernandez-Eugenio, G., M.L. Fardeau, J.L. Cayol, B.K. Patel, P. Thomas, H. Macarie, J.L. Garcia and B. Ollivier. 2002b. *Sporanaerobacter acetigenes* gen. nov., sp. nov., a novel acetogenic, facultatively sulfur-reducing bacterium. Int. J. Syst. Evol. Microbiol. *52*: 1217–1223.

Hernandez-Eugenio, G., M.L. Fardeau, B.K.C. Patel, H. Macarie, J.L. Garcia and B. Ollivier. 2000. *Desulfovibrio mexicanus* sp. nov., a sulfate-reducing bacterium isolated from an upflow anaerobic sludge blanket (UASB) reactor treating cheese wastewaters. Anaerobe *6*: 305–312.

Hernandez-Eugenio, G., M.L. Fardeau, B.K.C. Patel, H. Macarie, J.L. Garcia and B. Ollivier. 2001. *In* Validation of publication of new names and new combinations previously effectively published outside the IJSEM. List No. 79. Int. J. Syst. Evol. Microbiol. *51*: 263–265.

Hertel, C., W. Ludwig, M. Obst, R.F. Vogel, W.P. Hammes and K.H. Schleifer. 1991. 23S ribosomal RNA-targeted oligonucleotide probes for the rapid identification of meat lactobacilli. Syst. Appl. Microbiol. *14*: 173–177.

Heuer, H., M. Krsek, P. Baker, K. Smalla and E.M.H. Wellington. 1997. Analysis of actinomycete communities by specific amplification of genes encoding 16S rRNA and gel-electrophoretic separation in denaturing gradients. Appl. Environ. Microbiol. *63*: 3233–3241.

Heulin, T., M. Barakat, R. Christen, M. Lesourd, L. Sutra, G. De Luca and W. Achouak. 2003. *Ramlibacter tataouinensis* gen. nov., sp. nov., and *Ramlibacter henchirensis* sp. nov., cyst-producing bacteria isolated from subdesert soil in Tunisia. Int. J. Syst. Evol. Microbiol. *53*: 589–594.

Heyrman, J., A. Balcaen, P. De Vos, P. Schumann and J. Swings. 2002a. *Brachybacterium fresconis* sp. nov. and *Brachybacterium sacelli* sp. nov., isolated from deteriorated parts of a medieval wall painting of the chapel of Castle Herberstein (Austria). Int. J. Syst. Evol. Microbiol. *52*: 1641–1646.

Heyrman, J., A. Balcaen, P. De Vos and J. Swings. 2002b. *Halomonas muralis* sp. nov., isolated from microbial biofilms colonizing the walls and murals of the Saint-Catherine chapel (Castle Herberstein, Austria). Int. J. Syst. Evol. Microbiol. *52*: 2049–2054.

Heyrman, J., A. Balcaen, M. Rodriguez-Diaz, N.A. Logan, J. Swings and P. De Vos. 2003a. *Bacillus decolorationis* sp. nov., isolated from biodeteriorated parts of the mural paintings at the Servilia tomb (Roman necropolis of Carmona, Spain) and the Saint-Catherine chapel (Castle Herberstein, Austria). Int. J. Syst. Evol. Microbiol. *53*: 459–463.

Heyrman, J., N.A. Logan, H.J. Busse, A. Balcaen, L. Lebbe, M. Rodriguez-Diaz, J. Swings and P. De Vos. 2003b. *Virgibacillus carmonensis* sp. nov., *Virgibacillus necropolis* sp. nov. and *Virgibacillus picturae* sp. nov., three novel species isolated from deteriorated mural paintings, transfer of the species of the genus *Salibacillus* to *Virgibacillus*, as *Virgibacillus marismortui* comb. nov. and *Virgibacillus salexigens* comb. nov., and emended description of the genus *Virgibacillus*. Int. J. Syst. Evol. Microbiol. *53*: 501–511.

Hezayen, F.F., B.H. Rehm, B.J. Tindall and A. Steinbuchel. 2001. Transfer of *Natrialba asiatica* B1T to *Natrialba taiwanensis* sp. nov. and description of *Natrialba aegyptiaca* sp. nov., a novel extremely halophilic, aerobic, non-pigmented member of the *Archaea* from Egypt that produces extracellular poly(glutamic acid). Int. J. Syst. Evol. Microbiol. *51*: 1133–1142.

Hezayen, F.F., B.J. Tindall, A. Steinbuchel and B.H. Rehm. 2002. Characterization of a novel halophilic archaeon, *Halobiforma haloterrestris* gen. nov., sp. nov., and transfer of *Natronobacterium nitratireducens* to *Halobiforma nitratireducens* comb. nov. Int. J. Syst. Evol. Microbiol. *52*: 2271–2280.

Hicks, R.E., R.I. Amann and D.A. Stahl. 1992. Dual staining of natural bacterioplankton with 4′,6-diamidino-2-phenylindole and fluorescent oligonucleotide probes targeting kingdom-level 16S rRNA sequences. Appl. Environ. Microbiol. *58*: 2158–2163.

Hilario, E. and J.P. Gogarten. 1993. Horizontal transfer of ATPase genes—the tree of life becomes a net of life. Biosystems. *31*: 111–119.

Hilario, E. and J.P. Gogarten. 1998. The prokaryote-to-eukaryote transition reflected in the evolution of the V/F/A-ATPase catalytic and proteolipid subunits. J. Mol. Evol. *46*: 703–715.

Hiorns, W.D., R.C. Hastings, I.M. Head, A.J. McCarthy, J.R. Saunders, R.W. Pickup and G.H. Hall. 1995. Amplification of 16S ribosomal RNA genes of autotrophic ammonia-oxidizing bacteria demonstrates the ubiquity of *nitrosospiras* in the environment. Microbiology *141*: 2793–2800.

Hippe, H. 2000. *Leptospirillum* gen. nov. (ex Markosyan 1972), nom. rev., including *Leptospirillum ferrooxidans* sp. nov. (ex Markosyan 1972), nom. rev. and *Leptospirillum thermoferrooxidans* sp. nov. (Golovacheva et al. 1992). Int. J. Syst. Evol. Microbiol. *50*: 501–503.

Hippe, H., M. Vainshtein, G.I. Gogotova and E. Stackebrandt. 2003. Reclassification of *Desulfobacterium macestii* as *Desulfomicrobium macestii* comb. nov. Int. J. Syst. Evol. Microbiol. *53*: 1127–1130.

Hiraishi, A. 1994. Phylogenetic affiliations of *Rhodoferax fermentans* and related species of phototrophic bacteria as determined by automated 16S DNA sequencing. Curr. Microbiol. *28*: 25–28.

Hiraishi, A., Y. Hoshino and H. Kitamura. 1984. Isoprenoid quinone composition in the classification of *Rhodospirillaceae*. J. Gen. Appl. Microbiol. *30*: 197–210.

Hiraishi, A., Y. Hoshino and T. Satoh. 1991. *Rhodoferax fermentans* gen. nov., sp. nov., a phototrophic purple nonsulfur bacterium previously referred to as the "*Rhodocyclus gelatinosus*-like" group. Arch. Microbiol. *155*: 330–336.

Hiraishi, A. and H. Kitamura. 1984. Distribution of phototrophic purple nonsulfur bacteria in activated sludge systems and other aquatic environments. Bull. Jpn. Soc. Sci. Fish. *50*: 1929–1938.

Hiraishi, A., H. Kuraishi and K. Kawahara. 2000a. Emendation of the description of *Blastomonas natatoria* (Sly 1985) Sly and Cahill 1997 as an aerobic photosynthetic bacterium and reclassification of *Erythromonas ursincola* Yurkov et al. 1997 as *Blastomonas ursincola* comb. nov. Int. J. Syst. Evol. Microbiol. *50*: 1113–1118.

Hiraishi, A., Y. Matsuzawa, T. Kanbe and N. Wakao. 2000b. *Acidisphaera rubrifaciens* gen. nov., sp. nov., an aerobic bacteriochlorophyll-containing bacterium isolated from acidic environments. Int. J. Syst. Bacteriol. *50*: 1539–1546.

Hiraishi, A., K. Muramatsu and Y. Ueda. 1996. Molecular genetic analyses of *Rhodobacter azotoformans* sp. nov. and related species of phototrophic bacteria. Syst. Appl. Microbiol. *19*: 168–177.

Hiraishi, A., K.V.P. Nagashima, K. Matsuura, K. Shimada, S. Takaichi, N. Wakao and Y. Katayama. 1998. Phylogeny and photosynthetic features of *Thiobacillus acidophilus* and related acidophilic bacteria: its transfer to the genus *Acidiphilium* as *Acidiphilium acidophilum* comb. nov. Int. J. Syst. Bacteriol. *48*: 1389–1398.

Hiraishi, A. and K. Shimada. 2001. Aerobic anoxygenic photosynthetic bacteria with zinc-bacteriochlorophyll. J. Gen. Appl. Microbiol. *47*: 161–180.

Hiraishi, A. and Y. Ueda. 1994. Intrageneric structure of the genus *Rhodobacter*: transfer of *Rhodobacter sulfidophilus* and related marine species to the genus *Rhodovulum* gen. nov. Int. J. Syst. Bacteriol. *44*: 15–23.

Hiraishi, A. and Y. Ueda. 1995. Isolation and characterization of *Rhodovulum strictum* sp. nov., and some other purple nonsulfur bacteria from colored blooms in tidal and seawater pools. Int. J. Syst. Bacteriol. *45*: 319–326.

Hiraishi, A., Y. Yonemitsu, M. Matsushita, Y.K. Shin, H. Kuraishi and K. Kawahara. 2002a. Characterization of *Porphyrobacter sanguineus* sp. nov., an aerobic bacteriochlorophyll-containing bacterium capable of degrading biphenyl and dibenzofuran. Arch. Microbiol. *178*: 45–52.

Hiraishi, A., Y. Yonemitsu, M. Matsushita, Y.K. Shin, H. Kuraishi and K. Kawahara. 2002b. *In* Validation of publication of new names and new combinations previously effectively published outside the IJSEM. List No. 88. Int. J. Syst. Evol. Microbiol. *52*: 1915–1916.

Hirschler-Réa, A., R. Matheron, C. Riffaud, S. Moune, C. Eatock, R.A. Herbert, J.C. Willison and P. Caumette. 2003. Isolation and characterization of spirilloid purple phototrophic bacteria forming red layers in microbial mats of Mediterranean salterns: description of *Halorhodospira neutriphila* sp. nov. and emendation of the genus *Halorhodospira*. Int. J. Syst. Evol. Microbiol. *53*: 153–163.

Höfle, M.G. 1990. Transfer RNA as genotypic fingerprints of eubacteria. Arch. Microbiol. *153*: 299–304.

Höfle, M.G. 1991. Rapid genotyping of pseudomonads by using low-molecular-weight RNA profiles. *In* Galli, Silver and Witholt (Editors), *Pseudomonas* Molecular Biology and Biotechnology, American Society for Microbiology, Washington D.C. 116–126.

Hofstad, T., I. Olsen, E.R. Eribe, E. Falsen, M.D. Collins and P.A. Lawson. 2000. *Dysgonomonas* gen. nov. to accommodate *Dysgonomonas gadei* sp. nov., an organism isolated from a human gall bladder, and *Dysgonomonas capnocytophagoides* (formerly CDC group DF-3). Int. J. Syst. Evol. Microbiol. *50*: 2189–2195.

Holben, W.E. and D. Harris. 1995. DNA-based monitoring of total bacterial community structure in environmental samples. Mol. Ecol. *4*: 627–631.

Holben, W.E., K. Noto, T. Sumino and Y. Suwa. 1998. Molecular analysis of bacterial communities in a three compartment granular activated sludge system indicates community level control by incompatible nitrification processes. Appl. Environ. Microbiol. *64*: 2528–2532.

Holley, R.A., T.Y. Guan, M. Peirson and C.K. Yost. 2002. *Carnobacterium viridans* sp. nov., an alkaliphilic, facultative anaerobe isolated from refrigerated, vacuum-packed bologna sausage. Int. J. Syst. Evol. Microbiol. *52*: 1881–1885.

Hollocher, T.C., M.E. Tate and D.J. Nicholas. 1981. Oxidation of ammonia by *Nitrosomonas europaea*. Definite ^{18}O-tracer evidence that hydroxylamine formation involves a monooxygenase. J. Biol. Chem. *256*: 10834–10836.

Holt, J.G. (Editor). 1977. Shorter Bergey's Manual of Determinative Bacteriology, The Williams & Wilkins Co., Baltimore.

Holt, J.G. (Editor). 1984–1989. Bergey's Manual of Systematic Bacteriology, 1st Ed., Vols. 1–4, The Williams & Wilkins Co., Baltimore.

Holt, J.G., M.A. Bruns, B.J. Caldwell and C.D. Pease (Editors). 1992. Stedman's Bergey's Bacteria Words, The Williams & Wilkins Co., Baltimore.

Holt, J.G., N.R. Krieg, P.H.A. Sneath, J.T. Staley and S.T. Williams (Editors). 1994. Bergey's Manual of Determinative Bacteriology, 9th Ed., The Williams & Wilkins Co., Baltimore.

Hönerlage, W., D. Hahn and J. Zeyer. 1995. Detection of mRNA of *nprM* in *Bacillus megaterium* ATCC 14581 grown in soil by whole-cell hybridization. Arch. Microbiol. *163*: 235–241.

Hooper, A.B. 1969. Biochemical basis of obligate autotrophy in *Nitrosomonas europaea*. J. Bacteriol. *97*: 776–779.

Hooper, A.B. 1984. Ammonia oxidation and energy transduction in nitrifying bacteria. *In* Strohl and Tuovinen (Editors), Microbial Chemoautotrophy, Ohio State University Press, Columbus. 133–167.

Hooper, A.B. 1989. Biochemistry of the nitrifying lithoautotrophic bacteria. *In* Schlegel and Bowien (Editors), Autotrophic Bacteria, Sci. Tech. Publishers, Madison. pp. 239–265.

Hooper, A.B., J. Hansen and R. Bell. 1967. Characterization of glutamate dehydrogenase from the ammonia-oxidizing chemoautotroph *Nitrosomonas europaea*. J. Biol. Chem. *242*: 288–296.

Hooper, A.B. and K.R. Terry. 1973. Specific inhibitors of ammonia oxidation in *Nitrosomonas*. J. Bacteriol. *115*: 480–485.

Hooper, A.B. and K.R. Terry. 1974. Photoinactivation of ammonia oxidation in *Nitrosomonas*. J. Bacteriol. *119*: 899–906.

Hooper, A.B., T. Vannelli, D.J. Bergmann and D.M. Arciero. 1997. Enzymology of the oxidation of ammonia to nitrite by bacteria. Antonie Leeuwenhoek. *71*: 59–67.

Hoover, R.B., E.V. Pikuta, A.K. Bej, D. Marsic, W.B. Whitman, J. Tang and P. Krader. 2003. *Spirochaeta americana* sp. nov., a new haloalkaliphilic, obligately anaerobic spirochaete isolated from soda Mono Lake in California. Int. J. Syst. Evol. Microbiol. *53*: 815–821.

Horn, M., T.R. Fritsche, T. Linner, R.K. Gautom, M.D. Harzenetter and M. Wagner. 2002. Obligate bacterial endosymbionts of *Acanthamoeba* spp. related to the β-*Proteobacteria*: proposal of *Procabacter acanthamoebae* gen. nov., sp. nov. Int. J. Syst. Evol. Microbiol. *52*: 599–605.

Horn, M., M. Wagner, K.D. Muller, E.N. Schmid, T.R. Fritsche, K.H. Schleifer and R. Michel. 2000. *Neochlamydia hartmannellae* gen. nov., sp. nov. (*Parachlamydiaceae*), an endoparasite of the amoeba *Hartmannella vermiformis*. Microbiology-Sgm. *146*: 1231–1239.

Horn, M., M. Wagner, K.D. Muller, E.N. Schmid, T.R. Fritsche, K.H. Schleifer and R. Michel. 2001. *In* Validation of publication of new names and new combinations previously effectively published outside the IJSEM. List No. 81. Int. J. Syst. Evol. Microbiol. *51*: 1229.

Hougardy, A., B.J. Tindall and J.H. Klemme. 2000. *Rhodopseudomonas rhenobacensis* sp. nov., a new nitrate-reducing purple non-sulfur bacterium. Int. J. Syst. Evol. Microbiol. *50*: 985–992.

Hovanec, T.A. and E.F. DeLong. 1996. Comparative analysis of nitrifying bacteria associated with freshwater and marine aquaria. Appl. Environ. Microbiol. *62*: 2888–2896.

Hovanec, T.A., L.T. Taylor, A. Blakis and E.F. Delong. 1998. *Nitrospira*-like bacteria associated with nitrite oxidation in freshwater aquaria. Appl. Environ. Microbiol. *64*: 258–264.

Hoyles, L., E. Falsen, G. Foster and M.D. Collins. 2002a. *Actinomyces coleocanis* sp. nov., from the vagina of a dog. Int. J. Syst. Evol. Microbiol. *52*: 1201–1203.

Hoyles, L., E. Falsen, G. Foster, C. Pascual, C. Greko and M.D. Collins. 2000a. *Actinomyces canis* sp. nov., isolated from dogs. Int. J. Syst. Evol. Microbiol. *50*: 1547–1551.

Hoyles, L., E. Falsen, G. Foster, F. Rogerson and M.D. Collins. 2002b. *Arcanobacterium hippocoleae* sp. nov., from the vagina of a horse. Int. J. Syst. Evol. Microbiol. *52*: 617–619.

Hoyles, L., E. Falsen, G. Holmstrom, A. Persson, B. Sjoden and M.D. Collins. 2001a. *Actinomyces suimastitidis* sp. nov., isolated from pig mastitis. Int. J. Syst. Evol. Microbiol. *51*: 1323–1326.

Hoyles, L., E. Falsen, C. Pascual, B. Sjoden, G. Foster, D. Henderson and M.D. Collins. 2001b. *Actinomyces catuli* sp. nov., from dogs. Int. J. Syst. Evol. Microbiol. *51*: 679–682.

Hoyles, L., G. Foster, E. Falsen and M.D. Collins. 2000b. Characterization of a *Gemella*-like organism isolated from an abscess of a rabbit: description of *Gemella cunicula* sp. nov. Int. J. Syst. Evol. Microbiol. *50*: 2037–2041.

Hoyles, L., G. Foster, E. Falsen, L.F. Thomson and M.D. Collins. 2001c. *Facklamia miroungae* sp. nov., from a juvenile southern elephant seal (*Mirounga leonina*). Int. J. Syst. Evol. Microbiol. *51*: 1401–1403.

Hoyles, L., E. Inganas, E. Falsen, M. Drancourt, N. Weiss, A.L. McCartney and M.D. Collins. 2002c. *Bifidobacterium scardovii* sp. nov., from human sources. Int. J. Syst. Evol. Microbiol. *52*: 995–999.

Hoyles, L., P.A. Lawson, G. Foster, E. Falsen, M. Ohlen, J.M. Grainger and M.D. Collins. 2000c. *Vagococcus fessus* sp. nov., isolated from a seal and a harbour porpoise. Int. J. Syst. Evol. Microbiol. *50*: 1151–1154.

Hoyles, L., C. Pascual, E. Falsen, G. Foster, J.M. Grainger and M.D. Collins. 2001d. *Actinomyces marimammalium* sp. nov., from marine mammals. Int. J. Syst. Evol. Microbiol. *51*: 151–156.

Huang, C.Y., J.L. Garcia, B.K.C. Patel, J.L. Cayol, L. Baresi and R.A. Mah. 2000. *Salinivibrio costicola vallismortis* subsp. nov., a halotolerant facultative anaerobe from Death Valley, and emended description of *Salinivibrio costicola*. Int. J. Syst. Evol. Microbiol. *50*: 615–622.

Huang, Y., W. Qi, Z. Lu, Z. Liu and M. Goodfellow. 2001. *Amycolatopsis rubida* sp. nov., a new *Amycolatopsis* species from soil. Int. J. Syst. Evol. Microbiol. *51*: 1093–1097.

Huang, Y., L. Wang, Z. Lu, L. Hong, Z. Liu, G.Y. Tan and M. Goodfellow. 2002. Proposal to combine the genera *Actinobispora* and *Pseudonocardia* in an emended genus *Pseudonocardia*, and description of *Pseudonocardia zijingensis* sp. nov. Int. J. Syst. Evol. Microbiol. *52*: 977–982.

Huber, H., S. Burggraf, T. Mayer, I. Wyschkony, R. Rachel and K.O. Stetter. 2000a. *Ignicoccus* gen. nov., a novel genus of hyperthermophilic, chemolithoautotrophic *Archaea*, represented by two new species, *Ignicoccus islandicus* sp. nov. and *Ignicoccus pacificus* sp. nov. and *Ignicoccus pacificus* sp. nov. Int. J. Syst. Evol. Microbiol. *50*: 2093–2100.

Huber, H., S. Diller, C. Horn and R. Rachel. 2002. *Thermovibrio ruber* gen. nov., sp. nov., an extremely thermophilic, chemolithoautotrophic, nitrate-reducing bacterium that forms a deep branch within the phylum *Aquificae*. Int. J. Syst. Evol. Microbiol. *52*: 1859–1865.

Huber, H. and K.O. Stetter. 2001a. Class I. *Deferribacteres* class. nov. *In* Boone, Castenholz and Garrity (Editors), Bergey's Manual of Systematic Bacteriology, 2nd Edition, Volume 1. The archaea and the deeply branching and phototrophic bacteria. Springer, New York. 465.

Huber, H. and K.O. Stetter. 2001b. Family I. *Archaeoglobaceae* Stetter 1989, 2216. *In* Boone, Castenholz and Garrity (Editors), Bergey's Manual of Systematic Bacteriology, 2nd Edition, Volume 1. The archaea and the deeply branching and phototrophic bacteria. Springer, New York. 349.

Huber, H. and K.O. Stetter. 2001c. Family I. *Deferribacteraceae* 465–466. *In* Boone, Castenholz and Garrity (Editors), Bergey's Manual of Systematic Bacteriology, 2nd Edition, Volume 1. The archaea and the deeply branching and phototrophic bacteria. Springer, New York.

Huber, H. and K.O. Stetter. 2001d. Family I. *Methanopyraceae* fam. nov. *In* Boone, Castenholz and Garrity (Editors), Bergey's Manual of Systematic Bacteriology, 2nd Edition, Volume 1. The archaea and the deeply branching and phototrophic bacteria. Springer, New York. 353.

Huber, H. and K.O. Stetter. 2001e. Order I. *Archaeoglobales* ord. nov. Stetter 1989, 2216. *In* Boone, Castenholz and Garrity (Editors), Bergey's Manual of Systematic Bacteriology, 2nd Edition, Volume 1. The archaea and the deeply branching and phototrophic bacteria. Springer, New York. 349.

Huber, H. and K.O. Stetter. 2001f. Order I. *Deferribacterales* ord. nov. *In* Boone, Castenholz and Garrity (Editors), Bergey's Manual of Systematic Bacteriology, 2nd Edition, Volume 1. The archaea and the deeply branching and phototrophic bacteria. Springer, New York. 465.

Huber, H. and K.O. Stetter. 2001g. Order I. *Methanopyrales* ord. nov. *In* Boone, Castenholz and Garrity (Editors), Bergey's Manual of Systematic Bacteriology, 2nd Edition, Volume 1. The archaea and the deeply branching and phototrophic bacteria. Springer, New York. 353.

Huber, H. and K.O. Stetter. 2001h. Order II. *Desulfurococcales* ord. nov. *In* Boone, Castenholz and Garrity (Editors), Bergey's Manual of Systematic Bacteriology, 2nd Edition, Volume 1. The archaea and the deeply branching and phototrophic bacteria. Springer, New York. 179–180.

Huber, H. and K.O. Stetter. 2002. *In* Validation of publication of new names and new combinations previously effectively published outside the IJSEM. List No. 85. Int. J. Syst. Evol. Microbiol. *52*: 685–690.

Huber, R., S. Burggraf, T. Mayer, S.M. Barns, P. Rossnagel and K.O. Stetter. 1995. Isolation of a hyperthermophilic archaeum predicted by *in situ* RNA analysis. Nature *376*: 57–58.

Huber, R., W. Eder, S. Heldwein, G. Wanner, H. Huber, R. Rachel and K.O. Stetter. 1998. *Thermocrinis ruber* gen. nov., sp. nov., a pink-filament-forming hyperthermophilic bacterium isolated from Yellowstone National Park. Appl. Environ. Microbiol. *64*: 3576–3583.

Huber, R., M. Sacher, A. Vollmann, H. Huber and D. Rose. 2000b. Respiration of arsenate and selenate by hyperthermophilic archaea. Syst. Appl. Microbiol. *23*: 305–314.

Huber, R., M. Sacher, A. Vollmann, H. Huber and D. Rose. 2001. *In* Validation of publication of new names and new combinations previously effectively published outside the IJSEM. List No. 79. Int. J. Syst. Evol. Microbiol. *51*: 263–265.

Huber, R. and K.O. Stetter. 2001i. Genus 1. Aquifex. *In* Boone, Castenholz and Garrity (Editors), Bergey's Manual of Systematic Bacteriology, 2nd Edition, Volume 1. The archaea and the deeply branching and phototrophic bacteria. Springer, New York. 360–362.

Hugenholtz, P., B.M. Goebel and N.R. Pace. 1998a. Impact of culture-independent studies on the emerging phylogenetic view of bacterial diversity. J. Bacteriol. *180*: 4765–4774.

Hugenholtz, P., C. Pitulle, K.L. Hershberger and N.R. Pace. 1998b. Novel division level bacterial diversity in a Yellowstone hot spring. J. Bacteriol. *180*: 366–376.

Hugenholtz. P., B.M. Goebel and N.R. Pace. 1998. Impact of culture-independent studies on the emerging phylogenetic view of bacterial diversity. J. Bacteriol. *180*: 4765–4774.

Hugo, C.J., P. Segers, B. Hoste, M. Vancanneyt and K. Kersters. 2003. *Chryseobacterium joostei* sp. nov., isolated from the dairy environment. Int. J. Syst. Evol. Microbiol. *53*: 771–777.

Humphry, D.R., G.W. Black and S.P. Cummings. 2003. Reclassification of *'Pseudomonas fluorescens* subsp. *cellulosa'* NCIMB 10462 (Ueda et al. 1952) as *Cellvibrio japonicus* sp. nov. and revival of *Cellvibrio vulgaris* sp. nov., nom. rev. and *Cellvibrio fulvus* sp. nov., nom. rev. Int. J. Syst. Evol. Microbiol. *53*: 393–400.

Humphry, D.R., A. George, G.W. Black and S.P. Cummings. 2001. *Flavobacterium frigidarium* sp. nov., an aerobic, psychrophilic, xylanolytic and laminarinolytic bacterium from Antarctica. Int. J. Syst. Evol. Microbiol. *51*: 1235–1243.

Hungate, R.E. 1979. Evolution of a microbial ecologist. Annu. Rev. Microbiol. *33*: 1–20.

Hunik, J.H., H.J.G. Meijer and J. Tramper. 1992. Kinetics of *Nitrosomonas europaea* at extreme substrate, product and salt concentrations. Appl. Microbiol. Biotechnol. *37*: 802–807.

Huys, G., M. Cnockaert, J.M. Janda and J. Swings. 2003a. *Escherichia albertii* sp. nov., a diarrhoeagenic species isolated from stool specimens of Bangladeshi children. Int. J. Syst. Evol. Microbiol. *53*: 807–810.

Huys, G., P. Kämpfer, M.J. Albert, I. Kuhn, R. Denys and J. Swings. 2002. *Aeromonas hydrophila* subsp. *dhakensis* subsp. nov., isolated from children with diarrhoea in Bangladesh, and extended description of *Aeromonas hydrophila* subsp. *hydrophila* (Chester 1901) Stanier 1943 (Approved Lists 1980). Int. J. Syst. Evol. Microbiol. *52*: 705–712.

Huys, G., M. Pearson, P. Kämpfer, R. Denys, M. Cnockaert, V. Inglis and J. Swings. 2003b. *Aeromonas hydrophila* subsp. *ranae* subsp. nov., isolated from septicaemic farmed frogs in Thailand. Int. J. Syst. Evol. Microbiol. *53*: 885–891.

Hyman, M.R. and D.J. Arp. 1992. $^{14}C_2H_2$ and $^{14}CO_2$ labeling studies of the de novo synthesis of polypeptides by *Nitrosomonas europaea* during recovery from acetylene and light inactivation of ammonia monooxygenase. J. Biol. Chem. *267*: 1534–1545.

Hyman, M.R., I.B. Murton and D.J. Arp. 1988. Interaction of ammonia monooxygenase from *Nitrosomonas europaea* with alkanes, alkenes, and alkynes. Appl. Environ. Microbiol. *54*: 3187–3190.

Hyman, M.R. and P.M. Wood. 1983. Methane oxidation by *Nitrosomonas europaea*. Biochem. J. *212*: 31–37.

Hyman, M.R. and P.M. Wood. 1984a. Bromocarbon oxidations by *Nitrosomonas europaea*. *In* Crawford and Hanson (Editors), Microbial Growth on C_1 Compounds, American Society for Microbiology, Washington, DC. 49–52.

Hyman, M.R. and P.M. Wood. 1984b. Ethylene oxidation by *Nitrosomonas europaea*. Arch. Microbiol. *137*: 155–158.

Hyman, M.R. and P.M. Wood. 1985. Suicidal inactivation and labelling of ammonia mono-oxygenase by acetylene. Biochem. J. *227*: 719–725.

Hyypiä, T., A. Jalava, S.H. Larsen, P. Terho and V. Hukkanen. 1985. Detection of *Chlamydia trachomatis* in clinical specimens by nucleic acid spot hybridization. J. Gen. Microbiol. *131*: 975–978.

Igarashi, N., H. Moriyama, T. Fujiwara, Y. Fukumori and N. Tanaka. 1997. The 2.8 Å structure of hydroxylamine oxidoreductase from a nitrifying chemoautotrophic bacterium, *Nitrosomonas europaea*. Nat. Struct. Biol. *4*: 276–284.

Iizuka, T., Y. Jojima, R. Fudou, A. Hiraishi, J.W. Ahn and S. Yamanaka. 2003a. *Plesiocystis pacifica* gen. nov., sp. nov., a marine myxobacterium that contains dihydrogenated menaquinone, isolated from the Pacific coasts of Japan. Int. J. Syst. Evol. Microbiol. *53*: 189–195.

Iizuka, T., Y. Jojima, R. Fudou, M. Tokura, A. Hiraishi and S. Yamanaka. 2003b. *Enhygromyxa salina* gen. nov., sp. nov., a slightly halophilic myxobacterium isolated from the coastal areas of Japan. Syst. Appl. Microbiol. *26*: 189–196.

Iizuka, T., Y. Jojima, R. Fudou, M. Tokura, A. Hiraishi and S. Yamanaka. 2003c. *In* Validation of publication of new names and new combinations previously effectively published outside the IJSEM. List No. 93. Int. J. Syst. Evol. Microbiol. *53*: 1219–1220.

Ikuta, S., K. Takagi, R.B. Wallace and K. Itakura. 1987. Dissociation kinetics of 19 base paired oligonucleotide-DNA duplexes containing different single mismatched base pairs. Nucleic Acids Res. *15*: 797–811.

Imachi, H., Y. Sekiguchi, Y. Kamagata, S. Hanada, A. Ohashi and H. Harada. 2002. *Pelotomaculum thermopropionicum* gen. nov., sp. nov., an anaerobic, thermophilic, syntrophic propionate-oxidizing bacterium. Int. J. Syst. Evol. Microbiol. *52*: 1729–1735.

Imamura, Y., M. Ikeda, S. Yoshida and H. Kuraishi. 2000. *Janibacter brevis* sp. nov., a new trichloroethylene-degrading bacterium isolated from polluted environments. Int. J. Syst. Evol. Microbiol. *50*: 1899–1903.

Imhoff, J.F. 1982. Occurrence and evolutionary significance of two sulfate assimilation pathways in the *Rhodospirillaceae*. Arch. Microbiol. *132*: 197–203.

Imhoff, J.F. 1984. Quinones of phototrophic purple bacteria. FEMS Microbiol. Lett. *256*: 85–89.

Imhoff, J.F. 1991. Polar lipids and fatty acids in the genus *Rhodobacter*. Syst. Appl. Microbiol. *14*: 228–234.

Imhoff, J.F. 1992. Taxonomy, phylogeny, and general ecology of anoxygenic phototrophic bacteria. *In* Mann and Carr (Editors), Biotechnology Handbooks: Photosynthetic Prokaryotes, Vol. 6, Plenum Press, New York. pp. 53–92.

Imhoff, J.F. 1995. Taxonomy and physiology of phototrophic purple bacteria and green sulfer bacteria. *In* Blankenship, Madigan and Bauer (Editors), Advances in Photosynthesis: Anoxygenic Photosynthetic Bacteria, Vol. 2, Kluwer Academic Publishers, Dordrecht. pp. 1–15.

Imhoff, J.F. 1999. A phylogenetically oriented taxonomy of anoxygenic phototrophic bacteria. *In* Pescheck and Schmetterer (Editors), The Phototrophic Prokaryotes, Kluwer/Plenum, New York.

Imhoff, J.F. 2001a. The anoxygenic phototrophic purple bacteria. *In* Boone, Castenholz and Garrity (Editors), Bergey's Manual of Systematic Bacteriology, 2nd Ed., Vol. 1, Springer-Verlag, New York. 631–637.

Imhoff, J.F. 2001b. Transfer of *Pfennigia purpurea* Tindall 1999 (*Amoebobacter purpureus* Eichler and Pfennig 1988) to the genus *Lamprocystis* as *Lamprocystis purpurea* comb. nov. Int. J. Syst. Evol. Microbiol. *51*: 1699–1701.

Imhoff, J. F. 2001c. Transfer of *Rhodopseudomonas acidophila* to the new genus *Rhodoblastus* as *Rhodoblastus acidophilus* gen. nov., comb. nov. Int. J. Syst. Evol. Microbiol. *51*: 1863–1866.

Imhoff, J.F. 2003. Phylogenetic taxonomy of the family *Chlorobiaceae* on the basis of 16S rRNA and *fmo* (Fenna-Matthews-Olson protein) gene sequences. Int. J. Syst. Evol. Microbiol. *53*: 941–951.

Imhoff, J.F. and U. Bias-Imhoff. 1995. Lipids, quinones and fatty acids of anoxygenic phototrophic bacteria. *In* Blankenship, Madigan and Bauer (Editors), Anoxygenic Photosynthetic Bacteria, Kluwer Academic Publishing, The Netherlands. pp. 179–205.

Imhoff, J.F., R. Petri and J. Süling. 1998. Reclassification of species of the spiral-shaped phototrophic purple non-sulfur bacteria of the α-Proteobacteria: description of the new genera *Phaeospirillum* gen. nov., *Rhodovibrio* gen. nov., *Rhodothalassium* gen. nov. and *Roseospira* gen. nov. as well as transfer of *Rhodospirillum fulvum* to *Phaeospirillum fulvum* comb. nov., of *Rhodospirillum molischianum* to *Phaeospirillum molischianum* comb. nov., of *Rhodospirillum salinarum* to *Rhodovibrio salinarum* comb. nov., of *Rhodospirillum sodomense* to *Rhodovibrio sodomensis* comb. nov., of *Rhodospirillum salexigens* to *Rhodothalassium salexigens* comb. nov. and of *Rhodospirillum mediosalinum* to *Roseospira mediosalina* comb. nov. Int. J. Syst. Bacteriol. *48*: 793–798.

Imhoff, J.F. and N. Pfennig. 2001. *Thioflavicoccus mobilis* gen. nov., sp. nov., a novel purple sulfur bacterium with bacteriochlorophyll *b*. Int. J. Syst. Evol. Microbiol. *51*: 105–110.

Imhoff, J.F. and H.G. Trüper. 1989. Purple nonsulfur bacteria. *In* Staley, Bryant, Pfennig and Holt (Editors), Bergey's Manual of Systematic Bacteriology, 1st Ed., Vol. 3rd, The Williams & Wilkins Co., Baltimore. pp. 1658–1661.

Imhoff, J.F., H.G. Trüper and N. Pfennig. 1984. Rearrangement of the species and genera of the phototrophic "purple nonsulfur bacteria". Int. J. Syst. Bacteriol. *34*: 340–343.

Inagaki, F., K. Takai, H. Kobayashi, K.H. Nealson and K. Horikoshi. 2003. *Sulfurimonas autotrophica* gen. nov., sp. nov., a novel sulfur-oxidizing ε-proteobacterium isolated from hydrothermal sediments in the Mid-Okinawa Trough. Int. J. Syst. Evol. Microbiol. *53*: 1801–1805.

Inoue, K., K. Sugiyama, Y. Kosako, R. Sakazaki and S. Yamai. 2000. *Enterobacter cowanii* sp. nov., a new species of the family *Enterobacteriaceae*. Curr. Microbiol. *41*: 417–420.

Inoue, K., K. Sugiyama, Y. Kosako, R. Sakazaki and S. Yamai. 2001. *In* Validation of the publication of new names and new combinations previously effectively published outside the IJSB, List No. 82. Int. J. Syst. Evol. Microbiol. *51*: 1619.

International Committee on Systematic Bacteriology. 1997. VIIth International Congress of Microbiology and Applied Bacteriology. Minutes of the Meetings, 17, 18, and 22 August 1996, Jerusalem, Israel. Int. J. Syst. Bacteriol. *47*: 597–600.

International Committee on Systematic Bacteriology Subcommittee on the Taxonomy of *Mollicutes*. 1979. Proposal of miminal standards for descriptions of new species of the class *Mollicutes*. Int. J. Syst. Bacteriol. *29*: 172–180.

Ishikawa, M., S. Ishizaki, Y. Yamamoto and K. Yamasato. 2002. *Paraliobacillus ryukyuensis* gen. nov., sp. nov., a new Gram-positive, slightly halophilic, extremely halotolerant, facultative anaerobe isolated from a decomposing marine alga. J. Gen. Appl. Microbiol. *48*: 269–279.

Ishikawa, M., S. Ishizaki, Y. Yamamoto and K. Yamasato. 2003a. *In* Validation of publication of new names and new combinations previously effectively published outside the IJSEM. List No. 91. Int. J. Syst. Evol. Microbiol. *53*: 627–628.

Ishikawa, M., K. Nakajima, M. Yanagi, Y. Yamamoto and K. Yamasato. 2003b. *Marinilactibacillus psychrotolerans* gen. nov., sp. nov., a halophilic and alkaliphilic marine lactic acid bacterium isolated from marine organisms in temperate and subtropical areas of Japan. Int. J. Syst. Evol. Microbiol. *53*: 711–720.

Isnansetyo, A. and Y. Kamei. 2003. *Pseudoalteromonas phenolica* sp. nov., a novel marine bacterium that produces phenolic anti-methicillin-resistant *Staphylococcus aureus* substances. Int. J. Syst. Evol. Microbiol. *53*: 583–588.

Istock, C.A., J.A. Bell, N. Ferguson and N.L. Istock. 1996. Bacterial species and evolution: theoretical and practical perspectives. J. Ind. Microbiol. *17*: 137–150.

Itoh, T., K. Suzuki and T. Nakase. 2002. *Vulcanisaeta distributa* gen. nov., sp. nov., and *Vulcanisaeta souniana* sp. nov., novel hyperthermophilic, rod-shaped crenarchaeotes isolated from hot springs in Japan. Int. J. Syst. Evol. Microbiol. *52*: 1097–1104.

Itoh, T., K. Suzuki, P.C. Sanchez and T. Nakase. 2003. *Caldisphaera lagunensis* gen. nov., sp. nov., a novel thermoacidophilic crenarchaeote isolated from a hot spring at Mt. Maquiling, Philippines. Int. J. Syst. Evol. Microbiol. *53*: 1149–1154.

Ivanova, E.P., J. Chun, L.A. Romanenko, M.E. Matte, V.V. Mikhailov, G.M. Frolova, A. Huq and R.R. Colwell. 2000a. Reclassification of *Alteromonas distincta* Romanenko et al. 1995 as *Pseudoalteromonas distincta* comb. nov. Int. J. Syst. Evol. Microbiol. *50*: 141–144.

Ivanova, E.P., N.M. Gorshkova, T. Sawabe, K. Hayashi, N.I. Kalinovskaya, A.M. Lysenko, N.V. Zhukova, D.V. Nicolau, T.A. Kuznetsova, V.V. Mikhailov and R. Christen. 2002a. *Pseudomonas extremorientalis* sp. nov., isolated from a drinking water reservoir. Int. J. Syst. Evol. Microbiol. *52*: 2113–2120.

Ivanova, E.P. and V.V. Mikhailov. 2001a. A new family, *Alteromonadaceae* fam. nov., including marine proteobacteria of the genera *Alteromonas, Pseudoalteromonas, Idiomarina,* and *Colwellia*. Microbiology *70*: 10–17.

Ivanova, E.P. and V.V. Mikhailov. 2001b. *In* Validation of the publication of new names and new combinations previously effectively published outside the IJSEM. List No. 81. Int. J. Syst. Evol. Microbiol. *51*: 1229.

Ivanova, E.P., O.I. Nedashkovskaya, J. Chun, A.M. Lysenko, G.M. Frolova, V.I. Svetashev, M.V. Vysotskii, V.V. Mikhailov, A. Huq and R.R. Colwell. 2001a. *Arenibacter* gen. nov., new genus of the family *Flavobacteriaceae* and description of a new species, *Arenibacter latericius* sp. nov. Int. J. Syst. Evol. Microbiol. *51*: 1987–1995.

Ivanova, E.P., O.I. Nedashkovskaya, N.V. Zhukova, D.V. Nicolau, R. Christen and V.V. Mikhailov. 2003a. *Shewanella waksmanii* sp. nov., isolated from a sipuncula (*Phascolosoma japonicum*). Int. J. Syst. Evol. Microbiol. *53*: 1471–1477.

Ivanova, E.P., L.A. Romanenko, J. Chun, M.H. Matte, G.R. Matte, V.V. Mikhailov, V.I. Svetashev, A. Huq, T. Maugel and R.R. Colwell. 2000b. *Idiomarina* gen. nov., comprising novel indigenous deep-sea bacteria from the Pacific Ocean, including descriptions of two species, *Idiomarina abyssalis* sp. nov. and *Idiomarina zobellii* sp. nov. Int. J. Syst. Evol. Microbiol. *50*: 901–907.

Ivanova, E.P., L.A. Romanenko, M.H. Matté, G.R. Matté, A.M. Lysenko, U. Simidu, K. Kita-Tsukamoto, T. Sawabe, M.V. Vysotskii, G.M. Frolova, V. Mikhailov, R. Christen and R.R. Colwell. 2001b. Retrieval of the species *Alteromonas tetraodonis* Simidu et al. 1990 as *Pseudoalteromonas tetraodonis* comb. nov. and emendation of description. Int. J. Syst. Evol. Microbiol. *51*: 1071–1078.

Ivanova, E.P., T. Sawabe, Y.V. Alexeeva, A.M. Lysenko, N.M. Gorshkova, K. Hayashi, N.V. Zukova, R. Christen and V.V. Mikhailov. 2002b. *Pseudoalteromonas issachenkonii* sp. nov., a bacterium that degrades the thallus of the brown alga *Fucus evanescens*. Int. J. Syst. Evol. Microbiol. *52*: 229–234.

Ivanova, E.P., T. Sawabe, N.M. Gorshkova, V.I. Svetashev, V.V. Mikhailov, D.V. Nicolau and R. Christen. 2001c. *Shewanella japonica* sp. nov. Int. J. Syst. Evol. Microbiol. *51*: 1027–1033.

Ivanova, E.P., T. Sawabe, K. Hayashi, N.M. Gorshkova, N.V. Zhukova, O.I. Nedashkovskaya, V.V. Mikhailov, D.V. Nicolau and R. Christen. 2003b. *Shewanella fidelis* sp. nov., isolated from sediments and sea water. Int. J. Syst. Evol. Microbiol. *53*: 577–582.

Ivanova, E.P., T. Sawabe, A.M. Lysenko, N.M. Gorshkova, K. Hayashi, N.V. Zhukova, D.V. Nicolau, R. Christen and V.V. Mikhailov. 2002c. *Pseudoalteromonas translucida* sp. nov. and *Pseudoalteromonas paragorgicola* sp. nov., and emended description of the genus. Int. J. Syst. Evol. Microbiol. *52*: 1759–1766.

Ivanova, E.P., L.S. Shevchenko, T. Sawabe, A.M. Lysenko, V.I. Svetashev, N.M. Gorshkova, M. Satomi, R. Christen and V.V. Mikhailov. 2002d. *Pseudoalteromonas maricaloris* sp. nov., isolated from an Australian sponge, and reclassification of [*Pseudoalteromonas aurantia*] NCIMB 2033 as *Pseudoalteromonas flavipulchra* sp. nov. Int. J. Syst. Evol. Microbiol. *52*: 263–271.

Ivanova, T.L., T.P. Turova and A.S. Antonov. 1988. DNA-DNA hybridization studies on some purple nonsulfur bacteria. Syst. Appl. Microbiol. *10*: 259–263.

Iverson, T.M., D.M. Arciero, B.T. Hsu, M.S. Logan, A.B. Hooper and D.C. Rees. 1998. Heme packing motifs revealed by the crystal structure of the tetra-heme cytochrome c554 from *Nitrosomonas europaea*. Nat. Struct. Biol. *5*: 1005–1012.

Iwabe, N., K. Kuma, M. Hasegawa, S. Osawa and T. Miyata. 1989. Evolutionary relationship of archaebacteria, eubacteria, and eukaryotes inferred from phylogenetic trees of duplicated genes. Proc. Natl. Acad. Sci. U.S.A. *86*: 9355–9359.

Jablonski, E., E.W. Moomaw, R.H. Tullis and J.L. Ruth. 1986. Preparation of oligodeoxynucleotide-alkaline phosphatase conjugates and their use as hybridization probes. Nucleic Acids Res. *14*: 6115–6128.

Jackson, B.E., V.K. Bhupathiraju, R.S. Tanner, C.R. Woese and M.J. McInerney. 1999. *Syntrophus aciditrophicus* sp. nov., a new anaerobic bacterium that degrades fatty acids and benzoate in syntrophic association with hydrogen-using microorganisms. Arch. Microbiol. *171*: 107–114.

Jackson, B.E., V.K. Bhupathiraju, A.C. Tanner, C.R. Woese and B.V. McInerney. 2001. *In* Validation of the publication of new names and new combinations previously effectively published outside the IJSB, List No. 80. Int. J. Syst. Bacteriol. *51*: 793–794.

Jahn, E. 1924. Beitrage zur botanischen Protistologie I. Die Polyangiden, Verlag Gebruder Borntraeger, Leipzig. 107 pp. + 102 plates.

Jain, R.K., R.S. Burlage and G.S. Sayler. 1988. Methods for detecting recombinant DNA in the environment. Crit. Rev. Biotechnol. *8*: 33–84.

Jain, R., M.C. Rivera and J.A. Lake. 1999. Horizontal gene transfer among genomes: The complexity hypothesis. Proc. Nat. Acad. Sci. USA. *96*: 3801–3806.

Jain, R.K., G.S. Sayler, J.T. Wilson, L. Houston and D. Pacia. 1987. Maintenance and stability of introduced genotypes in groundwater aquifer material. Appl. Environ. Microbiol. *53*: 996–1002.

Jang, S.S., J.M. Donahue, A.B. Arata, J. Goris, L.M. Hansen, D.L. Earley, P.A.R. Vandamme, P.J. Timoney and D.C. Hirsh. 2001. *Taylorella asinigenitalis* sp. nov., a bacterium isolated from the genital tract of male donkeys (*Equus asinus*). Int. J. Syst. Evol. Microbiol. *51*: 971–976.

Janssen, P.H., W. Liesack and B. Schink. 2002. *Geovibrio thiophilus* sp. nov., a novel sulfur-reducing bacterium belonging to the phylum *Deferribacteres*. Int. J. Syst. Evol. Microbiol. *52*: 1341–1347.

Jarvis, B.D., H.L. Downer and J.P. Young. 1992. Phylogeny of fast-growing soybean-nodulating rhizobia support synonymy of *Sinorhizobium* and *Rhizobium* and assignment to *Rhizobium fredii*. Int. J. Syst. Bacteriol. *42*: 93–96.

Jeanthon, C., S. L'Haridon, V. Cueff, A. Banta, A.L. Reysenbach and D. Prieur. 2002. *Thermodesulfobacterium hydrogeniphilum* sp. nov., a thermophilic, chemolithoautotrophic, sulfate-reducing bacterium isolated from a deep-sea hydrothermal vent at Guaymas Basin, and emendation of the genus *Thermodesulfobacterium*. Int. J. Syst. Evol. Microbiol. *52*: 765–772.

Jeanthon, C., A.L. Reysenbach, S. L'Haridon, A. Gambacorta, N.R. Pace, P. Glenat and D. Prieur. 2000. *In* Validation of publication of new names and new combinations previously effectively published outside the IJSEM. List No. 76. Int. J. Syst. Evol. Microbiol. *50*: 1699–1700.

Jeanthon, C., A.L. Reysenbach, S. L'Haridon, A. Gambacorta, N.R. Pace, P. Glenat and D. Prieur. 1995. *Thermotoga subterranea* sp. nov., a new thermophilic bacterium isolated from a continental oil reservoir. Arch. Microbiol. *164*: 91–97.

Jeffrey, C. 1977. Biological Nomenclature, 2nd Ed., Arnold, London.

Jendrossek, D. 2001. Transfer of *[Pseudomonas] lemoignei*, a Gram-negative rod with restricted catabolic capacity, to *Paucimonas* gen. nov. with one species, *Paucimonas lemoignei* comb. nov. Int. J. Syst. Evol. Microbiol. *51*: 905–908.

Jensen, K., N.P. Revsbech and L.P. Nielsen. 1993. Microscale distribution of nitrification activity in sediment determined with a shielded microsensor for nitrate. Appl. Environ. Microbiol. *59*: 3287–3296.

Jetten, M.S.M., M. Strous, K T. van de Pas-Schoonen, J. Schalk, U.G.I.M. van Dongen, A.A. van de Graaf, S. Logemann, G. Muyzer, M.C.M. van Loosdrecht and J.G. Kuenen. 1998. The anaerobic oxidation of ammonium. FEMS Microbiol. Rev. *22*: 421–437.

Jian, W. and X. Dong. 2002. Transfer of *Bifidobacterium inopinatum* and *Bifidobacterium denticolens* to *Scardovia inopinata* gen. nov., comb. nov., and *Parascardovia denticolens* gen. nov., comb. nov., respectively. Int. J. Syst. Evol. Microbiol. *52*: 809–812.

Johnson, J.L. 1985. DNA reassociation and RNA hybridization of bacterial nucleic acids. *In* Gottschalk (Editor), Methods in Microbiology, Vol. 18, Academic Press, New York. pp. 33–74.

Johnson, P.W. and J.M. Sieburth. 1976. *In situ* morphology of nitrifying-like bacteria in aquaculture systems. Appl. Environ. Microbiol. *31*: 423–432.

Joklik, W.K. (Editor). 1999. Microbiology: A Centenary Perspective, American Society for Microbiology, Washington, DC.

Jolivet, E., S. L'Haridon, E. Corre, P. Forterre and D. Prieur. 2003. *Thermococcus gammatolerans* sp. nov., a hyperthermophilic archaeon from a deep-sea hydrothermal vent that resists ionizing radiation. Int. J. Syst. Evol. Microbiol. *53*: 847–851.

Jones, R.D. and R.Y. Morita. 1983a. Carbon monoxide oxidation by chemolithotrophic ammonium oxidizers. Can. J. Microbiol. *29*: 1545–1551.

Jones, R.D. and R.Y. Morita. 1983b. Methane oxidation by *Nitrosococcus oceanus* and *Nitrosomonas europaea*. Appl. Environ. Microbiol. *45*: 401–410.

Jones, R.D. and R.Y. Morita. 1985. Low temperature growth and whole cell kinetics of a marine ammonium oxidizer. Mar. Ecol.-Prog. Ser. *21*: 239–243.

Joulian, C., B.K. Patel, B. Ollivier, J.L. Garcia and P.A. Roger. 2000. *Methanobacterium oryzae* sp. nov., a novel methanogenic rod isolated from a Philippines ricefield. Int. J. Syst. Evol. Microbiol. *50*: 525–528.

Joyce, E.A., K. Chan, N.R. Salama and S. Falkow. 2002. Redefining bacterial populations: A post-genomic reformation. Nat. Rev. Genet. *3*: 462–473.

Jukes, T.H. and R.R. Cantor. 1969. Evolution of protein molecules. *In* Munzo (Editor), Mammalian Protein Metabolism, Academic Press, New York. pp. 21–132.

Juretschko, S., G. Timmermann, M. Schmid, K.H. Schleifer, A. Pommerening-Röser, H.P. Koops and M. Wagner. 1998. Combined molecular and conventional analysis of nitrifying bacterium diversity in activated sludge: *Nitrosococcus mobilis* and *Nitrospira*-like bacteria as dominant populations. Appl. Environ. Microbiol. *64*: 3042–3051.

Jürgens, K., J. Pernthaler, S. Schalla and R.I. Amann. 1999. Morphological and compositional changes in a planktonic bacterial community in response to protozoal grazing. Appl. Environ. Microbiol. *65*: 1241–1250.

Kafatos, F.C., C.W. Jones and A. Efstratiadis. 1979. Determination of nucleic acid sequence homologies and relative concentrations by a dot hybridization procedure. Nucleic Acids Res. *7*: 1541–1552.

Kageyama, A. and Y. Benno. 2000a. *Catenibacterium mitsuokai* gen. nov., sp. nov., a Gram-positive anaerobic bacterium isolated from human faeces. Int. J. Syst. Evol. Microbiol. *50*: 1595–1599.

Kageyama, A. and Y. Benno. 2000b. *Coprobacillus catenaformis* gen. nov., sp. nov., a new genus and species isolated from human feces. Microbiol Immunol. *44*: 23–28.

Kageyama, A. and Y. Benno. 2000c. Emendation of genus *Collinsella* and proposal of *Collinsella stercoris* sp. nov. and *Collinsella intestinalis* sp. nov. Int. J. Syst. Evol. Microbiol. *50*: 1767–1774.

Kageyama, A. and Y. Benno. 2000d. Phylogenic and phenotypic characterization of some *Eubacterium*-like isolates from human feces: Description of *Solobacterium moorei* gen. nov., sp. nov. Microbiol Immunol. *44*: 223–227.

Kageyama, A. and Y. Benno. 2000e. *In* Validation of publication of new names and new combinations previously effectively published outside the IJSEM. List No. 74. Int. J. Syst. Evol. Microbiol. *50*: 949–950.

Kageyama, A. and Y. Benno. 2000f. *In* Validation of publication of new names and new combinations previously effectively published outside the IJSEM. List No. 75. Int. J. Syst. Evol. Microbiol. *50*: 1415–1417.

Kakinuma, Y., K. Igarashi, K. Konishi and I. Yamato. 1991. Primary structure of the alpha-subunit of vacuolar-type $Na^{(+)}$-ATPase in *Enterococcus hirae*. Amplification of a 1000-bp fragment by polymerase chain reaction. FEBS Lett. *292*: 64–68.

Kalmbach, S., W. Manz and U. Szewzyk. 1997. Dynamics of biofilm formation in drinking water: phylogenetic affiliation and metabolic potential of single cells assessed by formazan reduction and *in situ* hybridization. FEMS Microbiol. Ecol. *22*: 265–279.

Kaluzhnaya, M., V. Khmelenina, B. Eshinimaev, N. Suzina, D. Nikitin, A. Solonin, J.L. Lin, I. McDonald, C. Murrell and Y. Trotsenko. 2001. Taxonomic characterization of new alkaliphilic and alkalitolerant methanotrophs from soda lakes of the Southeastern Transbaikal region and description of *Methylomicrobium buryatense* sp. nov. Syst. Appl. Microbiol. *24*: 166–176.

Kalyuzhnaya, M., V. Khmelenina, B. Eshinimaev, N. Suzina, D. Nikitin, A. Solonin, J.L. Lin, I. McDonald, C. Murrell and Y. Trotsenko. 2001. *In* Validation of publication of new names and new combinations previously effectively published outside the IJSEM. List No. 83. Int. J. Syst. Evol. Microbiol. *51*: 1945.

Kalyuzhnaya, M.G., V.N. Khmelenina, S. Kotelnikova, L. Holmquist, K. Pedersen and Y.A. Trotsenko. 1999. *Methylomonas scandinavica* sp. nov., a new methanotrophic psychrotrophic bacterium isolated from deep igneous rock ground water of Sweden. Syst. Appl. Microbiol. *22*: 565–572.

Kalyuzhnaya, M.G., V.N. Khmelenina, S. Kotelnikova, L. Holmquist, K. Pedersen and Y.A. Trotsenko. 2000. *In* Validation of publication of new names and new combinations previously effectively published outside the IJSEM. List No. 74. Int. J. Syst. Evol. Microbiol. *50*: 949–950.

Kämpfer, P., F.A. Rainey, M.A. Andersson, E.L. Nurmiaho Lassila, U. Ulrych, H.J. Busse, N. Weiss, R. Mikkola and M. Salkinoja-Salonen. 2000. *Frigoribacterium faeni* gen. nov., sp. nov., a novel psychrophilic genus of the family *Microbacteriaceae*. Int. J. Syst. Evol. Microbiol. *50*: 355–363.

Kämpfer, P., A. Albrecht, S. Buczolits and H.J. Busse. 2002a. *Psychrobacter faecalis* sp. nov., a new species from a bioaerosol originating from pigeon faeces. Syst. Appl. Microbiol. *25*: 31–36.

Kämpfer, P., A. Albrecht, S. Buczolits and H.J. Busse. 2002b. *In* Validation of the publication of new names and new combinations previously effectively published outside the IJSEM. List No. 87. Int. J. Syst. Evol. Microbiol. *52*: 1437–1438.

Kämpfer, P., M.A. Andersson, U. Jackel and M. Salkinoja-Salonen. 2003a. *Teichococcus ludipueritiae* gen. nov. sp. nov., and *Muricoccus roseus* gen. nov. sp. nov. representing two new genera of the α-1 subclass of the *Proteobacteria*. Syst. Appl. Microbiol. *26*: 319–319.

Kämpfer, P., M.A. Andersson, U. Jackel and M. Salkinoja-Salonen. 2003b. *In* Validation of publication of new names and new combinations previously effectively published outside the IJSEM. List No. 92. Int. J. Syst. Evol. Microbiol. *53*: 935–937.

Kämpfer, P., S. Buczolits, A. Albrecht, H.J. Busse and E. Stackebrandt. 2003c. Towards a standardized format for the description of a novel species (of an established genus): *Ochrobactrum gallinifaecis* sp. nov. Int. J. Syst. Evol. Microbiol. *53*: 893–896.

Kämpfer , P., H.J. Busse and F.A. Rainey. 2002. *Nocardiopsis compostus* sp. nov., from the atmosphere of a composting facility. Int. J. Syst. Evol. Microbiol. *52*: 621–627.

Kämpfer, P., U. Dreyer, A. Neef, W. Dott and H.J. Busse. 2003d. *Chryseobacterium defluvii* sp. nov., isolated from wastewater. Int. J. Syst. Evol. Microbiol. *53*: 93–97.

Kämpfer, P., R. Erhart, C. Beimfohr, J. Böhringer, M. Wagner and R.I. Amann. 1996. Characterization of bacterial communities from activated sludge: culture-dependent numerical identification versus *in situ* identification using group- and genus-specific rRNA-targeted oligonucleotide probes. Microb. Ecol. *32*: 101–121.

Kämpfer, P., R. Witzenberger, E.B.M. Denner, H.J. Busse and A. Neef. 2002c. *Novosphingobium hassiacum* sp. nov., a new species isolated from an aerated sewage pond. Syst. Appl. Microbiol. *25*: 37–45.

Kämpfer, P., R. Witzenberger, E.B. Denner, H.J. Busse and A. Neef. 2002d. *Sphingopyxis witflariensis* sp. nov., isolated from activated sludge. Int. J. Syst. Evol. Microbiol. *52*: 2029–2034.

Kämpfer, P., R. Witzenberger, E.B.M. Denner, H.J. Busse and A. Neef. 2002e. *In* Validation of publication of new names and new combinations previously effectively published outside the IJSEM. List No. 87. Int. J. Syst. Evol. Microbiol. *52*: 1437–1438.

Kane, M.D., L.K. Poulsen and D.A. Stahl. 1993. Monitoring the enrichment and isolation of sulfate-reducing bacteria by using oligonucleotide hybridization probes designed from environmentally derived 16S rRNA sequences. Appl. Environ. Microbiol. *59*: 682–686.

Kanso, S., A.C. Greene and B.K. Patel. 2002. *Bacillus subterraneus* sp. nov., an iron- and manganese-reducing bacterium from a deep subsurface Australian thermal aquifer. Int. J. Syst. Evol. Microbiol. *52*: 869–874.

Kanso, S. and B.K.C. Patel. 2003. *Microvirga subterranea* gen. nov., sp. nov., a moderate thermophile from a deep subsurface Australian thermal aquifer. Int. J. Syst. Evol. Microbiol. *53*: 401–406.

Karl, D.M. 1980. Cellular nucleotide measurements and applications in microbial ecology. Microbiol. Rev. *44*: 739–796.

Karlin, S., G.M. Weinstock and V. Brendel. 1995. Bacterial classifications derived from *recA* protein sequence comparisons. J. Bacteriol. *177*: 6881–6893.

Kasai, H., T. Tamura and S. Harayama. 2000. Intrageneric relationships among *Micromonospora* species deduced from *gyrB*-based phylogeny and DNA relatedness. Int. J. Syst. Evol. Microbiol. *50*: 127–134.

Kashefi, K., J.M. Tor, D.E. Holmes, C.V. Gaw Van Praagh, A.L. Reysenbach and D.R. Lovley. 2002. *Geoglobus ahangari* gen. nov., sp. nov., a novel hyperthermophilic archaeon capable of oxidizing organic acids and growing autotrophically on hydrogen with Fe(III) serving as the sole electron acceptor. Int. J. Syst. Evol. Microbiol. *52*: 719–728.

Kato, Y., R.M. Sakala, H. Hayashidani, A. Kiuchi, C. Kaneuchi and M. Ogawa. 2000. *Lactobacillus algidus* sp. nov., a psychrophilic lactic acid bacterium isolated from vacuum-packaged refrigerated beef. Int. J. Syst. Evol. Microbiol. *50*: 1143–1149.

Katsura, K., H. Kawasaki, W. Potacharoen, S. Saono, T. Seki, Y. Yamada, T. Uchimura and K. Komagata. 2001. *Asaia siamensis* sp. nov., an acetic acid bacterium in the alpha-*Proteobacteria*. Int. J. Syst. Evol. Microbiol. *51*: 559–563.

Kattar, M.M., B.T. Cookson, L.C. Carlson, S.K. Stiglich, M.A. Schwartz, T.T. Nguyen, R. Daza, C.K. Wallis, S.L. Yarfitz and M.B. Coyle. 2001. *Tsukamurella strandjordae* sp. nov., a proposed new species causing sepsis. J. Clin. Microbiol. *39*: 1467–1476.

Kattar, M.M., B.T. Cookson, L.C. Carlson, S.K. Stiglich, M.A. Schwartz, T.T. Nguyen, R. Daza, C.K. Wallis, S.L. Yarfitz and M.B. Coyle. 2002. *In* Validation of publication of new names and new combinations previously effectively published outside the IJSEM. List No. 86. Int. J. Syst. Evol. Microbiol. *52*: 1075–1076.

Kawasaki, H., Y. Hoshino and K. Yamasato. 1993. Phylogenetic diversity of phototrophic purple non-sulfur bacteria in the *alpha proteobacteria* group. FEMS Microbiol. Lett. *112*: 61–66.

Kawasaki, K., Y. Nogi, M. Hishinuma, Y. Nodasaka, H. Matsuyama and I. Yumoto. 2002. *Psychromonas marina* sp. nov., a novel halophilic, facultatively psychrophilic bacterium isolated from the coast of the Okhotsk Sea. Int. J. Syst. Evol. Microbiol. *52*: 1455–1459.

Kayton, I. 1982. Copyright in living genetically engineered works. Geo. Wash.L. Rev. *50*: 191–218.

Keen, G.A. and J.I. Prosser. 1987. Interrelationship between pH and surface growth of *Nitrobacter*. Soil Biol. Biochem. *19*: 665–672.

Keis, S., R. Shaheen and D.T. Jones. 2001. Emended descriptions of *Clostridium acetobutylicum* and *Clostridium beijerinckii*, and descriptions of *Clostridium saccharoperbutylacetonicum* sp. nov. and *Clostridium saccharobutylicum* sp. nov. Int. J. Syst. Evol. Microbiol. *51*: 2095–2103.

Kelly, D.P. 1971. Autotrophy: concepts of lithotrophic bacteria and their organic metabolism. Annu. Rev. Microbiol. *25*: 177–210.

Kelly, D.P., I.R. McDonald and A.P. Wood. 2000. Proposal for the reclassification of *Thiobacillus novellus* as *Starkeya novella* gen. nov., comb. nov., in the alpha-subclass of the *Proteobacteria*. Int. J. Syst. Evol. Microbiol. *50*: 1797–1802.

Kelly, D.P. and A.P. Wood. 2000. Reclassification of some species of *Thiobacillus* to the newly designated genera *Acidithiobacillus* gen. nov., *Halothiobacillus* gen. nov. and *Thermithiobacillus* gen. nov. Int. J. Syst. Evol. Microbiol. *50*: 511–516.

Kengen, S.W., G.B. Rikken, W.R. Hagen, C.G. van Ginkel and A.J. Stams. 1999. Purification and characterization of (per)chlorate reductase from the chlorate-respiring strain GR-1. J. Bacteriol. *181*: 6706–6711.

Keppen, O.I., T.P. Tourova, B.B. Kuznetsov, R.N. Ivanovsky and V.M. Gorlenko. 2000. Proposal of *Oscillochloridaceae* fam. nov. on the basis of a phylogenetic analysis of the filamentous anoxygenic phototrophic bacteria, and emended description of *Oscillochloris* and *Oscillochloris trichoides* in comparison with further new isolates. Int. J. Syst. Evol. Microbiol. *50*: 1529–1537.

Kersters, K. and J. De Ley. 1984. Genus III. *Agrobacterium*. *In* Krieg and Holt (Editors), Bergey's Manual of Systematic Bacteriology, 1st. Ed., Vol. 1, The Williams & Wilkins Co., Baltimore. pp. 244–254.

Kersters, K., W. Ludwig, M. Vancanneyt, P. De Vos, M. Gillis and K.-H. Schleifer. 1996. Recent changes in the classification of the pseudomonads: an overview. Syst. Appl. Microbiol. *19*: 465–477.

Kessler, C. 1991. The digoxigenin:anti-digoxigenin (DIG) technology - a survey on the concept and realization of a novel bioanalytical indicator system. Mol. Cell. Probes *5*: 161–205.

Kessler, C. 1994. Non-radioactive analysis of biomolecules. J. Biotechnol. *35*: 165–189.

Kester, R.A., W. deBoer and H.J. Laanbroek. 1997. Production of NO and N_2O by pure cultures of nitrifying and denitrifying bacteria during changes in aeration. Appl. Environ. Microbiol. *63*: 3872–3877.

Khan, A.A., S.J. Kim, D.D. Paine and C.E. Cerniglia. 2002. Classification of a polycyclic aromatic hydrocarbon-metabolizing bacterium, *Mycobacterium* sp. strain PYR-1, as *Mycobacterium vanbaalenii* sp. nov. Int. J. Syst. Evol. Microbiol. *52*: 1997–2002.

Khan, S.T. and A. Hiraishi. 2002. *Diaphorobacter nitroreducens* gen. nov., sp. nov., a poly(3-hydroxybutyrate)-degrading denitrifying bacterium isolated from activated sludge. J. Gen. Appl. Microbiol. *48*: 299–308.

Khan, S.T. and A. Hiraishi. 2003. *In* Validation of publication of new names and new combinations previously effectively published outside the IJSEM. List No. 92. Int. J. Syst. Evol. Microbiol. *53*: 935–937.

Killham, K. 1986. Heterotrophic nitrification. *In* Prosser (Editor), Nitrification, Vol. 36, IRL Press, Oxford. pp. 117–126.

Kim, B., A.M. al-Tai, S.B. Kim, P. Somasundaram and M. Goodfellow. 2000a. *Streptomyces thermocoprophilus* sp. nov., a cellulase-free endo-xylanase-producing streptomycete. Int. J. Syst. Evol. Microbiol. *50*: 505–509.

Kim, B.C., R. Grote, D.W. Lee, G. Antranikian and Y.R. Pyun. 2001. *Thermoanaerobacter yonseiensis* sp. nov., a novel extremely thermophilic, xylose-utilizing bacterium that grows at up to 85°C. Int. J. Syst. Evol. Microbiol. *51*: 1539–1548.

Kim, B., J. Lee, J. Jang, J. Kim and H. Han. 2003a. *Leuconostoc inhae* sp. nov., a lactic acid bacterium isolated from kimchi. Int. J. Syst. Evol. Microbiol. *53*: 1123–1126.

Kim, B., N. Sahin, G.Y. Tan, J. Zakrzewska-Czerwinska and M. Goodfellow. 2002a. *Amycolatopsis eurytherma* sp. nov., a thermophilic actinomycete isolated from soil. Int. J. Syst. Evol. Microbiol. *52*: 889–894.

Kim, J., J. Chun and H.U. Han. 2000b. *Leuconostoc kimchii* sp. nov., a new species from kimchi. Int. J. Syst. Evol. Microbiol. *50*: 1915–1919.

Kim, K.K., C.S. Lee, R.M. Kroppenstedt, E. Stackebrandt and S.T. Lee. 2003b. *Gordonia sihwensis* sp. nov., a novel nitrate-reducing bacterium isolated from a wastewater-treatment bioreactor. Int. J. Syst. Evol. Microbiol. *53*: 1427–1433.

Kim, K.K., A. Roth, S. Andrees, S.T. Lee and R.M. Kroppenstedt. 2002b. *Nocardia pseudovaccinii* sp. nov. Int. J. Syst. Evol. Microbiol. *52*: 1825–1829.

Kim, S.B., R. Brown, C. Oldfield, S.C. Gilbert, S. Iliarionov and M. Goodfellow. 2000c. *Gordonia amicalis* sp. nov., a novel dibenzothiophene-desulphurizing actinomycete. Int. J. Syst. Evol. Microbiol. *50*: 2031–2036.

Kim, S.J., J. Chun, K.S. Bae and Y.C. Kim. 2000d. Polyphasic assignment of an aromatic-degrading *Pseudomonas* sp., strain DJ77, in the genus *Sphingomonas* as *Sphingomonas chungbukensis* sp. nov. Int. J. Syst. Evol. Microbiol. *50*: 1641–1647.

Kim, S.B. and M. Goodfellow. 2002a. *Streptomyces avermitilis* sp. nov., nom. rev., a taxonomic home for the avermectin-producing streptomycetes. Int. J. Syst. Evol. Microbiol. *52*: 2011–2014.

Kim, S.B. and M. Goodfellow. 2002b. *Streptomyces thermospinisporus* sp. nov., a moderately thermophilic carboxydotrophic streptomycete isolated from soil. Int. J. Syst. Evol. Microbiol. *52*: 1225–1228.

Kim, S.B., J. Lonsdale, C.N. Seong and M. Goodfellow. 2003c. *Streptacidiphilus* gen. nov., acidophilic actinomycetes with wall chemotype I and emendation of the family *Streptomycetaceae* (Waksman and Henrici (1943)(AL)) emend. Rainey et al. 1997. Antonie Leeuwenhoek Int. J. of Gen. Molec. Microbiol. *83*: 107–116.

Kim, S.B., J. Lonsdale, C.N. Seong and M. Goodfellow. 2003d. *In* Validation of publication of new names and new combinations previously effectively published outside the IJSEM. List No. 93. Int. J. Syst. Evol. Microbiol. *53*: 1219–1220.

Kinoshita, N., Y. Homma, M. Igarashi, S. Ikeno, M. Hori and M. Hamada. 2001. *Nocardia vinacea* sp. nov. Actinomycetologica. *15*: 1–5.

Kinoshita, N., Y. Homma, M. Igarashi, S. Ikeno, M. Hori and M. Hamada. 2002. *In* Validation of publication of new names and new combinations previously effectively published outside the IJSEM. List No. 84. Int. J. Syst. Evol. Microbiol. *52*: 3–4.

Kinoshita, N., M. Igarashi, S. Ikeno, M. Hori and M. Hamada. 1999. *Saccharothrix tangerinus* sp. nov., the producer of the new antibiotic formamicin: Taxonomic studies. Actinomycetologica. *13*: 20–31.

Kinoshita, N., M. Igarashi, S. Ikeno, M. Hori and M. Hamada. 2000. *In* Validation of publication of new names and new combinations previously effectively published outside the IJSEM. List No. 74. Int. J. Syst. Evol. Microbiol. *50*: 949–950.

Kirchhof, G., B. Eckert, M. Stoffels, J.I. Baldani, V.M. Reis and A. Hartmann. 2001. *Herbaspirillum frisingense* sp. nov., a new nitrogen-fixing bacterial species that occurs in C4 fibre plants. Int. J. Syst. Evol. Microbiol. *51*: 157–168.

Kirsop, B.E. 1996. The Convention on Biological Diversity: some implications for microbiology and microbial culture collections. J. Ind. Microbiol. Biotechnol. *17*: 505–511.

Kirstein, K.O., E. Bock, D.J. Miller and D.J.D. Nicholas. 1986. Membrane-bound *b*-type cytochromes in *Nitrobacter*. FEMS Microbiol. Lett. *36*: 63–67.

Kitahara, M., F. Takamine, T. Imamura and Y. Benno. 2000. Assignment of *Eubacterium* sp. VPI 12708 and related strains with high bile acid 7-α-dehydroxylating activity to *Clostridium scindens* and proposal of *Clostridium hylemonae* sp. nov., isolated from human faeces. Int. J. Syst. Evol. Microbiol. *50*: 971–978.

Kitahara, M., F. Takamine, T. Imamura and Y. Benno. 2001. *Clostridium hiranonis* sp. nov., a human intestinal bacterium with bile acid 7-α-dehydroxylating activity. Int. J. Syst. Evol. Microbiol. *51*: 39–44.

Klenk, H.-P., T.D. Meier, P. Durovic, V. Schwass, F. Lottspeich, P.P. Dennis and W. Zillig. 1999. RNA polymerase of *Aquifex pyrophilus*: implications for the evolution of the bacterial *rpoBC*-operon and the extreme thermophilic bacteria. J. Mol. Evol. *48*: 528–541.

Klenk, H.-P., P. Palm and W. Zillig. 1994. DNA-dependent RNA polymerases as phylogenetic marker molecules. Syst. Appl. Microbiol. *16*: 638–647.

Klenk, H.-P., L. Zhou and J.C. Venter. 1997. Understanding life on this planet in the age of genomics. Proceedings of SPIE. *3111*: 306–317.

Kloos, K., U.M. Husgen and H. Bothe. 1998. DNA-probing for genes coding for denitrification, N_2 fixation and nitrification in bacteria isolated from different soils. Z. Naturforsch. [C]. *53*: 69–81.

Klotz, M.G., J. Alzerreca and J.M. Norton. 1997. A gene encoding a membrane protein exists upstream of the *amoA/amoB* genes in ammonia oxidizing bacteria: a third member of the *amo* operon? FEMS Microbiol. Lett. *150*: 65–73.

Kluyver, A.J. and C.B. van Niel. 1936. Prospects for a natural system of classification of bacteria. Zentbl. Bakteriol. Parasitenkd. Infektkrankh. Hyg. Abt. II *94*: 369–403.

Kobayashi, T., C. Imada, A. Hiraishi, H. Tsujibo, K. Miyamoto, Y. Inamori, N. Hamada and E. Watanabe. 2003. *Pseudoalteromonas sagamiensis* sp. nov., a marine bacterium that produces protease inhibitors. Int. J. Syst. Evol. Microbiol. *53*: 1807–1811.

Kobayashi, T., B. Kimura and T. Fujii. 2000. *Haloanaerobium fermentans* sp. nov., a strictly anaerobic, fermentative halophile isolated from fermented puffer fish ovaries. Int. J. Syst. Evol. Microbiol. *50*: 1621–1627.

Kogure, K., U. Simidu and N. Taga. 1979. A tentative direct microscopic method for counting living marine bacteria. Can. J. Microbiol. *25*: 415–420.

Kogure, K., U. Simidu and N. Taga. 1984. An improved direct viable count method for aquatic bacteria. Arch. Hydrobiol. *102*: 117–122.

Koike, I. and A. Hattori. 1978. Simultaneous determinations of nitrification and nitrate reduction in coastal sediments by an N15 dilution technique. Appl. Environ. Microbiol. *35*: 853–857.

Kolber, Z.S., F.G. Plumley, A.S. Lang, J.T. Beatty, R.E. Blankenship, C.L. VanDover, C. Vetriani, M. Koblizek, C. Rathgeber and P.G. Falkowski. 2001. Contribution of aerobic photoheterotrophic bacteria to the carbon cycle in the ocean. Science *292*: 2492–2495.

Komagata, K. 1989. Taxonomy of facultative methylotrophs. *In* Harashima, Shiba and Murata (Editors), Aerobic Photosynthetic Bacteria, Springer–Verlag, Berlin. 25–38.

Koonin, E.V., R.L. Tatusov and M.Y. Galperin. 1998. Beyond complete genomes: from sequence to structure and function. Curr. Opin. Struct. Biol. *8*: 355–363.

Koops, H.P., B. Böttcher, U.C. Möller, A. Pommerening-Röser and G. Stehr. 1990. Description of a new species of *Nitrosococcus*. Arch. Microbiol. *154*: 244–248.

Koops, H.P., B. Böttcher, U.C. Möller, A. Pommerening-Röser and G. Stehr. 1991. Classification of 8 new species of ammonia-oxidizing bacteria *Nitrosomonas communis* sp. nov., *Nitrosomonas ureae* sp. nov., *Nitrosomonas aestuarii* sp. nov., *Nitrosomonas marina* sp. nov., *Nitrosomonas nitrosa* sp. nov., *Nitrosomonas eutropha* sp. nov., *Nitrosomonas oligotropha* sp. nov., and *Nitrosomonas halophila* sp. nov. J. Gen. Microbiol. *137*: 1689–1699.

Koops, H.P.G., B. Böttcher, U.C. Möller, A. Pommerening-Röser and G. Stehr. 2001. *In* Validation of publication of new names and new combinations previously effectively published outside the IJSEM. List No. 83. Int. J. Syst. Evol. Microbiol. *51*: 1945.

Koops, H.P. and H. Harms. 1985. Deoxyribonucleic acid homologies among 96 strains of ammonia-oxidizing bacteria. Arch. Microbiol. *141*: 214–218.

Koops, H.P., H. Harms and H. Wehrmann. 1976. Isolation of a moderate halophilic ammonia-oxidizing bacterium, *Nitrosococcus mobilis* nov. sp. Arch. Microbiol. *107*: 277–282.

Koops, H.P. and U.C. Möller. 1992. The lithotrophic ammonia-oxidizing bacteria. *In* Balows, Trüper, Dworkin, Harder and Schleifer (Editors), The Prokaryotes: a Handbook on the Biology of Bacteria: Ecophysiology, Isolation, Identification, Applications, 2nd ed., Springer-Verlag, New York. 2625–2637.

Koops, H.P. and A. Pommerening-Röser. 2001. Distribution and ecophysiology of the nitrifying bacteria emphasizing cultured species. FEMS Microbiol. Ecol. *37*: 1–9.

Kopecny, J., M. Zorec, J. Mrazek, Y. Kobayashi and R. Marinsek-Logar. 2003. *Butyrivibrio hungatei* sp. nov. and *Pseudobutyrivibrio xylanivorans* sp. nov., butyrate-producing bacteria from the rumen. Int. J. Syst. Evol. Microbiol. *53*: 201–209.

Korolik, V., P.J. Coloe and V. Krishnapillai. 1988. A specific DNA probe for the identification of *Campylobacter jejuni*. J. Gen. Microbiol. *134*: 521–530.

Kosako, Y., E. Yabuuchi, T. Naka, N. Fujiwara and K. Kobayashi. 2000a. Proposal of *Sphingomonadaceae* fam. nov., consisting of *Sphingomonas* Yabuuchi et al. 1990, *Erythrobacter* Shiba and Shimidu 1982, *Erythromicrobium* Yurkov et al. 1994, *Porphyrobacter* Fuerst et al. 1993, *Zymomonas* Kluyver and van Niel 1936, and *Sandaracinobacter* Yurkov et al. 1997, with the type genus *Sphingomonas* Yabuuchi et al. 1990. Microbiol. Immunol. *44*: 563–575.

Kosako, Y., E. Yabuuchi, T. Naka, N. Fujiwara and K. Kobayashi. 2000b. *In* Validation of the publication of new names and new combinations previously effectively published outside the IJSEM. List No. 77. Int. J. Syst. Evol. Bacteriol. *50*: 1953.

Kosko, B. 1994. Fuzzy Thinking, Harper Collins Publishers, London.

Koussémon, M., Y. Combet-Blanc, B.K. Patel, J.L. Cayol, P. Thomas, J.L. Garcia and B. Ollivier. 2001. *Propionibacterium microaerophilum* sp. nov., a microaerophilic bacterium isolated from olive mill wastewater. Int. J. Syst. Evol. Microbiol. *51*: 1373–1382.

Kowalchuk, G.A., Z.S. Naoumenko, P.J. Derikx, A. Felske, J.R. Stephen and I.A. Arkhipchenko. 1999. Molecular analysis of ammonia-oxidizing bacteria of the beta subdivision of the class *Proteobacteria* in compost and composted materials. Appl. Environ. Microbiol. *65*: 396–403.

Kowalchuk, G.A., J.R. Stephen, W. De Boer, J.I. Prosser, T.M. Embley and J.W. Woldendorp. 1997. Analysis of ammonia-oxidizing bacteria of the beta subdivision of the class *Proteobacteria* in coastal sand dunes by denaturing gradient gel electrophoresis and sequencing of PCR-amplified 16S ribosomal DNA fragments. Appl. Environ. Microbiol. *63*: 1489–1497.

Krekeler, D., P. Sigalevich, A. Teske, H. Cypionka and Y. Cohen. 1997. A sulfate-reducing bacterium from the oxic layer of a microbial mat from Solar Lake (Sinai), *Desulfovibrio oxyclinae* sp. nov. Arch. Microbiol. *167*: 369–375.

Krekeler, D., P. Sigalevich, A. Teske, H. Cypionka and Y. Cohen. 2000. *In* Validation of the publication of new names and new combinations previously effectively published outside the IJSB. List no. 76. Int. J. Syst. Evol. Microbiol. *50*: 1699–1700.

Krichevsky, M.I. and L.M. Norton. 1974. Storage and manipulation of data by computers for determinative bacteriology. Int. J. Syst. Bacteriol. *24*: 525–531.

Krieg, N.R. and J.G. Holt (Editors). 1984. Bergey's Manual of Systematic Bacteriology, 1st Ed., Vol. 1, The Williams & Wilkins Co., Baltimore.

Kröckel, L., U. Schillinger, C.M. Franz, A. Bantleon and W. Ludwig. 2003. *Lactobacillus versmoldensis* sp. nov., isolated from raw fermented sausage. Int. J. Syst. Evol. Microbiol. *53*: 513–517.

Krooneman, J., F. Faber, A.C. Alderkamp, S.J. Elferink, F. Driehuis, I. Cleenwerck, J. Swings, J.C. Gottschal and M. Vancanneyt. 2002. *Lactobacillus diolivorans* sp. nov., a 1,2-propanediol-degrading bacterium isolated from aerobically stable maize silage. Int. J. Syst. Evol. Microbiol. *52*: 639–646.

Krumholz, L.R., J.P. McKinley, G.A. Ulrich and J.M. Suflita. 1997. Confined subsurface microbial communities in Cretaceous rock. Nature *386*: 64–66.

Krümmel, A. and H. Harms. 1982. Effect of organic matter on growth and cell yield of ammonia-oxidizing bacteria. Arch. Microbiol. *133*: 50–54.

Kuever, J., M. Konneke, A. Galushko and O. Drzyzga. 2001. Reclassification of *Desulfobacterium phenolicum* as *Desulfobacula phenolica* comb. nov. and description of strain SaxT as *Desulfotignum balticum* gen. nov., sp. nov. Int. J. Syst. Evol. Microbiol. *51*: 171–177.

Kuhner, C.H., C. Matthies, G. Acker, M. Schmittroth, A.S. Gossner and H.L. Drake. 2000. *Clostridium akagii* sp. nov. and *Clostridium acidisoli* sp. nov.: acid-tolerant, N_2-fixing clostridia isolated from acidic forest soil and litter. Int. J. Syst. Evol. Microbiol. *50*: 873–881.

Kuhner, C.H., S.S. Smith, K.M. Noll, R.S. Tanner and R.S. Wolfe. 1991. 7-Mercaptoheptanoylthreonine phosphate substitutes for heat-stable factor (mobile factor) for growth of *Methanomicrobium mobile*. Appl. Environ. Microbiol. *57*: 2891–2895.

Kukolya, J., I. Nagy, M. Laday, E. Toth, O. Oravecz, K. Marialigeti and L. Hornok. 2002. *Thermobifida cellulolytica* sp. nov., a novel lignocellulose-decomposing actinomycete. Int. J. Syst. Evol. Microbiol. *52*: 1193–1199.

Kurosawa, N., Y.H. Itoh and T. Itoh. 2003. Reclassification of *Sulfolobus hakonensis* Takayanagi et al. 1996 as *Metallosphaera hakonensis* comb. nov. based on phylogenetic evidence and DNA G+C content. Int. J. Syst. Evol. Microbiol. *53*: 1607–1608.

Kurtzman, C.P. 1986. The ARS Culture Collection: present status and new directions. Enzyme Microb. Technol. *8*: 328–333.

Kushmaro, A., E. Banin, Y. Loya, E. Stackebrandt and E. Rosenberg. 2001. *Vibrio shiloi* sp. nov., the causative agent of bleaching of the coral *Oculina patagonica*. Int. J. Syst. Evol. Microbiol. *51*: 1383–1388.

Kwon, S.W., J.S. Kim, I.C. Park, S.H. Yoon, D.H. Park, C.K. Lim and S.J. Go. 2003. *Pseudomonas koreensis* sp. nov., *Pseudomonas umsongensis* sp. nov. and *Pseudomonas jinjuensis* sp. nov., novel species from farm soils in Korea. Int. J. Syst. Evol. Microbiol. *53*: 21–27.

La Scola, B., R.J. Birtles, M.N. Mallet and D. Raoult. 1998. *Massilia timonae* gen. nov., sp. nov., isolated from blood of an immunocompromised patient with cerebellar lesions. J. Clin. Microbiol. *36*: 2847–2852.

La Scola, B., R.J. Birtles, M.N. Mallet and D. Raoult. 2000. *In* Validation of publication of new names and new combinations previously effectively published outside the IJSEM. List No. 73. Int. J. Syst. Evol. Microbiol. *50*: 423–424.

La Scola, B., F. Fenollar, P.E. Fournier, M. Altwegg, M.N. Mallet and D. Raoult. 2001. Description of *Tropheryma whipplei* gen. nov., sp. nov., the Whipple's disease bacillus. Int. J. Syst. Evol. Microbiol. *51*: 1471–1479.

La Scola, B., M.N. Mallet, P.A. Grimont and D. Raoult. 2002. Description of *Afipia birgiae* sp. nov. and *Afipia massiliensis* sp. nov. and recognition of *Afipia felis* genospecies A. Int. J. Syst. Evol. Microbiol. *52*: 1773–1782.

La Scola, B., M.N. Mallet, P.A. Grimont and D. Raoult. 2003. *Bosea eneae* sp. nov., *Bosea massiliensis* sp. nov. and *Bosea vestrisii* sp. nov., isolated from hospital water supplies, and emendation of the genus *Bosea* (Das et al. 1996). Int. J. Syst. Evol. Microbiol. *53*: 15–20.

Labeda, D.P. 2001. *Crossiella* gen. nov., a new genus related to *Streptoalloteichus*. Int. J. Syst. Evol. Microbiol. *51*: 1575–1579.

Labeda, D.P., J.M. Donahue, N.M. Williams, S.F. Sells and M.M. Henton. 2003. *Amycolatopsis kentuckyensis* sp. nov., *Amycolatopsis lexingtonensis* sp. nov. and *Amycolatopsis pretoriensis* sp. nov., isolated from equine placentas. Int. J. Syst. Evol. Microbiol. *53*: 1601–1605.

Labeda, D.P., K. Hatano, R.M. Kroppenstedt and T. Tamura. 2001. Revival of the genus *Lentzea* and proposal for *Lechevalieria* gen. nov. Int. J. Syst. Evol. Microbiol. *51*: 1045–1050.

Labeda, D.P. and R.M. Kroppenstedt. 2000. Phylogenetic analysis of *Saccharothrix* and related taxa: proposal for *Actinosynnemataceae* fam. nov. Int. J. Syst. Evol. Microbiol. *50*: 331–336.

Labrenz, M., M.D. Collins, P.A. Lawson, B.J. Tindall, P. Schumann and P. Hirsch. 1999. *Roseovarius tolerans* gen. nov., sp. nov., a budding bacterium with variable bacteriochlorophyll *a* production from hypersaline Ekho Lake. Int. J. Syst. Bacteriol. *49*: 137–147.

Labrenz, M., P.A. Lawson, B.J. Tindall, M.D. Collins and P. Hirsch. 2003. *Saccharospirillum impatiens* gen. nov., sp. nov., a novel γ-Proteobacterium isolated from hypersaline Ekho Lake (East Antarctica). Int. J. Syst. Evol. Microbiol. *53*: 653–660.

Labrenz, M., B.J. Tindall, P.A. Lawson, M.D. Collins, P. Schumann and P. Hirsch. 2000. *Staleya guttiformis* gen. nov., sp. nov. and *Sulfitobacter brevis* sp. nov., α-3-*Proteobacteria* from hypersaline, heliothermal and meromictic antarctic Ekho Lake. Int. J. Syst. Evol. Microbiol. *50*: 303–313.

Lack, J.G., S.K. Chaudhuri, R. Chakraborty, L.A. Achenbach and J.D. Coates. 2002a. Anaerobic biooxidation of Fe(II) by *Dechlorosoma suillum*. Microb. Ecol. *43*: 424–431.

Lack, J.G., S.K. Chaudhuri, S.D. Kelly, K.M. Kemner, S.M. O'Connor and J.D. Coates. 2002b. Immobilization of radionuclides and heavy metals through anaerobic bio-oxidation of Fe(II). Appl. Environ. Microbiol. *68*: 2704–2710.

Laffineur, K., V. Avesani, G. Cornu, J. Charlier, M. Janssens, G. Wauters and M. Delmee. 2003a. Bacteremia due to a novel *Microbacterium* species in a patient with leukemia and description of *Microbacterium paraoxydans* sp. nov. J. Clin. Microbiol. *41*: 2242–2246.

Laffineur, K., V. Avesani, G. Cornu, J. Charlier, M. Janssens, G. Wauters and M. Delmee. 2003b. *In* Validation of publication of new names and new combinations previously effectively published outside the IJSEM. List No. 92. Int. J. Syst. Evol. Microbiol. *53*: 935–937.

Lai, M.C. and S.C. Chen. 2001. *Methanofollis aquaemaris* sp. nov., a methanogen isolated from an aquaculture fish pond. Int. J. Syst. Evol. Microbiol. *51*: 1873–1880.

Lai, M.C., S.C. Chen, C.M. Shu, M.S. Chiou, C.C. Wang, M.J. Chuang, T.Y. Hong, C.C. Liu, L.J. Lai and J.J. Hua. 2002. *Methanocalculus taiwanensis* sp. nov., isolated from an estuarine environment. Int. J. Syst. Evol. Microbiol. *52*: 1799–1806.

Lake, J.A., R. Jain and M.C. Rivera. 1999. Mix and match in the tree of life. Science *283*: 2027–2028.

Lan, G.Q., Y.W. Ho and N. Abdullah. 2002. *Mitsuokella jalaludinii* sp. nov., from the rumens of cattle in Malaysia. Int. J. Syst. Evol. Microbiol. *52*: 713–718.

Langendijk, P.S., E.M. Kulik, H. Sandmeier, J. Meyer and J.S. van der Hoeven. 2001. Isolation of *Desulfomicrobium orale* sp. nov. and *Desulfovibrio* strain NY682, oral sulfate-reducing bacteria involved in human periodontal disease. Int. J. Syst. Evol. Bacteriol. *51*: 1035–1044.

Langer, P.R., A.A. Waldrop and D.C. Ward. 1981. Enzymatic synthesis of biotin-labeled polynucleotides: novel nucleic acid affinity probes. Proc. Natl. Acad. Sci. U.S.A. *78*: 6633–6637.

Lapage, S.P., S. Bascomb, W.R. Willcox and M.A. Curtis. 1973. Identification of bacteria by computer. General aspects and perspectives. J. Gen. Microbiol. *77*: 273–290.

Lapage, S.P., P.H.A. Sneath, E.F. Lessel, Jr., V.B.D. Skerman, H.P.R. Seeliger and W.A. Clark (Editors). 1975. International Code of Nomenclature of Bacteria, 1976 revision, American Society for Microbiology, Washington, DC.

Lapage, S.P., P.H.A. Sneath, E.F. Lessel Jr., V.B.D. Skerman, H.P.R. Seeliger and W.A. Clark (Editors). 1992. International Code of Nomenclature of Bacteria (1990) Revision. Bacteriological Code, American Society for Microbiology, Washington, DC.

Larsen, N., R. Overbeek, S. Pramanik, T.M. Schmidt, E.E. Selkov, O. Strunk, J.M. Tiedje and J.W. Urbance. 1997. Towards microbial data integration. J. Ind. Microbiol. Biotechnol. *18*: 68–72.

Lathe, R. 1985. Synthetic oligonucleotide probes deduced from amino acid sequence data: theoretical and practical considerations. J. Mol. Biol. *183*: 1–12.

Law-Brown, J. and P.R. Meyers. 2003. *Enterococcus phoeniculicola* sp. nov., a novel member of the enterococci isolated from the uropygial gland of the Red-billed Woodhoopoe, *Phoeniculus purpureus*. Int. J. Syst. Evol. Microbiol. *53*: 683–685.

Lawrence, J.G. and H. Ochman. 1998. Molecular archaeology of the *Escherichia coli* genome. Proc. Natl. Acad. Sci. U.S.A. *95*: 9413–9417.

Lawson, A.J., S.L. On, J.M. Logan and J. Stanley. 2001a. *Campylobacter hominis* sp. nov., from the human gastrointestinal tract. Int. J. Syst. Evol. Microbiol. *51*: 651–660.

Lawson, P.A., M.D. Collins, P. Schumann, B.J. Tindall, P. Hirsch and M. Labrenz. 2000a. New LL-diaminopimelic acid-containing actinomycetes from hypersaline, heliothermal and meromictic Antarctic Ekho Lake: *Nocardioides aquaticus* sp. nov. and *Friedmanniella lacustris* sp. nov. Syst. Appl. Microbiol. *23*: 458–458.

Lawson, P.A., M.D. Collins, P. Schumann, B.J. Tindall, P. Hirsch and M. Labrenz. 2000b. *In* Validation of publication of new names and new combinations previously effectively published outside the IJSEM. List No. 77. Int. J. Syst. Evol. Microbiol. *50*: 1953.

Lawson, P.A., E. Falsen, G. Foster, E. Eriksson, N. Weiss and M.D. Collins. 2001b. *Arcanobacterium pluranimalium* sp. nov., isolated from porpoise and deer. Int. J. Syst. Evol. Microbiol. *51*: 55–59.

Lawson, P.A., E. Falsen, E. Inganas, R.S. Weyant and M.D. Collins. 2002a. *Dysgonomonas mossii* sp. nov., from human sources. Syst. Appl. Microbiol. *25*: 194–197.

Lawson, P.A., E. Falsen, E. Inganas, R.S. Weyant and M.D. Collins. 2002b. *In* Validation of publication of new names and new combinations previously effectively published outside the IJSEM. List No. 88. Int. J. Syst. Evol. Microbiol. *52*: 1915–1916.

Lawson, P.A., E. Falsen, M. Ohlen and M.D. Collins. 2001c. *Aerococcus urinaehominis* sp. nov., isolated from human urine. Int. J. Syst. Evol. Microbiol. *51*: 683–686.

Lawson, P.A., E. Falsen, K. Truberg-Jensen and M.D. Collins. 2001d. *Aerococcus sanguicola* sp. nov., isolated from a human clinical source. Int. J. Syst. Evol. Microbiol. *51*: 475–479.

Lawson, P.A., G. Foster, E. Falsen, M. Ohlen and M.D. Collins. 2000c. *Atopobacter phocae* gen. nov., sp. nov., a novel bacterium isolated from common seals. Int. J. Syst. Evol. Microbiol. *50*: 1755–1760.

Lawson, P.A., N. Nikolaitchouk, E. Falsen, K. Westling and M.D. Collins. 2001e. *Actinomyces funkei* sp. nov., isolated from human clinical specimens. Int. J. Syst. Evol. Microbiol. *51*: 853–855.

Lawson, P.A., P. Papademas, C. Wacher, E. Falsen, R. Robinson and M.D. Collins. 2001f. *Lactobacillus cypricasei* sp. nov., isolated from Halloumi cheese. Int. J. Syst. Evol. Microbiol. *51*: 45–49.

Lawson, P.A., C. Wacher, I. Hansson, E. Falsen and M.D. Collins. 2001g. *Lactobacillus psittaci* sp. nov., isolated from a hyacinth macaw (*Anodorhynchus hyacinthinus*). Int. J. Syst. Evol. Microbiol. *51*: 967–970.

Layton, A.C., C.A. Lajoie, J.P. Easter, R. Jernigan, J. Sanseverino and G.S. Sayler. 1994. Molecular diagnostics and chemical analysis for assessing biodegradation of polychlorinated biphenyls in contaminated soils. J. Industrial Microbiol. Biotechnol. *13*: 392–401.

Le Minor, L. 1984. Genus II. *Salmonella*. *In* Kreig and Holt (Editors), Bergey's Manual of Systematic Bacteriology, Vol. 1, Williams and Wilkins, Baltimore. 427–458.

Le Minor, L. and M.Y. Popoff. 1987. Designation of *Salmonella enterica* sp. nov. nom. rev. as the type and only species of the genus *Salmonella*. Int. J. Syst. Bacteriol. *37*: 465–468.

Lebuhn, M., W. Achouak, M. Schloter, O. Berge, H. Meier, M. Barakat, A. Hartmann and T. Heulin. 2000. Taxonomic characterization of *Ochrobactrum* sp. isolates from soil samples and wheat roots, and description of *Ochrobactrum tritici* sp. nov. and *Ochrobactrum grignonense* sp. nov. Int. J. Syst. Evol. Microbiol. *50*: 2207–2223.

Lee, C.W., I.W. Wilkie, K.M. Townsend and A.J. Frost. 2000a. The demonstration of *Pasteurella multocida* in the alimentary tract of chickens after experimental oral infection. Vet. Microbiol. *72*: 47–55.

Lee, H.K., J. Chun, E.Y. Moon, S.H. Ko, D.S. Lee, H.S. Lee and K.S. Bae. 2001a. *Hahella chejuensis* gen. nov., sp. nov., an extracellular-polysaccharide-producing marine bacterium. Int. J. Syst. Evol. Microbiol. *51*: 661–666.

Lee, J.S., K.C. Lee, J.S. Ahn, T.I. Mheen, Y.R. Pyun and Y.H. Park. 2002a. *Weissella koreensis* sp. nov., isolated from kimchi. Int. J. Syst. Evol. Microbiol. *52*: 1257–1261.

Lee, J.S., K.C. Lee, Y.H. Chang, S.G. Hong, H.W. Oh, Y.R. Pyun and K.S. Bae. 2002b. *Paenibacillus daejeonensis* sp. nov., a novel alkaliphilic bacterium from soil. Int. J. Syst. Evol. Microbiol. *52*: 2107–2111.

Lee, J.S., K.C. Lee, Y.R. Pyun and K.S. Bae. 2003. *Arthrobacter koreensis* sp. nov., a novel alkalitolerant bacterium from soil. Int. J. Syst. Evol. Microbiol. *53*: 1277–1280.

Lee, J.S., Y.K. Shin, J.H. Yoon, M. Takeuchi, Y.R. Pyun and Y.H. Park. 2001b. *Sphingomonas aquatilis* sp. nov., *Sphingomonas koreensis* sp. nov., and *Sphingomonas taejonensis* sp. nov., yellow-pigmented bacteria isolated from natural mineral water. Int. J. Syst. Evol. Microbiol. *51*: 1491–1498.

Lee, S.D., M. Goodfellow and Y.C. Hah. 1999. A phylogenetic analysis of the genus *Catellatospora* based on 16S ribosomal DNA sequences, including transfer of *Catellatospora matsumotoense* to the genus *Micromonospora* as *Micromonospora matsumotoense* comb. nov. FEMS Microbiol Lett. *178*: 349–354.

Lee, S.D., M. Goodfellow and Y.C. Hah. 2000b. *In* Validation of publication of new names and new combinations previously effectively published outside the IJSEM. List No. 72. Int. J. Syst. Evol. Microbiol. *50*: 3–4.

Lee, S.D. and Y.C. Hah. 2001. *Amycolatopsis albidoflavus* sp. nov. Int. J. Syst. Evol. Microbiol. *51*: 645–650.

Lee, S.D. and Y.C. Hah. 2002. Proposal to transfer *Catellatospora ferruginea* and *'Catellatospora ishikariense'* to *Asanoa* gen. nov. as *Asanoa ferruginea* comb. nov. and *Asanoa ishikariensis* sp. nov., with emended description of the genus *Catellatospora*. Int. J. Syst. Evol. Microbiol. *52*: 967–972.

Lee, S.D., S.O. Kang and Y.C. Hah. 2000c. *Catellatospora koreensis* sp. nov., a novel actinomycete isolated from a gold-mine cave. Int. J. Syst. Evol. Microbiol. *50*: 1103–1111.

Lee, S.D., S.O. Kang and Y.C. Hah. 2000d. *Hongia* gen. nov., a new genus of the order *Actinomycetales*. Int. J. Syst. Evol. Microbiol. *50*: 191–199.

Lee, S.D., E.S. Kim, S.O. Kang and Y.C. Hah. 2002c. *Pseudonocardia spinosispora* sp. nov., isolated from Korean soil. Int. J. Syst. Evol. Microbiol. *52*: 1603–1608.

Lee, S.D., E.S. Kim, K.L. Min, W.Y. Lee, S.O. Kang and Y.C. Hah. 2001c. *Pseudonocardia kongjuensis* sp. nov., isolated from a gold mine cave. Int. J. Syst. Evol. Microbiol. *51*: 1505–1510.

Lee, S.D., E.S. Kim, J.H. Roe, J. Kim, S.O. Kang and Y.C. Hah. 2000e. *Saccharothrix violacea* sp. nov., isolated from a gold mine cave, and *Saccharothrix albidocapillata* comb. nov. Int. J. Syst. Evol. Microbiol. *50*: 1315–1323.

Lee, S., C. Malone and P.F. Kemp. 1993. Use of multiple 16S rRNA-targeted fluorescent probes to increase the signal strength and measure cellular RNA from natural planktonic bacteria. Mar. Ecol. Prog. Ser. *101*: 193–201.

Lehmann, K.B. and R. Neumann. 1896. Atlas and Grundress der Bakteriologie und Lehrbuch der speciellen bakteriologischen Diagnostik, 1st Ed., J.F. Lehmann, Munich.

Leisner, J.J., M. Vancanneyt, J. Goris, H. Christensen and G. Rusul. 2000. Description of *Paralactobacillus selangorensis* gen. nov., sp. nov., a new lactic acid bacterium isolated from chili bo, a Malaysian food ingredient. Int. J. Syst. Evol. Microbiol. *50*: 19–24.

Leisner, J.J., M. Vancanneyt, K. Lefebvre, K. Vandemeulebroecke, B. Hoste, N.E. Vilalta, G. Rusul and J. Swings. 2002. *Lactobacillus durianis* sp. nov., isolated from an acid-fermented condiment (tempoyak) in Malaysia. Int. J. Syst. Evol. Microbiol. *52*: 927–931.

Lerat, E., V. Daubin and N. Moran. 2003. From gene trees to organismal phylogeny in prokaryotes: the case of the *γ-Proteobacteria.* PLOS Biology. *1*: 1–9.

Levi, M.H., J. Bartell, L. Gandolfo, S.C. Smole, S.F. Costa, L.M. Weiss, L.K. Johnson, G. Osterhout and L.H. Herbst. 2003a. Characterization of *Mycobacterium montefiorense* sp. nov., a novel pathogenic mycobacterium from moray eels that is related to *Mycobacterium triplex.* J. Clin. Microbiol. *41*: 2147–2152.

Levi, M.H., J. Bartell, L. Gandolfo, S.C. Smole, S.F. Costa, L.M. Weiss, L.K. Johnson, G. Osterhout and L.H. Herbst. 2003b. *In* Validation of publication of new names and new combinations previously effectively published outside the IJSEM. List No. 94. Int. J. Syst. Evol. Microbiol. *53*: 1701–1702.

Li, M., J.H. Badger, X. Chen, S. Kwong, P. Kearney and H.Y. Zhang. 2001. An information-based sequence distance and its application to whole mitochondrial genome phylogeny. Bioinformatics. *17*: 149–154.

Li, M.G., W.J. Li, P. Xu, X.L. Cui, L.H. Xu and C.L. Jiang. 2003a. *Nocardiopsis xinjiangensis* sp. nov., a halophilic actinomycete isolated from a saline soil sample in China. Int. J. Syst. Evol. Microbiol. *53*: 317–321.

Li, W., B. Lanoot, Y. Zhang, M. Vancanneyt, J. Swings and Z. Liu. 2002a. *Streptomyces scopiformis* sp. nov., a novel streptomycete with fastigiate spore chains. Int. J. Syst. Evol. Microbiol. *52*: 1629–1633.

Li, W.J., S.K. Tang, E. Stackebrandt, R.M. Kroppenstedt, P. Schumann, L.H. Xu and C.L. Jiang. 2003b. *Saccharomonospora paurometabolica* sp. nov., a moderately halophilic actinomycete isolated from soil in China. Int. J. Syst. Evol. Microbiol. *53*: 1591–1594.

Li, W.J., P. Xu, S.K. Tang, L.H. Xu, R.M. Kroppenstedt, E. Stackebrandt and C.L. Jiang. 2003c. *Prauserella halophila* sp. nov. and *Prauserella alba* sp. nov., moderately halophilic actinomycetes from saline soil. Int. J. Syst. Evol. Microbiol. *53*: 1545–1549.

Li, W.J., P. Xu, L.P. Zhang, S.K. Tang, X.L. Cui, P.H. Mao, L.H. Xu, P. Schumann, E. Stackebrandt and C.L. Jiang. 2003d. *Streptomonospora alba* sp. nov., a novel halophilic actinomycete, and emended description of the genus *Streptomonospora* Cui et al. 2001. Int. J. Syst. Evol. Microbiol. *53*: 1421–1425.

Li, W.J., L.P. Zhang, P. Xu, X.L. Cui, Z.T. Lu, L.H. Xu and C.L. Jiang. 2002b. *Streptomyces beijiangensis* sp. nov., a psychrotolerant actinomycete isolated from soil in China. Int. J. Syst. Evol. Microbiol. *52*: 1695–1699.

Li, W.J., L.P. Zhang, P. Xu, X.L. Cui, L.H. Xu, Z. Zhang, P. Schumann, E. Stackebrandt and C.L. Jiang. 2003e. *Agromyces aurantiacus* sp. nov., isolated from a Chinese primeval forest. Int. J. Syst. Evol. Microbiol. *53*: 303–307.

Li, Z., Y. Kawamura, O. Shida, S. Yamagata, T. Deguchi and T. Ezaki. 2002c. *Bacillus okuhidensis* sp. nov., isolated from the Okuhida spa area of Japan. Int. J. Syst. Evol. Microbiol. *52*: 1205–1209.

Lie, T.J., M.L. Clawson, W. Godchaux and E.R. Leadbetter. 1999. Sulfidogenesis from 2-aminoethanesulfonate (taurine) fermentation by a morphologically unusual sulfate-reducing bacterium, *Desulforhopalus singaporensis* sp. nov. Appl. Environ. Microbiol. *65*: 3328–3334.

Lie, T.J., M.L. Clawson, W. Godchaux and E.R. Leadbetter. 2000. *In* Validation of the publication of new names and new combinations previously published outside the IJSB. List No. 76. Int. J. Syst. Bacteriol. *50*: 1699–1700.

Lilburn, T.G. and G.M. Garrity. 2004. Exploring prokaryotic taxonomy. Int. J. Syst. Evol. Microbiol. *54*: 7–13.

Lim, Y.W., K.S. Baik, S.K. Han, S.B. Kim and K.S. Bae. 2003. *Burkholderia sordidicola* sp. nov., isolated from the white-rot fungus *Phanerochaete sordida.* Int. J. Syst. Evol. Microbiol. *53*: 1631–1636.

Link, H.F. 1809. Observationes in ordines plantarum naturales. Dissertatio prima complectens anandrarum ordines Epiphytas, Mucedines, Gastromycos et Fungos. Ges. Nat. *3*: 3–42.

Linos, A., M.M. Berekaa, A. Steinbuchel, K.K. Kim, C. Sproer and R.M. Kroppenstedt. 2002. *Gordonia westfalica* sp. nov., a novel rubber-degrading actinomycete. Int. J. Syst. Evol. Microbiol. *52*: 1133–1139.

Lipschultz, F., O.C. Zafiriou, S.C. Wofsy, M.B. McElroy, F.W. Valois and S.W. Watson. 1981. Production of NO and N_2O by soil nitrifying bacteria. Nature (Lond.) *294*: 641–643.

Lipscomb, J.D., K.K. Andersson, E. Münck, T.A. Kent and A.B. Hooper. 1982. Resolution of multiple heme centers of hydroxylamine oxidoreductase from *Nitrosomonas.* II. Mossbauer spectroscopy. Biochemistry *21*: 3973–3976.

Lipski, A., E. Spieck, A. Makolla and K. Altendorf. 2001. Fatty acid profiles of nitrite-oxidizing bacteria reflect their phylogenetic heterogeneity. Syst. Appl. Microbiol. *24*: 377–384.

Lisdiyanti, P., H. Kawasaki, T. Seki, Y. Yamada, T. Uchimura and K.J. Komagata. 2000. Systematic study of the genus *Acetobacter* with descriptions of *Acetobacter indonesiensis* sp. nov., *Acetobacter tropicalis* sp. nov., *Acetobacter orleanensis* (Henneberg 1906) comb. nov., *Acetobacter lovaniensis* (Frateur 1950) comb. nov., and *Acetobacter estunensis* (Carr 1958) comb. nov. Gen. Appl. Microbiol. *46*: 147–165.

Lisdiyanti, P., H. Kawasaki, T. Seki, Y. Yamada, T. Uchimura and K.J. Komagata. 2001a. Identification of *Acetobacter* strains isolated from Indonesian sources, and proposals of *Acetobacter syzygii* sp. nov., *Acetobacter cibinongensis* sp. nov., and *Acetobacter orientalis* sp. nov. Gen. Appl. Microbiol. *47*: 119–131.

Lisdiyanti, P., H. Kawasaki, T. Seki, Y. Yamada, T. Uchimura and K. Komagata. 2001b. *In* Validation of the publication of new names and new combinations previously effectively published outside the IJSEM. List No. 51. Int. J. Syst. Evol. Microbiol. *51*: 263–265.

Lisdiyanti, P., H. Kawasaki, T. Seki, Y. Yamada, T. Uchimura and K. Komagata. 2002a. *In* Validation of the publication of new names and new combinations previously effectively published outside the IJSEM. List No. 84. Int. J. Syst. Evol. Microbiol. *52*: 3–4.

Lisdiyanti, P., H. Kawasaki, Y. Widyastuti, S. Saono, T. Seki, Y. Yamada, T. Uchimura and K. Komagata. 2002b. *Kozakia baliensis* gen. nov., sp. nov., a novel acetic acid bacterium in the *α-Proteobacteria.* Int. J. Syst. Evol. Microbiol. *52*: 813–818.

L'Haridon, S.L., M.L. Miroshnichenko, H. Hippe, M.L. Fardeau, E. Bonch-Osmolovskaya, E. Stackebrandt and C. Jeanthon. 2001. *Thermosipho geolei* sp. nov., a thermophilic bacterium isolated from a continental petroleum reservoir in Western Siberia. Int. J. Syst. Evol. Microbiol. *51*: 1327–1334.

L'Haridon, S., M.L. Miroshnichenko, H. Hippe, M.L. Fardeau, E.A. Bonch-Osmolovskaya, E. Stackebrandt and C. Jeanthon. 2002. *Petrotoga olearia* sp. nov. and *Petrotoga sibirica* sp. nov., two thermophilic bacteria isolated from a continental petroleum reservoir in Western Siberia. Int. J. Syst. Evol. Microbiol. *52*: 1715–1722.

L'Haridon, S., A.L. Reysenbach, A. Banta, P. Messner, P. Schumann, E. Stackebrandt and C. Jeanthon. 2003. *Methanocaldococcus indicus* sp. nov., a novel hyperthermophilic methanogen isolated from the Central Indian Ridge. Int. J. Syst. Evol. Microbiol. *53*: 1931–1935.

Liston, J., W. Weibe and R.R. Colwell. 1963. Quantitative approach to the study of bacterial species. J. Bacteriol. *85*: 1061–1070.

Liu, B. and X. Dong. 2002. *Lactobacillus pantheris* sp. nov., isolated from faeces of a jaguar. Int. J. Syst. Evol. Microbiol. *52*: 1745–1748.

Liu, H., Y. Xu, Y. Ma and P. Zhou. 2000. Characterization of *Micrococcus antarcticus* sp. nov., a psychrophilic bacterium from Antarctica. Int. J. Syst. Evol. Microbiol. *50*: 715–719.

Liu, J.R., R.S. Tanner, P. Schumann, N. Weiss, C.A. McKenzie, P.H. Janssen, E.M. Seviour, P.A. Lawson, T.D. Allen and R.J. Seviour. 2002a. Emended description of the genus *Trichococcus*, description of *Trichococcus collinsii* sp. nov., and reclassification of *Lactosphaera pasteurii* as *Trichococcus pasteurii* comb. nov. and of *Ruminococcus palustris* as *Tri-*

chococcus palustris comb. nov. in the low-G+C Gram-positive bacteria. Int. J. Syst. Evol. Microbiol. *52*: 1113–1126.

Liu, W.T., S. Hanada, T.L. Marsh, Y. Kamagata and K. Nakamura. 2002b. *Kineosphaera limosa* gen. nov., sp. nov., a novel Gram-positive polyhydroxyalkanoate-accumulating coccus isolated from activated sludge. Int. J. Syst. Evol. Microbiol. *52*: 1845–1849.

Liu, W.-T., T.L. Marsh, H. Cheng and L.J. Forney. 1997. Characterization of microbial diversity by determining terminal restriction fragment length polymorphisms of genes encoding 16S rRNA. Appl. Environ. Microbiol. *63*: 4516–4522.

Lizama, C., M. Monteoliva-Sanchez, A. Suarez-Garcia, R. Rosello-Mora, M. Aguilera, V. Campos and A. Ramos-Cormenzana. 2002. *Halorubrum tebenquichense* sp. nov., a novel halophilic archaeon isolated from the Atacama Saltern, Chile. Int. J. Syst. Evol. Microbiol. *52*: 149–155.

Llobet-Brossa, E., R. Rosselló-Mora and R.I. Amann. 1998. Microbial community composition of Wadden Sea sediments as revealed by fluorescence *in situ* hybridization. Appl. Environ. Microbiol. *64*: 2691–2696.

Lo Presti, F., S. Riffard, H. Meugnier, M. Reyrolle, Y. Lasne, P.A. Grimont, F. Grimont, R.F. Benson, D.J. Brenner, A.G. Steigerwalt, J. Etienne and J. Freney. 2001. *Legionella gresilensis* sp. nov. and *Legionella beliardensis* sp. nov., isolated from water in France. Int. J. Syst. Evol. Microbiol. *51*: 1949–1957.

Lockhart, W.R. and J. Liston. 1970. Methods for Numerical Taxonomy, American Society for Microbiology, Washington, DC.

Logan, B. 1998. A review of chlorate- and perchlorate-respiring microorganisms. Bioremed. J. *2*: 69–79.

Logan, B.E., H. Zhang, P. Mulvaney, M.G. Milner, I.M. Head and R.F. Unz. 2001. Kinetics of perchlorate- and chlorate-respiring bacteria. Appl. Environ. Microbiol. *67*: 2499–2506.

Logan, J.M.J., A. Burnens, D. Linton, A.J. Lawson and J. Stanley. 2000a. *Campylobacter lanienae* sp. nov., a new species isolated from workers in an abattoir. Int. J. Syst. Evol. Microbiol. *50*: 865–872.

Logan, N.A. 1994. Bacterial Systematics, Blackwell Scientific Publications, Oxford.

Logan, N.A., G. Forsyth, L. Lebbe, J. Goris, M. Heyndrickx, A. Balcaen, A. Verhelst, E. Falsen, A. Ljungh, H.B. Hansson and P. De Vos. 2002a. Polyphasic identification of *Bacillus* and *Brevibacillus* strains from clinical, dairy and industrial specimens and proposal of *Brevibacillus invocatus* sp. nov. Int. J. Syst. Evol. Microbiol. *52*: 953–966.

Logan, N.A., L. Lebbe, B. Hoste, J. Goris, G. Forsyth, M. Heyndrickx, B.L. Murray, N. Syme, D.D. Wynn-Williams and P. De Vos. 2000b. Aerobic endospore-forming bacteria from geothermal environments in northern Victoria Land, Antarctica, and Candlemas Island, South Sandwich archipelago, with the proposal of *Bacillus fumarioli* sp. nov. Int. J. Syst. Evol. Microbiol. *50*: 1741–1753.

Logan, N.A., L. Lebbe, A. Verhelst, J. Goris, G. Forsyth, M. Rodriguez-Diaz, M. Heyndrickx and P. De Vos. 2002b. *Bacillus luciferensis* sp. nov., from volcanic soil on Candlemas Island, South Sandwich archipelago. Int. J. Syst. Evol. Microbiol. *52*: 1985–1989.

López-López, A. , Pujalte, M.J. , Benlloch, S. , Mata-Roig, M., Rosselló-Mora, R., Garay, E. and Rodríguez-Valera, F.. 2002. *Thalassospira lucentensis* gen. nov., sp. nov., a new marine member of the *α-Proteobacteria*. Int. J. Syst. Evol. Microbiol. *52*: 1277–1283.

Loubinoux, J., F.M.A. Valente, I.A.C. Pereira, A. Costa, P.A.D. Grimont and A.E. Le Faou. 2002. Reclassification of the only species of the genus *Desulfomonas*, *Desulfomonas pigra*, as *Desulfovibrio piger* comb. nov. Int. J. Syst. Evol. Microbiol. *52*: 1305–1308.

Loveland-Curtze, J., P.P. Sheridan, K.R. Gutshall and J.E. Brenchley. 1999. Biochemical and phylogenetic analyses of psychrophilic isolates belonging to the *Arthrobacter* subgroup and description of *Arthrobacter psychrolactophilus*, sp. nov. Arch. Microbiol. *171*: 355–363.

Loveland-Curtze, J., P.P. Sheridan, K.R. Gutshall and J.E. Brenchley. 2000. *In* Validation of publication of new names and new combinations previously effectively published outside the IJSEM. List No. 72. Int. J. Syst. Evol. Microbiol. *50*: 3–4.

Lu, J. , Y. Nogi and H. Takami. 2001a. *Oceanobacillus iheyensis* gen. nov., sp. nov., a deep-sea extremely halotolerant and alkaliphilic species isolated from a depth of 1050 m on the Iheya Ridge. FEMS Microbiol. Lett. *205*: 291–297.

Lu, J. , Y. Nogi and H. Takami. 2002. *In* Validation of publication of new names and new combinations previously effectively published outside the IJSEM. List No. 85. Int. J. Syst. Evol. Microbiol. *52*: 685–690.

Lu, Z., Z. Liu, L. Wang, Y. Zhang, W. Qi and M. Goodfellow. 2001b. *Saccharopolyspora flava* sp. nov. and *Saccharopolyspora thermophila* sp. nov., novel actinomycetes from soil. Int. J. Syst. Evol. Microbiol. *51*: 319–325.

Lu, Z., L. Wang, Y. Zhang, Y. Shi, Z. Liu, E.T. Quintana and M. Goodfellow. 2003. *Actinomadura catellatispora* sp. nov. and *Actinomadura glauciflava* sp. nov., from a sewage ditch and soil in southern China. Int. J. Syst. Evol. Microbiol. *53*: 137–142.

Ludwig, W. 1995. Sequence databases. *In* Akkermans, van Elsas and de Bruijn (Editors), Molecular Microbial Ecology Manual, Vol. 3.3.5, Kluwer Academic Publishers, Dordrecht. pp. 1–22.

Ludwig, W., R.I. Amann, E. Martinez-Romero, W. Schönhuber, S. Bauer, A. Neef and K.H. Schleifer. 1998a. rRNA based identification and detection systems for rhizobia and other bacteria. Plant Soil *204*: 1–19.

Ludwig, W., S.H. Bauer, M. Bauer, I. Held, G. Kirchhof, R. Schulze, I. Huber, S. Spring, A. Hartmann and K.H. Schleifer. 1997. Detection and *in situ* identification of representatives of a widely distributed new bacterial phylum. FEMS Microbiol. Lett. *153*: 181–190.

Ludwig, W. and H.-P., Klenk. 2001. Overview: A phylogenetic backbone and taxonomic framework for procaryotic systematics. *In* Boone, Castenholz and Garrity (Editors), Bergey's Manual of Systematic Bacteriology, 2nd Edition, Volume 1. The archaea and the deeply branching and phototrophic bacteria. Springer, New York. 49–65.

Ludwig, W., J. Neumaier, N. Klugbauer, E. Brockmann, C. Roller, S. Jilg, K. Reetz, I. Schachtner, A. Ludvigsen, M. Bachleitner, U. Fischer and K.-H. Schleifer. 1993. Phylogenetic relationships of *Bacteria* based on comparative sequence analysis of elongation factor Tu and ATP-synthase β-subunit genes. Antonie Leeuwenhoek *64*: 285–305.

Ludwig, W., R. Rosselló-Mora, R. Aznar, S. Klugbauer, S. Spring, K. Reetz, C. Beimfohr, E. Brockmann, G. Kirchhof, S. Dorn, M. Bachleitner, N. Klugbauer, N. Springer, D. Lane, R. Nietupsky, M. Weiznegger and K.H. Schleifer. 1995. Comparative sequence analysis of 23S rRNA from *Proteobacteria*. Syst. Appl. Microbiol. *18*: 164–188.

Ludwig, W. and K.H. Schleifer. 1994. Bacterial phylogeny based on 16S and 23S rRNA sequence analysis. FEMS Microbiol. Rev. *15*: 155–173.

Ludwig, W., O. Strunk, S. Klugbauer, N. Klugbauer, M. Weizenegger, J. Neumaier, M. Bachleitner and K.H. Schleifer. 1998b. Bacterial phylogeny based on comparative sequence analysis. Electrophoresis *19*: 554–568.

Luijten, M.L., J. de Weert, H. Smidt, H.T. Boschker, W.M. de Vos, G. Schraa and A.J. Stams. 2003. Description of *Sulfurospirillum halorespirans* sp. nov., an anaerobic, tetrachloroethene-respiring bacterium, and transfer of *Dehalospirillum multivorans* to the genus *Sulfurospirillum* as *Sulfurospirillum multivorans* comb. nov. Int. J. Syst. Evol. Microbiol. *53*: 787–793.

Lunder, T., H. Sørum, G. Holstad, A.G. Steigerwalt, P. Mowinckel and D.J. Brenner. 2000. Phenotypic and genotypic characterization of *Vibrio viscosus* sp. nov. and *Vibrio wodanis* sp. nov. isolated from Atlantic salmon (*Salmo salar*) with "winter ulcer". Int. J. Syst. Evol. Microbiol. *50*: 427–450.

Lyimo, T.J., A. Pol, H.J. Op den Camp, H.R. Harhangi and G.D. Vogels. 2000. *Methanosarcina semesiae* sp. nov., a dimethylsulfide-utilizing methanogen from mangrove sediment. Int. J. Syst. Evol. Microbiol. *50*: 171–178.

MacAdoo, T.O. 1993. Nomenclatural literacy. *In* Goodfellow and O'Donnell (Editors), Handbook of New Bacterial Systematics, Academic Press, London. pp. 339–360.

Macario, A.J.L., F.A. Visser, J.B. van Lier and E. Conway de Macario. 1991. Topography of methanogenic subpopulations in a microbial consortium adapting to thermophilic conditions. J. Gen. Microbiol. *137*: 2179–2190.

Macgregor, B.J., D.P. Moser, E.W. Alm, K.H. Nealson and D.A. Stahl.

1997. Crenarchaeota in Lake Michigan sediment. Appl. Environ. Microbiol. *63*: 1178–1181.

Macián, M.C., W. Ludwig, R. Aznar, P.A. Grimont, K.H. Schleifer, E. Garay and M.J. Pujalte. 2001a. *Vibrio lentus* sp. nov., isolated from Mediterranean oysters. Int. J. Syst. Evol. Microbiol. *51*: 1449–1456.

Macián, M.C., W. Ludwig, K.H. Schleifer, E. Garay and M.J. Pujalte. 2001b. *Thalassomonas viridans* gen. nov., sp. nov., a novel marine γ-proteobacterium. Int. J. Syst. Evol. Microbiol. *51*: 1283–1289.

Macián, M.C., W. Ludwig, K.H. Schleifer, M.J. Pujalte and E. Garay. 2001c. *Vibrio agarivorans* sp. nov., a novel agarolytic marine bacterium. Int. J. Syst. Evol. Microbiol. *51*: 2031–2036.

Macián, M.C., M.J. Pujalte, M.C. Marquez, W. Ludwig, A. Ventosa, E. Garay and K.H. Schleifer. 2002. *Gelidibacter mesophilus* sp. nov., a novel marine bacterium in the family *Flavobacteriaceae*. Int. J. Syst. Evol. Microbiol. *52*: 1325–1329.

Macy, J.M., K. Nunan, K.D. Hagen, D.R. Dixon, P.J. Harbour, M. Cahill and L.I. Sly. 1996. *Chrysiogenes arsenatis* gen. nov., sp. nov., a new arsenate-respiring bacterium isolated from gold mine wastewater. Int. J. Syst. Bacteriol. *46*: 1153–1157.

Madigan, M.T., D.O. Jung, C.R. Woese and L.A. Achenbach. 2000. *Rhodoferax antarcticus* sp. nov., a moderately psychrophilic purple nonsulfur bacterium isolated from an Antarctic microbial mat. Arch. Microbiol. *173*: 269–277.

Madigan, M.T., D.O. Jung, C.R. Woese and L.A. Achenbach. 2001. *In* Validation of publication of new names and new combinations previously effectively published outside the IJSEM. List No. 80. Int. J. Syst. Evol. Microbiol. *51*: 793–794.

Magnusson, J., H. Jonsson, J. Schnurer and S. Roos. 2002. *Weissella soli* sp. nov., a lactic acid bacterium isolated from soil. Int. J. Syst. Evol. Microbiol. *52*: 831–834.

Maidak, B.L., J.R. Cole, C.T. Parker, Jr., G.M. Garrity, N. Larsen, B. Li, T.G. Lilburn, M.J. McCaughey, G.J. Olsen, R. Overbeek, S. Pramanik, T.M. Schmidt, J.M. Tiedje and C.R. Woese. 1999. A new version of the RDP (Ribosomal Database Project). Nucleic Acids Res. *27*: 171–173.

Maldonado, L., J.V. Hookey, A.C. Ward and M. Goodfellow. 2000. The *Nocardia salmonicida* clade, including descriptions of *Nocardia cummidelens* sp. nov., *Nocardia fluminea* sp. nov. and *Nocardia soli* sp. nov. Antonie Leeuwenhoek Int. J. Gen. Molec. Microbiol. *78*: 367–377.

Maldonado, L., J.V. Hookey, A.C. Ward and M. Goodfellow. 2001. *In* Validation of publication of new names and new combinations previously effectively published outside the IJSEM. List No. 82. Int. J. Syst. Evol. Microbiol. *51*: 1619–1620.

Maldonado, L.A., F.M. Stainsby and A.C. Ward. 2003a. *In* Validation of publication of new names and new combinations previously effectively published outside the IJSEM. List No. 93. Int. J. Syst. Evol. Microbiol. *53*: 1219–1220.

Maldonado, L.A., F.M. Stainsby, A.C. Ward and M. Goodfellow. 2003b. *Gordonia sinesedis* sp. nov., a novel soil isolate. Antonie Leeuwenhoek Int. J. Gen. Molec. Microbiol. *83*: 75–80.

Malik, K.A. and D. Claus. 1987. Bacterial culture collections: their importance to biotechnology and microbiology. *In* Russell (Editor), Biotechnology and Genetic Engineering Reviews, Vol. 5, Intercept, Ltd., Dorset. pp. 137–198.

Malmqvist, A., T. Welander, E. Moore, A. Ternström, G. Molin and I.-M. Stenström. 1994. *Ideonella dechloratans* gen. nov., sp. nov., a new bacterium capable of growing anaerobically with chlorate as an electron acceptor. Syst. Appl. Microbiol. *17*: 58–64.

Manachini, P.L., D. Mora, G. Nicastro, C. Parini, E. Stackebrandt, R. Pukall and M.G. Fortina. 2000. *Bacillus thermodenitrificans* sp. nov., nom. rev. Int. J. Syst. Evol. Microbiol. *50*: 1331–1337.

Manaia, C.M. and E.R. Moore. 2002. *Pseudomonas thermotolerans* sp. nov., a thermotolerant species of the genus *Pseudomonas sensu stricto*. Int. J. Syst. Evol. Microbiol. *52*: 2203–2209.

Manaia, C.M., B. Nogales and O.C. Nunes. 2003a. *Tepidiphilus margaritifer* gen. nov., sp. nov., isolated from a thermophilic aerobic digester. Int. J. Syst. Evol. Microbiol. *53*: 1405–1410.

Manaia, C.M., O.C. Nunes and B. Nogales. 2003b. *Caenibacterium thermophilum* gen. nov., sp. nov., isolated from a thermophilic aerobic digester of municipal sludge. Int. J. Syst. Evol. Microbiol. *53*: 1375–1382.

Mandel, M. 1969. New approaches to bacterial taxonomy: perspective and prospects. Annu. Rev. Microbiol. *23*: 239–274.

Manfio, G.P., E. Atalan, J. Zakrzewska-Czerwinska, M. Mordarski, C. Rodriguez, M.D. Collins and M. Goodfellow. 2003a. Classification of novel soil streptomycetes as *Streptomyces aureus* sp. nov., *Streptomyces laceyi* sp. nov. and *Streptomyces sanglieri* sp. nov. Antonie Leeuwenhoek Int. J. Gen. Molec.Microbiol. *83*: 245–255.

Manfio, G.P., E. Atalan, J. Zakrzewska-Czerwinska, M. Mordarski, C. Rodriguez, M.D. Collins and M. Goodfellow. 2003b. *In* Validation of publication of new names and new combinations previously effectively published outside the IJSEM. List No. 93. Int. J. Syst. Evol. Microbiol. *53*: 1219–1220.

Mannerova, S., R. Pantucek, J. Doskar, P. Svec, C. Snauwaert, M. Vancanneyt, J. Swings and I. Sedlacek. 2003. *Macrococcus brunensis* sp. nov., *Macrococcus hajekii* sp. nov. and *Macrococcus lamae* sp. nov., from the skin of llamas. Int. J. Syst. Evol. Microbiol. *53*: 1647–1654.

Männistö , M.K., P. Schumann, F.A. Rainey, P. Kämpfer, I. Tsitko, M.A. Tiirola and M.S. Salkinoja-Salonen. 2000. *Subtercola boreus* gen. nov., sp. nov. and *Subtercola frigoramans* sp. nov., two new psychrophilic actinobacteria isolated from boreal groundwater. Int. J. Syst. Evol. Microbiol. *50*: 1731–1739.

Mansch, R. and E. Bock. 1998. Biodeterioration of natural stone with special reference to nitrifying bacteria. Biodegradation *9*: 47–64.

Manz, W., R. Amann, W. Ludwig, M. Wagner and K.H. Schleifer. 1992. Phylogenetic oligodeoxynucleotide probes for the major subclasses of *Proteobacteria*: Problems and solutions. System. Appl. Microbiol. *15*: 593–600.

Manz, W., R.I. Amann, R. Szewzyk, U. Szewzyk, T.A. Stenstrom, P. Hutzler and K.H. Schleifer. 1995. *In situ* identification of *Legionellaceae* using 16S rRNA-targeted oligonucleotide probes and confocal laser scanning microscopy. Microbiology (Reading) *141*: 29–39.

Manz, W., U. Szewzyk, P. Ericsson, R.I. Amann, K.H. Schleifer and T.A. Stenström. 1993. *In situ* identification of bacteria in drinking water and adjoining biofilms by hybridization with 16S and 23S rRNA-directed fluorescent oligonucleotide probes. Appl. Environ. Microbiol. *59*: 2293–2298.

Manz, W., M. Wagner, R.I. Amann and K.H. Schleifer. 1994. *In situ* characterization of the microbial consortia active in two wastewater treatment plants. Water Res. *28*: 1715–1723.

Marchandin, H., E. Jumas-Bilak, B. Gay, C. Teyssier, H. Jean-Pierre, M.S. de Buochberg, C. Carriere and J.P. Carlier. 2003. Phylogenetic analysis of some *Sporomusa* sub-branch members isolated from human clinical specimens: description of *Megasphaera micronuciformis* sp. nov. Int. J. Syst. Evol. Microbiol. *53*: 547–553.

Marchesi, J.R., T. Sato, A.J. Weightman, T.A. Martin, J.C. Fry, S.J. Hiom and W.G. Wade. 1998. Design and evaluation of useful bacterium-specific PCR primers that amplify genes coding for bacterial 16S rRNA. Appl. Environ. Microbiol. *64*: 795–799.

Mardia, K.V., J.T. Kent and J.M. Bibby. 1979. Multivariate Analysis, Academic Press, London.

Margesin, R., C. Sproer, P. Schumann and F. Schinner. 2003. *Pedobacter cryoconitis* sp. nov., a facultative psychrophile from alpine glacier cryoconite. Int. J. Syst. Evol. Microbiol. *53*: 1291–1296.

Marmur, J. and P. Doty. 1961. Thermal renaturation of DNA. J. Mol. Biol. *3*: 584–594.

Marmur, J. and D. Lane. 1960. Strand separation and specific recombination in deoxyribonucleic acids: biological studies. Proc. Natl. Acad. Sci. U.S.A. *46*: 453–461.

Martin, S., M.C. Marquez, C. Sanchez-Porro, E. Mellado, D.R. Arahal and A. Ventosa. 2003. *Marinobacter lipolyticus* sp. nov., a novel moderate halophile with lipolytic activity. Int. J. Syst. Evol. Microbiol. *53*: 1383–1387.

Martin, S.M. and V.B.D. Skerman (Editors). 1972. World Directory of Collections of Cultures of Microorganisms, Wiley-Interscience, New York.

Martin, W., H. Brinkmann, C. Savonna and R. Cerff. 1993. Evidence for a chimeric nature of nuclear genomes: eubacterial origin of eukaryotic glyceraldehyde-3-phosphate dehydrogenase genes. Proc. Natl. Acad. Sci. U.S.A. *90*: 8692–8696.

Martin, W. and R. Cerff. 1986. Prokaryotic features of a nucleus-encoded enzyme. cDNA sequences for chloroplast and cytosolic glyceraldehyde-3-phosphate dehydrogenases from mustard (*Sinapis alba*). Eur. J. Biochem. *159*: 323–331.

Martinec, T., M. Kocur and I. Habetova. 1966–1967. František Král, founder of the first collection of microorganisms. Publ. Fac. Sci. Univ. J.E. Purkyně Brno. no. 475. *K38*: 261–265.

Martiny, H. and H.P. Koops. 1982. Incorporation of organic compounds into cell protein by lithotrophic, ammonia-oxidizing bacteria. Antonie van Leeuwenhoek. *48*: 327–336.

Maruyama, A., D Honda, H. Yamamoto, K. Kitamura and T. Higashihara. 2000. Phylogenetic analysis of psychrophilic bacteria isolated from the Japan Trench, including a description of the deep-sea species *Psychrobacter pacificensis* sp. nov. Int. J. Syst. Evol. Microbiol. *50*: 835–846.

Masuzawa, T., N. Takada, M. Kudeken, T. Fukui, Y. Yano, F. Ishiguro, Y. Kawamura, Y. Imai and T. Ezaki. 2001. *Borrelia sinica* sp. nov., a lyme disease-related *Borrelia* species isolated in China. Int. J. Syst. Evol. Microbiol. *51*: 1817–1824.

Maszenan, A.M. , R.J. Seviour, B.K.C. Patel and P. Schumann. 2002. *Quadricoccus australiensis* gen. nov., sp. nov., a β-proteobacterium from activated sludge biomass. Int. J. Syst. Evol. Microbiol. *52*: 223–228.

Maszenan, A.M., R.J. Seviour, B.K. Patel, P. Schumann, J. Burghardt, Y. Tokiwa and H.M. Stratton. 2000. Three isolates of novel polyphosphate-accumulating Gram-positive cocci, obtained from activated sludge, belong to a new genus, *Tetrasphaera* gen. nov., and description of two new species, *Tetrasphaera japonica* sp. nov. and *Tetrasphaera australiensis* sp. nov. Int. J. Syst. Evol. Microbiol. *50*: 593–603.

Matin, A. 1978. Organic nutrition of chemolithotrophic bacteria. Annu. Rev. Microbiol. *32*: 433–468.

Matsubara, H., K. Goto, T. Matsumura, K. Mochida, M. Iwaki, M. Niwa and K. Yamasato. 2002. *Alicyclobacillus acidiphilus* sp. nov., a novel thermo-acidophilic, ω-alicyclic fatty acid-containing bacterium isolated from acidic beverages. Int. J. Syst. Evol. Microbiol. *52*: 1681–1685.

Matsumoto, A., Y. Takahashi, T. Kudo, A. Seino, Y. Iwai and S. Omura. 2000. *Actinoplanes capillaceus* sp. nov., a new species of the genus *Actinoplanes.* Antonie Leeuwenhoek Int. J. Gen. Molec. Microbiol. *78*: 107–115.

Matsumoto, A., Y. Takahashi, T. Kudo, A. Seino, Y. Iwai and S. Omura. 2001. *In* Validation of publication of new names and new combinations previously effectively published outside the IJSEM. List No. 80. Int. J. Syst. Evol. Microbiol. *51*: 793–794.

Matsumoto, A., Y. Takahashi, M. Shinose, A. Seino, Y. Iwai and S. Omura. 2003. *Longispora albida* gen. nov., sp. nov., a novel genus of the family *Micromonosporaceae.* Int. J. Syst. Evol. Microbiol. *53*: 1553–1559.

Matsuyama, H., I. Yumoto, T. Kudo and O. Shida. 2003. *Rhodococcus tukisamuensis* sp. nov., isolated from soil. Int. J. Syst. Evol. Microbiol. *53*: 1333–1337.

Matthies, C., S. Evers, W. Ludwig and B. Schink. 2000a. *Anaerovorax odorimutans* gen. nov., sp. nov., a putrescine-fermenting, strictly anaerobic bacterium. Int. J. Syst. Evol. Microbiol. *50*: 1591–1594.

Matthies, C., C.H. Kuhner, G. Acker and H.L. Drake. 2001. *Clostridium uliginosum* sp. nov., a novel acid-tolerant, anaerobic bacterium with connecting filaments. Int. J. Syst. Evol. Microbiol. *51*: 1119–1125.

Matthies, C., N. Springer, W. Ludwig and B. Schink. 2000b. *Pelospora glutarica* gen. nov., sp. nov., a glutarate-fermenting, strictly anaerobic, spore-forming bacterium. Int. J. Syst. Evol. Microbiol. *50*: 645–648.

Matulewich, V.A., P.F. Strom and M.S. Finstein. 1975. Length of incubation for enumerating nitrifying bacteria present in various environments. Appl. Microbiol. *29*: 265–268.

Maxam, A.M. and W. Gilbert. 1980. Sequencing end-labeled DNA with base-specific chemical cleavages. Methods Enzymol. *65*: 499–560.

Maymó-Gatell, X., Y.T. Chien, J.M. Gossett and S.H. Zinder. 1997. Isolation of a bacterium that reductively dechlorinates tetrachloroethene to ethene. Science *276*: 1568–1571.

Maynard Smith, J. 1995. Do bacteria have population genetics? *In* Baumberg, Young, Wellington and Saunders (Editors), Population Genetics of Bacteria, Cambridge University Press, Cambridge. 1–12.

Mayr, E. 1942. Systematics and the Origin of Species, Columbia University Press, New York.

Mayr, E. 1998. Two empires or three? Proc. Natl. Acad. Sci. U.S.A. *95*: 9720–9723.

McCaig, A.E., T.M. Embley and J.I. Prosser. 1994. Molecular analysis of enrichment cultures of marine ammonia oxidizers. FEMS Microbiol. Lett. *120*: 363–367.

McCammon, S.A. and J.P. Bowman. 2000. Taxonomy of Antarctic *Flavobacterium* species: description of *Flavobacterium gillisiae* sp. nov., *Flavobacterium tegetincola* sp. nov., and *Flavobacterium xanthum* sp. nov., nom. rev. and reclassification of *[Flavobacterium] salegens* as *Salegentibacter salegens* gen. nov., comb. nov. Int. J. Syst. Evol. Microbiol. *50*: 1055–1063.

McDonald, I.R., N.V. Doronina, Y.A. Trotsenko, C. McAnulla and J.C. Marrell. 2001. *Hyphomicrobium chloromethanicum* sp. nov. and *Methylobacterium chloromethanicum* sp. nov., chloromethane-utilizing bacteria isolated from a polluted environment. Int. J. Syst. Evol. Microbiol. *51*: 119–122.

McGowan, V.F. and V.B.D. Skerman (Editors). 1982. World Directory of Collections of Cultures of Microorganisms, 2nd Ed., World Data Center for Microorganisms, Brisbane.

McLean, R.J.C., M. Whiteley, D.J. Stickler and W.C. Fuqua. 1997. Evidence of autoinducer activity in naturally occurring biofilms. FEMS Microbiol. Lett. *154*: 259–263.

McTavish, H., J.A. Fuchs and A.B. Hooper. 1993a. Sequence of the gene coding for ammonia monooxygenase in *Nitrosomonas europaea.* J. Bacteriol. *175*: 2436–2444.

McTavish, H., F. LaQuier, D. Arciero, M. Logan, G. Mundfrom, J.A. Fuchs and A.B. Hooper. 1993b. Multiple copies of genes coding for electron transport proteins in the bacterium *Nitrosomonas europaea.* J. Bacteriol. *175*: 2445–2447.

Mechichi, T., M.L. Fardeau, M. Labat, J.L. Garcia, F. Verhe and B.K. Patel. 2000a. *Clostridium peptidivorans* sp. nov., a peptide-fermenting bacterium from an olive mill wastewater treatment digester. Int. J. Syst. Evol. Microbiol. *50*: 1259–1264.

Mechichi, T., M. Labat, J.L. Garcia, P. Thomas and B.K.C. Patel. 199 Characterization of a new xylanolytic bacterium, *Clostridium xyla vorans* sp. nov. Syst. Appl. Microbiol. *22*: 366–371.

Mechichi, T., M. Labat, T.H.S. Woo, P. Thomas, J.L. Garcia and B Patel. 1998. *Eubacterium aggregans* sp. nov., a new homoaceto bacterium from olive mill wastewater treatment digestor. Ana *4*: 283–291.

Mechichi, T., M. Labat, T.H.S. Woo, P. Thomas, J.L. Garcia an Patel. 2000b. *In* Validation of publication of new names combinations previously effectively published outside the I No. 72. Int. J. Syst. Evol. Microbiol. *50*: 3–4.

Mechichi, T., M. Labat, T.H.S. Woo, P. Thomas, J.L. Garcia Patel. 2000c. *In* Validation of publication of new nam combinations previously effectively published outside th No. 76. Int. J. Syst. Evol. Microbiol. *50*: 1699–1700.

Mechichi, T., E. Stackebrandt and G. Fuchs. 2003. *Alicy cans* gen. nov., sp. nov., a cyclohexanol-degrading, ni proteobacterium. Int. J. Syst. Evol. Microbiol. *53*: 1

Mechichi, T., E. Stackebrandt, N. Gad'on and G. Fu genetic and metabolic diversity of bacteria degrad pounds under denitrifying conditions, and des *phenylacetica* sp. nov., *Thauera aminoaromatica* sp *buckelii* sp. nov. Arch. Microbiol. *178*: 26–35.

Mechichi, T., E. Stackebrandt, N. Gad'on and G. dation of publication of new names and new c effectively published outside the IJSEM. List Microbiol. *52*: 1437–1438.

Meehan, C., A.J. Bjourson and G. McMullan.

ucens sp. nov., a synthetic azo dye decolorizing bacterium from industrial wastewater. Int. J. Syst. Evol. Microbiol. *51*: 1681–1685.

Meincke, M., E. Krieg and E. Bock. 1989. *Nitrosovibrio* spp., the dominant ammonia oxidizing bacteria in building sandstone. Appl. Environ. Microbiol. *55*: 2108–2110.

Menes, R.J. and L. Muxí. 2002. *Anaerobaculum mobile* sp. nov., a novel anaerobic, moderately thermophilic, peptide-fermenting bacterium that uses crotonate as an electron acceptor, and emended description of the genus *Anaerobaculum*. Int. J. Syst. Evol. Microbiol. *52*: 157–164.

Meredith, R. 1997. Winning the race to invent. Nat. Biotechnol. *15*: 283–284.

Mergaert, J., M.C. Cnockaert and J. Swings. 2002. *Fulvimonas soli* gen. nov., sp. nov., a γ-proteobacterium isolated from soil after enrichment on acetylated starch plastic. Int. J. Syst. Evol. Microbiol. *52*: 1285–1289.

Mergaert, J., M.C. Cnockaert and J. Swings. 2003a. *Thermomonas fusca* sp. nov. and *Thermomonas brevis* sp. nov., two mesophilic species isolated from a denitrification reactor with poly(ε-caprolactone) plastic granules as fixed bed, and emended description of the genus *Thermomonas*. Int. J. Syst. Evol. Microbiol. *53*: 1961–1966.

Mergaert, J., D. Lednicka, J. Goris, M.C. Cnockaert, P. De Vos and J. Swings. 2003b. Taxonomic study of *Cellvibrio* strains and description of *Cellvibrio ostraviensis* sp. nov., *Cellvibrio fibrivorans* sp. nov. and *Cellvibrio gandavensis* sp. nov. Int. J. Syst. Evol. Microbiol. *53*: 465–471.

Messick, J.B., P.G. Walker, W. Raphael, L. Berent and X. Shi. 2002. *Mycoplasma haemodidelphidis* sp. nov., *Mycoplasma haemolamae* sp. nov. and *Mycoplasma haemocanis* comb. nov., haemotrophic parasites from a naturally infected opossum (*Didelphis virginiana*), alpaca (*Lama pacos*) and dog (*Canis familiaris*): phylogenetic and secondary structural relatedness of their 16S rRNA genes to other mycoplasmas. Int. J. Syst. Evol. Microbiol. *52*: 693–698.

Mevs, U., E. Stackebrandt, P. Schumann, C.A. Gallikowski and P. Hirsch. 2000. *Modestobacter multiseptatus* gen. nov., sp. nov., a budding actinomycete from soils of the Asgard Range (Transantarctic Mountains). Int. J. Syst. Evol. Microbiol. *50*: 337–346.

…ers, P.R., D.S. Porter, C. Omorogie, J.M. Pule and T. Kwetane. 2003. …eptomyces speibonae sp. nov., a novel streptomycete with blue substrate …elium isolated from South African soil. Int. J. Syst. Evol. Microbiol. …01–805.

…u, U., L.A. Achenbach and J.D. Coates. 2000. Isolation and …ization of two novel (per)chlorate-reducing bacteria from … lagoons. *In* Urbansky (Editor), Perchlorate in the Envi… …wer Academic/Plenum, New York. pp. 271–283.

…n, S., G. Bourg, E. Jumas-Bilak, P. Guigue-Talet, A. … D. O'Callaghan and M. Ramuz. 1997. Genome struc… …y in the genus *Brucella*. J. Bacteriol. *179*: 3244–

…nkunde für Landwirte, Berlin.

… neues System der Bakterien. Arb. Bakteriol. …38.

…etes (Bacteria, Bakterien). *In* Engler and …familien, Tiel I, Abt. 1a, W. Englemann,

…kterien, Vol. 1, Gustav Fischer, Jena.

…terien, Vol. 2, Gustav Fischer, Jena.

…e, F.S. Colwell and D.R. Boone. 2003a. … deep marine sediments that contain …on of *Methanoculleus submarinus* sp. … 3311–3316.

…S. Colwell and D.R. Boone. 2003b. … names and new combinations …de the IJSEM. List No. 94. Int.

Mil…nd M.T. Madigan. 2000. *Rho*… …an alkaliphilic purple non… …oda lakes. Arch. Microbiol.

… M.T. Madigan. 2001. *In*

Validation of the publication of new names and new combinations previously effectively published outside the IJSB. List No. 80. Int. J. Syst. Bacteriol. *51*: 793–794.

Miller, D.J. and D.J.D. Nicholas. 1985. Characterization of a soluble cytochrome oxidase-nitrite reductase from *Nitrosomonas europaea*. J. Gen. Microbiol. *131*: 2851–2854.

Miller, D.J., P.M. Wood and D.J.D. Nicholas. 1984. Further characterization of cytochrome P_{460} in *Nitrosomonas europaea*. J. Gen. Microbiol. *130*: 3049–3054.

Miller, J.M. and C.M. O'Hara. 1995. Substrate utilization systems for the identifcation of bacteria and yeasts. *In* Murray, Baron, Pfaller, Tenover and Yolken (Editors), Manual of Clinical Microbiology, 6th Ed., American Society for Microbiology, Washington, D.C. pp. 103–109.

Miller, T.L. and C. Lin. 2002. Description of *Methanobrevibacter gottschalkii* sp. nov., *Methanobrevibacter thaueri* sp. nov., *Methanobrevibacter woesei* sp. nov. and *Methanobrevibacter wolinii* sp. nov. Int. J. Syst. Evol. Microbiol. *52*: 819–822.

Miranda-Tello, E., M.L. Fardeau, J. Sepulveda, L. Fernandez, J.L. Cayol, P. Thomas and B. Ollivier. 2003. *Garciella nitratireducens* gen. nov., sp. nov., an anaerobic, thermophilic, nitrate- and thiosulfate-reducing bacterium isolated from an oilfield separator in the Gulf of Mexico. Int. J. Syst. Evol. Microbiol. *53*: 1509–1514.

Miroshnichenko, M.L., H. Hippe, E. Stackebrandt, N.A. Kostrikina, N.A. Chernyh, C. Jeanthon, T.N. Nazina, S.S. Belyaev and E.A. Bonch-Osmolovskaya. 2001a. Isolation and characterization of *Thermococcus sibiricus* sp. nov. from a Western Siberia high-temperature oil reservoir. Extremophiles *5*: 85–91.

Miroshnichenko, M.L., H. Hippe, E. Stackebrandt, N.A. Kostrikina, N.A. Chernyh, C. Jeanthon, T.N. Nazina, S.S. Belyaev and E.A. Bonch-Osmolovskaya. 2001b. *In* Validation of publication of new names and new combinations previously effectively published outside the IJSEM. List No. 82. Int. J. Syst. Evol. Microbiol. *51*: 1619–1620.

Miroshnichenko, M.L., N.A. Kostrikina, N.A. Chernyh, N.V. Pimenov, T.P. Tourova, A.N. Antipov, S. Spring, E. Stackebrandt and E.A. Bonch-Osmolovskaya. 2003a. *Caldithrix abyssi* gen. nov., sp. nov., a nitrate-reducing, thermophilic, anaerobic bacterium isolated from a Mid-Atlantic Ridge hydrothermal vent, represents a novel bacterial lineage. Int. J. Syst. Evol. Microbiol. *53*: 323–329.

Miroshnichenko, M.L., N.A. Kostrikina, S. L'Haridon, C. Jeanthon, H. Hippe, E. Stackebrandt and E.A. Bonch-Osmolovskaya. 2002. *Nautilia lithotrophica* gen. nov., sp. nov., a thermophilic sulfur-reducing epsilonproteobacterium isolated from a deep-sea hydrothermal vent. Int. J. Syst. Evol. Microbiol. *52*: 1299–1304.

Miroshnichenko, M.L., S. L'Haridon, C. Jeanthon, A.N. Antipov, N.A. Kostrikina, B.J. Tindall, P. Schumann, S. Spring, E. Stackebrandt and E.A. Bonch-Osmolovskaya. 2003b. *Oceanithermus profundus* gen. nov., sp. nov., a thermophilic, microaerophilic, facultatively chemolithoheterotrophic bacterium from a deep-sea hydrothermal vent. Int. J. Syst. Evol. Microbiol. *53*: 747–752.

Miroshnichenko, M.L., S. L'Haridon, O. Nercessian, A.N. Antipov, N.A. Kostrikina, B.J. Tindall, P. Schumann, S. Spring, E. Stackebrandt, E.A. Bonch-Osmolovskaya and C. Jeanthon. 2003c. *Vulcanithermus mediatlanticus* gen. nov., sp. nov., a novel member of the family *Thermaceae* from a deep-sea hot vent. Int. J. Syst. Evol. Microbiol. *53*: 1143–1148.

Miroshnichenko, M.L., A.I. Slobodkin, N.A. Kostrikina, S. L'Haridon, O. Nercessian, S. Spring, E. Stackebrandt, E.A. Bonch-Osmolovskaya and C. Jeanthon. 2003d. *Deferribacter abyssi* sp. nov., an anaerobic thermophile from deep-sea hydrothermal vents of the Mid-Atlantic Ridge. Int. J. Syst. Evol. Microbiol. *53*: 1637–1641.

Mirzabekov, A.D. 1994. DNA sequencing by hybridization: a megasequencing method and a diagnostic tool? Trends Biotechnol. *12*: 27–32.

Miyamoto, Y. and K. Itoh. 2000. *Bacteroides acidifaciens* sp. nov., isolated from the caecum of mice. Int. J. Syst. Evol. Microbiol. *50*: 145–148.

Mobarry, B.K., M. Wagner, V. Urbain, B.E. Rittmann and D.A. Stahl. 1996. Phylogenetic probes for analyzing abundance and spatial organization of nitrifying bacteria. Appl. Environ. Microbiol. *62*: 2156–2162.

Mohn, W.W. and J.M. Tiedje. 1992. Microbial reductive dehalogenation. Microbiol. Rev. *56*: 482–507.

Molisch, H. 1907. Die Purpurbakterien Nach Neuen Untersuchungen, G. Fischer, Jena.

Monciardini, P., L. Cavaletti, P. Schumann, M. Rohde and S. Donadio. 2003. *Conexibacter woesei* gen. nov., sp. nov., a novel representative of a deep evolutionary line of descent within the class *Actinobacteria*. Int. J. Syst. Evol. Microbiol. *53*: 569–576.

Monserrate, E., S.B. Leschine and E. Canale-Parola. 2001. *Clostridium hungatei* sp. nov., a mesophilic, N_2-fixing cellulolytic bacterium isolated from soil. Int. J. Syst. Evol. Microbiol. *51*: 123–132.

Montalvo-Rodríguez, R., J. Lopez-Garriga, R.H. Vreeland, A. Oren, A. Ventosa and M. Kamekura. 2000. *Haloterrigena thermotolerans* sp. nov., a halophilic archaeon from Puerto Rico. Int. J. Syst. Evol. Microbiol. *50*: 1065–1071.

Montgomery, L., B. Flesher and D. Stahl. 1988. Transfer of *Bacteroides succinogenes* (Hungate) to *Fibrobacter*, gen. nov. as *Fibrobacter succinogenes*, comb. nov., and description of *Fibrobacter intestinalis*, sp. nov. Int. J. Syst. Bacteriol. *38*: 430–435.

Moore, W.E.C. and Moore, L.V.H. (Editors). 1989. Index of the Bacterial and Yeast Nomenclatural Changes Published in the *International Journal of Systematic Bacteriology* since the 1980 Approved Lists of Bacterial Names (January 1, 1980 to January 1, 1989), American Society for Microbiology, Washington, DC.

Mora, D., M. Scarpellini, L. Franzetti, S. Colombo and A. Galli. 2003. Reclassification of *Lactobacillus maltaromicus* (Miller et al. 1974) DSM 20342(T) and DSM 20344 and *Carnobacterium piscicola* (Collins et al. 1987) DSM 20730(T) and DSM 20722 as *Carnobacterium maltaromaticum* comb. nov. Int. J. Syst. Evol. Microbiol. *53*: 675–678.

Moreira, C., F.A. Rainey, M.F. Nobre, M.T. da Silva and M.S. da Costa. 2000. *Tepidimonas ignava* gen. nov., sp. nov., a new chemolithoheterotrophic and slightly thermophilic member of the *β-Proteobacteria*. Int. J. Syst. Evol. Microbiol. *50*: 735–742.

Mori, K., S. Hanada, A. Maruyama and K. Marumo. 2002. *Thermanaeromonas toyohensis* gen. nov., sp. nov., a novel thermophilic anaerobe isolated from a subterranean vein in the Toyoha Mines. Int. J. Syst. Evol. Microbiol. *52*: 1675–1680.

Mori, K., H. Yamamoto, Y. Kamagata, M. Hatsu and K. Takamizawa. 2000. *Methanocalculus pumilus* sp. nov., a heavy-metal-tolerant methanogen isolated from a waste-disposal site. Int. J. Syst. Evol. Microbiol. *50*: 1723–1729.

Mormile, M.R., M.F. Romine, M.T. Garcia, A. Ventosa, T.J. Bailey and B.M. Peyton. 1999. *Halomonas campisalis* sp. nov., a denitrifying, moderately haloalkaliphilic bacterium. Syst. Appl. Microbiol. *22*: 551–558.

Mormile, M.R., M.F. Romine, M.T. Garcia, A. Ventosa, T.J. Bailey and B.M. Peyton. 2000. *In* Validation of the publication of new names and new combinations previously effectively published outside the IJSEM, List No. 74. Int. J. Syst. Evol. Microbiol. *50*: 949–950.

Morotomi, M., N. Yuki, Y. Kado, A. Kushiro, T. Shimazaki, K. Watanabe and T. Yuyama. 2002. *Lactobacillus equi* sp. nov., a predominant intestinal *Lactobacillus* species of the horse isolated from faeces of healthy horses. Int. J. Syst. Evol. Microbiol. *52*: 211–214.

Moseley, S.L., P. Echeverria, J. Seriwatana, C. Tirapat, W. Chaicumpa, T. Sakuldaipeara and S. Falkow. 1982. Identification of enterotoxigenic *Escherichia coli* by colony hybridization using three enterotoxin gene probes. J. Infect. Dis. *145*: 863–869.

Moshkovski, S.D. 1945. Cytotropic inducers of infection and the classification of the *Rickettsiae* with *Chlamydozoa* (in Russian, English summary). Adv. Mod. Biol. (Moscow). *19*: 1–44.

Mouné, S., C. Eatock, R. Matheron, J.C. Willison, A. Hirschler, R. Herbert and P. Caumette. 2000. *Orenia salinaria* sp. nov., a fermentative bacterium isolated from anaerobic sediments of Mediterranean salterns. Int. J. Syst. Evol. Microbiol. *50*: 721–729.

Mukai, T., K. Arihara, A. Ikeda, K. Nomura, F. Suzuki and H. Ohori. 2003. *Lactobacillus kitasatonis* sp. nov., from chicken intestine. Int. J. Syst. Evol. Microbiol. *53*: 2055–2059.

Mukamolova, G.V., A.S. Kaprelyants, D.I. Young, M. Young and D.B. Kell. 1998. A bacterial cytokine. Proc. Natl. Acad. Sci. U.S.A. *95*: 8916–8921.

Mulder, A., A.A. Vandegraaf, L.A. Robertson and J.G. Kuenen. 1995. Anaerobic ammonium oxidation discovered in a denitrifying fluidized-bed reactor. FEMS Microbiol. Ecol. *16*: 177–183.

Müller, M.R., M.A. Ehrmann and R.F. Vogel. 2000. *Lactobacillus frumenti* sp. nov., a new lactic acid bacterium isolated from rye-bran fermentations with a long fermentation period. Int. J. Syst. Evol. Microbiol. *50*: 2127–2133.

Müller, O.F. 1773. Vermium Terrestrium et Fluviatilium, seu Animalium Infusoriorum, Helminthicorum et Testaceorum, non Marinorum. Succincta Historia. *1*: 1–135.

Müller, O.F. 1786. Animalcula Infusoria Fluviatilia et Marina, quae Detexit, Systematice Descripsit et ad Vivum Delineari Curavit,

Munsch, P., T. Alatossava, N. Marttinen, J.M. Meyer, R. Christen and L. Gardan. 2002. *Pseudomonas costantinii* sp. nov., another causal agent of brown blotch disease, isolated from cultivated mushroom sporophores in Finland. Int. J. Syst. Evol. Microbiol. *52*: 1973–1983.

Murdoch, D.A. and H.N. Shah. 1999. Reclassification of *Peptostreptococcus magnus* (Prevot 1933) Holdeman and Moore 1972 as *Finegoldia magna* comb. nov. and *Peptostreptococcus micros* (Prevot 1933) Smith 1957 as *Micromonas micros* comb. nov. Anaerobe *5*: 555–559.

Murdoch, D.A. and H.N. Shah. 2000. *In* Validation of publication of new names and new combinations previously effectively published outside the IJSEM. List No. 75. Int. J. Syst. Evol. Microbiol. *50*: 1415–1417.

Murray, R.G.E. and K.-H., Schleifer. 1994. Taxonomic notes: a proposal for recording the properties of putative taxa of procaryotes. Int. J. Syst. Bacteriol. *44*: 174–176.

Murray, R.G.E. and E. Stackebrandt. 1995. Taxonomic note: implementation of the provisional status *Candidatus* for incompletely described procaryotes. Int. J. Syst. Bacteriol. *45*: 186–187.

Muyzer, G. and K. Smalla. 1998. Application of denaturing gradient gel electrophoresis (DGGE) and temperature gradient gel electrophoresis (TGGE) in microbial ecology. Antonie Leeuwenhoek *73*: 127–141.

Myhr, S. and T. Torsvik. 2000. *Denitrovibrio acetiphilus*, a novel genus and species of dissimilatory nitrate-reducing bacterium isolated from a[n] oil reservoir model column. Int. J. Syst. Evol. Microbiol. *50*: 161[1]–1619.

Mylvaganam, S. and P.P. Dennis. 1992. Sequence heterogeneity betw[een] the two genes encoding 16S rRNA from the halophilic archae[bac]terium *Haloarcula marismortui*. Genetics. *130*: 399–410.

Nakagawa, S., K. Takai, K. Horikoshi and Y. Sako. 2003. *Persep[honella] hydrogeniphila* sp. nov., a novel thermophilic, hydrogen-oxidizi[ng bac]terium from a deep-sea hydrothermal vent chimney. Int. J. Sy[st. Evol.] Microbiol. *53*: 863–869.

Nakamura, L.K., O. Shida, H. Takagi and K. Komagata. 200[2]. [Bacillus] *pycnus* sp. nov. and *Bacillus neidei* sp. nov., round-spored ba[cteria from] soil. Int. J. Syst. Evol. Microbiol. *52*: 501–505.

Nakazawa, F., S.E. Poco, Jr., M. Sato, T. Ikeda, S. Kalfas, G. S[…] E. Hoshino. 2002. Taxonomic characterization of *Mog*[…] *ersum* sp. nov. and *Mogibacterium neglectum* sp. nov., iso[…] man oral cavities. Int. J. Syst. Evol. Microbiol. *52*: 11[…]

Nakazawa, F., M. Sato, S.E. Poco, T. Hashimura, T. Ike[…] Sundqvist and E. Hoshino. 2000. Description of *Mog*[…] gen. nov., sp. nov. and *Mogibacterium vescum* gen. […] reclassification of *Eubacterium timidum* (Holdema[…] *gibacterium timidum* gen. nov., comb. nov. Int. J. S[…] *50*: 679–688.

Nam, S.W., J. Chun, S. Kim, W. Kim, J. Zakrzews[…] Goodfellow. 2003a. *Tsukamurella spumae* sp. no[…] associated with foaming in activated sludge […] biol. *26*: 367–375.

Nam, S.W., J. Chun, S. Kim, W. Kim, J. Zakrz[…] Goodfellow. 2003b. *In* Validation of publ[…] new combinations previously effectively p[…] List No. 94. Int. J. Syst. Evol. Microbiol. […]

Navarro, E., P. Simonet, P. Normand and[…]

zation of natural populations of *Nitrobacter* spp. using PCR/RFLP analysis of the ribosomal intergenic spacer. Arch. Microbiol. *157*: 107–115.

Nazina, T.N., T.P. Tourova, A.B. Poltaraus, E.V. Novikova, A.A. Grigoryan, A.E. Ivanova, A.M. Lysenko, V.V. Petrunyaka, G.A. Osipov, S.S. Belyaev and M.V. Ivanov. 2001. Taxonomic study of aerobic thermophilic bacilli: descriptions of *Geobacillus subterraneus* gen. nov., sp. nov. and *Geobacillus uzenensis* sp. nov. from petroleum reservoirs and transfer of *Bacillus stearothermophilus, Bacillus thermocatenulatus, Bacillus thermoleovorans, Bacillus kaustophilus, Bacillus thermodenitrificans* to *Geobacillus* as the new combinations *G. stearothermophilus, G. thermocatenulatus, G. thermoleovorans, G. kaustophilus, G. thermoglucosidasius* and *G. thermodenitrificans.* Int. J. Syst. Evol. Microbiol. *51*: 433–446.

Nedashkovskaya, O.I., S.B. Kim, S.K. Han, A.M. Lysenko, M. Rohde, N.V. Zhukova, E. Falsen, G.M. Frolova, V.V. Mikhailov and K.S. Bae. 2003a. *Mesonia algae* gen. nov., sp. nov., a novel marine bacterium of the family *Flavobacteriaceae* isolated from the green alga *Acrosiphonia sonderi* (Kutz) Kornm. Int. J. Syst. Evol. Microbiol. *53*: 1967–1971.

Nedashkovskaya, O.I., M. Suzuki, M.V. Vysotskii and V.V. Mikhailov. 2003b. *Reichenbachia agariperforans* gen. nov., sp. nov., a novel marine bacterium in the phylum *Cytophaga-Flavobacterium-Bacteroides.* Int. J. Syst. Evol. Microbiol. *53*: 81–85.

Nedashkovskaya, O.I., M. Suzuki, M.V. Vysotskii and V.V. Mikhailov. 2003c. *Vitellibacter vladivostokensis* gen. nov., sp. nov., a new member of the phylum *Cytophaga-Flavobacterium-Bacteroides.* Int. J. Syst. Evol. Microbiol. *53*: 1281–1286.

Neef, A., A. Zaglauer, H. Meier, R.I. Amann, H. Lemmer and K.H. Schleifer. 1996. Population analysis in a denitrifying sand filter: conventional and *in situ* identification of *Paracoccus* spp. in methanol-fed biofilms. Appl. Environ. Microbiol. *62*: 4329–4339.

ei, M. and S. Kumar. 2000. Molecular Evolution and Phylogenetics, Oxford University Press, Oxford.

hardt, F.C. , R. Curtiss, J. Ingraham, E. Lin, K. Brooks Low, B. Ma- anik, W. Rfznikopp, M. Riley, M. Schaechter and H.E. Umbarger ors). 1996. *Escherichia coli* and *Salmonella* — Cellular and Molec- iology, ASM Press, Washington, D.C.

., K.E. Johansson, Y. Rikihisa and J.G. Tully. 2002. Revision trophic *Mycoplasma* species names. Int. J. Syst. Evol. Micro- 3.

e Baere, I. Tjernberg, M. Vaneechoutte, T.J. van der Dijkshoorn. 2001. *Acinetobacter ursingii* sp. nov. and *dleri* sp. nov., isolated from human clinical speci- vol. Microbiol. *51*: 1891–1899.

n, I. Cleenwerck, T. De Baere, D. Janssens, T.J. zek and M. Vaneechoutte. 2003. *Acinetobacter* colony-forming species isolated from human . Syst. Evol. Microbiol. *53*: 1563–1567.

W.F. Doolittle. 2001. Defining the core of ic genes: The euryarchaeal core. J. Mol.

sel, E.-J. Finke and H. Meyer. 2000a. *In* ew names and new combinations pre- tside the IJSEM. List No. 75. Int. J. 417.

E.-J. Finke and H. Meyer. 2000b. e types belong to the same geno- Ne oups. Int. J. Med. Microbiol. *290*:

n Untereinheit der V-Typ- und Neu Technical University, Munich.

O. Stetter. 1990. *Thermococcus* ly thermophilic marine ar- Mi 07.

Newma tetter. 2001. *In* Validation nations previously effec- 82. Int. J. Syst. Evol.

ann, D.J. Ellis, D.R.

Lovley and F.M.M. Morel. 1977. Dissimilatory arsenate and sulfate reduction in *Desulfotomaculum auripigmentum* sp. nov. Arch. Microbiol. *168*: 380–388.

Newman, D.K., E.K. Kennedy, J.D. Coates, D. Ahmann, D.J. Ellis, D.R. Lovley and F.M.M. Morel. 2000. *In* Validation of publication of new names and new combinations previously effectively published outside the IJSEM. List No. 75. Int. J. Syst. Evol. Microbiol. *50*: 1415–1417.

Niamsup, P., I.N. Sujaya, M. Tanaka, T. Sone, S. Hanada, Y. Kamagata, S. Lumyong, A. Assavanig, K. Asano, F. Tomita and A. Yokota. 2003. *Lactobacillus thermotolerans* sp. nov., a novel thermotolerant species isolated from chicken faeces. Int. J. Syst. Evol. Microbiol. *53*: 263–268.

Nicolaus, B., R. Improta, M.C. Manca, L. Lama, E. Esposito and A. Gambacorta. 1998. Alicyclobacilli from an unexplored geothermal soil in Antarctica: Mount Rittmann. Polar Biol. *19*: 133–141.

Nicolaus, B., R. Improta, M.C. Manca, L. Lama, E. Esposito and A. Gambacorta. 2002. *In* Validation of publication of new names and new combinations previously effectively published outside the IJSEM. List No. 84. Int. J. Syst. Evol. Microbiol. *52*: 3–4.

Nicolaus, B., L. Lama, E. Esposito, M.C. Manca, G. diPrisco and A. Gambacorta. 1996. *"Bacillus thermoantarcticus"* sp. nov., from Mount Melbourne, Antarctica: A novel thermophilic species. Polar Biol. *16*: 101–104.

Nielsen, L.P. 1992. Denitrification in sediment determined from nitrogen isotope pairing. FEMS Microbiol. Ecol. *86*: 357–362.

Nielsen, P.H., K. Andreasen, M. Wagner, L.L. Blackall, H. Lemmer and R. Seviour. 1997. Variability of type 021N in activated sludge as determined by *in situ* substrate uptake pattern and *in situ* hybridization with fluorescent rRNA targeted probes. 2nd International Conference on Microorganisms in Activated Sludge and Biofilm Processes, Berkeley, California. IAWQ. pp. 255–262.

Niemann, S., E. Richter and S. Rusch-Gerdes. 2002. Biochemical and genetic evidence for the transfer of *Mycobacterium tuberculosis* subsp. *caprae* Aranaz et al. 1999 to the species *Mycobacterium bovis* Karlson and Lessel 1970 (approved lists 1980) as *Mycobacterium bovis* subsp. *caprae* comb. nov. Int. J. Syst. Evol. Microbiol. *52*: 433–436.

Nikolaitchouk, N., L. Hoyles, E. Falsen, J.M. Grainger and M.D. Collins. 2000. Characterization of *Actinomyces* isolates from samples from the human urogenital tract: description of *Actinomyces urogenitalis* sp. nov. Int. J. Syst. Evol. Microbiol. *50*: 1649–1654.

Nikolaitchouk, N., C. Wacher, E. Falsen, B. Andersch, M.D. Collins and P.A. Lawson. 2001. *Lactobacillus coleohominis* sp. nov., isolated from human sources. Int. J. Syst. Evol. Microbiol. *51*: 2081–2085.

Nishimori, E., K. Kita-Tsukamoto and H. Wakabayashi. 2000. *Pseudomonas plecoglossicida* sp. nov., the causative agent of bacterial haemorrhagic ascites of ayu, *Plecoglossus altivelis.* Int. J. Syst. Evol. Microbiol. *50*: 83–89.

Nishimura, Y., M. Kano, T. Ino, H. Iizuka, Y. Kosako and T. Kaneko. 1987. Deoxyribonucleic acid relationship among the radiation-resistant *Acinetobacter* and other *Acinetobacter.* J. Gen. Appl. Microbiol. *33*: 371–376.

Nogi, Y., C. Kato and K. Horikoshi. 2002. *Psychromonas kaikoae* sp. nov., a novel piezophilic bacterium from the deepest cold-seep sediments in the Japan Trench. Int. J. Syst. Evol. Microbiol. *52*: 1527–1532.

Nomoto, T., Y. Fukumori and T. Yamanaka. 1993. Membrane-bound cytochrome *c* is an alternative electron donor for cytochrome aa3 in *Nitrobacter winogradskyi.* J. Bacteriol. *175*: 4400–4404.

Norton, J.M. and M.G. Klotz. 1991. Homology among ammonia monooxygenase genes from ammonia-oxidizing bacteria. *In* Abstracts of the 95th General Meeting of the American Society for Microbiology, American Society for Microbiology, Washington, DC. 564, K–164.

Norton, J.M., J.M. Low and M.G. Klotz. 1996. The gene encoding ammonia monooxygenase subunit A exists in three nearly identical copies in *Nitrosospira* sp. NpAV. FEMS Microbiol. Lett. *139*: 181–188.

Ntougias, S. and N.J. Russell. 2001. *Alkalibacterium olivoapovliticus* gen. nov., sp. nov., a new obligately alkaliphilic bacterium isolated from edible-olive wash-waters. Int. J. Syst. Evol. Microbiol. *51*: 1161–1170.

Nübel, U., B. Engelen, A. Felske, J. Snaidr, A. Weishuber, R.I. Amann, W. Ludwig and H. Backhaus. 1996. Sequence heterogeneities of genes encoding 16S rRNAs in *Paenibacillus polymyxa* detected by temperature gradient gel electrophoresis. J. Bacteriol. *178*: 5636–5643.

Nübel, U., F. Garcia-Pichel and G. Muyzer. 2000. The halotolerance and phylogeny of cyanobacteria with tightly coiled trichomes (*Spirulina* Turpin) and the description of *Halospirulina tapeticola* gen. nov., sp. nov. Int. J. Syst. Evol. Microbiol. *50*: 1265–1277.

Numata, M., T. Saito, T. Yamazaki, Y. Fukumori and T. Yamanaka. 1990. Cytochrome P-460 of *Nitrosomonas europaea*: further purification and further characterization. J. Biochem. (Tokyo) *108*: 1016–1021.

Nunoura, T., S. Akihara, K. Takai and Y. Sako. 2002a. *Thermaerobacter nagasakiensis* sp. nov., a novel aerobic and extremely thermophilic marine bacterium. Arch. Microbiol. *177*: 339–344.

Nunoura, T., S. Akihara, K. Takai and Y. Sako. 2002b. *In* Validation of publication of new names and new combinations previously effectively published outside the IJSEM. List No. 86. Int. J. Syst. Evol. Microbiol. *52*: 1075–1076.

Nüsslein, K. and J.M. Tiedje. 1998. Characterization of the dominant and rare members of a young Hawaiian soil bacterial community with small-subunit ribosomal DNA amplified from DNA fractionated on the basis of its guanine and cytosine composition. Appl. Environ. Microbiol. *64*: 1283–1289.

Ochi, K. 1995. Comparative ribosomal-protein sequence analyses of a phylogenetically defined genus, *Pseudomonas*, and its relatives. Int. J. Syst. Bacteriol. *45*: 268–273.

Ochiai, S., Y. Adachi and K. Mori. 1997. Unification of the genera *Serpulina* and *Brachyspira*, and proposals of *Brachyspira hyodysenteriae* comb. nov., *Brachyspira innocens* comb. nov. and *Brachyspira pilosicoli* comb. nov. Microbiol. Immunol. *41*: 445–452.

O'Connor, S.M. and J.D. Coates. 2002. Universal immunoprobe for (per)chlorate-reducing bacteria. Appl. Environ. Microbiol. *68*: 3108–3113.

O'Hara, C.M., F.W. Brenner, A.G. Steigerwalt, B.C. Hill, B. Holmes, P.A. Grimont, P.M. Hawkey, J.L. Penner, J.M. Miller and D.J. Brenner. 2000. Classification of *Proteus vulgaris* biogroup 3 with recognition of *Proteus hauseri* sp. nov., nom. rev. and unnamed *Proteus* genomospecies 4, 5 and 6. Int. J. Syst. Evol. Microbiol. *50*: 1869–1875.

Ohno, M., H. Shiratori, M.J. Park, Y. Saitoh, Y. Kumon, N. Yamashita, A. Hirata, H. Nishida, K. Ueda and T. Beppu. 2000. *Symbiobacterium thermophilum* gen. nov., sp. nov., a symbiotic thermophile that depends on co-culture with a *Bacillus* strain for growth. Int. J. Syst. Evol. Microbiol. *50*: 1829–1832.

Okamura, K., F. Mitsumori, O. Ito, K.I. Takamiya and M. Nishimura. 1986. Photophosphorylation and oxidative phosphorylation in intact cells and chromatophores of an aerobic photosynthetic bacterium *Erythrobacter* sp. strain Och-114. J. Bacteriol. *168*: 1142–1146.

Olsen, G.J. and C.R. Woese. 1993. Ribosomal RNA: a key to phylogeny. FASEB J. *7*: 113–123.

Olsen, G.J. and C.R. Woese. 1997. Archaeal genomics: an overview. Cell *89*: 991–994.

Olsen, G.J., C.R. Woese and R. Overbeek. 1994. The winds of (evolutionary) change: breathing new life into microbiology. J. Bacteriol. *176*: 1–6.

Olson, T.C. and A.B. Hooper. 1983. Energy coupling in the bacterial oxidation of small molecules: an extracytoplasmic dehydrogenase in *Nitrosomonas*. FEMS Microbiol. Lett. *19*: 47–50.

Oltmann, L.F., W.N.M. Reijnders and A.H. Stouthamer. 1976. Characterization of purified nitrate reductase a and chlorate reductase c from *Proteus mirabilis*. Arch. Microbiol. *111*: 25–35.

Omel'chenko, M.V., L.V. Vasil'eva, G.A. Zavarzin, N.D. Savel'eva, A.M. Lysenko, L.L. Mityushina, V.N. Khmelenina and Y.A. Trotsenko. 1996. A novel psychrophilic methanotroph of the genus *Methylobacter*. Microbiology *65*: 339–343.

Omel'chenko, M.V., L.V. Vasil'eva, G.A. Zavarzin, N.D. Savel'eva, A.M. Lysenko, L.L. Mityushina, V.N. Khmelenina and Y.A. Trotsenko. 2000. *In* Validation of the publication of new names and new combinations previously effectively published outside the IJSEM. List No. 50. Int. J. Syst. Evol. Microbiol. *50*: 423–424.

Onderdonk, A.B. and M. Sasser. 1995. Gas-liquid and high-performance chromatographic methods for the identification of micoorganisms. *In* Murray, Baron, Pfaller, Tenover and Yolken (Editors), Manual of Clinical Microbiology, 6th Ed., American Society for Microbiology, Washington, D.C. pp. 123–129.

O'Neill, J.G. and J.F. Wilkinson. 1977. Oxidation of ammonia by methane-oxidizing bacteria and effects of ammonia on methane oxidation. J. Gen. Microbiol. *100*: 407–412.

Oren, A., R. Elevi, S. Watanabe, K. Ihara and A. Corcelli. 2002. *Halomicrobium mukohataei* gen. nov., comb. nov., and emended description of *Halomicrobium mukohataei*. Int. J. Syst. Evol. Microbiol. *52*: 1831–1835.

Orla-Jensen, S. 1909. Die Hauptlinien des natürlichen Bakterien-systems. Zentbl. Bakteriol. Parasitenkd. Infektkrankh. Hyg. Abt. II *22*: 97–98 and 305–346.

Orla-Jensen, S. 1919. The Lactic Acid Bacteria, Høst, Copenhagen.

Orla-Jensen, S. 1921. The main lines of the bacterial system. J. Bacteriol. *6*: 263–273.

Otoguro, M., M. Hayakawa, T. Yamazaki, T. Tamura, K. Hatano and Y. Imai. 2001. Numerical phenetic and phylogenic analyses of *Actinokineospora* isolates, with a description of *Actinokineospora auranticolor* sp. nov. and *Actinokineospora enzanensis* sp. nov. Actinomycetologica. *15*: 30–39.

Otoguro, M., M. Hayakawa, T. Yamazaki, T. Tamura, K. Hatano and Y. Imai. 2003. *In* Validation of publication of new names and new combinations previously effectively published outside the IJSEM. List No. 89. Int. J. Syst. Evol. Microbiol. *53*: 1–2.

Ouattara, A.S., E.A. Assih, S. Thierry, J.L. Cayol, M. Labat, O. Monroy and H. Macarie. 2003. *Bosea minatitlanensis* sp. nov., a strictly aerobic bacterium isolated from an anaerobic digester. Int. J. Syst. Evol. Microbiol. *53*: 1247–1251.

Ouverney, C.C. and J.A. Fuhrman. 1997. Increase in fluorescence intensity of 16S rRNA *in situ* hybridization in natural samples treated with chloramphenicol. Appl. Environ. Microbiol. *63*: 2735–2740.

Pace, N.R. 1997. A molecular view of microbial diversity and the biosphere. Science *276*: 734–740.

Painter, H.A. 1986. Nitrification in the treatment of sewage and waste-waters. *In* Prosser (Editor), Nitrification, IRL Press, Oxford. pp. 185–211.

Palleroni, N.J., R. Kunisawa, R. Contopoulou and M. Doudoroff. 1973. Nucleic acid homologies in the genus *Pseudomonas*. Int. J. Syst. Bacteriol. *23*: 333–339.

Palmisano, M.M., L.K. Nakamura, K.E. Duncan, C.A. Istock and F.M. Cohan. 2001. *Bacillus sonorensis* sp. nov., a close relative of *Bacillus licheniformis*, isolated from soil in the Sonoran Desert, Arizona. Int. J. Syst. Evol. Microbiol. *51*: 1671–1679.

Pandey, K.K., S. Mayilraj and T. Chakrabarti. 2002. *Pseudomonas indica* sp. nov., a novel butane-utilizing species. Int. J. Syst. Evol. Microbiol. *52*: 1559–1567.

Park, D.H., J.S. Kim, S.W. Kwon, C. Wilson, Y.M. Yu, J.H. Hur and C.K. Lim. 2003a. *Streptomyces luridiscabiei* sp. nov., *Streptomyces puniciscabiei* sp. nov. and *Streptomyces niveiscabiei* sp. nov., which cause potato common scab disease in Korea. Int. J. Syst. Evol. Microbiol. *53*: 2049–2054.

Park, M.Y., K.S. Ko, H.K. Lee, M.S. Park and Y.H. Kook. 2003b. *Legionella busanensis* sp. nov., isolated from cooling tower water in Korea. Int. J. Syst. Evol. Microbiol. *53*: 77–80.

Parshina, S.N., R. Kleerebezem, J.L. Sanz, G. Lettinga, A.N. Nozhevnikova, N.A. Kostrikina, A.M. Lysenko and A.J. Stams. 2003. *Soehngenia saccharolytica* gen. nov., sp. nov. and *Clostridium amygdalinum* sp. nov., two novel anaerobic, benzaldehyde-converting bacteria. Int. J. Syst. Evol. Microbiol. *53*: 1791–1799.

Patterson, M.M., M.D. Schrenzel, Y. Feng, S. Xu, F.E. Dewhirst, B.J. Paster, S.A. Thibodeau, J. Versalovic and J.G. Fox. 2000. *Helicobacter aurati* sp. nov., a urease-positive *Helicobacter* species cultured from gastrointestinal tissues of Syrian hamsters. J. Clin. Microbiol. *38*: 3722–3728.

Patterson, M.M., M.D. Schrenzel, Y. Feng, S. Xu, F.E. Dewhirst, B.J. Paster, S.A. Thibodeau, J. Versalovic and J.G. Fox. 2002. *In* Validation of new names and new combinations previously effectively published outside the IJSEM. List No. 84. Int. J. Syst. Evol. Microbiol. *52*: 3–4.

Pavan, M.E., S.L. Abbott, J. Zorzópulos and J.M. Janda. 2000. *Aeromonas salmonicida* subsp. *pectinolytica* subsp. nov., a new pectinase-positive subspecies isolated from a heavily polluted river. Int. J. Syst. Evol. Microbiol. *50*: 1119–1124.

Pease, A.C., D. Solas, E.J. Sullivan, M.T. Cronin, C.P. Holmes and S.P.A. Fodor. 1994. Light-generated oligonucleotide arrays for rapid DNA sequence analysis. Proc. Natl. Acad. Sci. U.S.A. *91*: 5022–5026.

Peix, A., R. Rivas, P.F. Mateos, E. Martinez-Molina, C. Rodriguez-Barrueco and E. Velazquez. 2003. *Pseudomonas rhizosphaerae* sp. nov., a novel species that actively solubilizes phosphate in vitro. Int. J. Syst. Evol. Microbiol. *53*: 2067–2072.

Peltola, J.S.P., M.A. Andersson, P. Kämpfer, G. Auling, R.M. Kroppenstedt, H.J. Busse, M.S. Salkinoja-Salonen and F.A. Rainey. 2001. Isolation of toxigenic *Nocardiopsis* strains from indoor environments and description of two new *Nocardiopsis* species, *N. exhalans* sp. nov. and *N. umidischolae* sp. nov. Appl. Environ. Microbiol. *67*: 4293–4304.

Peltola, J.S.P., M.A. Andersson, P. Kämpfer, G. Auling, R.M. Kroppenstedt, H.J. Busse, M.S. Salkinoja-Salonen and F.A. Rainey. 2002. *In* Validation of publication of new names and new combinations previously effectively published outside the IJSEM. List No. 84. Int. J. Syst. Evol. Microbiol. *52*: 3–4.

Pernthaler, J., F.-O. Glöckner, S. Unterholzner, A. Alfreider, R. Psenner and R.I. Amann. 1998. Seasonal community and population dynamics of pelagic bacteria and archaea in a high mountain lake. Appl. Environ. Microbiol. *64*: 4299–4306.

Pernthaler, J., T. Posch, K. Simek, J. Vrba, R.I. Amann and R. Psenner. 1997. Contrasting bacterial strategies to coexist with a flagellate predator in an experimental microbial assemblage. Appl. Environ. Microbiol. *63*: 596–601.

Petrosyan, P., M. Garcia-Varela, A. Luz-Madrigal, C. Huitron and M.E. Flores. 2003. *Streptomyces mexicanus* sp. nov., a xylanolytic micro-organism isolated from soil. Int. J. Syst. Evol. Microbiol. *53*: 269–273.

Pettersson, B., S.K. de Silva, M. Uhlen and F.G. Priest. 2000. *Bacillus siralis* sp. nov., a novel species from silage with a higher order structural attribute in the 16S rRNA genes. Int. J. Syst. Evol. Microbiol. *50*: 2181–2187.

Pfennig, N. 1969. *Rhodospirillum tenue* sp. n., a new species of the purple nonsulfur bacteria. J. Bacteriol. *99*: 619–620.

Pfennig, N. 1978. *Rhodocyclus purpureus* gen. nov. and sp. nov. a ring-shaped, vitamin B_{12}-requiring member of the family *Rhodospirillaceae*. Int. J. Syst. Bacteriol. *28*: 283–288.

Pfennig, N. and H.G. Trüper. 1971. Higher taxa of the phototrophic bacteria. Int. J. Syst. Bacteriol. *21*: 17–18.

Pfennig, N. and H.G. Trüper. 1974. The phototrophic bacteria. *In* Buchanan and Gibbons (Editors), Bergey's Manual of Determinative Bacteriology, 8th Ed., The Williams & Wilkins Co., Baltimore. pp. 24–60.

Phillips, C.J., Z. Smith, T.M. Embley and J.I. Prosser. 1999. Phylogenetic differences between particle associated and planktonic ammonia-oxidizing bacteria of the beta subdivision of the class *Proteobacteria* in the northwestern Mediterranean Sea. Appl. Environ. Microbiol. *65*: 779–786.

Phillips, R.W., J. Wiegel, C.J. Berry, C. Fliermans, A.D. Peacock, D.C. White and L.J. Shimkets. 2002. *Kineococcus radiotolerans* sp. nov., a radiation-resistant, Gram-positive bacterium. Int. J. Syst. Evol. Microbiol. *52*: 933–938.

Picard, C., C. Ponsonnet, E. Paget, X. Nesme and P. Simonet. 1992. Detection and enumeration of bacteria in soil by direct DNA extraction and polymerase chain reaction. Appl. Environ. Microbiol. *58*: 2717–2722.

Pickup, R.W. and J.R. Saunders (Editors). 1996. Molecular approaches to environmental microbiology, Ellis Horwood, London.

Pidiyar, V., A. Kaznowski, N.B. Narayan, M. Patole and Y.S. Shouche. 2002. *Aeromonas culicicola* sp. nov., from the midgut of *Culex quinquefasciatus*. Int. J. Syst. Evol. Microbiol. *52*: 1723–1728.

Pikuta, E.V., R.B. Hoover, A.K. Bej, D. Marsic, E.N. Detkova, W.B. Whitman and P. Krader. 2003a. *Tindallia californiensis* sp. nov., a new anaerobic, haloalkaliphilic, spore-forming acetogen isolated from Mono Lake in California. Extremophiles *7*: 327–334.

Pikuta, E.V., R.B. Hoover, A.K. Bej, D. Marsic, E.N. Detkova, W.B. Whitman and P. Krader. 2003b. *In* Validation of publication of new names and new combinations previously effectively published outside the IJSEM. List No. 94. Int. J. Syst. Evol. Microbiol. *53*: 1701–1702.

Pikuta, E.V., R.B. Hoover, A.K. Bej, D. Marsic, W.B. Whitman, D. Cleland and P. Krader. 2003c. *Desulfonatronum thiodismutans* sp. nov., a novel alkaliphilic, sulfate-reducing bacterium capable of lithoautotrophic growth. Int. J. Syst. Evol. Microbiol. *53*: 1327–1332.

Pikuta, E., A. Lysenko, N. Chuvilskaya, U. Mendrock, H. Hippe, N. Suzina, D. Nikitin, G. Osipov and K. Laurinavichius. 2000a. *Anoxybacillus pushchinensis* gen. nov., sp. nov., a novel anaerobic, alkaliphilic, moderately thermophilic bacterium from manure, and description of *Anoxybacillus flavitherms* comb. nov. Int. J. Syst. Evol. Microbiol. *50*: 2109–2117.

Pikuta, E., A. Lysenko, N. Suzina, G. Osipov, B. Kuznetsov, T. Tourova, V. Akimenko and K. Laurinavichius. 2000b. *Desulfotomaculum alkaliphilum* sp. nov., a new alkaliphilic, moderately thermophilic, sulfate-reducing bacterium. Int. J. Syst. Evol. Microbiol. *50*: 25–33.

Pinck, C., C. Coeur, P. Potier and E. Bock. 2001. Polyclonal antibodies recognizing the AmoB protein of ammonia oxidizers of the β-subclass of the class *Proteobacteria*. Appl. Environ. Microbiol. *67*: 118–124.

Place, R.B., D. Hiestand, S. Burri and M. Teuber. 2002. *Staphylococcus succinus* subsp. *casei* subsp nov., a dominant isolate from a surface ripened cheese. Syst. Appl. Microbiol. *25*: 353–359.

Place, R.B., D. Hiestand, S. Burri and M. Teuber. 2003a. *In* Validation of publication of new names and new combinations previously effectively published outside the IJSEM. List No. 89. Int. J. Syst. Evol. Microbiol. *53*: 1–2.

Place, R.B., D. Hiestand, H.R. Gallmann and M. Teuber. 2003b. *Staphylococcus equorum* subsp. *linens*, subsp. nov., a starter culture component for surface ripened semi-hard cheeses. Syst. Appl. Microbiol. *26*: 30–37.

Place, R.B., D. Hiestand, H.R. Gallmann and M. Teuber. 2003c. *In* Validation of publication of new names and new combinations previously effectively published outside the IJSEM. List No. 93. Int. J. Syst. Evol. Microbiol. *53*: 1219–1220.

Plugge, C.M., M. Balk and A.J.M. Stams. 2002a. *Desulfotomaculum thermobenzoicum* subsp. *thermosyntrophicum* subsp. nov., a thermophilic, syntrophic, propionate-oxidizing, spore-forming bacterium. Int. J. Syst. Evol. Microbiol. *52*: 391–399.

Plugge, C.M., M. Balk, E.G. Zoetendal and A.J. Stams. 2002b. *Gelria glutamica* gen. nov., sp. nov., a thermophilic, obligately syntrophic, glutamate-degrading anaerobe. Int. J. Syst. Evol. Microbiol. *52*: 401–407.

Plugge, C.M., E.G. Zoetendal and A.J. Stams. 2000. *Caloramator coolhaasii* sp. nov., a glutamate-degrading, moderately thermophilic anaerobe. Int. J. Syst. Evol. Microbiol. *50*: 1155–1162.

Polz, M.F. and C.M. Cavanaugh. 1998. Bias in template-to-product ratios in multitemplate PCR. Appl. Environ. Microbiol. *64*: 3724–3730.

Pommerening-Röser, A. 1993. Untersuchen zur phylogenie ammoniak oxidierender bakterien, University of Hamburg

Pommerening-Röser, A., G. Rath and H.P. Koops. 1996. Phylogenetic diversity within the genus *Nitrosomonas*. System. Appl. Microbiol. *19*: 344–351.

Porter, J.R. 1976. The world view of culture collections. *In* Colwell (Editor), The Role of Culture Collections in the Era of Molecular Biology, American Society for Microbiology, Washington, D.C. pp. 62–72.

Porter, R.W. and Y.S. Feig. 1980. The use of DAPI for identifying and counting aquatic microflora. Limnol. Oceanogr. *25*: 943–948.

Poth, M. 1986. Dinitrogen production from nitrite by a *Nitrosomonas* isolate. Appl. Environ. Microbiol. *52*: 957–959.

Poth, M. and D.D. Focht. 1985. ^{15}N Kinetic analysis of N_2O production by *Nitrosomonas europaea* an examination of nitrifier denitrification. Appl. Environ. Microbiol. *49*: 1134–1141.

Poulsen, L.K., G. Ballard and D.A. Stahl. 1993. Use of rRNA fluorescence *in situ* hybridization for measuring the activity of single cells in young and established biofilms. Appl. Environ. Microbiol. *59*: 1354–1360.

Poyart, C., G. Quesne and P. Trieu-Cuot. 2002. Taxonomic dissection of the *Streptococcus bovis* group by analysis of manganese-dependent superoxide dismutase gene (*sodA*) sequences: reclassification of '*Streptococcus infantarius* subsp. *coli*' as *Streptococcus lutetiensis* sp. nov. and of *Streptococcus bovis* biotype 11.2 as *Streptococcus pasteurianus* sp. nov. Int. J. Syst. Evol. Microbiol. *52*: 1247–1255.

Preston, C.M., K.Y. Wu, T.F. Molinski and E.F. DeLong. 1996. A psychrophilic crenarchaeon inhabits a marine sponge: *Crenarchaeum symbiosum* gen. nov., sp. nov. Proc. Natl. Acad. Sci. U.S.A. *93*: 6241–6246.

Priest, F. and B. Austin. 1993. Modern Bacterial Taxonomy, 2nd Ed., Chapman and Hall, London.

Prokofeva, M.I., M.L. Miroshnichenko, N.A. Kostrikina, N.A. Chernyh, B.B. Kuznetsov, T.P. Tourova and E.A. Bonch-Osmolovskaya. 2000. *Acidilobus aceticus* gen. nov., sp. nov., a novel anaerobic thermoacidophilic archaeon from continental hot vents in Kamchatka. Int. J. Syst. Evol. Microbiol. *50*: 2001–2008.

Prosser, J.I. 1989. Autotrophic nitrification in bacteria. *In* Rose and Tempest (Editors), Microbial Physiology, Academic Press, London. pp. 125–181.

Provorov, N.A. 1998. Coevolution of rhizobia with legumes: Facts and hypotheses. Symbiosis *24*: 337–367.

Prowe, S.G. and G. Antranikian. 2001. *Anaerobranca gottschalkii* sp. nov., a novel thermoalkaliphilic bacterium that grows anaerobically at high pH and temperature. Int. J. Syst. Evol. Microbiol. *51*: 457–465.

Puchkova, N.N., J.F. Imhoff and V.M. Gorlenko. 2000. *Thiocapsa litoralis* sp. nov., a new purple sulfur bacterium from microbial mats from the White Sea. Int. J. Syst. Evol. Microbiol. *50*: 1441–1447.

Pukall, R., M. Laroche, R.M. Kroppenstedt, P. Schumann, E. Stackebrandt and R. Ulber. 2003. *Paracoccus seriniphilus* sp. nov., an L-serine-dehydratase-producing coccus isolated from the marine bryozoan *Bugula plumosa*. Int. J. Syst. Evol. Microbiol. *53*: 443–447.

Purkhold, U., A. Pommerening-Röser, S. Juretschko, M.C. Schmid, H.P. Koops and M. Wagner. 2000. Phylogeny of all recognized species of ammonia oxidizers based on comparative 16S rRNA and *amoA* sequence analysis: implications for molecular diversity surveys. Appl. Environ. Microbiol. *66*: 5368–5382.

Purkhold, U. , M. Wagner, G. Timmermann, A. Pommerening-Roser and H.P. Koops. 2003. 16S rRNA and amoA-based phylogeny of 12 novel betaproteobacterial ammonia-oxidizing isolates: extension of the dataset and proposal of a new lineage within the nitrosomonads. Int. J. Syst. Evol. Microbiol. *53*: 1485–1494.

Qiu, L.Y., Y. Sekiguchi, H. Imachi, Y. Kamagata, I.C. Tseng, S.S. Cheng, A. Ohashi and H. Harada. 2003a. *Sporotomaculum syntrophicum* sp. nov., a novel anaerobic, syntrophic benzoate-degrading bacterium isolated from methanogenic sludge treating wastewater from terephthalate manufacturing. Arch. Microbiol. *179*: 242–249.

Qiu, L.Y., Y. Sekiguchi, H. Imachi, Y. Kamagata, I.C. Tseng, S.S. Cheng, A. Ohashi and H. Harada. 2003b. *In* Validation of publication of new names and new combinations previously effectively published outside the IJSEM. List No. 92. Int. J. Syst. Evol. Microbiol. *53*: 935–937.

Rabus, R., R. Nordhaus, W. Ludwig and F. Widdel. 1993. Complete oxidation of toluene under strictly anoxic conditions by a new sulfate-reducing bacterium. Appl. Environ. Microbiol. *59*: 1444–1451.

Rabus, R., R. Nordhaus, W. Ludwig and F. Widdel. 2000. *In* Validation of the publication of new names and new combinations previously effectively published outside the IJSEM. List No. 75. Int. J. Syst. Evol. Microbiol. *50*: 1415–1417.

Rainey, F.A. and M.S. da Costa. 2001. Order II. *Thermales* ord. nov. *In* Boone, Castenholz and Garrity (Editors), Bergey's Manual of Systematic Bacteriology, 2nd Edition, Volume 1. The archaea and the deeply branching and phototrophic bacteria. Springer, New York. 403.

Rainey, F.A. and M.S. da Costa. 2002. *In* Validation of publication of new names and new combinations previously effectively published outside the IJSEM. List No. 85. Int. J. Syst. Evol. Microbiol. *52*: 685–690.

Rainey, F.A., J. Silva, M.F. Nobre, M.T. Silva and M.S. da Costa. 2003. *Porphyrobacter cryptus* sp. nov., a novel slightly thermophilic, aerobic, bacteriochlorophyll *a*-containing species. Int. J. Syst. Evol. Microbiol. *53*: 35–41.

Rainey, F.A., N.L. Ward-Rainey, P.H. Janssen, H. Hippe and E. Stackebrandt. 1996. *Clostridium paradoxum* DSM 7308^T contains multiple 16S rRNA genes with heterogeneous intervening sequences. Microbiology (Reading) *142*: 2087–2095.

Ramsing, N.B., H. Fossing, T.G. Ferdelman, F. Andersen and B. Thamdrup. 1996. Distribution of bacterial populations in a stratified Fjord (Mariager Fjord, Denmark) quantified by *in situ* hybridization and related to chemical gradients in the water column. Appl. Environ. Microbiol. *62*: 1391–1404.

Ramsing, N.B., M. Kühl and B.B. Jørgensen. 1993. Distribution of sulfate-reducing bacteria, O_2, and H_2S in photosynthetic biofilms determined by oligonucleotide probes and microelectrodes. Appl. Environ. Microbiol. *59*: 3840–3849.

Rasche, M.E., R.E. Hicks, M.R. Hyman and D.J. Arp. 1990a. Oxidation of monohalogenated ethanes and n-chlorinated alkanes by whole cells of *Nitrosomonas europaea*. J. Bacteriol. *172*: 5368–5373.

Rasche, M.E., M.R. Hyman and D.J. Arp. 1990b. Biodegradation of halogenated hydrocarbon fumigants by nitrifying bacteria. Appl. Environ. Microbiol. *56*: 2568–2571.

Raskin, L., L.K. Poulsen, D.R. Noguera, B.E. Rittmann and D.A. Stahl. 1994a. Quantification of methanogenic groups in anaerobic biological reactors by oligonucleotide probe hybridization. Appl. Environ. Microbiol. *60*: 1241–1248.

Raskin, L., J.M. Stromley, B.E. Rittmann and D.A. Stahl. 1994b. Group-specific 16S rRNA hybridization probes to describe natural communities of methanogens. Appl. Environ. Microbiol. *60*: 1232–1240.

Ratcliff, R. 1981. Terminal deoxynucleotidyltransferase. *In* Boyer (Editor), The Enzymes, Vol. 14, Academic Press, New York. pp. 105–118.

Rautio, M., E. Eerola, M.L. Vaisanen-Tunkelrott, D. Molitoris, P. Lawson, M.D. Collins and H. Jousimies-Somer. 2003a. Reclassification of *Bacteroides putredinis* (Weinberg et al., 1937) in a new genus *Alistipes* gen. nov., as *Alistipes putredinis* comb. nov., and description of *Alistipes finegoldii* sp. nov., from human sources. Syst. Appl. Microbiol. *26*: 182–188.

Rautio, M., E. Eerola, M.L. Vaisanen-Tunkelrott, D. Molitoris, P. Lawson, M.D. Collins and H. Jousimies-Somer. 2003b. *In* Validation of publication of new names and new combinations previously effectively published outside the IJSEM. List No. 94. Int. J. Syst. Evol. Microbiol. *53*: 1701–1702.

Ravin, A.W. 1963. Experimental approaches to the study of bacterial phylogeny. Am. Natur. *97*: 307–318.

Raymond, J., O. Zhaxybayeva, J.P. Gogarten, S.Y. Gerdes and R.E. Blankenship. 2002. Whole-genome analysis of photosynthetic prokaryotes. Science (Wash. D. C.) *298*: 1616–1620.

Reddy, G.S., R.K. Aggarwal, G.I. Matsumoto and S. Shivaji. 2000. *Arthrobacter flavus* sp. nov., a psychrophilic bacterium isolated from a pond in McMurdo Dry Valley, Antarctica. Int. J. Syst. Evol. Microbiol. *50*: 1553–1561.

Reddy, G.S., G.I. Matsumoto and S. Shivaji. 2003a. *Sporosarcina macmurdoensis* sp. nov., from a cyanobacterial mat sample from a pond in the McMurdo Dry Valleys, Antarctica. Int. J. Syst. Evol. Microbiol. *53*: 1363–1367.

Reddy, G.S., J.S. Prakash, G.I. Matsumoto, E. Stackebrandt and S. Shivaji. 2002a. *Arthrobacter roseus* sp. nov., a psychrophilic bacterium isolated from an antarctic cyanobacterial mat sample. Int. J. Syst. Evol. Microbiol. *52*: 1017–1021.

Reddy, G.S., J.S. Prakash, V. Prabahar, G.I. Matsumoto, E. Stackebrandt and S. Shivaji. 2003b. *Kocuria polaris* sp. nov., an orange-pigmented psychrophilic bacterium isolated from an Antarctic cyanobacterial mat sample. Int. J. Syst. Evol. Microbiol. *53*: 183–187.

Reddy, G.S., J.S. Prakash, R. Srinivas, G.I. Matsumoto and S. Shivaji. 2003c. *Leifsonia rubra* sp. nov. and *Leifsonia aurea* sp. nov., psychrophiles from a pond in Antarctica. Int. J. Syst. Evol. Microbiol. *53*: 977–984.

Reddy, G.S.N., J.S.S. Prakash, M. Vairamani, S. Prabhakar, G.I. Matsumoto and S. Shivaji. 2002b. *Planococcus antarcticus* and *Planococcus psychro-*

philus spp. nov. isolated from cyanobacterial mat samples collected from ponds in Antarctica. Extremophiles *6*: 253–261.

Reddy, G.S.N., J.S.S. Prakash, M. Vairamani, S. Prabhakar, G.I. Matsumoto and S. Shivaji. 2002c. *In* Validation of publication of new names and new combinations previously effectively published outside the IJSEM. List No. 87. Int. J. Syst. Evol. Microbiol. *52*: 1437–1438.

Rees, G.N., C.G. Harfoot, P.H. Janssen, L. Schoenborn, J. Kuever and H. Lunsdorf. 2002. *Thiobaca trueperi* gen. nov., sp. nov., a phototrophic purple sulfur bacterium isolated from freshwater lake sediment. Int. J. Syst. Evol. Microbiol. *52*: 671–678.

Rees, G.N. and B.K.C. Patel. 2001. *Desulforegula conservatrix* gen. nov., sp. nov., a long-chain fatty acid-oxidizing, sulfate-reducing bacterium isolated from sediments of a freshwater lake. Int. J. Syst. Evol. Bacteriol. *51*: 1911–1916.

Reeves, M.W., G.M. Evins, A.A. Heiba, B.D. Plikaytis and J.J. Farmer, III. 1989. Clonal nature of *Salmonella typhi* and its genetic relatedness to other salmonellae as shown by multilocus enzyme electrophoresis, and proposal of *Salmonella bongori* comb. nov. J. Clin. Microbiol. *27*: 313–320.

Reinhold-Hurek, B. and T. Hurek. 2000. Reassessment of the taxonomic structure of the diazotrophic genus *Azoarcus sensu lato* and description of three new genera and new species, *Azovibrio restrictus* gen. nov., sp. nov., *Azospira oryzae* gen. nov., sp. nov. and *Azonexus fungiphilus* gen. nov., sp. nov. Int. J. Syst. Evol. Microbiol. *50*: 649–659.

Relman, D.A. 1999. The search for unrecognized pathogens. Science *284*: 1308–1310.

Renaud, F.N., D. Aubel, P. Riegel, H. Meugnier and C. Bollet. 2001. *Corynebacterium freneyi* sp. nov., α-glucosidase-positive strains related to *Corynebacterium xerosis*. Int. J. Syst. Evol. Microbiol. *51*: 1723–1728.

Rennie, R.J. and E.L. Schmidt. 1977. Immunofluorescence studies of *Nitrobacter* populations in soils. Can. J. Microbiol. *23*: 1011–1017.

Reva, O.N., V.V. Smirnov, B. Pettersson and F.G. Priest. 2002. *Bacillus endophyticus* sp. nov., isolated from the inner tissues of cotton plants (*Gossypium* sp.). Int. J. Syst. Evol. Microbiol. *52*: 101–107.

Reysenbach, A.-L. 2001a. Class I. *Aquificae* class. nov. *In* Boone, Castenholz and Garrity (Editors), Bergey's Manual of Systematic Bacteriology, 2nd Edition, Volume 1. The archaea and the deeply branching and phototrophic bacteria. Springer, New York. 359.

Reysenbach, A.-L. 2001b. Class I. *Thermoprotei* class. nov. *In* Boone, Castenholz and Garrity (Editors), Bergey's Manual of Systematic Bacteriology, 2nd Edition, Volume 1. The archaea and the deeply branching and phototrophic bacteria. Springer, New York. 169.

Reysenbach, A.-L. 2001c. Class I. *Thermotogae* class. nov. *In* Boone, Castenholz and Garrity (Editors), Bergey's Manual of Systematic Bacteriology, 2nd Edition, Volume 1. The archaea and the deeply branching and phototrophic bacteria. Springer, New York. 369.

Reysenbach, A.-L. 2001d. Class IV. *Thermoplasmata* class. nov. *In* Boone, Castenholz and Garrity (Editors), Bergey's Manual of Systematic Bacteriology, 2nd Edition, Volume 1. The archaea and the deeply branching and phototrophic bacteria. Springer, New York. 335.

Reysenbach, A.-L. 2001e. Family I. *Aquificaceae* fam. nov. *In* Boone, Castenholz and Garrity (Editors), Bergey's Manual of Systematic Bacteriology, 2nd Edition, Volume 1. The archaea and the deeply branching and phototrophic bacteria. Springer, New York. 360.

Reysenbach, A.-L. 2001f. Family I. *Thermoplasmataceae* fam. nov. *In* Boone, Castenholz and Garrity (Editors), Bergey's Manual of Systematic Bacteriology, 2nd Edition, Volume 1. The archaea and the deeply branching and phototrophic bacteria. Springer, New York. 335.

Reysenbach, A.-L. 2001g. Family I. *Thermotogaceae* fam. nov. *In* Boone, Castenholz and Garrity (Editors), Bergey's Manual of Systematic Bacteriology, 2nd Edition, Volume 1. The archaea and the deeply branching and phototrophic bacteria. Springer, New York. 370.

Reysenbach, A.-L. 2001h. Order I. Aquificales ord. nov. *In* Boone, Castenholz and Garrity (Editors), Bergey's Manual of Systematic Bacteriology, 2nd Edition, Volume 1. The archaea and the deeply branching and phototrophic bacteria. Springer, New York. 359.

Reysenbach, A.-L. 2001i. Order I. *Thermoplasmatales* ord. nov. *In* Boone, Castenholz and Garrity (Editors), Bergey's Manual of Systematic Bacteriology, 2nd Edition, Volume 1. The archaea and the deeply branching and phototrophic bacteria. Springer, New York. 335.

Reysenbach, A.-L. 2001j. Order I. *Thermotogales* ord. nov. *In* Boone, Castenholz and Garrity (Editors), Bergey's Manual of Systematic Bacteriology, 2nd Edition, Volume 1. The archaea and the deeply branching and phototrophic bacteria. Springer, New York. 369.

Reysenbach, A.-L. 2001k. Phylum BI. *Aquaficae* phyl. nov. *In* Boone, Castenholz and Garrity (Editors), Bergey's Manual of Systematic Bacteriology, 2nd Edition, Volume 1. The archaea and the deeply branching and phototrophic bacteria. Springer, New York. 359.

Reysenbach, A.-L. 2001l. Phylum BII. *Thermotogae* phyl. nov. *In* Boone, Castenholz and Garrity (Editors), Bergey's Manual of Systematic Bacteriology, 2nd Edition, Volume 1. The archaea and the deeply branching and phototrophic bacteria. Springer, New York. 369.

Reysenbach, A.-L. 2002. *In* Validation of publication of new names and new combinations previously effectively published outside the IJSEM. List No. 85. Int. J. Syst. Evol. Microbiol. *52*: 685–690.

Reysenbach, A.L., G.S. Wickham and N.R. Pace. 1994. Phylogenetic analysis of the hyperthermophilic pink filament community in Octopus Spring, Yellowstone National Park. Appl. Environ. Microbiol. *60*: 2113–2119.

Rhodes, M.W., H. Kator, S. Kotob, P. van Berkum, I. Kaattari, W. Vogelbein, F. Quinn, M.M. Floyd, W.R. Butler and C.A. Ottinger. 2003. *Mycobacterium shottsii* sp. nov., a slowly growing species isolated from Chesapeake Bay striped bass (*Morone saxatilis*). Int. J. Syst. Evol. Microbiol. *53*: 421–424.

Richter, E., S. Niemann, F.O. Gloeckner, G.E. Pfyffer and S. Rusch-Gerdes. 2002. *Mycobacterium holsaticum* sp. nov. Int. J. Syst. Evol. Microbiol. *52*: 1991–1996.

Rijkenberg, M.J.A., R. Kort and K.J. Hellingwerf. 2001. *Alkalispirillum mobile* gen. nov., spec. nov., an alkaliphilic non-phototrophic member of the *Ectothiorhodospiraceae*. Arch. Microbiol. *175*: 369–375.

Rijkenberg, M.J.A., R. Kort and K.J. Hellingwerf. 2002. *In* Validation of publication of new names and new combinations previously effectively published outside the IJSEM. List No. 86. Int. J. Syst. Evol. Microbiol. *52*: 1075–1076.

Rikken, G., A. Kroon and C.v. Ginkel. 1996. Transformation of (per)chlorate into chloride by a newly isolated bacterium: reduction and dismutation. Appl. Microbiol. Biotechnol. *45*: 420–426.

Rippka, R., T. Coursin, W. Hess, C. Lichtle, D.J. Scanlan, K.A. Palinska, I. Iteman, F. Partensky, J. Houmard and M. Herdman. 2000. *Prochlorococcus marinus* Chisholm et al. 1992 subsp. *pastoris* subsp. nov. strain PCC 9511, the first axenic chlorophyll a(2)/b(2)-containing cyanobacterium (*Oxyphotobacteria*). Int. J. Syst. Evol. Microbiol. *50*: 1833–1847.

Rippka, R., T. Coursin, W. Hess, C. Lichtle, D.J. Scanlan, K.A. Palinska, I. Iteman, F. Partensky, J. Houmard and M. Herdman. 2001. *In* Validation of publication of new names and new combinations previously effectively published outside the IJSEM. List No. 79. Int. J. Syst. Evol. Microbiol. *51*: 263–265.

Risatti, J.B., W.C. Capman and D.A. Stahl. 1994. Community structure of a microbial mat: the phylogenetic dimension. Proc. Natl. Acad. Sci. U.S.A. *91*: 10173–10177.

Rivas, R., M. Sanchez, M.E. Trujillo, J.L. Zurdo-Pineiro, P.F. Mateos, E. Martinez-Molina and E. Velazquez. 2003a. *Xylanimonas cellulosilytica* gen. nov., sp. nov., a xylanolytic bacterium isolated from a decayed tree (*Ulmus nigra*). Int. J. Syst. Evol. Microbiol. *53*: 99–103.

Rivas, R., A. Willems, N.S. Subba-Rao, P.F. Mateos, F.B. Dazzo, R.M. Kroppenstedt, E. Martinez-Molína, M. Gillis and E. Velázquez. 2003b. *In* Validation of publication of new names and new combinations previously effectively published outside the IJSEM. List No. 92. Int. J. Syst. Evol. Microbiol. *53*: 935–937.

Rivas, R., A. Willems, N.S. Subba-Rao, P.F. Mateos, F.B. Dazzo, R.M. Kroppenstedt, E. Martinez-Molina, M. Gillis and E. Velazquez. 2003c. Description of *Devosia neptuniae* sp. nov. that nodulates and fixes nitrogen in symbiosis with *Neptunia natans*, an aquatic legume from India. Syst. Appl. Microbiol. *26*: 47–53.

Robertson, B.R., J. O'Rourke, P. Vandamme, S.L.W. On and A. Lee.

2001a. *Helicobacter ganmani* sp. nov., a urease-negative anaerobe isolated from the intestines of laboratory mice. Int. J. Syst. Bacteriol. *51*: 1881–1889.

Robertson, J.A., G.W. Stemke, J.W. Davis, Jr., R. Harasawa, D. Thirkell, F. Kong, M.C. Shepard and D.K. Ford. 2002. Proposal of *Ureaplasma parvum* sp. nov. and emended description of *Ureaplasma urealyticum* (Shepard et al. 1974) Robertson et al. 2001. Int. J. Syst. Evol. Microbiol. *52*: 587–597.

Robertson, W.J., J.P. Bowman, P.D. Franzmann and B.J. Mee. 2001b. *Desulfosporosinus meridiei* sp. nov., a spore-forming sulfate-reducing bacterium isolated from gasolene-contaminated groundwater. Int. J. Syst. Evol. Microbiol. *51*: 133–140.

Roldan, M.D., F. Reyes, C. Moreno-Vivian and F. Castillo. 1994. Chlorate and nitrate reduction in the phototrophic bacteria *Rhodobacter capsulatus* and *Rhodobacter sphaeroides*. Curr. Microbiol. *29*: 241–245.

Roller, C., M. Wagner, R.I. Amann, W. Ludwig and K.H. Schleifer. 1994. *In situ* probing of Gram-positive bacteria with high DNA G + C content using 23S rRNA-targeted oligonucleotides: environmental application of nucleic acid hybridization. Microbiology (Reading) *140*: 2849–2858.

Romanenko, L.A., P. Schumann, M. Rohde, A.M. Lysenko, V.V. Mikhailov and E. Stackebrandt. 2002a. *Psychrobacter submarinus* sp. nov. and *Psychrobacter marincola* sp. nov., psychrophilic halophiles from marine environments. Int. J. Syst. Evol. Microbiol. *52*: 1291–1297.

Romanenko, L.A., P. Schumann, M. Rohde, V.V. Mikhailov and E. Stackebrandt. 2002b. *Halomonas halocynthiae* sp. nov., isolated from the marine ascidian *Halocynthia aurantium*. Int. J. Syst. Evol. Microbiol. *52*: 1767–1772.

Romanenko, L.A., P. Schumann, N.V. Zhukova, M. Rohde, V.V. Mikhailov and E. Stackebrandt. 2003a. *Oceanisphaera litoralis* gen. nov., sp. nov., a novel halophilic bacterium from marine bottom sediments. Int. J. Syst. Evol. Microbiol. *53*: 1885–1888.

Romanenko, L.A., M. Uchino, E. Falsen, N.V. Zhukova, V.V. Mikhailov and T. Uchimura. 2003b. *Rheinheimera pacifica* sp. nov., a novel halotolerant bacterium isolated from deep sea water of the Pacific. Int. J. Syst. Evol. Microbiol. *53*: 1973–1977.

Romanenko, L.A., M. Uchino, V.V. Mikhailov, N.V. Zhukova and T. Uchimura. 2003c. *Marinomonas primoryensis* sp. nov., a novel psychrophile isolated from coastal sea-ice in the Sea of Japan. Int. J. Syst. Evol. Microbiol. *53*: 829–832.

Romanenko, L.A., N.V. Zhukova, A.M. Lysenko, V.V. Mikhailov and E. Stackebrandt. 2003d. Assignment of *'Alteromonas marinoglutinosa'* NCIMB 1770 to *Pseudoalteromonas mariniglutinosa* sp. nov., nom. rev., comb. nov. Int. J. Syst. Evol. Microbiol. *53*: 1105–1109.

Romanenko, L.A., N.V. Zhukova, M. Rohde, A.M. Lysenko, V.V. Mikhailov and E. Stackebrandt. 2003e. *Glaciecola mesophila* sp. nov., a novel marine agar-digesting bacterium. Int. J. Syst. Evol. Microbiol. *53*: 647–651.

Romanenko, L.A., N.V. Zhukova, M. Rohde, A.M. Lysenko, V.V. Mikhailov and E. Stackebrandt. 2003f. *Pseudoalteromonas agarivorans* sp. nov., a novel marine agarolytic bacterium. Int. J. Syst. Evol. Microbiol. *53*: 125–131.

Romanenko, V.I., V.N. Korenkov and S.I. Kuznetsov. 1976. Bacterial decomposition of ammonium perchlorate. Mikrobiologiya *45*: 204–209.

Romano, I., A. Giordano, L. Lama, B. Nicolaus and A. Gambacorta. 2003a. *Planococcus rifietensis* sp. nov., isolated from algal mat collected from a sulfurous spring in Campania (Italy). Syst. Appl. Microbiol. *26*: 357–366.

Romano, I., A. Giordano, L. Lama, B. Nicolaus and A. Gambacorta. 2003b. *In* Validation of publication of new names and new combinations previously effectively published outside the IJSEM. List No. 94. Int. J. Syst. Evol. Microbiol. *53*: 1701–1702.

Rondon, M.R., S.J. Raffel, R.M. Goodman and J. Handelsman. 1999. Toward functional genomics in bacteria: analysis of gene expression in *Escherichia coli* from a bacterial artificial chromosome library of *Bacillus cereus*. Proc. Natl. Acad. Sci. U.S.A. *96*: 6451–6455.

Rooney-Varga, J.N., R. Devereux, R.S. Evans and M.E. Hines. 1997. Seasonal changes in the relative abundance of uncultivated sulfate-reducing bacteria in a salt marsh sediment and in the rhizosphere of *Spartina alterniflora*. Appl. Environ. Microbiol. *63*: 3895–3901.

Roos, S., F. Karner, L. Axelsson and H. Jonsson. 2000. *Lactobacillus mucosae* sp. nov., a new species with in vitro mucus-binding activity isolated from pig intestine. Int. J. Syst. Evol. Microbiol. *50*: 251–258.

Rosemarin, A., K. K. Lehtinen and M. M. Notini. 1990. Effects of treated and untreated softwood pulp mill effluents on Baltic Sea algae and invertebrates in model ecosystems. Nord. Pulp Paper Res. J. *2*: 83–87.

Rosselló-Mora, R.A., B. Thamdrup, H. Schäfer, R. Weller and R.I. Amann. 1999. Marine sediment microbial community response to organic carbon amendation under anaerobic conditions. Syst. Appl. Microbiol. *22*: 237–248.

Rosselló-Mora, R.A., M. Wagner, R.I. Amann and K.H. Schleifer. 1995. The abundance of *Zoogloea ramigera* in sewage treatment plants. Appl. Environ. Microbiol. *6*: 702–707.

Roth, A., U. Reischl, N. Schonfeld, L. Naumann, S. Emler, M. Fischer, H. Mauch, R. Loddenkemper and R.M. Kroppenstedt. 2000. *Mycobacterium heckeshornense* sp. nov., a new pathogenic slowly growing *Mycobacterium* sp. causing cavitary lung disease in an immunocompetent patient. J. Clin. Microbiol. *38*: 4102–4107.

Roth, A., U. Reischl, N. Schonfeld, L. Naumann, S. Emler, M. Fischer, H. Mauch, R. Loddenkemper and R.M. Kroppenstedt. 2001. *In* Validation of publication of new names and new combinations previously effectively published outside the IJSEM. List No. 79. Int. J. Syst. Evol. Microbiol. *51*: 263–265.

Rotthauwe, J.H., W. De Boer and W. Liesack. 1995. Comparative analysis of gene sequences encoding ammonia monooxygenase of *Nitrosospira* sp. AHB1 and *Nitrosolobus multiformis* C-71. FEMS Microbiol. Lett. *133*: 131–135.

Rotthauwe, J.H., K.P. Witzel and W. Liesack. 1997. The ammonia monooxygenase structural gene *amoA* as a functional marker: molecular fine-scale analysis of natural ammonia-oxidizing populations. Appl. Environ. Microbiol. *63*: 4704–4712.

Rurangirwa, F.R., C.A. Teitzel, J. Cui, D.M. French, P.L. McDonough and T. Besser. 2000. *Streptococcus didelphis* sp. nov., a streptococcus with marked catalase activity isolated from opossums (*Didelphis virginiana*) with suppurative dermatitis and liver fibrosis. Int. J. Syst. Evol. Microbiol. *50*: 759–765.

Saenger, W. 1984. Principles of Nucleic Acid Structure, Springer-Verlag, Berlin.

Sahin, N., K. Isik, A.U. Tamer and M. Goodfellow. 2000a. Taxonomic position of "*Pseudomonas oxalaticus*" strain Ox1^T (DSM 1105^T) (Khambata and Bhat, 1953) and its description in the genus *Ralstonia* as *Ralstonia oxalatica* comb. nov. Syst. Appl. Microbiol. *23*: 206–209.

Sahin, N., K. Isik, A.U. Tamer and M. Goodfellow. 2000b. *In* Validation of the publication of new names and new combinations previously effectively published outside the IJSEM. List No. 77. Int. J. Syst. Evol. Microbiol. *50*: 1953.

Sahm, K., B.J. Macgregor, B.B. Jørgensen and D.A. Stahl. 1999. Sulphate reduction and vertical distribution of sulphate-reducing bacteria quantified by rRNA slot-blot hybridization in a coastal marine sediment. Environ. Microbiol. *1*: 65–74.

Saiki, R.K., D.H. Gelfand, S. Stoffel, S.J. Scharf, R. Higuchi, G.T. Horn, K.B. Mullis and H.A. Erlich. 1988. Primer-directed enzymatic amplification of DNA with a thermostable DNA polymerase. Science *239*: 487–491.

Saiki, R.K., P.S. Walsh, C.H. Levenson and H.A. Erlich. 1989. Genetic analysis of amplified DNA with immobilized sequence-specific oligonucleotide probes. Proc. Natl. Acad. Sci. U.S.A. *86*: 6230–6234.

Saintpierre, D., H. Amir, R. Pineau, L. Sembiring and M. Goodfellow. 2003a. *Streptomyces yatensis* sp. nov., a novel bioactive streptomycete isolated from a New Caledonian ultramafic soil. Antonie Leeuwenhoek Int. J. Gen. Molec.Microbiol. *83*: 21–26.

Saintpierre, D., H. Amir, R. Pineau, L. Sembiring and M. Goodfellow. 2003b. *In* Validation of publication of new names and new combinations previously effectively published outside the IJSEM. List No. 93. Int. J. Syst. Evol. Microbiol. *53*: 1219–1220.

Saitoh, S., T. Suziki and Y. Nishimura. 1998. Proposal of *Craurococcus roseus* gen. nov., sp. nov. and *Paracraurococcus ruber* gen. nov., sp. nov., novel aerobic bacteriochlorophyll *a*-containing bacteria from soil. Int. J. Syst. Bacteriol. *48*: 1043–1047.

Saitou, N. and M. Nei. 1987. The neighbor-joining method: a new method for reconstructing phylogenetic trees. Mol. Biol. Evol. *4*: 406–425.

Sakaguchi, T., A. Arakaki and T. Matsunaga. 2002. *Desulfovibrio magneticus* sp. nov., a novel sulfate-reducing bacterium that produces intracellular single-domain-sized magnetite particles. Int. J. Syst. Evol. Microbiol. *52*: 215–221.

Sakala, R.M., Y. Kato, H. Hayashidani, M. Murakami, C. Kaneuchi and M. Ogawa. 2002. *Lactobacillus fuchuensis* sp. nov., isolated from vacuum-packaged refrigerated beef. Int. J. Syst. Evol. Microbiol. *52*: 1151–1154.

Sakamoto, M., M. Suzuki, M. Umeda, L. Ishikawa and Y. Benno. 2002. Reclassification of *Bacteroides forsythus* (Tanner et al. 1986) as *Tannerella forsythensis* corrig., gen. nov., comb. nov. Int. J. Syst. Evol. Microbiol. *52*: 841–849.

Sako, Y., S. Nakagawa, K. Takai and K. Horikoshi. 2003. *Marinithermus hydrothermalis* gen. nov., sp. nov., a strictly aerobic, thermophilic bacterium from a deep-sea hydrothermal vent chimney. Int. J. Syst. Evol. Microbiol. *53*: 59–65.

Sako, Y., T. Nunoura and A. Uchida. 2001. *Pyrobaculum oguniense* sp. nov., a novel facultatively aerobic and hyperthermophilic archaeon growing at up to 97°C. Int. J. Syst. Evol. Microbiol. *51*: 303–309.

Sambrook, J., E.F. Fritsch and T. Maniatis. 1989. Molecular cloning: a laboratory manual, Cold Spring Harbor Press, Cold Spring Harbor.

Sanford, R.A., J.R. Cole, F.E. Löffler and J.M. Tiedje. 1996. Characterization of *Desulfitobacterium chlororespirans* sp. nov., which grows by coupling the oxidation of lactate to the reductive dechlorination of 3-chloro-4-hydroxybenzoate. Appl. Environ. Microbiol. *62*: 3800–3808.

Sanford, R.A. , J.R. Cole, F.E. Löffler and J.M. Tiedje. 2001. *In* Validation of publication of new names and new combinations previously effectively published outside the IJSEM. List No. 80. Int. J. Syst. Evol. Microbiol. *51*: 793–794.

Sanford, R.A., J.R. Cole and J.M. Tiedje. 2002a. Characterization and description of *Anaeromyxobacter dehalogenans* gen. nov., sp. nov., an aryl-halorespiring facultative anaerobic myxobacterium. Appl. Environ. Microbiol. *68*: 893–900.

Sanford, R.A., J.R. Cole and J.M. Tiedje. 2002b. *In* Validation of publication of new names and new combinations previously effectively published outside the IJSEM. List No. 86. Int. J. Syst. Evol. Microbiol. *52*: 1075–1076.

Sarathchandra, S.U. 1978. Nitrification activities and changes in populations of nitrifying bacteria in soil perfused at 2 different H-ion concentrations. Plant Soil *50*: 99–111.

Sass, A., H. Rutters, H. Cypionka and H. Sass. 2002a. *Desulfobulbus mediterraneus* sp. nov., a sulfate-reducing bacterium growing on mono- and disaccharides. Arch. Microbiol. *177*: 468–474.

Sass, A., H. Rutters, H. Cypionka and H. Sass. 2002b. *In* Validation of publication of new names and new combinations previously effectively published outside the IJSEM. List No. 87. Int. J. Syst. Evol. Microbiol. *52*: 1437–1438.

Sato, K. 1978. Bacteriochlorophyll formation by facultative methylotrophs, *Protaminobacter ruber* and *Pseudomonas* AM 1. FEBS Lett. *85*: 207–210.

Sato, K., K. Hagiwara and S. Shimizu. 1985. Effect of cultural conditions on tetrapyrrole formation, especially bacteriochlorophyll formation in a facultative methylotroph, *Protaminobacter ruber*. Agric Biol Chem. *49*: 1–5.

Sato, K., T. Shiba and Y. Shioi. 1989. Regulation of the biosynthesis of bacteriochlorophyll. *In* Harashima, Shima and Murata (Editors), Aerobic Photosynthetic Bacteria, Japan Scientific Societies Press, Springer Verlag, Tokyo, Berlin, Heidelberg, New York, London, Paris. 95–124.

Sato, K. and S. Shimizu. 1979. The conditions for bacteriochlorophyll formation and the ultrastructure of a methanol-utilizing bacterium, *Protaminobacter ruber*, classified as a nonphotosynthetic bacterium. Agric. Biol. Chem. *43*: 1669–1676.

Satoh, T., Y. Hoshina and H. Kitamura. 1976. *Rhodopseudomonas sphaeroides* forma subsp. *denitrificans*, a denitrifying strain as a subspecies of *Rhodopseudomonas sphaeroides*. Arch. Microbiol. *108*: 265–269.

Satomi, M. , B. Kimura, T. Hamada, S. Harayama and T. Fujii. 2002. Phylogenetic study of the genus *Oceanospirillum* based on 16S rRNA and *gyrB* genes: emended description of the genus *Oceanospirillum*, description of *Pseudospirillum* gen. nov., *Oceanobacter* gen. nov. and *Terasakiella* gen. nov. and transfer of *Oceanospirillum jannaschii* and *Pseudomonas stanieri* to *Marinobacterium* as *Marinobacterium jannaschii* comb. nov. and *Marinobacterium stanieri* comb. nov. Int. J. Syst. Evol. Microbiol. *52*: 739–747.

Satomi, M., H. Oikawa and Y. Yano. 2003. *Shewanella marinintestina* sp. nov., *Shewanella schlegeliana* sp. nov. and *Shewanella sairae* sp. nov., novel eicosapentaenoic-acid-producing marine bacteria isolated from sea-animal intestines. Int. J. Syst. Evol. Microbiol. *53*: 491–499.

Savant, D.V., Y.S. Shouche, S. Prakash and D.R. Ranade. 2002. *Methanobrevibacter acididurans* sp. nov., a novel methanogen from a sour anaerobic digester. Int. J. Syst. Evol. Microbiol. *52*: 1081–1087.

Sawabe, T., R. Tanaka, M.M. Iqbal, K. Tajima, Y. Ezura, E.P. Ivanova and R. Christen. 2000. Assignment of *Alteromonas elyakovii* KMM162^{T} and five strains isolated from spot-wounded fronds of *Laminaria japonica* to *Pseudoalteromonas elyakovii* comb. nov. and the extended description of the species. Int. J. Syst. Evol. Bacteriol. *50*: 265–271.

Sayavedra-Soto, L.A., N.G. Hommes and D.J. Arp. 1994. Characterization of the gene encoding hydroxylamine oxidoreductase in *Nitrosomonas europaea*. J. Bacteriol. *176*: 504–510.

Sayler, G.S. and A.C. Layton. 1990. Environmental application of nucleic acid hybridization. Annu. Rev. Microbiol. *44*: 625–628.

Sayler, G.S., M.S. Shields, E.T. Tedford, A. Breen, S.W. Hooper, K.M. Sirotkin and J.W. Davis. 1985. Application of DNA–DNA colony hybridization to the detection of catabolic genotypes in environmental samples. Appl. Environ. Microbiol. *49*: 1295–1303.

Schaechter, M.O., O. Maaløe and N.O. Kjeldgaard. 1958. Dependency on medium and temperature of cell size and chemical composition during balanced growth of *Salmonella typhimurium*. J. Gen. Microbiol. *19*: 592–606.

Schaefer, J.K., K.D. Goodwin, I.R. McDonald, J.C. Murrell and R.S. Oremland. 2002. *Leisingera methylohatidivorans* gen. nov., sp. nov., a marine methylotroph that grows on methyl bromide. Int. J. Syst. Evol. Microbiol. *52*: 851–859.

Schink, B. 1997. Energetics of syntrophic cooperation in methanogenic degradation. Microbiol. Mol. Biol. Rev. *61*: 262–280.

Schink, B., V. Thiemann, H. Laue and M.W. Friedrich. 2002a. *Desulfotignum phosphitoxidans* sp. nov., a new marine sulfate reducer that oxidizes phosphite to phosphate. Arch. Microbiol. *177*: 381–391.

Schink, B., V. Thiemann, H. Laue and M.W. Friedrich. 2002b. *In* Validation of publication of new names and new combinations previously effectively published outside the IJSEM. List No. 87. Int. J. Syst. Evol. Microbiol. *52*: 1437–1438.

Schinsky, M.F., M.M. McNeil, A.M. Whitney, A.G. Steigerwalt, B.A. Lasker, M.M. Floyd, G.G. Hogg, D.J. Brenner and J.M. Brown. 2000. *Mycobacterium septicum* sp. nov., a new rapidly growing species associated with catheter-related bacteraemia. Int. J. Syst. Evol. Microbiol. *50*: 575–581.

Schippers, A., K. Bosecker, S. Willscher, C. Sproer, P. Schumann and R.M. Kroppenstedt. 2002. *Nocardiopsis metallicus* sp. nov., a metal-leaching actinomycete isolated from an alkaline slag dump. Int. J. Syst. Evol. Microbiol. *52*: 2291–2295.

Schlegel, L., F. Grimont, E. Ageron, P.A. Grimont and A. Bouvet. 2003. Reappraisal of the taxonomy of the *Streptococcus bovis/Streptococcus equinus* complex and related species: description of *Streptococcus gallolyticus* subsp. *gallolyticus* subsp. nov., *Streptococcus gallolyticus* subsp. *macedonicus* subsp. nov. and *Streptococcus gallolyticus* subsp. *pasteurianus* nov. Int. J. Syst. Evol. Microbiol. *53*: 631–645.

Schlegel, L., F. Grimont, M.D. Collins, B. Regnault, P.A. Grimont and A. Bouvet. 2000. *Streptococcus infantarius* sp. nov., *Streptococcus infantarius* subsp. *infantarius* subsp. nov. and *Streptococcus infantarius* subsp. *coli* subsp. nov., isolated from humans and food. Int. J. Syst. Evol. Microbiol. *50*: 1425–1434.

Schleifer, K.H., M. Ehrmann, C. Beimfohr, E. Brockmann, W. Ludwig and R.I. Amann. 1995. Application of molecular methods for the classification and identification of lactic acid bacteria. Int. Dairy J. *5*: 1081–1094.

Schleifer, K.H., W. Ludwig and R. Amann. 1993. Nucleic acid probes. *In* Goodfellow and O'Donnell (Editors), Handbook of new bacterial systematics, Academic Press Limited, London. 464–512.

Schleifer, K.H. and E. Stackebrandt. 1983. Molecular systematics of prokaryotes. Annu. Rev. Microbiol. *37*: 143–187.

Schlesner, H., P.A. Lawson, M.D. Collins, N. Weiss, U. Wehmeyer, H. Volker and M. Thomm. 2001. *Filobacillus milensis* gen. nov., sp. nov., a new halophilic spore-forming bacterium with Orn-D-Glu-type peptidoglycan. Int. J. Syst. Evol. Microbiol. *51*: 425–431.

Schlesner, H. and E. Stackebrandt. 1986. Assignment of the genera *Planctomyces* and *Pirella* to a new family *Planctomycetaceae* fam. nov. and description of the order *Planctomycetales* ord. nov. Syst. Appl. Microbiol. *8*: 174–176.

Schmidhuber, S., W. Ludwig and K.H. Schleifer. 1988. Construction of a DNA probe for the specific identification of *Streptococcus oralis*. J. Clin. Microbiol. *26*: 1042–1044.

Schmidt, E.L. 1982. Nitrification in soil. *In* Stevenson (Editor), Nitrogen in Agricultural Soils, American Society of Agronomy, Madison. pp. 253–288.

Schmidt, I. 1997. Anaerobe Ammoniakoxidation von *Nitrosomonas eutropha*, University of Hamburg

Schmidt, I. and E. Bock. 1997. Anaerobic ammonia oxidation with nitrogen dioxide by *Nitrosomonas eutropha*. Arch. Microbiol. *167*: 106–111.

Schmidt, I. and E. Bock. 1998. Anaerobic ammonia oxidation by cell-free extracts of *Nitrosomonas eutropha*. Antonie Leeuwenhoek *73*: 271–278.

Schmidt, K. 1978. Biosynthesis of carotenoids. *In* Clayton and Sistrom (Editors), The Photosynthetic Bacteria, Plenum Press, New York. pp. 729–750.

Scholz-Muramatsu, H., A. Neumann, M. Messmer, E. Moore and G. Diekert. 1995. Isolation and characterization of *Dehalospirillum multivorans* gen. nov., sp. nov, a tetrachloroethene utilizing, strictly anaerobic bacterium. Arch. Microbiol. *163*: 48–56.

Scholz-Muramatsu, H., A. Neumann, M. Messmer, E. Moore and G. Diekert. 2002. *In* Validation of publication of new names and new combinations previously effectively published outside the IJSEM. List No. 88. Int. J. Syst. Evol. Microbiol. *52*: 1915–1916.

Schönhuber, W., B. Fuchs, S. Juretschko and R.I. Amann. 1997. Improved sensitivity of whole-cell hybridization by the combination of horseradish peroxidase-labeled oligonucleotides and tyramide signal amplification. Appl. Environ. Microbiol. *63*: 3268–3273.

Schramm, A., D. de Beer, J.C. van den Heuvel, S. Ottengraf and R. Amann. 1999. Microscale distribution of populations and activities of *Nitrosospira* and *Nitrospira* spp. along a macroscale gradient in a nitrifying bioreactor: Quantification by *in situ* hybridization and the use of microsensors. Appl. Environ. Microbiol. *65*: 3690–3696.

Schramm, A., D. De Beer, M. Wagner and R. Amann. 1998. Identification and activities *in situ* of *Nitrosospira* and *Nitrospira* spp. as dominant populations in a nitrifying fluidized bed reactor. Appl. Environ. Microbiol. *64*: 3480–3485.

Schramm, A., L.H. Larsen, N.P. Revsbech, N.B. Ramsing, R. Amann and K.H. Schleifer. 1996. Structure and function of a nitrifying biofilm as determined by *in situ* hybridization and the use of microelectrodes. Appl. Environ. Microbiol. *62*: 4641–4647.

Schroeter, J. 1885–1889. Kryptogamenflora von Schlesien. Bd. 3 Heft 3, Pilze. *In* Cohn (Editor), Breslau, J.U. Kern's Verlag.

Schroll, G., H.J. Busse, G. Parrer, S. Rölleke, W. Lubitz and E.B.M. Denner. 2001a. *Alcaligenes faecalis* subsp. *parafaecalis* subsp. nov., a bacterium accumulating poly-beta-hydroxybutyrate from acetone-butanol bioprocess residues. Syst. Appl. Microbiol. *24*: 37–43.

Schroll, G., H.J. Busse, G. Parrer, S. Rolleke, W. Lubitz and E.B.M. Denner. 2001b. *In* Validation of the publication of new names and new combinations previously effectively published outside the IJSEM. List No. 82. Int. J. Syst. Evol. Microbiol. *51*: 1619–1620.

Schüller, G., C. Hertel and W.P. Hammes. 2000. *Gluconacetobacter entanii* sp. nov., isolated from submerged high-acid industrial vinegar fermentations. Int. J. Syst. Evol. Microbiol. *50*: 2013–2020.

Schumann, P., U. Behrendt, A. Ulrich and K. Suzuki. 2003. Reclassification of *Subtercola pratensis* Behrendt et al. 2002 as *Agreia pratensis* comb. nov. Int. J. Syst. Evol. Microbiol. *53*: 2041–2044.

Schumann, P., N. Weiss and E. Stackebrandt. 2001. Reclassification of *Cellulomonas cellulans* (Stackebrandt and Keddie 1986) as *Cellulosimicrobium cellulans* gen. nov., comb. nov. Int. J. Syst. Evol. Microbiol. *51*: 1007–1010.

Schwiertz, A., G.L. Hold, S.H. Duncan, B. Gruhl, M.D. Collins, P.A. Lawson, H.J. Flint and M. Blaut. 2002a. *Anaerostipes caccae* gen. nov., sp. nov., a new saccharolytic, acetate-utilising, butyrate-producing bacterium from human faeces. Syst. Appl. Microbiol. *25*: 46–51.

Schwiertz, A., G.L. Hold, S.H. Duncan, B. Gruhl, M.D. Collins, P.A. Lawson, H.J. Flint and M. Blaut. 2002b. *In* Validation of publication of new names and new combinations previously effectively published outside the IJSEM. List No. 87. Int. J. Syst. Evol. Microbiol. *52*: 1437–1438.

Sekiguchi, Y., Y. Kamagata, K. Nakamura, A. Ohashi and H. Harada. 2000. *Syntrophothermus lipocalidus* gen. nov., sp. nov., a novel thermophilic, syntrophic, fatty-acid-oxidizing anaerobe which utilizes isobutyrate. Int. J. Syst. Evol. Microbiol. *50*: 771–779.

Sekiguchi, Y., T. Yamada, S. Hanada, A. Ohashi, H. Harada and Y. Kamagata. 2003. *Anaerolinea thermophila* gen. nov., sp. nov. and *Caldilinea aerophila* gen. nov., sp. nov., novel filamentous thermophiles that represent a previously uncultured lineage of the domain *Bacteria* at the subphylum level. Int. J. Syst. Evol. Microbiol. *53*: 1843–1851.

Sembiring, L., A.C. Ward and M. Goodfellow. 2000. Selective isolation and characterisation of members of the *Streptomyces violaceusniger* clade associated with the roots of *Paraserianthes falcataria*. Antonie Leeuwenhoek Int. J. Gen. Molec.Microbiol. *78*: 353–366.

Sembiring, L., A.C. Ward and M. Goodfellow. 2001. *In* Validation of publication of new names and new combinations previously effectively published outside the IJSEM. List No. 82. Int. J. Syst. Evol. Microbiol. *51*: 1619–1620.

Seyfried, M., D. Lyon, F.A. Rainey and J. Wiegel. 2002. *Caloramator viterbensis* sp. nov., a novel thermophilic, glycerol-fermenting bacterium isolated from a hot spring in Italy. Int. J. Syst. Evol. Microbiol. *52*: 1177–1184.

Sheridan, P.P., J. Loveland-Curtze, V.I. Miteva and J.E. Brenchley. 2003. *Rhodoglobus vestalii* gen. nov., sp. nov., a novel psychrophilic organism isolated from an Antarctic Dry Valley lake. Int. J. Syst. Evol. Microbiol. *53*: 985–994.

Shi, B.H., V. Arunpairojana, S. Palakawong and A. Yokota. 2002. *Tistrella mobilis* gen. nov., sp. nov., a novel polyhydroxyalkanoate-producing bacterium belonging to α-*Proteobacteria*. J. Gen. Appl. Microbiol. *48*: 335–343.

Shi, B.H., V. Arunpairojana, S. Palakawong and A. Yokota. 2003. *In* Validation of publication of new names and new combinations previously effectively published outside the IJSEM. List No. 92. Int. J. Syst. Evol. Microbiol. *53*: 935–937.

Shiba, T. 1984. Utilization of light energy by the strictly aerobic bacterium *Erythrobacter* sp. och 114. J. Gen. Appl. Microbiol. *30*: 239–244.

Shiba, T. 1989. Overview of the aerobic photosynthetic bacteria. *In* Hirashima, Shiba and Murata (Editors), Aerobic Photosynthetic Bacteria, Springer-Verlag, Berlin. 1–8.

Shiba, T. 1991. *Roseobacter litoralis* gen. nov., sp. nov., and *Roseobacter denitrificans* sp. nov., aerobic pink-pigmented bacteria which contain bacteriochlorophyll *a*. Syst. Appl. Microbiol. *14*: 140–145.

Shiba, T. and K. Abe. 1987. An aerobic bacterium containing bacteriochlorophyll-proteins showing absorption maxima of 802, 844 and 862 nm in the near infrared region. Agric. Biol. Chem. *51*: 945–946.

Shiba, T., Y. Shioi, K. Takamiya, D.C. Sutton and C.R. Wilkinson. 1991. Distribution and physiology of aerobic-bacteria containing bacterio-

maoka et al. 1987) gen. nov., comb. nov. Int. J. Syst. Bacteriol. *49*: 567–576.

Wernegreen, J.J. and M.A. Riley. 1999. Comparison of the evolutionary dynamics of symbiotic and housekeeping loci: a case for the genetic coherence of rhizobial lineages. Mol. Biol. Evol. *16*: 98–113.

Werner, F.C. (Editor). 1972. Wortelemente Lateinisch-Griechischer Fachausdrücke in den Biologischen Wissenschaften, 3rd Ed., Suhrkamp Taschenbuch Verlag, Berlin.

Wertz, J.E., C. Goldstone, D.M. Gordon and M.A. Riley. 2003. A molecular phylogeny of enteric bacteria and implications for a bacterial species concept. J. Evol. Biol. *16*: 1236–1248.

Wery, N., F. Lesongeur, P. Pignet, V. Derennes, M.A. Cambon-Bonavita, A. Godfroy and G. Barbier. 2001a. *Marinitoga camini* gen. nov., sp. nov., a rod-shaped bacterium belonging to the order *Thermotogales*, isolated from a deep-sea hydrothermal vent. Int. J. Syst. Evol. Microbiol. *51*: 495–504.

Wery, N., J.M. Moricet, V. Cueff, J. Jean, P. Pignet, F. Lesongeur, M.A. Cambon-Bonavita and G. Barbier. 2001b. *Caloranaerobacter azorensis* gen. nov., sp. nov., an anaerobic thermophilic bacterium isolated from a deep-sea hydrothermal vent. Int. J. Syst. Evol. Microbiol. *51*: 1789–1796.

Westerberg, K., A.M. Elvang, E. Stackebrandt and J.K. Jansson. 2000. *Arthrobacter chlorophenolicus* sp. nov., a new species capable of degrading high concentrations of 4-chlorophenol. Int. J. Syst. Evol. Microbiol. *50*: 2083–2092.

Wetmur, J.G., D.M. Wong, B. Ortiz, J. Tong, F. Reichert and D.H. Gelfand. 1994. Cloning, sequencing, and expression of RecA proteins from three distantly related thermophilic eubacteria. J. Biol. Chem. *269*: 25928–25935.

Weyant, R.S., C.W. Moss, R.E. Weaver, D.G. Hollis, J.G. Jordan, E.C. Cook and M.I. Daneshvar (Editors). 1996. Identification of Unusual Pathogenic Gram-negative Aerobic and Facultatively Anaerobic Bacteria, 2nd Ed,, The Williams & Wilkins Co., Baltimore.

Whitford, M.F., L.J. Yanke, R.J. Forster and R.M. Teather. 2001. *Lachnobacterium bovis* gen. nov., sp. nov., a novel bacterium isolated from the rumen and faeces of cattle. Int. J. Syst. Evol. Microbiol. *51*: 1977–1981.

Whitman, W.B. 2001a. Genus I. *Methanocaldococcus. In* Boone, Castenholz and Garrity (Editors), Bergey's Manual of Systematic Bacteriology, 2nd Edition, Vol. 1, Springer, New York. 243–245.

Whitman, W.B. 2001b. Genus II. *Methanothermococcus. In* Boone, Castenholz and Garrity (Editors), Bergey's Manual of Systematic Bacteriology, 2nd Edition, Vol. 1, Springer, New York. 241–242.

Whitman, W.B. 2001c. Genus II. *Methanotorris. In* Boone, Castenholz and Garrity (Editors), Bergey's Manual of Systematic Bacteriology, 2nd Edition, Vol. 1, Springer, New York. 245–246.

Whitman, W.B. 2002. *In* Validation of publication of new names and new combinations previously effectively published outside the IJSEM. List No. 85. Int. J. Syst. Evol. Microbiol. *52*: 685–690.

Whitman, W.B., D.R. Boone and Y. Koga. 2001. Family II. *Methanocaldococcaceae* fam. nov. *In* Boone, Castenholz and Garrity (Editors), Bergey's Manual of Systematic Bacteriology, 2nd Edition, Volume 1. The archaea and the deeply branching and phototrophic bacteria. Springer, New York. 242–243.

Whitman, W.B., D.R. Boone and Y. Koga. 2002. *In* Validation of publication of new names and new combinations previously effectively published outside the IJSEM. List No. 85. Int. J. Syst. Evol. Microbiol. *52*: 685–690.

Whitman, W.B., D.C. Coleman and W.J. Wiebe. 1998. Prokaryotes: the unseen majority. Proc. Natl. Acad. Sci. U.S.A. *95*: 6578–6583.

Whittaker, M., D. Bergmann, D. Arciero and A.B. Hooper. 2000. Electron transfer during the oxidation of ammonia by the chemolithotrophic bacterium *Nitrosomonas europaea.* Biochim. Biophys. Acta *1459*: 346–355.

Widdel, F., S. Schnell, S. Heising, A. Ehrenreich, B. Assmus and B. Schink. 1993. Ferrous iron oxidation by anoxygenic phototrophic bacteria. Nature (Lond.) *362*: 834–836.

Willcox, M.D., H. Zhu and K.W. Knox. 2001. *Streptococcus australis* sp. nov., a novel oral streptococcus. Int. J. Syst. Evol. Microbiol. *51*: 1277–1281.

Willcox, W.R., S.P. Lapage and B. Holmes. 1980. A review of numerical methods in bacterial identification. Antonie Leeuwenhoek *46*: 233–299.

Willems, A., J. De Ley, M. Gillis and K. Kersters. 1991a. *Comamonadaceae*, a new family encompassing the acidovorans ribosomal RNA complex, including *Variovorax paradoxus*, gen. nov., comb. nov., for *Alcaligenes paradoxus* (Davis 1969). Int. J. Syst. Bacteriol. *41*: 445–450.

Willems, A., M. Fernandez-Lopez, E. Munoz-Adelantado, J. Goris, P. De Vos, E. Martinez-Romero, N. Toro and M. Gillis. 2003. Description of new *Ensifer* strains from nodules and proposal to transfer *Ensifer adhaerens* Casida 1982 to *Sinorhizobium* as *Sinorhizobium adhaerens* comb. nov. Request for an opinion. Int. J. Syst. Evol. Microbiol. *53*: 1207–1217.

Willems, A., H. Gilhaus, W. Beer, H. Mietke, H.R. Gelderblom, B. Burghardt, W. Voigt and R. Reissbrodt. 2002. *Brackiella oedipodis* gen. nov., sp. nov., Gram-negative, oxidase-positive rods that cause endocarditis of cotton-topped tamarin (*Saguinus oedipus*). Int. J. Syst. Evol. Microbiol. *52*: 179–186.

Willems, A., M. Gillis and J. De Ley. 1991b. Transfer of *Rhodocyclus gelatinosus* to *Rubrivivax gelatinosus* gen. nov., comb. nov., and phylogenetic relationships with *Leptothrix, Sphaerotilus natans, Pseudomonas saccharophila* and *Alcaligenes latus.* Int. J. Syst. Bacteriol. *41*: 65–73.

Willems, A., B. Pot, E. Falsen, P. Vandamme, M. Gillis, K. Kersters and J. De Ley. 1991c. Polyphasic taxonomic study of the emended genus *Comamonas*: Relationship to *Aquaspirillum aquaticum*, E. Falsen group 10, and other clinical isolates. Int. J. Syst. Bacteriol. *41*: 427–444.

Williams, P.J. and S.W. Watson. 1968. Autotrophy in *Nitrosocystis oceanus.* J. Bacteriol. *96*: 1640–1648.

Williams, S.T., M.E. Sharpe and J.G. Holt (Editors). 1989. Bergey' s Manual of Systematic Bacteriology, 1st Ed., Vol. 4, The Williams & Wilkins Co., Baltimore.

Willumsen, P., U. Karlson, E. Stackebrandt and R.M. Kroppenstedt. 2001. *Mycobacterium frederiksbergense* sp. nov., a novel polycyclic aromatic hydrocarbon-degrading *Mycobacterium* species. Int. J. Syst. Evol. Microbiol. *51*: 1715–1722.

Wilmotte, A. and M. Herdman. 2001. Phylogenetic relationships among the cyanobacteria based on 16S rRNA sequences. *In* Boone, Castenholz and Garrity (Editors), Bergey's Manual of Systematic Bacteriology, 2nd Edition, Volume 1. The archaea and the deeply branching and phototrophic bacteria. Springer, New York. 487–493.

Wilson, E.O. 1994. Naturalist, Island Press, Washington, DC.

Wilson, I.G. 1997. Inhibition and facilitation of nucleic acid amplification. Appl. Environ. Microbiol. *63*: 3741–3751.

Wilson, R.W., V.A. Steingrube, E.C. Bottger, B. Springer, B.A. Brown-Elliott, V. Vincent, K.C. Jost, Jr., Y. Zhang, M.J. Garcia, S.H. Chiu, G.O. Onyi, H. Rossmoore, D.R. Nash and R.J. Wallace, Jr.. 2001. *Mycobacterium immunogenum* sp. nov., a novel species related to *Mycobacterium abscessus* and associated with clinical disease, pseudo-outbreaks and contaminated metalworking fluids: an international cooperative study on mycobacterial taxonomy. Int. J. Syst. Evol. Microbiol. *51*: 1751–1764.

Wink, J.M., R.M. Kroppenstedt, B.N. Ganguli, S.R. Nadkarni, P. Schumann, G. Seibert and E. Stackebrandt. 2003a. Three new antibiotic producing species of the genus *Amycolatopsis, Amycolatopsis balhimycina* sp. nov., *A. tolypomycina* sp. nov., *A. vancoresmycina* sp. nov., and description of *Amycolatopsis keratiniphila* subsp. *keratiniphila* subsp. nov. and *A. keratiniphila* subsp. *nogabecina* subsp. nov. Syst. Appl. Microbiol. *26*: 38–46.

Wink, J.M., R.M. Kroppenstedt, B.N. Ganguli, S.R. Nadkarni, P. Schumann, G. Seibert and E. Stackebrandt. 2003b. *In* Validation of publication of new names and new combinations previously effectively published outside the IJSEM. List No. 92. Int. J. Syst. Evol. Microbiol. *53*: 935–937.

Wink, J., R.M. Kroppenstedt, G. Seibert and E. Stackebrandt. 2003c. *Actinomadura namibiensis* sp. nov. Int. J. Syst. Evol. Microbiol. *53*: 721–724.

chlorophyll *a* on the east and west coasts of Australia. Appl. Environ. Microbiol. *57*: 295–300.

Shiba, T., U. Simidu and N. Taga. 1979. Distribution of aerobic bacteria which contain bacteriochlorophyll *a*. Appl. Environ. Microbiol. *38*: 43–45.

Shieh, W.Y., Y.W. Chen, S.M. Chaw and H.H. Chiu. 2003a. *Vibrio ruber* sp. nov., a red, facultatively anaerobic, marine bacterium isolated from sea water. Int. J. Syst. Evol. Microbiol. *53*: 479–484.

Shieh, W.Y., A.L. Chen and H.H. Chiu. 2000. *Vibrio aerogenes* sp. nov., a facultatively anaerobic marine bacterium that ferments glucose with gas production. Int. J. Syst. Evol. Microbiol. *50*: 321–329.

Shieh, W.Y., W.D. Jean, Y.T. Lin and M. Tseng. 2003b. *Marinobacter lutaoensis* sp. nov., a thermotolerant marine bacterium isolated from a coastal hot spring in Lutao, Taiwan. Can. J. Microbiol. *49*: 244–252.

Shieh, W.Y., W.D. Jean, Y.T. Lin and M. Tseng. 2003c. *In* Validation of publication of new names and new combinations previously effectively published outside the IJSEM. List No. 94. Int. J. Syst. Evol. Microbiol. *53*: 1701–1702.

Shigematsu, T., K. Yumihara, Y. Ueda, M. Numaguchi, S. Morimura and K. Kida. 2003. *Delftia tsuruhatensis* sp. nov., a terephthalate-assimilating bacterium isolated from activated sludge. Int. J. Syst. Evol. Microbiol. *53*: 1479–1483.

Shimada, K. 1995. Aerobic anoxygenic phototrophs. *In* Blankenship, Madigan and Bauer (Editors), Anoxygenic Photosynthetic Bacteria, Kluwer Academic Publishers, Dordrecht. 105–122.

Shintani, T., W.T. Liu, S. Hanada, Y. Kamagata, S. Miyaoka, T. Suzuki and K. Nakamura. 2000. *Micropruina glycogenica* gen. nov., sp. nov., a new Gram-positive glycogen-accumulating bacterium isolated from activated sludge. Int. J. Syst. Evol. Microbiol. *50*: 201–207.

Shioi, Y. 1986. Growth characteristics and substrate specificity of aerobic photosynthetic bacterium *Erythrobacter* sp. Och-114. Plant Cell Physiol. *27*: 567–572.

Shioi, Y. and M. Doi. 1988. Control of bacteriochlorophyll accumulation by light in an aerobic photosynthetic bacterium *Erythrobacter* sp. Och 114. Arch. Biochem. Biophys. *266*: 470–477.

Shively, J.M., E. Bock, K. Westphal and G.C. Cannon. 1977. Icosahedral inclusions (carboxysomes) of *Nitrobacter agilis*. J. Bacteriol. *132*: 673–675.

Shojaei, H., J.G. Magee, R. Freeman, M. Yates, N.U. Horadagoda and M. Goodfellow. 2000. *Mycobacterium elephantis* sp. nov., a rapidly growing non-chromogenic *Mycobacterium* isolated from an elephant. Int. J. Syst. Evol. Microbiol. *50*: 1817–1820.

Sibley, C.G. and J.E. Ahlquist. 1987. DNA hybridization evidence of hominoid phylogeny: results from an expanded data set. J. Mol. Evol. *26*: 99–121.

Sibley, C.G., J.A. Comstock and J.E. Ahlquist. 1990. DNA hybridization evidence of hominoid phylogeny: a reanalysis of the data. J. Mol. Evol. *30*: 202–236.

Siefert, E. and V.B. Koppenhagen. 1982. Studies on the vitamin B_{12} auxotrophy of *Rhodocyclus purpureus* and two other vitamin B_{12}-requiring purple nonsulfur bacteria. Arch. Microbiol. *132*: 173–178.

Sievers, M., W. Ludwig and M. Teuber. 1994. Phylogentic positioning of *Acetobacter*, *Gluconobacter*, *Rhodopila* and *Acidiphilium* species as a branch of acidophilic bacteria in the alpha-subclass of *Proteobacteria* based on 16S ribosomal DNA sequences. Syst. Appl. Microbiol. *17*: 189–196.

Sievert, S.M., T. Heidorn and J. Kuever. 2000. *Halothiobacillus kellyi* sp. nov., a mesophilic, obligately chemolithoautotrophic, sulfur-oxidizing bacterium isolated from a shallow-water hydrothermal vent in the Aegean Sea, and emended description of the genus *Halothiobacillus*. Int. J. Syst. Evol. Microbiol. *50*: 1229–1237.

Sievert, S.M. and J. Kuever. 2000. *Desulfacinum hydrothermale* sp. nov., a thermophilic, sulfate-reducing bacterium from geothermally heated sediments near Milos Island (Greece). Int. J. Syst. Evol. Microbiol. *50*: 1239–1246.

Sikorski, J., E. Stackebrandt and W. Wackernagel. 2001. *Pseudomonas kilonensis* sp. nov., a bacterium isolated from agricultural soil. Int. J. Syst. Evol. Microbiol. *51*: 1549–1555.

Silvestri, L., M. Turri, L.R. Hill and E. Gilardi. 1962. A quantitative approach to the systematics of *Actinomycetales* based on overall similarity. Microbial Classification, Symp. Soc. Gen. Microbiol., *12*: 333–360.

Simankova, M.V., O.R. Kotsyurbenko, E. Stackebrandt, N.A. Kostrikina, A.M. Lysenko, G.A. Osipov and A.N. Nozhevnikova. 2000. *Acetobacterium tundrae* sp. nov., a new psychrophilic acetogenic bacterium from tundra soil. Arch. Microbiol. *174*: 440–447.

Simankova, M.V., O.R. Kotsyurbenko, E. Stackebrandt, N.A. Kostrikina, A.M. Lysenko, G.A. Osipov and A.N. Nozhevnikova. 2001a. *In* Validation of publication of new names and new combinations previously effectively published outside the IJSEM. List No. 80. Int. J. Syst. Evol. Microbiol. *51*: 793–794.

Simankova, M.V., S.N. Parshina, T.P. Tourova, T.V. Kolganova, A.J.B. Zehnder and A.N. Nozhevnikova. 2001b. *Methanosarcina lacustris* sp. nov., a new psychrotolerant methanogenic archaeon from anoxic lake sediments. Syst. Appl. Microbiol. *24*: 362–367.

Simankova, M.V., S.N. Parshina, T.P. Tourova, T.V. Kolganova, A.J.B. Zehnder and A.N. Nozhevnikova. 2002. *In* Validation of publication of new names and new combinations previously effectively published outside the IJSEM. List No. 85. Int. J. Syst. Evol. Microbiol. *52*: 685–690.

Simmering, R., D. Taras, A. Schwiertz, G. Le Blay, B. Gruhl, P.A. Lawson, M.D. Collins and M. Blaut. 2002a. *Ruminococcus luti* sp. nov., isolated from a human faecal sample. Syst. Appl. Microbiol. *25*: 189–193.

Simmering, R., D. Taras, A. Schwiertz, G. Le Blay, B. Gruhl, P.A. Lawson, M.D. Collins and M. Blaut. 2002b. *In* Validation of publication of new names and new combinations previously effectively published outside the IJSEM. List No. 88. Int. J. Syst. Evol. Microbiol. *52*: 1915–1916.

Simmons, J.H., L.K. Riley, C.L. Besch-Williford and C.L. Franklin. 2000a. *Helicobacter mesocricetorum* sp. nov., a novel *helicobacter* isolated from the feces of Syrian hamsters. J. Clin. Microbiol. *38*: 1811–1817.

Simmons, J.H., L.K. Riley, C.L. Besch-Williford and C.L. Franklin. 2000b. *In* Validation of the publication of new names and new combinations previously effectively published outside the IJSB. List No. 76. Int. J. Syst. Evol. Bacteriol. *50*: 1699–1700.

Simpson, K.L., B. Pettersson and F.G. Priest. 2001. Characterization of lactobacilli from Scotch malt whisky distilleries and description of *Lactobacillus ferintoshensis* sp. nov., a new species isolated from malt whisky fermentations. Microbiology-Sgm. *147*: 1007–1016.

Simpson, K.L., B. Pettersson and F.G. Priest. 2002. *In* Validation of publication of new names and new combinations previously effectively published outside the IJSEM. List No. 86. Int. J. Syst. Evol. Microbiol. *52*: 1075–1076.

Singleton, D.R., M.A. Furlong, A.D. Peacock, D.C. White, D.C. Coleman and W.B. Whitman. 2003. *Solirubrobacter pauli* gen. nov., sp. nov., a mesophilic bacterium within the *Rubrobacteridae* related to common soil clones. Int. J. Syst. Evol. Microbiol. *53*: 485–490.

Sinigalliano, C.D., D.N. Kuhn and R.D. Jones. 1995. Amplification of the *amoA* gene from diverse species of ammonium-oxidizing bacteria and from an indigenous bacterial population from seawater. Appl. Environ. Microbiol. *61*: 2702–2706.

Sistrom, W.R. 1960. A requirement for sodium in the growth of *Rhodopseudomonas sphaeroides*. J. Gen. Microbiol. *22*: 778–782.

Skerman, V.B.D. 1967. A Guide for the Identification of the Genera of Bacteria, 2nd Ed., The Williams & Wilkins Co., Baltimore.

Skerman, V.B.D., V. McGowan and P.H.A. Sneath. 1980. Approved lists of bacterial names. Int. J. Syst. Bacteriol. *30*: 225–420.

Skerratt, J.H., J.P. Bowman and P.D. Nichols. 2002. *Shewanella olleyana* sp. nov., a marine species isolated from a temperate estuary which produces high levels of polyunsaturated fatty acids. Int. J. Syst. Evol. Microbiol. *52*: 2101–2106.

Skinner, F.E., R.C.T. Jones and J.E.A. Mollison. 1952. Comparison of a direct- and a plate-counting technique for the quantitative estimation of soil microorganisms. J. Gen. Microbiol. *32*: 261–271.

Slobodkin, A.I., T.P. Tourova, N.A. Kostrikina, N.A. Chernyh, E.A. Bonch-Osmolovskaya, C. Jeanthon and B.E. Jones. 2003. *Tepidibacter thalassicus* gen. nov., sp. nov., a novel moderately thermophilic, anaerobic, fermentative bacterium from a deep-sea hydrothermal vent. Int. J. Syst. Evol. Microbiol. *53*: 1131–1134.

Smibert, R.M. and N.R. Krieg. 1994. Phenotypic characterization. *In* Gerhardt, Murray, Wood and Krieg (Editors), Methods for General and Molecular Bacteriology, American Society for Microbiology, Washington, D.C. pp. 607–654.

Smith, A.J. and D.S. Hoare. 1977. Specialist phototrophs, lithotrophs, and methylotrophs: a unity among a diversity of procaryotes? Bacteriol. Rev. *41*: 419–448.

Smith, E., P. Leeflang and K. Wernars. 1997. Detection of shifts in microbial community structure and diversity in soil caused by copper contamination using amplified ribosomal DNA restriction analysis. FEMS Microbiol. Ecol. *23*: 249–261.

Smorczewski, W.T. and E.L. Schmidt. 1991. Numbers, activities, and diversity of autotrophic ammonia oxidizing bacteria in a fresh water, eutrophic lake sediment. Can. J. Microbiol. *37*: 828–833.

Snaidr, J., R.I. Amann, I. Huber, W. Ludwig and K.H. Schleifer. 1997. Phylogenetic analysis and *in situ* identification of bacteria in activated sludge. Appl. Environ. Microbiol. *63*: 2884–2896.

Snaidr, J., B.M. Fuchs, G. Wallner, M. Wagner and K.H. Schleifer. 1999. Phylogeny and *in situ* identification of a morphologically conspicuous bacterium, *Candidatus* Magnospira bakii, present in very low frequency in activated sludge. Environ. Microbiol. *1*: 125–136.

Sneath, P.H.A. 1972. Computer taxonomy. *In* Norris and Ribbons (Editors), Methods in Microbiology, Academic Press, London. 29–98.

Sneath, P.H.A. 1974. Phylogeny of microorganisms. Symp. Soc. Gen. Microbiol. *24*: 1–39.

Sneath, P.H.A. 1977. A method for testing the distinctness of clusters: a test for the disjunction of two clusters in euclidean space as measured by their overlap. J. Int. Assoc. Math. Geol. *9*: 123–143.

Sneath, P.H.A. 1978a. Classification of microorganisms. *In* Norris and Richmond (Editors), Essays in Microbiology, John Wiley, Chichester. 9/1–9/31.

Sneath, P.H.A. 1978b. Identification of microorganisms. *In* Norris and Richmond (Editors), Essays in Microbiology, John Wiley, Chichester. 10/1–10/32.

Sneath, P.H.A. 1979a. BASIC program for a significance test for clusters in UPGMA dendrograms obtained from square euclidean distances. Comput. Geosci. *5*: 127–137.

Sneath, P.H.A. 1979b. BASIC program for a significance test for two clusters in euclidean space as measured by their overlap. Comput. Geosci. *5*: 143–155.

Sneath, P.H.A. 1992. International Code of Nomenclature of Bacteria (1990 Revision), American Society for Microbiology, Washington, DC.

Sneath, P.H. 1993. Evidence from *Aeromonas* for genetic crossing-over in ribosomal sequences [letter]. Int. J. Syst. Bacteriol. *43*: 626–629.

Sneath, P.H.A., N.S. Mair, M.E. Sharpe and J.G. Holt (Editors). 1986. Bergey's Manual of Systematic Bacteriology, 1st Ed., Vol. 2, The Williams & Wilkins Co., Baltimore.

Sneath, P.H.A. and Sokal, R.R.. 1973. Numerical Taxonomy: The Principles and Practice of Numerical Classification, W.H. Freeman and Co., San Francisco.

Sohn, K., S.G. Hong, K.S. Bae and J. Chun. 2003. Transfer of *Hongia koreensis* Lee et al. 2000 to the genus *Kribbella* Park et al. 1999 as *Kribbella koreensis* comb. nov. Int. J. Syst. Evol. Microbiol. *53*: 1005–1007.

Sokal, R.R. and P.H.A. Sneath. 1963. Principles of Numerical Taxonomy, W.H. Freeman and Co., San Francisco.

Sokolova, T.G., J.M. Gonzalez, N.A. Kostrikina, N.A. Chernyh, T.P. Tourova, C. Kato, E.A. Bonch-Osmolovskaya and F.T. Robb. 2001. *Carboxydobrachium pacificum* gen. nov., sp. nov., a new anaerobic, thermophilic, CO-utilizing marine bacterium from Okinawa Trough. Int. J. Syst. Evol. Microbiol. *51*: 141–149.

Sokolova, T.G., N.A. Kostrikina, N.A. Chernyh, T.P. Tourova, T.V. Kolganova and E.A. Bonch-Osmolovskaya. 2002. *Carboxydocella thermautotrophica* gen. nov., sp. nov., a novel anaerobic, CO-utilizing thermophile from a Kamchatkan hot spring. Int. J. Syst. Evol. Microbiol. *52*: 1961–1967.

Song, B., N.J. Palleroni, L.J. Kerkhof and M.M. Häggblom. 2001. Characterization of halobenzoate-degrading, denitrifying *Azoarcus* and *Thauera* isolates and description of *Thauera chlorobenzoica* sp. nov. Int. J. Syst. Evol. Microbiol. *51*: 589–602.

Song, Y.L., C.X. Liu, D.R. Molitoris, T.J. Tomzynski, P.A. Lawson, M.D. Collins and S.M. Finegold. 2003a. *Clostridium bolteae* sp. nov., isolated from human sources. Syst. Appl. Microbiol. *26*: 84–89.

Song, Y.L., C.X. Liu, D.R. Molitoris, T.J. Tomzynski, P.A. Lawson, M.D. Collins and S.M. Finegold. 2003b. *In* Validation of publication of new names and new combinations previously effectively published outside the IJSEM. List No. 92. Int. J. Syst. Evol. Microbiol. *53*: 935–937.

Sonne-Hansen, J. and B.K. Ahring. 1999. *Thermodesulfobacterium hveragerdense* sp. nov., and *Thermodesulfovibrio islandicus* sp. nov., two thermophilic sulfate reducing bacteria isolated from a Icelandic hot spring. Syst. Appl. Microbiol. *22*: 559–564.

Sonne-Hansen, J. and B.K. Ahring. 2000. *In* Validation of publication of new names and new combinations previously effectively published outside the IJSEM. List No. 74. Int. J. Syst. Evol. Microbiol. *50*: 949–950.

Sørheim, R., V.L. Torsvik and J. Goksøyr. 1989. Phenotypical divergences between populations of soil bacteria isolated on different media. Microb. Ecol. *17*: 181–192.

Sorokin, D.Y., V.M. Gorlenko, T.P. Tourova, A.I. Tsapin, K.H. Nealson and G.J. Kuenen. 2002a. *Thioalkalimicrobium cyclicum* sp. nov. and *Thioalkalivibrio jannaschii* sp. nov., novel species of haloalkaliphilic, obligately chemolithoautotrophic sulfur-oxidizing bacteria from hypersaline alkaline Mono Lake (California). Int. J. Syst. Evol. Microbiol. *52*: 913–920.

Sorokin, D.Y., A.M. Lysenko, L.L. Mityushina, T.P. Tourova, B.E. Jones, F.A. Rainey, L.A. Robertson and G.J. Kuenen. 2001a. *Thioalkalimicrobium aerophilum* gen. nov., sp. nov. and *Thioalkalimicrobium sibericum* sp. nov., and *Thioalkalivibrio versutus* gen. nov., sp. nov., *Thioalkalivibrio nitratis* sp. nov. and *Thioalkalivibrio denitrificans* sp. nov., novel obligately alkaliphilic and obligately chemolithoautotrophic sulfur-oxidizing bacteria from soda lakes. Int. J. Syst. Evol. Microbiol. *51*: 565–580.

Sorokin, D.Y., G. Muyzer, T. Brinkhoff, J.G. Kuenen and M.S.M. Jetten. 1998. Isolation and characterization of a novel facultatively alkaliphilic *Nitrobacter* species, *N. alkalicus* sp. nov. Arch. Microbiol. *170*: 345–352.

Sorokin, D.Y. , G. Muyzer, T. Brinkhoff, J.G. Kuenen and M.S.M. Jetten. 2001b. *In* Validation of publication of new names and new combinations previously effectively published outside the IJSEM. List No. 78. Int. J. Syst. Evol. Microbiol. *51*: 1–2.

Sorokin, D.Y., T.P. Tourova, T.V. Kolganova, K.A. Sjollema and J.G. Kuenen. 2002b. *Thioalkalispira microaerophila* gen. nov., sp. nov., a novel lithoautotrophic, sulfur-oxidizing bacterium from a soda lake. Int. J. Syst. Evol. Microbiol. *52*: 2175–2182.

Sorokin, D.Y., T.P. Tourova, B.B. Kuznetsov, I.A. Bryantseva and V.M. Gorlenko. 2000a. *Roseinatronobacter thiooxidans* gen. nov., sp. nov., a new alkaliphilic aerobic bacteriochlorophyll *a*-containing bacterium isolated from a soda lake. Mikrobiologiya *69*: 89–97.

Sorokin, D.Y., T.P. Tourova, B.B. Kuznetsov, I.A. Bryantseva and V.M. Gorlenko. 2000b. *In* Validation of the publication of new names and new combinations previously effectively published outside the IJSEM. List No. 75. Int. J. Syst. Evol. Microbiol. *50*: 1415.

Sorokin, D.Y., T.P. Tourova, A.M. Lysenko and J.G. Kuenen. 2001c. Microbial thiocyanate utilization under highly alkaline conditions. Appl. Environ. Microbiol. *67*: 528–538.

Sorokin, D.Y., T.P. Tourova, A.M. Lysenko, L.L. Mityushina and J.G. Kuenen. 2002c. *Thioalkalivibrio thiocyanoxidans* sp. nov. and *Thioalkalivibrio paradoxus* sp. nov., novel alkaliphilic, obligately autotrophic, sulfur-oxidizing bacteria capable of growth on thiocyanate, from soda lakes. Int. J. Syst. Evol. Microbiol. *52*: 657–664.

Sorokin, D. , T. Tourova, M.C. Schmid, M. Wagner, H.P. Koops, J.G. Kuenen and M. Jetten. 2001d. Isolation and properties of obligately chemolithoautotrophic and extremely alkali-tolerant ammonia-oxidizing bacteria from Mongolian soda lakes. Arch. Microbiol. *176*: 170–177.

Sorokin, D.Y., T.P. Tourova, K.A. Sjollema and J.G. Kuenen. 2003. *Thialkalivibrio nitratireducens* sp. nov., a nitrate-reducing member of an autotrophic denitrifying consortium from a soda lake. Int. J. Syst. Evol. Microbiol. *53*: 1779–1783.

Sowitzki, D. 1994. Untersuchungen zur urease von *Notrosospira* sp. Nsp1 klonierung, teilsequenzierung und expression in *E. coli*, University of Hamburg.

Spanevello, M.D., H. Yamamoto and B.K. Patel. 2002. *Thermaerobacter subterraneus* sp. nov., a novel aerobic bacterium from the Great Artesian Basin of Australia, and emendation of the genus *Thermaerobacter*. Int. J. Syst. Evol. Microbiol. *52*: 795–800.

Speksnijder, A.G., G.A. Kowalchuk, K. Roest and H.J. Laanbroek. 1998. Recovery of a *Nitrosomonas*-like 16S rDNA sequence group from freshwater habitats. Syst. Appl. Microbiol. *21*: 321–330.

Spergser, J., M. Wieser, M. Taubel, R.A. Rosselló-Mora, R. Rosengarten and H.J. Busse. 2003. *Staphylococcus nepalensis* sp. nov., isolated from goats of the Himalayan region. Int. J. Syst. Evol. Microbiol. *53*: 2007–2011.

Spieck, E., J. Aamand, S. Bartosch and E. Bock. 1996a. Immunocytochemical detection and location of the membrane-bound nitrite oxidoreductase in cells of *Nitrobacter* and *Nitrospira*. FEMS Microbiol. Lett. *139*: 71–76.

Spieck, E., S. Ehrich, J. Aamand and E. Bock. 1998. Isolation and immunocytochemical location of the nitrite-oxidizing system in *Nitrospira moscoviensis*. Arch. Microbiol. *169*: 225–230.

Spieck, E., S. Muller, A. Engel, E. Mandelkow, H. Patel and E. Bock. 1996b. Two-dimensional structure of membrane-bound nitrite oxidoreductase from *Nitrobacter hamburgensis*. J. Struct. Biol. *117*: 117–123.

Sprenger, W.W., M.C. van Belzen, J. Rosenberg, J.H. Hackstein and J.T. Keltjens. 2000. *Methanomicrococcus blatticola* gen. nov., sp. nov., a methanol- and methylamine-reducing methanogen from the hindgut of the cockroach *Periplaneta americana*. Int. J. Syst. Evol. Microbiol. *50*: 1989–1999.

Spring, S. , P. Kämpfer and K.-H. Schleifer. 2001. *Limnobacter thiooxidans* gen. nov., sp. nov., a novel thiosulfate-oxidizing bacterium isolated from freshwater lake sediment. Int. J. Syst. Evol. Microbiol. *51*: 1463–1470.

Spring, S., B. Merkhoffer, N. Weiss, R.M. Kroppenstedt, H. Hippe and E. Stackebrandt. 2003. Characterization of novel psychrophilic clostridia from an Antarctic microbial mat: description of *Clostridium frigoris* sp. nov., *Clostridium lacusfryxellense* sp. nov., *Clostridium bowmanii* sp. nov. and *Clostridium psychrophilum* sp. nov. and reclassification of *Clostridium laramiense* as *Clostridium estertheticum* subsp. *laramiense* subsp. nov. Int. J. Syst. Evol. Microbiol. *53*: 1019–1029.

Springer, N., R. Amann, W. Ludwig, K.H. Schleifer and H. Schmidt. 1996. *Polynucleobacter necessarius*, an obligate bacterial endosymbiont of the hypotrichous ciliate *Euplotes aediculatus*, is a member of the beta-subclass of *Proteobacteria*. FEMS Microbiol. Lett. *135*: 333–336.

Squartini, A., P. Struffi, H. Doring, S. Selenska-Pobell, E. Tola, A. Giacomini, E. Vendramin, E. Velazquez, P.F. Mateos, E. Martinez-Molina, F.B. Dazzo, S. Casella and M.P. Nuti. 2002. *Rhizobium sullae* sp. nov. (formerly '*Rhizobium hedysari*'), the root-nodule microsymbiont of *Hedysarum coronarium* L. Int. J. Syst. Evol. Microbiol. *52*: 1267–1276.

Stackebrandt, E. 1988. Phylogenetic relationships vs. phenotypic diversity: how to achieve a phylogenetic classification system of the eubacteria. Can. J. Microbiol. *34*: 552–556.

Stackebrandt, E., S. Breymann, U. Steiner, H. Prauser, N. Weiss and P. Schumann. 2002a. Re-evaluation of the status of the genus *Oerskovia*, reclassification of *Promicromonospora enterophila* (Jager et al. 1983) as *Oerskovia enterophila* comb. nov. and description of *Oerskovia jenensis* sp. nov. and *Oerskovia paurometabola* sp. nov. Int. J. Syst. Evol. Microbiol. *52*: 1105–1111.

Stackebrandt, E., W. Frederiksen, G.M. Garrity, P.A.D. Grimont, P. Kämpfer, M.C.J. Maiden, X. Nesme, R. Rosselló-Mora, J. Swings, H.G. Trüper, L. Vauterin, A.C. Ward and W.B. Whitman. 2002b. Report of the ad hoc committee for the re-evaluation of the species definition in bacteriology. Int. J. Syst. Evol. Microbiol. *52*: 1043–1047.

Stackebrandt, E. and B.M. Goebel. 1994. Taxonomic note: A place for DNA–DNA reassociation and 16S rRNA sequence analysis in the present species definition in bacteriology. Int. J. Syst. Bacteriol. *44*: 846–849.

Stackebrandt, E., R.G.E. Murray and H.G. Trüper. 1988. *Proteobacteria* classis nov., a name for the phylogenetic taxon that includes the "purple bacteria and their relatives". Int. J. Syst. Bacteriol. *38*: 321–325.

Stackebrandt, E., F.A. Rainey and N.L. Ward-Rainey. 1997. Proposal for a new hierarchic classification system, *Actinobacteria* classis nov. Int. J. Syst. Bacteriol. *47*: 479–491.

Stackebrandt, E. and P. Schumann. 2000. Description of *Bogoriellaceae* fam. nov., *Dermacoccaceae* fam. nov., *Rarobacteraceae* fam. nov. and *Sanguibacteraceae* fam. nov. and emendation of some families of the suborder *Micrococcineae*. Int. J. Syst. Evol. Microbiol. *50*: 1279–1285.

Stackebrandt, E., P. Schumann, K.P. Schaal and N. Weiss. 2002c. *Propionimicrobium* gen. nov., a new genus to accommodate *Propionibacterium lymphophilum* (Torrey 1916) Johnson and Cummins 1972, 1057^{AL} as *Propionimicrobium lymphophilum* comb. nov. Int. J. Syst. Evol. Microbiol. *52*: 1925–1927.

Stackebrandt, E., P. Schumann, E. Schuler and H. Hippe. 2003. Reclassification of *Desulfotomaculum auripigmentum* as *Desulfosporosinus auripigmenti* corrig., comb. nov. Int. J. Syst. Evol. Microbiol. *53*: 1439–1443.

Stackebrandt, E., J. Wink, U. Steiner and R.M. Kroppenstedt. 2001. *Nonomuraea dietzii* sp. nov. Int. J. Syst. Evol. Microbiol. *51*: 1437–1441.

Stahl, D.A. 1986. Evolution, ecology, and diagnosis: unity in variety. Bio/Technol. *4*: 623–628.

Stahl, D.A. and R.I. Amann. 1991. Development and application of nucleic acid probes in bacterial systematics. *In* Stackebrandt and Goodfellow (Editors), Nucleic Acid Techniques in Bacterial Systematics, John Wiley & Sons, Chichester. pp. 205–248.

Stahl, D.A., B. Flesher, H.R. Mansfield and L. Montgomery. 1988. Use of phylogenetically based hybridization probes for studies of ruminal microbial ecology. Appl. Environ. Microbiol. *54*: 1079–1084.

Staley, J.T. 1997. Biodiversity: are microbial species threatened? Curr. Opin. Biotechnol. *8*: 340–345.

Staley, J.T. 1999. Bacterial biodiversity: a time for place. ASM News. *65*: 681–687.

Staley, J.T., M.P. Bryant, N. Pfennig and J.G. Holt (Editors). 1989. Bergey's Manual of Systematic Bacteriology, 1st Ed., Vol. 3, The Williams & Wilkins Co., Baltimore.

Staley, J.T. and A. Konopka. 1985. Measurement of *in situ* activities of nonphotosynthetic microorganisms in aquatic and terrestrial habitats. Annu. Rev. Microbiol. *39*: 321–346.

Stams, A.J.M., E.M. Flameling and E.C.L. Marnette. 1990. The importance of autotrophic versus heterotrophic oxidation of atmospheric ammonium in forest ecosystems with acid soil. FEMS Microbiol. Ecol. *74*: 337–344.

Stan-Lotter, H., M. Pfaffenhuemer, A. Legat, H.J. Busse, C. Radax and C. Gruber. 2002. *Halococcus dombrowskii* sp. nov., an archaeal isolate from a Permian alpine salt deposit. Int. J. Syst. Evol. Microbiol. *52*: 1807–1814.

Stanier, R.Y. and C.B. Van Niel. 1962. The concept of a bacterium. Arch. Mikrobiol. *42*: 17–35.

Stanley, P.M. and E.L. Schmidt. 1981. Serological diversity of *Nitrobacter* spp. from soil and aquatic habitats. Appl. Environ. Microbiol. *41*: 1069–1071.

Starkey, R.L. 1948. Family I. *Nitrobacteriaceae* Buchanan, 1917. *In* Breed, Murray and Hitchens (Editors), Bergey's Manual of Determinative Bacteriology, 6th ed., The Williams & Wilkins Co., Baltimore. 69–81.

Starkey, R.G. 1957. Family I. *Nitrobacteriaceae* Buchanan, 1917. *In* Breed, Murray and Smith (Editors), Bergey's Manual of Determinative Bacteriology, 7th ed., The Williams & Wilkins Co., Baltimore. 68–73.

Steer, T., M.D. Collins, G.R. Gibson, H. Hippe and P.A. Lawson. 2001. *Clostridium hathewayi* sp. nov., from human faeces. Syst. Appl. Microbiol. *24*: 353–357.

Steer, T., M.D. Collins, G.R. Gibson, H. Hippe and P.A. Lawson. 2002. *In* Validation of publication of new names and new combinations previously effectively published outside the IJSEM. List No. 85. Int. J. Syst. Evol. Microbiol. *52*: 685–690.

Stehr, G. 1996. Untersuchungen zur ökologie ammoniak oxidierender bakterien, University of Hamburg

Stehr, G., B. Bottcher, P. Dittberner, G. Rath and H.P. Koops. 1995a. The ammonia-oxidizing nitrifying population of the river Elbe estuary. FEMS Microbiol. Ecol. *17*: 177–186.

Stehr, G., S. Zörner, B. Böttcher and H.P. Koops. 1995b. Exopolymers: an ecological characteristic of a floc-attached, ammonia-oxidizing bacterium. Microb. Ecol. *30*: 115–126.

Stenklo, K., H.D. Thorell, H. Bergius, R. Aasa and T. Nilsson. 2001. Chlorite dismutase from *Ideonella dechloratans*. J. Biol. Inorg. Chem. *6*: 601–607.

Stepanyuk, V., G. Smirnova, T. Klyushnikova, N. Kanyuk, L. Panchenko, T. Nogina and V. Prima. 1992. New species of the *Acinetobacter* genus, *Acinetobacter thermotoleranticus* sp. nov. Mikrobiologiya *61*: 347–356.

Stephen, J.R., G.A. Kowalchuk, M.A.V. Bruns, A.E. McCaig, C.J. Phillips, T.M. Embley and J.I. Prosser. 1998. Analysis of beta-subgroup proteobacterial ammonia oxidizer populations in soil by denaturing gradient gel electrophoresis analysis and hierarchical phylogenetic probing. Appl. Environ. Microbiol. *64*: 2958–2965.

Stephen, J.R., A.E. McCaig, Z. Smith, J.I. Prosser and T.M. Embley. 1996. Molecular diversity of soil and marine 16S rRNA gene sequences related to β-subgroup ammonia-oxidizing bacteria. Appl. Environ. Microbiol. *62*: 4147–4154.

Stetter, K.O. 1995. Microbial life in hyperthermal environments. ASM News. *61*: 285–290.

Stetter, K.O. 2001. Genus VII. *Thermodiscus* gen. nov. *In* Boone, Castenholz and Garrity (Editors), Bergey's Manual of Systematic Bacteriology, 2nd Edition, Vol. 1, Springer, New York. 189–190.

Stetter, K.O. 2003. *In* Validation of publication of new names and new combinations previously effectively published outside the IJSEM. List No. 89. Int. J. Syst. Evol. Microbiol. *53*: 1–2.

Stewart, V. 1988. Nitrate respiration in relation to facultative metabolism in enterobacteria. Microbiol. Rev. *52*: 190–232.

Stöhr, R., A. Waberski, W. Liesack, H. Völker, U. Wehmeyer and M. Thomm. 2001a. *Hydrogenophilus hirschii* sp. nov., a novel thermophilic hydrogen-oxidizing bega-proteobacterium isolated from Yellowstone National Park. Int. J. Syst. Evol. Microbiol. *51*: 481–488.

Stöhr, R., A. Waberski, H. Volker, B.J. Tindall and M. Thomm. 2001b. *Hydrogenothermus marinus* gen. nov., sp. nov., a novel thermophilic hydrogen-oxidizing bacterium, recognition of *Calderobacterium hydrogenophilum* as a member of the genus *Hydrogenobacter* and proposal of the reclassification of *Hydrogenobacter acidophilus* as *Hydrogenobaculum acidophilum* gen. nov., comb. nov., in the phylum *'Hydrogenobacter/Aquifex'*. Int. J. Syst. Evol. Microbiol. *51*: 1853–1862.

Stolz, A., C. Schmidt-Maag, E.B.M. Denner, H.-J. Busse, T. Egli and P. Kämpfer. 2000. Description of *Sphingomonas xenophaga* sp. nov. for strains $BN6^T$ and N,N which degrade xenobiotic aromatic compounds. Int. J. Syst. Evol. Microbiol. *50*: 35–41.

Storms, V., L.A. Devriese, R. Coopman, P. Schumann, F. Vyncke and M. Gillis. 2003. *Arthrobacter gandavensis* sp. nov., for strains of veterinary origin. Int. J. Syst. Evol. Microbiol. *53*: 1881–1884.

Straub, K.L. and B.E. Buchholz-Cleven. 2001. *Geobacter bremensis* sp. nov. and *Geobacter pelophilus* sp. nov., two dissimilatory ferric-iron-reducing bacteria. Int. J. Syst. Evol. Microbiol. *51*: 1805–1808.

Straub, K.L., F.A. Rainey and F. Widdel. 1999. *Rhodovulum iodosum* sp. nov. and *Rhodovulum robiginosum* sp. nov., two new marine phototrophic ferrous-iron-oxidizing purple bacteria. Int. J. Syst. Bacteriol. *49*: 729–735.

Strecker, M., E. Sickinger, R.S. English, S. J.M. and E. Bock. 1994. Calvin cycle genes in *Nitrobacter vulgaris* T3. FEMS Microbiol. Lett. *120*: 45–50.

Strömpl, C., G.L. Hold, H. Lunsdorf, J. Graham, S. Gallacher, W.R. Abraham, E.R. Moore and K.N. Timmis. 2003. *Oceanicaulis alexandrii* gen. nov., sp. nov., a novel stalked bacterium isolated from a culture of the dinoflagellate *Alexandrium tamarense* (Lebour) Balech. Int. J. Syst. Evol. Microbiol. *53*: 1901–1906.

Strömpl, C., B.J. Tindall, H. Lunsdorf, T.Y. Wong, E.R. Moore and H. Hippe. 2000. Reclassification of *Clostridium quercicolum* as *Dendrosporobacter quercicolus* gen. nov., comb. nov. Int. J. Syst. Evol. Microbiol. *50*: 101–106.

Strous, M., J.A. Fuerst, E.H.M. Kramer, S. Logemann, G. Muyzer, K.T. Van de Pas-Schoonen, R. Webb, J.G. Kuenen and M.S.M. Jetten. 1999. Missing lithotroph identified as new planctomycete. Nature (Lond.) *400*: 446–449.

Stuart, G.W. , K. Moffett and S. Baker. 2002. Integrated gene and species phylogenies from unaligned whole genome protein sequences. Bioinformatics. *18*: 100–108.

Stupperich, E. and H.J. Eisinger. 1989. Biosynthesis of *p*-cresolyl cobamide in *Sporomusa ovata*. Arch. Microbiol. *151*: 372–377.

Sugawara, H., S. Miyazaki, J. Shimura and Y. Ichiyanagi. 1996. Bioinformatics tools for the study of microbial diversity. J. Ind. Microbiol. *17*: 490–497.

Suggs, S.V., T. Hirose, T. Miyake, E.H. Kawashima, M.J. Johnson, K. Itakura and R.B. Wallace. 1981. Use of synthetic oligonucleotides for the isolation of specific cloned DNA sequences. *In* Brown and Fox (Editors), Developmental Biology Using Purified Genes, Academic Press, New York. pp. 683–693.

Sun, B.L., J.R. Cole, R.A. Sanford and J.M. Tiedje. 2000. Isolation and characterization of *Desulfovibrio dechloracetivorans* sp. nov., a marine dechlorinating bacterium growing by coupling the oxidation of acetate to the reductive dechlorination of 2-chlorophenol. Appl. Environ. Microbiol. *66*: 2408–2413.

Sun, B.L., J.R. Cole, R.A. Sanford and J.M. Tiedje. 2001a. *In* Validation of publication of new names and new combinations previously effectively published outside the IJSEM. List No. 78. Int. J. Syst. Evol. Microbiol. *51*: 1–2.

Sun, B., J.R. Cole and J.M. Tiedje. 2001b. *Desulfomonile limimaris* sp. nov., an anaerobic dehalogenating bacterium from marine sediments. Int. J. Syst. Evol. Microbiol. *51*: 365–371.

Sundermeyer, H. and E. Bock. 1981a. Characterization of the nitrite-oxidizing system in *Nitrobacter*. *In* Bothe and Trebst (Editors), Biology of Inorganic Nitrogen and Sulfur, Springer-Verlag, Berlin. pp. 317–324.

Sundermeyer, H. and E. Bock. 1981b. Energy metabolism of autotrophically and heterotrophically grown cells of *Nitrobacter winogradskyi*. Arch. Microbiol. *130*: 250–254.

Sung, M.H., H. Kim, J.W. Bae, S.K. Rhee, C.O. Jeon, K. Kim, J.J. Kim, S.P. Hong, S.G. Lee, J.H. Yoon, Y.H. Park and D.H. Baek. 2002. *Geobacillus toebii* sp. nov., a novel thermophilic bacterium isolated from hay compost. Int. J. Syst. Evol. Microbiol. *52*: 2251–2255.

Suominen, I., C. Sproer, P. Kämpfer, F.A. Rainey, K. Lounatmaa and M. Salkinoja-Salonen. 2003. *Paenibacillus stellifer* sp. nov., a cyclodextrin-producing species isolated from paperboard. Int. J. Syst. Evol. Microbiol. *53*: 1369–1374.

Surkov, A.V., G.A. Dubinina, A.M. Lysenko, F.O. Glockner and J. Kuever. 2001. *Dethiosulfovibrio russensis* sp. nov., *Dethosulfovibrio marinus* sp. nov. and *Dethosulfovibrio acidaminovorans* sp. nov., novel anaerobic, thiosulfate- and sulfur-reducing bacteria isolated from *'Thiodendron'* sulfur mats in different saline environments. Int. J. Syst. Evol. Microbiol. *51*: 327–337.

Sutra, L., R. Christen, C. Bollet, P. Simoneau and L. Gardan. 2001. *Samsonia erythrinae* gen. nov., sp. nov., isolated from bark necrotic lesions of *Erythrina* sp., and discrimination of plant-pathogenic *Enterobacteriaceae* by phenotypic features. Int. J. Syst. Evol. Microbiol. *51*: 1291–1304.

Suwa, Y., Y. Imamura, T. Suzuki, T. Tashiro and Y. Urushigawa. 1994. Ammonia-oxidizing bacteria with different sensitivities to $(NH_4)_2SO_4$ in activated sludges. Water Res. *28*: 1523–1532.

Suwa, Y., T. Sumino and K. Noto. 1997. Phylogenetic relationships of activated sludge isolates of ammonia oxidizers with different sensitivities to ammonium sulfate. J. Gen. Appl. Microbiol. *43*: 373–379.

Suyama, T., T. Shigematsu, S. Takaichi, Y. Nodasaka, S. Fujikawa, H. Hosoya, Y. Tokowa, T. Kanagawa and S. Hanada. 1999. *Roseateles depolymerans* gen. nov., sp. nov. a new bacteriochlorophyll *a*-containing obligate aerobe belonging to the β-subclass of the *Proteobacteria*. Int. J. Syst. Bacteriol. *49*: 449–457.

Suzuki, I. 1984. Oxidation of inorganic nitrogen compounds. *In* Crawford and Hanson (Editors), Microbial Growth on C_1 Compounds, American Society for Microbiology, Washington, DC. 42–52.

Suzuki, I., U. Dular and S.C. Kwok. 1974. Ammonia or ammonium ion as substrate for oxidation by *Nitrosomonas europaea* cells and extracts. J. Bacteriol. *120*: 556–558.

Suzuki, I., S.C. Kwok and U. Dular. 1976. Competitive inhibition of ammonia oxidation in *Nitrosomonas europaea* by methane, carbon monoxide, or methanol. FEBS Lett. *72*: 117–120.

Suzuki, I., S.C. Kwok, U. Dular and D.C. Tsang. 1981. Cell free ammonia-oxidizing system of *Nitrosomonas europaea*: general conditions and properties. Can. J. Biochem. *59*: 477–483.

Suzuki, K., M. Suzuki, J. Sasaki, Y.H. Park and K. Komagata. 1999a. *Leifsonia* gen. nov., a genus for 2,4-diaminobutyric acid-containing actinomycetes to accommodate *"Corynebacterium aquaticum"* Leifson 1962 and *Clavibacter xyli* subsp. *cynodontis* Davis et al. 1984. J. Gen. Appl. Microbiol. *45*: 253–262.

Suzuki, K., M. Suzuki, J. Sasaki, Y.H. Park and K. Komagata. 2000a. *In* Validation of publication of new names and new combinations previously effectively published outside the IJSEM. List No. 75. Int. J. Syst. Evol. Microbiol. *50*: 1415–1417.

Suzuki, M., Y. Nakagawa, S. Harayama and S. Yamamoto. 2001. Phylogenetic analysis and taxonomic study of marine *Cytophaga*-like bacteria: proposal for *Tenacibaculum* gen. nov. with *Tenacibaculum maritimum* comb. nov. and *Tenacibaculum ovolyticum* comb. nov., and description of *Tenacibaculum mesophilum* sp. nov. and *Tenacibaculum amylolyticum* sp. nov. Int. J. Syst. Evol. Microbiol. *51*: 1639–1652.

Suzuki, M., M.S. Rappe and S.J. Giovannoni. 1998. Kinetic bias in estimates of coastal picoplankton community structure obtained by measurements of small-subunit rRNA gene PCR amplicon length heterogeneity. Appl. Environ. Microbiol. *64*: 4522–4529.

Suzuki, T., T. Iwasaki, T. Uzawa, K. Hara, N. Nemoto, T. Kon, T. Ueki, A. Yamagishi and T. Oshima. 2002a. *Sulfolobus tokodaii* sp. nov. (f. *Sulfolobus* sp. strain 7), a new member of the genus *Sulfolobus* isolated from Beppu Hot Springs, Japan. Extremophiles *6*: 39–44.

Suzuki, T., T. Iwasaki, T. Uzawa, K. Hara, N. Nemoto, T. Kon, T. Ueki, A. Yamagishi and T. Oshima. 2002b. *In* Validation of publication of new names and new combinations previously effectively published outside the IJSEM. List No. 87. Int. J. Syst. Evol. Microbiol. *52*: 1437–1438.

Suzuki, T., Y. Muroga, M. Takahama and Y. Nishimura. 1999b. *Roseivivax halodurans* gen. nov., sp. nov. and *Roseivivax halotolerans* sp. nov., aerobic bacteriochlorophyll-containing bacteria isolated from a saline lake. Int. J. Syst. Bacteriol. *49*: 629–634.

Suzuki, T., Y. Muroga, M. Takahama and Y. Nishimura. 2000b. *Roseibium denhamense* gen. nov., sp. nov. and *Roseibium hamelinense* sp. nov., aerobic bacteriochlorophyll-containing bacteria isolated from the east and west coasts of Australia. Int. J. Syst. Evol. Microbiol. *50*: 2151–2156.

Suzuki, T., Y. Muroga, M. Takahama, T. Shiba and Y. Nishimura. 1999c. *Rubrimonas cliftonensis* gen. nov., sp. nov., an aerobic bacteriochlorophyll-containing bacterium isolated from a saline lake. Int. J. Syst. Bacteriol. *49*: 201–205.

Svec, P., L.A. Devriese, I. Sedlacek, M. Baele, M. Vancanneyt, F. Haesebrouck, J. Swings and J. Doskar. 2001. *Enterococcus haemoperoxidus* sp. nov. and *Enterococcus moraviensis* sp. nov., isolated from water. Int. J. Syst. Evol. Microbiol. *51*: 1567–1574.

Swezey, J.L., L.K. Nakamura, T.P. Abbott and R.E. Peterson. 2000. *Lactobacillus arizonensis* sp. nov., isolated from jojoba meal. Int. J. Syst. Evol. Microbiol. *50*: 1803–1809.

Switzer Blum, J., A.B. Bindi, J. Buzzelli, J.F. Stolz and R.S. Oremland. 1998. *Bacillus arsenicoselenatis*, sp. nov., and *Bacillus selenitireducens*, sp. nov.: two haloalkaliphiles from Mono Lake, California that respire oxyanions of selenium and arsenic. Arch. Microbiol. *171*: 19–30.

Switzer Blum, J., A. Burns Bindi, J. Buzzelli, J.F. Stolz and R.S. Oremland. 2001a. *In* Validation of publication of new names and new combinations previously effectively published outside the IJSEM. List No. 80. Int. J. Syst. Evol. Microbiol. *51*: 793–794.

Switzer Blum, J., J.F. Stolz, A. Oren and R.S. Oremland. 2001b. *Selenihalanaerobacter shriftii* gen. nov., sp. nov., a halophilic anaerobe from Dead Sea sediments that respires selenate. Arch. Microbiol. *175*: 208–219.

Switzer Blum, J., J.F. Stolz, A. Oren and R.S. Oremland. 2001c. *In* Validation of publication of new names and new combinations previously effectively published outside the IJSEM. List No. 81. Int. J. Syst. Evol. Microbiol. *51*: 1229.

Swofford, D.L., G.J. Olsen, P.J. Waddell and D.M. Hillis. 1996. Phylogenetic inference. *In* Hillis, Moritz and Mable (Editors), Molecular Systematics, 2nd Ed., Sinauer Associates, Inc., Sunderland. pp. 407–514.

Szostak, J.W., J.I. Stiles, B.K. Tye, P. Chiu, F. Sherman and R. Wu. 1979. Hybridization with synthetic oligonucleotides. Methods Enzymol. *68*: 419–428.

Tajima, K., Y. Takahashi, A. Seino, Y. Iwai and S. Omura. 2001. Description of two novel species of the genus *Kitasatospora* Omura et al. 1982, *Kitasatospora cineracea* sp. nov. and *Kitasatospora niigatensis* sp. nov. Int. J. Syst. Evol. Microbiol. *51*: 1765–1771.

Takahashi, Y., A. Matsumoto, A. Seino, J. Ueno, Y. Iwai and S. Omura. 2002. *Streptomyces avermectinius* sp. nov., an avermectin-producing strain. Int. J. Syst. Evol. Microbiol. *52*: 2163–2168.

Takahata, Y., M. Nishijima, T. Hoaki and T. Maruyama. 2001. *Thermotoga petrophila* sp. nov. and *Thermotoga naphthophila* sp. nov., two hyperthermophilic bacteria from the Kubiki oil reservoir in Niigata, Japan. Int. J. Syst. Evol. Microbiol. *51*: 1901–1909.

Takai, K. and K. Horikoshi. 2000a. *Thermosipho japonicus* sp. nov., an extremely thermophilic bacterium isolated from a deep-sea hydrothermal vent in Japan. Extremophiles *4*: 9–17.

Takai, K. and K. Horikoshi. 2000b. *In* Validation of publication of new names and new combinations previously effectively published outside the IJSEM. List No. 76. Int. J. Syst. Evol. Microbiol. *50*: 1699–1700.

Takai, K., A. Inoue and K. Horikoshi. 2002. *Methanothermococcus okinawensis* sp. nov., a thermophilic, methane-producing archaeon isolated from a Western Pacific deep-sea hydrothermal vent system. Int. J. Syst. Evol. Microbiol. *52*: 1089–1095.

Takai, K., H. Kobayashi, K.H. Nealson and K. Horikoshi. 2003a. *Deferribacter desulfuricans* sp. nov., a novel sulfur-, nitrate- and arsenate-reducing thermophile isolated from a deep-sea hydrothermal vent. Int. J. Syst. Evol. Microbiol. *53*: 839–846.

Takai, K., H. Kobayashi, K.H. Nealson and K. Horikoshi. 2003b. *Sulfurihydrogenibium subterraneum* gen. nov., sp. nov., from a subsurface hot aquifer. Int. J. Syst. Evol. Microbiol. *53*: 823–827.

Takai, K., T. Komatsu and K. Horikoshi. 2001a. *Hydrogenobacter subterraneus* sp. nov., an extremely thermophilic, heterotrophic bacterium unable to grow on hydrogen gas, from deep subsurface geothermal water. Int. J. Syst. Evol. Microbiol. *51*: 1425–1435.

Takai, K., D.P. Moser, T.C. Onstott, N. Spoelstra, S.M. Pfiffner, A. Dohnalkova and J.K. Fredrickson. 2001b. *Alkaliphilus transvaalensis* gen. nov., sp. nov., an extremely alkaliphilic bacterium isolated from a deep South African gold mine. Int. J. Syst. Evol. Microbiol. *51*: 1245–1256.

Takai, K., S. Nakagawa, Y. Sako and K. Horikoshi. 2003c. *Balnearium lithotrophicum* gen. nov., sp. nov., a novel thermophilic, strictly anaerobic, hydrogen-oxidizing chemolithoautotroph isolated from a black smoker chimney in the Suiyo Seamount hydrothermal system. Int. J. Syst. Evol. Microbiol. *53*: 1947–1954.

Takai, K., A. Sugai, T. Itoh and K. Horikoshi. 2000. *Palaeococcus ferrophilus* gen. nov., sp. nov., a barophilic, hyperthermophilic archaeon from a deep-sea hydrothermal vent chimney. Int. J. Syst. Evol. Microbiol. *50*: 489–500.

Takaichi, S., N. Wakao, A. Hiraishi, S. Itoh and K. Shimada. 1999. Nomenclature of metal-substituted (bacterio)chlorophylls in natural photosynthesis: Metal-(bacterio)chlorophyll and M-(B)Chl. Photosynth. Res. *59*: 255–256.

Takamiya, K.I. and K. Okamura. 1984. Photochemical activities and photosynthetic ATP formation in membrane preparation from a facultative methylotroph *Protaminobacter ruber* strain Nr-1. Arch. Microbiol. *140*: 21–26.

Takamiya, K.I., Y. Shioi, H. Shimada and H. Arata. 1992. Inhibition of accumulation of bacteriochlorophyll and carotenoids by blue light in an aerobic photosynthetic bacterium, *Roseobacter denitrificans*, during anaerobic respiration. Plant Cell Physiol. *33*: 1171–1174.

Takeda, M., Y. Kamagata, W.C. Ghiorse, S. Hanada and J. Koizumi. 2002a. *Caldimonas manganoxidans* gen. nov., sp. nov., a poly(3-hydroxybutyrate)-degrading, manganese-oxidizing thermophile. Int. J. Syst. Evol. Microbiol. *52*: 895–900.

Takeda, M., Y. Kamagata, S. Shinmaru, T. Nishiyama and J. Koizumi. 2002b. *Paenibacillus koleovorans* sp. nov., able to grow on the sheath of *Sphaerotilus natans*. Int. J. Syst. Evol. Microbiol. *52*: 1597–1601.

Takeuchi, M., K. Hamana and A. Hiraishi. 2001. Proposal of the genus *Sphingomonas sensu stricto* and three new genera, *Sphingobium*, *Novosphingobium* and *Sphingopyxis*, on the basis of phylogenetic and chemotaxonomic analyses. Int. J. Syst. Evol. Bacteriol. *51*: 1405–1417.

Takeuchi, M. and K. Hatano. 2001. *Agromyces luteolus* sp. nov., *Agromyces rhizospherae* sp. nov. and *Agromyces bracchium* sp. nov., from the mangrove rhizosphere. Int. J. Syst. Evol. Microbiol. *51*: 1529–1537.

Takeuchi, M., K. Hatano, I. Sedlacek and Z. Pacova. 2002. *Rhodococcus jostii* sp. nov., isolated from a medieval grave. Int. J. Syst. Evol. Microbiol. *52*: 409–413.

Tamaki, H., S. Hanada, Y. Kamagata, K. Nakamura, N. Nomura, K. Nakano and M. Matsumura. 2003. *Flavobacterium limicola* sp. nov., a psychrophilic, organic-polymer-degrading bacterium isolated from freshwater sediments. Int. J. Syst. Evol. Microbiol. *53*: 519–526.

Tamer, A.U. , M. Aragno and N. Sahin. 2002. Isolation and characterization of a new type of aerobic, oxalic acid utilizing bacteria, and proposal of *Oxalicibacterium flavum* gen. nov., sp. nov. Syst. Appl. Microbiol. *25*: 513–519.

Tamer, A.U. , M. Aragno and N. Sahin. 2003. *In* Validation of the publication of new names and new combinations previously effectively published outside the IJSEM. List no. 91. Int. J. Syst. Evol. Microbiol. *53*: 627–628.

Tamura, T. and K. Hatano. 2001. Phylogenetic analysis of the genus *Actinoplanes* and transfer of *Actinoplanes minutisporangius* Ruan et al. 1986 and *'Actinoplanes aurantiacus'* to *Cryptosporangium minutisporangium* comb. nov. and *Cryptosporangium aurantiacum* sp. nov. Int. J. Syst. Evol. Microbiol. *51*: 2119–2125.

Tamura, T., M. Hayakawa and K. Hatano. 2001. A new genus of the order *Actinomycetales*, *Virgosporangium* gen. nov., with descriptions of *Virgosporangium ochraceum* sp. nov. and *Virgosporangium aurantiacum* sp. nov. Int. J. Syst. Evol. Microbiol. *51*: 1809–1816.

Tamura, T., S. Suzuki and K. Hatano. 2000a. *Acrocarpospora* gen. nov., a new genus of the order *Actinomycetales*. Int. J. Syst. Evol. Microbiol. *50*: 1163–1171.

Tamura, T., L. Zhiheng, Z. Yamei and K. Hatano. 2000b. *Actinoalloteichus cyanogriseus* gen. nov., sp. nov. Int. J. Syst. Evol. Microbiol. *50*: 1435–1440.

Tan, Z.Y., F.L. Kan, G.X. Peng, E.T. Wang, B. Reinhold-Hurek and W.X. Chen. 2001. *Rhizobium yanglingense* sp. nov., isolated from arid and semi-arid regions in China. Int. J. Syst. Evol. Microbiol. *51*: 909–914.

Tanaka, K., E. Stackebrandt, S. Tohyama and T. Eguchi. 2000. *Desulfovirga adipica* gen. nov., sp. nov. an adipate-degrading, Gram-negative, sulfate-reducing bacterium. Int. J. Syst. Evol. Bacteriol. *50*: 639–644.

Tanasupawat, S., O. Shida, S. Okada and K. Komagata. 2000. *Lactobacillus acidipiscis* sp. nov. and *Weissella thailandensis* sp. nov., isolated from fermented fish in Thailand. Int. J. Syst. Evol. Microbiol. *50*: 1479–1485.

Taras, D., R. Simmering, M.D. Collins, P.A. Lawson and M. Blaut. 2002. Reclassification of *Eubacterium formicigenerans* Holdeman and Moore 1974 as *Dorea formicigenerans* gen. nov., comb. nov., and description of *Dorea longicatena* sp. nov., isolated from human faeces. Int. J. Syst. Evol. Microbiol. *52*: 423–428.

Tarlera, S. and E.B.M. Denner. 2003. *Sterolibacterium denitrificans* gen. nov., sp. nov., a novel cholesterol-oxidizing, denitrifying member of the β-*Proteobacteria*. Int. J. Syst. Evol. Microbiol. *53*: 1085–1091.

Tatusov, R.L., E.V. Koonin and D.J. Lipman. 1997. A genomic perspective on protein families. Science *278*: 631–637.

Täubel, M., P. Kämpfer, S. Buczolits, W. Lubitz and H.J. Busse. 2003. *Bacillus barbaricus* sp. nov., isolated from an experimental wall painting. Int. J. Syst. Evol. Microbiol. *53*: 725–730.

Tchan, Y.T., J. Pochon and A.R. Prévot. 1948. Études de systématique bactérienne. VIII. Essai de classification des *Cytophaga*. Ann. Inst. Pasteur. *74*: 394–400.

Tee, W., P. Midolo, P.H. Janssen, T. Kerr and M.L. Dyall-Smith. 2001. Bacteremia due to *Leptotrichia trevisanii* sp. nov. Eur. J. Clin. Microbiol Infect. Dis. *20*: 765–769.

Tee, W., P. Midolo, P.H. Janssen, T. Kerr and M.L. Dyall-Smith. 2002. *In* Validation of publication of new names and new combinations previously effectively published outside the IJSEM. List No. 85. Int. J. Syst. Evol. Microbiol. *52*: 685–690.

Teixeira, L.M., M.G. Carvalho, M.M. Espinola, A.G. Steigerwalt, M.P. Douglas, D.J. Brenner and R.R. Facklam. 2001. *Enterococcus porcinus* sp. nov. and *Enterococcus ratti* sp. nov., associated with enteric disorders in animals. Int. J. Syst. Evol. Microbiol. *51*: 1737–1743.

Teske, A., E. Alm, J.M. Regan, S. Toze, B.E. Rittmann and D.A. Stahl. 1994. Evolutionary relationships among ammonia- and nitrite-oxidizing bacteria. J. Bacteriol. *176*: 6623–6630.

Teske, A., P. Sigalevich, Y. Cohen and G. Muyzer. 1996. Molecular identification of bacteria from a coculture by denaturing gradient gel electrophoresis of 16S ribosomal DNA fragments as a tool for isolation in pure cultures. Appl. Environ. Microbiol. *62*: 4210–4215.

Thompson, F.L., B. Hoste, C.C. Thompson, J. Goris, B. Gomez-Gil, L. Huys, P. De Vos and J. Swings. 2002. *Enterovibrio norvegicus* gen. nov., sp. nov., isolated from the gut of turbot (*Scophthalmus maximus*) larvae: a new member of the family *Vibrionaceae*. Int. J. Syst. Evol. Microbiol. *52*: 2015–2022.

Thompson, F.L., B. Hoste, K. Vandemeulebroecke and J. Swings. 2003a. Reclassification of *Vibrio hollisae* as *Grimontia hollisae* gen. nov., comb. nov. Int. J. Syst. Evol. Microbiol. *53*: 1615–1617.

Thompson, F.L., Y. Li, B. Gomez-Gil, C.C. Thompson, B. Hoste, K. Vandemeulebroecke, G.S. Rupp, A. Pereira, M.M. De Bem, P. Sorgeloos and J. Swings. 2003b. *Vibrio neptunius* sp. nov., *Vibrio brasiliensis* sp. nov. and *Vibrio xuii* sp. nov., isolated from the marine aquaculture environment (bivalves, fish, rotifers and shrimps). Int. J. Syst. Evol. Microbiol. *53*: 245–252.

Thompson, F.L., C.C. Thompson, B. Hoste, K. Vandemeulebroecke, M. Gullian and J. Swings. 2003c. *Vibrio fortis* sp. nov. and *Vibrio hepatarius* sp. nov., isolated from aquatic animals and the marine environment. Int. J. Syst. Evol. Microbiol. *53*: 1495–1501.

Thompson, F.L., C.C. Thompson, Y. Li, B. Gomez-Gil, J. Vandenberghe, B. Hoste and J. Swings. 2003d. *Vibrio kanaloae* sp. nov., *Vibrio pomeroyi* sp. nov. and *Vibrio chagasii* sp. nov., from sea water and marine animals. Int. J. Syst. Evol. Microbiol. *53*: 753–759.

Thompson, F.L., C.C. Thompson and J. Swings. 2003e. *In* Validation of publication of new names and new combinations previously effectively published outside the IJSEM. List No. 94. Int. J. Syst. Evol. Microbiol. *53*: 1701–1702.

Thompson, F.L., C.C. Thompson and J. Swings. 2003f. *Vibrio tasmaniensis* sp. nov., isolated from Atlantic Salmon (*Salmo salar* L.). Syst. Appl. Microbiol. *26*: 65–69.

Tighe, S.W., P. de Lajudie, K. Dipietro, K. Lindstrom, G. Nick and B.D. Jarvis. 2000. Analysis of cellular fatty acids and phenotypic relationships of *Agrobacterium*, *Bradyrhizobium*, *Mesorhizobium*, *Rhizobium* and *Sinorhizobium* species using the Sherlock Microbial Identification System. Int. J. Syst. Evol. Microbiol. *50*: 787–801.

Tong, H., X. Gao and X. Dong. 2003. *Streptococcus oligofermentans* sp. nov., a novel oral isolate from caries-free humans. Int. J. Syst. Evol. Microbiol. *53*: 1101–1104.

Topley, W.W.C. and G.S. Wilson. 1929. The Principles of Bacteriology and Immunity, Edward Arnold and Co., London.

Topley, W.W.C. and G.S. Wilson. 1964. Topley and Wilson's Principles of Bacteriology and Immunity, The Williams & Wilkins Co., Baltimore.

Torkko, P., S. Suomalainen, E. Iivanainen, M. Suutari, E. Tortoli, L. Paulin and M.L. Katila. 2000. *Mycobacterium xenopi* and related organisms isolated from stream waters in Finland and description of *Mycobacterium botniense* sp. nov. Int. J. Syst. Evol. Microbiol. *50*: 283–289.

Torkko, P., S. Suomalainen, E. Iivanainen, E. Tortoli, M. Suutari, J. Seppanen, L. Paulin and M.L. Katila. 2002. *Mycobacterium palustre* sp. nov., a potentially pathogenic, slowly growing mycobacterium isolated from clinical and veterinary specimens and from Finnish stream waters. Int. J. Syst. Evol. Microbiol. *52*: 1519–1525.

Torsvik, V., J. Goksøyr and F.L. Daae. 1990a. High diversity in DNA of soil bacteria. Appl. Environ. Microbiol. *56*: 782–787.

Torsvik, V., K. Salte, R. Sørheim and J. Goksøyr. 1990b. Comparison of phenotypic diversity and DNA heterogeneity in a population of soil bacteria. Appl. Environ. Microbiol. *56*: 776–781.

Tortoli, E., C. Piersimoni, R.M. Kroppenstedt, J.I. Montoya-Burgos, U. Reischl, A. Giacometti and S. Emler. 2001. *Mycobacterium doricum* sp. nov. Int. J. Syst. Evol. Microbiol. *51*: 2007–2012.

Tóth, E., G. Kovács, P. Schumann, A.L. Kovacs, U. Steiner, A. Halbritter and K. Márialigeti. 2001. *Schineria larvae* gen. nov., sp. nov., isolated from the 1st and 2nd larval stages of *Wohlfahrtia magnifica* (Diptera: Sarcophagidae). Int. J. Syst. Evol. Microbiol. *51*: 401–407.

Totten, P.A., K.K. Holmes, H.H. Handsfield, J.S. Knapp, P.L. Perine and S. Falkow. 1983. DNA hybridization technique for the detection of *Neisseria gonorrhoeae* in men with urethritis. J. Infect. Dis. *148*: 462–471.

Touzel, J.P., M. O'Donohue, P. Debeire, E. Samain and C. Breton. 2000. *Thermobacillus xylanilyticus* gen. nov., sp. nov., a new aerobic thermophilic xylan-degrading bacterium isolated from farm soil. Int. J. Syst. Evol. Microbiol. *50*: 315–320.

Trébaol, G., L. Gardan, C. Manceau, J.L. Tanguy, Y. Tirilly and S. Boury. 2000. Genomic and phenotypic characterization of *Xanthomonas cynarae* sp. nov., a new species that causes bacterial bract spot of artichoke (*Cynara scolymus* L.). Int. J. Syst. Evol. Microbiol. *50*: 1471–1478.

Trevisan, V. 1887. Sul micrococco della rabbia e sulla possibilità di riconoscere durante il periode d'incubazione, dell'esame del sangue della persona moricata, de ha contratta l'infezione rabbica. Rend. Ist. Lombardo (Ser.2) *20*: 88–105.

Trevisan, V. 1889. I generi e le specie delle batteriacee, Zanaboni and Gabuzzi, Milan. 1–35.

Trüper, H.G. 1992. Prokaryotes: an overview with respect to biodiversity and environmental importance. Biodivers. Conserv. *1*: 227–236.

Trüper, H.G. 1996. Help! Latin! How to avoid the most common mistakes while giving Latin names to newly discovered prokaryotes. Microbiol. SEM. *12*: 473–475.

Trüper, H.G. and L. De'Clari. 1997. Taxonomic note: necessary correction of specific epithets formed as substantives (nouns) "in apposition". Int. J. Syst. Bacteriol. *47*: 908–909.

Trüper, H.G. and L. De'Clari. 1998. Taxonomic note: erratum and correction of further specific epithets formed as substantives (nouns) "in apposition". Int. J. Syst. Bacteriol. *48*: 615.

Tsang, D.C. and I. Suzuki. 1982. Cytochrome c554 as a possible electron donor in the hydroxylation of ammonia and carbon monoxide in *Nitrosomonas europaea*. Can. J. Biochem. *60*: 1018–1024.

Tsien, H.C. and H. Laudelout. 1968. Minimal size of *Nitrobacter* membrane fragments retaining nitrite oxidizing activity. Arch. Mikrobiol. *61*: 280–291.

Tsukamoto, T., M. Takeuchi, O. Shida, H. Murata and A. Shirata. 2001. Proposal of *Mycetocola* gen. nov. in the family *Microbacteriaceae* and three new species, *Mycetocola saprophilus* sp. nov., *Mycetocola tolaasinivorans* sp. nov. and *Mycetocola lacteus* sp. nov., isolated from cultivated mushroom, *Pleurotus ostreatus*. Int. J. Syst. Evol. Microbiol. *51*: 937–944.

Tsuruoka, N., Y. Isono, O. Shida, H. Hemmi, T. Nakayama and T. Nishino. 2003. *Alicyclobacillus sendaiensis* sp. nov., a novel acidophilic, slightly thermophilic species isolated from soil in Sendai, Japan. Int. J. Syst. Evol. Microbiol. *53*: 1081–1084.

Tsutsumi, S., K. Denda, K. Yokoyama, T. Oshima, T. Date and M. Yoshida. 1991. Molecular cloning of genes encoding two major subunits of a eubacterial V-type ATPase from *Thermus thermophilus*. Biochim. Biophys. Acta *1098*: 13–20.

Turenne, C., P. Chedore, J. Wolfe, F. Jamieson, G. Broukhanski, K. May and A. Kabani. 2002. *Mycobacterium lacus* sp. nov., a novel slowly growing, non-chromogenic clinical isolate. Int. J. Syst. Evol. Microbiol. *52*: 2135–2140.

Turner, S.L. and J.P.W. Young. 2000. The glutamine synthetases of rhizobia: Phylogenetics and evolutionary implications. Mol. Biol. Evol. *17*: 309–319.

Tyrrell, G.J., L. Turnbull, L.M. Teixeira, J. Lefebvre, M.D.S. Carvalho, R.R. Facklam and M. Lovgren. 2002a. *Enterococcus gilvus* sp. nov. and *Enterococcus pallens* sp. nov. isolated from human clinical specimens. J. Clin. Microbiol. *40*: 1140–1145.

Tyrrell, G.J., L. Turnbull, L.M. Teixeira, J. Lefebvre, M.D.S. Carvalho, R.R. Facklam and M. Lovgren. 2002b. *In* Validation of publication of new names and new combinations previously effectively published outside the IJSEM. List No. 86. Int. J. Syst. Evol. Microbiol. *52*: 1075–1076.

Uchino, M., O. Shida, T. Uchimura and K. Komagata. 2001. Recharacterization of *Pseudomonas fulva* Lizuka and Komagata 1963, and proposals of *Pseudomonas parafulva* sp. nov. and *Pseudomonas cremoricolorata* sp. nov. J. Gen. Appl. Microbiol. *47*: 247–261.

Uchino, M., O. Shida, T. Uchimura and K. Komagata. 2002a. *In* Validation of publication of new names and new combinations previously effectively published outside the IJSEM. List No. 85. Int. J. Syst. Evol. Microbiol. *52*: 685–690.

Uchino, Y., T. Hamada and A. Yokota. 2002b. Proposal of *Pseudorhodobacter ferrugineus* gen. nov., comb. nov., for a non-photosynthetic marine bacterium, *Agrobacterium ferrugineum*, related to the genus *Rhodobacter*. J. Gen. Appl. Microbiol. *48*: 309–319.

Uchino, Y., T. Hamada and A. Yokota. 2003. *In* Validation of the publication of new names and new combinations previously effectively published outside the IJSEM. List no. 92. Int. J. Syst. Evol. Microbiol. *53*: 935–937.

Ueda, K., T. Seki, T. Kudo, T. Yoshida and M. Kataoka. 1999. Two distinct mechanisms cause heterogeneity of 16S rRNA. J. Bacteriol. *181*: 78–82.

Uetanabaro, A.P., C. Wahrenburg, W. Hunger, R. Pukall, C. Sproer, E. Stackebrandt, V.P. de Canhos, D. Claus and D. Fritze. 2003. *Paenibacillus agarexedens* sp. nov., nom. rev., and *Paenibacillus agaridevorans* sp. nov. Int. J. Syst. Evol. Microbiol. *53*: 1051–1057.

Urakami, T. and K. Komagata. 1984. *Protomonas*, new genus of facultatively methylotrophic bacteria. Int. J. Syst. Bacteriol. *34*: 188–201.

Urbance, J.W., B.J. Bratina, S.F. Stoddard and T.M. Schmidt. 2001. Taxonomic characterization of *Ketogulonigenium vulgare* gen. nov., sp. nov. and *Ketogulonigenium robustum* sp. nov., which oxidize L-sorbose to 2-keto-L-gulonic acid. Int. J. Syst. Evol. Microbiol. *51*: 1059–1070.

Urbanski, T. 1984a. Composite propellants. *In* Urbanski (Editor), Chemistry and Technology of Explosives, Pergamon Press, New York. pp. 602–620.

Urbanski, T. 1984b. Salts of nitric acid and of oxy-acids of chlorine. *In* Urbanski (Editor), Chemistry and Technology of Explosives, Pergamon Press, New York. pp. 444–461.

Urbansky, E.T., M.L. Magnuson, C.A. Kelty, B. Gu and G.M. Brown. 2000. Comment on "Perchlorate identification in fertilizers" and the subsequent addition/correction. Environ. Sci. Technol. *34*: 4452–4453.

Urcì, C., P. Salamone, P. Schumann and E. Stackebrandt. 2000. *Marmoricola aurantiacus* gen. nov., sp. nov., a coccoid member of the family *Nocardioidaceae* isolated from a marble statue. Int. J. Syst. Evol. Microbiol. *50*: 529–536.

Urdea, M.S., B.D. Warner, J.A. Running, M. Stempien, J. Clyne and T. Horn. 1988. A comparison of non-radioisotopic hybridization assay methods using fluorescent, chemiluminescent and enzyme labeled synthetic oligodeoxyribonucleotide probes. Nucleic Acids Res. *16*: 4937–4956.

Ushiba, Y., Y. Takahara and H. Ohta. 2003. *Sphingobium amiense* sp. nov., a novel nonylphenol-degrading bacterium isolated from a river sediment. Int. J. Syst. Evol. Microbiol. *53*: 2045–2048.

Utåker, J.B., L. Bakken, Q.Q. Jiang and I.F. Nes. 1995. Phylogentic analysis of seven new isolates of ammonia-oxidizing bacteria based on 16S rRNA gene sequences. System. Appl. Microbiol. *18*: 549–559.

Utåker, J.B. and I.F. Nes. 1998. A qualitative evaluation of the published oligonucleotides specific for the 16S rRNA gene sequences of the ammonia-oxidizing bacteria. Syst. Appl. Microbiol. *21*: 72–88.

Valverde, A., E. Velazquez, C. Gutierrez, E. Cervantes, A. Ventosa and J.M. Igual. 2003. *Herbaspirillum lusitanum* sp. nov., a novel nitrogen-fixing bacterium associated with root nodules of *Phaseolus vulgaris*. Int. J. Syst. Evol. Microbiol. *53*: 1979–1983.

van Berkum, P., Z. Terefework, L. Paulin, S. Suomalainen, K. Lindstrom and B.D. Eardly. 2003. Discordant phylogenies within the *rrn* loci of rhizobia. J. Bacteriol. *185*: 2988–2998.

Van de Graaf, A.A., A. Mulder, P. De Bruijn, M.S. Jetten, L.A. Robertson and J.G. Kuenen. 1995. Anaerobic oxidation of ammonium is a biologically mediated process. Appl. Environ. Microbiol. *61*: 1246–1251.

van de Peer, Y., E. Robbrecht, S. de Hoog, A. Caers, P. de Rijk and R. de Wachter. 1999. Database on the structure of small subunit ribosomal RNA. Nucleic Acids Res. *27*: 179–183.

van der Maarel, M., A. Veen and D.J. Wijbenga. 2000. *Paenibacillus granivorans* sp. nov., a new *Paenibacillus* species which degrades native potato starch granules. Syst. Appl. Microbiol. *23*: 344–348.

van der Maarel, M., A. Veen and D.J. Wijbenga. 2001. *In* Validation of publication of new names and new combinations previously effectively published outside the IJSEM. List No. 79. Int. J. Syst. Evol. Microbiol. *51*: 263–265.

van der Wielen, P.W., G.M. Rovers, J.M. Scheepens and S. Biesterveld. 2002. *Clostridium lactatifermentans* sp. nov., a lactate-fermenting anaerobe isolated from the caeca of a chicken. Int. J. Syst. Evol. Microbiol. *52*: 921–925.

van Ginkel, C.G., G.B. Rikken, A.G. Kroon and S.W. Kengen. 1996. Purification and characterization of chlorite dismutase: a novel oxygen-generating enzyme. Arch. Microbiol. *166*: 321–326.

Van Trappen, S., J. Mergaert and J. Swings. 2003. *Flavobacterium gelidilacus* sp. nov., isolated from microbial mats in Antarctic lakes. Int. J. Syst. Evol. Microbiol. *53*: 1241–1245.

Vancanneyt, M., F. Schut, C. Snauwaert, J. Goris, J. Swings and J.C. Gottschal. 2001a. *Sphingomonas alaskensis* sp. nov., a dominant bacterium from a marine oligotrophic environment. Int. J. Syst. Evol. Microbiol. *51*: 73–79.

Vancanneyt, M., C. Snauwaert, I. Cleenwerck, M. Baele, P. Descheemaeker, H. Goossens, B. Pot, P. Vandamme, J. Swings, F. Haesebrouck and L.A. Devriese. 2001b. *Enterococcus villorum* sp. nov., an enteroadherent bacterium associated with diarrhoea in piglets. Int. J. Syst. Evol. Microbiol. *51*: 393–400.

Vandamme, P. 1998. Speciation. *In* Williams, Ketley and Salmond (Editors), Methods in Microbiology, Vol. 27. Bacterial Pathogenesis, Academic Press, London. pp. 51–56.

Vandamme, P., E. Falsen, R. Rossau, B. Hoste, P. Segers, R. Tytgat and J. De Ley. 1991. Revision of *Campylobacter*, *Helicobacter*, and *Wolinella* taxonomy: emendation of generic descriptions and proposal of *Arcobacter* gen. nov. Int. J. Syst. Bacteriol. *41*: 88–103.

Vandamme, P., J. Goris, W.M. Chen, P. de Vos and A. Willems. 2002a. *Burkholderia tuberum* sp. nov. and *Burkholderia phymatum* sp. nov., nodulate the roots of tropical legumes. Syst. Appl. Microbiol. *25*: 507–512.

Vandamme, P., J. Goris, W.M. Chen, P. de Vos and A. Willems. 2003a. *In* Validation of publication of new names and new combinations previously effectively published outside the IJSEM. List No. 91. Int. J. Syst. Evol. Microbiol. *53*: 627–628.

Vandamme, P., D. Henry, T. Coenye, S. Nzula, M. Vancanneyt, J.J. LiPuma, D.P. Speert, J.R.W. Govan and E. Mahenthiralingam. 2002b. *Burkholderia anthina* sp. nov. and *Burkholderia pyrrocinia*, two additional *Burkholderia cepacia* complex bacteria, may confound results of new molecular diagnostic tools. FEMS Immunology and Medical Microbiology *33*: 143–149.

Vandamme, P., D. Henry, T. Coenye, S. Nzula, M. Vancanneyt, J.J. LiPuma, D.P. Speert, J.R.W. Govan and E. Mahenthiralingam. 2002c. *In* Validation of publication of new names and new combinations previously effectively published outside the IJSEM. List No. 87. Int. J. Syst. Evol. Microbiol. *52*: 1437–1438.

Vandamme, P., B. Holmes, T. Coenye, J. Goris, E. Mahenthiralingam, J.J. LiPuma and J.R.W. Govan. 2003b. *Burkholderia cenocepacia* sp. nov.–a new twist to an old story. Res. Microbiol. *154*: 91–96.

Vandamme, P., B. Holmes, T. Coenye, J. Goris, E. Mahenthiralingam, J.J. LiPuma and J.R.W. Govan. 2003c. *In* Validation of publication of new names and new combinations previously effectively published outside the IJSEM. List No. 92. Int. J. Syst. Evol. Microbiol. *53*: 935–937.

Vandamme, P., J. Hommez, C. Snauwaert, B. Hoste, I. Cleenwerck, K. Lefebvre, M. Vancanneyt, J. Swings, L.A. Devriese and F. Haesebrouck. 2001. *Globicatella sulfidifaciens* sp. nov., isolated from purulent infections in domestic animals. Int. J. Syst. Evol. Microbiol. *51*: 1745–1749.

Vandamme, P., E. Mahenthiralingam, B. Holmes, T. Coenye, B. Hoste, P. De Vos, D. Henry and D.P. Speert. 2000a. Identification and population structure of *Burkholderia stabilis* sp. nov. (formerly *Burkholderia cepacia* genomovar IV). J. Clin. Microbiol. *38*: 1042–1047.

Vandamme, P., E. Mahenthiralingam, B. Holmes, T. Coenye, B. Hoste, P. De Vos, D. Henry and D.P. Speert. 2000b. *In* Validation of publication of new names and new combinations previously effectively published outside the IJSEM. List No. 75. Int. J. Syst. Evol. Microbiol. *50*: 1415–1417.

Vandamme, P., B. Pot, M. Gillis, P. de Vos, K. Kersters and J. Swings. 1996a. Polyphasic taxonomy, a consensus approach to bacterial systematics. Microbiol. Rev. *60*: 407–438.

Vandamme, P., M. Vancanneyt, A. Van Belkum, P. Segers, W.G.V. Quint, K. Kersters, B.J. Paster and F.E. Dewhirst. 1996b. Polyphasic analysis of strains of the genus *Capnocytophaga* and Centers for Disease Control group DF-3. Int. J. Syst. Bacteriol. *46*: 782–791.

Vandekerckhove, T.T., A. Willems, M. Gillis and A. Coomans. 2000. Occurrence of novel verrucomicrobial species, endosymbiotic and associated with parthenogenesis in *Xiphinema americanum*-group species (Nematoda, *Longidoridae*). Int. J. Syst. Evol. Microbiol. *50*: 2197–2205.

Vannelli, T., M. Logan, D.M. Arciero and A.B. Hooper. 1990. Degradation of halogenated aliphatic compounds by the ammonia- oxidizing bacterium *Nitrosomonas europaea*. Appl. Environ. Microbiol. *56*: 1169–1171.

Vauterin, L., B. Hoste, K. Kersters and J. Swings. 1995. Reclassification of *Xanthomonas*. Int. J. Syst. Bacteriol. *45*: 472–489.

Vela, A.I., M.D. Collins, M.V. Latre, A. Mateos, M.A. Moreno, R. Hutson, L. Dominguez and J.F. Fernandez-Garayzabal. 2003a. *Psychrobacter pulmonis* sp. nov., isolated from the lungs of lambs. Int. J. Syst. Evol. Microbiol. *53*: 415–419.

Vela, A.I., E. Fernandez, P.A. Lawson, M.V. Latre, E. Falsen, L. Dominguez, M.D. Collins and J.F. Fernandez-Garayzabal. 2002. *Streptococcus entericus* sp. nov., isolated from cattle intestine. Int. J. Syst. Evol. Microbiol. *52*: 665–669.

Vela, A.I., A. Mateos, M.D. Collins, V. Briones, R.A. Hutson, L. Dominguez and J.F. Fernandez-Garayzabal. 2003b. *Corynebacterium suicordis* sp. nov., from pigs. Int. J. Syst. Evol. Microbiol. *53*: 2027–2031.

Velázquez, E., J.M. Igual, A. Willems, M.P. Fernández, E. Muñoz, P.F. Mateos, A. Abril, N. Toro, P. Normand, E. Cervantes, M. Gillis and E. Martínez-Molina. 2001. *Mesorhizobium chacoense* sp. nov., a novel species that nodulates *Prosopis alba* in the Chaco Arido region (Argentina). Int. J. Syst. Evol. Bacteriol. *51*: 1011–1021.

Venables, W.N. and B.D. Ripley. 1994. Modern Applied Statistics with S-Plus, Springer-Verlag, New York.

Venkateswaran, K. and N. Dohmoto. 2000. *Pseudoalteromonas peptidolytica* sp. nov., a novel marine mussel-thread-degrading bacterium isolated from the Sea of Japan. Int. J. Syst. Evol. Microbiol. *50*: 565–574.

Venkateswaran, K., M. Kempf, F. Chen, M. Satomi, W. Nicholson and R. Kern. 2003. *Bacillus nealsonii* sp. nov., isolated from a spacecraft-assembly facility, whose spores are γ-radiation resistant. Int. J. Syst. Evol. Microbiol. *53*: 165–172.

Verger, J.M., F. Grimont, P.A.D. Grimont and M. Grayon. 1985. *Brucella*, a monospecific genus as shown by deoxyribonucleic acid hybridization. Int. J. Syst. Bacteriol. *35*: 292–295.

Vernozy-Rozand, C., C. Mazuy, H. Meugnier, M. Bes, Y. Lasne, F. Fiedler, J. Etienne and J. Freney. 2000. *Staphylococcus fleurettii* sp. nov., isolated from goat's milk cheeses. Int. J. Syst. Evol. Microbiol. *50*: 1521–1527.

Viale, A.M., A. Arakaki, F.C. Soncini and R.G. Ferreyra. 1994. Evolutionary relationships among eubacterial groups as inferred from GroEL (chaperonin) sequence comparisons. Int. J. Syst. Bacteriol. *44*: 527–533.

Vijgenboom, E., L.P. Woudt, P.W. Heinstra, K. Rietveld, J. van Haarlem, G.P. van Wezel, S. Shochat and L. Bosch. 1994. Three *tuf*-like genes in the kirromycin producer *Streptomyces ramocissimus*. Microbiology *140*: 983–998.

Vinga, S. and J. Almeida. 2003. Alignment-free sequence comparison – a review. Bioinformatics. *19*: 513–523.

von der Weid, I., G.F. Duarte, J.D. van Elsas and L. Seldin. 2002. *Paenibacillus brasilensis* sp. nov., a novel nitrogen-fixing species isolated from the maize rhizosphere in Brazil. Int. J. Syst. Evol. Microbiol. *52*: 2147–2153.

von Klein, D., H. Arab, H. Volker and M. Thomm. 2002a. *Methanosarcina baltica*, sp. nov., a novel methanogen isolated from the Gotland Deep of the Baltic Sea. Extremophiles *6*: 103–110.

von Klein, D., H. Arab, H. Volker and M. Thomm. 2002b. *In* Validation of publication of new names and new combinations previously effectively published outside the IJSEM. List No. 85. Int. J. Syst. Evol. Microbiol. *52*: 685–690.

von Wintzingerode, F., U.B. Gobel, R.A. Siddiqui, U. Rosick, P. Schumann, A. Fruhling, M. Rohde, R. Pukall and E. Stackebrandt. 2001a. *Salana multivorans* gen. nov., sp. nov., a novel actinobacterium isolated from an anaerobic bioreactor and capable of selenate reduction. Int. J. Syst. Evol. Microbiol. *51*: 1653–1661.

von Wintzingerode, F., U.B. Gobel and E. Stackebrandt. 1997. Determination of microbial diversity in environmental samples: pitfalls of PCR-based rRNA analysis. FEMS Microbiol. Rev. *21*: 213–229.

von Wintzingerode, F., A. Schattke, R.A. Siddiqui, U. Rosick, U.B. Gobel and R. Gross. 2001b. *Bordetella petrii* sp. nov., isolated from an anaerobic bioreactor, and emended description of the genus *Bordetella*. Int. J. Syst. Evol. Microbiol. *51*: 1257–1265.

Voordouw, G., J.K. Voordouw, T.R. Jack, J. Foght, P.M. Fedorak and D.W.S. Westlake. 1992. Identification of distinct communities of sulfate-reducing bacteria in oil fields by reverse sample genome probing. Appl. Environ. Microbiol. *58*: 3542–3552.

Voordouw, G., J.K. Voordouw, R.R. Karkoff-Schweizer, P.M. Fedorak and D.W.S. Westlake. 1991. Reverse sample genome probing, a new technique for identification of bacteria in environmental samples by DNA hybridization, and its application to the identification of sulfate-reducing bacteria in oil field samples. Appl. Environ. Microbiol. *57*: 3070–3078.

Voytek, M.A. 1996. Relative abundance and species diversity of autotrophic ammonia-oxidizing bacteria in aquatic systems (nitrifying bacteria), University of California, Santa Cruz.

Voytek, M.A. and B.B. Ward. 1995. Detection of ammonium-oxidizing bacteria of the beta-subclass of the class *Proteobacteria* in aquatic samples with PCR. Appl. Environ. Microbiol. *61*: 1444–1450.

Voytek, M.A., B.B. Ward and J.C. Priscu. 1998. The abundance of ammonia-oxidizing bacteria in Lake Bonney, Antarctica determined by immunofluorescence, PCR, and in situ hybridization. *In* Priscu (Editor), Ecosystem Dynamics in a Polar Desert: the McMurdo Dry Valleys, Antarctica, Antarctica Research Series 72, American Geophysical Union, Washington, D.C. 217–228.

Vreeland, R.H., S. Straight, J. Krammes, K. Dougherty, W.D. Rosenzweig and M. Kamekura. 2002. *Halosimplex carlsbadense* gen. nov., sp. nov., a unique halophilic archaeon, with three 16S rRNA genes, that grows only in defined medium with glycerol and acetate or pyruvate. Extremophiles *6*: 445–452.

Vreeland, R.H., S. Straight, J. Krammes, K. Dougherty, W.D. Rosenzweig and M. Kamekura. 2003. *In* Validation of publication of new names and new combinations previously effectively published outside the IJSEM. List No. 92. Int. J. Syst. Evol. Microbiol. *53*: 935–937.

Wagner, D., E. Spieck, E. Bock and E.M. Pfeiffer. 2001. Microbial life in terrestrial permafrost: Methanogenesis and nitrification in gelisols as potentials for exobiological processes. *In* Horneck and Baumstark-Khan (Editors), Astrobiology - The Quest for the Conditions of Life, Springer-Verlag, Berlin; New York. pp. 143–159.

Wagner, M., R.I. Amann, H. Lemmer and K.H. Schleifer. 1993. Probing activated sludge with oligonucleotides specific for proteobacteria: inadequacy of culture-dependent methods for describing microbial community structure. Appl. Environ. Microbiol. *59*: 1520–1525.

Wagner, M., R. Erhart, W. Manz, R. Amann, H. Lemmer, D. Wedi and K.H. Schleifer. 1994. Development of an rRNA-targeted oligonucleotide probe specific for the genus *Acinetobacter* and its application for *in situ* monitoring in activated sludge. Appl. Environ. Microbiol. *60*: 792–800.

Wagner, M., D.R. Noguera, S. Juretschko, G. Rath, H.P. Koops and K.H. Schleifer. 1998a. Combining fluorescent *in situ* hybridization (FISH) with cultivation and mathematical modeling to study population structure and function of ammonia-oxidizing bacteria in activated sludge. Water Sci. Technol. *37*: 441–449.

Wagner, M., G. Rath, R. Amann, H.P. Koops and K.H. Schleifer. 1995. *In situ* identification of ammonia-oxidizing bacteria. Syst. Appl. Microbiol. *18*: 251–264.

Wagner, M., G. Rath, H.P. Koops, J. Flood and R. Amann. 1996. *In situ* analysis of nitrifying bacteria in sewage treatment plants. Water Sci. Technol. *34*: 237–244.

Wagner, M., M. Schmid, S. Juretschko, T.K. Trebesius, A. Bubert, W. Goebel and K.H. Schleifer. 1998b. *In situ* detection of a virulence factor mRNA and 16S rRNA in *Listeria monocytogenes*. FEMS Microbiol. Lett. *160*: 159–168.

Wagner-Döbler, I. , H. Rheims, A. Felske, R. Pukall and B.J. Tindall. 2003. *Jannaschia helgolandensis* gen. nov., sp. nov., a novel abundant member of the marine *Roseobacter* clade from the North Sea. Int. J. Syst. Evol. Microbiol. *53*: 731–738.

Wahl, G.M., S.L. Berger and A.R. Kimmel. 1987. Molecular hybridization of immobilized nucleic acids: theoretical concepts and practical considerations. Methods Enzymol. *152*: 399–407.

Wainø, M., B.J. Tindall and K. Ingvorsen. 2000. *Halorhabdus utahensis* gen. nov., sp. nov., an aerobic, extremely halophilic member of the Archaea from Great Salt Lake, Utah. Int. J. Syst. Evol. Microbiol. *50*: 183–190.

Wakao, N., N. Yokoi, N. Isoyama, A. Hiraishi, K. Shimada, M. Kobayashi, H. Kise, M. Iwaki, S. Itoh, S. Takaichi and Y. Sakurai. 1996. Discovery of natural photosynthesis using Zn-containing bacteriochlorophyll in an aerobic bacterium *Acidiphilium rubrum*. Plant Cell Physiol. *37*: 889–893.

Walker, N. 1978. On the diversity of nitrifiers in nature. *In* Schlessinger (Editor), Microbiology 1978, American Society for Microbiology, Washington. 346–347.

Walker, N. and K.N. Wickramasinghe. 1979. Nitrification and autotrophic nitrifying bacteria in acid tea soils. Soil Biol. Biochem. *11*: 231–236.

Wallace, R.J., N. McKain, N.R. McEwan, E. Miyagawa, L.C. Chaudhary, T.P. King, N.D. Walker, J.H. Apajalahti and C.J. Newbold. 2003. *Eubacterium pyruvativorans* sp. nov., a novel non-saccharolytic anaerobe from the rumen that ferments pyruvate and amino acids, forms caproate and utilizes acetate and propionate. Int. J. Syst. Evol. Microbiol. *53*: 965–970.

Wallace, W., S.E. Knowles and D.J. Nicholas. 1970. Intermediary metabolism of carbon compounds by nitrifying bacteria. Arch. Mikrobiol. *70*: 26–42.

Wallace, W. and D.J. Nicholas. 1969. Glutamate dehydrogenase in *Nitrosomonas europaea* and the effect of hydroxylamine, oximes and related compounds on its activity. Biochim. Biophys. Acta *171*: 229–237.

Wallace, W., T. Ward, A. Breen and H. Attaway. 1996. Identification of an anaerobic bacterium which reduces perchlorate and chlorate as *Wolinella succinogenes*. J. Ind. Microbiol. *16*: 68–72.

Wallner, G., R.I. Amann and W. Beisker. 1993. Optimizing fluorescent *in situ* hybridization with rRNA-targeted oligonucleotide probes for flow cytometric identification of microorganisms. Cytometry. *14*: 136–143.

Wallner, G., R. Erhart and R.I. Amann. 1995. Flow cytometric analysis of activated sludge with rRNA-targeted probes. Appl. Environ. Microbiol. *61*: 1859–1866.

Wallner, G., B. Fuchs, S. Spring, W. Beisker and R.I. Amann. 1997. Flow cytometric sorting of microorganisms for molecular analysis. Appl. Environ. Microbiol. *63*: 4223–4231.

Wang, E.T., Z.Y. Tan, A. Willems, M. Fernandez-Lopez, B. Reinhold-Hurek and E. Martinez-Romero. 2002. *Sinorhizobium morelense* sp. nov., a *Leucaena leucocephala*-associated bacterium that is highly resistant to multiple antibiotics. Int. J. Syst. Evol. Microbiol. *52*: 1687–1693.

Wang, L., Y. Zhang, Z. Lu, Y. Shi, Z. Liu, L. Maldonado and M. Goodfellow. 2001a. *Nocardia beijingensis* sp. nov., a novel isolate from soil. Int. J. Syst. Evol. Microbiol. *51*: 1783–1788.

Wang, Y.M., Z.S. Zhang, X.L. Xu, J.S. Ruan and Y. Wang. 2001b. *Actinopolymorpha singaporensis* gen. nov., sp. nov., a novel actinomycete from the tropical rainforest of Singapore. Int. J. Syst. Evol. Microbiol. *51*: 467–473.

Ward, B.B. 1982. Oceanic distribution of ammonium-oxidizing bacteria determined by immunofluorescent assay. J. Mar. Res. *40*: 1155–1172.

Ward, B.B. 1986. Nitrification in marine environments. *In* Prosser (Editor), Nitrification, Vol. 20, Special publication for the Society of General Microbiology. IRL Press, Oxford; Washington, DC. 157–184.

Ward, B.B. and A.F. Carlucci. 1985. Marine ammonia-oxidizing and nitrite-oxidizing bacteria serological diversity determined by immunofluorescence in culture and in the environment. Appl. Environ. Microbiol. *50*: 194–201.

Ward, B.B., D.P. Martino, M.C. Diaz and S.B. Joye. 2000. Analysis of ammonia-oxidizing bacteria from hypersaline Mono Lake, California, on the basis of 16S rRNA sequences. Appl. Environ. Microbiol. *66*: 2873–2881.

Ward, B.B. and M.J. Perry. 1980. Immunofluorescent assay for the marine ammonium-oxidizing bacterium *Nitrosococcus oceanus*. Appl. Environ. Microbiol. *39*: 913–918.

Ward, B.B., M.A. Voytek and R.P. Witzel. 1997. Phylogenetic diversity of natural populations of ammonia oxidizers investigated by specific PCR amplification. Microb. Ecol. *33*: 87–96.

Ward, D.M., M.J. Ferris, S.C. Nold and M.M. Bateson. 1998. A natural view of microbial biodiversity within hot spring cyanobacterial mat communities. Microbiol. Mol. Biol. Rev. *62*: 1353–1370.

Warnick, T.A., B.A. Methe and S.B. Leschine. 2002. *Clostridium phytofermentans* sp. nov., a cellulolytic mesophile from forest soil. Int. J. Syst. Evol. Microbiol. *52*: 1155–1160.

Wasserfallen, A., J. Nolling, P. Pfister, J. Reeve and E. Conway de Macario. 2000. Phylogenetic analysis of 18 thermophilic *Methanobacterium* isolates supports the proposals to create a new genus, *Methanothermobacter* gen. nov., and to reclassify several isolates in three species, *Methanothermobacter thermautotrophicus* comb. nov., *Methanothermobacter wolfeii* comb. nov., and *Methanothermobacter marburgensis* sp. nov. Int. J. Syst. Evol. Microbiol. *50*: 43–53.

Wassill, L., W. Ludwig and K.H. Schleifer. 1998. Development of a modified subtraction hybridization technique and its application for the design of strain specific PCR systems for lactococci. FEMS Microbiol. Lett. *166*: 63–70.

Watson, S.W. 1965. Characteristics of a marine, nitrifying bacterium, *Nitrocystis oceanus* sp. nov. Limnol. Oceanogr. (Suppl.) *10*: R274–289.

Watson, S.W. 1971. Taxonomic considerations of the family *Nitrobacteraceae* Buchanan. Request for opinions. Int. J. Syst. Bacteriol. *21*: 254–270.

Watson, S.W. 1974. Family I. *Nitrobacteriaceae* Buchanan. *In* Buchanan and Gibbons (Editors), Bergey's Manual of Determinative Bacteriology, 8th ed., The Williams & Wilkins Co., Baltimore. 450–456.

Watson, S.W., E. Bock, H. Harms, H.P. Koops and A.B. Hooper. 1989. Nitrifying bacteria. *In* Staley, Bryant, Pfennig and Holt (Editors), Bergey's Manual of Systematic Bacteriology, 1st Ed., Vol. 3, The Williams & Wilkins Co., Baltimore. pp. 1808–1833.

Watson, S.W., E. Bock, F.W. Valois, J.B. Waterbury and U. Schlosser. 1986. *Nitrospira marina*, gen. nov. sp. nov.: a chemolithotrophic nitrite-oxidizing bacterium. Arch. Microbiol. *144*: 1–7.

Watson, S.W., L.B. Graham, C.C. Remsen and F.W. Valois. 1971. A lobular ammonia-oxidizing bacterium, *Nitrosolobus multiformis* nov. gen. sp. Arch. Mikrobiol. *76*: 183–203.

Watson, S.W. and M. Mandel. 1971. Comparison of the morphology and deoxyribonucleic acid composition of 27 strains of nitrifying bacteria. J. Bacteriol. *107*: 563–569.

Watson, S.W., F.W. Valois and J.B. Waterbury. 1981. The Family *Nitrobacteraceae. In* Starr, Stolp, Trüper, Balows and Schlegel (Editors), The Prokaryotes: a Handbook on Habitats, Isolation, and Identification of Bacteria, Springer-Verlag, Berlin; New York. 1005–1022.

Watson, S.W. and J.B. Waterbury. 1971. Characteristics of two marine nitrite oxidizing bacteria, *Nitrospina gracilis* nov. gen. nov. sp. and *Nitrococcus mobilis* nov. gen. nov. sp. Arch. Microbiol. *77*: 203–230.

Wattiau, P., M. Janssens and G. Wauters. 2000. *Corynebacterium simulans* sp. nov., a non-lipophilic, fermentative *Corynebacterium.* Int. J. Syst. Evol. Microbiol. *50*: 347–353.

Wauters, G., V. Avesani, K. Laffineur, J. Charlier, M. Janssens, B. Van Bosterhaut and M. Delmee. 2003a. *Brevibacterium lutescens* sp. nov., from human and environmental samples. Int. J. Syst. Evol. Microbiol. *53*: 1321–1325.

Wauters, G., J. Charlier, M. Janssens and M. Delmee. 2000a. Identification of *Arthrobacter oxydans, Arthrobacter luteolus* sp. nov., and *Arthrobacter albus* sp. nov., isolated from human clinical specimens. J. Clin. Microbiol. *38*: 2412–2415.

Wauters, G., J. Charlier, M. Janssens and M. Delmee. 2000b. *In* Validation of publication of new names and new combinations previously effectively published outside the IJSEM. List No. 76. Int. J. Syst. Evol. Microbiol. *50*: 1699–1700.

Wauters, G., J. Charlier, M. Janssens and M. Delmee. 2001. *Brevibacterium paucivorans* sp. nov., from human clinical specimens. Int. J. Syst. Evol. Microbiol. *51*: 1703–1707.

Wauters, G., T. De Baere, A. Willems, E. Falsen and M. Vaneechoutte. 2003b. Description of *Comamonas aquatica* comb. nov. and *Comamonas kerstersii* sp. nov. for two subgroups of *Comamonas terrigena* and emended description of *Comamonas terrigena.* Int. J. Syst. Evol. Microbiol. *53*: 859–862.

Wayne, L.G., D.J. Brenner, R.R. Colwell, P.A.D. Grimont, O. Kandler, M.I. Krichevsky, L.H. Moore, W.E.C. Moore, R.G.E. Murray, E. Stackebrandt, M.P. Starr and H.G. Trüper. 1987. Report of the ad hoc committee on reconciliation of approaches to bacterial systematics. Int. J. Syst. Bacteriol. *37*: 463–464.

Weckesser, J., H. Mayer and G. Shulz. 1995. Anoxygenic phototrophic bacteria: model organisms for studies on cell wall macromolecules. *In* Blankenship, Madigan and Bauer (Editors), Anoxygenic Photosynthetic Bacteria, Kluwer Academic Publishing, The Netherlands. pp. 207–230.

Wei, G.H., Z.Y. Tan, M.E. Zhu, E.T. Wang, S.Z. Han and W.X. Chen. 2003. Characterization of rhizobia isolated from legume species within the genera *Astragalus* and *Lespedeza* grown in the Loess Plateau of China and description of *Rhizobium loessense* sp. nov. Int. J. Syst. Evol. Microbiol. *53*: 1575–1583.

Wei, G.H., E.T. Wang, Z.Y. Tan, M.E. Zhu and W.X. Chen. 2002. *Rhizobium indigoferae* sp. nov. and *Sinorhizobium kummerowiae* sp. nov., respectively isolated from *Indigofera* spp. and *Kummerowia stipulacea.* Int. J. Syst. Evol. Microbiol. *52*: 2231–2239.

Weiner, R.M., M. Melick, K. O'Neill and E.J. Quintero. 2000. *Hyphomonas adhaerens* sp. nov., *Hyphomonas johnsonii* sp. nov. and *Hyphomonas rosenbergii* sp. nov., marine budding and prosthecate bacteria. Int. J. Syst. Evol. Microbiol. *50*: 459–469.

Welch, D.F., K.C. Carroll, E.K. Hofmeister, D.H. Persing, D.A. Robison, A.G. Steigerwalt and D.J. Brenner. 1999. Isolation of a new subspecies, *Bartonella vinsonii* arupensis, from a cattle rancher: identity with isolates found in conjunction with *Borrelia burgdorferi* and *Babesia microti* among naturally infected mice. J. Clin. Microbiol. *37*: 2598–2601.

Welch, D.F., K.C. Carroll, E.K. Hofmeister, D.H. Persing, D.A. Robison, A.G. Steigerwalt and D.J. Brenner. 2000. *In* Validation of the publication of new names and new combinations previously effectively published outside the IJSB. List No. 72. Int. J. Syst. Evol. Microbiol. *50*: 3–4.

Wen, A., M. Fegan, C. Hayward, S. Chakraborty and L.I. Sly. 1999. Phylogenetic relationships among members of the *Comamonadaceae*, and description of *Delftia acidovorans* (den Dooren de Jong 1926 and Ta-

Winogradsky, H. 1935a. On the number and variety of nitrifying organisms. Int. Congr. Soil. Sci. *1*: 138–140.

Winogradsky, H. 1935b. Sur la microflore nitrificatrice des boues activees de Paris. R. C. Acad. Sci. *200*: 1886–1888.

Winogradsky, H. 1937. Contributions a l'étude de la microflore nitrificatrice des boues activees de Paris. Ann. Inst. Pasteur (Paris) *58*: 326–340.

Winogradsky, S. 1890a. Recherches sure les organismes de la nitrification. Ann. Inst. Pasteur (Paris) *4*: 257–275.

Winogradsky, S. 1890b. Sur les organismes de la nitrification. C R Acad. Sci. (Paris). *110*: 1013–1016.

Winogradsky, S. 1891. Recherches sur les organismes de la nitrification. Ann. Inst. Pasteur (Paris) *5*: 577–616.

Winogradsky, S. 1892. Contributions a la morphologie des organismes de la nitrification. Arch. Sci. Biol. (St. Petersb.) *1*: 86–137.

Winogradsky, S. 1904. Die Nitrifikation. *In* Handbuch de Technischen Mykologie, Lafar, Jena. 132–181.

Winogradsky, S. 1930. Microbes de la nitrification. Travaux recents. Bull. Inst. Pasteur. (Paris). *28*: 683–687.

Winogradsky, S. 1931. Noucvelles recherershes sur les microbes de la nitrification. C R Acad. Sci. (Paris). *192*: 1000–1004.

Winogradsky, S. 1935c. Travaux récents sur la nitrification. Revue critique. Bull. Inst. Pasteur. (Paris). *33*: 1073–1079.

Winogradsky, S. and H. Winogradsky. 1933. Études sur la microbiologie du sol. VII. Nouvelles recherches sur les organismes de la nitrification. Ann. Inst. Pasteur (Paris) *50*: 350–432.

Winslow, C.-E.A., J. Broadhurst, R.E. Buchanan, C.J. Krumwiede, L.A. Rogers and G.H. Smith. 1917. The families and genera of the bacteria. Preliminary report of the Committee of the Society of American Bacteriologists on characterization and classification of bacterial types. J. Bacteriol. *2*: 506–566.

Winslow, C.-E.A., J. Broadhurst, R.E. Buchanan, C.J. Krumwiede, L.A. Rogers and G.H. Smith. 1920. The families and genera of the Bacteria. Final report of the Committee of the Society of American Bacteriologists on characterization and classification of bacterial types. J. Bacteriol. *5*: 191–229.

Winslow, C.-E.A. and A. Winslow. 1908. The Systematic Relationships of the *Coccaceae*, John Wiley and Sons, New York.

Wise, M.G., J.V. McArthur and L.J. Shimkets. 1997. Bacterial diversity of a Carolina bay as determined by 16S rRNA gene analysis: confirmation of novel taxa. Appl. Environ. Microbiol. *63*: 1505–1514.

Wise, M.G., J.V. McArthur and L.J. Shimkets. 2001. *Methylosarcina fibrata* gen. nov., sp. nov. and *Methylosarcina quisquiliarum* sp. nov., novel type I methanotrophs. Int. J. Syst. Evol. Microbiol. *51*: 611–621.

Woese, C.R. 1987. Bacterial evolution. Microbiol. Rev. *51*: 221–271.

Woese, C.R., O. Kandler and M.L. Wheelis. 1990. Towards a natural system of organisms: proposal for the domains *Archaea, Bacteria*, and *Eucarya*. Proc. Natl. Acad. Sci. U.S.A. *87*: 4576–4579.

Woese, C.R., E. Stackebrandt, W. Weisburg, B.J. Paster, M.T. Madigan, V.J. Fowler, C.M. Hahn, P. Blanz, R. Gupta, K.H. Nealson and G.E. Fox. 1984a. The phylogeny of purple bacteria: the alpha subdivision. Syst. Appl. Microbiol. *5*: 315–326.

Woese, C.R., W.G. Weisburg, C.M. Hahn, B.J. Paster, L.B. Zablen, B.J. Lewis, T.J. Macke, W. Ludwig and E. Stackebrandt. 1985. The phylogeny of purple bacteria: the gamma subdivision. Syst. Appl. Microbiol. *6*: 25–33.

Woese, C.R., W.G. Weisburg, B.J. Paster, C.M. Hahn, R.S. Tanner, N.R. Krieg, H.P. Koops, H. Harms and E. Stackebrandt. 1984b. The phylogeny of purple bacteria: the beta subdivision. Syst. Appl. Microbiol. *5*: 327–336.

Wolf, A., A. Fritze, M. Hagemann and G. Berg. 2002a. *Stenotrophomonas rhizophila* sp. nov., a novel plant-associated bacterium with antifungal properties. Int. J. Syst. Evol. Microbiol. *52*: 1937–1944.

Wolf, Y.I., I.B. Rogozin, N.V. Grishin and E.V. Koonin. 2002b. Genome trees and the Tree of Life. Trends Genet. *18*: 472–479.

Wolf, Y.I., I.B. Rogozin, N.V. Grishin, R.L. Tatusov and E.V. Koonin. 2001. Genome trees constructed using five different approaches suggest new major bacterial clades. BioMed Central Evol. Biol. *1*: 8.

Wolterink, A.F., A.B. Jonker, S.W. Kengen and A.J. Stams. 2002. *Pseudomonas chloritidismutans* sp. nov., a non-denitrifying, chlorate-reducing bacterium. Int. J. Syst. Evol. Microbiol. *52*: 2183–2190.

Woo, P.C.Y., D.M.W. Tam, K.W. Leung, S.K.P. Lau, J.L.L. Teng, M.K.M. Wong and K.Y. Yuen. 2002a. *Streptococcus sinensis* sp. nov., a novel species isolated from a patient with infective endocarditis. J. Clin. Microbiol. *40*: 805–810.

Woo, P.C.Y., D.M.W. Tam, K.W. Leung, S.K.P. Lau, J.L.L. Teng, M.K.M. Wong and K.Y. Yuen. 2002b. *In* Validation of publication of new names and new combinations previously effectively published outside the IJSEM. List No. 87. Int. J. Syst. Evol. Microbiol. *52*: 1437–1438.

Wood, P.M. 1986. Nitrification as a bacterial energy source. *In* Prosser (Editor), Nitrification, IRL Press, Oxford. pp. 39–62.

Wood, P.M. 1988. Chemolithotrophy. *In* Anthony (Editor), Bacterial Energy Transduction, Academic Press, London; San Diego. 183–230.

Wu, J., R.F. Unz, H. Zhang and B.E. Logan. 2001. Persistence of perchlorate and the relative numbers of perchlorate- and chlorate-respiring microorganisms in natural waters, soils, and wastewater. Bioremed. J. *5*: 119–130.

Wullenweber, M., H.P. Koops and H. Harms. 1977. Polyhedral inclusion bodies in cells of *Nitrosomonas* spec. Arch. Microbiol. *112*: 69–72.

Wyss, C., F.E. Dewhirst, R. Gmur, T. Thurnheer, Y. Xue, P. Schupbach, B. Guggenheim and B.J. Paster. 2001. *Treponema parvum* sp. nov., a small, glucoronic or galacturonic acid-dependent oral spirochaete from lesions of human periodontitis and acute necrotizing ulcerative gingivitis. Int. J. Syst. Evol. Microbiol. *51*: 955–962.

Xia, Y., T.M. Embley and A.G. O'Donnell. 1994. Phylogenetic analysis of *Azospirillum* by direct sequencing of PCR amplified 16S rDNA. Syst. Appl. Microbiol. *17*: 197–201.

Xie, Q., Y. Wang, Y. Huang, Y. Wu, F. Ba and Z. Liu. 2002. Description of *Lentzea flaviverrucosa* sp. nov. and transfer of the type strain of *Saccharothrix aerocolonigenes* subsp. *staurosporea* to *Lentzea albida*. Int. J. Syst. Evol. Microbiol. *52*: 1815–1820.

Xin, H., T. Itoh, P. Zhou, K. Suzuki, M. Kamekura and T. Nakase. 2000. *Natrinema versiforme* sp. nov., an extremely halophilic archaeon from Aibi salt lake, Xinjiang, China. Int. J. Syst. Evol. Microbiol. *50*: 1297–1303.

Xin, H., T. Itoh, P. Zhou, K. Suzuki and T. Nakase. 2001. *Natronobacterium nitratireducens* sp. nov., a haloalkaliphilic archaeon isolated from a soda lake in China. Int. J. Syst. Evol. Microbiol. *51*: 1825–1829.

Xu, Y., Y. Nogi, C. Kato, Z.Y. Liang, H.J. Ruger, D. De Kegel and N. Glansdorff. 2003. *Psychromonas profunda* sp. nov., a psychropiezophilic bacterium from deep Atlantic sediments. Int. J. Syst. Evol. Microbiol. *53*: 527–532.

Xu, Y., Z. Wang, Y. Xue, P. Zhou, Y. Ma, A. Ventosa and W.D. Grant. 2001. *Natrialba hulunbeirensis* sp. nov. and *Natrialba chahannaoensis* sp. nov., novel haloalkaliphilic archaea from soda lakes in Inner Mongolia Autonomous Region, China. Int. J. Syst. Evol. Microbiol. *51*: 1693–1698.

Xue, Y., X. Sun, P. Zhou, R. Liu, F. Liang and Y. Ma. 2003. *Gordonia paraffinivorans* sp. nov., a hydrocarbon-degrading actinomycete isolated from an oil-producing well. Int. J. Syst. Evol. Microbiol. *53*: 1643–1646.

Xue, Y., Y. Xu, Y. Liu, Y. Ma and P. Zhou. 2001. *Thermoanaerobacter tengcongensis* sp. nov., a novel anaerobic, saccharolytic, thermophilic bacterium isolated from a hot spring in Tengcong, China. Int. J. Syst. Evol. Microbiol. *51*: 1335–1341.

Yabuuchi, E. and T. Ezaki. 2000. Arguments against the replacement of type species of the genus *Salmonella* from *Salmonella choleraesuis* to '*Salmonella enterica*' and the creation of the term 'neotype species', and for conservation of *Salmonella choleraesuis*. Int. J. Syst. Evol. Microbiol. *50*: 1693–1694.

Yabuuchi, E., Y. Kawamura, T. Ezaki, M. Ikedo, S. Dejsirilert, N. Fujiwara, T. Naka and K. Kobayashi. 2000a. *Burkholderia uboniae* sp. nov., L-arabinose-assimilating but different from *Burkholderia thailandensis* and *Burkholderia vietnamiensis*. Microbiol Immunol. *44*: 307–317.

Yabuuchi, E., Y. Kawamura, T. Ezaki, M. Ikedo, S. Dejsirilert, N. Fujiwara, T. Naka and K. Kobayashi. 2000b. *In* Validation of publication of new names and new combinations previously effectively published outside the IJSEM. List No. 75. Int. J. Syst. Evol. Microbiol. *50*: 1415–1417.

Yabuuchi, E., Y. Kosako, N. Fujiwara, T. Naka, I. Matsunaga, H. Ogura and K. Kobayashi. 2002. Emendation of the genus *Sphingomonas* Yabuuchi et al. 1990 and junior objective synonymy of the species of three genera, *Sphingobium*, *Novosphingobium* and *Sphingopyxis*, in conjunction with *Blastomonas ursincola*. Int. J. Syst. Evol. Microbiol. *52*: 1485–1496.

Yabuuchi, E., Y. Kosako, T. Naka, S. Suzuki and I. Yano. 1999. Proposal of *Sphingomonas suberifaciens* (van Bruggen, Jochimsen and Brown 1990) comb. nov., *Sphingomonas natatoria* (Sly 1985) comb. nov., *Sphingomonas ursincola* (Yurkov et al. 1997) comb. nov., and emendation of the genus *Sphingomonas*. Microbiol. Immunol. *43*: 339–349.

Yabuuchi, E., H. Yamamoto, S. Terakubo, N. Okamura, T. Naka, N. Fujiwara, K. Kobayashi, Y. Kosako and A. Hiraishi. 2001. Proposal of *Sphingomonas wittichii* sp. nov. for strain RW1^T, known as a dibenzo-*p*-dioxin metabolizer. Int. J. Syst. Evol. Microbiol. *51*: 281–292.

Yabuuchi, E., I. Yano, H. Oyaizu, Y. Hashimoto, T. Ezaki and H. Yamamoto. 1990. Proposals of *Sphingomonas paucimobilis* gen. nov. and comb. nov., *Sphingomonas parapaucimobilis* sp. nov., *Sphingomonas yanoikuyae* sp. nov., *Sphingomonas adhaesiva* sp. nov., *Sphingomonas capsulata* comb. nov., and two genospecies of the genus *Sphingomonas*. Microbiol. Immunol. *34*: 99–119.

Yakimov, M.M., L. Giuliano, T.N. Chernikova, G. Gentile, W.R. Abraham, H. Lunsdorf, K.N. Timmis and P.N. Golyshin. 2001. *Alcalilimnicola halodurans* gen. nov., sp. nov., an alkaliphilic, moderately halophilic and extremely halotolerant bacterium, isolated from sediments of soda-depositing Lake Natron, East Africa Rift Valley. Int. J. Syst. Evol. Microbiol. *51*: 2133–2143.

Yakimov, M.M. , L. Giuliano, G. Gentile, E. Crisafi, T.N. Chernikova, W.R. Abraham, H. Lunsdorf, K.N. Timmis and P.N. Golyshin. 2003a. *Oleispira antarctica* gen. nov., sp. nov., a novel hydrocarbonoclastic marine bacterium isolated from Antarctic coastal sea water. Int.J. Syst. Evol. Microbiol. *53*: 779–785.

Yakimov, M.M., H. Lünsdorf and P.N. Golyshin. 2003b. *Thermoleophilum album* and *Thermoleophilum minutum* are culturable representatives of group 2 of the *Rubrobacteridae* (*Acinetobacter*). Int. J. Syst. Evol. Microbiol. *53*: 377–380.

Yamada, Y. 1983. *Acetobacter xylinus* sp. nov., nom. rev., for the cellulose-forming and cellulose-less acetate-oxidising acetic acid bacteria with the Q-10 system. J. Gen. Appl. Microbiol. *29*: 417–420.

Yamada, Y. 2000. Transfer of *Acetobacter oboediens* Sokollek et al. 1998 and *Acetobacter intermedius* Boesch et al. 1998 to the genus *Gluconacetobacter* as *Gluconacetobacter oboediens* comb. nov. and *Gluconacetobacter intermedius* comb. nov. Int. J. Syst. Evol. Microbiol. *50*: 2225–2227.

Yamada, Y., K. Katsura, H. Kawasaki, Y. Widyastuti, S. Saono, T. Seki, T. Uchimura and K. Komagata. 2000. *Asaia bogorensis* gen. nov., sp. nov., an unusual acetic acid bacterium in the alpha-*Proteobacteria*. Int. J. Syst. Evol. Microbiol. *50*: 823–829.

Yamanaka, T. and Y. Fukumori. 1988. The nitrite oxidizing system of *Nitrobacter winogradskyi*. FEMS Microbiol. Rev. *4*: 259–270.

Yamanaka, T. and M. Shinra. 1974. Cytochrome c-552 and cytochrome c-554 derived from *Nitrosomonas europaea*. Purification, properties, and their function in hydroxylamine oxidation. J. Biochem. (Tokyo) *75*: 1265–1273.

Yao, Z.Y., F.L. Kan, E.T. Wang, G.H. Wei and W.X. Chen. 2002. Characterization of rhizobia that nodulate legume species of the genus *Lespedeza* and description of *Bradyrhizobium yuanmingense* sp. nov. Int. J. Syst. Evol. Microbiol. *52*: 2219–2230.

Yap, W.H., Z.S. Zhang and Y. Wang. 1999. Distinct types of rRNA operons exist in the genome of the actinomycete *Thermomonospora chromogena* and evidence for horizontal transfer of an entire rRNA operon. J. Bacteriol. *181*: 5201–5209.

Yassin, A.F., R.M. Kroppenstedt and W. Ludwig. 2003a. *Corynebacterium glaucum* sp. nov. Int. J. Syst. Evol. Microbiol. *53*: 705–709.

Yassin, A.F., F.A. Rainey, J. Burghardt, H. Brzezinka, M. Mauch and K.P. Schaal. 2000a. *Nocardia paucivorans* sp. nov. Int. J. Syst. Evol. Microbiol. *50*: 803–809.

Yassin, A.F., F.A. Rainey, U. Mendrock, H. Brzezinka and K.P. Schaal. 2000b. *Nocardia abscessus* sp. nov. Int. J. Syst. Evol. Microbiol. *50*: 1487–1493.

Yassin, A.F., F.A. Rainey and U. Steiner. 2001a. *Nocardia cyriacigeorgici* sp. nov. Int. J. Syst. Evol. Microbiol. *51*: 1419–1423.

Yassin, A.F., F.A. Rainey and U. Steiner. 2001b. *Nocardia ignorata* sp. nov. Int. J. Syst. Evol. Microbiol. *51*: 2127–2131.

Yassin, A.F., U. Steiner and W. Ludwig. 2002a. *Corynebacterium appendicis* sp. nov. Int. J. Syst. Evol. Microbiol. *52*: 1165–1169.

Yassin, A.F., U. Steiner and W. Ludwig. 2002b. *Corynebacterium aurimucosum* sp. nov. and emended description of *Corynebacterium minutissimum* Collins and Jones (1983). Int. J. Syst. Evol. Microbiol. *52*: 1001–1005.

Yassin, A.F., B. Straubler, P. Schumann and K.P. Schaal. 2003b. *Nocardia puris* sp. nov. Int. J. Syst. Evol. Microbiol. *53*: 1595–1599.

Yershov, G., V. Barsky, A. Belgovsky, E. Kirillov, E. Kreindlin, I. Ivanov, S. Parinov, D. Guschin, A. Drobishev, S. Dubiley and A. Mirzabekov. 1996. DNA analysis and diagnostics on oligonucleotide microchips. Proc. Natl. Acad. Sci. U.S.A. *93*: 4913–4918.

Yi, H., Y.H. Chang, H.W. Oh, K.S. Bae and J. Chun. 2003. *Zooshikella ganghwensis* gen. nov., sp. nov., isolated from tidal flat sediments. Int. J. Syst. Evol. Microbiol. *53*: 1013–1018.

Yoon, J.H., Y.G. Cho, S.S. Kang, S.B. Kim, S.T. Lee and Y.H. Park. 2000a. *Rhodococcus koreensis* sp. nov., a 2,4-dinitrophenol-degrading bacterium. Int. J. Syst. Evol. Microbiol. *50*: 1193–1201.

Yoon, J.H., S.H. Choi, K.C. Lee, Y.H. Kho, K.H. Kang and Y.H. Park. 2001a. *Halomonas marisflavae* sp. nov., a halophilic bacterium isolated from the Yellow Sea in Korea. Int. J. Syst. Evol. Microbiol. *51*: 1171–1177.

Yoon, J.H., S.S. Kang, Y.G. Cho, S.T. Lee, Y.H. Kho, C.J. Kim and Y.H. Park. 2000b. *Rhodococcus pyridinivorans* sp. nov., a pyridine-degrading bacterium. Int. J. Syst. Evol. Microbiol. *50*: 2173–2180.

Yoon, J.H., S.S. Kang, K.C. Lee, Y.H. Kho, S.H. Choi, K.H. Kang and Y.H. Park. 2001b. *Bacillus jeotgali* sp. nov., isolated from jeotgal, Korean traditional fermented seafood. Int. J. Syst. Evol. Microbiol. *51*: 1087–1092.

Yoon, J.H., S.S. Kang, K.C. Lee, E.S. Lee, Y.H. Kho, K.H. Kang and Y.H. Park. 2001c. *Planomicrobium koreense* gen. nov., sp. nov., a bacterium isolated from the Korean traditional fermented seafood jeotgal, and transfer of *Planococcus okeanokoites* (Nakagawa et al. 1996) and *Planococcus mcmeekinii* (Junge et al. 1998) to the genus *Planomicrobium*. Int. J. Syst. Evol. Microbiol. *51*: 1511–1520.

Yoon, J.H., S.S. Kang, T.I. Mheen, J.S. Ahn, H.J. Lee, T.K. Kim, C.S. Park, Y.H. Kho, K.H. Kang and Y.H. Park. 2000c. *Lactobacillus kimchii* sp. nov., a new species from kimchi. Int. J. Syst. Evol. Microbiol. *50*: 1789–1795.

Yoon, J.H., K.H. Kang and Y.H. Park. 2002a. *Lentibacillus salicampi* gen. nov., sp. nov., a moderately halophilic bacterium isolated from a salt field in Korea. Int. J. Syst. Evol. Microbiol. *52*: 2043–2048.

Yoon, J.H., K.H. Kang and Y.H. Park. 2003a. *Halobacillus salinus* sp. nov., isolated from a salt lake on the coast of the East Sea in Korea. Int. J. Syst. Evol. Microbiol. *53*: 687–693.

Yoon, J.H., K.H. Kang and Y.H. Park. 2003b. *Psychrobacter jeotgali* sp. nov., isolated from jeotgal, a traditional Korean fermented seafood. Int. J. Syst. Evol. Microbiol. *53*: 449–454.

Yoon, J.H., I.G. Kim, K.H. Kang, T.K. Oh and Y.H. Park. 2003c. *Alteromonas marina* sp. nov., isolated from sea water of the East Sea in Korea. Int. J. Syst. Evol. Microbiol. *53*: 1625–1630.

Yoon, J.H., I.G. Kim, K.H. Kang, T.K. Oh and Y.H. Park. 2003d. *Bacillus marisflavi* sp. nov. and *Bacillus aquimaris* sp. nov., isolated from sea water of a tidal flat of the Yellow Sea in Korea. Int. J. Syst. Evol. Microbiol. *53*: 1297–1303.

Yoon, J.H., H. Kim, K.H. Kang, T.K. Oh and Y.H. Park. 2003e. Transfer of *Pseudomonas elongata* Humm 1946 to the genus *Microbulbifer* as *Microbulbifer elongatus* comb. nov. Int. J. Syst. Evol. Microbiol. *53*: 1357–1361.

Yoon, J.H., H. Kim, I.G. Kim, K.H. Kang and Y.H. Park. 2003f. *Erythrobacter flavus* sp. nov., a slight halophile from the East Sea in Korea. Int. J. Syst. Evol. Microbiol. *53*: 1169–1174.

Yoon, J.H., I.G. Kim, D.Y. Shin, K.H. Kang and Y.H. Park. 2003g. *Microbulbifer salipaludis* sp. nov., a moderate halophile isolated from a Korean salt marsh. Int. J. Syst. Evol. Microbiol. *53*: 53–57.

Yoon, J.H., K.C. Lee, S.S. Kang, Y.H. Kho, K.H. Kang and Y.H. Park. 2000d. *Janibacter terrae* sp. nov., a bacterium isolated from soil around a wastewater treatment plant. Int. J. Syst. Evol. Microbiol. *50*: 1821–1827.

Yoon, J.H., J.J. Lee, S.S. Kang, M. Takeuchi, Y.K. Shin, S.T. Lee, K.H. Kang and Y.H. Park. 2000e. *Gordonia nitida* sp. nov., a bacterium that degrades 3-ethylpyridine and 3-methylpyridine. Int. J. Syst. Evol. Microbiol. *50*: 1203–1210.

Yoon, J.H., K.C. Lee, Y.H. Kho, K.H. Kang, C.J. Kim and Y.H. Park. 2002b. *Halomonas alimentaria* sp. nov., isolated from jeotgal, a traditional Korean fermented seafood. Int. J. Syst. Evol. Microbiol. *52*: 123–130.

Yoon, J.H., K.C. Lee, N. Weiss, K.H. Kang and Y.H. Park. 2003h. *Jeotgalicoccus halotolerans* gen. nov., sp. nov. and *Jeotgalicoccus psychrophilus* sp. nov., isolated from the traditional Korean fermented seafood jeotgal. Int. J. Syst. Evol. Microbiol. *53*: 595–602.

Yoon, J.H., K.C. Lee, N. Weiss, Y.H. Kho, K.H. Kang and Y.H. Park. 2001d. *Sporosarcina aquimarina* sp. nov., a bacterium isolated from seawater in Korea, and transfer of *Bacillus globisporus* (Larkin and Stokes 1967), *Bacillus psychrophilus* (Nakamura 1984) and *Bacillus pasteurii* (Chester 1898) to the genus *Sporosarcina* as *Sporosarcina globispora* comb. nov., *Sporosarcina psychrophila* comb. nov. and *Sporosarcina pasteurii* comb. nov., and emended description of the genus *Sporosarcina*. Int. J. Syst. Evol. Microbiol. *51*: 1079–1086.

Yoon, J.H., H.M. Oh, B.D. Yoon, K.H. Kang and Y.H. Park. 2003i. *Paenibacillus kribbensis* sp. nov. and *Paenibacillus terrae* sp. nov., bioflocculants for efficient harvesting of algal cells. Int. J. Syst. Evol. Microbiol. *53*: 295–301.

Yoon, J.H., W.T. Seo, Y.K. Shin, Y.H. Kho, K.H. Kang and Y.H. Park. 2002c. *Paenibacillus chinjuensis* sp. nov., a novel exopolysaccharide-producing bacterium. Int. J. Syst. Evol. Microbiol. *52*: 415–421.

Yoon, J.H., D.Y. Shin, I.G. Kim, K.H. Kang and Y.H. Park. 2003j. *Marinobacter litoralis* sp. nov., a moderately halophilic bacterium isolated from sea water from the East Sea in Korea. Int. J. Syst. Evol. Microbiol. *53*: 563–568.

Yoon, J.H., N. Weiss, K.H. Kang, T.K. Oh and Y.H. Park. 2003k. *Planococcus maritimus* sp. nov., isolated from sea water of a tidal flat in Korea. Int. J. Syst. Evol. Microbiol. *53*: 2013–2017.

Yoon, J.H., N. Weiss, K.C. Lee, I.S. Lee, K.H. Kang and Y.H. Park. 2001e. Jeotgalibacillus alimentarius gen. nov., sp. nov., a novel bacterium isolated from jeotgal with L-lysine in the cell wall, and reclassification of *Bacillus marinus* Ruger 1983 as *Marinibacillus marinus* gen nov., comb. nov. Int. J. Syst. Evol. Microbiol. *51*: 2087–2093.

Young, J.M. 2003. The genus name *Ensifer* Casida 1982 takes priority over *Sinorhizobium* Chen et al. 1988, and *Sinorhizobium morelense* Wang et al. 2002 is a later synonym of *Ensifer adhaerens* Casida 1982. Is the combination *"Sinorhizobium adhaerens"* (Casida 1982) Willems et al. 2003 legitimate? Request for an Opinion. Int. J. Syst. Evol. Microbiol. *53*: 2107–2110.

Young, J.P.W. and K.E. Haukka. 1996. Diversity and phylogeny of rhizobia. New Phytol. *133*: 87–94.

Young, J.M., L.D. Kuykendall, E. Martínez-Romero, A. Kerr and H. Sawada. 2001. A revision of *Rhizobium* Frank 1889, with an emended description of the genus, and the inclusion of all species of *Agrobacterium* Conn 1942 and *Allorhizobium undicola* de Lajudie et al. 1998 as new combinations: *Rhizobium radiobacter, R. rhizogenes, R. rubi, R. undicola* and *R. vitis*. Int. J. Syst. Evol. Microbiol. *51*: 89–103.

Yu, Z.G. , V. Anh and K.S. Lau. 2001. Measure representation and multifractal analysis of complete genomes. Physical Review E. *6403*: DOI 10.1103/PhysRevE.64.03193.

Yuen, K.Y., P.C. Woo, J.L. Teng, K.W. Leung, M.K. Wong and S.K. Lau. 2001. *Laribacter hongkongensis* gen. nov., sp. nov., a novel gram-negative bacterium isolated from a cirrhotic patient with bacteremia and empyema. J. Clin. Microbiol. *39*: 4227–4232.

Yuen, K.Y. ., P.C. Woo, J.L. Teng, K.W. Leung, M.K. Wong and S.K. Lau. 2002. *In* Validation of publication of new names and new combinations previously effectively published outside the IJSEM. List No. 87. Int. J. Syst. Evol. Microbiol. *52*: 1437–1438.

Yumoto, I., K. Hirota, Y. Sogabe, Y. Nodasaka, Y. Yokota and T. Hoshino. 2003a. *Psychrobacter okhotskensis* sp. nov., a lipase-producing facultative psychrophile isolated from the coast of the Okhotsk Sea. Int. J. Syst. Evol. Microbiol. *53*: 1985–1989.

Yumoto, I., T. Kusano, T. Shingyo, Y. Nodasaka, H. Matsuyama and H. Okuyama. 2001a. Assignment of *Pseudomonas* sp. strain E-3 to *Pseudomonas psychrophila* sp. nov., a new facultatively psychrophilic bacterium. Extremophiles *5*: 343–349.

Yumoto, I., T. Kusano, T. Shingyo, Y. Nodasaka, H. Matsuyama and H. Okuyama. 2002a. *In* Validation of publication of new names and new combinations previously effectively published outside the IJSEM. List No. 85. Int. J. Syst. Evol. Microbiol. *52*: 685–690.

Yumoto, I., A. Nakamura, H. Iwata, K. Kojima, K. Kusumoto, Y. Nodasaka and H. Matsuyama. 2002b. *Dietzia psychralcaliphila* sp. nov., a novel, facultatively psychrophilic alkaliphile that grows on hydrocarbons. Int. J. Syst. Evol. Microbiol. *52*: 85–90.

Yumoto, I., S. Yamaga, Y. Sogabe, Y. Nodasaka, H. Matsuyama, K. Nakajima and A. Suemori. 2003b. *Bacillus krulwichiae* sp. nov., a halotolerant obligate alkaliphile that utilizes benzoate and *m*-hydroxybenzoate. Int. J. Syst. Evol. Microbiol. *53*: 1531–1536.

Yumoto, I., K. Yamazaki, M. Hishinuma, Y. Nodasaka, A. Suemori, K. Nakajima, N. Inoue and K. Kawasaki. 2001b. *Pseudomonas alcaliphila* sp. nov., a novel faculatatively psychrophilic alkaliphile isolated from seawater. Int. J. Syst. Evol. Microbiol. *51*: 349–355.

Yun, N.R., Y.K. Shin, S.Y. Hwang, H. Kuraishi, J. Sugiyama and K. Kawahara. 2000a. Chemotaxonomic and phylogenetic analyses of *Sphingomonas* strains isolated from ears of plants in the family Gramineae and a proposal of *Sphingomonas roseoflava* sp. nov. J. Gen. Appl. Microbiol. *46*: 9–18.

Yun, N.R., Y.K. Shin, S.Y. Hwang, H. Kuraishi, J. Sugiyama and K. Kawahara. 2000b. *In* Validation of the publication of new names and new combinations previously effectively published outside the IJSEM. List No. 75. Int. J. Syst. Evol. Microbiol. *50*: 1415–1417.

Yurkov, V.V. and J.T. Beatty. 1998a. Anoxygenic aerobic phototrophic bacteria. Microbiol. Mol. Biol. Rev. *62*: 695–724.

Yurkov, V.V. and J.T. Beatty. 1998b. Isolation of aerobic anoxygenic photosynthetic bacteria from black smoker plume wates of the Juan de Fuca Ridge in the Pacific Ocean. Appl. Environ. Microbiol. *64*: 337–341.

Yurkov, V.V. and V.M. Gorlenko. 1992a. A new genus of freshwater aerobic, bacteriochlorophyll *a*-containing bacteria, *Roseococcus* gen. nov. Microbiology *60*: 628–632.

Yurkov, V.V. and V.M. Gorlenko. 1992b. New species of aerobic bacteria from the genus *Erythromicrobium* containing bacteriochlorophyll *a*. Mikrobiologiya *61*: 163–168.

Yurkov, V.V., V.M. Gorlenko and E.I. Kompantseva. 1992. A new genus of orange-coloured bacteria containing bacteriochlorophyll *a*: *Erythromicrobium* gen. nov. Mikrobiologiya *61*: 256–260.

Yurkov, V.V., S. Krieger, E. Stackebrandt and T. Beatty. 1999. *Citromicrobium bathyomarinum*, a novel aerobic bacterium isolated from deep-sea hydrothermal vent plume waters that contains photosynthetic pigment-protein complexes. J. Bacteriol. *181*: 4517–4525.

Yurkov, V.V., A.M. Lysenko and V.M. Gorlenko. 1991. Hybridization analysis of the classification of bacteriochlorophyll *a*-containing freshwater aerobic bacteria. Microbiology *60*: 362–366.

Yurkov, V., E. Stackebrandt, O. Buss, A. Vermeglio, V. Gorlenko and J.T. Beatty. 1997. Reorganization of the genus *Erythromicrobium*: description of *"Erythromicrobium sibiricum"* as *Sandaracinobacter sibiricus* gen. nov., sp. nov., and of *"Erythromicrobium ursincola"* as *Erthyromonas ursincola* gen. nov., sp. nov. Int. J. Syst. Bacteriol. *47*: 1172–1178.

Yurkov, V., E. Stackebrandt, A. Holmes, J.A. Fuerst, P. Hugenholtz, J. Golecki, N. Gad'on, V.M. Gorlenko, E.I. Kompantseva and G. Drews. 1994. Phylogenetic positions of novel aerobic, bacteriochlorophyll *a*-containing bacteria and description of *Roseococcus thiosulfatophilus* gen.

nov., sp. nov., *Erythromicrobium ramosum* gen. nov., sp. nov., and *Erythrobacter litoralis* sp. nov. Int. J. Syst. Bacteriol. *44*: 427–434.

Zarda, B., D. Hahn, A. Chatzinotas, W. Schönhuber, A. Neef, R.I. Amann and J. Zeyer. 1997. Analysis of bacterial community structure in bulk soil by *in situ* hybridization. Arch. Microbiol. *168*: 185–192.

Zart, D. 1997. Entwicklung eines verfahrens zur entferung von ammonium-stickoff mit hilfe einer reinkultur von *Nitrosomonas eutropha*, University of Hamburg.

Zart, D. and E. Bock. 1998. High rate of aerobic nitrification and denitrification by *Nitrosomonas eutropha* grown in a fermentor with complete biomass retention in the presence of gaseous NO_2 or NO. Arch. Microbiol. *169*: 282–286.

Zavarzina, D.G., T.P. Tourova, B.B. Kuznetsov, E.A. Bonch-Osmolovskaya and A.I. Slobodkin. 2002. *Thermovenabulum ferriorganovorum* gen. nov., sp. nov., a novel thermophilic, anaerobic, endospore-forming bacterium. Int. J. Syst. Evol. Microbiol. *52*: 1737–1743.

Zavarzina, D.G., T.N. Zhilina, T.P. Tourova, B.B. Kuznetsov, N.A. Kostrikina and E.A. Bonch-Osmolovskaya. 2000. *Thermanaerovibrio velox* sp. nov., a new anaerobic, thermophilic, organotrophic bacterium that reduces elemental sulfur, and emended description of the genus *Thermanaerovibrio*. Int. J. Syst. Evol. Microbiol. *50*: 1287–1295.

Zhang, D., H. Yang, Z. Huang, W. Zhang and S.J. Liu. 2002a. *Rhodopseudomonas faecalis* sp. nov., a phototrophic bacterium isolated from an anaerobic reactor that digests chicken faeces. Int. J. Syst. Evol. Microbiol. *52*: 2055–2060.

Zhang, D., H. Yang, W. Zhang, Z. Huang and S.J. Liu. 2003a. *Rhodocista pekingensis* sp. nov., a cyst-forming phototrophic bacterium from a municipal wastewater treatment plant. Int. J. Syst. Evol. Microbiol. *53*: 1111–1114.

Zhang, H., S. Hanada, T. Shigematsu, K. Shibuya, Y. Kamagata, T. Kanagawa and R. Kurane. 2000. *Burkholderia kururiensis* sp. nov., a trichloroethylene (TCE)- degrading bacterium isolated from an aquifer polluted with TCE. Int. J. Syst. Evol. Microbiol. *50*: 743–749.

Zhang, H., Y. Sekiguchi, S. Hanada, P. Hugenholtz, H. Kim, Y. Kamagata and K. Nakamura. 2003b. *Gemmatimonas aurantiaca* gen. nov., sp. nov., a Gram-negative, aerobic, polyphosphate-accumulating micro-organism, the first cultured representative of the new bacterial phylum *Gemmatimonadetes* phyl. nov. Int. J. Syst. Evol. Microbiol. *53*: 1155–1163.

Zhang, J., Z. Liu and M. Goodfellow. 2003c. *Nocardia caishijiensis* sp. nov., a novel soil actinomycete. Int. J. Syst. Evol. Microbiol. *53*: 999–1004.

Zhang, J., Y. Zhang, C. Xiao, Z. Liu and M. Goodfellow. 2002b. *Rhodococcus maanshanensis* sp. nov., a novel actinomycete from soil. Int. J. Syst. Evol. Microbiol. *52*: 2121–2126.

Zhang, L.P., C.L. Jiang and W.X. Chen. 2002c. *Streptosporangium subroseum* sp. nov., an actinomycete with an unusual phospholipid pattern. Int. J. Syst. Evol. Microbiol. *52*: 1235–1238.

Zhang, Q., W.J. Li, X.L. Cui, M.G. Li, L.H. Xu and C.L. Jiang. 2003d. *Streptomyces yunnanensis* sp. nov., a mesophile from soils in Yunnan, China. Int. J. Syst. Evol. Microbiol. *53*: 217–221.

Zhang, W.Z., Y.F. Xue, Y.H. Ma, W.D. Grant, A. Ventosa and P.J. Zhou. 2002d. *Marinospirillum alkaliphilum* sp. nov., a new alkaliphilic helical bacterium from Haoji soda lake in Inner Mongolia Autonomous Region of China. Extremophiles *6*: 33–37.

Zhang, W.Z., Y.F. Xue, Y.H. Ma, W.D. Grant, A. Ventosa and P.J. Zhou. 2002e. *In* Validation of publication of new names and new combinations previously effectively published outside the IJSEM. List No. 87. Int. J. Syst. Evol. Microbiol. *52*: 1437–1438.

Zhang, W.Z., Y.F. Xue, Y.H. Ma, P. Zhou, A. Ventosa and W.D. Grant. 2002f. *Salinicoccus alkaliphilus* sp. nov., a novel alkaliphile and moderate halophile from Baer Soda Lake in Inner Mongolia Autonomous Region, China. Int. J. Syst. Evol. Microbiol. *52*: 789–793.

Zhang, Z., T. Kudo, Y. Nakajima and Y. Wang. 2001. Clarification of the relationship between the members of the family *Thermomonosporaceae* on the basis of 16S rDNA, 16S-23S rRNA internal transcribed spacer and 23S rDNA sequences and chemotaxonomic analyses. Int. J. Syst. Evol. Microbiol. *51*: 373–383.

Zhilina, T.N., E.S. Garnova, T.P. Tourova, N.A. Kostrikina and G.A. Zavarzin. 2001a. *Amphibacillus fermentum* sp. nov. and *Amphibacillus tropicus* sp. nov., new alkaliphilic, facultatively anaerobic, saccharolytic bacilli from Lake Magadi. Microbiology *70*: 711–722.

Zhilina, T.N., E.S. Garnova, T.P. Tourova, N.A. Kostrikina and G.A. Zavarzin. 2001b. *Halonatronum saccharophilum* gen. nov., sp. nov.: A new haloalkaliphilic bacterium of the order *Haloanaerobiales* from Lake Magadi. Microbiology *70*: 64–72.

Zhilina, T.N., E.S. Garnova, T.P. Tourova, N.A. Kostrikina and G.A. Zavarzin. 2001c. *In* Validation of publication of new names and new combinations previously effectively published outside the IJSEM. List No. 79. Int. J. Syst. Evol. Microbiol. *51*: 263–265.

Zhilina, T.N., E.S. Garnova, T.P. Tourova, N.A. Kostrikina and G.A. Zavarzin. 2002. *In* Validation of publication of new names and new combinations previously effectively published outside the IJSEM. List No. 85. Int. J. Syst. Evol. Microbiol. *52*: 685–690.

Zhilina, T.N., T.P. Tourova, B.B. Kuznetsov, N.A. Kostrikina and A.M. Lysenko. 1999. *Orenia sivashensis* sp. nov., a new moderately halophilic anaerobic bacterium from Lake Sivash lagoons. Microbiology *68*: 452–459.

Zhilina, T.N., T.P. Tourova, B.B. Kuznetsov, N.A. Kostrikina and A.M. Lysenko. 2000. *In* Validation of publication of new names and new combinations previously effectively published outside the IJSEM. List No. 72. Int. J. Syst. Evol. Microbiol. *50*: 3–4.

Zhu, F., S. Wang and P. Zhou. 2003a. *Flavobacterium xinjiangense* sp. nov. and *Flavobacterium omnivorum* sp. nov., novel psychrophiles from the China No. 1 glacier. Int. J. Syst. Evol. Microbiol. *53*: 853–857.

Zhu, H., M.D. Willcox and K.W. Knox. 2000. A new species of oral *Streptococcus* isolated from Sprague-Dawley rats, *Streptococcus orisratti* sp. nov. Int. J. Syst. Evol. Microbiol. *50*: 55–61.

Zhu, L., W. Li and X. Dong. 2003b. Species identification of genus *Bifidobacterium* based on partial HSP60 gene sequences and proposal of *Bifidobacterium thermacidophilum* subsp. *porcinum* subsp. nov. Int. J. Syst. Evol. Microbiol. *53*: 1619–1623.

Zillig, W. 1988. *In* Validation of the publication of new names and new combinations previously effectively published outside the IJSB. List No. 24. Int. J. Syst. Bacteriol. *38*: 136–137.

Zillig, W., H.-P. Klenk, P. Palm, G. Pühler, F. Gropp, R.A. Garrett and H. Leffers. 1989. The phylogenetic relations of DNA-dependent RNA polymerases of archaebacteria, eukaryotes, and eubacteria. Can. J. Microbiol. *35*: 73–80.

Zillig, W. and A.-L., Reysenbach. 2001. Class V. *Thermococci* class. nov. *In* Boone, Castenholz and Garrity (Editors), Bergey's Manual of Systematic Bacteriology, 2nd Edition, Volume 1. The archaea and the deeply branching and phototrophic bacteria. Springer, New York. 341.

Zillig, W. and A.-L., Reysenbach. 2002. *In* Validation of publication of new names and new combinations previously effectively published outside the IJSEM. List No. 85. Int. J. Syst. Evol. Microbiol. *52*: 685–690.

Zillig, W. and K.O. Stetter. 1983. *In* Validation of the publication of new names and new combinations previously effectively published outside the IJSB. List No. 10. Int. J. Syst. Bacteriol. *33*: 438–440.

Zlamala, C., P. Schumann, P. Kämpfer, R. Rosselló-Mora, W. Lubitz and H.J. Busse. 2002a. *Agrococcus baldri* sp. nov., isolated from the air in the 'Virgilkapelle' in Vienna. Int. J. Syst. Evol. Microbiol. *52*: 1211–1216.

Zlamala, C., P. Schumann, P. Kämpfer, M. Valens, R. Rosselló-Mora, W. Lubitz and H.J. Busse. 2002b. *Microbacterium aerolatum* sp. nov., isolated from the air in the 'Virgilkapelle' in Vienna. Int. J. Syst. Evol. Microbiol. *52*: 1229–1234.

Zoetendal, E.G., C.M. Plugge, A.D. Akkermans and W.M. de Vos. 2003. *Victivallis vadensis* gen. nov., sp. nov., a sugar-fermenting anaerobe from human faeces. Int. J. Syst. Evol. Microbiol. *53*: 211–215.

Zopf, W. 1885. Die Spaltpilze, 3rd Ed., Edward Trewendt, Breslau.

Zuckerkandl, E. and L. Pauling. 1965. Molecules as documents of evolutionary history. J. Theor. Biol. *8*: 357–366.

Zweifel, U.L. and A. Hagström. 1995. Total counts of marine bacteria include a large fraction of non-nucleoid-containing bacteria (ghosts). Appl. Environ. Microbiol. *61*: 2180–2185.

Zwickl, D.J. and D.M. Hillis. 2002. Increased taxon sampling greatly reduces phylogenetic error. Syst. Biol. *51*: 588–598.

Index of Scientific Names of *Archaea* and *Bacteria*

Key to the fonts and symbols used in this index:

Nomenclature Lower case, Roman	Genera, species, and subspecies of bacteria. Every bacterial name mentioned in the *Manual* is listed in the index. Specific epithets are listed individually and also under the genus.*
CAPITALS, ROMAN:	Names of taxa higher than genus (tribes, families, orders, classes, divisions, kingdoms).
Pagination Roman:	Pages on which taxa are mentioned.
Boldface:	Indicates page on which the description of a taxon is given.†

* Infrasubspecific names, such as serovars, biovars, and pathovars, are not listed in the index.

† A description may not necessarily be given in the *Manual* for a taxon that is considered as *incertae sedis* or that is listed in an addendum or note added in proof; however, the page on which the complete citation of such a taxon is given is indicated in boldface type.

Index of Scientific Names of *Archaea* and *Bacteria*